DESIGN AND CONTROL OF CONCRETE MIXTURES

SEVENTEENTH EDITION

by Michelle L. Wilson and Paul D. Tennis

Since 1916

America's Cement Manufacturers™

Portland Cement Association
200 Massachusetts Ave NW, Suite 200
Washington, D.C. 20001-5655
cement.org
shapedbyconcrete.com
cementprogress.com

The Portland Cement Association (PCA) is the premier policy, research, education, and market intelligence organization serving America's cement manufacturers. PCA supports sustainability, innovation, and safety, while fostering continuous improvement in cement manufacturing, distribution, infrastructure, and economic growth.

shaped
BY CONCRETE

KEYWORDS: admixtures, aggregates, air-entrained concrete, batching, cement, cold weather, curing, durability, fibers, finishing, high-performance concrete, hot weather, innovations, imperfections, mixing, mixing water, mixture proportioning, paving, placing, portland cement concrete, properties, standards, structures, supplementary cementing materials, sustainability, tests, and volume changes.

ABSTRACT: The fresh and hardened properties of concrete, such as workability, strength, volume change, and durability, are presented. All concrete ingredients (cementing materials, water, aggregates, chemical admixtures, and fibers) are reviewed for their optimal use in designing and proportioning concrete mixtures, including examples of high-performance concrete. The use of concrete from design to batching, mixing, transporting, placing, consolidating, finishing, and curing is addressed, and highlights the specific contribution of cement and concrete to sustainability of the built environment. Applications for concrete including pavements and structures are also reviewed.

REFERENCE: Wilson, Michelle L. and Tennis, Paul D., *Design and Control of Concrete Mixtures*, EB001, 17th edition, Portland Cement Association, Skokie, Illinois, USA, 2021, 600 pages.

The authors of this engineering bulletin are:

Michelle L. Wilson
Senior Director of Cement and Concrete Technology
Portland Cement Association

Paul D. Tennis
Senior Director of Research and Product Standards
Portland Cement Association

COVER PHOTOGRAPH: Marina City, Chicago, Illinois, designed by Bertrand Goldberg.

Seventeenth Edition Print History
First Printing 2021
Second Printing rev. 2022- pages 29, 35, 36,45, 46, and 423
Third Printing rev. 2023- pages 1, 5, 6, 15, 21, 24, 29, 35, 44, 45, 51, 54, 55, 58, 102, 123, 245, 246, 248, 249, 256, 257, 258, 259, 423, 540, 541, 543, 544, 559, and 571

WARNING: Contact with wet (unhardened) concrete, mortar, cement, or cement mixtures can cause SKIN IRRITATION, SEVERE CHEMICAL BURNS (THIRD-DEGREE), or SERIOUS EYE DAMAGE. Frequent exposure may be associated with irritant and/or allergic contact dermatitis. Wear waterproof gloves, a long-sleeved shirt, full-length trousers, and proper eye protection when working with these materials. If you have to stand in wet concrete, use waterproof boots that are high enough to keep concrete from flowing into them. Wash wet concrete, mortar, cement, or cement mixtures from your skin immediately. Flush eyes with clean water immediately after contact. Indirect contact through clothing can be as serious as direct contact, so promptly rinse out wet concrete, mortar, cement, or cement mixtures from clothing. Seek immediate medical attention if you have persistent or severe discomfort.

TABLE OF CONTENTS

PREFACE AND ACKNOWLEDGMENTS

Preface

Design and Control of Concrete Mixtures has been the cement and concrete industry's primary reference on concrete knowledge for almost 100 years. Since the first edition was published in 1924, the U.S. version has been updated 16 times to reflect advances in cement and concrete technology and to meet the growing needs of architects, engineers, builders, concrete producers, concrete technologists, instructors, and students.

This fully revised 17th edition was written to provide a concise, current reference on concrete, including the many developments that occurred since the last edition was published in 2016. The text is backed by over 100 years of research by the Portland Cement Association and other industry groups. It reflects the latest guidance on standards, specifications, and test methods of ASTM International (ASTM), the American Association of State Highway and Transportation Officials (AASHTO), and the American Concrete Institute (ACI).

The 17th edition includes an in-depth restructuring of the existing content, presenting a 40% increase in new information over the prior edition within the previous chapters. Many advances have been made in concrete technology over the decades ranging from new materials and testing methods, to improved concrete properties and construction practices.

This book addresses the properties of concrete needed in construction applications, including strength and durability and highlights the contributions of cement and concrete to sustainability. It provides guidance on all aspects of concrete technology from selection of suitable materials to designing and proportioning a concrete mixture. The requirements for batching, mixing, transporting, placing, consolidating, finishing, and curing concrete are covered. Applications for concrete including pavements and structures are also reviewed. This edition also has added two new chapters on imperfections in concrete, and innovations in concrete.

More than 3 million copies of past editions of the book have been distributed, making this book a primary reference on concrete technology.

Acknowledgments

The authors wish to acknowledge contributions made by many individuals and organizations who provided valuable assistance in the writing and publishing of the 17th edition. We are particularly grateful for the assistance of the individuals listed below. This acknowledgment does not necessarily imply approval of the text by these individuals since the entire handbook was not reviewed by all those listed and since final editorial prerogatives have necessarily been exercised by the Portland Cement Association.

A special thanks to the following PCA staff for extensive technical recommendations, contributions, photography, and text edits; Kaitlin Beer (market intelligence), Richard Bohan (sustainability), Jamie Farny (building marketing), Nick Ferrari (communications), Greg Halsted (paving), David Orense (standards and references), Libby Pritchard (safety), Brian Schmidt (market intelligence), Trevor Storck (market intelligence), Louisa Verma (library services), Mike Zande (communications and publishing), and Dave Zwicke (market intelligence).

Additional thanks for technical assistance, references, photography, and editorial reviews goes to the following individuals and organizations: American Concrete Institute (ACI); American Concrete Pavement Association (ACPA); Paul Aubee, WRI; Michael Ayers, Global Pavement Consultants; Baker Concrete Construction; Katie Bartojay, Bureau of Reclamation; Giuseppe Basilico, Italcementi; Randy Bass, Schnabel Engineering; Nick Beristain, Votorantim Cimentos; Jeff Bowman, Kryton; Jared Brewe, PCI; Drew Burns, SCA; Scott Campbell, NRMCA; Nicholas Carino; Joseph Catella; Concrete Canvas; Concrete Reinforcing Steel Institute (CRSI); Mary Christiansen, University of Minnesota-Duluth; Boyd Clark, CTLGroup; Eamon Connolly, McHugh Engineering Group; James Cornell, JN Cornell Associates; Tim Cost, VT Cost Consulting; Matthew D'Ambrosia, MJ2 Consulting; Davis Colors; Anthony DeCarlo Jr., TWC Concrete Construction; Doka USA; Rex Donahey, ACI; Thano Drimalas, University of Texas at Austin; Eric Ferrebee, ACPA; Federal Highway Administration (FHWA); Walter Flood IV, Flood Testing Laboratories; John Fleck;

Fiber Reinforced Concrete Association; John Gajda, MJ2 Consulting; Sabrina Garber, Transtec COMMAND Center; Tom Gepford, Mitsubishi Cement; Gomaco Corporation; Robin Graves, Cemex; Brian Green, USACE; Thomas Greene, GCP Applied Technologies; Jeremy Gregory, MIT Concrete Sustainability Hub; Groundheaters, Inc; Jim Grove, FHWA; Dale Harrington, Snyder-Associates; John Hausfeld, Baker Concrete Construction; HiCon Inc.; R. Doug Hooton, University of Toronto; Kenneth Hover, Cornell University; G. Terry Harris, GCP Applied Technologies; ICON; Jason Ideker, Oregon State University; Al Innis; Burkan Isgor, Oregon State University; Jazz; Ara Jeknavorian; Victoria Jennings, MJ2 Consulting; Cecil Jones, Diversified Engineering Services; Josie; Maria Juenger, University of Texas at Austin; Amanda Kaminsky, Building Product Ecosystems; J. Scott Keim, USBR; Neel Khosa, AMSYSCO; Eric Koehler, Titan America; Ron Kozikowski, NorthStarr Concrete; Steven Kosmatka, University of Wisconsin-Milwaukee; Frank Kozeliski, Michele's Ready Mix; Eric Kreiger, USACE; Charles Lambert, Ernst Concrete of Georgia; David Lange, University of Illinois; Sang Y. Lee, CTLGroup; Tyler Ley, Oklahoma State University; Victor Li, University of Michigan; Colin Lobo; NRMCA Lucem; Kevin MacDonald, Beton Consulting Engineers; Jim Mack, CEMEX; Michael Mahoney, Euclid Chemical Company; Nicholas Marks, Continental Cement; Marshalll Concrete Products, Inc.; Clayton McCabe, Argos USA; Kirk McDonald, CalPortland; John Melander; Richard Meininger, FHWA; Michael Morrison, ACI; John Milano; National Concrete Pavement Technology Center (NCPTC); Natural Pozzolan Association (NPA); NCC/Dukane Precast; Theodore Neff, GTI; Charles Nmai, Master Builders Solutions USA; Karthik Obla, NRMCA; Jan Olek, Purdue University; Scott Orthey, ASTM; Claudia Ostertag, University of California, Berkeley; Thomas Palansky, CTLGroup; Bill Palmer, Informa; Stephen Parkes, Votorantim Cimentos; PCA Product Standards and Technology Committee; Claus Petersen, Germann Instruments; Nicholas Popoff, Votorantim Cimentos; Precast-Prestressed Concrete Institute (PCI); Henry Prenger, LafargeHolcim; Cheng Qi, Ash Grove; Klaus-Alexander Rieder, GCP Applied Technologies; Yuisa Rios, FEMA; Thomas Rewerts, Thos. Rewerts & Co.; Larry Rowland, Lehigh White Cement Company; Anton Schindler, Auburn University; Michael Schneider, Baker Concrete Construction; David Scott, Concrete Engineering Consultants; Matthew Sheehan, 3Con; Skidmore, Owings, and Merrill; Konstantine Sobolev, University of Wisconsin-Milwaukee; Dave Suchorski, Ash Grove Cement; Larry Sutter, Michigan Technological University; Laszlo Szabo, Zimmerman Industries; Scott Tarr, NorthStarr Concrete; Peter Taylor, NCPTC; Glenda Tennis; Joseph Thomas, NPA; Michael Thomas, University of New Brunswick; Tilt-up Concrete Association; Daniel Toon, United Forming; William Turley, Construction and Demolition Recycling Association; John Turner, WJE; Urban Mining; Tom Van Dam, NCE; Jan Vosahlik, CTLGroup; W. Jason Weiss, Oregon State University; Nathan Westin, CRSI; Jay Whitt, Lehigh Hanson; Stephen Wilcox, Argos, USA; Wilco; Wire Reinforcement Institute (WRI); James Wolsiefer, Ferroglobe; Kari Yuers, Kryton; Zhaozhou Zhang, Purdue University; and numerous others who provided reviews, photographs, and technical material for EB001 over the past several years.

In addition, a special thanks goes to George Seegebrecht, Concrete Consulting Engineers, and Aimee Pergalsky, Red Brick Farm, for their meticulous technical edits; and to Adam Frint, Hilde Neumayer, and George Hayes for their professional expertise in the design layout and artwork for this publication.

Thanks also goes to ASTM, AASHTO, and ACI for the use of their material and documents referenced throughout the book.

FORWARD

shaped

BY CONCRETE

Modern society has been built using cement and concrete. These materials are used throughout our daily lives to shape the world around us. It is used in highways, streets, driveways, sidewalks, parking lots, garages, bridges, tunnels, dams, high-rise buildings, and residential homes, and numerous other applications. From the tallest building in the world (Burj Khalifa in Dubai, U.A.E.) to the largest dam in the world (Three Gorges Dam in Yichang City, China), concrete is the building material of choice.

Concrete is highly recognized for its architectural appeal that is exhibited in the unique and iconic Chicago high rise, Marina City, displayed on the cover of this book.

Designed by Bertrand Goldberg, Marina City fascinated the world when it opened in 1962. At that time, the 65-story towers were the tallest residential building in the world, and the tallest concrete structure in the world. Today, it remains a key feature of Chicago architecture and a testament to concrete's beauty, durability, and resilience almost 60 years later.

For over 100 years, the Portland Cement Association (PCA) has provided technical guidance to the concrete practitioner on how to design and control concrete mixtures.

Many advances have been made in concrete technology over the decades ranging from new materials and testing methods, to improved concrete properties and construction practices, contributing to a more sustainable and resilient future.

Cityscape view of Chicago skyscrapers from the Chicago River, with Marina Towers featured in the center.

PCA's Product Standards and Technology Committee ensures that concrete continues to evolve to become more durable, economical, and sustainable. This book supports these goals. On behalf of the Committee, we hope that you will find this 17th edition of *Design and Control of Concrete Mixtures* to be an essential resource.

Nicholas Popoff, Votorantim Cimentos
Stephen Wilcox, Argos USA

Co-Chairs
PCA Product Standards and Technology Committee

INTRODUCTION TO CONCRETE

FIGURE 1-1. Concrete components: cement, water, coarse aggregate, fine aggregate, supplementary cementing materials, and chemical admixtures.

Concrete's versatility, durability, sustainability, and economy have made it the world's most widely used construction material, and the most widely produced man-made material, accounting for more than half of all man-made materials every year by mass (Ashby 2009). About 4.5 metric tons (4.9 tons) of concrete are produced per person per year worldwide and about 2.2 metric tons (2.5 tons) per person in the United States (PCA 2023).

The term *concrete* refers to a mixture of aggregates held together by a binder of cementitious paste. The aggregates are usually sand and gravel or crushed stone. The paste is typically made up of hydraulic cements (portland or blended cement), with or without supplementary cementitious materials (SCMs, such as fly ash or slag cement), water, and may also contain chemical admixtures (used to impart special properties), and possibly entrained air (Figure 1-1).

FUNDAMENTALS OF QUALITY CONCRETE

The performance of concrete is related to material characteristics, mixture proportions, workmanship, adequacy of curing, and its service environment. Every step in the production process is important to ensuring that concrete performs as expected, or better. Quality concrete involves a variety of materials and a number of different processes, including:

• using suitable materials: the production and testing of raw materials;

• determining the desired properties of concrete;

• proportioning of concrete constituents to meet the design requirements;

• batching, mixing, and handling to achieve consistency;

• proper placement, finishing, and adequate consolidation to ensure uniformity; and

• proper curing, maintenance of moisture and temperature conditions, to promote strength gain and durability.

When these factors are not carefully controlled, they may adversely affect the performance of the fresh and hardened properties.

Suitable Materials

Concrete is basically a mixture of two components: aggregates and paste. The paste, essentially composed of hydraulic cement and water, provides workability to fresh concrete, then binds the aggregates into a rocklike mass as the paste hardens from the chemical reactions between cement and water. Supplementary cementitious materials and chemical admixtures may also be included in the paste and will impact the chemical reactions and fresh and hardened concrete properties.

The paste may also contain entrapped air or purposely entrained air. As shown in Figure 1-2, the paste typically constitutes about 25% to 40% of the total volume of concrete. The volume of cement is usually between 7% and 15% and the volume of water between 14% and 21%. Air content in air-entrained concrete ranges from about 4% to 8% by volume.

Aggregates are divided into two groups: fine and coarse. Fine aggregates consist of natural or manufactured sand with particle sizes ranging up to 9.5 mm (⅜ in.); coarse aggregates are particles retained on the 1.18-mm (No. 16) sieve and ranging up to 150 mm (6 in.) in size. The maximum size of coarse aggregate is typically 19 mm or 25 mm (¾ in. or 1 in.). An intermediate-sized aggregate, around 9.5 mm (⅜ in.), is sometimes added to improve the overall aggregate gradation.

Since aggregates make up about 60% to 75% of the total volume of concrete, their selection is important. Aggregates should consist of particles with adequate strength and resistance to exposure conditions and should not contain materials that will cause deterioration of the concrete. A continuous gradation of aggregate particle sizes is desirable for efficient use of the paste.

The freshly mixed (plastic) and hardened properties of concrete may be changed by adding chemical admixtures to the concrete during batching. Admixtures are commonly used to: adjust setting time or hardening, reduce water demand, increase workability, intentionally entrain air, and adjust other fresh or hardened concrete properties.

The quality of the concrete depends upon the quality of the paste and aggregate and the bond between the two. In properly made concrete, each particle of aggregate is completely coated with paste and all of the spaces between aggregate particles are completely filled with paste, as illustrated in Figure 1-3.

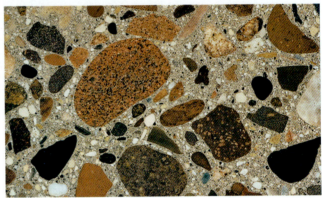

FIGURE 1-3. Cross section of hardened concrete. Cementitious materials and water (paste) completely coat each aggregate particle and fills all space between particles.

The selection of suitable materials is a very important step in the proportioning of a concrete mixture. The properties of the individual components and how they interact with each other can vary. In addition to incompatibilities within the mixture components, some materials may not be suitable for the environment in which the concrete is expected to perform. For more on materials for use in concrete, see Chapters 2 to 7.

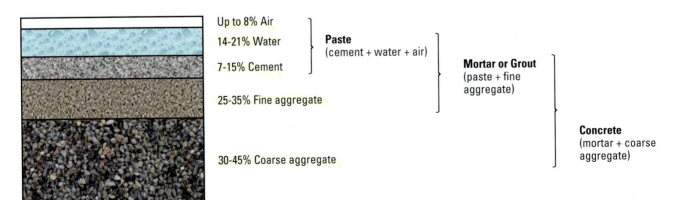

FIGURE 1-2. Concrete is a mixture of paste (cementitious materials, water, and air) and aggregates. Range in proportions of materials used in concrete, are shown by volume.

Water-Cementitious Materials Ratio

The water-to-cement ratio (w/c, or water-cementitious materials ratio, w/cm) is the ratio of the mass of water to the mass of cement (or cementitious materials). In 1918, Abrams published data that showed for a given set of concrete materials, the strength of concrete is dependent on the water-to-cement ratio, which is known as *Abram's Law* (Figure 1-4). Abrams noted that the compressive strength of concrete does not depend on the maximum density of aggregate or maximum density of concrete; rather it is dependent on the water-cement ratio, the degree of hydration (age of the concrete), and the curing and environmental conditions. This allows the 28-day strength of concrete to be predicted.

The equation of the curve is of the form,

$$S = \frac{A}{B^x} \quad \dots \dots \dots \dots \dots \dots \dots \dots (1)$$

where S is the compressive strength of concrete and x is the ratio of the volume of water to the volume of cement in the batch. A and B are constants whose values depend on the quality of the cement used, the age of the concrete, curing conditions, etc.

This equation expresses the law of strength of concrete so far as the proportions of materials are concerned. It is seen that for given concrete materials the strength depends on only one factor—the ratio of water to cement. Equations which have been proposed in the past for this purpose contain terms which take into account such factors as quantity of cement, proportions of fine and coarse aggregate, voids in aggregate, etc., but they have uniformly omitted the only term

FIGURE 1-4. Abrams' Law (Abrams 1918).

In current practice, w/cm is typically used and *cm* represents the mass of cementing materials, which includes portland or blended cement plus any supplementary cementing materials.

Unnecessarily high water contents dilute the cement paste (essentially the glue of concrete) and increases the volume of the concrete produced (Figure 1-5). This increases the porosity of the paste, which decreases its density, strength, and durability.

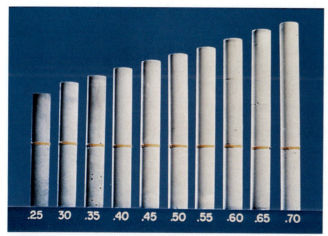

FIGURE 1-5. Ten paste cylinders with water-cement ratios from 0.25 to 0.70. The band indicates that each cylinder contains the same amount of cement. Increased water dilutes the cement paste, increasing volume, reducing density, and lowering strength.

Reducing the water content of concrete, and thereby reducing the w/cm, leads to increased strength and stiffness, and reduced creep. The drying shrinkage and associated risk of cracking will also be reduced. The concrete will have a lower permeability and increased water tightness that render it more resistant to weathering and the action of aggressive chemicals, improving durability. The lower water-to-cementitious materials ratio also improves the bond between the concrete and embedded steel reinforcement.

Some advantages of reducing water content include:

• increased compressive and flexural strength,

• lower permeability and increased watertightness,

• increased durability and resistance to weathering,

• better bond between concrete and reinforcement,

• reduced drying shrinkage and cracking, and

• less volume change from wetting and drying.

The less water used, the better the quality of the concrete provided the mixture can still be consolidated properly. Smaller amounts of mixing water can result in stiffer mixtures; with vibration, stiffer mixtures can often be easily placed. Certain chemical admixtures can also improve workability, allowing easier placement of lower water content mixtures.

Design-Workmanship-Environment

Ultimately, the quality of the final product depends on the quality of the design and workmanship, and knowledge of the environment it is to be placed (Figure 1-6). It is essential that the designers and workforce be adequately trained for this purpose.

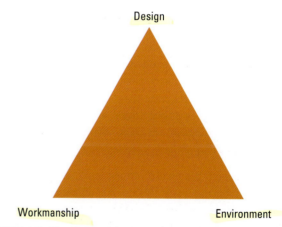

FIGURE 1-6. The quality of the final product depends on the quality of the design and workmanship, and knowledge of the environment it is to be placed.

Concrete structures are designed and built to withstand a variety of loads and may be exposed to many different environments, such as seawater, deicing salts, sulfate-bearing soils, abrasion, and cyclic wetting and drying.

Developing a proper mix design for the intended application and environment is critical to ensure long-term performance of the concrete. The materials and proportions used to produce concrete will depend on the loads it is required to carry and the environment to which it will be exposed. Properly designed and built concrete structures are strong and durable throughout their service life. For more information on designing concrete mixtures see Chapter 13.

There are many steps between mixing the concrete and having a finished concrete element. Each of these steps must be executed properly to ensure a quality product. The method of transportation must be suitable for the type of mixture to avoid early stiffening, segregation, or loss of material. Placement and consolidation methods must be controlled to fully fill the formwork with a homogeneous concrete material. Finishing is used to establish the final appearance of any concrete surface and proper curing is necessary to ensure satisfactory hydration and hardening of the cementitious materials. For more information on mixing and transportation, placing and finishing, and curing concrete, see Chapters 14 to 17.

Lastly, the environment in which the concrete is to be placed and expected to perform must be considered when specifying the mixture design, choosing suitable materials, and determining appropriate handling, placing, finishing, and curing techniques. This includes planning for temperatures at the time of placement, exposure to freeze-thaw, various forms of chemical attack, or marine exposure. For more information on concrete durability, see Chapter 11.

SUSTAINABLE DEVELOPMENT

Concrete is the basis of much of civilization's infrastructure and much of its physical development. More concrete is used throughout the world than all other building materials combined. It is a fundamental building material to municipal and transportation infrastructure, office buildings, and homes. And, while cement manufacturing is resource- and energy-intensive, the characteristics of concrete make it a low environmental-impact construction material, from an environmental and sustainability perspective. In fact, many applications for concrete directly contribute to sustainability.

America's cement producers' strong culture of innovation that has led to new sustainable manufacturing practices that continually lessen environmental impacts. Over the last 40 years, U.S. cement manufacturers have reduced the energy used to produce a metric ton of cement by roughly 40%. More-efficient equipment has led to faster, safer projects. Company-driven improvements have led to increased use of alternative fuels, which represent about 15% of total cement plant energy consumption in the U.S. This reduces the flow of industrial byproducts to landfills. For more information on the sustainability of cement and concrete see Chapter 12.

CODES AND REFERENCED STANDARDS

Codes and standards provide a set of rules and requirements for the design, construction, and operations of buildings. The building codes enacted at the state or local level becomes the minimum legal requirements to which buildings are designed and constructed.

The model building codes in the United States are developed by the International Code Council (ICC). The main codes in the U.S. are the International Building Code (IBC) or International Residential Code (IRC). In Canada, national model codes are published by the National Research Council of Canada.

The national model codes may be adopted by state and local jurisdictions with or without modifications or amendments, depending on their needs. States and municipalities typically reserve the right to amend the model codes to assure that the requirements for design and construction of buildings are appropriate for the climatic, geographical, geological, political, and economic conditions within their jurisdiction. Each state has adopted some form of the IBC.

The U.S. Department of Transportation (DOT) and the Federal Highway Administration (FHWA) cover state and local government public works. This includes requirements for the construction, maintenance, and supervision of the nation's highways, bridges, and tunnels.

Many technical aspects of the model buildings codes and state and federal agencies are addressed in reference publications developed by national consensus standards development organizations (SDO). These standards are typically referenced in the appropriate sections of the building code and include standards from the following organizations:

- American Concrete Institute (ACI),
- ASTM International (formerly known as American Society for Testing and Materials [ASTM]),
- American Association of State Highway and Transportation Officials (AASHTO), and
- The Masonry Society (TMS).

Concrete specifications and design standards, and also non-mandatory materials and construction guides, are available through ACI.

Specifications for concrete materials and testing are available from ASTM and AASHTO. ASTM Committee C01 standards support cement quality and performance, while Committee C09 addresses standards for concrete and concrete aggregates. There are many other ASTM committees that are involved in concrete materials specification and testing including those for masonry (C12 and C15), sustainability (E60), and road and

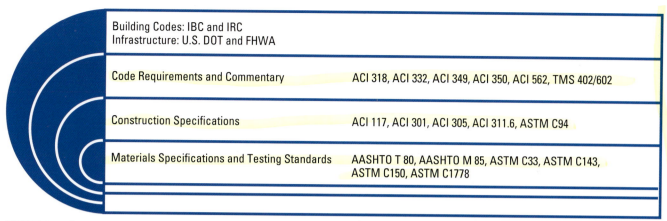

Building Codes: IBC and IRC Infrastructure: U.S. DOT and FHWA	
Code Requirements and Commentary	ACI 318, ACI 332, ACI 349, ACI 350, ACI 562, TMS 402/602
Construction Specifications	ACI 117, ACI 301, ACI 305, ACI 311.6, ASTM C94
Materials Specifications and Testing Standards	AASHTO T 80, AASHTO M 85, ASTM C33, ASTM C143, ASTM C150, ASTM C1778

FIGURE 1-7. Codes and standards umbrella depicting relationships between codes, specifications, and other construction standards.

paving materials (D18). AASHTO represents highway and transportation departments in the 50 states, the District of Columbia, and Puerto Rico. It represents all transportation modes including air, highways, public transportation, active transportation, rail, and water.

Masonry construction is generally covered by a code and specification co-published by The Masonry Society (TMS), American Concrete Institute (ACI), and the American Society of Civil Engineers (ASCE).

The relationship between codes, specifications, and other construction standards are depicted in Figure 1-7. It is important to distinguish between mandatory language codes and standards versus other industry guides that are in non-mandatory language. Codes and standards should not reference non-mandatory documents.

INDUSTRY TRENDS

The cement industry is essential to the nation's construction industry. Few construction projects are viable without utilizing cement-based products.

Cement Consumption

U.S. cement production is dispersed with the operation of 90 cement plants in 34 states. The top five companies collectively operate around 60% of US clinker capacity (PCA 2023). The United States consumed 108.4 million metric tons (120 million short tons) of portland cement in 2022.

Cement consumption varies based on the time of year and prevalent weather conditions. Nearly two-thirds of U.S. cement consumption occurs in the six-month period between May and October. The seasonal nature of the industry can result in large swings in cement and clinker (unfinished raw material) inventories at cement plants over the course of a year. Cement

producers will typically build up inventories during the winter and then ship them during the summer.

The domestic U.S. cement industry is regional in nature. The logistics of shipping cement limits distribution over long distances. As a result, customers traditionally purchase cement from local sources. About 97% of U.S. cement is shipped to customers by truck. Barge and rail account for the remaining distribution modes.

The majority of cement shipments are sent to ready-mixed concrete producers. The remainder are shipped to manufacturers of concrete related products, contractors, materials dealers, oil well/mining/drilling companies, as well as government entities. Portland cement consumption in the U.S. by user groups is defined in Figure 1-8.

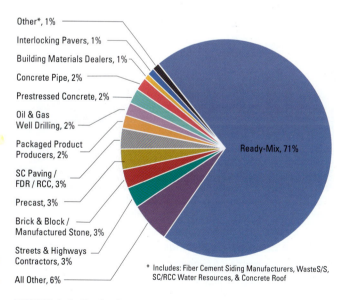

Other*, 1%
Interlocking Pavers, 1%
Building Materials Dealers, 1%
Concrete Pipe, 2%
Prestressed Concrete, 2%
Oil & Gas Well Drilling, 2%
Packaged Product Producers, 2%
SC Paving / FDR / RCC, 3%
Precast, 3%
Brick & Block / Manufactured Stone, 3%
Streets & Highways Contractors, 3%
All Other, 6%
Ready-Mix, 71%

* Includes: Fiber Cement Siding Manufacturers, WasteS/S, SC/RCC Water Resources, & Concrete Roof

FIGURE 1-8. Portland cement consumption in U.S. by user groups (PCA 2022).

Primary Markets and Applications

Concrete is used in highways, streets, parking lots, parking garages, bridges, high-rise buildings, dams, homes, floors, sidewalks, driveways, and numerous other applications as shown in Figure 1-9 (PCA 2023). The primary markets of paving and structures (Figure 1-10) are described in further detail in Chapter 21 and Chapter 22.

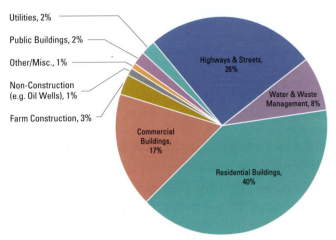

FIGURE 1-9. Apparent use of portland cement in the U.S. by market segment (PCA 2023).

FIGURE 1-10. Concrete is used as a building material for many applications including paving (top) and buildings (bottom).

There are concrete products for all types of applications and construction needs. As shown in Figure 1-8, ready-mixed concrete for cast-in-place construction is the most common type of concrete produced in the US. Other popular applications for concrete include precast concrete products and concrete masonry units. Specialty applications for cement and concrete include packaged products, shotcrete, roller-compacted concrete, pervious concrete, soil cement, and controlled low strength material (flowable fill).

Cast-in-place concrete. Cast-in-place concrete is transported in a plastic state as ready-mixed concrete and deposited in forms on site. Ready-mixed concrete is batched in plants that proportion and mix concrete (Figure 1-11), and then transport it to construction sites in trucks with revolving drums or in dump trucks. In some cases, ready-mixed concrete is produced by temporary plants set up on a construction site, minimizing transportation distances. The United States uses about 267 million cubic meters (349 million cubic yards) of ready-mixed concrete each year (PCA 2023).

FIGURE 1-11. Ready-mixed concrete is conveniently delivered to jobsites in trucks with revolving drums.

Precast Concrete. Precast concrete is produced by casting plastic concrete in a reusable mold or form which is then moist cured in a controlled environment, stripped from the forms, and ultimately transported to the construction site and lifted (or erected) into place. By producing precast concrete in a environment where temperature and humidity are controlled precast concrete is afforded the opportunity to cure and mature, while being closely monitored.

Typically, large precast concrete elements include tunnel segments, beams and columns, and other structural elements (Figure 1-12). Concrete pipe, pavement sections, walls, roofing tiles, pavers, railroad cross ties, and concrete block are also precast concrete. Precast concrete offers efficient, economical construction in all weather conditions and provides the long clear spans and open spaces needed, for example in parking structures and commercial buildings.

FIGURE 1-12. Precast concrete bridge girder being delivered to construction site.

Concrete Masonry Units. Concrete masonry units (CMUs) are used in all types of low-rise buildings, from residential to educational to commercial and industrial. Concrete masonry units are manufactured from very dry, stiff concrete mixtures into a wide variety of shapes and sizes (Figure 1-13). The no-slump or low-slump material is placed into molds, vibrated and compacted, and demolded quickly. Units are stiff enough to hold their shape as they enter the curing chamber. The final product is often referred to as concrete block or concrete brick, and is typically joined together into an integral structure with masonry mortar, and may also be filled with grout depending on load, reinforcement and load-bearing properties.

FIGURE 1-13. Concrete masonry units are manufactured in a wide variety of sizes and shapes (courtesy of Besser Company).

THE BEGINNING OF AN INDUSTRY

The oldest concrete discovered dates from around 7000 BC. It was found in 1985 when a concrete floor was uncovered during the construction of a road at Yiftah El in Galilee, Israel. It consisted of a lime concrete, made from burning limestone to produce quicklime, which when mixed with water and stone, hardened to form concrete (Brown 1996 and Auburn 2000).

A cementing material was used between the stone blocks in the construction of the Great Pyramid at Giza in ancient Egypt around 2500 BC. Some reports say it was a lime mortar while others say the cementing material was made from burnt gypsum. By 500 BC, the art of making lime-based mortar arrived in ancient Greece. The Greeks used lime-based materials as a binder between stone and brick and as a rendering material over porous limestones commonly used in the construction of their temples and palaces.

Natural pozzolans have been used for centuries. The term "pozzolan" comes from a volcanic ash mined at Pozzuoli, a village near Naples, Italy. Sometime during the second century BC the Romans quarried a volcanic ash near Pozzuoli. Believing that the material was sand, they mixed it with lime and found the mixture to be a much stronger material than previously produced. This discovery was to have a significant effect on construction. The material was not sand, but a fine volcanic ash containing silica and alumina. When combined chemically with lime, this material produced what became known as pozzolanic cement. However, the use of volcanic ash and calcined clay dates back to 2000 BC and earlier in other cultures. Many of the Roman, Greek, Indian, and Egyptian pozzolan concrete structures can still be seen today. The longevity of these structures attests to the durability of these materials.

Examples of early Roman concrete have been found dating back to 300 BC. The very word concrete is derived from the Latin word *concretus* meaning grown together or compounded. The Romans perfected the use of pozzolan as a cementing material. This material was used by builders of the famous Roman walls, aqueducts, and other historic structures including the Theatre at Pompeii, Pantheon, and Colliseum in Rome (Figure 1-14). Building practices were much less refined in the Middle Ages and the quality of cementing materials deteriorated.

The practice of burning lime and the use of pozzolan was lost until the 1300s. In the 18th century, John Smeaton concentrated his work to determine why some limes possess hydraulic properties while others (those made from essentially pure limestones) did not. He found that an impure, soft limestone containing clay minerals made the best hydraulic cement.

FIGURE 1-14. Theatre at Pompeii (left), completed in 70 BC, and the Coliseum in Rome (right), completed in 80 AD, were both constructed of concrete. Much of each still stands today (courtesy of J. Catella).

This hydraulic cement, combined with a pozzolan imported from Italy, was used in the reconstruction of the Eddystone Lighthouse in the English Channel, southwest of Plymouth, England (Figure 1-15).

FIGURE 1-15. Eddystone lighthouse constructed of natural cement by John Smeaton.

The project began operation in 1759 and took three years to complete. It was recognized as a turning point in the development of the cement industry. A number of discoveries followed as efforts within a growing natural cement industry were now directed to the production of a consistent quality material. Natural cement was manufactured in Rosendale, New York, in the early 1800s (White 1820). One of the first uses of natural cement was to build the Erie Canal in 1818 (Snell and Snell 2000).

The development of portland cement was the result of persistent investigation by science and industry to produce a superior quality natural cement. The invention of portland cement is generally credited to Joseph Aspdin, an English mason. Aspdin pulverized limestone and clay, burned the mixture to form pebble-sized lumps called clinker, then ground the clinker into powder. To make the process faster and easier, Aspdin reportedly used limestone from local roads that had already been pulverized by traffic – and was fined for stealing the limestone! In 1824, he obtained a patent for a product he named portland cement (Figure 1-16). When hardened, Aspdin's product resembled the color of the natural limestone quarried on the Isle of Portland in the English Channel (Figure 1-17) (Aspdin 1824). The name has endured and is now used throughout the world, with many manufacturers adding their own trade or brand names.

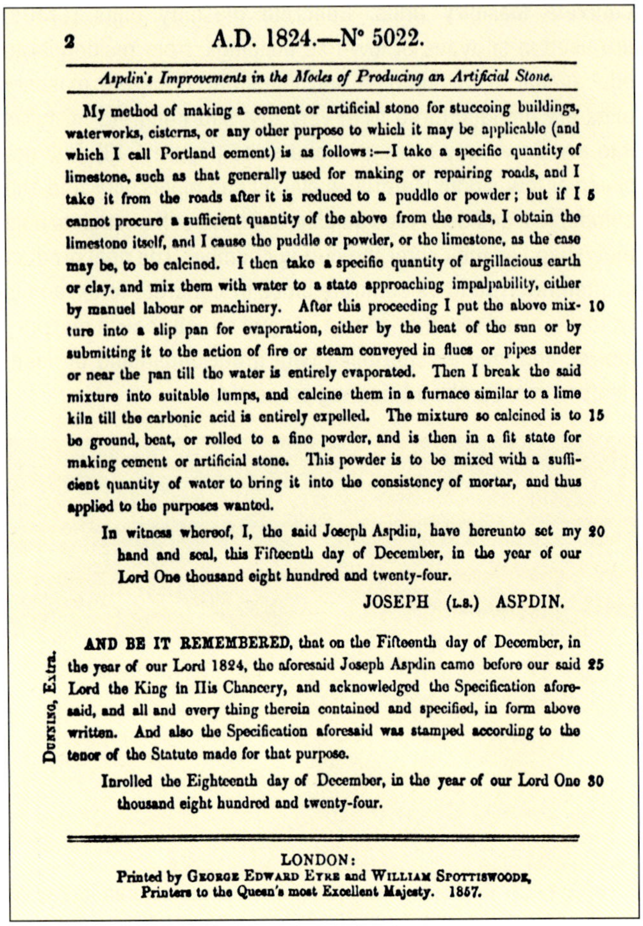

FIGURE 1-16. Aspdin's patent for portland cement.

FIGURE 1-17. Isle of Portland quarry stone (after which portland cement was named) next to a cylinder of modern concrete.

In 1845, I. C. Johnson, of White and Sons, Swanscombe, England, improved upon Asdin's recipe. He burned the cement raw materials at higher temperatures (closer to the temperatures used today), until the mass was nearly vitrified, producing a portland cement as we now know it. This cement became the popular choice during the middle of the 19th century and was

 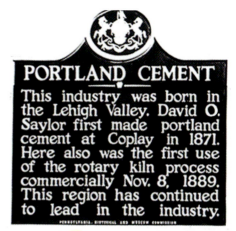

FIGURE 1-18. The first portland cement manufactured in the United States was produced at a plant in Coplay, Pennsylvania, in 1871.

exported from England throughout the world. Production also began in Belgium, France, and Germany about the same time and export of these products from Europe to North America began about 1865. The first recorded shipment of portland cement to the United States was in 1868.

The first portland cement manufactured in the United States was produced at a plant in Coplay, Pennsylvania, by David O. Saylor in 1871 (Figure 1-18). Three years later, Saylor and his associates founded the Coplay Cement Company near the Lehigh Valley town of the same name, launching America's portland cement industry.

With rapid U.S. expansion and building came a burgeoning demand for nearby domestic sources of cement, and other plants followed. The Lehigh Valley, with abundant raw materials, remained the nucleus of the cement industry at the time: Universal Atlas Portland Cement (1889), Alpha Portland Cement Company (1895), and Lehigh Cement Company (1897).

The rotary kiln was introduced to the United States by Jose de Navarro in 1889, who in the same year founded the Keystone Cement Company. His first kilns, built next to Saylor's Coplay works in the Lehigh Valley, were 7.6 m long and 1.5 m in diameter and their operating temperature approached that of today's kilns: 1370°C to 1650°C (2500°F to 3000°F). Navarro's Keystone Cement Company pioneered the use of pulverized coal for firing the kilns.

Other pioneers in cement-making technology included Thomas A. Edison. In the 1890s, Edison developed new crushing equipment for a scheme to recover iron ore from low-grade rock. When his plan proved unworkable, he adapted this equipment to cement making and in 1902 founded the Edison Portland Cement Works at New Village, New Jersey. There, he introduced the first long rotary kilns (more than 45 m [150 ft]) to realize dramatic gains in efficiency. Paralleling these kiln innovations were improvements in crushing and grinding methods.

Cement making was no longer a trial-and-error industry conducted by a handful of isolated entrepreneurs. Production methods were well established, if imperfect by today's standards. A thriving domestic industry had supplanted imported cement as the material of choice. In just one generation, an entire industry was born and brought to fruition.

As the industry continued to expand, there were needs for reliable technical information, research, and uniform test methods and standards. The Portland Cement Association was founded in Chicago in 1916.

The growth of the US cement and concrete industries since 1900 is highlighted in Figure 1-19, which shows the volume of cement produced and made into concrete. The ups and downs over the last more than 100 years parallels significant economic events in US history: World Wars I and II, the Great Depression, the 1970s energy crises, and the recession of 2006 to 2008. These parallels highlight the signification contribution to the US economy and the development of modern society.

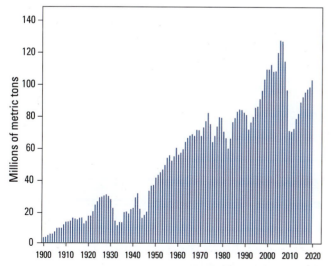

FIGURE 1-19. US portland cement consumption from 1900 to 2019 parallels many major events in US history, highlighting the relationship of concrete to modern society.

HISTORICAL CONTRIBUTIONS – 100 PLUS YEARS OF CONCRETE KNOWLEDGE

Since the early 1900s, PCA has been a leader in education and research on cement and concrete technology. That includes a broad mix of publications that were developed to increase the knowledge and understanding of cement and concrete. Over one hundred years ago, PCA published its first book on concrete mixtures, *Proportioning Concrete Mixtures and Mixing and Placing Concrete* (1916) (Figure 1-20). The book proportioned concrete for different applications using simple bulk volume relationships such as a 1:2:3 mix (1 part cement: 2 parts sand: 3 parts gravel).

From 1918 to 1925 the Lewis Institute in Chicago, under sponsorship of PCA, presented concrete research in a bulletin series. These are labeled by PCA as the Lewis, or LS series, and are also listed as Bulletins by the Lewis Institute.

From 1939 to 1969 PCA Research Department Bulletins were issued to cover the fields of cement and concrete materials technology. These are indexed as the RX series.

From 1953 to 1969, PCA Development Department Bulletins were issued covering research in concrete products, structures, and engineering, indexed as the DX series.

From 1969 to present, PCA Research and Development bulletins have been issued as the RD series. It comprises bulletins relating to cement and concrete materials research and to structural, transportation, fire, and building construction development programs.

In addition to the research documents listed above, Engineering Bulletins (EB), Special Publications (SP), Market Intelligence (MI), and Information Sheets (IS) are referenced throughout this publication.

FIGURE 1-20. PCA's first book, *Proportioning Concrete Mixtures and Mixing and Placing Concrete* (1916).

Table 1-1 provides a list of selected research reports and achievements in the cement and concrete industry over a span of more than 100 years. This does not even remotely try to cover the entire catalogue of PCA research and technology publications, but rather highlights some of the more notable works from each decade that have greatly influenced the development of cement and concrete technology (Figure 1-21). Summaries of many of PCA research report series are included in publications RX228 (PCA 1969), DX147 (PCA 1970), and RD100 (PCA 1990).

FIGURE 1-21. PCA central research laboratory was built in 1948, and structural laboratory in 1957, in Skokie, IL.

TABLE 1-1. Selection of Significant PCA Research in the Cement and Concrete Industry from 1916-2020

ID	YEAR	TITLE	AUTHOR	SUMMARY OF FINDINGS
LS001	1918	*Design of Concrete Mixtures*	D.A. Abrams	The first comprehensive, researched-based guide to designing concrete mixtures in the U.S. Includes recognition of the important factors of water-cement ratio and aggregate gradation on consistency and strength of concrete.
LS013	1924	*Calcium Chloride as an Admixture in Concrete*	D.A. Abrams	Research results indicated that $CaCl_2$ in concrete in appropriate amounts could lead to strength increases of 0.7 MPa to 1.4 MPa (100 psi to 200 psi). In addition, $CaCl_2$ contents above about 3% by weight of cement did not improve strength.
LS016	1925	*Effect of Size and Shape of Test Specimen on Compressive Strength of Concrete*	H.F. Gonnerman	Compression tests were made at 7 days to 1-year on 1755 concrete specimens to study the impacts of various specimen shapes. These included cylinders, ranging from 33 mm to 300 mm (1⅓ in. to 12 in.) in diameter, and varying in length; 150-mm × 200-mm (6-in. × 8-in.) cubes; and prisms, 150 mm × 300 mm (6 in. × 12 in.), and 200 mm × 400 mm (8 in. × 16 in.). The size, grading of aggregate, mixture consistency, and age were additional variables. The 150-mm × 300-mm (6-in. × 12-in.) cylinder proved to be a satisfactory specimen.
LS017	1925	*Studies of Bond between Concrete and Steel*	D.A. Abrams	Study to determine relation between bond to steel and compressive strength of concrete. Used a pull-out test and found correlation between bond and compressive strength. Tests included 735 pullout tests in 25-mm (1-in.) plain round steel bars embedded in 200-mm × 200-mm (8-in. × 8-in.) concrete cylinders and 735 parallel compression tests on 150-mm × 300-mm (6-in. × 12-in.) cylinders at ages of 7 days to 1 year. Study provided foundational research to reinforced concrete design, and established the pull-out test for bond strength.
Major Series - 171, 209, and 210	1928	*Flexure and Tension Tests of Plain Concrete*	H. F. Gonnerman and E. C. Shuman	Summarizes many tests by the Structural Materials Research Laboratory and PCA Research Laboratory from about 1922 to 1928. Data allows evaluations of the effects of aggregate type and gradation, concrete curing, age, and water content (among other variables), on the relationship between flexural and compressive strengths.
SN0038	1929	*Calculation of Compounds in Portland Cement*	R.H. Bogue	The basis for a calculation of the phases present in portland cement from chemical analyses are presented. The formulas that came to be known as the 'Bogue Equations' are given and diagrammatical methods are also provided.
RX001	1938	*Estimation of Phase Composition of Clinker in the System 3CaO • SiO₂ – 2CaO • SiO₂ – 3CaO • Al₂O₃ – 4CaO • Al₂O₃ • Fe₂O₃ at Clinkering Temperatures*	L. A. Dahl	Methods are given for calculating the percentages of solid phases and liquid at various stages of the heating or cooling process, and for calculating clinker composition based on equilibrium calculations of oxide analyses.
RX002	1939	*The Bleeding of Portland Cement Paste, Mortar and Concrete Treated as a Special Case of Sedimentation*	T.C. Powers and L.A. Dahl	In-depth study on bleeding of portland cement paste, mortar, and concrete. Bleeding is shown to be a process of subsidence or sedimentation. The rate of bleeding follows a general law derived from Poiseuille's law of capillary flow. A technique for measuring bleeding is given.
RX012	1946	*The Influence of Gypsum on the Hydration and Properties of Portland Cement Pastes*	W. Lerch	Foundational research study exploring the effects of gypsum content and fineness on cement hydration, setting, and early strength development. Study found that proper regulation of the early hardening reactions has impacts beyond regulating the time of setting.
RX019	1947	*Procedures for Determining the Air Content of Freshly-Mixed Concrete by the Rolling and Pressure Methods*	C.A. Menzel	Apparatus and test methods were developed for determining air content of freshly mixed concrete based on volumetric and pressure methods. Both laboratory and field experience with different methods indicated that the pressure method is most practical for field tests.
RX022	1948	*Studies of the Physical Properties of Hardened Portland Cement Paste*	T.C. Powers and T.L. Brownyard	Study on hardened cement paste characteristics, including adsorption, specific surface, porosity, gel-water and capillary water, volume change, moisture diffusion, hydration limits, and freezing of water.
RX026	1948	*Long-Time Study of Cement Performance in Concrete*	F.R. McMillan, I.L.Tyler, W.C. Hansen, W. Lerch, C.L. Ford,and L.S. Brown	In 1941, PCA installed a long-term field study of cement performance on concrete. The durability of these structures under the various exposures provided the principal basis for evaluating the performance of the cements. The initial summary of cement chemical and physical properties were published in 1948 and other findings were reported in later decades.

ID	YEAR	TITLE	AUTHOR	SUMMARY OF FINDINGS
RX033	1949	The Air Requirement of Frost-Resistant Concrete	T.C. Powers	Findings showed the effectiveness of entrained air depends on air void spacing. The actual maximum spacing factor for certain frost-resistant concretes was estimated to be about 0.01 in.
RX053	1954	Permeability of Portland Cement Paste	T.C. Powers, L.E. Copeland, J.C. Hayes, and H.M. Mann	Apparatus and methods for measuring the permeability of portland cement pastes. Test results show effects of curing, cement content, cement composition, and cement fineness.
RX054	1955	Some Observations on the Mechanics of Alkali-Aggregate Reaction	L.S. Brown	Fundamental work in establishing the physical damage from ASR. Describes visual features of alkali-aggregate reaction in concrete whereby reaction may be identified not only as having occurred, but particularly as having been a cause of physical damage.
IS015.01	1955	Hot Cement and Hot Weather Concrete Tests	W. Lerch	Reviews the data on hot cement and hot weather concrete tests and describes procedures for controlling the temperature of the freshly mixed concrete during hot weather. The temperature of the cement, because of its, low specific heat and its relatively small quantity in the mix, has only a minor impact on the temperature of the freshly mixed concrete.
RX083	1956	Studies of Salt Scaling of Concrete	G. Verbeck and P. Klieger	Investigation of mechanism that de-icers cause scaling on concrete surface. Provide information on effect of type and concentration of de-icer, curing condition, air entrainment, and other variables. Lead to better understanding and development of remediation measures.
RX087	1958	Carbonation of Hydrated Portland Cement	G.J. Verbeck	Carbonation of portland cement hydration products discussed. Effects of humidity on carbonation. Phenolphthalein color test detailed.
RX090	1958	The Physical Structure and Engineering Properties of Concrete	T.C. Powers	Research suggested that several components of cement hydrate at the same rate because of the identical composition of hydration products regardless of age at which they are formed. The strength of the paste as related to gel-space ratio is discussed.
DX031	1959	Ultimate Strength Criteria for Reinforced Concrete	L.B. Kriz	Derivation of ultimate strength requirements for structural members based on analytical determination of strains in concrete. The ultimate strengths derived were in agreement with tests, and found to be a function of only the stress at the extreme compression edge and properties of the cross-section involved. For members governed by tension, the design equations of the 1956 ACI Building Code were substantiated with this research.
RX119	1960	Concrete Mix Water – How Impure Can It Be?	H.H. Steinour	Findings showed that natural, fresh water rarely contains more than 2000 ppm (0.2%) of dissolved solids and is generally suitable as mixing water. Higher amounts of impurities present in natural waters, and some industrial wastewaters, can also be tolerated, except for alkali carbonates and bicarbonates which may have significant effects even at 2000 ppm.
DX089	1965	Moisture Migration – Concrete Slab-On-Ground Construction	H.W. Brewer	In-depth study covering a wide range of concrete quality topics related to moisture. Included exposure conditions varying from water-in-contact drying only, as well as the effect of admixtures, vapor barriers, and gravel capillary breaks. Good correlation was found between moisture migration and water-cement ratio. The study found that flow increased directly with increased w/c.
DX103	1966	Influence of Size and Shape of Member on the Shrinkage and Creep of Concrete	T.C. Hansen and A.H. Mattock	The influence of size and shape of member on the shrinkage and creep of concrete. Conclusions showed that the volume-surface ratio is a suitable parameter for use in structural design when estimating the influence of size and shape of member on shrinkage and creep formations.
DX107	1966	Seismic Resistance of Reinforced Concrete – A Laboratory Test Rig	N.W. Hanson and H.W. Conner	The seismic resistance of reinforced concrete was simulated. Earthquake loads and displacements were applied to joints between beams and columns in reinforced concrete building frames.
RD004	1970	Fire Resistance of Lightweight Insulating Concretes	A.H. Gustaferro, M.S. Abrams, and A. Litvin	Data presented on fire resistance of lightweight concretes. Relationships between slab thickness and fire endurance are presented. Design charts for estimation of fire endurance of specimens are made.
RD009	1971	Fire Resistance of Prestressed Concrete Beams	A.H. Gustaferro, M.S. Abrams, and E.A.B. Salse	Investigation of fire resistance of simply supported prestressed concrete beams. Fire endurance as well as ultimate capacity and deflection are tested. Thermal and creep strains were considered. Found feasible to calculate fire endurance of reinforced and prestressed beams.

ID	YEAR	TITLE	AUTHOR	SUMMARY OF FINDINGS
RD047	1976	*Characteristics and Utilization of Coarse Aggregates Associated with D-Cracking*	D. Stark	Coarse aggregate response to freezing and thawing and resistance to D-cracking. Studies indicated that nondurable material may reach critical saturation when the concrete is in contact only with capillary-held water. Sorption properties and freeze-thaw tests differentiate between durable and nondurable materials. Reducing maximum aggregate particle sizes, as indicated by laboratory freeze-thaw tests, is the most feasible method of improving durability.
RD086	1982	*Longtime Study of Concrete Durability in Sulfate Soils*	D. Stark	Report presents 11-year performance data for concrete beams stored in sulfate bearing soil in Sacramento, California. Results reaffirm the importance of cement type (the tricalcium aluminate content) and, more significantly, cement factor with attendant change in water-cement ratio on resistance to sulfate attack.
RD089	1986	*Effect of Fly Ash on Some of the Physical Properties of Concrete*	S.H. Gebler and P. Klieger	Tests of portland cement concretes containing Class F and Class C fly ashes were examined. At early ages, compressive strength of concretes with fly ash, regardless of class, was essentially unaffected by moisture availability. Abrasion resistance of control concretes and concretes containing fly ash was dependent on compressive strength. Drying shrinkage and absorption of the concretes were generally unaffected by the use of fly ash.
RD098	1989	*Influence of Design and Materials on Corrosion Resistance of Steel in Concrete*	D. Stark	Investigated the influence of design and materials on corrosion resistance of steel in concrete. The effects of w/c ratio, cover over steel, use of galvanized steel, fly ash replacements for cement, cracks extending from the surface to the reinforcing steel, and mixed steel in reinforcing mats on the corrosion resistance in concrete were investigated.
RD105	1992	*Optimization of Sulfate Form and Content*	F.J. Tang	The study suggested that early aluminate hydration reactions have a profound effect on cement paste flow and strength development. Chemical and physical forms of sulfates in cement can have a strong influence on these early reactions. Sulfate level for strength development at all ages, the sulfate level that showed highest one-day strengths did not necessarily provide best strengths at 28 days.
RD107	1992	*Effects of Conventional and High-Range Water Reducers on Concrete Properties*	D. Whiting and W. Dziedzic	Use of chemical admixtures to reduce both water and cement contents resulted in accelerated rates of slump loss and shorter working times compared to controls. When used to produce flowing concretes, working times were equivalent to those for mixes not containing the admixtures. In general, setting times were increased from 1 to 2 h at 23°C (73°F) and by lesser amounts at 32°C (90°F). Bleeding of flowing concretes was greater than that of control mixtures, especially when a copolymer-type HRWR was used. Rate of air loss was significantly greater in cement-reduced mixtures containing HRWR compared with controls; however, air loss rates in flowing concretes were roughly equivalent for all mixtures tested.
RD104	1994	*Engineering Properties of Commercially Available High-Strength Concretes (Including Three-Year Data)*	R.G. Burg and B.W. Ost	Six concretes in the range of 69 MPa to 138 MPa (10,000 psi to 20,000 psi) were studied. Properties of temperature rise, compressive strength, static modulus, tensile strength, modulus of rupture, thermal expansion, drying shrinkage, freeze-thaw durability, relative humidity, specific heat, thermal conductivity, and water absorption were tested. Specimens were also tested for creep, drying shrinkage, rapid chloride permeability, resistivity, corrosion rate, and petrographic analysis.
RD119	2001	*Long-Term Performance of Plain and Reinforced Concrete in Seawater Environments*	D. Stark	Summary on the long-term performance of plain and reinforced concrete in seawater. Observations of 37- to 39-year exposures revealed excellent performance of the concrete mixtures. Of singular importance, however, was longitudinal cracking due to corrosion of embedded reinforcing steel. This cracking was more severe in concrete beams located above high-tide range than companion beams immersed in seawater. It was evident that both chloride ion and dissolved oxygen in seawater induced corrosion and cracking in the beams.
RD122	2001	*Frost and Scaling Resistance of High-Strength Concrete*	R.C.A. Pinto and K.C. Hover	Work assessed the effect of air entrainment and time of surface finishing operations on the frost durability and scaling resistance of high-strength concrete. No SCMs were used. Frost resistance was investigated as an interior concrete property, via modified ASTM C666, and as a surface property, via ASTM C672. Mixtures studied with w/c of 0.25 and no intentionally entrained air were shown to be frost resistant. Further, properly air-entrained mixtures with w/c of 0.50 were frost resistant.

ID	YEAR	TITLE	AUTHOR	SUMMARY OF FINDINGS
SN2526	2005	Chemical Path of Ettringite Formation in Heat-cured Mortar and Its Relationship to Expansion	Y. Shimada	Study conducted at Northwestern University on delayed ettringite formation (DEF) within mortar systems with different changes. Research concluded that normal aging process of the cementitious systems involves dissolution of uniformly distributed fine ettringite crystals within the hardened cement paste and subsequent recrystallization as innocuous crystals in the largest accessible spaces. This process, known as Ostwald ripening, facilitates relaxation of expansion pressure and is impeded in well compacted dense microstructures. Therefore, alteration of the mortar parameters to enhance microstructure for improved mechanical strength increases the risk of DEF-related expansion. The potential for later expansion depends on monosulfate availability, diffusion of ettringite constituents, and the degree of supersaturation of pore solution with respect to ettringite.
SN3011	2007	Life Cycle Inventory of Portland Cement Concrete	M.L. Marceau, M.A. Nisbet, and M.G. VanGeem	This report presents results of an Life Cycle Inventory (LCI) of three concrete products: ready mixed concrete, concrete masonry, and precast concrete. More accurate data availability results in LCI results that are lower for most of the flows reported in previous reports.
SN3174	2011	Methods, Impacts, and Opportunities in the Concrete Pavement Life Cycle	N. Santero, A. Loijos, M. Akbarian, and J. Ochsendorf	Life-cycle assessment approaches for pavements are developed in order to quantify current greenhouse gas (GHG) emissions, identify opportunities for improvement, and estimate the cost-effectiveness of potential reduction strategies. A model for pavement structural properties that relate to vehicle fuel consumption is developed, highlighting the use-phase impacts of pavement selection.
SN3119	2012	Life-Cycle Evaluation of Concrete Building Construction as a Strategy for Sustainable Cities	E. Masanet, A. Stadel, and P. Gursel	Report provides a concise overview of the development and use of a life-cycle assessment (LCA) model for structural materials in U.S. commercial buildings—the Berkeley Lab Building Materials Pathways (B-PATH) model. The report summarizes literature review findings, methods development, model use, and recommendations for future work in the area of LCA for commercial buildings.
SN3240	2012	Mechanistic Approach to Pavement–Vehicle Interaction and Its Impact on Life-Cycle Assessment	M. Akbarian, S.S. Moeini-Ardakani, F.J. Ulm, and M. Nazzal	The deflection response of a pavement system and its relationship to pavement structure and material (Pavement Vehicle Interaction or PVI) can be linked to a change in fuel consumption with empirical studies. Calibration and validation of the deflection model were performed against the falling weight deflectometer (FWD) time history data recorded by FHWA's Long-Term Pavement Performance program. Report highlights the importance of considering the use phase impacts, as concrete pavement's stiffness and longer service life can result in lower overall greenhouse gas emissions.
SN3256	2014	Carbon Sequestration in Old and New Portland Cement Concrete Pavement Interiors	L. Hasselback and A. Alam	Pavement samples that were from one to 85 years old were sliced and paste/mortar specimens were chipped from different depths. CO_2 sequestration was estimated using a thermo-gravimetric analysis (TGA) for the carbonate levels and X-ray fluorescence analysis (XRF) for elemental analysis. The results indicate that the amount of carbon sequestration in the interior of these pavements is in the early years of life around 10% of that released during calcinations increasing to 25% or more in arid climates.
SN3315	2017	Calcium Oxychloride Formation Potential in Cementitious Pastes Exposed to Blends of Deicing Salt	P. Suraneni, J. Monical, E. Unal, Y. Farnam, and J. Weiss	Suggests two strategies to mitigate the amount of calcium oxychloride that may be formed in cementitious matrix: reduction in the amount of calcium hydroxide in the pastes through use of supplementary cementitious materials, and the use of deicing salt blends that include lower amounts of calcium chloride. A model is developed to estimate the amount of calcium oxychloride formed in mixtures, given the calcium hydroxide and calcium chloride contents.
SN3342	2020	Evaluation of Cement Soundness using the ASTM C151 Autoclave Expansion Test	H. Kabir and R.D. Hooton	Study noting lack of correlation between the autoclave expansion test and concrete soundness issues in the field. Research concluded that the AET provides an unrealistic basis for assessing volume stability as well as for rejection of portland cements for use in concrete.

REFERENCES

PCA's online catalog includes links to PDF versions of many of our research reports and other classic publications. Visit: cement.org/library/catalog.

Abrams, D.A., *Design of Concrete Mixtures*, Lewis Institute, Structural Materials Research Laboratory, Bulletin No. 1, PCA LS001, 1918, 20 pages.

Abrams, D.A., *Calcium Chloride as an Admixture in Concrete*, Lewis Institute, Structural Materials Research Laboratory, Bulletin No. 13, PCA LS013, Chicago, IL, 1924, 57 pages.

Abrams, D.A., *Studies of Bond Between Concrete and Steel*, Lewis Institute, Structural Materials Research Laboratory, Bulletin No. 17, PCA LS017, Chicago, IL, 1925.

Akbarian, M.; Moeini-Ardakani, S.S.; Ulm, F.J.; and Nazzal, M., "Mechanistic Approach to Pavement-Vehicle Interaction and Its Impact on Life-Cycle Assessment," *Transportation Research Record: Journal of the Transportation Research Board*, No. 2306, 2012, pages 171 to 179.

Ashby, M.F., *Materials and the Environment: Eco-informed Material Choice*, Butterworth-Heinemann, Burlington, MA, 2009, 400 pages.

Aspdin, J., *Artificial Stone*, British Patent No. 5022, December 15, 1824, 2 pages.

Auburn, *Historical Timeline of Concrete*, AU BSC 314, Auburn University, June 2000.

Bogue, R.H., *Calculation of Compounds in Portland Cement*, Portland Cement Association Fellowship, National Bureau of Standards, Paper No. 21, October, 1929, 8 pages.

Brewer, H.W., *Moisture Migration – Concrete Slab-On-Ground Construction*, Development Department Bulletin DX089, Portland Cement Association, 1965.

Brown, G.E., *Analysis and History of Cement*, Gordon E. Brown Associates, Keswick, Ontario, 1996, 259 pages.

Brown, L.S., *Some Observations on the Mechanics of Alkali-Aggregate Reaction*, Research Department Bulletin RX054, Portland Cement Association, 1955.

Burg, R.G.; and Ost, B.W., *Engineering Properties of Commercially Available High-Strength Concrete (Including Three-Year Data)*, Research and Development Bulletin RD104, Portland Cement Association, 1994, 58 pages.

Dahl, L.A., *Estimation of Phase Composition of Clinker in the System $3CaO \bullet SiO_2$-$2CaO \bullet SiO_2$-$3CaO \bullet Al_2O_3$-$4CaO \bullet Al_2O_3 \bullet Fe_2O_3$ at Clinkering Temperatures*, Research Department Bulletin RX001, Portland Cement Association, 1938

Gebler, S.H.; and Klieger, P., *Effect of Fly Ash on Some of the Physical Properties of Concrete*, RD089, Portland Cement Association, 1986.

Gonnerman, H.F., *Effect of Size and Shape of Test Specimen on Compressive Strength of Concrete*, Lewis Institute, Structural Materials Research Laboratory, Bulletin No. 16, PCA LS016, Chicago, IL, 1925, 16 pages.

Gonnerman, H.F.; and Shuman, E.C., "Flexure and Tension Tests of Plain Concrete," Major Series 171, 209, and 210, *Report of the Director of Research*, Portland Cement Association, November 1928, pages 137 to 200.

Gustaferro, A.H.; Abrams, M.S.; and Litvin, Albert, *Fire Resistance of Lightweight Insulating Concretes*, Research and Development Bulletin RD004, Portland Cement Association, 1970.

Gustaferro, A.H.; Abrams, M.S.; and Salse, E.A.B., *Fire Resistance of Prestressed Concrete Beams, Study C: Structural Behavior During Fire Test*, Research and Development Bulletin RD009, Portland Cement Association, 1971.

Hansen, T.C.; and Mattock, A.H., *Influence of Size and Shape of Member on the Shrinkage and Creep of Concrete*, Development Department Bulletin DX103, Portland Cement Association, Skokie, IL, 1966, 42 pages.

Hanson, N.W.; and Conner, H.W., *Seismic Resistance of Reinforced Concrete – A Laboratory Test Rig*, Development Department Bulletin DX107, Portland Cement Association, 1966.

Haselbach, L.; and Alam, A., "Carbon Sequestration in Old and New Portland Cement Concrete Pavement Interiors," *ASCE Construction Institute: Innovative Materials and Design for Sustainable Transportation Infrastructure*, 2015, pages 71 to 82.

Kabir, H.; and Hooton, R.D., *Evaluation of Cement Soundness Using the ASTM C151 Autoclave Expansion Test*, SN3342, Portland Cement Association, Skokie, IL, 2020, 34 pages.

Kriz, L.B., *Ultimate Strength Criteria For Reinforced Concrete*, Development Department Bulletin DX031, Portland Cement Association, 1959.

Lerch, W., *The Influence of Gypsum on the Hydration and Properties of Portland Cement Pastes*, Research Department Bulletin RX012, Portland Cement Association, 1946.

Lerch, W., *Hot Cement and Hot Weather Concrete*, IS015.01T, Portland Cement Association, Skokie, IL, 1955, 11 pages.

Marceau, M.L.; Nisbet, M.A.; and VanGeem, M.G., *Life Cycle Inventory of Portland Cement Concrete*, SN3011, Portland Cement Association, Skokie, IL, 2007, 69 pages.

Masanet, E.; Stadel, A.; and Gursel, P., *Life-Cycle Evaluation of Concrete Building Construction as a Strategy for Sustainable Cities*, SN3119, Portland Cement Association, Skokie, IL, 2012, 87 pages.

McMillan, F.R.; Tyler, I.L.; Hansen, W.C.; Lerch, W.; Ford, C.L.; and Brown, L.S., *Long-Time Study of Cement Performance in Concrete*, Research Department Bulletin RX026, Portland Cement Association, 1948.

Menzel, C.A., *Procedure for Determining the Air Content of Freshly-Mixed Concrete by the Rolling and Pressure Methods*, Research Department Bulletin RX019, Portland Cement Association, 1947.

PCA, *Index of Research Department Bulletins 1-227: Annotated List with Author and Subject Index*, RX228, Portland Cement Association, Skokie, IL, July 1969, 130 pages.

PCA, *Index of Development Department Bulletins D1-D146: Annotated List with Author and Subject Index*, DX147, Portland Cement Association, Skokie, IL, December 1970, 65 pages.

PCA, *Index of Research and Development Bulletins 1939-1989, Including Lewis Institute Series*, 1918-1925, RD100, Portland Cement Association, Skokie, IL, 1990, 134 pages.

PCA, *Survey of Portland Cement Consumption by User Group*, Portland Cement Association, Fourth Quarter 2022, 43 pages.

PCA, *2023 U.S. Cement Industry Annual Yearbook*, Portland Cement Association, Skokie, IL, 2023, 54 pages.

Pinto, R.C.A.; and Hover, K.C., *Frost and Scaling Resistance of High-Strength Concrete*, Research and Development Bulletin RD122, Portland Cement Association, 2001, 70 pages.

Powers, T.C., *The Air Requirements of Frost-Resistant Concrete*, Research Department Bulletin RX033, Portland Cement Association, 1949.

Powers, T.C., *The Physical Structure and Engineering Properties of Concrete*, Research Department Bulletin RX090, Portland Cement Association, 1958.

Powers, T.C.; and Brownyard, T.L., *Studies of the Physical Properties of Hardened Portland Cement Paste*, Research Department Bulletin RX022, Portland Cement Association, 1947.

Powers, T.C.; Copeland, L.E.; Hayes, J.C.; and Mann, H.M., *Permeability of Portland Cement Pastes*, Research Department Bulletin RX053, Portland Cement Association, 1954.

Powers, T.C.; Dahl, L.H., *The Bleeding of Portland Cement Paste, Mortar and Concrete Treated As a Special Case of Sedimentation*, Research Department Bulletin RX002, Portland Cement Association, 1939, 171 pages.

Shimada, Y.E., *Chemical Path of Ettringite Formation in Heat Cured Mortar and Its Relationship to Expansion*, Ph.D. Thesis, Northwestern University, Evanston, IL, 2005 [PCA SN2526].

Snell, L.M.; and Snell, B.G., *The Erie Canal – America's First Concrete Classroom*, 2000.

Stark, D., *Characteristics and Utilization of Coarse Aggregates Associated with D-Cracking*, RD047, Portland Cement Association, Skokie, IL, 1976, 9 pages.

Stark, D., *Longtime Study of Concrete Durability in Sulfate Soils*, RD086, Portland Cement Association, Skokie, IL, 1982, 15 pages.

Stark, D., *Influence of Design and Materials on Corrosion Resistance of Steel in Concrete*, Research and Development Bulletin RD098, Portland Cement Association, 1989, 44 pages.

Stark, D., *Long-Term Performance of Plain and Reinforced Concrete in Seawater Environments*, Research and Development Bulletin RD119, Portland Cement Association, 2001, 14 pages.

Steinour, H.H., *Concrete Mix Water – How Impure Can It Be?*, Research Department Bulletin RX119, Portland Cement Association, 1960, 20 pages.

Suraneni, P.; Monical, J.; Unal, E.; Farnam, Y.; and Weiss, J., "Calcium Oxychloride Formation Potential in Cementitious Pastes Exposed to Blends of Deicing Salt," *ACI Materials Journal*, Vol. 114, No. 4, July-August 2017, pages 631 to 641.

Tang, F.J., *Optimization of Sulfate Form and Content*, Research and Development Bulletin RD105, Portland Cement Association, 1992, 44 pages.

Verbeck, G.J., *Carbonation of Hydrated Portland Cement*, Research Department Bulletin RX087, Portland Cement Association, 1958, 21 pages.

Verbeck, G.; and Klieger, P., *Studies of "Salt" Scaling of Concrete*, Research Department Bulletin RX083, Portland Cement Association, 1956, 14 pages.

White, C., *Hydraulic Cement*, U. S. patent, 1820.

Whiting, D.; and Dziedzic, W., *Effects of Conventional and High-Range Water Reducers on Concrete Properties*, Research and Development Bulletin RD107, Portland Cement Association, 1992, 25 pages.

PORTLAND, BLENDED, AND OTHER HYDRAULIC CEMENTS

FIGURE 2-1. Aerial view of a cement plant.

Portland cement is a key ingredient in most concrete for general construction and is composed primarily of hydraulic calcium silicates. Hydraulic cements set and harden by chemically reacting with water and maintain their stability underwater. Portland cement and blended cement are common hydraulic cements and the focus of this chapter.

MANUFACTURE OF PORTLAND CEMENT

The past century has seen dramatic developments in nearly every aspect of cement manufacturing (Bhatty and others 2011). Process improvements have greatly improved the uniformity and quality of portland cement. One particularly significant improvement is the rate of strength gain of the finished product (Bhatty and Tennis 2008). Quality control (Moore 2007) has continued to improve throughout the history of portland cement production. Commercial manufacturers now offer a wider range of cements, including blended cements, suitable for numerous applications. The cement industry has focused resources towards reducing the environmental impact of cement production, particularly carbon dioxide generation and energy efficiency, while maintaining performance characteristics, as described in Chapter 12.

Figure 2-1 shows an aerial view of a modern cement manufacturing plant. Portland cement clinker is manufactured by combining precise proportions of raw materials. These materials are then fired in a rotary kiln at high temperatures to form new chemical compounds which are hydraulic in nature. These compounds (phases) are formed from oxides of calcium, silica, alumina, and iron. Clinker is then ground with other ingredients to make cement.

Stages in the manufacture of cement are illustrated in Figure 2-2.

OVERVIEW OF CEMENT PLANT OPERATIONS

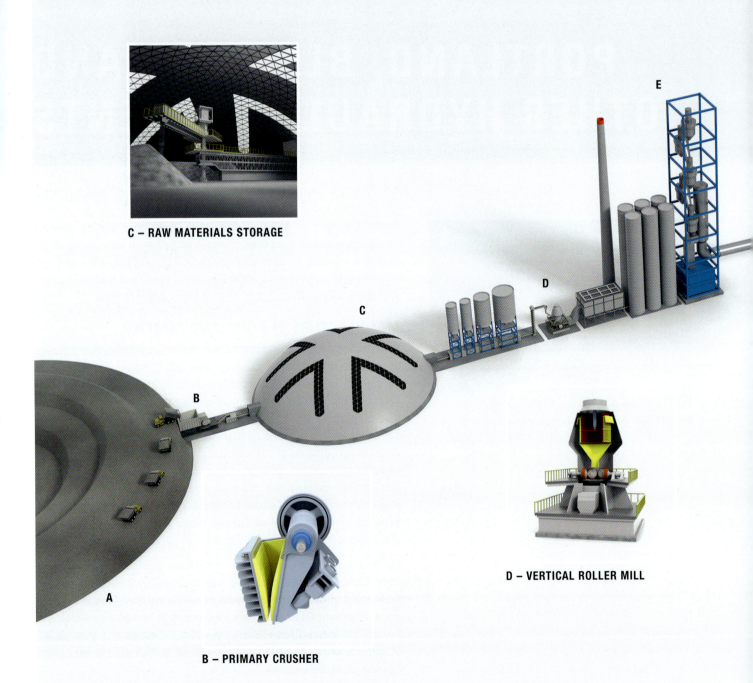

C – RAW MATERIALS STORAGE

D – VERTICAL ROLLER MILL

B – PRIMARY CRUSHER

FIGURE 2-2. Overview of typical cement plant operations: (A) quarrying of primary raw materials; (B) crushing of raw materials; (C) raw materials storage and blending; (D) raw material grinding and blending; (E) preheating/precalcining; (F) rotary kiln processing; (G) clinker cooling; (H) clinker storage; (I) blending and finish grinding; (J) cement storage and transportation to concrete production facilities; and (K) baghouse systems collect dust at several points in the process. For an interactive version see: cement.org/cement-concrete/how-cement-is-made.

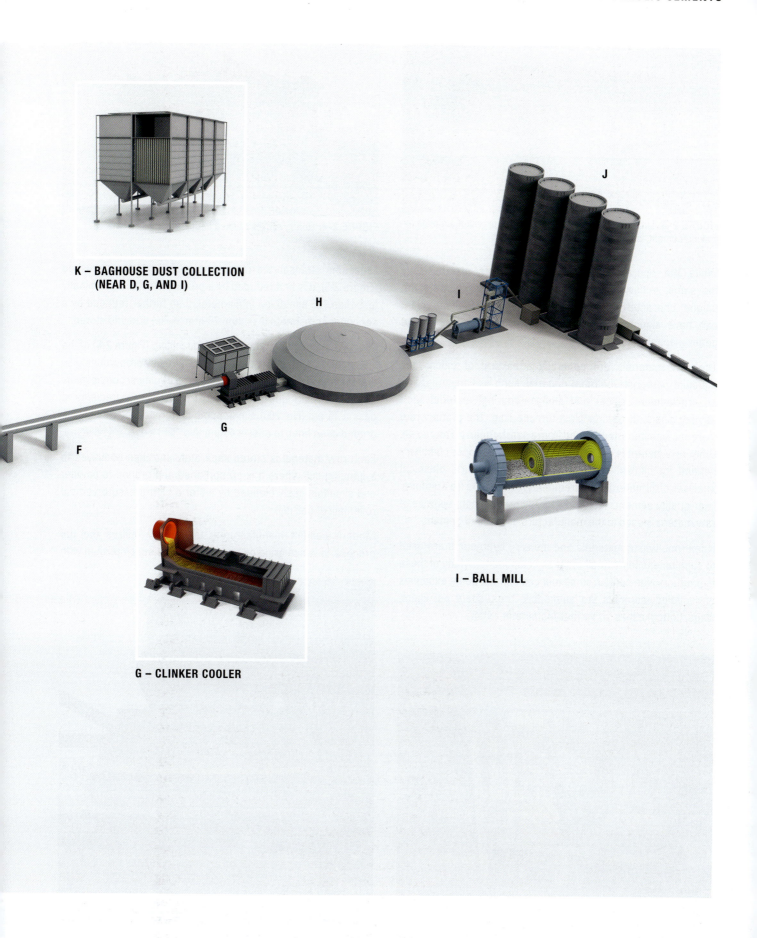

K – BAGHOUSE DUST COLLECTION
(NEAR D, G, AND I)

J

H

I

G

F

I – BALL MILL

G – CLINKER COOLER

FIGURE 2-3. Limestone, a primary raw material providing calcium for making cement, is quarried near the cement plant.

FIGURE 2-4. Quarry rock is trucked to the primary crusher.

While the operations of all cement plants are basically the same, a flow diagram cannot adequately illustrate the unique characteristics of each cement plant. Every plant can have significant differences in layout, equipment, and general appearance.

The raw materials are generally a mixture of calcareous (calcium carbonate-bearing) material, such as limestone, and an argillaceous (silica- and alumina-bearing) material such as clay, shale, fly ash, or blast-furnace slag. The primary raw material, limestone, is obtained from a nearby quarry (Figure 2-3). Other raw materials are sourced to obtain the proper chemistry needed for clinker formation. During manufacture, chemical analyses of all materials are frequently made to ensure a uniform, high-quality cement. Table 2-1 lists the predominant sources of raw materials used in the manufacture of portland cement.

Numerous waste materials and industry byproducts are used to supplement some or all of these components. Many of these byproducts also contain some fuel component. Some examples of recycled materials are automobile tires, spent pot liners, slags, bottom ashes, or various incinerator ashes.

Limestone is transported from the quarry (Figure 2-4) and crushed (Figure 2-5). It is first reduced by a primary crusher to nominally less than 125 mm (5 in.) in diameter, then further reduced by a secondary crusher to less than 20-mm (¾-in.) size and stored.

The next stage of manufacture is final milling (Figure 2-6) of the limestone and other raw materials, followed by proportioning so that the resulting mixture has the desired chemical composition. The final milling step reduces most of the raw ingredients to pass a 75-μm (No. 200) sieve. Some materials may need to be ground even finer, to pass a 45-μm (No. 325) sieve.

Each raw material is stored separately and then conveyed to a grinding mill where it is proportioned and ground to powder and simultaneously blended by either a high-pressure vertical roller mill or ball mill.

Modern cement manufacturing technology utilizes the dry process, in which grinding and blending are accomplished with

FIGURE 2-5. The first stage of comminution is primary crushing, where quarried rock is reduced to less than 125 mm (5 in.) in diameter. Secondary crushing results in further reduction to less than 20 mm (¾ in.) in diameter.

FIGURE 2-6. Final raw feed milling reduces the raw feed to pass a 75-μm (No. 200) sieve.

 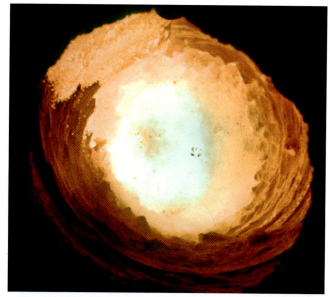

FIGURE 2-7. (left) Rotary kiln (furnace) for manufacturing portland cement clinker. (right) View inside the kiln where material temperatures reach to 1400°C to 1550°C (2550°F to 2800°F).

dry materials. In a wet process, the raw material grinding and blending operations are performed with the materials mixed with water to form a slurry. In other respects, the dry and wet processes are similar.

Pyroprocessing is a process in which the raw materials are subjected to high temperatures in order to bring about a chemical or physical change. In cement manufacturing, pyroprocessing requires kiln temperatures that can reach 1550°C (2800°F) or higher.

After blending, the ground raw materials are fed into the upper or back end of a kiln (Figure 2-7). The raw mix passes through the kiln at a rate controlled by the slope and rotational speed of the kiln. Fuel (pulverized coal, petroleum coke, new or recycled petroleum oil, natural gas, tire-derived fuel, or other byproduct fuel) is forced into the lower end of the kiln where it ignites and generates material temperatures of 1400°C to 1550°C (2550°F to 2800°F). Most modern kilns (post 1980) have increased energy efficiencies by preheating raw mix prior to introduction to the kiln via a series of vertical heat exchange devices known as preheater cyclones and precalciner vessels (Figure 2-8). Rotary kilns in the newest plants are only used for the final burning zone reactions.

As the raw materials are heated in the kiln, a number of chemical reactions occur during calcination, changing the raw material into portland cement clinker. Figure 2-9 illustrates the clinker production process from raw feed to the final product.

In the final state of pyroprocessing, the clinker is rapidly cooled (Figure 2-10). During clinker cooling, ambient temperature air is passed over the clinker as it is transported over a grate system. This allows the air to be preheated and recycles some of the energy back into the process. The rapid cooling of the clinker quenches in the cement phases formed at high temperature. The clinker is often stored before final grinding.

In the final stage of manufacture the clinker and other cement ingredients are proportioned and then pulverized into a fine material, typically by ball mills (Figure 2-11). During the finish grinding stage, small amounts of calcium sulfate (typically gypsum), are added to regulate the setting time of the cement and to improve shrinkage and strength development properties (Lerch 1946 and Tang 1992). Limestone may be added, in amounts up to 5% by mass for ASTM C150 portland cements, and up to 15% by mass for ASTM C595 Type IL, portland-limestone cements. Inorganic processing additions can be added to portland and blended in amounts up to 5% if qualified by ASTM C465. Organic processing additions may be added in amounts up to 1% by mass.

FIGURE 2-8. A 5-stage preheater cyclone tower, often used with precalciner. These technologies have improved the energy efficiency of cement production.

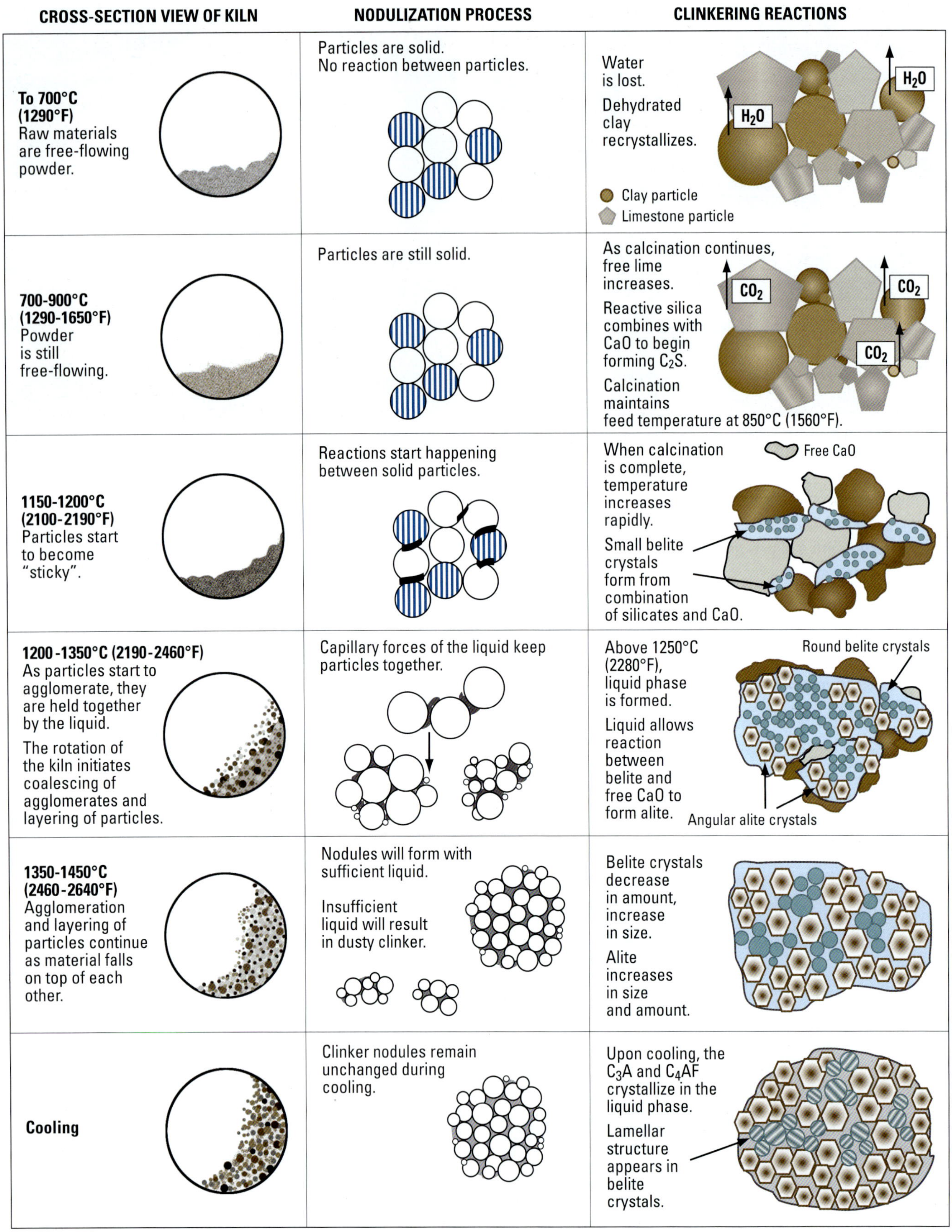

CROSS-SECTION VIEW OF KILN	NODULIZATION PROCESS	CLINKERING REACTIONS
To 700°C (1290°F) Raw materials are free-flowing powder.	Particles are solid. No reaction between particles.	Water is lost. Dehydrated clay recrystallizes. ⬤ Clay particle ⬡ Limestone particle
700-900°C (1290-1650°F) Powder is still free-flowing.	Particles are still solid.	As calcination continues, free lime increases. Reactive silica combines with CaO to begin forming C_2S. Calcination maintains feed temperature at 850°C (1560°F).
1150-1200°C (2100-2190°F) Particles start to become "sticky".	Reactions start happening between solid particles.	When calcination is complete, temperature increases rapidly. Small belite crystals form from combination of silicates and CaO. Free CaO
1200-1350°C (2190-2460°F) As particles start to agglomerate, they are held together by the liquid. The rotation of the kiln initiates coalescing of agglomerates and layering of particles.	Capillary forces of the liquid keep particles together.	Above 1250°C (2280°F), liquid phase is formed. Liquid allows reaction between belite and free CaO to form alite. Round belite crystals Angular alite crystals
1350-1450°C (2460-2640°F) Agglomeration and layering of particles continue as material falls on top of each other.	Nodules will form with sufficient liquid. Insufficient liquid will result in dusty clinker.	Belite crystals decrease in amount, increase in size. Alite increases in size and amount.
Cooling	Clinker nodules remain unchanged during cooling.	Upon cooling, the C_3A and C_4AF crystallize in the liquid phase. Lamellar structure appears in belite crystals.

FIGURE 2-9. Process of clinker production from raw feed to the final product (after Hills 2000).

TABLE 2-1. Sources of Raw Materials Used in Manufacture of Portland Cement

CALCIUM	IRON	SILICA	ALUMINA	SULFATE
Alkali waste	Blast-furnace flue dust	Calcium silicate	Aluminum-ore refuse*	Anhydrite
Aragonite*	Clay*	Cement rock	Bauxite	Calcium sulfate
Calcite*	Iron ore*	Clay*	Cement rock	Gypsum*
Cement-kiln dust	Mill scale*	Fly ash	Clay*	Synthetic gypsum
Cement rock	Ore washings	Fuller's earth	Copper slag	
Chalk	Pyrite cinders	Limestone	Fly ash*	
Clay	Shale	Loess	Fuller's earth	
Fuller's earth		Marl*	Granodiorite	
Limestone*		Ore washings	Limestone	
Marble		Quartzite	Loess	
Marl*		Rice-hull ash	Ore washings	
Seashells		Sand*	Shale*	
Shale*		Sandstone	Slag	
Slag		Shale*	Staurolite	
		Slag		
		Traprock		

*Most common sources.
Note: Many industrial byproducts have potential as raw materials for the manufacture of portland cement.

FIGURE 2-10. A clinker cooler rapidly reduces the clinker temperature. (inset) Heated clinker prior to cooling.

FIGURE 2-11. A ball mill used for finish grinding at a cement plant. (inset) Grinding media in first comparment of ball mill.

These additions provide significant environmental benefits as well as manufacturing improvements (Hawkins and others 2005 and Taylor 2008).

The clinker and all other ingredients are ground so fine that nearly all of the material passes through a 45-µm (No. 325) sieve. This extremely fine gray powder is now portland cement (Figure 2-12). If supplementary cementing materials (SCM) (for example, slag cement, fly ash, silica fume, or other pozzolans) or additional limestone are added to portland cement, either by blending or during the finish grinding, a blended hydraulic cement is produced. The finished cement is stored in silos until

FIGURE 2-12. Finish grinding of hydraulic cements results in almost all of the material passing a 45-µm (No. 325) sieve.

transported by ship, rail, or truck to a concrete plant. Some cement is bagged prior to transport as well. See **Handling Cement** for more information.

HYDRAULIC CEMENT INGREDIENTS

Although the goal of the quarrying, raw material grinding, proportioning, pyroprocessing, and cooling processes is to produce portland cement clinker, other ingredients are included during the finish grinding, or ground separately and blended, to produce a final hydraulic cement. The major categories of cement ingredients are described below.

Portland Cement Clinker

The foundation of most hydraulic cements for general construction is portland cement clinker: the grayish black pellets predominantly the size of marbles or coarse sand (Figure 2-13). This clinker consists predominantly of four chemical phases (compounds): tricalcium silicate, or alite (abbreviated C_3S); dicalcium silicate, or belite (C_2S); tricalcium aluminate (C_3A); and tetracalcium aluminoferrite (C_4AF). These phases are discussed in further detail under **Portland Cement Clinker Phases (Compounds)**.

FIGURE 2-13. Portland cement clinker is formed by burning calcium and siliceous raw materials in a kiln. This particular clinker averages about 20 mm (¾ in.) in diameter.

Gypsum

Gypsum (calcium sulfate dihydrate, Figure 2-14), is the predominant source of sulfate used in cement. Sulfate is added to react with C_3A to form ettringite (calcium trisulfoaluminate hydrate). This controls the rapid hydration of C_3A. Without sulfate, cement would set much too rapidly and normal property development would be interrupted, and concrete placement would be more difficult. In addition to controlling setting and early strength gain, the sulfate also helps control drying shrinkage and can influence strength through 28 days (Lerch 1946 and Tang 1992). Typically, about 4% to 8% by mass of cement is calcium sulfate, with higher amounts sometimes required for finer cements. Finish grinding mill temperatures

are often controlled to produce a balance of calcium sulfate hemihydrate (also called hemihydrate, bassanite, or plaster of paris) and gypsum.

In addition to natural forms of calcium sulfate, synthetic gypsum (also known as syngyp) is commonly used in the production of cements. These are often produced as byproducts of other industries. For example, flue gas desulfurization (FGD) gypsum is produced while scrubbing stack gases of coal fired power plants with calcium carbonate.

FIGURE 2-14. Gypsum, a form of calcium sulfate, is interground with portland cement clinker to form portland cement. It helps control setting, drying shrinkage properties, and strength development.

Limestone

Limestone may be used as an ingredient in portland cement in amounts up to 5% by mass, and between 5% and 15% in a blended Type IL cement. This limestone is not calcined in the cement kiln, but is interground (or finely ground and blended) with the portland cement clinker and other ingredients. Since this limestone is not pyroprocessed, energy requirements to produce cement are reduced and other environmental benefits are achieved for the cement, without loss of performance. Although limestone is relatively inert, a small part of it does react chemically. Limestone typically broadens the particle size distribution of a cement, improving the particle packing at a fine level.

In 2014, about 60% of U.S. portland cements included limestone as an ingredient, with an average amount of 3.1% (Tennis 2016). For Type IL blended cements, the limestone content in 2014 averaged 11.1% (see Table 2-2). See Tennis and others (2011), Hooton and Thomas (2010), Matschei and Glasser (2006), Hawkins and others (2005), and Bharadwaj and others (2021) for more information on use of limestone in hydraulic cements.

Supplementary Cementitious Materials

Supplementary cementitious materials (SCMs) used in blended cements include slag cement and pozzolans. Pozzolans used in the U.S. include fly ash, silica fume, and natural pozzolans. These materials generally react slower than portland cement phases, but contribute to long-term strength and durability of concretes. Many SCMs are byproducts of other industries and thus their beneficial reuse in blended cements also improves the sustainability characteristics of blended cements as well. Supplementary cementitious materials contents vary depending on the specific material (See **Types of Blended Hydraulic Cements**). Chapter 3 includes more information on SCMs and Chapter 12 discusses sustainability in detail.

Processing Additions

Processing additions are added during the finish grinding of cements to assist with manufacture. Both organic and inorganic processing additions may be used. A typical organic processing addition is a grinding aid and is typically used in amounts of a few tenths of percent by mass or less. Grinding aids improve efficiency in the cement production process by satisfying surface charges on particles, minimizing their tendency to coat grinding media. Inorganic processing additions can be included in amounts up to 5% by mass. A typical inorganic processing addition is granulated blast-furnace slag. Since it is harder than clinker, it also helps grinding efficiency, acting as a sort of small-scale grinding media. ASTM C465, *Standard Specification for Processing Additions for Use in the Manufacture of Hydraulic Cements* (AASHTO M 327), requirements must be met for organic processing additions and for inorganic processing additions used in amounts greater than 1% by mass in portland cements.

Functional Additions

Functional additions can be interground during final grinding to beneficially change the properties of hydraulic cement. These additions must meet the requirements of ASTM C226, *Standard Specification for Air-Entraining Additions for Use in the Manufacture of Air-Entraining Hydraulic Cement*, or ASTM C688, *Standard Specification for Functional Additions for Use in Hydraulic Cements*. ASTM C226 addresses air-entraining additions while ASTM C688 addresses the following types of additions: water-reducing, retarding, accelerating, water-reducing and retarding, water-reducing and accelerating, and set-control additions. These additions can be used to enhance the performance of the cement for normal or special concrete construction, grouting, and other applications. For more on admixtures for use in concrete see Chapter 6.

TYPES OF PORTLAND CEMENTS

Different types of portland cement are manufactured to meet various physical and chemical requirements for specific purposes. Portland cements are manufactured to meet the requirements of ASTM C150, *Standard Specification for Portland Cement* (AASHTO M 85). The requirements of AASHTO M 85 and ASTM C150 are harmonized.

ASTM C150 and AASHTO M 85 provides for six basic types of portland cement:

Type I	General use
Type II	Moderate sulfate resistance
Type II(MH)	Moderate heat of hydration and moderate sulfate resistance
Type III	High early strength
Type IV	Low heat of hydration
Type V	High sulfate resistance

Some cements are also labeled with more than one type designation, for example Type I/II. This simply means that such a cement meets the requirements of Type I and Type II cement.

A detailed review of ASTM C150 (AASHTO M 85) portland cement types follows.

Type I

Type I portland cement is a general-purpose cement suitable for all uses where the special properties of other cement types are not required. Its uses in concrete include pavements, floors, reinforced concrete buildings, bridges, tanks, reservoirs, pipe, masonry units, and precast concrete products (Figure 2-15).

FIGURE 2-15. Typical uses for normal or general use cements include (left to right) highway pavements, floors, bridges, and buildings.

FIGURE 2-16. Moderate sulfate resistant cements and high sulfate resistant cements improve the sulfate resistance of concrete elements, such as (left to right) slabs on ground, pipe, and concrete posts exposed to high-sulfate soils.

Type II

Type II portland cement is used where protection against moderate sulfate attack is necessary. It is used in structures or elements exposed to soil or ground waters where sulfate concentrations are higher than normal, but not unusually severe (Figure 2-16). Type II cement has moderate sulfate resistant properties because it contains no more than 8% tricalcium aluminate (C_3A).

Sulfates in soil or water may enter the concrete and react with hydrated C_3A phases, resulting in expansion, scaling, and cracking of concrete. Some sulfate compounds, such as magnesium sulfate, directly attack calcium silicate hydrate.

To effectively control sulfate attack, the use of Type II (or Type V) cement in concrete must be accompanied by the use of a low water-to-cementitious materials ratio (to achieve low permeability). Figure 2-17 illustrates the improved sulfate

resistance of Type II cement over Type I cement. Chapter 11 includes more detailed guidance on designing concrete for sulfate exposure applications.

Concrete exposed to seawater is often made with Type II cement. Seawater contains significant amounts of sulfates and chlorides. Although sulfates in seawater are capable of attacking concrete, the presence of chlorides in seawater inhibits the expansive reaction characteristic of sulfate attack. Thus, a marine environment is an application requiring a *moderate* sulfate exposure class. Observations from a number of sources show that the performance of concretes in seawater with portland cements having C_3A contents as high as 10% have demonstrated satisfactory durability, provided the permeability of the concrete is low and the reinforcing steel has adequate cover (Zhang and others 2003). Chapter 11 provides more information on sulfate attack.

Type II(MH)

Type II(MH) cements are manufactured to generate heat at a slower rate than Type I or most Type II cements by limiting the heat index to a maximum of 100. The heat index is a function of the clinker phases equal to $C_3S + 4.75 \, C_3A$ (% by mass). This requirement is roughly equivalent to 3-day heat of hydration measurements of 315 kJ/kg using ASTM C1702, *Standard Test Method for Measurement of Heat of Hydration of Hydraulic Cementitious Materials Using Isothermal Conduction Calorimetry*. As an alternate to the heat index requirement, an optional requirement can be specified limiting the heat of hydration to 335 kJ/kg at 3 days determined by ASTM C1702. Type II(MH) (or Type IV, if available) may often be used in mass concrete structures, such as large piers, large foundations, and thick retaining walls. Using MH cements can reduce temperature rise and peak temperature, and minimize temperature related cracking. Thermal control is also especially important when concrete is placed in warm weather (see Chapter 18).

Type II(MH) cements are also moderately sulfate resistant, as the maximum tricalcium aluminate content (C_3A) is limited to a maximum of 8% by mass.

Cement content = 390 kg/m³ (658 lbs/yd³)

- ● ASTM Type II, 8% C_3A w/c = 0.38
- ■ ASTM Type I, 13% C_3A w/c = 0.39
- ▲ ASTM Type V, 4% C_3A w/c = 0.37

FIGURE 2-17. Performance of concretes made with different cements in sulfate soil. Type II and Type V cements have lower C_3A contents that improve sulfate resistance when used in appropriate concrete mixtures (Stark 2002).

FIGURE 2-18. High early strength cements are used where early concrete strength is needed, such as in (left to right) cold weather concreting, fast track paving to minimize traffic congestion, and rapid form removal for precast concrete.

Type III

Type III portland cement provides strength at an earlier age than other cement types. For example, design strength may be achieved in a matter of days as compared to the typical requirements of 28 days. Type III is typically chemically similar to Type I cement, except that its particles have been ground finer. It is used to achieve high-early strengths for early form removal and where a structure must be put into service quickly. In cold weather its use permits a reduction in the length of the curing period (Figure 2-18).

Type IV

Type IV portland cement was developed for use in applications where the rate and amount of heat generated from hydration must be minimized. It develops heat of hydration and strength at a slower rate than other cement types. Type IV cement is intended for use in massive concrete structures, such as large gravity dams (Figure 2-19), where the temperature rise resulting from heat generated during hardening must be minimized (see Chapter 18). Type IV cement is not commonly manufactured in North America as other measures for controlling heat rise in concrete are now more commonly available, including the use of blended cements and SCMs (see Chapter 3).

FIGURE 2-19. Moderate heat and low heat cements minimize heat generation in massive elements or structures such as (left) very thick bridge supports, and (right) dams. Hoover Dam, shown here, used a Type IV cement to control temperature rise.

Type V

Type V portland cement is used in concrete exposed to severe sulfate environments – principally where soils or groundwaters have a high sulfate content. It tends to gain strength more slowly than Type I cement. (See Chapter 11 for sulfate exposures requiring the use of Type V cement.) The high sulfate resistance of Type V cement is attributed to a low tricalcium aluminate (C_3A) content, not more than 5%. The use of a low water to cementitious materials ratio and low permeability are critical to the performance of any concrete exposed to sulfates. Even Type V cement concrete cannot withstand a severe sulfate exposure if the concrete has a high water-cementitious materials ratio (Figure 2-20 left). Type V cement, like other portland cements, is not resistant to acids and other highly corrosive substances.

FIGURE 2-20. Specimens used in the outdoor sulfate test plot in Sacramento, California, are 150 × 150 × 760-mm (6 × 6 × 30-in.) beams. A comparison of deterioration ratings is illustrated: (left) a rating of 5 for 12-year old concretes made with Type V cement and a water-to-cement ratio of 0.65; and (right) a rating of 2 for 16-year old concretes made with Type V cement and a water-to-cement ratio of 0.37 (Stark 2002).

ASTM C150 (AASHTO M 85) allows both a chemical approach (for example, limiting C_3A) and a physical approach (ASTM C452, *Standard Test Method for Potential Expansion of Portland-Cement Mortars Exposed to Sulfate*) to assure the sulfate resistance of Type V cement. Either the chemical or the physical approach can be specified, but not both.

Air-Entraining Portland Cements

Specifications for four types of air-entraining portland cement (Types IA, IIA, II(MH)A, and IIIA) are given in ASTM C150 (AASHTO M 85). They correspond in composition to ASTM C150 Types I, II, II(MH), and III, respectively, except that small quantities of air-entraining additions are interground with the clinker during manufacture. These cements can produce concrete with improved resistance to freezing and thawing. When mixed with proper intensity and for an appropriate duration, such concrete contains fine, well-distributed, and completely separated air bubbles. Air entrainment for most concretes is achieved through the use of an air-entraining admixture added when batching concrete, rather than through the use of air-entraining cements (see Chapter 6). Therefore, it is advisable to check on local availability of air-entraining portland cements before specifying them.

White Portland Cements

White portland cement differs from gray portland cement chiefly in color. It is made to conform to the requirements of ASTM C150 (AASHTO M 85), usually Type I or Type III. The manufacture of white cement is accomplished by limiting the amount of iron and manganese in the raw materials. These two elements are responsible for portland cement's characteristic gray color.

Special processing, such as water quenching of clinker and use of ceramic grinding media in finish milling, might also be required in the manufacture of white cement. White portland cement is used primarily for architectural purposes in structural walls, precast, and glass fiber reinforced concrete (GFRC) facing panels, terrazzo surfaces, stucco, cement paint, tile grout, and decorative concrete (Figure 2-21). Its use is recommended wherever white or colored concrete, grout, or mortar is desired and should be specified as white portland cement meeting the requirements of ASTM C150 (AASHTO M 85), Type I, II, III, or V. See Farny (2001) for more information on white cement.

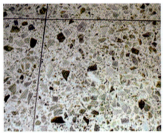

FIGURE 2-21. White portland cement is used in white or light-colored architectural concrete, ranging from (left to right) terrazzo for floors shown here with white cement and green granite aggregate, to decorative and structural precast and cast-in-place building (courtesy of Condell Medical Center).

TYPES OF BLENDED HYDRAULIC CEMENTS

ASTM C595, *Standard Specification for Blended Hydraulic Cements* (AASHTO M 240) recognizes four primary classes of blended cements as follows:

Type IL(X)	Portland-limestone cement
Type IP(X)	Portland-pozzolan cement
Type IS(X)	Portland blast-furnace slag cement
Type IT(AX)(BY)	Ternary blended cement

The letters "X" and "Y" represent the nominal mass percentage of the ingredient included in the blended cement with the remaining mass percentage being portland cement. For example, a cement designated as Type IS(50) contains 50% by mass of slag cement (and 50% portland cement). For ternary blended cements, "A" is the type of ingredient present in the largest amount ("X") while "B" is the other ingredient type. "A" or "B" may be either S for slag cement, P for pozzolan, or L for limestone. "X" and "Y" are the corresponding percentages of those materials. As an example, Type IT(L10)(P10) indicates a ternary blended cement with 10% limestone and 10% pozzolan. In this example, portland cement accounts for 80% of the total cementitious content. Table 2-2 lists the average amounts of limestone, slag cement, and pozzolans for U.S. blended cements.

TABLE 2-2. Nominal Amount of Ingredient in U.S. Blended Cements Reported for 2014 (Tennis 2016).

INGREDIENT/CEMENT TYPE		CONTENT, % by mass
Limestone/Type IL	mean	11.1
	range	10–14
Pozzolan/Type IP	mean	23.5
	range	9–32
Slag cement/Type IS	mean	30.3
	range	20–40

*Data based on 11 Type IL cements, 6 Type IP and 6 Type IS cements. No data for Type IT cements was provided in the survey.

Blended hydraulic cements are produced by intimately and uniformly intergrinding or blending two or more types of fine materials. The primary materials are portland cement, limestone, slag cement, fly ash, silica fume, calcined clay, other pozzolans, hydrated lime, and preblended combinations of these materials (Figure 2-22). The requirements of AASHTO M 240 are nearly identical to those in ASTM C595.

FIGURE 2-22. Blended cements use a combination of portland cement or clinker and gypsum blended or interground with fly ash or other pozzolans, slag, or limestone. Shown is blended cement (center) surrounded by (right and clockwise) clinker, gypsum, portland cement, fly ash, slag, silica fume, and calcined clay.

Blended cements' share of total U.S. cement consumption have increased substantially since 2021 (Figure 2-23). Use of portland-limestone cement (Type IL) has increased from less than 5% in 2021 to over 50% as of July 2023. USGS has estimated that Type IL accounts for more than 95% of blended cements in 2023.

Blended cements are used in all aspects of concrete construction in the same manner as portland cements. Blended cements can be used as the sole cementitious material in concrete or they can be used in combination with SCMs added at the concrete plant. In general, blended cements can provide an advantage over using portland cements with SCMs added during concrete batching in that the finished cement undergoes quality control checks at the cement plant. Blended cements optimized for fineness and sulfate content, for example, minimize incompatibility issues. See Klieger and Isberner 1967 and PCA 1995 for additional information on blended cements.

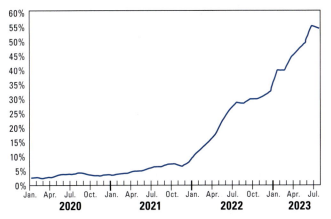

FIGURE 2-23. Blended cements' share of total U.S. cement consumption have increased substantially since 2021 (USGS 2023).

An overview of ASTM C595 and AASHTO M 240 blended cement types follows.

Type IL

Portland-limestone cements are blended cements which contain more than 5% but less than or equal to 15% by mass of finely ground limestone as an ingredient. These cements perform similarly to Type I cements and offer sustainability advantages (Bushi and Meil 2014). Through additional testing, Type IL cements may be qualified having special properties (see **Blended Cements with Special Properties**). For more information on portland-limestone cements, see Tennis and others (2011) and Bharadwaj and others (2021).

Type IP

Portland-pozzolan cements are designated as Type IP. Type IP cement may be used for general construction. The pozzolan content of these cements is up to 40% by mass. Performance of concrete made with Type IP cement as a group is similar to or better than that of Type I cement concrete. Type IP may be produced with special properties (see **Blended Cements with Special Properties**).

Type IS

Portland blast-furnace slag cement is classified into two categories depending on its slag cement content: Type IS(< 70) has less than 70% by mass of slag cement and is suitable for general concrete construction, while Type IS(≥70) has 70% slag cement or more by mass, may contain hydrated lime, and is used in conjunction with portland cement in making concrete or with lime in making mortar, but is not used alone for structural concrete applications. ASTM C595 (AASHTO M 240) includes subcategories for optional special properties for Type IS(< 70) cements (see **Blended Cements with Special Properties**).

Type IT

Ternary blended cements include, along with portland cement, two of the following: limestone, pozzolan, or slag cement. Ternary cements can sometimes offer advantages over binary blended cements. The combinations of different cementing materials provided by ternary blended cements may optimize the overall performance of these cements for a specific application. For example, a rapidly reacting pozzolan like silica fume might be combined with a slower reacting material such as fly ash to enhance both early and longer-term strength development. Other concrete properties might be improved as well, such as reduced permeability and resistance to ASR.

Type IT(S≥70) is limited to a maximum limestone content of 15% and can contain hydrated lime. All other Type IT cements are limited to a maximum of 70% total pozzolan, slag cement, and limestone, content with pozzolan limited to 40% and limestone limited to 15%. Type IT ternary blended cements meet the same chemical and physical requirements as for binary blended cements for the non-portland cement ingredient present in the largest amount. For instance, if slag cement is present in the highest amount in a Type IT(S<70) cement, the provisions of Type IS(<70) apply, while if a pozzolan is present in the highest amount, then the requirements of Type IP apply. The provisions of Type IP also apply if the pozzolan and slag cement content are the same.

Blended Cements with Special Properties

Special properties may be specified for blended cement types IL, IP, IS(<70), and IT(S<70), and are indicated by suffixes as follows:

A – air-entraining cements
MS – moderate sulfate resistance
HS – high sulfate resistance
MH – moderate heat of hydration
LH – low heat of hydration
HE – high early strength

For example, an air-entraining portland blast-furnace slag cement with 40% slag that has high sulfate resistance would be designated as Type IS(40)(HS)(A).

Requirements for these special properties are identified in ASTM C595 (AASHTO M 240). Availability should be verified before specifying, as cements with these special properties are not produced in all areas.

Historic Nomenclature for Blended Cements

Older literature may refer to blended cement Types I(PM) or I(SM), which were referred to as "pozzolan-modified portland cement" or "slag-modified portland cement," respectively. Type I(PM) contained less than 15% pozzolan and Type I(SM) contained less than 25% slag cement. Likewise, Type P and Type S are no longer defined in ASTM C595 (AASHTO M 240). These references became obsolete with revisions to ASTM C595 (AASHTO M 240) in 2006 to include the amount of pozzolan or slag in the nomenclature for the cement type. References to these cement types should be avoided in favor of current, and more descriptive, cement type designations.

TYPES OF PERFORMANCE-BASED HYDRAULIC CEMENTS

All portland and blended cements are hydraulic cements. "Hydraulic cement" is merely a broader term indicating the formation of solid hydrates following reaction with water. ASTM C1157, *Standard Performance Specification for Hydraulic Cement,* is a performance-based specification and can apply to both portland cement and blended hydraulic cements. Cements specified under ASTM C1157 meet physical performance test requirements, as opposed to prescriptive restrictions on ingredients or cement chemistry as found in other cement specifications. ASTM C1157 provides for six types of hydraulic cement as follows:

Type GU	General use
Type HE	High early strength
Type MS	Moderate sulfate resistance
Type HS	High sulfate resistance
Type MH	Moderate heat of hydration
Type LH	Low heat of hydration

In addition, these cements can also include an Option R – Low Reactivity with Alkali-Reactive Aggregates – specified to help control expansion due to alkali-silica reactivity (see Chapter 11). An "R" is appended to the cement type designation: for example, Type GU-R is a general use hydraulic cement that has low reactivity when used with alkali-reactive aggregates. Requirements for air-entraining cements are also defined in ASTM C1157 and are indicated with an "A" appended to the cement type.

A detailed review of ASTM C1157 cements follows:

Type GU

Type GU is a general-purpose cement suitable for all applications where the special properties of other types are not required. Its uses in concrete include pavements, floors, reinforced concrete buildings, bridges, pipe, precast concrete products, and other applications where Type I is used.

Type HE

Type HE cement provides higher strengths at an early age, usually a week or less. It is used in the same manner as Type III portland cement and has the same 1- and 3-day minimum strength requirements.

Type MS

Type MS cement is used where precaution against moderate sulfate attack is important. Applications may include drainage structures where sulfate concentrations in ground waters are higher than normal but not unusually severe (see Chapter 11). It is used in the same manner as Type II and II(MH) portland cement (although Type MS cements do not necessarily have moderate heat of hydration). Like Type II and II(MH), concrete made with Type MS cement must be designed and placed with a low water-cementitious materials ratio to provide adequate sulfate resistance. ASTM C1157 also requires that mortar bar specimens demonstrate resistance to expansion while exposed to solution containing a high concentration of sulfates, as per ASTM C1012, *Standard Test Method for Length Change of Hydraulic-Cement Mortars Exposed to a Sulfate Solution.*

Type HS

Type HS cement is used in concrete exposed to severe sulfate action – principally where soils or ground waters have a high sulfate content (see Chapter 11). It is used in the same manner as Type V portland cement. As noted previously, low water-cementitious materials ratios are critically important to assure performance of concrete exposed to sulfate exposures. Similar to MS cements, ASTM C1157 also requires that mortar bar specimens demonstrate resistance to expansion while exposed to solution containing a high concentration of sulfates per ASTM C1012.

Type MH

Type MH cement is used in applications requiring a moderate heat of hydration and a controlled temperature rise. Type MH cement is used in the same manner as Type II(MH) portland cement, or MH-designated blended cements where moderate heat of hydration is desired.

Type LH

Type LH cement is used where the rate and amount of heat generated from hydration must be minimized. It develops strength at a slower rate than other cement types. Type LH cement is intended for use in massive concrete structures where the temperature rise resulting from heat generated during hardening must be minimized. It is used as an alternative to Type IV portland cement or in the same manner as a LH-designated blended cement.

Masonry and Mortar Cements

Masonry cements and mortar cements are hydraulic cements designed for use in mortar for masonry construction. They consist of a mixture of portland cement or blended hydraulic cement and plasticizing materials (such as limestone, hydrated lime, or hydraulic lime), together with other materials introduced to enhance one or more properties such as setting time, workability, water retention, and durability. These components are proportioned and packaged at a cement plant under controlled conditions to assure uniformity of performance.

Masonry cements meet the requirements of ASTM C91, *Standard Specification for Masonry Cement*. ASTM C91 classifies masonry cements as Type N, Type S, and Type M. White masonry cement and colored masonry cements meeting ASTM C91 are also available in some areas.

The increased use of masonry in demanding structural applications, such as high seismic areas, resulted in the development of ASTM C1329, *Standard Specification for Mortar Cement*. ASTM C91 and ASTM C1329 requirements for compressive strength and air content limits of masonry cement and mortar cements are identical. However, ASTM C1329 also includes bond strength performance criteria for mortar cements.

Masonry and mortar cements are further classified as Type N, Type S, and Type M. A brief description of each type follows:

- Type N masonry cement and Type N mortar cement are used to produce ASTM C270, *Standard Specification for Mortar for Unit Masonry*, Type N and Type O mortars. They may also be used with portland or blended cements to produce Type S and Type M mortars.
- Type S masonry cement and Type S mortar cement are used to produce ASTM C270, Type S mortar. They may also be used with portland or blended cements to produce Type M mortar.
- Type M masonry cement and Type M mortar cement are used to produce ASTM C270, Type M mortar without the addition of other cements or hydrated lime.

Types N, S, and M generally have increasing levels of portland cement and higher strength with Type M providing the highest strength. Type N masonry and mortar cements are most commonly used for above grade masonry.

The workability, strength, and color of masonry cements and mortar cements remain at a high level of uniformity through the use of modern manufacturing controls. In addition to mortar for masonry construction, masonry cements and mortar cements are commonly used for parging. Masonry cements are also used in portland cement-based plaster or stucco (Figure 2-24) construction (see ASTM C926, *Standard Specification for Application of Portland Cement-Based Plaster*). Masonry cement and mortar cement are not suitable for use in concrete production. For more information on concrete masonry and mortar cements, refer to PCA's *Concrete Masonry Handbook* (Farny and others 2008).

Plastic Cements

Plastic cement is a hydraulic cement that meets the requirements of ASTM C1328, *Standard Specification for Plastic (Stucco) Cement*. It is used to produce portland cement-based plaster or stucco (ASTM C926) (Figure 2-24). Plastic cements consist of a mixture of portland or blended hydraulic cement and materials such as limestone, hydrated or hydraulic lime, which act as plasticizing agents. Other ingredients may be added to enhance one or more properties such as setting time, workability, water retention, and durability.

FIGURE 2-24. Masonry cement and plastic cement are used to make plaster or stucco for commercial, institutional, and residential buildings. Shown is a church with a stucco exterior.

ASTM C1328 defines separate requirements for Type M and Type S plastic cement with Type M having higher strength requirements. The International Building Code (IBC) does not classify plastic cement into different types but does specify that it must meet the requirements of ASTM C1328. When plastic cement is used, lime or any other plasticizer may not be added to the plaster at the time of mixing.

The term "plastic" in plastic cement does not refer to the inclusion of any organic compounds in the cement. Instead, "plastic" refers to the ability of the cement to introduce a higher degree of workability ("plasticity") to the plaster. Plaster using this cement must remain workable long enough for it to be reworked to obtain the desired densification and texture. Plastic cement should not be used to make concrete. For more information on the use of plastic cement and plaster, see PCA's *Portland Cement Plaster/Stucco Manual* (Melander and others 2003).

SPECIAL CEMENTS

Special cements are produced for specific applications. Table 2-3 summarizes the special cements discussed below. See ACI ITG-10.1R, *Report on Alternative Cements* (2018b), Odler (2000) and Klemm (1998) for more information.

Calcium Aluminate Cements

Calcium aluminate cement is not portland cement based. It is used in special applications for early strength gain, resistance to high temperatures, and resistance to sulfates, weak acids, and seawater. Portland cement and calcium aluminate cement combinations have been used to make rapid setting concretes and mortars. Typical applications for calcium aluminate cement concrete include: chemically resistant, heat resistant, and corrosion resistant industrial floors; refractory castables; and repair applications. Standards addressing these cements include ASTM C1600, *Standard Specification for Rapid Hardening Hydraulic Cement*, and European Standard EN 14647, *Calcium Aluminate Cement: Composition, Specifications and Conformity Criteria*.

Calcium aluminate cement concrete must be used at low water-to-cement ratios (less than 0.40); this minimizes the potential for conversion of less stable hexagonal calcium aluminate hydrate (CAH_{10}) to the stable cubic tricalcium aluminate hydrate (C_3AH_6), hydrous alumina (AH_3), and water. Given sufficient time and particular moisture conditions and temperatures, this conversion causes a 53% decrease in volume of hydrated material. However, this internal volume change occurs without a dramatic alteration of the overall dimensions of a concrete element, resulting in increased paste porosity and decreased compressive strength. At low water-to-cement ratios, there is insufficient space for all the calcium aluminate to react and form CAH_{10}. The released water produced by the conversion reacts with additional calcium aluminate, partially compensating for the effects of conversion. Concrete design strength must therefore be based on the converted strength. Because of the unique concerns associated with the conversion phenomenon, calcium aluminate cement is generally used in nonstructural applications and is rarely, if ever used within structural applications (Taylor 1997).

Carbonated Calcium Silicate Cements

Although carbonated calcium silicate cements (CCSC) are non-hydraulic cements that harden through the reaction of carbon dioxide with calcium silicates, they are mixed with aggregates and water and formed like conventional concrete. Water is then removed from the mixture while the concrete is in a closed chamber filled with carbon dioxide. Although this is an emerging technology, it is most suited for some types of precast products and some have been manufactured commercially. See ACI ITG-10R, *Practitioner's Guide for Alternative Cements* (2018a) and ACI ITG-10.1R (2018b) for more information.

TABLE 2-3. Applications for Special Cements

SPECIAL CEMENTS	TYPE	APPLICATION
Calcium aluminate cement		Repair, chemical resistance, high temperature exposures
Carbonated calcium silicates		Some precast products (not a hydraulic cement)
Cements with functional additions, ASTM C595 (AASHTO M 240), ASTM C1157		General concrete construction needing special characteristics such as: water-reducing, retarding, air entraining, set control, and accelerating properties
Ettringite cements		Waste stabilization*
Expansive cements, ASTM C845	E-1(K), E-1(M), E-1(S)	Shrinkage compensating concrete
Finely ground (ultrafine) cement		Geotechnical grouting*
Geopolymer cement		General construction, repair, waste stabilization*
Magnesium phosphate cement		Repair and chemical resistance
Natural cement, ASTM C10	Natural cement, Quick-setting natural cement	Historic restoration of natural cement mortar, cement plaster, grout, whitewash, and concrete
Rapid hardening hydraulic cement, ASTM C1600	URH, VRH, MRH, GRH, RH-CAC	General paving where very rapid (about 4 hours) strength development is required
Regulated-set cements		Early strength and repair*
Sulfur cements		Repair and chemical resistance
Water-repellent cements		Tile grout, paint, and stucco finish coats
Well cements, API 10A	A, B, C, D, G, H, K, L	Grouting wells

* Portland and blended hydraulic cements are also used for these applications.

Cements with Functional Additions

Functional additions can be interground with cement clinker to beneficially change the properties of hydraulic cement. These additions must meet the requirements of ASTM C226, *Standard Specification for Air-Entraining Additions for Use in the Manufacture of Air-Entraining Hydraulic Cement*, or ASTM C688, *Standard Specification for Functional Additions for Use in Hydraulic Cements*. ASTM C226 addresses air-entraining additions while ASTM C688 addresses the following types of additions: water-reducing, retarding, accelerating, water-reducing and retarding, water-reducing and accelerating, and set-control additions. Cement specifications ASTM C595, AASHTO M 240, and some other specifications allow functional additions. These additions can be used to enhance the performance of the cement for normal or special concrete construction, grouting, and other applications.

Ettringite Cements

Ettringite cements are calcium sulfoaluminate cements that are specially formulated for particular uses, such as the stabilization of waste materials (Klemm 1998). They can be formulated to form large amounts of ettringite to stabilize particular metallic ions within the ettringite structure. Ettringite cements have also been used in rapid setting applications, including use in coal mines. Also see **Expansive Cements**.

Expansive Cements

Expansive cement is a hydraulic cement that expands slightly during the early hardening period after initial set. It must meet the requirements of ASTM C845, *Standard Specification for Expansive Hydraulic Cement*, where it is designated as Type E-1. Three varieties of expansive cement are recognized in ASTM C845. They are designated as K, M, and S. These designations are added as a suffix to the type. Type E-1(K) contains portland cement, tetracalcium trialuminosulfate, calcium sulfate, and uncombined calcium oxide (lime). Type E-1(M) contains portland cement, calcium aluminate cement, and calcium sulfate. Type E-1(S) contains portland cement with a high tricalcium aluminate content and calcium sulfate.

Expansive cement may also contain formulations other than those noted. The expansive properties of each type can be varied over a considerable range – refer to ACI 223R, *Guide for the Use of Shrinkage-Compensating Concrete* (2010).

When expansion is restrained, for example by reinforcing steel, expansive cement concrete (also called shrinkage compensating concrete) can be used to:

- compensate for the volume decrease associated with drying shrinkage,
- induce tensile stress in reinforcement (post-tensioning), and
- stabilize the long-term dimensions of post-tensioned concrete structures in comparison to the original design dimensions.

Expansive cement can control and reduce drying shrinkage cracks commonly associated with the use of portland cement. Figure 2-25 illustrates the length change (early expansion and drying shrinkage) history of shrinkage-compensating concrete and conventional portland cement concrete. For more information see ACI 223R (2010), Odler (2000), Russell (1978), and Pfeifer and Perenchio (1973).

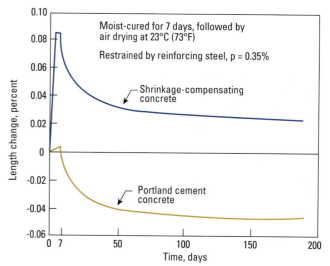

FIGURE 2-25. Length-change history of shrinkage compensating concrete containing Type E-1(S) cement and Type I portland cement concrete (Pfeifer and Perenchio 1973).

Finely-Ground Cements (Ultrafine Cements)

Finely-ground cements, also referred to as ultrafine cements, are hydraulic cements that are ground very fine for use in grouting into fine soil or thin rock fissures (Figure 2-26). The cement particles are less than 10 µm in diameter with 50% of particles less than 5 µm. Blaine surface area of these cements often exceeds 800 m²/kg. These very fine cements consist of portland cement, slag cement, and other mineral additives.

FIGURE 2-26. A slurry of finely ground cement and water can be injected into the ground, as shown here, to stabilize in-place materials, to provide strength for foundations, or to chemically retain contaminants in soil.

Geopolymer Cements

Geopolymer cements are inorganic hydraulic cements that are based on the polymerization of minerals (Davidovits and others 1999). The term more specifically refers to alkali-activated aluminosilicate cements, also called zeolitic or polysialate cements. These cements often contain industrial byproducts, such as fly ash. They have been used in general construction, high-early strength applications, and waste stabilization. These cements do not contain organic polymers or plastics.

Magnesium-Based Cements

Magnesium-based cements include magnesium phosphate and magnesium oxychloride (also known as Sorel cement). Magnesium phosphate cement is a rapid setting, early strength gain cement. It is usually used for special applications, such as repair of pavements and concrete structures, or for resistance to certain aggressive chemicals. It does not contain portland cement. Magnesium oxychloride cements are rapid strength cement with high early and late strengths, but show limited resistance to water.

Natural Cements

Natural cements were used in the 1800s before portland cement became popular. Natural cement is a coarse ground hydraulic cement with slow strength gain properties and was used in mortar, plaster, whitewashing, and concrete. It was used extensively in canal, bridge, and building construction. It is still available for use in restoring historic structures and must meet ASTM C10, *Standard Specification for Natural Cement.*

Rapid Hardening Cements

Rapid hardening, high-early strength, hydraulic cement is used in construction applications, such as fast-track paving, where fast strength development is required (design or load-carrying strength in about four hours). Rapid hardening cements are specified by performance according to ASTM C1600, and classified as Type URH – ultra rapid hardening, Type VRH – very rapid hardening, Type MRH – medium rapid hardening, Type GRH – general rapid hardening, and Type RH-CAC – rapid hardening incorporating calcium aluminate.

Regulated-Set Cements

Regulated-set cement is a calcium fluoroaluminate hydraulic cement that can be formulated and controlled to produce concrete with setting times ranging from a few minutes to one hour and with corresponding rapid early strength development (Greening and others 1971). It is a portland-based cement with functional additions that can be manufactured in the same kiln used to manufacture conventional portland cement. Regulated-set cement incorporates set control and early-strength-development components. The final physical properties of the resulting concrete are in most respects similar to comparable concretes made with portland cement.

Sulfur Cements

Sulfur cement is used with conventional aggregates to make sulfur cement concrete for repairs and chemically resistant applications. Sulfur cement melts at temperatures between 113°C and 121°C (235°F and 250°F). Sulfur concrete is maintained at temperatures around 130°C (270°F) during mixing and placing. The material gains strength quickly as it cools and is resistant to acids and aggressive chemicals. Sulfur cement does not contain portland or hydraulic cement.

Water-Repellent Cements

Water-repellent cements, sometimes called waterproofed cements, are usually made by adding a small amount of water-repellent additive such as stearate (sodium, aluminum, or other) to cement clinker during final grinding (Lea 1971). Chapter 6 provides detail on some of these additives. Manufactured in either white or gray color, water-repellent cements reduce capillary water transmission provided there is little to no hydrostatic pressure. However, they do not stop water-vapor transmission. Water-repellent cements are used in tile grouts, paint, stucco finish coats, and in the manufacture of specialty precast units.

Well Cements

Well cements (sometimes referred to as oil-well cements) are used for grouting oil wells and other types of wells. (This procedure is often called oil-well cementing.) Well cements are usually made from portland cement clinker or from blended hydraulic cements. Generally, they must be slow-setting and resistant to high temperatures and pressures. The American Petroleum Institute's *Specification for Cements and Materials for Well Cementing* (API Specification 10A) includes requirements for eight classes of well cements (Classes A, B, C, D, G, H, K, or L) and three grades (Grades O – ordinary, MSR – moderate sulfate resistant, and HSR – high sulfate resistant). Each class is applicable for use at a certain range of well depths, temperatures, pressures, and sulfate environments. The petroleum industry also uses conventional types of portland cement with suitable cement-modifiers. (ASTM C150 Type II or Type V cements may meet requirements for API Class G and H cements, while ASTM C150 Types I, II, and III cements are similar to API Class A, Class B, and Class C cements, respectively.) Class K well cements are combinations of portland cement and finely ground silica, while Class L well cements are portland cement and fly ash or pozzolan blends. See API Specification 10A (2015) for more detail. Expansive cements have also performed adequately as well cements.

SELECTING AND SPECIFYING CEMENTS

When specifying cements for a project, the availability of cement types should be verified. Specifications should allow flexibility in cement selection. Limiting a project to only one cement type, one brand, or one standard cement specification can result in project delays and it may not allow for the best use of local materials. Cements with special properties should not be required unless special characteristics are necessary. In addition, the use of SCMs should not inhibit the use of any particular portland or blended cement. The project specifications should focus on the needs of the concrete structure and allow use of a variety of materials to accomplish those needs. A typical specification may call for portland cements meeting ASTM C150 (AASHTO M 85), blended cements meeting ASTM C595 (AASHTO M 240), or for performance-based hydraulic cements meeting ASTM C1157.

If no special properties (such as low heat generation or sulfate resistance) are required, all general use cements should be permitted, including: Types I, GU, IS(<70), IP, IL, and IT(S<70). Also, it should be noted that some cement types meet more than one specification requirement. For example, nearly all Type II cements meet the requirements of Type I, but not all Type I cements meet the requirements of Type II. Some cements are marketed as Type I/II cements, indicating that they meet all of the specification requirements for both Type I and Type II.

Table 2-4 provides a matrix of commonly used cements and their typical applications in concrete construction. Often, cements with special properties can also be used in general purpose applications and cements with high sulfate resistance can be used in moderate sulfate exposures. For example, at Type HS cement can be used in moderate sulfate exposures.

Availability of Cements

Some types of cement may not be readily available in all areas of the United States. ASTM C150 (AASHTO M 85) Type II portland cement was historically the most widely available in the U.S. Type I cement is also very common. As of July 2023, Type IL presents over 50% of the cement produced in the United States. Type III cement and white cement are usually available in larger metropolitan areas and represent about 4% of cement produced in the U.S. Type IV cement is not readily available.

Type V cement comprised about 15% of cement produced in the U.S. in 2021 and is typically manufactured only in regions of the country having high sulfate environments. Air-entraining cements use has decreased as the popularity of air-entraining admixtures has increased. See Sullivan and others (2023) for statistics on cement use.

If a given cement type is not available, comparable results may be obtained with another cement type that is available. For example, high-early-strength concrete can be made using concrete mixtures with higher cement contents and lower water-to-cement ratios. Also, the effects of heat of hydration can be minimized using leaner mixtures, placing smaller lifts, artificial cooling, or by adding an SCM to the concrete. Because of the range of phase composition permitted in cement specifications, a Type II cement may also meet the requirements of a Type V. Options for producing concrete for moderate or high sulfate exposures also include blended cements with MS or HS designations or using suitable types and quantities of SCMs with portland cements, although additional testing may be required for this latter option (see Chapters 3 and 11). In all cases, a low water-to-cementitious materials ratio and adequate curing is also necessary to achieve low permeability.

Blended cements can be obtained throughout most of the United States. However, certain types of blended cement may not be available in some areas. When ASTM C595 (AASHTO M 240) blended cements are desired, but are not available, similar properties can often be obtained by adding pozzolans meeting requirements of ASTM C618, *Standard Specification for Coal Ash and Raw or Calcined Natural Pozzolan for Use in Concrete* (AASHTO M 295); slag cement meeting requirements of ASTM C989, *Standard Specification for Slag Cement for Use in Concrete and Mortars* (AASHTO M 302); silica fume meeting ASTM C1240, *Standard Specification for Silica Fume Used in Cementitious Mixtures* (AASHTO M 307); a ground glass pozzolan meeting ASTM C1866, *Standard Specification for Ground-Glass Pozzolan for Use in Concrete*; or a blended

TABLE 2-4. Applications for Hydraulic Cements Used in Concrete Construction*

CEMENT SPECIFICATION	GENERAL PURPOSE	MODERATE HEAT OF HYDRATION	HIGH EARLY STRENGTH	LOW HEAT OF HYDRATION	MODERATE SULFATE RESISTANCE	HIGH SULFATE RESISTANCE	AIR-ENTRAINING
ASTM C150 portland cements	I	II(MH)	III	IV	II, II(MH)	V	IA, IIA, II(MH)A, IIIA
ASTM C595 blended hydraulic cements	IL, IP, IS(<70), IT(S<70)	IL(MH), IP(MH), IS(<70)(MH), IT(S<70)(MH)	IL(HE), IP(HE), IS(<70)(HE), IT(S<70)(HE)	IL(LH), IP(LH), IS(<70)(LH), IT(S<70)(LH)	IL(MS), IP(MS), IS(<70)(MS), IT(S<70)(MS)	IL(HS), IP(HS), IS(<70)(HS), IT(S<70)(HS)	Option A[†]
ASTM C1157 hydraulic cements	GU	MH	HE	LH	MS	HS	Option A[†]

* Check the local availability of specific cement types as some cements are not available everywhere.
† Any cement in the columns to the left can be used to make air-entraining cement.

SCM meeting ASTM C1697, *Standard Specification for Blended Supplementary Cementitious Materials*, when batching concrete along with ordinary portland cement. Like any concrete mixture, these mixes should be tested for time of set, strength gain, durability, and other properties prior to their use in construction (see Chapter 3).

The availability of all special cements should be verified before specifying.

Drinking Water Applications. Concrete has demonstrated decades of safe use in drinking water applications. Materials in contact with drinking water must meet special requirements to control chemicals entering the water supply. Some localities may require that cement or concrete meet the special requirements of the NSF International/American National Standards Institute/Standards Council of Canada standard NSF/ANSI/CAN 61, *Drinking Water System Components – Health Effects*. Availability of cement meeting NSF/ANSI/CAN 61 requirements of should be verified before specifying.

Specifying International Cement Standards

In some instances, projects in the United States designed by engineering firms from other countries refer to cement standards other than those in ASTM or AASHTO. There is typically no direct equivalency between ASTM and international cement standards because of differences in test methods and limits on required properties. However, imported cements can be tested to confirm conformance with ASTM and AASHTO specifications.

Canadian Cement Specifications. In Canada, portland cements are manufactured to meet the specifications of the Canadian Standards Association (CSA) Standard A3001 *Cementitious Materials for Use in Concrete*. CSA A3001 includes portland cements, blended cements, portland-limestone cements, and portland-limestone blended cements as different categories of cements. Provisions of the specifications and referenced test methods are largely similar to ASTM standards, but not completely identical.

CSA A3001 designates portland cements similar to ASTM C1157 with requirements similar to ASTM C150. Four types of portland cement are specified in CSA A3001 and are identified as follows:

- GU General use hydraulic cement
- MS Moderate sulfate-resistant hydraulic cement
- HE High early-strength hydraulic cement
- HS High sulfate-resistant hydraulic cement

The nomenclature for blended hydraulic cements is a three-letter descriptive designation (GUb, MSb, MHb, HEb, LHb, and HSb) to address its equivalent performance to portland cements with up to three supplementary cementing materials. Upon request, the designations for blended cements can also provide information on the composition of blended hydraulic cements. The designations then follow the form: BHb-Axx/Byy/Czz, where BHb is the blended hydraulic cement type, and xx, yy, and zz are the supplementary materials used in the cement in proportions A, B, and C, respectively.

In addition, CSA A3001 includes provisions for portland-limestone cements (PLC) with 5% to 15% by mass of limestone. Four types of portland-limestone cements are defined and are designated by an L suffix: Type GUL, Type HEL, Type HSL, and Type MSL. CSA A3001 also includes provisions for portland-limestone blended cements, which contain between 5% and 15% limestone, as well as SCMs. Type designations are GULb, HELb, MSLb, and HSLb. See Chapter 3 for more information on SCMs defined in CSA A3001.

European Cement Specifications. The European cement standard, EN 197-1, *Cement-Part 1: Composition, Specifications and Conformity Criterial for Common Cements*, may be specified for use in contract documents. EN 197-1 defines 27 types of cement and uses base types of CEM I, II, III, IV, and V. These types do *not* correspond to the cement types in ASTM C150, nor can ASTM cements be substituted for EN specified cement without the designer's approval. CEM I is a portland cement and CEM II through V are blended cements (called composite cements in Europe).

The test methods and requirements referenced in EN 197-1 and US specifications are different. EN 197-1 includes strength classes, 32.5, 42.5, and 52.5, which correspond to limits on 28-d strength ranges of 32.5 MPa to 52.5 MPa, 42.5 MPa to 62.5 MPa and greater than 52.5 MPa, respectively. The strengths are defined using a test method found in EN 196-1, *Methods of Testing Cement – Part 1: Determination of Strength*, that differs from ASTM C109, *Standard Test Method for Compressive Strength of Hydraulic Cement (Using 2-in. or [50-mm] Cube Specimens (AASHTO T 106)*, in several details (sand gradation, water content, compaction method, specimen size, curing condition, etc.), so the values cannot be directly compared. Cements meeting EN 197-1 and other international specifications are usually not available in the United States; therefore, the best approach is to inform the designer about locally available cements and request changes in the project specifications that allow use of an ASTM or AASHTO cement.

Mexican Cement Specifications. Mexican Specification NMX-C-414-ONNCCE-2017, *Building Industry – Hydraulic Cements – Specifications and Testing Methods*, details the requirements for hydraulic cements used in Mexico.

It defines requirements for six basic cement types as follows:

- Type CPO   Ordinary portland cement
- Type CPP   Pozzolanic portland cement (between 6% and 50% pozzolan)
- Type CPEG Portland cement with blast-furnace slag (between 6% and 60% slag cement)
- Type CPC   Portland composite cement (at least 50% portland cement clinker and sulfate, and
  - 6% to 35% slag cement,
  - 6 % to 35% pozzolan,
  - 1% to 10% silica fume, and/or
  - 6% to 35% limestone.
- Type CPS   Portland cement with silica fume
- Type CEG   Granulated blast-furnace slag cement

All cement types can include up to 5% minor constituents, such as limestone. Any of the six basic types of cement may be specified with one of five strength classes. Strength classes 20, 30, and 40 are defined by 28-day strengths of between 20 MPa and 40 MPa, 30 MPa and 50 MPa, and at least 40 MPa, respectively, using test method NMX-C-061-ONNCCE, *Determination of the Compressive Strength of Hydraulic Cements*. There are also strength classes 30 R and 40 R requiring minimum 3-day strengths of 20 MPa and 30 MPa respectively (in addition to 28-day requirements). As with European and other international cement standards, directly equivalence with US cement specifications cannot be assumed due to differences in test methods and compositional requirements.

Specifying Low-Alkali Cements

Portland cement specification ASTM C150 (AASHTO M 85) historically has included an optional limit on equivalent alkalies for cement. Job specifications should not reference this outdated requirement. This optional limit was intended for use with ASR-susceptible aggregates; however, that optional limit was removed because low-alkali cements are not always effective on their own in mitigating ASR. Higher cement contents with low-alkali cements can provide the same alkali loading as concrete mixtures with lower cement contents with higher alkali levels. The alkali loading of the concrete mixture provides a much better approach to mitigating ASR (see **Alkalies** later in this chapter, Chapter 11, and ACI 225 2020). Portland cement mill test reports routinely report equivalent alkali contents to permit alkali loading calculations when necessary.

CHEMICAL CHARACTERISTICS OF CEMENTS

Portland Cement Clinker Phases (Compounds)

During the firing operation in the manufacture of portland cement clinker, calcium combines with the other components of the raw mix to form four principal phases that make up about 85% to 90% of portland cement by mass. Gypsum, or other forms of calcium sulfate, limestone, and grinding aids or other processing additions are also added during finish grinding.

Cement chemists use the following chemical shorthand (abbreviations):

$$A = Al_2O_3, C = CaO, \bar{C} = CO_2, F = Fe_2O_3,$$
$$H = H_2O, M = MgO, S = SiO_2, \text{ and } \bar{S} = SO_3.$$

The term "phases" is generally used to describe the components of clinker, because the word "compounds" generally refers to specific, ideal compositions, while the compositions of the phases in cement often include trace elements. Following are the four primary phases in portland cement, their approximate chemical formulas (oxide notation), and abbreviations:

Tricalcium silicate -	$3CaO \bullet SiO_2 = C_3S$
Dicalcium silicate -	$2CaO \bullet SiO_2 = C_2S$
Tricalcium aluminate -	$3CaO \bullet Al_2O_3 = C_3A$
Tetracalcium aluminoferrite -	$4CaO \bullet Al_2O_3 \bullet Fe_2O_3 = C_4AF$

C_3S and C_2S in clinker are also referred to as alite and belite, respectively. Alite generally constitutes 50% to 70% of the clinker, whereas belite accounts for only about 10% to 25% (Figure 2-27). Aluminate phases constitute up to about 10% of the clinker, and ferrite compounds generally up to as much as 15% (Bhatty and Tennis 2008). These and other phases may

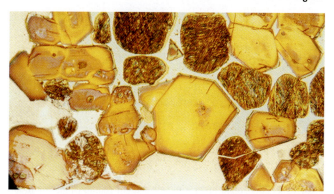

FIGURE 2-27. (left) Polished thin-section examination of portland cement clinker shows alite (C_3S) as light, angular crystals. The darker, rounded crystals are belite (C_2S) (field width approximately 350 µm). (right) Scanning electron microscope (SEM) micrograph of alite (C_3S) crystals in portland cement clinker (field width approximately 70 µm) (Campbell 1999).

be observed and analyzed through the use of microscopy (ASTM C1356, *Standard Test Method for Quantitative Determination of Phases in Portland Cement Clinker by Microscopical Point-Count Procedure,* and Campbell 1999) or quantitative x-ray diffraction analysis (QXRDA, see ASTM C1365, *Standard Test Method for Determination of the Proportion of Phases in Portland Cement and Portland-Cement Clinker Using X-Ray Powder Diffraction Analysis*).

The primary portland cement phases have the following properties:

Tricalcium Silicate, C_3S, hydrates and hardens rapidly and is largely responsible for initial set and early strength. In general, the early strength of portland cement concrete is higher with increased percentages of C_3S.

Dicalcium Silicate, C_2S, hydrates and hardens slowly and contributes primarily to strength increase at ages beyond one week.

Tricalcium Aluminate, C_3A, liberates a large amount of heat during the first few days of hydration and hardening. It also contributes slightly to early strength development. Cements with low percentages of C_3A are more resistant to soils and waters containing sulfates.

Tetracalcium Aluminoferrite, C_4AF, is the product resulting from the use of iron and aluminum raw materials to reduce the clinkering temperature during cement manufacture. It contributes little to strength. Most color effects that give cement its characteristic gray color are due to the presence of C_4AF and its hydrates.

Oxides

Chemical analyses of portland and other hydraulic cements typically are reported as a series of oxides. Reporting is traditionally in the order SiO_2, CaO, Al_2O_3, Fe_2O_3, MgO, and SO_3, (see Table 2-5) which was the order of wet chemical analysis. More rapid methods are overwhelmingly used today, predominantly x-ray fluorescence (XRF) spectroscopy. This procedure provides an elemental analysis, which is then converted to oxides. Although elements are reported as oxides for consistency, they are usually not found in that oxide form in the cement. For example, sulfur is reported as SO_3 (sulfur trioxide), however, cement does not contain any sulfur trioxide, but sulfur is found predominantly in various forms of calcium sulfate, like gypsum. Similarly, cement does not include around 20% SiO_2 as silica, but about 20% by mass of the cement is accounted for by silicon (and its associated oxygen) in the calcium silicates (alite and belite) in the cement. The amount of calcium, silica, alumina, and iron establish the amounts of the primary phases in the cement and significantly impact the properties of hydrated cement (see **Bogue Phase Estimates** in this chapter. See also Bhatty 1995 and PCA 1992 for more information.).

Bogue Phase Estimates

The percentage of each cement phase in a portland cement can be estimated from an oxide analysis, ASTM C114 (AASHTO T 105), *Standard Test Methods for Chemical Analysis of Hydraulic Cement,* or of the unhydrated cement using the Bogue equations, a form of which are provided in ASTM C150 (AASHTO M 85). X-ray diffraction (XRD) techniques (ASTM C1365) can also be used to determine phase composition. Since the production of clinker depends on natural raw materials, there can be a significant level of trace elements in the major phases, which are not accounted for in the Bogue equations. Bogue calculations also assume a "perfect combination" (complete equilibrium) in the high temperature chemical reactions. X-ray diffraction can eliminate this bias, but equipment is not generally suitable for production laboratories. Table 2-6 shows typical oxide and phase composition and fineness for each of the principal types of portland cement. Bogue phases are not generally estimated for blended cements, as the oxide analyses of blended cements are confounded by limestone and SCMs present.

Calcium Sulfate

Gypsum and other forms of calcium sulfate are used in cement to control the otherwise rapid reaction of tricalcium aluminate (C_3A) discussed above. Finish grinding temperature is often controlled to produce a balance of calcium sulfate hemihydrate (also called hemihydrate, bassanite, or plaster of paris) and gypsum. The heat generated during finish grinding will dehydrate a portion of the gypsum to hemihydrate. The dehydration and rehydration can be shown graphically in Figure 2-28.

The forms of calcium sulfate, their chemical formulas, and abbreviations are:

Calcium sulfate dihydrate (gypsum)
$$CaSO_4 \cdot 2H_2O = CaO \cdot SO_3 \cdot 2H_2O = C\bar{S}H_2$$

Calcium sulfate hemihydrate
$$CaSO_4 \cdot \tfrac{1}{2}H_2O = CaO \cdot SO_3 \cdot \tfrac{1}{2}H_2O = C\bar{S}H_{\frac{1}{2}}$$

Anhydrous calcium sulfate (anhydrite)
$$CaSO_4 = CaO \cdot SO_3 = C\bar{S}$$

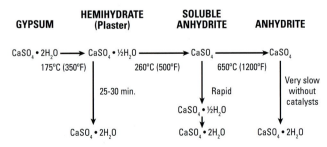

FIGURE 2-28. Dehydration and rehydration relationships for forms of calcium sulfate. Temperatures are approximate (after Hansen and others 1988).

TABLE 2-5. Chemical Analyses of U.S. Portland Cements *

TYPE OF PORTLAND CEMENT		CHEMICAL COMPOSITION†, % BY MASS							
		SiO$_2$	Al$_2$O$_3$	Fe$_2$O$_3$	CaO	MgO	SO$_3$	Na$_2$Oeq	Chloride
I	mean	19.8	5.1	2.5	63.3	2.3	3.3	0.62	0.020
	range	18.8-20.6	4.4-5.7	1.3-3.6	60.7-64.6	0.8-3.0	2.5-4.1	0.26-1.04	0.005-0.037
II	mean	20.1	4.6	3.4	63.6	2.0	3.1	0.55	0.014
	range	18.9-21.7	2.8-5.4	2.4-4.5	60.9-65.3	0.7-4.4	2.5-3.8	0.20-1.03	0.003-0.050
III	mean	20.0	4.7	3.1	63.3	2.1	3.6	0.56	0.014
	range	18.7-21.2	3.7-6.2	1.2-3.9	60.6-65.2	0.7-4.5	2.8-5.2	0.29-1.04	0.003-0.045
IV	mean	22.2	4.6	5.0	62.5	1.9	2.2	0.36	–
	range	21.5-22.8	3.5-5.3	3.7-5.9	62.0-63.4	1.0-3.8	1.7-2.5	0.29-0.42	–
V	mean	20.8	4.0	3.8	63.7	2.2	2.6	0.45	0.008
	range	19.7-22.0	3.5-4.4	3.1-5.0	62.7-64.6	1.1-4.5	2.1-3.4	0.29-0.57	0.002-0.019
White	mean	22.7	4.1	0.3	66.7	0.9	2.7	0.18	–
	range	22.0-24.4	2.2-4.0	0.5-0.6	63.9-68.7	0.3-1.4	2.3-3.1	0.09-0.38	–

* Data for Types I, II, III, and V, are a summary of a PCA survey of cement characteristics (Tennis 2016). Data for 8 white cements and 4 Type IV cement reported in Gebhardt 1995. Values generally represent 10 Type I cements, 56 Type II cements (including Type I /II cements), 47 Type III cements and 19 Type V cements.

† Chemical composition is typically based on elemental analysis and converted to oxides for consistency in reporting, but the elements are not generally found in the form of these oxides in cement. Thus, cement does not include around 20% SiO$_2$ as silica, but about 20% by mass of the cement is accounted for by silicon (and its associated oxygen) in calcium silicates (alite and belite) in the cement.

—Not enough data available to provide summary statistics.

TABLE 2-6. Bogue Potential Phase Estimates of U.S. Portland Cements*

TYPE OF PORTLAND CEMENT		POTENTIAL PHASE COMPOSITION, % BY MASS			
		C$_3$S	C$_2$S	C$_3$A	C$_4$AF
I	mean	57.5	12.7	9.3	7.3
	range	49-62	9-16	7-11	4-11
II	mean	59.1	12.7	6.5†	10.2
	range	51-68	7-20	3†-8	7-13
III	mean	58.0	13.5	7.3	9.1
	range	49-66	7-20	4-14	4-12
IV	mean	42.2	31.7	3.7	15.1
	range	37-49	27-36	3-4	11-18
V	mean	59.2	14.6	4.1	11.7
	range	52-63	8-22	2-5	9-15
White	mean	62.7	17.8	10.4	1.0
	range	50-72	9-25	5-13	1-2

* Data for Types I, II, III, and V, are a summary of a PCA survey of cement characteristics (Tennis 2016). Data for 8 white cements and 4 Type IV cement reported in Gebhardt 1995. Values generally represent 10 Type I cements, 56 Type II cements (including Type I /II cements), 47 Type III cements and 19 Type V cements. Potential phase composition is based on Bogue calculations as given in ASTM C150 (AASHTO M 85).

† One Type II cement had an A/F ratio of less than 0.64, which results in a potential C$_3$A content of 0, as a solid solution forms with a composition between C$_4$AF and C$_2$F. The mean and range shown in the table exclude that cement.

Bogue Equations

Bogue (1929) developed equations to estimate the amounts of the primary phases of a portland cement from an oxide analysis, based on assumptions of ideal phase compositions, and complete kiln reactions. Although only approximate, this approach has been useful to roughly characterize portland cements, including for use in specification ASTM C150 (AASHTO M 85):

$\% \: C_3S = 4.071 \times \% \: CaO - 7.600 \times \% \: SiO_2 - 6.718 \times \% \: Al_2O_3 - 1.430 \times \% \: Fe_2O_3 - 2.852 \times \% \: SO_3$

$\% \: C_2S = 2.867 \times \% \: SiO_2 - 0.7544 \times \% \: C_3S$

$\% \: C_3A = 2.650 \times \% \: Al_2O_3 - 1.692 \times \% \: Fe_2O_3$

$\% \: C_4AF = 3.043 \times \% \: Fe_2O_3$

While these equations work reasonably well for the majority of portland cements made with clinker and gypsum, portland cements with an Al_2O_3/Fe_2O_3 ratio of less than 0.64 do not form C_3A, but form a solid solution of calcium aluminoferrite (usually shown as ss[C_4AF+C_2F]), estimated by:

$\% \: ss(C_4AF+C_2F) = 2.100 \times \% Al_2O_3 + 1.702 \times \% \: Fe_2O_3$

and in this case, the C_3S content is estimated by:

$\% \: C_3S = 4.071 \times \% \: CaO - 7.600 \times \% \: SiO_2 - 4.479 \times \% \: Al_2O_3 - 2.859 \times \% \: Fe_2O_3 - 2.852 \times \% \: SO_3.$

ASTM C150 (AASHTO M 85) includes adjustments for ingredients such as limestone and inorganic processing additions, which are not pyroprocessed, but impact the oxide analyses.

The form and fineness of a calcium sulfate source will influence is dissolution rate and its availability to participate in controlling the rapid C_3A reactions (see Figure 2-29).

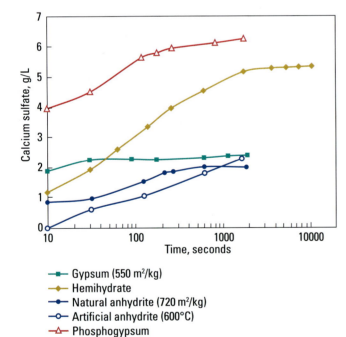

FIGURE 2-29. Rate of solubility of several forms of sulfate. (Adapted from Hansen and others 1988 and Frigione 1983.)

Alkalies

The dominant alkalies in portland and blended cements are sodium (Na) and potassium (K), which are naturally present in raw materials that make clinker. The equivalent alkalies (Na_2Oeq) are a convenient way of reporting the alkali content of portland cement and is simply the sum of the oxides analyzed with an adjustment to the potassium to account for different molecular weights:

$$Na_2Oeq = Na_2O + 0.658 \: K_2O$$

This combination is meaningful because sodium and potassium are mostly in phases (compounds) that rapidly dissolve when the cement is mixed with water, and their action is similar: both serve to raise the internal pH of the concrete. This high pH has several impacts on concrete: it improves corrosion resistance for steel reinforcement by creating a passivation layer (see Chapter 11); it stabilizes C-S-H, the main binding phase in concrete; it helps activate pozzolans and slag cement in concrete; and it can improve the effectiveness of some air-entraining agents.

Alkalies in some concrete mixtures can also have an adverse impact: if the alkali loading is too high for the reactivity level of the aggregate and exposure conditions, premature deterioration due to alkali-silica reactivity (ASR) can result, unless proper mitigation measures are employed. Alkali loading is calculated as the portland cement (or portland cement portion of a blended cement) content of a concrete mixture times the portland cement equivalent alkali content.

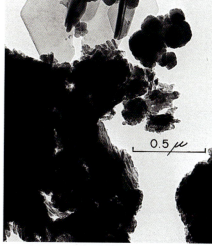

FIGURE 2-30. Electron micrographs of (left) (middle) calcium silicate hydrate (C-S-H), and (right) hydrated portland cement. Note the fibrous nature of the calcium silicate hydrates. Broken fragments of angular calcium hydroxide crystallites are also present (right). The aggregation of fibers and adhesion of the hydration particles are responsible for the strength development of portland cement paste. Reference (left and middle) Brunauer (1962) and (right) Copeland and Schulz (1962).

See also ASTM C1778, *Standard Guide for Reducing the Risk of Deleterious Alkali-Aggregate Reaction in Concrete* (AASHTO R 80), and Chapter 11 for more information on ASR, and discussion under **Specifying "Low-Alkali" Cements** in this chapter for more information on cement alkalies.

HYDRATION OF CEMENT

Chemical Reactions of Primary Phases

In the presence of water, the primary phases hydrate (chemically react with water) to form new solids that are the infrastructure of hardened cement paste in concrete (Figure 2-30). The cement particles slowly dissolve and new solid phases form, filling in part of the space occupied by the mixing water. The portion of the mixing water not filled in by solids remain as capillary pores. The calcium silicates, C_3S and C_2S, hydrate to form calcium silicate hydrate (C-S-H) and calcium hydroxide. Hydrated portland cement paste typically contains 15% to 25% calcium hydroxide and about 50% calcium silicate hydrate by mass. The strength and other properties of hydrated cement are due

primarily to calcium silicate hydrate (Figures 2-30 and 2-31). C_3A, sulfates (gypsum, anhydrite, or other sulfate source), and water combine to form ettringite (calcium trisulfoaluminate hydrate), calcium monosulfoaluminate, and other related compounds. C_4AF reacts with water to form calcium aluminoferrite hydrates. These basic reactions are shown in Table 2-7, and Figure 2-32A illustrates how the amounts of materials change over time as hydration proceeds. Powers and Brownyard (1947), Brunauer (1957), Copeland and others (1960), Powers (1961), Lea (1971), Taylor (1997), and Brouwers (2011), address the pore structure and chemistry of cement paste. Figure 2-32B shows estimates of the relative volumes of the phases in portland cement pastes as hydration proceeds, and Figure 2-33 provides estimates of how fast each of the primary cement phases reacts. The National Institute of Standards and Technology (NIST) has compiled a monograph (Garboczi and others 2019) of detailed modelling techniques to estimate various properties of hydrated cement pastes, mortars, and concrete. Additionally, Lothenbach and others (2019a, 2019b) have developed a thermodynamic

FIGURE 2-31. Scanning-electron micrographs of hardened cement paste at two magnifications (approximate field widths [left] 180 μm and [right] 85 μm).

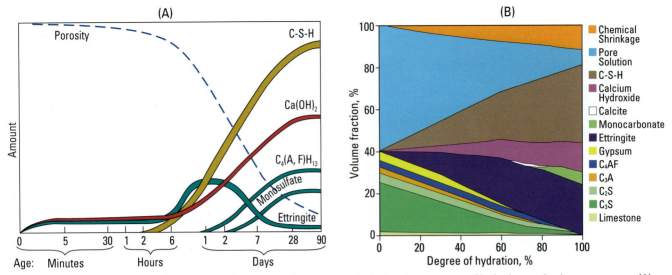

FIGURE 2-32. Relative volumes of the cement phases and hydration products in the microstructure of hydrating portland cement pastes: (A) as a function of time (adapted from Locher and others 1976); and (B) as a function of the degree of hydration as estimated by a computer model for a water-to-cement ratio of 0.485. For (B) a typical Type II cement composition was used: C_3S=59%, C_2S=13%, C_3A=6.5% and C4AF=10%. (Courtesy of B. Isgor, W.J. Weiss, and K. Bharadwaj using methods discussed in Jafari Azad and others 2017.)

modelling system to model cement hydration reactions. The Concrete Microscopy Library (Lange and Stutzman 2019) provides an interesting collection of micrographs related to cement hydration and concrete.

Water (Evaporable and Nonevaporable)

Water is a key ingredient of pastes, mortars, and concretes (see Chapter 4). The phases in portland cement must chemically react with water to develop strength. The amount of water added to a mixture impacts the durability as well. The space initially occupied by water in a cementitious mixture is either partially or completely replaced over time as the hydration reactions proceed (Table 2-7). If more than about 35% water by mass of cement (a water-to-cement ratio of 0.35) is used, then more space is available than needed for cement hydration products to form and some porosity in the hardened material will remain,

even after complete hydration. This is called capillary porosity. Figure 2-34 shows that cement pastes with high and low water-to-cement ratios have equal masses after drying (thus indicating that evaporable water was removed). The cement consumed the same amount of water in both pastes resulting in more bulk volume in the higher water-to-cement ratio paste. As the water-to-cement ratio increases, the capillary porosity increases, and the strength decreases. Also, transport properties such as permeability and diffusivity are increased. This degradation in transport properties allows detrimental chemicals to more readily attack the concrete or reinforcing steel.

Water is observed in cementitious materials in different forms. Free water includes mixing water that has not reacted with the cement phases. Bound water is chemically combined in the solid phases or physically bound to the solid surfaces.

TABLE 2-7. Portland Cement Phase Hydration Reactions (Oxide Notation)

2 (3CaO • SiO₂) Tricalcium silicate	+ 11 H₂O Water	→	3CaO • 2SiO₂ • 8H₂O Calcium silicate hydrate (C-S-H)	+ 3 (CaO • H₂O) Calcium hydroxide
2 (2CaO • SiO₂) Dicalcium silicate	+ 9 H₂O Water	→	3CaO • 2SiO₂ • 8H₂O Calcium silicate hydrate (C-S-H)	+ CaO • H₂O Calcium hydroxide
3CaO • Al₂O₃ Tricalcium aluminate	+ 3 (CaO • SO₃ • 2H₂O) Gypsum	+ 26 H₂O Water	→	6CaO • Al₂O₃ • 3SO₃ • 32H₂O Ettringite
2 (3CaO • Al₂O₃) Tricalcium aluminate	+ 6CaO • Al₂O₃ • 3SO₃ • 32H₂O Ettringite	+ 4 H₂O Water	→	3 (4CaO • Al₂O₃ • SO₃ • 12H₂O) Calcium monosulfoaluminate)
3CaO • Al₂O₃ Tricalcium aluminate	+ CaO • H₂O Calcium hydroxide	+ 12 H₂O Water	→	4CaO • Al₂O₃ • 13H₂O Tetracalcium aluminate hydrate
4CaO • Al₂O₃ • Fe₂O₃ Tetracalcium aluminoferrite	+ 2 (CaO • H₂O) Calcium hydroxide	+ 10 H₂O Water	→	6CaO • Al₂O₃ • Fe₂O₃ • 12H₂O Calcium aluminoferrite hydrate

Note: This table includes only primary reactions and not several additional minor reactions. The composition of calcium silicate hydrate (C-S-H) is not stoichiometric (Tennis and Jennings 2000).

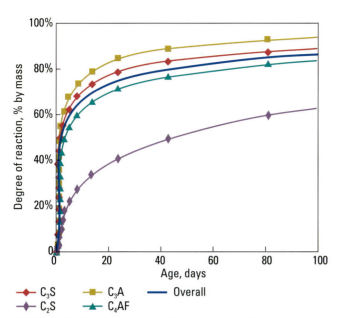

FIGURE 2-33. Relative reactivity of cement phases. The curve labeled "Overall" has a composition of 58% C_3S, 13% C_2S, 9% C_3A, and 7% C_4AF, a typical Type I cement composition (adapted from Tennis and Jennings 2000).

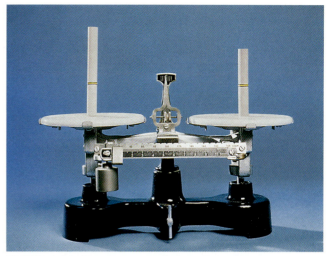

FIGURE 2-34. Cement paste cylinders of equal mass and equal cement content, but mixed with different water-to-cement ratios, after all water has evaporated from the cylinders. Cylinder on the left has much higher residual porosity.

A reliable separation of the chemically combined water from the physically adsorbed water is not possible. Accordingly, Powers (1949) distinguished between evaporable and nonevaporable water. The nonevaporable water is the amount retained by a sample after it has been subjected to a drying procedure intended to remove all the free water (traditionally, by heating to 105°C [221°F]). Evaporable water was originally considered to be free water, but it is now recognized that some bound water is also lost as the sample is heated. All nonevaporable water is bound water, yet not all bound water is nonevaporable.

For complete hydration of portland cement to occur, only about 40% water (a water-to-cement ratio of 0.40) is required. If a water-to-cement ratio greater than about 0.40 is used, the excess water not needed for cement hydration remains in the capillary pores or evaporates. If a concrete mixture has a water-cement ratio less than about 0.40, some cement will remain unhydrated due to the lack of water necessary for the hydration reactions.

To estimate the degree of hydration of a hydrated material, the nonevaporable water content is often used. To convert the measured nonevaporable water into degrees of hydration, it is necessary to know the value of nonevaporable water-to-cement ratio (w_n/c) at complete hydration. This can be experimentally determined by preparing a high water-to-cement ratio cement paste (for example, 1.0) and continuously grinding the paste sample while it hydrates in a roller mill. In this procedure, complete hydration of the cement will typically be achieved after 28 days.

Alternatively, an estimate of the quantity of nonevaporable water-to-cement ratio (w_n/c) at complete hydration can be obtained from the potential Bogue composition of the cement. Nonevaporable water contents for the major phases of portland cement are provided in Table 2-8. For a typical Type I cement, these coefficients will generally result in a calculated w_n/c for completely hydrated cement somewhere between 0.22 and 0.25.

TABLE 2-8. Nonevaporable Water Contents for Fully Hydrated Major Cement Phases (Molina 1992)

CEMENT PHASE	NONEVAPORABLE (COMBINED) WATER CONTENT (g water/g cement paste)
C_3S	0.24
$C2_S$	0.21
C_3A	0.40
C_4AF	0.37
Free lime (CaO)	0.33

PROPERTIES OF CEMENTS

The properties of hydraulic cements are governed by their chemical and physical characteristics: the amounts of the different Bogue phases, the form and amount of sulfate, the fineness and particle size distribution, and the effects of other ingredients, particularly limestone and SCMs in blended cements. Specifications for cements place limits on their physical properties and often their chemical composition (see Tables 2-9, 2-10, and 2-11). An understanding of the significance of some of the physical properties is helpful in interpreting results of cement tests (Johansen and others 2006). More information on testing cement can be found in Chapter 20.

TABLE 2-9. Compositional Requirements in ASTM Specifications C150 and C595* (% by mass)

OXIDE, ANALYTE, OR PHASE	TYPES	LIMIT, % BY MASS (maximum, unless otherwise noted)	NOTES
Al_2O_3	II, II(MH)	6.0	
Fe_2O_3	II, II(MH)	6.0	Does not apply if optional heat of hydration or optional sulfate resistance requirements are specified.
	IV	6.5	
MgO	I, II, II(MH), III, IV, V, IP, IT(P≥S), IT(P≥L)	6.0	
Sulfate, SO_3†	IV, V	2.3	
	I, II, II(MH) IL, IS(<70), IT(P<S<70), IT(L≥S); IT(L≥P), IT(L<S<70)	3.0	3.5% for Type I if C_3A content is >8.0%.
	III	3.5	4.5% for Type III if C_3A content is >8.0%.
	IP, IS(≥70), IT(P≥S), IT(P≥L), IT(S≥70)	4.0	
Sulfide, S^{2-}	IS, IT(P<S<70), IT(S≥70), IT(L<S<70)	2.0	
Loss on ignition (LOI)	IV	2.5	
	I, II, II(MH), III, V, IS(<70), IT(P<S<70)	3.0	3.5% for Types I, II II(MH), III, IV, V when limestone is an ingredient in the cement.
	IS(≥70), IT(S≥70)	4.0	
	IP, IT(P≥S)	5.0	
	IL, IT(L≥S) IT(L≥P)	10.0	
Insoluble residue (IR)	IS, IT(P<S<70), IT(S≥70), IT(L<S<70)	1.0	
	I, II, II(MH), III, IV, V	1.5	
Tricalcium silicate, C_3S	IV	35	Does not apply if optional heat of hydration requirements are specified.
Dicalcium silicate, C_2S	IV	40 minimum	Does not apply if optional heat of hydration requirements are specified.
Tricalcium aluminate, C_3A	V, III‡	5.0	For high sulfate resistance.
	IV	7.0	Does not apply if optional heat of hydration requirements are specified.
	II, II(MH), III‡	8.0	For moderate sulfate resistance.
	III	15.0	
$C_3S+4.75(C_3A)$	II(MH)	100	Referred to as the "heat index;" Not applicable if the cement meets a heat of hydration limit.
$C_4AF+2(C_3A)$ (or solid solution C_2F+C_3A)	V	25	Does not apply if optional sulfate resistance requirements are specified.
Limestone	IL, IT(L≥S) IT(L≥P)	5.0 minimum, 15.0 maximum	
	I, II, II(MH), III, IV, V	5.0	
Organic processing additions	I, II, II(MH), III, IV, V	1.0	Qualification by ASTM C465 required.
Inorganic processing addition	I, II, II(MH), III, IV, V	5.0	Qualification by ASTM C465 required for amounts >1%.

*As of October 2023. Specification requirements are updated frequently. For complete and current details, review the most recent editions. Requirements in ASTM C150 and AASHTO M 85 are the same, as are requirements in ASTM C595 and AASHTO M 240.

† Default SO_3 limits can be exceeded if ASTM C1038 expansion test results do not exceed 0.020% at 14 days.

‡ This is an optional requirement that applies only when specifically requested.

TABLE 2-10. Select Physical Requirements for Cements in ASTM C150, ASTM C595, and ASTM C1157

TEST METHOD	ASTM SPECIFICATION	TYPES	LIMIT	NOTES*
ASTM C109 Water requirement	C595	Cements with LH option	64%	
ASTM C185 Air content	C150	I, II, II(MH), III, IV, V,	12% maximum	Air entraining cements are required to contain between 16% and 22% air.
	C595†	IL, IP, IS, IT		
	C1157	GU, HE, HS, MS, MH, LH		
ASTM C188 Density	C595 C1157	IL, IP, IS, IT	–	No requirement, but data is required to be reported.
ASTM C191 Initial setting time	C150	I, II, II(MH), III, IV, V	Between 45 minutes and 375 minutes	Vicat method.
	C595†	IL, IP, IS, IT	Between 45 minutes and 420 minutes	
	C1157	GU, HE, HS, MS, MH, LH		
ASTM C204 Fineness	C150	I, II, II(MH), IV, V	260 m²/kg minimum	Air permeability (Blaine) method.
		II(MH), IV	430 m²/kg maximum	
	C595†	IL, IP, IS, IT	–	No limit, but data is required to be reported.
	C1157	GU, HE, HS, MS, MH, LH		
ASTM C430 or ASTM C1891 Retained on 45-μm sieve	C595†	IL, IP, IS, IT	–	No requirement, but data is required to be reported.
	C1157	GU, HE, HS, MS, MH, LH		
ASTM C451 False set‡	C150	I, II, II(MH), III, IV, V	50% minimum	
	C1157	GU, HE, HS, MS, MH, LH		
ASTM C452 Sulfate resistance	C150	V‡	0.040% expansion maximum	C₃A, C₄AF+2C₃A, and Fe₂O₃ limits do not apply when this optional requirement is invoked. Cements meeting this high sulfate resistance requirement are deemed to meet moderate sulfate resistance requirements.
ASTM C1012 Sulfate resistance	C1157	MS	maximum of 0.10% expansion at 6 months	
	C595	Cements with MS option		
	C1157	HS	maximum of 0.05% expansion at 6 months or maximum of 0.10% expansion at 1 year	Testing at one year is not required if the 6 month limit is met. Cement is not rejected unless both the 6 month and 1 year limit are exceeded.
	C595	Cements with HS option		
ASTM C227 Mortar expansion‡	C1157	GU, HE, HS, MS, MH, LH	maximum of 0.020% at 14 days. and 0.060% at 56 days;	Optional requirement for cement that will be used with reactive aggregates.
ASTM C1038 Mortar bar expansion	C150	I, II, II(MH), III, IV, V	maximum of 0.020% at 14 days	Required for C150 and C595 cements only if default limits on SO₃ are exceeded.
	C595	IL, IP, IS, IT		
	C1157	GU, HE, HS, MS, MH, LH		
ASTM C1702 Heat of hydration	C1157	MH	maximum of 335 kJ/kg at 3 days	Heat index limit for Type II(MH) does not apply when this requirement is invoked. For Type IV, C₃S, C₂S, C₃A, and Fe₂O₃ limits do not apply when this requirement is invoked. C1702 data required to be reported for Type II(MH).
	C150	II(MH)‡		
	C595	Cements with MH option		
	C1157	LH	maximum of 200 kJ/kg at 3 days, and 225 kJ/kg at 7 days	
	C150	IV‡		
	C595	Cements with LH option		

* For complete details, review the current standards: ASTM C150 (AASHTO M 85), ASTM C595 (AASHTO M 240), and ASTM C1157

† Requirement also applies to blended cements with special properties.

‡ Optional requirement that only applies if specifically requested.

TABLE 2-11. Minimum Compressive Strength Requirements in ASTM and AASHTO Cement Specifications*

SPECIFICATION	TYPE	LIMITS, MPa (psi)			
		1 DAY	**3 DAYS**	**7 DAYS**	**28 DAYS**
ASTM C150 (AASHTO M 85)	I	—	12.0 (1740)	19.0 (2760)	28.0 (4060)†
	II, II(MH)	—	10.0 (1450)	17.0 (2470)	28.0 (4060)†
	III	12.0 (1740)	24.0 (3480)	—	—
	IV	—	—	7.0 (1020)	17.0 (2470)
	V	—	8.0 (1160)	15.0 (2180)	21.0 (3050)
ASTM C595 (AASHTO M 240)	IL, IP, IS(<70), IT(S<70), IL(MS), IP(MS), IS(<70)(MS), IT(S<70)(MS), IL(HS), IP(HS), IS(<70)(HS), IT(S<70)(HS)	—	13.0 (1890)	20.0 (2900)	25.0 (3620)
	IS(≥70), IT(S≥70)	—	—	5.0 (720)	11.0 (1600)
	IL(HE), IP(HE), IS(<70)(HE), IT(S<70)(HE)	12.0 (1740)	24.0 (3480)	—	—
	IL(MH), IP(MH), IS(<70)(MH), IT(S<70)(MH)	—	10.0 (1450)	17.0 (2470)	22.0 (3190)
	IL(LH), IP(LH), IS(<70)(LH), IT(S<70)(LH)	—	—	11.0 (1600)	21.0 (3050)
ASTM C1157	GU	—	13.0 (1890)	20.0 (2900)	28.0 (4060)
	HE	12.0 (1740)	24.0 (3480)	—	—
	MS	—	11.0 (1600)	18.0 (2610)	28.0 (4060)†
	HS	—	11.0 (1600)	18.0 (2610)	25.0 (3620)
	MH	—	5.0 (720)	11.0 (1600)	22.0 (3190)†
	LH	—	—	11.0 (1600)	21.0 (3050)

* For complete details, review the relevant standards. As determined by ASTM C109 (AASHTO T 106). Air-entraining cements are not included, but generally have somewhat lower requirements.

† Optional requirement that applies only if specifically requested.

— = no requirement at this age.

Compressive Strength

Compressive strength is influenced by the cement type, specifically: the phase composition, sulfate content, and fineness of the cement. Minimum strength requirements in cement specifications are typically exceeded comfortably.

Table 2-11 shows the minimum compressive strength requirements in U.S. specifications. In general, cement strengths (based on ASTM C109 [AASHTO T 106] mortar cube tests) cannot be used to predict concrete strengths with any degree of accuracy because of the many variables that influence concrete strength, including: aggregate characteristics, concrete mixture proportions, construction procedures, and environmental conditions in the field (Weaver and others 1974 and DeHayes 1990). Therefore, it should never be assumed that two types of cement meeting the same minimum requirements will produce the same strength of mortar or concrete without modification of mix proportions.

Figures 2-35 and 2-36 illustrate the strength development for standard mortars made with various types of portland cement.

Wood (1992) provides long-term strength properties of mortars and concretes made with portland and blended cements.

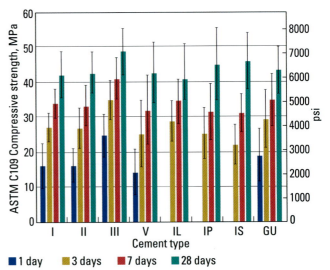

FIGURE 2-35. Average (mean) ASTM C109 (AASHTO T 106) mortar cube compressive strengths for portland, blended, and Type GU cements (Tennis 2016). Range of reported values indicated by error bars.

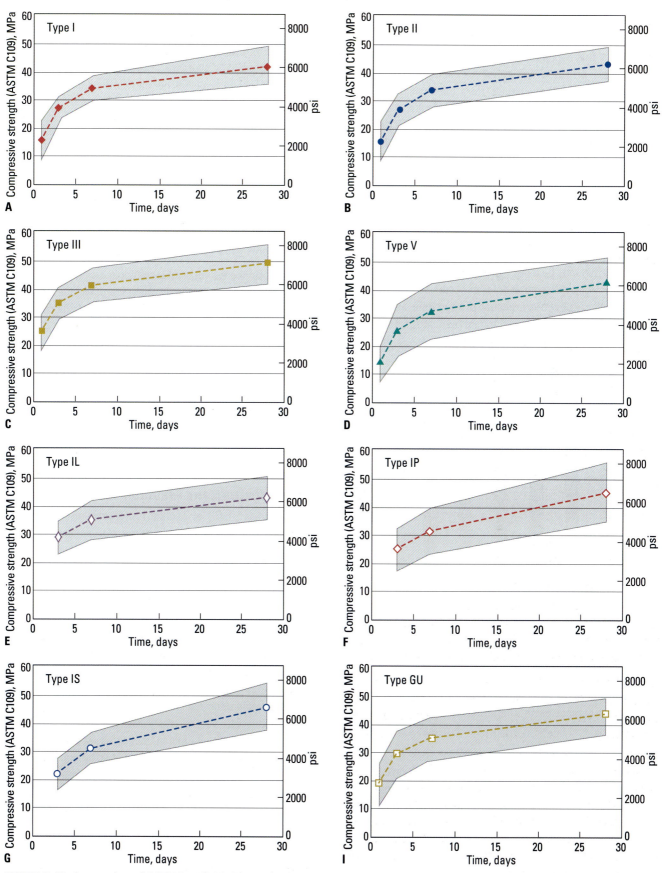

FIGURE 2-36. Average (mean) ASTM C109 (AASHTO T 106) mortar cube compressive strengths for: (A) 10 Type I portland cements; (B) 56 Type II portland cements; (C) 45 Type III portland cements; (D) 19 Type V portland cements; (E) 12 Type IL blended cements; (F) 8 Type IP blended cements; (G) 8 Type IS blended cements; and (H) 10 Type GU cements. *Shaded areas indicate two standard deviations around the mean values indicated by the dashed line* (adapted from Tennis 2016).

Setting Time

The setting properties of hydraulic cement are measured to ensure that the cement is hydrating normally. Two arbitrary setting times are used: initial set – which is considered to represent the time that elapses from the moment water is added until the paste ceases to be fluid and plastic, and final set – the time required for the paste to acquire a certain degree of hardness. Tests are performed in accordance with ASTM C191, *Standard Test Methods for Time of Setting of Hydraulic Cement by Vicat Needle* (AASHTO T 131).

The setting times indicate if a paste is undergoing normal hydration reactions. Sulfate (from gypsum or other sources) in the cement regulates setting time, but setting time is also affected by cement fineness, water-to-cement ratio, and any admixtures that may be used. Setting times of concretes do not correlate directly with setting times of pastes because of water loss to the air or substrate, presence of aggregate, and because of temperature differences in the field (as contrasted with the controlled temperature in a testing lab). Figure 2-37 illustrates mean set times for U.S. hydraulic cements.

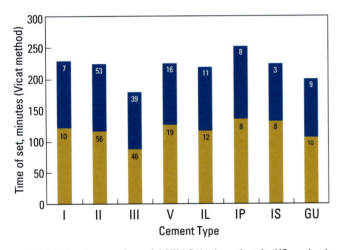

FIGURE 2-37. Average (mean) ASTM C191 time of set for US portland, blended, and Type GU cements. Numbers on the columns indicate the number of cements included in the average (Tennis 2016).

Early Stiffening (False Set and Flash Set). Early stiffening is the early development of stiffness in the working characteristics or plasticity of cement paste, mortar, or concrete. This includes both false set and flash set. Figure 2-38 provides a summary.

False set is evidenced by a significant loss of plasticity without the evolution of much heat shortly after mixing. False set occurs when a large portion of gypsum dehydrates in the cement finish mill to form plaster. Stiffening is caused by the rapid crystallization of interlocking needle-like secondary gypsum. Ettringite precipitation can also contribute to false set. Additional mixing without added water breaks up these crystals to restore workability without harming the concrete.

Flash set or quick set is evidenced by a rapid and early loss of workability in paste, mortar, or concrete. It is usually

accompanied by the evolution of considerable heat resulting primarily from the rapid reaction of aluminates. If the proper amount or form of calcium sulfate is not available to control tricalcium aluminate hydration, stiffening becomes apparent. Flash set cannot be dispelled, nor can the plasticity be regained by further mixing without the addition of water.

Proper stiffening results from the careful balance of the sulfate and aluminate compounds, as well as the temperature and fineness of the materials (which control the dissolution and precipitation rates). The amount of sulfate in the form of plaster has a significant effect. For example, with one particular cement, 2% plaster provided a 5-hour set time, while 1% plaster caused flash set to occur, and 3% demonstrated false set (Helmuth and others 1995).

Cements are tested for early stiffening using ASTM C451, *Standard Test Method for Early Stiffening of Hydraulic Cement (Paste Method)* (AASHTO T 186), and ASTM C359, *Standard Test Method for Early Stiffening of Hydraulic Cement (Mortar Method)* (AASHTO T 185), which use the penetration techniques of the Vicat apparatus. However, these tests do not address all of the variables that can influence early stiffening, which include mixing time, placing temperature, and field conditions. For example, concretes mixed for very short periods (less than one minute) tend to be more susceptible to early stiffening. These tests also do not address early stiffening caused by interactions with other concrete ingredients.

Loss on Ignition and Insoluble Residue

Loss on ignition (LOI) of cement is determined by heating a cement sample of known weight in accordance with ASTM C114 (AASHTO T 105). The weight loss of the sample is then determined. Loss-on-ignition values for portland cements typically range from 0% to 3.5%. Traditionally, a high LOI for portland cements has been an indication of prehydration and carbonation, which may have been caused by improper or prolonged storage, or contamination during transport. Modern cements may have an LOI over 2% due to the use of limestone as an ingredient in portland cement. Blended cements with limestone may have LOI values higher than 7% primarily due to limestone, while other blended cements (Type IP and Type IS), typically have LOI values under 3%. A modified procedure known as split loss-on-ignition measures weight loss on a sample of cement sequentially heated to two temperatures. The differences in weight between the two temperature ranges is generally due to CO_2 loss due to limestone.

The insoluble residue is determined also using procedures in ASTM C114 . This material is predominantly silica (quartz) that has not reacted in the kiln to form silicates, or from impurities in the gypsum source or limestone. Small variations in insoluble residue are normal and will have little impact on performance (ACI 225R 2019). ASTM C150 (AASHTO M 85) limits the insoluble residue for portland cements to a maximum of 1.5%,

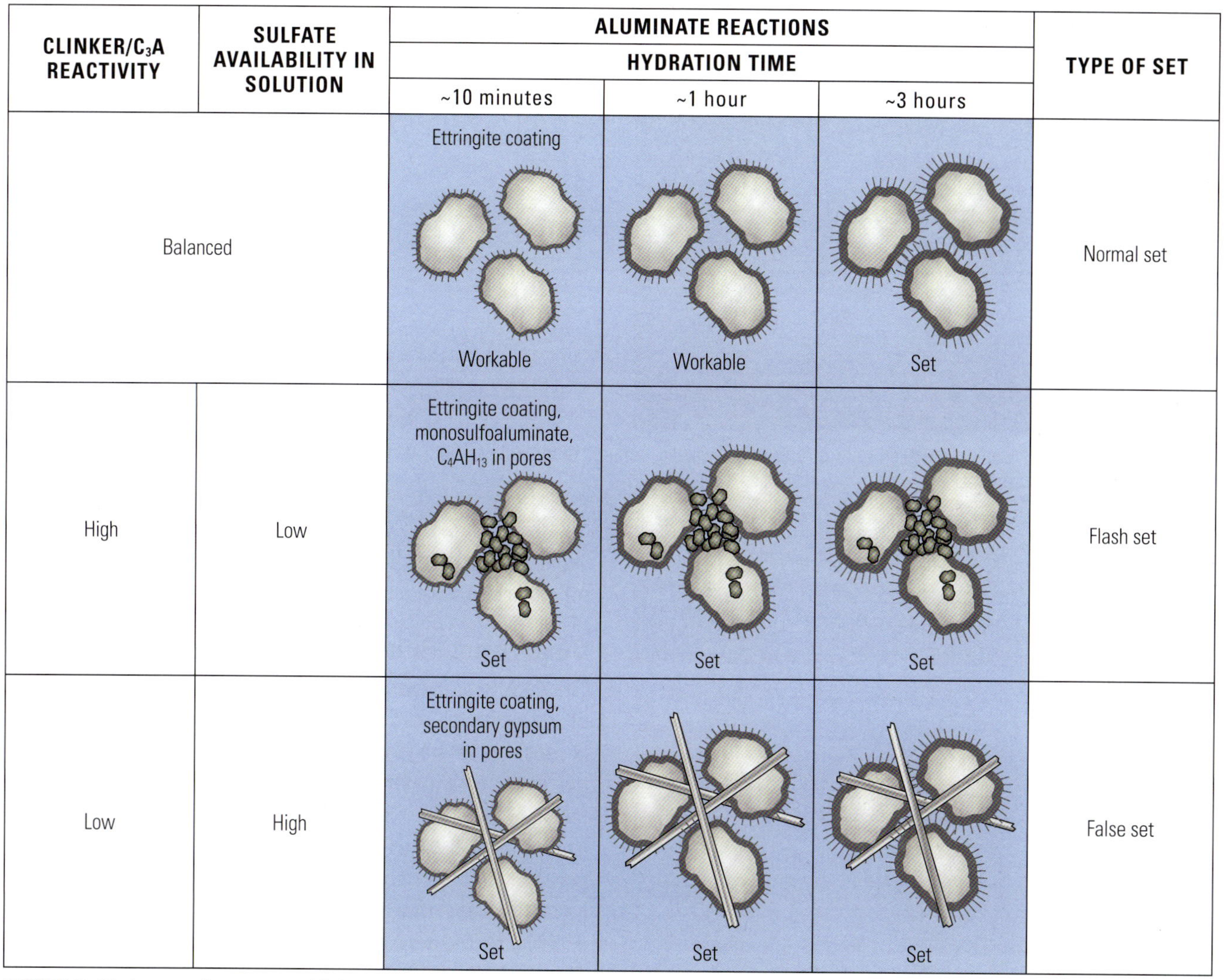

CLINKER/C₃A REACTIVITY	SULFATE AVAILABILITY IN SOLUTION	ALUMINATE REACTIONS			TYPE OF SET
		HYDRATION TIME			
		~10 minutes	~1 hour	~3 hours	
Balanced		Ettringite coating Workable	 Workable	 Set	Normal set
High	Low	Ettringite coating, monosulfoaluminate, C₄AH₁₃ in pores Set	 Set	 Set	Flash set
Low	High	Ettringite coating, secondary gypsum in pores Set	 Set	 Set	False set

FIGURE 2-38. Impact of gypsum and C_3A on setting of portland cements (after Hansen and others 1988).

by mass, and ASTM C595 (AASHTO M 240) limits the insoluble residue for blended cements made with slag cement to 1.0% by mass. More information on LOI and insoluble residue is provided in Chapter 20.

Particle Size and Fineness

Portland cement consists of individual angular particles with a range of sizes, the result of pulverizing clinker in the grinding mill (Figure 2-39 left). Approximately 95% of cement particles are smaller than 45 µm, with the average particle size of approximately 15 µm. Figure 2-39 (right) illustrates the particle size distribution for a portland cement. The fineness is, at best, a general indication of the particle size distribution of a cement. Finer cements, as indicated by higher Blaine fineness measurements, are typically indicative of smaller individual cement particles. The fineness of cement affects heat released and the rate of hydration. Greater cement fineness (smaller particle size) increases the rate at which cement hydrates and thus accelerates strength development. The effects of greater fineness on paste strength are manifested principally during the first seven days.

Fineness is usually measured by the Blaine air-permeability test, ASTM C204 (AASHTO T 153), *Standard Test Methods for Fineness of Hydraulic Cement by Air-Permeability Apparatus.* Cements with finer particles have more surface area in square meters per kilogram of cement. (The use of square centimeters per gram for reporting fineness is now considered archaic.) ASTM C150 and AASHTO M 85 have minimum Blaine fineness limits of 260 m²/kg for all cement types except Type III, and maximum limits on fineness for Type II(MH) and Type IV cements of 430 m²/kg. Blended cement specifications require reporting of the cement fineness with no limit. Due to the nature of the non-portland cement ingredients, blended cements tend to have higher finenesses than portland cements. Tests are also available that quantify the fineness of cement in the form of the amount of cement passing a 45-µm (No. 325) sieve (ASTM C430, *Standard Test Method for Fineness of Hydraulic Cement by the 45-µm [No. 325] Sieve* [AASHTO T 192], and ASTM C1891, *Standard Test Method for Fineness of Hydraulic Cement by Air-Jet Sieving at 45-µm [No. 325]),* and x-ray or laser particle size analyzers can also be used to determine fineness values (see Chapter 20).

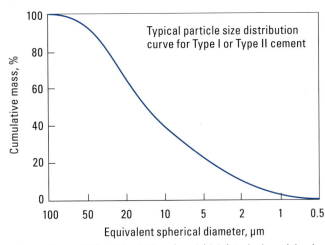

FIGURE 2-39. (left) Scanning-electron micrograph of powdered cement (approximate field width of 95 μm) and (right) typical particle size distribution of portland cement.

TABLE 2-12. Fineness of U.S. Hydraulic Cements*

TYPE OF PORTLAND CEMENT		BLAINE FINENESS m²/kg	PASSING A 45-μm (NO. 325) SIEVE %
I	mean	397.2	95.1
	range	375-440	91.2-97.9
II	mean	392.7	96.4
	range	305-471	85.7-99.9
III	mean	560.8	99.0
	range	365-723	95.1-99.9
IV	mean	339.5	–
	range	319-362	–
V	mean	401.1	95.9
	range	302-551	81.0-99.0
White	mean	482.4	–
	range	384-564	–
IL	mean	479	97.4
	range	409-660	95.4-99.1
IP	mean	524	96.7
	range	397-850	91.3-98.6
IS	mean	475	97.0
	range	391-539	93.6-99.2
GU	mean	514	97.5
	range	432-619	94.0-99.2

* Data for all but Type IV and white cements are from a PCA survey of cement characteristics (Tennis 2016). Data for 8 white cements and 4 Type IV cement reported in Gebhardt 1995. Values generally represent averages of 10 Type I cements, 56 Type II cements (including Type I /II cements), 45 Type III cements and 19 Type V cements, 12 Type IL cements, 8 Type IP cements, 8 Type IS cements and 10 Type GU cements.

Table 2-12 provides data on fineness of US hydraulic cements for general concrete construction.

Density and Relative Density (Specific Gravity)

The density of cement is defined as the mass of a unit volume of the solids or particles, excluding air between particles. It is reported as megagrams per cubic meter (Mg/m^3) or grams per cubic centimeter (g/cm^3). The numeric value is identical for both units. The particle density of portland cement ranges from 3.10 to 3.25, averaging 3.15 Mg/m^3. Portland-limestone, portland blast-furnace slag and portland-pozzolan cements have densities ranging from 2.95 to 3.15, averaging about 3.05 Mg/m^3. The density of a cement, determined by ASTM C188, *Standard Test Method for Density of Hydraulic Cement* (AASHTO T 133), is not an indication of the cement's quality; rather, its principal use is in mixture proportioning calculations.

For mixture proportioning, it may be more useful to express the density as relative density (also called specific gravity). The relative density is a dimensionless number determined by dividing the cement density by the density of water at 4°C, which is 1.0 Mg/m^3 (1.0 g/cm^3 or 1000 kg/m^3) or 62.4 lb/ft^3.

A relative density of 3.15 is assumed for portland cement in volumetric calculations of concrete mix proportioning. As noted above, the relative density of blended cements tends to be slightly lower; for this reason, density for blended cements is reported on mill test reports and should be used in volumetric calculations. As mix proportions list quantities of concrete ingredients in kilograms or pounds, the relative density must be multiplied by the density of water at 4°C to determine the particle density in kg/m^3 or lb/ft^3. This product is then divided into the mass or weight of cement to determine the absolute volume of cement in cubic meters or cubic feet. See Chapter 13 for examples.

Bulk Density

The bulk density of cement is defined as the mass of cement particles plus air between particles per unit volume. The bulk density of cement can vary considerably depending on how it is handled and stored. Portland cement that is aerated or loosely deposited may have a bulk density of only 830 kg/m³ (52 lb/ft³), but with consolidation caused by vibration, the same cement can have a bulk density as much as 1650 kg/m³ (103 lb/ft³) (Toler 1963). Thus, batching cement by volume can lead to significant inconsistencies, depending on how loosely the cement is packed. For this reason alone, good practice dictates that cement must be weighed for each batch of concrete produced, as opposed to using a volumetric measure (Figure 2-40).

FIGURE 2-40. Both 500-mL beakers contain 500 grams of dry powdered cement. On the left, cement was simply poured into the beaker. On the right, cement was slightly vibrated – imitating consolidation during transport or packing while stored in a silo. There is a 20% difference in bulk volume.

Soundness

Soundness refers to the ability of a hardened cement paste to retain its volume after setting. Lack of soundness or delayed destructive expansion can be caused by excessive amounts of hard-burned free lime or magnesia (periclase). Many specifications for hydraulic cement limit the magnesia (MgO) content to control soundness issues. Past versions of ASTM and AASHTO specifications for hydraulic cements included limits on expansion of cement paste bars as measured by ASTM C151, *Standard Test Method for Autoclave Expansion of Hydraulic Cement* (AASHTO T 107). However, research has demonstrated that the test does not relate to concrete performance, and the requirement was removed from current standards. Cases of abnormal expansion attributed to unsound cement are exceedingly rare, which may be due to better quality control and an improved understanding of the role of free lime and magnesia. See Gonnerman and others (1953), Helmuth and West (1998), Klemm (2005), and Kabir and Hooton (2020) for extensive reviews of the autoclave expansion test.

Heat of Hydration

Heat of hydration is the heat generated when cement and water react. The amount of heat generated is dependent chiefly upon the chemical composition of the cement, with C_3A and C_3S phases primarily responsible for high heat evolution. The water-to-cement ratio, fineness of the cement, certain chemical admixtures, and curing temperature are also factors. Increases in the fineness, cement content, or curing temperature generally increase the heat of hydration. For blended cements, limestone or SCMs can reduce heat generation. Although portland and blended cements can continue to react slowly for years, the rate of heat generation is greatest at early ages. A large amount of heat evolves within the first three days with the greatest rate of heat liberation usually occurring within the first 24 hours (Copeland and others 1960).

For many concrete elements, such as slabs and pavements, heat generation is not a concern because the heat is quickly dissipated into the environment. However, in structures of considerable mass, or high cement contents, the rate and amount of heat generated are critical. If this heat is not rapidly dissipated in massive elements, a significant rise in concrete temperature can occur. This may be undesirable, since, after hardening at an elevated temperature, non-uniform cooling of the concrete to ambient temperature may create excessive tensile stresses and result in cracking. In addition, if concrete temperatures during curing exceed about 70°C (158°F), a distress mechanism known as delayed ettringite formation (DEF) may be triggered, potentially leading to reduced service life for the concrete (see Chapter 11 for more information on DEF). On the other hand, a rise in concrete temperature caused by heat of hydration is often beneficial in cold weather, if it helps maintain favorable curing temperatures. The temperature differential and maximum temperature are two critical factors (see Chapter 18 and 19).

TABLE 2-13. ASTM C1702 Heat of Hydration for Selected Portland Cements in the U.S. in 2016, kJ/kg*

CEMENT TYPE	TYPE I		TYPE II		TYPE II(MH)		TYPE V		TYPE III	
Age	3 days	7 days	3 days	7 days	3 days	7 days	3 days	7 days	3 days	7 days
Mean	337	367	305	348	298	335	292	364	354	383
St. Dev.			22	22	13	1	9	30	18	10
N	1	1	33	24	7	2	2	4	2	2

*Table is based on limited data for several cement types. To convert to cal/g, divide by 4.184.
Note: No data for Type IV cements was available.

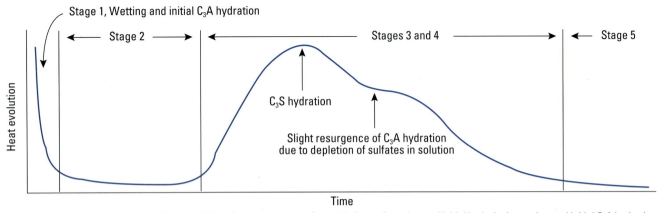

FIGURE 2-41. Heat evolution as a function of time for cement paste. Stage 1 is heat of wetting and initial hydrolysis wetting and initial C_3A hydration. Stage 2 is a dormant period related to initial set. Stage 3 is an accelerated reaction of the hydration products that determines rate of hardening and final set. Stage 4 decelerates formation of hydration products and determines the rate of early strength gain. Stage 5 is a slow, steady formation of hydration products establishing the rate of later strength gain.

Table 2-13 provides heat of hydration values for a variety of portland cements. Although limited data are available, it illustrates the expected trend that Type III cements generally have a higher heat of hydration than other types.

Cements do not generate heat at a constant rate. The heat output during hydration of a typical Type I portland cement is illustrated in Figure 2-41. The first peak shown in the heat profile is caused by heat generated from the initial hydration reactions of cement compounds such as tricalcium aluminate. Sometimes called the heat of wetting, this initial heat peak is followed by a period of slow thermal activity known as the induction or dormant period, when calcium sulfate in solution interacts with the C_3A to delay the onset of other reactions. After several hours, a broad second heat peak attributed to tricalcium silicate hydration emerges, signaling the onset of the paste hardening process. Finally, a minor third peak due to the renewed activity of tricalcium aluminate may be evident; its intensity and location are normally dependent on the amount of tricalcium aluminate and sulfate present in the cement.

In calorimetry testing (ASTM C1702), the first heat evolution can be measured immediately or even during mixing of the paste; as a result, often only the downward slope of the first peak is observed. The second peak (C_3S peak) generally occurs between 6 and 12 hours. The third peak (also known as the sulfate depletion peak) usually occurs within a few hours after the second peak, but may be difficult to distinguish with some cements (Tang 1992).

When it is desired to minimize heat generation in concrete, designers should choose a lower heat cement, such as an ASTM C150 (AASHTO M 85) Type II(MH) portland cement. Type IV cement can also be used to control temperature rise, but it is rarely available. Moderate-heat and low-heat cements are also available in ASTM C595 (AASHTO M 240; with (MH) or (LH) suffixes) and ASTM C1157 (Type MH or LH) specifications. Replacement of a significant portion of the cement with low-heat supplementary cementitious materials is often used to reduce temperature rise. For information on testing for heat of hydration see Chapter 20.

HANDLING CEMENT

Packaging

Cement is often bagged for convenient use at construction sites (Figure 2-42) and for small jobs. In the United States a bag of portland cement typically has a mass of 42 kg (92.6 lb) and a volume of 28 L (1 ft³). Traditionally, a U.S. bag contained 94 lb (42.6 kg) and had a volume of 1 ft³ (28 L). In Canada, a bag of portland cement has a mass of 40 kg, and in other countries a bag of cement may have a mass of 25 kg or 50 kg.

FIGURE 2-42. A small amount of cement is shipped in bags, primarily for masonry applications and for small projects.

Transportation

Tens of millions of metric tons of cement are shipped per year throughout the United States. Most portland cements are

FIGURE 2-43. Portland cements are shipped from the plant silos (left) to the user in bulk by railcar, cement truck (center), or ship (right).

shipped in bulk by rail, truck, barge, or ship (Figure 2-43). Pneumatic (pressurized air) loading and unloading of the transport vehicle is the most popular means of handling bulk cement. However, large sacks holding one to twelve tons of cement provide an alternative to bulk cement handling. Bulk cement is measured by the metric ton (1000 kg) or short ton (2000 lb).

Storage of Cement

Cement is a moisture-sensitive material; while dry, it will retain its quality indefinitely. Cement stored in contact with damp air or moisture can set more slowly and develop less strength than cement that is kept dry. At the cement plant and at ready mixed concrete facilities, bulk cement is stored in silos. The relative humidity in a warehouse or shed used to store bagged cement should be as low as possible. All cracks and openings in walls and roofs should be closed. Cement bags should not be stored on damp surfaces but should rest on raised platforms or pallets. Bags should be stacked closely together to reduce air circulation and should never be stacked against outside walls. Bags to be stored for long periods should be covered with tarpaulins or other waterproof covering (Figure 2-44).

FIGURE 2-44. When stored on the job, cement should be protected from moisture.

Cement stored for long periods may develop what is called warehouse pack. This can usually be corrected by rolling the bags on the floor. At the time of use, cement should be free-flowing and free of lumps. If lumps do not break up easily, the cement should be tested before it is used in important work. Standard strength tests or loss-on-ignition tests should be made whenever the quality of cement is in question.

Ordinarily, bulk cement does not remain in storage long, but it can be stored for long periods without deterioration. Bulk cement should be stored in weather tight concrete or steel bins or silos. Dry low-pressure aeration or vibration should be used in bins or silos to keep the cement flowable and avoid bridging. It is not uncommon for cement silos to hold only about 80% of their rated capacity because of loose packing of aerated cement increases volume.

Hot Cement

When cement clinker is pulverized in the grinding mill during cement manufacture, the friction of the particle grinding process generates heat. Freshly ground cement can be well above ambient temperatures, or hot when placed in storage silos at the cement plant. This heat dissipates slowly. Therefore, during summer months when demand is high, cement may still be hot when delivered to a ready mixed concrete plant or jobsite. Tests have shown that the effect of hot cement on the workability and strength development of concrete is not significant (Lerch 1955). Since cement makes up only about 7% to 15% by volume of a typical concrete mixture, the temperatures of the mixing water and aggregates play a much greater role in establishing the final temperature of the concrete as mixed (see Chapter 18).

REFERENCES

PCA's online catalog includes links to PDF versions of many of our research reports and other classic publications.
Visit: cement.org/library/catalog.

ACI Committee 223, *Guide for the Use of Shrinkage-Compensating Concrete*, ACI 223R-10, American Concrete Institute, Farmington Hills, MI, 2010.

ACI Committee 225, *Removal of the Optional Alkali Limit in ASTM C150, Standard Specifications for Portland Cement* ACI, 225.1T-20, Technical Note, American Concrete Institute, Farmington Hills, MI, 2020.

ACI Committee 225, *Guide to the Selection and Use of Hydraulic Cements*, ACI 225R-19, ACI Committee 225 Report, American Concrete Institute, Farmington Hills, MI, 2019.

ACI Innovation Task Group 10, *Practitioner's Guide for Alternative Cements*, ITG-10R-18, American Concrete Institute, Farmington Hills, MI, 2018a, 21 pages.

ACI Innovation Task Group 10, *Report on Alternative Cements*, ITG-10.1R-18, American Concrete Institute, Farmington Hills, MI, 2018b, 33 pages.

American Petroleum Institute (API), *Specification for Cements and Materials for Well Cementing*, API Specification 10A, 25th Ed., API Publishing Services, Washington, DC, March 2019.

Bhatty, J.I.; Miller, F.M.; Kosmatka, S.H.; and Bohan, R.P., editors, *Innovations in Portland Cement Manufacturing*, SP400, Portland Cement Association, Skokie, IL, 2011, 1734 pages.

Bhatty, J.I., *Role of Minor Elements in Cement Manufacture and Use*, Research and Development Bulletin RD109, Portland Cement Association, Skokie, IL, 1995, 48 pages.

Bhatty, J.I.; and Tennis, P.D., *U.S. and Canadian Cement Characteristics: 2004*, SN2879, Portland Cement Association, Skokie, IL, 2008, 67 pages.

Bharadwaj, K.; Chopperla, K.; Choudhary A.; Glosser, D.; Ghantous, R.; Vasudevan, G.; Ideker, J.; Isgor, O.; Trejo, D.; and Weiss, J., *Impact of the Use of Portland-Limestone Cement on Concrete Performance as Plain or Reinforced Material*, Oregon State University, Corvallis, Oregon, 2021, 320 pages.

Bogue, R.H., "Calculation of Compounds in Portland Cement," *Analytical Edition, Industrial and Engineering Chemistry*, Vol. 1, No. 4, October 15, 1929, pages 192 to 202.

Brouwers, H.J.H., *A Hydration Model of Portland Cement Using the Work of Powers and Brownyard*, Eindhoven University of Technology, Eindhoven, Netherlands, 2011, 168 pages.

Brunauer, S., *Some Aspects of the Physics and Chemistry of Cement*, Research Department Bulletin RX080, Portland Cement Association, Skokie, IL, 1957, 43 pages.

Brunauer, S., *Tobermorite Gel – The Heart of Concrete*, Research Department Bulletin RX138, Portland Cement Association, Skokie, IL, 1962, 20 pages.

British Standards Institute, *Calcium Aluminate Cement. Composition, Specifications and Conformity Criteria*, BS EN 14647, 2005.

Bushi, L. and Meil, J., *An Environmental Live Cycle Assessment of Portland-Limestone and Ordinary Portland Cements in Concrete*, Technical Brief, Athena Sustainable Materials Institute, January 2014, 10 pages.

Campbell, D.H., *Microscopical Examination and Interpretation of Portland Cement and Clinker*, SP030, Portland Cement Association, 1999, 214 pages.

Copeland, L.E.; Kantro, D.L.; and Verbeck, G., *Chemistry of Hydration of Portland Cement*, Research Department Bulletin RX153, Portland Cement Association, 1960.

Copeland, L.E.; and Schulz, E.G., *Electron Optical Investigation of the Hydration Products of Calcium Silicates and Portland Cement*, Research Department Bulletin RX135, Portland Cement Association, 1962.

CSA A3000, *Cementitious Materials Compendium*, A3000-18, CSA Group, Standards Council of Canada, Ottowa, Ontario, 2021, 253 pages.

Davidovits, J.; Davidovits, R.; and James, C., editors, *Geopolymer '99*, The Geopolymer Institute, Insset, Université de Picardie, Saint-Quentin, France, July 1999.

DeHayes, S.M., "C 109 vs. Concrete Strengths" *Proceedings of the Twelfth International Conference on Cement Microscopy*, International Cement Microscopy Association, Duncanville, Texas, 1990.

EN 197-1, European Standard for Cement – Part 1: *Composition, Specifications and Common Cements*, European Committee for Standardization (CEN), Brussels, 2000.

Farny, J.A., *White Cement Concrete*, EB217, Portland Cement Association, 2001, 32 pages.

Farny, J.A.; Melander, J.M.; and Panarese, W.C., *Concrete Masonry Handbook for Architects, Engineers, Builders*, EB008, 6th edition, Portland Cement Association, Skokie, IL, 2008, 308 pages.

Garboczi, E.J.; Bentz, D.P.; Snyder, K.A.; Martys, N.S.; Stutzman, P.E.; Ferraris, C.F.; and Bullard, J.W., *Concrete Materials Monograph*, National Institute of Standards and Technology, Gaithersburg, MD, 2019.

Gebhardt, R.F., "Survey of North American Portland Cements: 1994," *Cement, Concrete, and Aggregates, American Society for Testing and Materials*, West Conshohocken, PA, December 1995, pages 145 to 189.

Gonnerman, H.F.; Lerch, W.; and Whiteside, T.M., *Investigations of the Hydration Expansion Characteristics of Portland Cements*, Research Department Bulletin RX045, Portland Cement Association, Chicago, IL, 1953, 181 pages.

Greening, N.R.; Copeland, L.E.; and Verbeck, G.J., *Modified Portland Cement and Process*, United States Patent No. 3,628,973, Issued to Portland Cement Association, December 21, 1971.

Hansen, W.C.; Offutt, J.S.; Roy, D.M.; Grutzek, M.W.; and Shatzlein, K.J., *Gypsum and Anhydrite in Portland Cement*, 3rd edition, United States Gypsum Company, Chicago, IL, 1988, 101 pages.

Hawkins, P.; Tennis, P.D.; and Detwiler, R.J., *The Use of Limestone in Portland Cement: A State-of-the-Art Review*, EB227, Portland Cement Association, Skokie, IL, 2005, 44 pages.

Helmuth, R.; Hills, L.M.; Whiting, D.A.; and Bhattacharja, S., *Abnormal Concrete Performance in the Presence of Admixtures*, RP333, Portland Cement Association, Skokie, IL, 1995, 92 pages.

Helmuth, R.; and West, P.B., "Reappraisal of the Autoclave Expansion Test," *Cement, Concrete, and Aggregates*, CCAGDP, Vol. 20, No. 1, June 1998, pages 194 to 219.

Hills, L.M., "Clinker Formation and the Value of Microscopy," *Proceedings of the Twenty-Second International Conference on Cement Microscopy*, Montreal, 2000, pages 1 to 12.

Hooton, R.D.; and Thomas, M.D.A., *The Durability of Concrete Produced with Portland-Limestone Cement: Canadian Studies*, SN3142, Portland Cement Association, Skokie, IL, 2010, 28 pages.

Jafari Azad, V.; Suraneni, P.; Isgor, O. B.; and Weiss, W.J., "Interpreting the Pore Structure of Hydrating Cement Phases through a Synergistic Use of the Powers-Brownyard Model, Hydration Kinetics, and Thermodynamic Calculations," *ASTM Advances in Civil Engineering Materials*, Vol. 6, No. 1, 2017, pages 1 to 16.

Johansen, V.C.; Taylor, P.C.; and Tennis, P.D., *Effect of Cement Characteristics on Concrete Properties*, EB226, 2nd edition, Portland Cement Association, Skokie, IL, 2006, 48 pages.

Kabir, H.; and Hooton, R.D., *Evaluation of Cement Soundness using the ASTM C151 Autoclave Expansion Test*, SN3442, Portland Cement Association, Skokie, IL, 2020, 35 pages.

Klemm, W.A., *Ettringite and Oxyanion-Substituted Ettringites – Their Characterization and Applications in the Fixation of Heavy Metals: A Synthesis of the Literature*, Research and Development Bulletin RD116, Portland Cement Association, 1998, 80 pages.

Klemm, W.A., *Cement Soundness and the Autoclave Expansion Test: An Update of the Literature*, SN2651, Portland Cement Association, Skokie, IL, 2005, 20 pages.

Klieger, P.; and Isberner, A.W., *Laboratory Studies of Blended Cements – Portland Blast-Furnace Slag Cements*, Research Department Bulletin RX218, Portland Cement Association, 1967.

Lange, D. A.; and Stutzman, P. E., *The Concrete Microscopy Library*, http://publish.illinois.edu/concretemicroscopylibrary/, 2019.

Lea, F.M., *The Chemistry of Cement and Concrete*, 3rd ed., Chemical Publishing Co., Inc., New York, NY, 1971.

Lerch, W., *Hot Cement and Hot Weather Concrete Tests,* IS015.01T, Portland Cement Association, Chicago, IL, 1955.

Lerch, W., *The Influence of Gypsum on the Hydration and Properties of Portland Cement Pastes*, Research Department Bulletin RX012, Portland Cement Association, Chicago, IL, 1946.

Locher, F.W.; Richartz, W.; and Sprung, S., "Setting of Cement–Part I: Reaction and Development of Structure," *ZKG International*, Vol. 29, No. 10, 1976, pages 435 to 442.

Lothenbach, B.; Kulik, D.; Matschei, T.; Balonis, M.; Baquerizo, L.G.; Dilnesa, B.Z.; Miron, G.D.; and Myers, D., "Cemdata18: A Thermodynamic Database for Hydrated Portland Cements and Alkali-Activated Materials," *Cement and Concrete Research*, Vol. 115, January 2019a, pages 472 to 506.

Lothenbach, B.; and Zajac, M., "Application of Thermodynamic Modelling to Hydrated Cements," *Cement and Concrete Research*, Vol. 123, September 2019b, pages 1 to 21.

Matschei, T.; and Glasser, F.P., "The Influence of Limestone on Cement Hydration," *ZKG International*, Vol. 59, No. 12, 2006, pages 78 to 86.

Melander, J.M.; Farny, J.A.; and Isberner, A.W. Jr., *Portland Cement Plaster/Stucco Manual*, EB049.05, Portland Cement Association, Skokie, IL, 2003, 72 pages.

Molina, L., *On Predicting The Influence of Curing Conditions on the Degree of Hydration*, CBI Report 5:92, Swedish Cement and Concrete Research Institute, Stockholm, Sweden, 1992, 96 pages.

Moore, C.W., *Control of Portland Cement Quality*, EB121, Portland Cement Association, Skokie, IL, 2007, 170 pages.

NSF International/American National Standards Institute/Standards Council of Canada, *Drinking Water System Components – Health Effects*, NSF/ANSI/CAN 61, 2016, 292 pages.

Odler, I., *Special Inorganic Cements*, E&FN Spon, New York, NY, 2000, 420 pages.

PCA, *An Analysis of Selected Trace Metals in Cement and Kiln Dust*, SP109, Portland Cement Association, Skokie, IL, 1992, 60 pages.

PCA, *Emerging Technologies Symposium on Cements for the 21st Century*, SP206, Portland Cement Association, Skokie, IL, 1995, 140 pages.

Pfeifer, D.W.; and Perenchio, W.F., *Reinforced Concrete Pipe Made with Expansive Cements*, Research and Development Bulletin RD015, Portland Cement Association, Skokie, IL, 1973.

Poole, T., *Revision of Test Methods and Specifications for Controlling Heat of Hydration in Hydraulic Cement*, SN3007, Portland Cement Association, Skokie, IL, 2007, 42 pages.

Poole, T., "Predicting Seven-Day Heat of Hydration of Hydraulic Cement from Standard Test Properties," *Journal of ASTM International*, June 2009, 10 pages.

Powers, T.C., *Some Physical Aspects of the Hydration of Portland Cement*, Research Department Bulletin RX126, Portland Cement Association, Chicago, IL, 1961.

Powers, T.C., *The Nonevaporable Water Content of Hardened Portland-Cement Paste–Its Significance for Concrete Research and Its Method of Determination*, Research Department Bulletin RX029, Portland Cement Association, Chicago, IL, 1949.

Powers, T.C.; and Brownyard, T.L., *Studies of the Physical Properties of Hardened Portland Cement Paste*, Research Department Bulletin RX022, Portland Cement Association, Chicago, IL, 1947.

Russell, H.G., *Performance of Shrinkage-Compensating Concretes in Slabs*, Research and Development Bulletin RD057, Portland Cement Association, Skokie, IL, 1978.

Stark, D., *Performance of Concrete in Sulfate Environments*, Research and Development Bulletin RD129, Portland Cement Association, Skokie, IL, 2002.

Sullivan, E. J.; Zwicke, D.; Lan, D.; Schmidt, B.; Chiappe, J.T.; Storck, T.; Beer, K.; Tate, W.; and Jang, E., *U.S. Cement Industry Annual Yearbook: 2023*, Portland Cement Association, Skokie, IL, 2023, 54 pages.

Tang, F.J., *Optimization of Sulfate Form and Content*, Research and Development Bulletin RD105, Portland Cement Association, Skokie, IL, 44 pages, 1992.

Taylor, P., *Specifications and Protocols for Acceptance Tests on Processing Additions in Cement Manufacturing*, NCHRP Report 607, Transportation Research Board, Washington, D.C., 2008, 96 pages.

Taylor, H.F.W., *Cement Chemistry*, Thomas Telford Publishing, London, 1997, 477 pages.

Tennis, P.D.; and Jennings, H.M., "A Model for Two Types of Calcium Silicate Hydrate in the Microstructure of Portland Cement Pastes," *Cement and Concrete Research*, Vol. 30, No. 6, 2000, pages 855 to 863.

Tennis, P.D.; Thomas, M.D.A.; and Weiss, W.J., *State-of-the-Art Report on Use of Limestone in Cements at Levels of up to 15%*, SN3148, Portland Cement Association, Skokie, IL, 2011, 78 pages.

Tennis, P.D., *Chemical and Physical Characteristics of US Cements: 2014*, SN3284, Portland Cement Association, Skokie, IL, 2016, 32 pages.

Toler, H.R., *Flowability of Cement*, Research Department Report MP-106, Portland Cement Association, Chicago, IL, October 1963.

USGS, Mineral Industry Surveys, Cement in July 2023, Geological Survey (U.S.), Mineral Resource Surveys Program (U.S.), September 23, 2023, 28 pages.

Weaver, W.S.; Isabelle, H.L.; and Williamson, F., "A Study of Cement and Concrete Correlation," *Journal of Testing and Evaluation*, American Society for Testing and Materials, West Conshohocken, Pennsylvania, January 1974, pages 260 to 280.

Wood, S.L., *Evaluation of the Long-Term Properties of Concrete*, Research and Development Bulletin RD102, Portland Cement Association, Skokie, IL, 1992, 99 pages.

Zhang, M.-H.; Bremner, T.W.; Malhotra, V.M., "The Effect of Portland Cement Types on Performance," *Concrete International*, January 2003, pages 87 to 94.

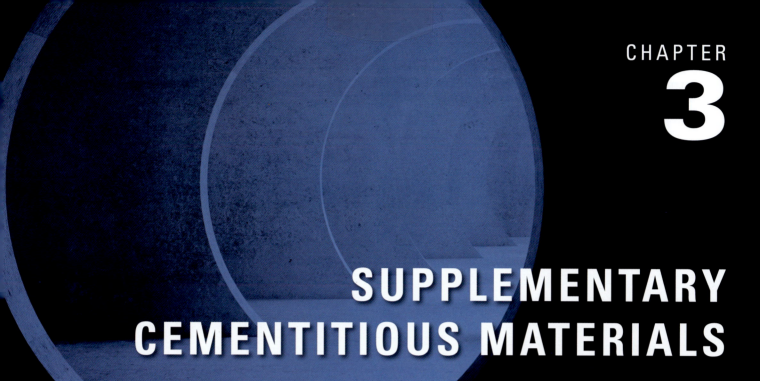

SUPPLEMENTARY CEMENTITIOUS MATERIALS

FIGURE 3-1. Supplementary cementitious materials. From left to right, fly ash (Class C), metakaolin (calcined clay), silica fume, fly ash (Class F), slag cement, and calcined shale.

Supplementary cementitious materials (SCMs), also called supplementary cementing materials, are materials that when used in conjunction with portland or blended cement, contribute to the properties of concrete through hydraulic or pozzolanic activity, or both. Hydraulic materials react chemically with water to form cementitious compounds. A pozzolan is a fine siliceous or aluminosiliceous material that chemically reacts with calcium hydroxide to form calcium silicate hydrate and other cementitious phases. Fly ash, slag cement, silica fume, natural pozzolans (such as volcanic ashes, pumiceous tephra, and metakaolin), and ground glass, may be classified as SCMs (Figure 3-1). Tables 3-1 and 3-2 list the applicable U.S. and Canadian specifications these materials are required to meet and show the wide range of classes and types of SCMs available. Table 3-3 provides typical chemical analyses and Table 3-4 gives selected properties of SCMs. Figure 3-2 shows typical ranges of SiO_2, Al_2O_3, and CaO for several SCMs on a ternary diagram.

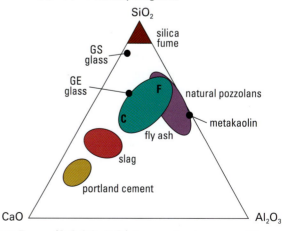

FIGURE 3-2. Ternary (CaO-SiO_2-Al_2O_3) diagram of various cementitious materials, in weight percent (adapted from Lothenbach and others 2011).

TABLE 3-1. US Standard Specifications* and Classes of Supplementary Cementitious Materials

STANDARD SPECIFICATION	TYPES OF SCM		CLASSIFICATION	PROPERTIES
ASTM C618, *Standard Specification for Coal Ash and Raw or Calcined Natural Pozzolan for Use in Concrete* (AASHTO M 295)	Natural pozzolans	Pumice/pumicte, Obsidian, Rhyolitic glassy tuff, Perlite, Volcanic ashes, Shales, Diatomaceous earths	Class N	Raw
		Calcined clay (including Metakaolin), Calcined shale		Calcined
	Coal ash	Fly ash, Bottom ash, Harvested ash	Class F	Pozzolanic properties
			Class C	Pozzolanic and cementitious properties
ASTM C989, *Standard Specification for Slag Cement for use in Concrete and Mortars* (AASHTO M 302)	Slag cement		Grade 80	Low activity index
			Grade 100	Moderate activity index
			Grade 120	High activity index
ASTM C1240, *Standard Specification for Silica Fume Used in Cementitious Mixtures* (AASHTO M 307)	Silica fume			Pozzolanic
ASTM C1697, *Standard Specification for Blended Supplementary Cementitious Materials*	Blended SCMs	xx, yy and zz refer to the amounts in mass %	A, B, and possibly C are the types of SCMs as listed below:	Type SCMb-Axx/Byy/Czz[†]
		Class N pozzolan	N	
		Class F fly ash	F	
		Class C fly ash	C	
		Silica fume	SF	
		Slag cement	S	
ASTM C1866, *Standard Specification for Ground-Glass Pozzolan for Use in Concrete*	Ground-glass pozzolan		Class GS	Ground soda-lime glass
			Class GE	Ground fiberglass
AASHTO M 321, *Standard Specification for High-Reactivity Pozzolans for Use in Hydraulic-Cement Concrete, Mortar, and Grout*	High-reactivity pozzolans	Metakaolin, Rice hull ash, Ultra-fine fly ash, and other materials		

* In addition, ASTM C1709, *Standard Guide for Evaluation of Alternative Supplementary Cementitious Materials (ASCM) for Use in Concrete*, is a protocol for evaluating potential SCMs not covered by standard specifications.

[†] For example, SCMb-75S/25F indicates a blended SCM with 75% slag cement and 25% Class F fly ash.

The practice of using SCMs in concrete mixtures has been growing in North America since the 1970s. Many of these materials are byproducts of industrial processes. Their judicious use provides performance benefits to concrete as well as contributing positively to sustainability characteristics. SCMs are used in about 65% of ready-mixed concrete (Obla and others 2012).

Supplementary cementitious materials may be used to improve a particular concrete property, such as controlling heat of hydration in mass concrete or resistance to alkali-silica reactivity. The quantity used is dependent on the properties of the specific materials and the desired effect on concrete performance. The appropriate amount to use should be based on field performance or established by testing to verify that fresh and hardened concrete properties are as anticipated.

Supplementary cementitious materials are added to concrete as part of the total cementitious system. Typical practice in the U.S. uses SCMs, or combinations of them, as additions to, or as a partial replacement for, hydraulic cements (including portland or blended cement) in concrete. Blended cements, which may already contain pozzolans, slag, or limestone, are designed to be used with or without additional SCMs (keeping in mind total SCM content limits for some applications). The use of these materials in blended cements is discussed in Chapter 2, and by Thomas (2013), Thomas and Wilson (2002), and Detwiler and others (1996).

TABLE 3-2. Canadian Classes of Supplementary Cementitious Materials*

TYPE OF SCM	CLASSIFICATION	PROPERTIES
Fly ash	Type F	low CaO content (≤15%)
	Type CI	intermediate CaO content (>15% and ≤20%)
	Type CH	high CaO content (>20%)
Ground glass	Type G$_H$	high alkali content (>4% and ≤13%)
	Type G$_L$	with low alkali content (≤4%)
Ground granulated blast-furnace slag	Type S	
Natural pozzolan	Type N	
Silica fume	Type SF	high SiO$_2$ content (≥85%)
	Type SFI	intermediate SiO$_2$ content (≥75%)
Blended SCMs	Type BMb	

* CSA A3001, *Cementitious Materials for Use in Concrete.*

FLY ASH

Fly ash, the most widely used SCM in concrete in the US, is a byproduct of the combustion of pulverized coal in electric power generating plants. During combustion, the coal's mineral impurities (such as clay, feldspar, quartz, and shale) fuse in suspension and are carried away from the combustion chamber by the exhaust gases. In the process, the fused material cools and solidifies into spherical glassy particles called fly ash. The fly ash is collected from the exhaust gases by electrostatic precipitators or bag filters as a finely divided powder (Figure 3-3). In some cases, fly ash stored in ponds or other impoundments has been reclaimed (also known as harvesting), processed, and used as an SCM in concrete (Diaz-Loya and others 2019).

FIGURE 3-3. Fly ash, a byproduct of the combustion of pulverized coal in electric power generating plants, has been used in concrete since the 1930s.

TABLE 3-3. Typical Chemical Analysis of Supplementary Cementitious Materials (% by mass)

	CLASS F FLY ASH	CLASS C FLY ASH	SLAG CEMENT	SILICA FUME	RAW NATURAL POZZOLAN	CALCINED CLAY	CALCINED SHALE	META-KAOLIN	TYPE GE GROUND GLASS	TYPE GC GROUND GLASS
SiO$_2$	52	35	35	90	71	58	50	53	60	71
Al$_2$O$_3$	23	18	12	0.4	15	29	20	43	13	1
Fe$_2$O$_3$	11	6	1	0.4	1	4	8	0.5	0.4	0.5
CaO	5	21	40	1.6	1.0	1	8	0.1	21	10
SO$_3$	0.8	4.1	2	0.4	0.1	0.5	0.4	0.1	<0.1	<0.1
Na$_2$O	1.0	5.8	0.3	0.5	3.7	0.2	—	0.05	0.8	13
K$_2$O	2.0	0.7	0.4	2.2	3.9	2	—	0.4	0.1	0.3
Na$_2$Oeq (total)	2.2	6.3	0.6	1.9	6.2	1.5	—	0.3	0.9	13.2

TABLE 3-4. Selected Properties of Supplementary Cementitious Materials

	CLASS F FLY ASH	CLASS C FLY ASH	SLAG CEMENT	SILICA FUME	RAW NATURAL POZZOLAN	CALCINED CLAY	CALCINED SHALE	META-KAOLIN	TYPE GE GROUND GLASS	TYPE GC GROUND GLASS
Loss on ignition, %	1.5	0.5	1.0	3.0	3.4	1.5	3.0	0.7	<0.1	0.1
Fineness*, m²/kg	420	420	400	20,000†	500	990	730	17,000†	400	400
Relative density	2.38	2.65	2.94	2.40	2.45	2.50	2.63	2.50	2.51	2.55

* Blaine
† Nitrogen absorption

The potential for using fly ash as an SCM in concrete has been known since the early 1900s (Anon 1914 and Davis and others 1937). However, fly ash from coal-burning electricity generating plants did not become widely available until the 1930s (ACI 232 2018). The first major use of fly ash in the U.S. was in the construction of the Hungry Horse Dam in Montana (USBR 1948). Today, fly ash is used in more than 50% of ready mixed concrete (Obla and others 2012). Class F fly ash is often used at dosages of 15% to 25% by mass of cementitious material and Class C fly ash is used at dosages of 15% to 40% by mass of cementitious material. High volume fly ash (HVFA) concrete contains more than 40% fly ash by mass of cementitious material. Dosage varies with the reactivity of the ash and the desired effects on concrete properties (Thomas 2007, Helmuth 1987, and ACI 232 2018).

In North America, the vast majority of the fly ash used in concrete is added at the mixer as a separate ingredient. A small proportion of concrete incorporates fly ash as an ingredient in Type IP blended cement. The use of blended cements containing fly ash (or other SCMs) is more common in European countries and others outside of North America.

ASTM C311, *Standard Test Methods for Sampling and Testing Fly Ash or Natural Pozzolans for Use in Portland-Cement Concrete*, provides test methods for fly ash (and natural pozzolans) for use as SCMs in concrete.

Beneficiation of Fly Ash

Fly ash is typically used directly from the precipitators with no additional processing before mixing in concrete. There are some sources of fly ash which do not meet the requirements for use in concrete; these materials might be beneficiated for improved characteristics by optimizing material fineness and reducing carbon content (ACI 232 2018). Particle size reductions have been achieved by air classifiers and grinding (Cornelissen and others 1995, and Obla and others 2001). There are also technologies available to reduce the carbon content of fly ash including triboelectric separation (Whitlock 1993), carbon burnout (Cochran and Boyd 1993), and froth flotation (Groppo 2001). Likewise, some harvested fly ash may need additional processing (for example drying, grinding, carbon content reduction) before use in concrete (Tritsch and others 2020).

Classification of Fly Ash

Fly ashes are divided into two classes in accordance with ASTM C618, *Standard Specification for Coal Ash and Raw or Calcined Natural Pozzolan for Use in Concrete* (AASHTO M 295): Class F and Class C.

Class F fly ash has primarily pozzolanic properties. When compared to Class F fly ash, Class C fly ash is less pozzolanic but is typically is more cementitious. Both classes of fly ash meet the applicable requirements provided in ASTM C618. Although both Class F and Class C fly ash are required to have

the contents of SiO_2, Al_2O_3, and Fe_2O_3 sum to 50% or more, as a primary distinguishing characteristic, the calcium oxide (CaO) content must be less than or equal to 18% for Class F fly ash and greater than 18% for Class C fly ash.

As shown in Table 3-2, CSA A3000 designations for fly ash are different, with calcium oxide (CaO) content requirements of 15% or less to define Type F fly ash, between 15 and 20% to define Type CI fly ash, and more than 20% to define Type C ash.

Physical Properties of Fly Ash

Most fly ash particles are fine solid spheres although some are hollow cenospheres (Figure 3-4) or plerospheres, hollow spherical particles containing smaller spheres. The particle sizes in fly ash vary from less than 1 μm to more than 100 μm with the average particle size measuring less than 20 μm (typically 10 μm). Typically, only 10% to 30% of the particles by mass are larger than 45 μm. The Blaine fineness is typically 300 m²/kg to 500 m²/kg, although some fly ashes can have surface areas as low as 200 m²/kg and some as high as 700 m²/kg. For fly ash without compaction, the bulk density (mass per unit volume including air between articles) can vary from 540 kg/m³ to 860 kg/m³ (34 lb/ft³ to 54 lb/ft³), whereas with close packed storage or vibration, the range can be 1120 kg/m³ to 1500 kg/m³ (70 lb/ft³ to 94 lb/ft³). The relative density (specific gravity) of fly ash generally ranges between 1.90 and 2.80 and the color ranges from gray to tan.

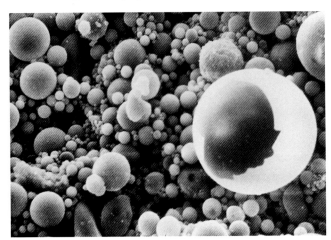

FIGURE 3-4. Scanning electron microscope (SEM) micrograph of fly ash particles. Field width approximately 235 μm.

Chemical Properties of Fly Ash

Fly ash is primarily aluminosilicate glass containing silica, alumina, iron, and calcium. Minor constituents are magnesium, sulfur, sodium, potassium, and carbon. The chemical composition of fly ash depends on the source of the coal as well as the combustion process. Because of the variety of coal types mined in North America, the composition of fly ash covers a wide range. Hard bituminous or anthracitic coals tend to produce ashes high in silica and alumina but low in calcium (Class F). Fly ashes from softer lignite of subbituminous coals tend to be higher in calcium – anywhere from a minimum of 18% to more than 30% CaO – and lower in silica and alumina (Class C).

The performance of fly ash in concrete is strongly influenced by its chemical composition. Fly ash from different sources can behave very differently. As the molten droplets of fly ash leave the burning zone in a coal plant, the surface cools rapidly, preventing the formation of well-ordered crystalline phases, and the fly ash is dominated by amorphous or glassy material. The glass content can typically be as low as 50% and as high as 90%. The interior of the droplets cools more slowly, especially in the larger particles; this permits the formation of crystalline material.

The abundance and composition of the glass in fly ash is important. In low-calcium fly ash, the crystalline phases that form – quartz, mullite, hematite, and magnetite – are inert in concrete; only the glass phase reacts and contributes to the properties of the concrete. The glass phase in low-calcium fly ash (typically Class F) is an aluminosilicate that will not react with water unless there is a source of calcium hydroxide or some other activator present. This material is pozzolanic.

As the calcium content in the fly ash increases, the fly ash becomes more hydraulic. Some of the calcium is incorporated into the glass (Hemmings and Berry 1988). This changes the structure of the glass and makes it more reactive. It is possible for the glass to become somewhat hydraulic (able to react directly with water). In most instances, Class C fly ash is both hydraulic as well as pozzolanic.

Refer to ACI 232.2R, *Report of the Use of Fly Ash in Concrete* (2018), Thomas (2007), and Thomas (2013) for more information on using fly ash in concrete.

SLAG CEMENT

Slag cement, previously known as ground, granulated blast-furnace slag (GGBFS), is the glassy material formed from molten slag produced in blast furnaces as a byproduct of the production of iron used in steel making. The molten slag is formed as the non-ferrous ingredients from the iron ore and fluxes used to make iron melt, at a temperature of about 1500°C (2730°F), and floats above the denser molten iron. In order to transform the molten slag into a cementitious material, it is rapidly quenched in water to form a glassy, sand-like, granulated material, then dried and ground into a fine powder (Figures 3-5 and 3-6). If the slag is allowed to cool slowly in air, it will form crystalline products that have no cementitious properties. Air-cooled slag is inert, and is used for other applications, such as aggregate for structural backfill.

The use of slag cement as a cementitious material in combination with slaked lime dates back to 1774 (Mather 1957). One of the first major uses of slag-lime cements was in the construction of the Paris underground in the late 1800s (Thomas 1979). Blended slag-portland cements were introduced in the

FIGURE 3-5. Ground granulated blast-furnace slag.

FIGURE 3-6. Scanning electron microscope micrograph of slag particles. Field width approximately 43 μm.

U.S. in 1896. When used in general purpose concrete in North America, slag cement commonly constitutes between 25% and 50% of the cementitious material in the mixture. Some concretes have a slag cement contents of 70% or more of the cementitious material by weight for applications such as mass concrete and marine environments. As with fly ash, the majority of slag cement used in the US is batched as a separate ingredient at the concrete plant, with 5% to 10% of ready-mixed concrete mixtures incorporating slag (Obla and others 2012).

Classification of Slag Cement

There are three grades of slag cement in accordance with ASTM C989, *Specification for Slag Cement for Use in Concrete and Mortars* (AASHTO M 302). Grade 80, Grade 100, and Grade 120. ASTM C989 (AASHTO M 302) defines the slag activity index (SAI) as the ratio of 28-day compressive strengths of mortars made with 50% of a reference portland cement and 50% slag cement and those for mortars made with the reference portland cement. Grade 80 requires a minimum average SAI of 75%, Grade 100 requires a minimum average SAI of 95% and Grade 120 requires a minimum average SAI of 115%. The average SAI limits listed are determined on the average of the last consecutive 5 samples. Limits on 28-day SAI values of 70, 90, or 110 for individual samples for Grades 80, 100, and 120, respectively, also apply. The 7-day slag activity index is also reported as additional information on slag cement mill test reports for Grades 100 and 120.

Physical Properties of Slag Cement

The rough and angular-shaped granulated slag is ground to produce slag cement, a fine powder material of approximately the same or greater fineness than portland cement. Most of the angular-shaped particles are ground to less than 45 μm and have a Blaine fineness of about 400 m²/kg to 600 m²/kg. The relative density (specific gravity) for slag cement is typically between 2.85 and 2.95. The bulk density varies from 1050 kg/m³ to 1375 kg/m³ (66 lb/ft³ to 86 lb/ft³). The appearance of the finished product resembles white cement.

Chemical Properties of Slag Cement

The chemical composition of slag cement depends mainly on the composition of the charge to the blast furnace – oxides of silica, calcium, alumina, magnesium, and iron generally make up more than 95% of the slag cement. Although the composition may vary between sources, the variation from an individual plant is generally low due to the requirements of the iron-making process.

There have been a number of attempts to relate chemistry of slag cements to their reactivity through various types of hydraulic moduli (Hooton and Emery 1983, Pal and others 2003). Due to the complexities of the interactions between slag cement glass content, primary oxides, fineness, and alkali loading of the concrete, ASTM C989 (AASHTO M 302) uses a performance-based approach, the slag activity index (SAI) to classify slag cements (see **Classification of Slag Cement**).

Slag cement in the presence of water and an activator such as NaOH, KOH, or Ca(OH)$_2$ (all of which are supplied by portland cement) hydrates and sets in a manner similar to portland cement. Slag cement will still hydrate and set in the absence of an activator, but the process is slow.

Refer to ACI 233R, *Guide to the Use of Slag Cement in Concrete and Mortar* (2017) for more information on slag cement.

SILICA FUME

Silica fume is the ultrafine non-crystalline silica produced in electric-arc furnaces as an industrial byproduct of the production of silicon and ferrosilicon alloys. Silica fume is also known as condensed silica fume or microsilica. In silicon metal production, a source of high purity silica (such as quartz or quartzite) together with wood chips and coal are heated in an electric arc furnace to remove the oxygen from the silica (reducing conditions). Silica fume rises as an oxidized vapor from the 2000°C (3630°F) furnaces. When it cools it condenses and is collected in baghouse filters. Silica fume is processed to remove impurities and to assure that the final produce meets applicable specifications.

Silica fume was first collected in Kristiansand, Norway, in 1947. Investigations into silica fume's properties continued for many years (Bernhardt 1952) and one of the first structural placements occurred in Norway in 1971 (Fiskaa and others 1971). In late 1983, the U.S. Army Corps of Engineers selected silica fume for its high abrasion resistance for use in the repair of Kinzua Dam in Pennsylvania (Holland and others 1986).

Silica fume is typically used in amounts between 5% and 10% by mass of the total cementitious material. It is used in applications where a high degree of impermeability is needed or high-strength concrete is required. In the U.S. the trend has been to add silica fume as a separate ingredient to the concrete mixer, but silica fume can also be used as a blended cement ingredient in ASTM C595 (AASHTO M 240) Type IP or Type IT.

Classification of Silica Fume

Silica fume for use as a pozzolanic material in concrete may be supplied in one of three forms in accordance with ASTM C1240, *Standard Specification for Silica Fume Used in Cementitious Mixtures* (AASHTO M 307); undensified, as a water-based slurry, or densified. Densified silica fume is by far the most common version used in the US in ready-mixed and precast concrete applications. Undensified silica fume is utilized to a much smaller degree, primarily in specialty applications such as Ultra High-Performance Concrete (see Chapter 23). The use of silica fume in slurry form is relatively rare. Both undensified silica fume and slurried silica fume can present challenges in transportation, handling, and storage.

Physical Properties of Silica Fume

Silica fume is composed of microscopic spherical amorphous (non-crystalline) particles. It is extremely fine with particles less than 1 μm in diameter and with an average diameter of about 0.1 μm, about 100 times smaller than average cement particles (Figures 3-7 and 3-8).

Silica fume has a surface area of about 20,000 m²/kg (nitrogen adsorption method). For comparison, tobacco smoke's surface area is about 10,000 m²/kg. Type I and Type III cements typically have surface areas between about 350 m²/kg to 450 m²/kg and 550 m²/kg to 700 m²/kg (Blaine), respectively.

FIGURE 3-7. Silica fume powder.

FIGURE 3-8. Scanning electron microscope micrograph of silica fume particles. Field width approximately 15 μm.

The bulk density of silica fume varies from 130 kg/m³ to 430 kg/m³ (8 lb/ft³ to 27 lb/ft³). The relative density (specific gravity) is usually around 2.20. The color of silica fume is gray, with the actual shade primarily dependent on the carbon content. Specially processed white silica fume is available when color is an important consideration.

Silica fume is sold more commonly in powder form but is also available internationally as a slurry. Densification is typically achieved by passing compressed air through a silo containing silica fume causing the particles to tumble and collide resulting in loosely bound agglomerations. Alternatively, densification can be achieved by mechanical means – such as passing the material through a screw auger or pressure densification. These methods can increase the bulk density of silica fume from 600 kg/m³ to 740 kg/m³ (38 lb/ft³ to 46 lb/ft³).

Chemical Properties of Silica Fume

Silica fume is predominantly composed of silicon dioxide, SiO_2, in noncrystalline, glassy (amorphous) form. When produced during the manufacture of silicon metal, the silica content is usually above 90% by mass and may even be as high as 99%. For silica fumes produced during ferrosilicon alloy production, the quantity of silica in the fume decreases as the amount of silicon in the alloy decreases. For alloys with a nominal silicon content of 75%, the resulting silica fume typically contains 85% silica or more. However, the manufacture of alloys containing only 50% silicon will generally produce silica fumes with less than 85% silica; these materials may be not be permitted by some specifications for use in concrete. ASTM C1240 (AASHTO M 307) requires a minimum of 85% SiO_2 content, while Canadian Standards Association (CSA) A3001 defines a silica fume Type SFI as having a minimum of 75% SiO_2, in addition to Type SF which requires a minimum 85% SiO_2.

For more information on silica fume, see ACI 234R, *Guide to the Use of Silica Fume in Concrete* (2006), and Holland (2005).

NATURAL POZZOLANS

Natural pozzolans are generally produced from natural deposits. The North American experience with natural pozzolans dates back to early 20th century public works projects, such as dams, where they were used to control temperature rise in mass concrete. In the 1930s and 1940s, a blended cement containing 25% of calcined Monterey shale interground with portland cement was produced in California and was used by the State Division of Highways in several structures including the Golden Gate Bridge. In addition to controlling temperature rise, natural pozzolans were used to improve resistance to sulfate attack and were among the first materials found to mitigate alkali-silica reaction.

Metakaolin. Metakaolin is a calcined clay which is produced by calcination of kaolin clay. The clay is purified by water processing prior to very carefully controlled thermal activation at relatively low temperature (650°C to 800°C [1202°F to 1472°F]) compared to other calcined clays.

Metakaolin is used in applications where very low permeability or very high strength is required. In these applications, metakaolin is used more as an additive to the concrete rather than a replacement of cement; typical additions are around 10% of the cement mass.

Classification of Natural Pozzolans

Natural pozzolans are classified by ASTM C618 (AASHTO M 295) as Class N pozzolans. The most common are "raw" natural pozzolans, derived from volcanic glasses, pumice, perlite, tuff, diatomaceous earth, tephra, basalt, and similar materials. Other natural pozzolans are derived from shales or clays, that are calcined, or heated to a sufficient temperature to transform the clay structure into a disordered amorphous aluminosilicate with pozzolanic properties (Figures 3-9 and 3-10). This process is often referred to as thermal activation. The optimum heat treatment varies for different materials. After heating the material is usually ground. Less common natural pozzolans include rice husk ash (RHA) and diatomaceous earth.

FIGURE 3-9. Scanning electron microscope micrograph of calcined shale particles. Field width is approximately 20 μm.

FIGURE 3-10. Scanning electron microscope micrograph of calcined clay particles. Field width is about 40 μm.

FIGURE 3-11. Metakaolin, a calcined clay.

Metakaolin has been marketed under the description "high-reactivity metakaolin" to distinguish it from the normally less reactive calcined clays. AASHTO M 321, *Standard Specification for High-Reactivity Pozzolans for Use in Hydraulic Cement Concrete, Mortar and Grout*, provides detailed requirements.

Physical Properties of Natural Pozzolans

The average particle size of natural pozzolans is generally between 1 μm and 15 μm. Raw natural pozzolans generally have a relative density (specific gravity) of between 2.35 and 2.50, with a specific surface area of between 500 m^2/kg and 1500 m^2/kg. Calcined clays have a relative density (specific gravity) of between 2.40 and 2.61 with Blaine fineness ranging from 650 m^2/kg to 1350 m^2/kg.

Metakaolin is ground to a very high fineness with an average particle size of about 1 μm to 3 μm (Figure 3-11). It has a high Hunter L whiteness value of about 90 (on a scale of 0 for black to 100 for maximum whiteness), which tends to lighten the color of concrete.

For normal SCM contents, water demand for concretes with raw natural pozzolans are generally comparable to those without SCMs. When natural pozzolans are used at 20% of cementitious materials contents, compressive strengths at 28 days are generally 85% to 105% of that of comparable concrete mixtures without SCMs.

Chemical Properties of Natural Pozzolans

Calcined shale may contain on the order of 5% to 10% calcium, which results in the material having some hydraulic properties. Because of the amount of residual calcite that is not fully calcined, and the bound water molecules in the clay minerals, calcined shale will typically have a loss on ignition (LOI) of approximately 1% to 5%, although it may be higher. The LOI value for raw and calcined natural pozzolans is a reflection of this bound water, and is not an indication of carbon content as would be the case for fly ash. The LOI value for raw and calcined natural pozzolans does not impact air-entraining admixtures dosages.

Metakaolin is composed of kaolin clay consisting predominantly of silica (SiO_2) and alumina (Al_2O_3). The sum of the silica and alumina content is generally greater than 95% by mass. The remaining portion contains small amounts of iron, calcium, and alkalies. Raw natural pozzolans are also typically low in calcium, and high in alumina and silica, which are responsible for its reactivity.

ACI 232.1R, *Use of Raw or Processed Natural Pozzolans in Concrete* (2012) provides additional detail on use of natural pozzolans in concrete.

GROUND GLASS

The use of finely ground glass (Figures 3-12 and 3-13) as a pozzolan for concrete is relatively new in the US. In general, these pozzolans have a relatively high silica content and when ground fine enough, act as a pozzolan. They do not have hydraulic properties due to low calcium content. Typical use in concrete are at between 10% and 40% of the cementitious materials by mass. Approximately 10 million metric tonnes (11 million tons) of waste glass is produced annually in the US, although more than 25% of it recycled (EPA 2019) into other glass products. Statistics on the use of ground-glass pozzolan in concrete in the US are not yet available, as a specification, ASTM C1866, *Standard Specification for Ground-Glass Pozzolan for Use in Concrete*, became available in 2020.

FIGURE 3-12. Ground glass, when ground fine enough, acts as a pozzolan (courtesy of Urban Mining).

FIGURE 3-13. Scanning electron microscope micrograph of ground-glass pozzolan. Field width approximately 120 µm (courtesy of L. Sutter).

Classification of Ground Glass

Ground glass is specified under ASTM C1866, *Standard Specification for Ground-Glass Pozzolan for Use in Concrete.* Two types of ground glass are covered: Type GS, ground soda-lime glass and Type GE, which is typically ground fiberglass. Soda-lime glass is the most commonly produced glass and used for beverage and food containers, as well as windowpanes. Fiberglass, most typically used for insulation and reinforcement of plastics and other polymers, is the source of Type GE ground-glass pozzolan.

Physical Properties of Ground Glass

Both types of ground-glass pozzolan are white powders, with finenesses limited by 45-µm (No. 325) sieve residue of 5.0% maximum. Specific surface areas are generally around 400 m²/kg (Blaine). Specific gravity of Type GE is about 2.60 while that of Type GS is about 2.51. As with other pozzolans, early strengths of concretes made with ground glass can be slightly lower initially, but by 28 days often is higher than a similar concrete made without pozzolans. Although ground-glass pozzolan particles are relatively angular, use of ground glass typically results in slight decreases in water demand or slight increases in workability.

Chemical Properties of Ground Glass

Both types of ground-glass pozzolan are predominantly silica, with Type GS typically more than 70% SiO_2 and Type GE around 60%. The silica is more than 95% amorphous rather than crystalline. A primary distinction between the two types is the alkali content: Type GS has an equivalent alkali content typically between 10% and 15%, while Type GE typically has less than 1% equivalent alkalies. Type GE glass also has a higher alumina content, typically 10% to 15%. The alkali content difference makes Type GE much more suitable for reducing expansion due to ASR. Typical chemistry is given in Table 3-3. The loss on ignition (LOI) is limited by specification to 0.5% maximum. Values above that amount, typically from label adhesives, can increase the entrained air in concrete.

A NOTE ON GROUND CALCIUM CARBONATE AND MINERAL FILLERS

Ground calcium carbonate and other mineral fillers are ingredients that are added to concrete to improve some characteristics of concrete. In some cases fresh properties, such as workability, may be enhanced, and mineral fillers are common ingredients in some self-consolidating concrete mixtures (see Chapter 23). It should be emphasized that mineral fillers are NOT supplementary cementitious materials. These ingredients are relatively inert, and do not chemically react to a significant extent with water, the high pH pore solution of concrete, or with calcium hydroxide. There are benefits to using fillers in some concrete mixtures, however, they do not provide improvements to strength and durability attributable to SCMs' pozzolanic and hydraulic reactions. More information on using mineral fillers in concrete can be found in ACI 211.7R, *Guide for Proportioning Concrete Mixtures with Ground Calcium Carbonate and Other Mineral Fillers* (2020) and Chapter 5.

ALTERNATIVE SCMS

Supplementary cementitious materials that meet their standard specifications will generally perform well in concrete mixtures when used in proper proportions for the designated application. In addition to these materials, there are SCMs that do not meet the standard specifications which may also perform well in some concrete mixtures and applications. Such materials, which fall outside the scope of ASTM C618, ASTM C989, ASTM C1240, and ASTM C1866, are referred to as alternative supplementary cementitious materials (ASCMs). As an example, a fly ash with a carbon content that exceeds that permitted in C618 may perform well in concretes that do not require entrained air. ASTM C1709, *Standard Guide for Evaluation of Alternative Supplementary Cementitious Materials for Use in Concrete*, provides guidance on testing such materials to assure that the fresh and hardened concrete properties are appropriate for the intended concrete application. This standard is not a specification, but a guide for testing such materials.

REACTIONS OF SCMS

Pozzolanic Reactions

As discussed in Chapter 2, calcium silicate hydrate (C-S-H), is the primary cementitious binding phase in hardened portland cement concrete and is often referred to as the glue that holds the aggregates together. Calcium hydroxide (CH) has little to no cementitious properties and contributes only slightly to the strength of the hydrated material. It is easily leached by water and attacked by chemical agents. However, CH does help maintain the high pH necessary to stabilize C-S-H.

A siliceous pozzolan reacts with calcium hydroxide formed by the hydration of portland cement and the reaction results in the production of more C-S-H. The general pozzolanic reaction can be illustrated as (Helmuth 1987):

$$xCH + yS + zH \rightarrow C_xS_yH_{x+z}$$

EQUATION 3.1

Where: CH is calcium hydroxide, S is silica, H is water, C-S-H is calcium silicate hydrate, and x, y, and z are molar coefficients. Typically, the ratio of x to y is slightly lower in pozzolanic C-S-H than that produced by alite and belite, but this does not adversely impact performance. For many SCMs, pozzolanic C-S-H forms primarily after the pore structure of the paste is established by the reaction of portland cement phases, and acts to reduce porosity and close off pores, reducing concrete's permeability and increasing its strength.

Many pozzolans and slag cements contain reactive alumina in addition to silica. This alumina reacts to produce various calcium-aluminate hydrates (C-A-H) and calcium aluminosilicate hydrates (C-A-S-H). The alumina may also participate in reactions with the various sulfate phases in the system.

The rate of the pozzolanic reaction, or pozzolanic activity, is influenced by a number of parameters (Massazza 1998). The greater the surface area of the pozzolan, the faster it can react. This is what makes silica fume an extremely reactive material that contributes to concrete properties at very early ages. The composition and amount of glass in the pozzolan also affects the rate of reaction. For instance, calcium is a glass modifier and renders the glassy phases of Class C fly ash more rapidly reactive than Class F fly ash. Like most chemical reactions, the pozzolanic reaction rate increases with temperature. However, the pozzolanic reaction appears to be more sensitive to temperature than the normal hydration of portland cement. The solubility of glass increases with pH. This increases its availability for reaction with CH. As a result, a pozzolan will tend to react more quickly when used in concrete with higher alkali loading.

Hydraulic Reactions

Slag cement is referred to as a latent hydraulic material. This means that it will hydrate when it is mixed solely with water but the process is typically slow. Latent hydraulic materials benefit from chemical activation to promote significant hydration: the most effective activators for slag cement are calcium hydroxide or alkali compounds. Hydration of portland cement releases both calcium and alkali hydroxides and serves as a very efficient activator for slag cement. There are also some high calcium fly ashes which have latent hydraulic properties.

When portland cement hydrates, it produces C-S-H and CH together with other aluminate and ferrite phases. The hydration of slag cement is similar except that no calcium hydroxide is produced. A small amount of the calcium hydroxide released by the portland cement may be consumed by the hydration of the slag cement; this could be considered to be a form of pozzolanic reaction. The hydration of a portland cement-slag cement blend will produce more C-S-H and less CH than an ordinary portland cement concrete mixture.

As with pozzolans, the rate of reaction of slag cement in concrete is dependent on many factors. Reactivity increases with the fineness and glass content of the slag cement, and with temperature (ambient and concrete). The composition of the slag cement is also important. Because slag cement requires activation by the portland cement, it will tend to react faster when the ratio of portland cement to slag cement is higher. Like pozzolans, the rate of dissolution of the glass in slag cement increases at higher pH. Slag cement will react faster in concrete with higher alkali loadings.

Slag cement hydration is accelerated when the temperature is elevated. Like pozzolans, the rate of slag cement reaction is generally more sensitive to temperature effects than the rate of portland cement hydration.

Table 3-5 provides a summary of the pozzolanic and hydraulic character of various SCMs. Note that this is not necessarily related to reactivity, which depends on the fineness and the mineralogy of the material, but related to the type of reactions that take place in concrete. As shown in Table 3-3, the pozzolanic behavior of an SCM is directly related to its calcium content.

TABLE 3-5. Pozzolanic and Hydraulic Reactions of SCMs*

SCM	POZZOLANIC BEHAVIOR	HYDRAULIC BEHAVIOR	CALCIUM CONTENT
Silica fume, metakaolin	◆◆◆◆◆		Low (<1%)
Low-CaO fly ash, natural pozzolans, Type GS ground glass	◆◆◆◆		
Type GE ground glass	◆◆◆	◆	
High-CaO fly ash	◆◆◆	◆◆	
Slag cement	◆	◆◆◆◆	High (> 30%)

*Adapted from Thomas and Wilson 2002.

EFFECTS ON FRESHLY MIXED CONCRETE

This section provides a brief review of the freshly mixed concrete properties that SCMs affect and their degree of influence (assuming everything else is equal). Table 3-6 summarizes general effects that these materials have on the fresh properties of concrete mixtures. It should be noted that effects vary considerably between, and often within, classifications of SCMs. Specified performance should be evaluated for specific mixtures.

TABLE 3-6. The Impact of SCM Characteristics on the Fresh Properties of Concrete*

| | FLY ASH | | SLAG CEMENT | SILICA FUME | NATURAL POZZOLANS | | | GROUND GLASS |
	CLASS F	CLASS C			RAW	CALCINED CLAY/SHALE	METAKAOLIN	
Water demand	↓	↓	↓	↑	↔	↔	↑	↓
Workability	↑	↑	↑	↓	↑	↑	↓	↑
Bleeding and segregation	↓	↓	↕	↓	↔	↔	↓	↓
Setting time	↑	↕	↑	↔	↔	↔	↔	↔
Air content	↓	↓	↔	↓	↔	↔	↓	↔
Heat of hydration	↓	↕	↓	↕	↓	↓	↔	↓

* The properties will change depending on the material composition, fineness, dosage, and other mixture parameters.
  Adapted from Thomas and Wilson (2002) and Omran and others (2018).

Key: ↓ Lowers   ↑ Increases   ↕ May increase or lower   ↔ No impact

Water Demand

Of the various SCMs, fly ash (Class C and F) has the most beneficial effect on reducing water demand. Concrete mixtures containing fly ash generally require less water (about 1% to 10% less at normal dosages) for a given slump than concrete containing only portland cement. Higher dosages can result in greater water reduction (Table 3-7). However, fly ashes with a high percentage of coarse particles (larger than 45 µm) are less efficient in reducing the water demand and in extreme cases may increase the amount of water required up to 5% (Gebler and Klieger 1986). Fly ash reduces water demand in a manner similar to liquid chemical water reducers (Helmuth 1987).

The use of slag cement can result in a reduction in the water demand of concrete although the impact is dependent on the slag fineness and is less marked than with fly ash. In general, water demand for a given slump in concrete mixtures with slag cement will be 3% to 5% lower than ordinary portland cement concrete.

The water demand of concrete containing silica fume increases with higher amounts of silica fume. Some mixtures may not experience an increase in water demand when only a small amount (less than 5%) of silica fume is present.

Calcined clays and calcined shales generally have little effect on water demand at normal dosages; however, other natural pozzolans can significantly increase or decrease water demand. The use of finely ground calcined clay, such as metakaolin, may lead to an increased water demand, especially when used at higher dosages. The increased requirement for water can be offset by the use of water-reducing chemical admixtures.

Use of ground-glass pozzolan tends to result in a slight reduction in water demand.

Workability

Generally, the use of fly ash, slag cement, and calcined clay and shale increase workability. This means that for a given slump, concrete containing these materials are generally easier to place, consolidate, and finish. The use of SCMs generally aids the pumpability of concrete. However, fly ashes with high percentages of coarse particles (retained on a 45-µm [No. 325] sieve) or high carbon content (indicated by a high loss on ignition) can increase the water demand and decrease the workability (Thomas 2013). Slag cement generally increases the workability of concrete mixtures, typically reducing the water demand by 3% to 5%. This is less than fly ash, which may reduce the water demand by 3% for every 10% fly ash added (Thomas 2013).

Most concretes containing SCMs will generally have equal or improved finishability compared to similar concrete mixtures without them. Mixtures that contain higher dosages of supplementary cementitious materials, especially silica fume in amounts greater than about 4%, can increase the stickiness of a concrete mixture. Adjustments, such as the use of high-range water reducers, may be required to maintain workability and permit proper compaction and finishing.

Bleeding and Segregation

In general, the finer the supplementary cementing material, the lower the bleed rate and bleeding capacity. Increasing the SCM content will also typically lower bleeding.

TABLE 3-7. Effect of Fly Ash on Mixing Water Requirements for Air-Entrained Concrete*

FLY ASH MIX IDENTIFICATION	CLASS OF FLY ASH	FLY ASH CONTENT, % by mass of cementitious material	CHANGE IN WATER REQUIREMENT COMPARED TO CONTROL, %
C1A	C	25	-6
C1D	F	25	-2
C1E	F	25	-6
C1F	C	25	-8
C1G	C	25	-6
C1J	F	25	-6
C2A	C	50	-18
C2D	F	50	-6
C2E	F	50	-14
C2F	C	50	-16
C2G	C	50	-12
C2J	F	50	-10

* All mixtures had cementitious materials contents of 335 kg/m^3 (564 lb/yd^3), a slump of 125 mm ± 25 mm (5 in. ± 1 in.), and an air content of 6% ± 1%. Water-to-cementitious materials ratios varied from 0.40 to 0.48 (Whiting 1989).

Concretes using fly ash generally exhibit less bleeding and segregation than plain concretes (Table 3-8). This effect makes the use of fly ash particularly valuable in concrete mixtures made with aggregates that are deficient in fines. The reduction in bleed water is primarily due to the reduced water demand of mortars and concretes using fly ash (Gebler and Klieger 1986).

The effect of slag cement on bleeding and segregation is generally dependent on its fineness. Concretes containing ground slags of comparable fineness to that of the portland cement tend to show an increased rate and amount of bleeding than plain concretes, but this appears to have no adverse effect on segregation. Slag cements ground finer than portland cement tend to reduce bleeding.

The incorporation of silica fume in concrete can have a very profound effect on bleeding. Concrete mixtures containing normal levels of silica fume (5% to 10%) and w/cm below 0.50 may not bleed. Because of this, special attention needs to be paid to placing, finishing, and curing operations when using silica fume.

Raw natural pozzolans and calcined clays and calcined shales have little effect on bleeding. The effect will depend on fineness and level of replacement. For instance, a very finely ground calcined clay such as metakaolin will reduce bleeding.

The bleeding period of concrete with SCMs is typically extended over a longer duration compared to plain concrete. This is an important factor in timing of finishing operations (see Chapter 15).

TABLE 3-8. Effect of Fly Ash on Bleeding of Concrete (ASTM C232, AASHTO T 158)*

FLY ASH MIXTURES		BLEEDING	
IDENTIFICATION	CLASS OF FLY ASH	PERCENT	mL/cm^{2†}
A	C	0.22	0.007
B	F	1.11	0.036
C	F	1.61	0.053
D	F	1.88	0.067
E	F	1.18	0.035
F	C	0.13	0.004
G	C	0.89	0.028
H	F	0.58	0.022
I	C	0.12	0.004
J	F	1.48	0.051
Class C, average Class F, average		0.34 1.31	0.011 0.044
Control mixture		1.75	0.059

* All mixtures had cementitious materials contents of 307 kg/m^3 (517 lb/yd^3), a slump of 75 mm ± 25 mm (3 in. ± 1 in.), and an air content of 6 ± 1%. Fly ash mixtures contained 25% ash by mass of cementitious material (Gebler and Klieger 1986).

† Volume of bleed water per surface area.

Plastic Shrinkage Cracking. Concrete containing silica fume may exhibit an increase in plastic shrinkage cracking because of its low bleeding characteristics. The problem may be avoided by ensuring that such concrete is protected against drying, before, during, and after finishing (see Chapter 17). Other pozzolans and slag cement generally have little effect on plastic shrinkage cracking, unless they significantly delay set time, which can increase the risk of plastic shrinkage cracking.

Setting Time

The use of SCMs will generally retard the setting time of concrete (Table 3-9). The extent of set retardation depends on many factors including the fineness and composition of the SCM, the level of replacement used, amount and composition of the portland cement or blended cement (particularly its alkali content), water-to-cementitious materials ratio (w/cm), and temperature of the concrete. Set retardation is an advantage during hot weather, allowing more time to place and finish the concrete. However, during cold weather, pronounced retardation can occur with some materials, especially when used in higher quantities, significantly delaying finishing operations. Accelerating admixtures can be used to decrease the setting time.

Lower-calcium Class F fly ashes, slag cement, and some natural pozzolans tend to delay the time of setting of concrete. At normal laboratory temperatures the extent of the retardation is generally in the range of 15 minutes to one hour for initial set and 30 minutes to two hours for final set. Longer delays in setting will be experienced when higher quantities of SCMs are

used or concrete is placed at lower temperatures, while little effect may be observed at higher temperatures.

The situation for higher-calcium Class C fly ashes is less clear. The use of some Class C ashes may lead to a slight acceleration, reducing the time to initial and final set, especially in hot weather. Other Class C ashes can cause retardation of a few hours or more, especially when used at relatively high replacement levels. It is not possible to predict the effect based on the chemistry of the fly ash. Individual cement-ash-admixture combinations must be tested.

In some cases, pollution control techniques at coal-burning power plants can lead to fly ashes with sulfate phases that can impact the setting of concrete (Zunino and others 2018).

Silica fume is generally used at relatively low replacement levels and has little significant impact on the setting behavior of concrete. Likewise, most raw and calcined natural pozzolans have little effect on setting time unless used at high replacement rates for portland cement.

Air Content

Supplementary cementing materials generally require an increase in the amount of air-entraining admixture necessary due to the increase in fineness of the cementitious materials content.

For fly ash concrete mixtures, the amount of air-entraining admixture required to achieve a certain air content in the concrete is a primarily a function of the carbon content, fineness, and alkali content of the fly ash. Increases in alkali contents decrease air-entraining admixture dosage requirements,

TABLE 3-9. Effect of Fly Ash on Setting Time of Concrete*

FLY ASH MIXTURES		SETTING TIME, HR:MIN		RETARDATION RELATIVE TO CONTROL, HR:MIN	
IDENTIFICATION	CLASS OF FLY ASH	INITIAL	FINAL	INITIAL	FINAL
A	C	4:30	5:35	0:15	0:05
B	F	4:40	6:15	0:25	0:45
C	F	4:25	6:15	0:10	0:45
D	F	5:05	7:15	0:50	1:45
E	F	4:25	5:50	0:10	0:20
F	C	4:25	6:00	0:10	0:30
G	C	4:55	6:30	0:40	1:00
H	F	5:10	7:10	0:55	1:40
I	C	5:00	6:50	0:45	1:20
J	F	5:10	7:40	0:55	2:10
Class C, average Class F, average		4:40 4:50	6:15 6:45	0:30 0:35	0:45 1:15
Control mixture		4:15	5:30	–	–

* Concretes had a cementitious materials content of 307 kg/m³ (517 lb/yd³). Fly ash mixtures contained 25% ash by mass of cementitious material. Water to cementitious materials ratio = 0.40 to 0.45. Tested at 23°C (73°F) (Gebler and Klieger 1986).

while increases in the fineness and carbon content typically require an increase in dosage requirements. The carbon content of fly ash is indicated by its loss on ignition (LOI). A Class F fly ash with relatively high carbon content (near the 6% maximum LOI permitted by C618) may increase the required admixture dosage by as much as 5 times compared to a portland cement concrete without fly ash.

The LOI provides a measure of the quantity of carbon in a fly ash, but different fly ashes can have a similar LOI content and affect air entrainment differently. For a given fly ash, it is the adsorption properties of the carbon that determine the fly ash's impact on air entrainment, in addition to the quantity of carbon present.

ASTM C1827, *Standard Test Method for Determination of the Air-Entraining Admixture Demand of a Cementitious Mixture,* commonly known as the Foam Index Test, provides a qualitative assessment of the adsorption properties of a fly ash and the related impact on the required dosage of air-entraining admixture for a specific combination of fly ash, air entrainer, and portland cement, relative to a portland cement-only mixture. This is known as the relative volume of AEA and, although a good relative indicator, it cannot be used to quantify an AEA dosage in a concrete mixture. Changes in the foam index value for a fly ash (or other SCM) can be used to anticipate the need to increase or decrease the dosage based on changes in materials characteristics (Dodson 1990).

Slag cements can have a small effect on the required dosage of air-entraining admixtures. The increase in admixture dosage required with slag cement is associated with the fine particle size of these materials as compared with portland cement.

Silica fume has a marked influence on the air-entraining admixture dosage requirements, which in most cases rapidly increase with an increase in the amount of silica fume used in the concrete, on the order of 25% to 50% higher for typical silica fume contents, due to the large surface area of the silica fume, which adsorbs part of the admixture.

Heat of Hydration

Concretes made with the majority of supplementary cementing materials (fly ash, natural pozzolans, slag cement, and ground glass) typically have a lower heat of hydration than portland cement concrete. Consequently, their use will reduce the amount of heat built up during hydration in a concrete structure. Supplementary cementitious materials are often used specifically to reduce the temperature rise of mass concrete (Gajda 2007). However, the impact of SCMs on the heat of hydration must be confirmed prior to use through methods like (semi-) adiabatic calorimetry.

When Class F fly ash is used in cementitious systems, the contribution to the heat of hydration by the fly ash is approximately 50% of that of portland cement (Gajda and

VanGeem 2002). This value is an accepted approximation. Since all cements and fly ashes are different, if specific data are needed, then the heat of hydration should be measured. Class C fly ash is less commonly used in mass concrete because its heat of hydration is higher than that of Class F. The heat of hydration of Class C fly ash depends on the CaO content. For low-CaO Class C fly ashes, the heat of hydration may be similar to that of Class F fly ash. However, for higher-CaO Class C fly ashes, the heat of hydration may be similar to that of cement.

When used in sufficient quantities, slag cement reduces the temperature rise of concrete (Gajda 2007). Figure 3-14 shows the effect of 70% slag cement on the heat of hydration. The amount of reduction is based on the proportion of slag in the concrete. Different combinations of slag cement and portland and blended cements will behave differently.

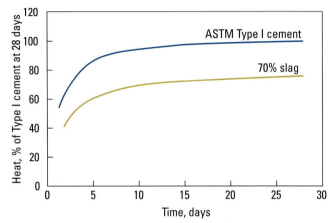

FIGURE 3-14. Effect of 70% slag cement on heat of hydration at 20°C (68°F) compared to a Type I cement.

Silica fume typically contributes a heat of hydration that is equal to or somewhat greater than that of portland cement depending on the dosage level. Unless specific testing is done to determine the heat of hydration of a specific mix design, an acceptable approximation is that the contribution to the heat of hydration of the cementitious system by silica fume is 100% to 120% that of portland cement (Pinto and Hover 1999). However, for a given strength, use of silica fume typically reduces total cementitious materials content which results in a lower heat of hydration.

At typical dosage rates, most calcined clays impart a heat of hydration similar to moderate heat cements (Barger and others 1997). However, metakaolin has a heat of hydration equal to or somewhat greater than that of portland cement. A good estimate is that the heat of hydration of metakaolin is 100% to 125% that of portland cement (Gajda 2007). To get a more exact determination of the heat of hydration, specific testing should be done.

Gajda (2007) and Detwiler and others (1996) provide reviews of the effect of pozzolans and slag cement on heat generation.

EFFECTS ON HARDENED CONCRETE

This section provides a summary of the hardened concrete properties that SCMs affect and their degree of influence, on the assumption that everything else is equal. These materials vary considerably in their effect on concrete mixtures as summarized in Table 3-10. The effect of different SCMs within a classification can also vary.

Strength

The extent to which strength development of concrete is influenced by supplementary cementitious materials will depend on many factors such as: composition and amount of SCM, alkali loading of the concrete, mixture proportions of the concrete, and temperature conditions during placement and curing.

In general, supplementary cementing materials (fly ash, slag cement, silica fume, calcined shale, calcined clay [including metakaolin], and ground glass) all contribute to the long-term strength gain of concrete (Figure 3-15).

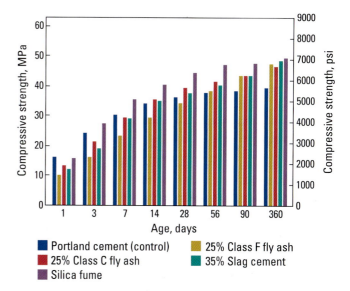

FIGURE 3-15. Strength development of a range of SCMs in air-entrained concretes with water-to-cementitious materials ratio of 0.45 (after Thomas 2013).

TABLE 3-10. The Impact of SCM Characteristics on the Hardened Properties of Concrete*

| | FLY ASH | | SLAG CEMENT | SILICA FUME | NATURAL POZZOLANS | | | GROUND GLASS |
	CLASS F	CLASS C			RAW	CALCINED CLAY/SHALE	METAKAOLIN	
Early age strength gain	↓	↔	↕	↑	↔	↔	↑	↓
Long term strength gain	↑	↑	↑	↑	↑	↑	↑	↑
Abrasion resistance	↔	↔	↔	↑	↔	↔	↔	↔
Drying shrinkage and creep	↔	↔	↔	↔	↔	↔	↔	↔
Permeability and absorption	↓	↓	↓	↓	↓	↓	↓	↓
Corrosion resistance	↑	↑	↑	↑	↑	↑	↑	↑
Alkali-silica reactivity	↓	↓	↓	↓	↓	↓	↓	↓
Sulfate resistance	↑	↕	↑	↑	↑	↑	↑	↑
Freezing and thawing	↔	↔	↔	↔	↔	↔	↔	↔
Deicer scaling resistance	↕̄	↕̄	↕̄	↕̄	↕̄	↕̄	↕̄	↕̄

* The properties will change depending on the material composition, fineness, dosage, other mixture parameters, as well as curing conditions, temperature, and other variables. These general trends may not apply to all materials and therefore testing should be performed to verify the impact. (Adapted from Thomas and Wilson 2002, and Omran and others 2018.)

Key: ↓ Lowers   ↑ Increases   ↕ May increase or lower   ↕̄ No impact   ↔ May lower or have no impact

Concretes made with certain highly reactive fly ashes (especially high-calcium Class C ashes) or slag cements can equal or exceed the strength a control concrete in 1 to 28 days. Some fly ashes and other pozzolans require 28 to 90 days to exceed control strength, depending on the mixture proportions, fineness, and curing conditions. Tensile, flexural, torsional, and bond strengths are affected in the same manner as compressive strength.

However, the strength of concrete containing these materials can be either higher or lower than the strength of concrete using only portland cement depending on the age and/or the ambient placement and curing temperatures. Figure 3-16 illustrates this effect for various fly ashes.

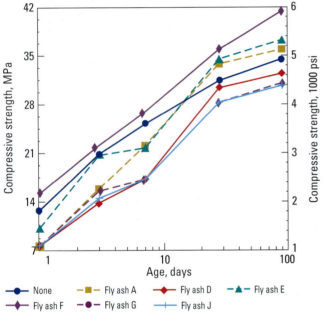

FIGURE 3-16. Compressive strength development at 1, 3, 7, 28, and 90 days of concrete mixtures containing 307 kg/m³ (517 lb/yd³) of cementitious materials with a fly ash dosage of 25% of the cementitious materials (Whiting 1989).

When a portion of portland cement or a blended cement is replaced by an SCM, the strength-w/cm relationship changes. The direction in which it changes depends upon the nature of the SCM, its dosage level, the age of test, and other factors. Provided the relationship is known, the desired strength can be obtained with almost any concrete mixture by simple selection of the appropriate w/cm for that particular mixture. Therefore, the lower early-age strength observed with some SCMs can be compensated for by using a lower w/cm mixture.

Because of its increased reactivity and cementitious behavior, Class C fly ash makes a greater contribution to early age concrete strength than Class F fly ash. Class C fly ash generally achieves the same or similar strength as the control after 3 to 7 days assuming comparable ambient temperature.

Slag cement at a 35% replacement can reach the same strength as a mixture without SCM after about 7 days assuming

comparable total cementitious material contents, w/cm, and ambient placement and curing temperature.

Because of its fine particle size, high surface area, and highly amorphous nature, silica fume reacts very rapidly in concrete and its contribution to strength is seen very early on. One-day strength using silica fume may be slightly increased or decreased depending on the nature of the silica fume and the alkali content of the portland cement. Between 3 and 28 days, a silica fume concrete invariably shows superior strength gain compared with a control, assuming comparable ambient placement and curing temperatures. Silica fume also aids in increasing the early strength gain of concretes made with fly ash or slag cement. Other very fine, highly reactive pozzolans, such as metakaolin, might be expected to show similar behavior.

Ground-glass pozzolans of sufficient fineness have been shown to increase the compressive strength of mortars and concrete (Figure 3-17).

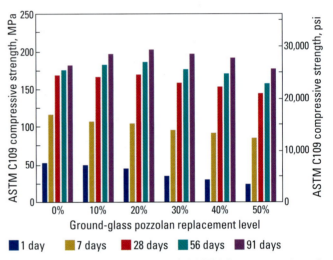

FIGURE 3-17. Compressive strength of ASTM C109 mortars through 91 days, using various percentages of ground-glass pozzolan (after Soliman and Tagnit-Hamou 2016).

Figure 3-18 illustrates the benefit of using fly ash as an addition instead of a partial cement replacement to improve strength development in cold weather. The control concrete had a cement content of 332 kg/m³ (560 lb/yd³) and w/c of 0.45. The fly ash curves show substitution for cement (S), partial (equal) substitution for cement and sand (P), and addition of fly ash by mass of cement (A). The use of partial cement substitution or addition of fly ash increases strength development comparable to the portland cement-only control, even in cold weather (Detwiler 2000). Mass concrete mixture design often takes advantage of the delayed strength gain of pozzolans, as these structures are often not put into full service immediately. With appropriate mixture adjustments, all SCMs can be used in all seasons and still provide the necessary strength (see Chapters 18 and 19).

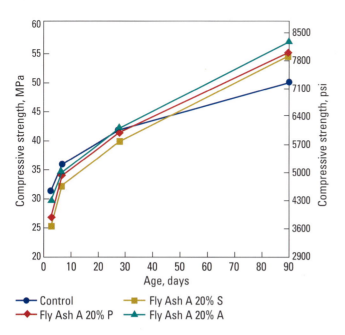

FIGURE 3-18. Compressive strengths for concretes cured at 23°C (73°F) for the first 24 hours and 4°C (40°F) for the remaining time. Letters in the caption refer to (S), substitution for cement, (P) partial (equal) substitution for cement and sand, and (A) addition of fly ash by mass of cement (Detwiler 2000).

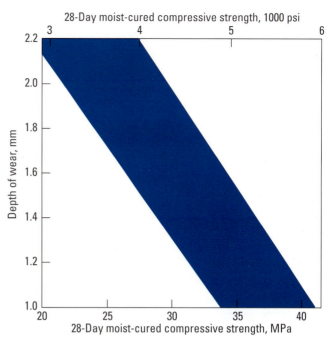

FIGURE 3-19. Comparison of abrasion resistance and compressive strength of various concretes with 25% fly ash. Abrasion resistance increases with strength (Gebler and Klieger 1986).

Supplementary cementitious materials are often used in the production of high-strength concrete. Fly ash has been used in production of concrete with strengths up to 100 MPa (15,000 psi). With silica fume, ready mix producers now have the ability to make concrete with strengths up to 140 MPa (20,000 psi), when used with high-range water reducers and appropriate aggregates (Burg and Ost 1994).

Because of the slow pozzolanic reaction of some SCMs, continuous wet curing and favorable curing temperatures may be required for periods longer than normal to ensure adequate strength gain. A 7-day moist cure or membrane cure should be adequate for concretes with normal dosages of most SCMs. As with portland cement concrete, low curing temperatures can reduce early-strength gain (Gebler and Klieger 1986).

Impact and Abrasion Resistance

The impact resistance and abrasion resistance of concrete are related to compressive strength and aggregate type. Supplementary cementing materials generally do not affect these properties beyond their influence on strength. Concretes containing fly ash are as abrasion resistant as portland cement concretes without fly ash (Gebler and Klieger 1986). Figure 3-19 illustrates that abrasion resistance of fly ash concrete is related to strength.

Drying Shrinkage and Creep

When used in low to moderate amounts, the effect of fly ash, slag cement, calcined clay, calcined shale, and silica fume

on the drying shrinkage and creep of concrete is generally small and not considered significant (Figure 3-20). There are conflicting data on the effects of SCMs on creep depending on testing regimes used (Thomas 2013). Some studies indicate that silica fume may reduce specific creep (Burg and Ost 1994).

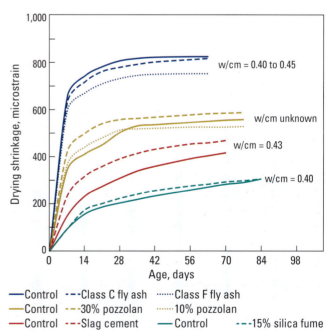

FIGURE 3-20. Drying shrinkage for a range of SCMs compared to controls at similar water-to-cementitious materials ratios. (After Thomas 2013, Gebler and Klieger 1986, Brooks and others 1992, Malhotra and others 1987, Mehta 1981.)

Permeability and Absorption

With appropriate concrete mixture design, control of w/cm, and adequate curing; fly ash, slag cement, and natural pozzolans generally reduce the permeability and absorption of concrete. Silica fume and metakaolin are especially effective in this regard. Silica fume and calcined clay can provide concrete with a chloride resistance of under 1000 coulombs using ASTM C1202, *Standard Test Method for Electrical Indication of Concrete's Ability to Resist Chloride Ion Penetration* (Barger and others 1997). Tests show that the permeability of concrete decreases as the quantity of hydrated cementitious materials increases and is age dependent. Fly ash and slag cement as well as ternary blends can provide ASTM C1202 values less than 1000 coulombs at later ages (Figure 3-21). The absorption of concrete (ASTM C1585, *Standard Test Method for Measurement of Rate of Adsorption of Water by Hydraulic-Cement Concretes*) containing fly ash is often about the same as concrete without ash, although some ashes can reduce absorption by 20% or more.

Corrosion Resistance

When concrete is properly cured, SCMs can help reduce reinforcing steel corrosion by reducing the permeability of concrete to water, air, and chloride ions.

Concrete with fly ash shows a slight improvement in the reduction to chloride ion ingress for concrete at an early age, it improves over time, reaching very low values at one year. By 56 days, concrete with slag cement and some other pozzolans generally exhibits lower chloride permeability compared to concretes without SCMs.

The incorporation of silica fume can have a dramatic effect, producing concrete with very low chloride permeability after just 28 days. Only small improvements at later ages are observed.

Concrete with metakaolin behaves in a manner similar to silica fume. Concrete containing silica fume or metakaolin is often used in overlays and full-depth slab placements on bridges and parking garages. These structures are particularly vulnerable to corrosion due to chloride-ion ingress.

Carbonation

The rate of carbonation is significantly increased in concrete with: a high water-cementitious materials ratio, low cement content, short curing period, low strength, and a highly permeable or porous paste. The depth of carbonation of good quality concrete is generally of little practical significance. At normal dosages, fly ash is generally reported to slightly increase the carbonation rate (Campbell and others 1991, and Stark and others 2002). However, it has been documented that concrete containing fly ash will carbonate at a similar rate compared with portland cement concrete of the same 28-day strength (Tsukayama 1980, Lewandowski 1983, Matthews 1984, Nagataki and others 1986, Hobbs 1988, and Dhir 1989). This means that fly ash increases the carbonation rate provided that the basis for comparison is an equal w/cm. It has also been shown that the increase due to fly ash is more pronounced at higher levels of replacement and in poorly-cured concrete of low strength (Thomas and Matthews 1992, and Thomas and Matthews 2000). Even when concretes are compared on the basis of equal strength, concrete with fly ash (especially at high levels of replacement) may carbonate more rapidly in poorly-cured, low strength concrete (Ho and Lewis 1983, Ho and Lewis 1997, Thomas and Matthews 1992, and Thomas and Matthews 2000).

Alkali-Silica Reactivity

Alkali-silica reactivity (ASR) can be generally be controlled through the appropriate use of SCMs (Figures 3-22 and 3-23). Supplementary cementitious materials provide additional calcium silicate hydrate which chemically binds some of the

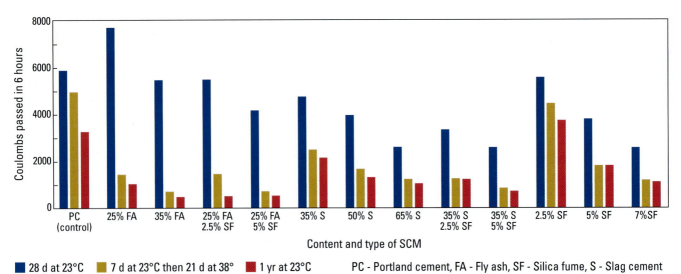

FIGURE 3-21. ASTM C1202 results for a range of SCM types and levels for binary and ternary concrete mixtures. Accelerated curing of specimens (7 days at 23°C then 21 days at 38°C) provides results for concretes with SCMs at 28 days that are generally comparable to those cured for 1 year at 23°C (after Lane and Ozyilidirim 1999).

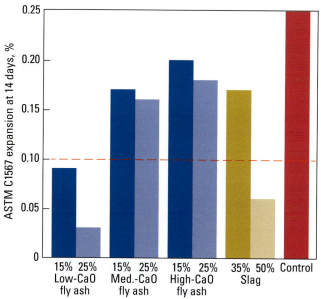

FIGURE 3-22. Effect of different fly ashes and slag cement on alkali-silica reactivity. Note that dosage of the ash or slag cement is critical. A highly reactive natural aggregate was used in this test. A less reactive aggregate would require less fly ash or slag cement to control the reaction (after Detwiler 2002). Dashed line indicates 0.10% expansion limit identified in ASTM C1778.

alkalies in concrete (Bhatty 1985 and Bhatty and Greening 1986) and reduce its permeability. When determining the optimum SCM content for ASR resistance, it is important to maximize the reduction in reactivity and to avoid dosages and materials that can aggravate reactivity. This latter effect is known as the pessimum effect (Figure 3-24).

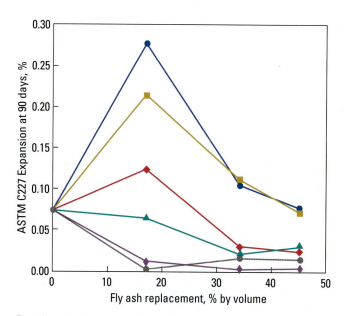

FIGURE 3-24. Illustration of the pessimum effect. ASTM C227 mortar bar expansion as a function of cement replacement level for highly reactive aggregate mixtures containing fly ash with greater than 1.5% alkali (Farbiarz and Carrasquillo 1987).

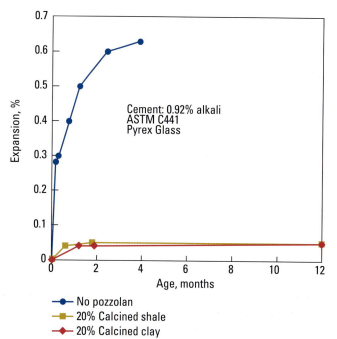

FIGURE 3-23. Reduction of expansion due to alkali-silica reactivity by calcined clay and calcined shale (Lerch 1950).

ASTM C1778, *Standard Guide for Reducing the Risk of Deleterious Alkali-Aggregate Reaction* (AASHTO R 80), provides prescriptive guidance on appropriate dosages of some SCMs and provides a performance-based approach to verify the SCM contents necessary to mitigate ASR for other SCMs. In the performance-based approach, dosage rates are verified by tests, such as ASTM C1567, *Standard Test Method for Determining the Potential Alkali-Silica Reactivity of Combinations of Cementitious Materials and Aggregate (Accelerated Mortar-Bar Method)*, or ASTM C1293, *Standard Test Method for Determination of Length Change of Concrete Due to Alkali-Silica Reaction*.

The effectiveness of Class F fly ash in controlling expansion due to ASR is well established (Thomas 2007). However, the effect of fly ash varies considerably as the composition of the ash varies. The calcium oxide (CaO) content of the fly ash is an indicator of how the material behaves with regard to controlling ASR. Low-calcium fly ash (Class F) is clearly the best at controlling ASR. The efficiency of fly ash in this role decreases as the CaO content increases above about 20% by mass (Figure 3-25). Low calcium Class F ashes result in reduced ASR-related expansion, in some cases up to 70% or more. At proper dosages, Class C ashes can often also reduce expansion due to ASR but to a lesser degree than most Class F ashes.

Fly ash can also contain significant quantities of soluble alkalies. This will tend to reduce its effectiveness in controlling ASR. Generally, the amount of fly ash required to control ASR will increase as any of the following parameters increase: calcium oxide or alkali content of the fly ash, reactivity of the aggregate, or amount of alkali available in the concrete.

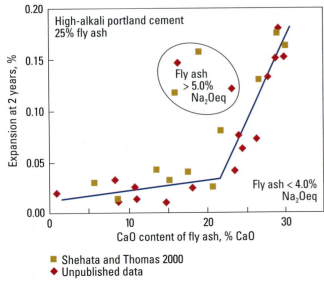

- ■ Shehata and Thomas 2000
- ◆ Unpublished data

FIGURE 3-25. Fly ashes with CaO contents below about 20% are generally effective in reducing expansion in ASTM C1293 concrete prisms made with a reactive aggregate at 25% replacement rate. However, some fly ashes with high alkali contents are not. Without fly ash, this aggregate expanded about 0.25% at 2 years (Thomas 2007).

ASR expansion decreases with use of slag cement. Generally, the amount of slag cement required to control expansion increases as either the reactivity of the aggregate or the amount of alkali in the mixture increases. As much as 65% slag cement may be needed to reduce the expansion to less than 0.04% in concrete with a highly-reactive aggregate in ASTM C1293. Typically, a minimum of 35% slag is required with moderately reactive aggregate.

The effect of silica fume on ASR resistance appears strongly dependent on the alkali available in the concrete system. Studies have shown that lower amounts of silica fume can be used when the alkali content of the concrete is lower than typically used in the concrete prism test (ASTM C1293, approximately 5.25 kg alkali/m³ or 8.8 lb alkali/yd³).

Metakaolin is also a highly reactive pozzolan and is nearly as effective as silica fume in improving ASR resistance – requiring a typical replacement level of somewhere between 10% to 15% to control expansion under ASTM C1293 test conditions.

However, supplementary cementitious materials that reduce alkali-silica reaction will not mitigate alkali-carbonate reaction (ACR), a type of reaction involving concrete alkalies and certain dolomitic limestones. The use of alkali-carbonate reactive aggregates should be avoided.

Chapter 11 provides more detailed information on ASR. Descriptions of aggregate testing and the measures needed to prevent deleterious alkali-aggregate reaction are discussed in ASTM C1778 (AASHTO R 80), Farny and Kerkhoff (2007), and Thomas and others (2008).

Sulfate Resistance

With proper proportioning and material selection, most supplementary cementing materials can improve the resistance of concrete to sulfate or seawater attack (Figure 3-26). This is done primarily by reducing permeability and the amount of reactive elements needed for expansive sulfate reactions. For improved sulfate resistance of lean concrete, one study (Stark 1989) showed that for a particular Class F fly ash, an adequate amount was approximately 20% of the cementitious materials content. This illustrates the need to determine optimum fly ash contents: With those cementitious materials in that mix design, higher fly ash contents were found to be detrimental.

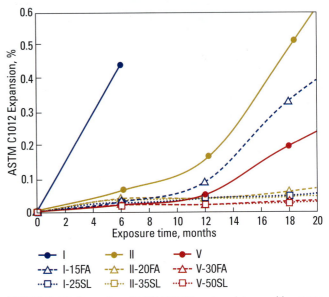

FIGURE 3-26. Expansion of ASTM C1012 mortar mixtures with a range of cements (I, II, V), with and without Class F fly ash (FA) and slag cement (SL) (after Obla and Lobo 2017).

Concretes with Class F fly ashes are generally more sulfate resistant than those made with Class C ashes. Some Class C ashes have been shown to reduce sulfate resistance at normal dosage rates (Thomas 2007).

Slag cement is generally considered beneficial in sulfate environments. Studies indicate that properly designed concrete with slag cement can provide sulfate resistance equal to or greater than concretes made with Type V sulfate-resistant portland cement (ACI 233 and Detwiler and others 1996). However, the Al_2O_3 content of the slag cement can influence its ability to mitigate sulfate attack (the higher the Al_2O_3 the less sulfate resistance). Adding additional sulfate to balance the alumina in the slag cement can restore its ability to improve sulfate resistance (ACI 233 2017).

Silica fume has been shown to provide sulfate resistance equivalent to slag cement and fly ash (Fidjestøl and Frearson 1994).

Calcined clay has been demonstrated to provide sulfate resistance greater than Type V cement (Barger and others 1997). Chapter 11 provides more information on sulfate resistance.

Requirements for sulfate resistance of hydraulic cements and cementitious materials based on 6-month or 12-month ASTM C1012 expansions for cementitious materials are given in ASTM C595, ASTM C1157, ASTM C618, ASTM C989, and ASTM C1240. ACI 318, *Building Code Requirements for Structural Concrete* (2019), includes requirements based on ASTM C1012 for S3 (very severe) exposures at 18 months. See Chapter 11 for more detail.

Delayed Ettringite Formation. Delayed ettringite formation (DEF), is a potential distress mechanism for concrete, mortar, or cement paste subjected to high early-age temperatures. Such early-age temperatures may result from heat treatment, or in other cases, from internal heat of hydration. It is generally accepted (Taylor 1994, ACI 201.2R (2016), Shimada and others 2005, and Gajda 2008) that concrete temperatures above about 70°C (158°F) are required for DEF to be an issue. Use of SCMs in blended cements or as concrete ingredients are recommended by ACI 301, *Specifications for Concrete Construction* (2020), and PCI 2013 to reduce the potential for DEF, if temperatures in the range of 158°F to 185°F (70°C to 85°C) are unavoidable. See Chapter 11 for further discussion on DEF, and for recommended SCM contents for reducing the potential for DEF.

Chemical Resistance

Supplementary cementitious materials often reduce chemical attack by reducing the permeability of concrete, making it more difficult for aggressive ions to penetrate and move through the concrete's pores. Although many of these materials may improve chemical resistance, they do not make concrete immune to chemical attack. Concrete in severe chemical exposure may need additional protection using barrier systems. Kerkhoff (2007) provides a discussion of methods and materials to protect concrete from aggressive chemicals and exposures.

Freeze-Thaw Resistance

As discussed in Chapters 6 and 11, the freeze-thaw resistance of concrete is dependent on the air void system of the paste, the strength of the concrete when exposed to freezing, the water-to-cementitious materials ratio, and the quality of aggregate relative to its freeze-thaw resistance. These properties in general are not influenced by SCMs with the exception of surface scaling.

Deicer-Scaling Resistance

Decades of field experience have demonstrated that air-entrained concretes containing normal dosages of fly ash, slag cement, silica fume, calcined clay, or calcined shale are resistant to the scaling caused by the application of deicing salts in a freeze-thaw environment (Figure 3-27). Laboratory tests also

FIGURE 3-27. View of concrete slabs in PCA outdoor test plot (Skokie, Illinois) containing (A) fly ash, (B) slag, (C) calcined shale, and (D) portland cement only, after 30 years of deicer and frost exposure. These samples demonstrate the durability of concretes containing various cementitious materials. Slabs made with Type V cement and SCMs indicated with 335 kg/m³ (564 lb/yd³) of cementitious materials and air entrainment. (Based on research described in Klieger 1963, and Stark and others 2002.)

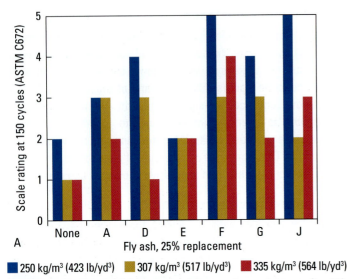

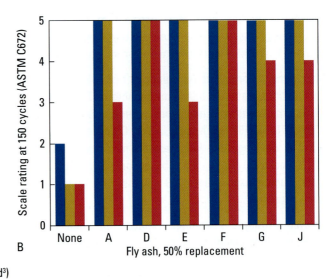

■ 250 kg/m³ (423 lb/yd³)   ■ 307 kg/m³ (517 lb/yd³)   ■ 335 kg/m³ (564 lb/yd³)

FIGURE 3-28. Relationship between deicer-scaling resistance after 150 cycles and dosage of fly ash for air-entrained concretes made with moderate to high water-cementitious materials ratios. Replacement of portland cement with fly ash: (A) 25% and (B) 50%. A scale rating of 0 is no scaling and 5 is severe scaling (Whiting 1989).

indicate that the deicer-scaling resistance of concrete made with SCMs is often equal to concrete made without SCMs.

Scaling resistance can decrease as the amount of certain SCMs increases. However, concretes that are properly designed, placed, and cured have demonstrated good scaling resistance even when made with high dosages of some SCMs.

Concrete containing supplementary cementing materials at appropriate dosages may be expected to be adequately resistant to the effects of deicing salts provided the concrete has a w/cm of ≤ 0.45 or below, an adequate air-void system is present, and proper finishing and curing is applied. This includes ensuring bleeding has ceased and initial set has occurred prior to final finishing, and the concrete is given ample opportunity to air-dry prior to first exposure to salts and freezing conditions (see Chapter 11 and Chapter 16 for more information on deicer scaling resistance).

The effect of high fly ash dosages and low cementing material contents is demonstrated in Figure 3-28. The importance of using a low water-cement ratio for scaling resistance is demonstrated in Figure 3-29. Gebler and Klieger (1986a) demonstrate that well designed, placed, and cured concretes with and without fly ash can be equally resistant to deicer scaling.

The ACI 318-19 building code states that the maximum dosage of fly ash, slag, and silica fume should be 25%, 50%, and 10% by mass of cementing materials, respectively, for deicer exposures. Total SCM content should not exceed 50% of the cementitious material. ACI 201.2R (2016) notes that these restrictions may apply only to surfaces finished by hand, rather than those finished by machine or cast against formwork. Dosages higher than these limits have been shown to be durable in some cases but not all. Materials may respond differently in different environments. The selection of materials and dosages should be based on local experience and the durability should be demonstrated by field history or laboratory performance.

Aesthetics

Supplementary cementitious materials may alter the color of hardened concrete; depending on the color and amount of the material used, the effect may be slight. Many SCMs resemble the color of portland cement and therefore have little effect on color of the hardened concrete. Some silica fumes may give concrete a slightly bluish or dark gray tint, and tan fly ash may impart a tan color to concrete when used in large quantities. Slag cement and metakaolin can make concrete whiter and may be used in place of white cement depending on dosage. Sometimes slag cements can initially impart a bluish or greenish undertone; this effect fades over time as concrete is exposed to oxygen in air.

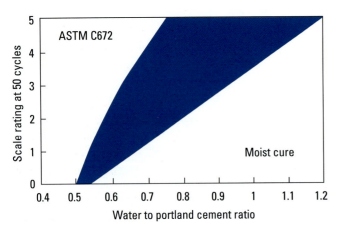

FIGURE 3-29. Relationship between deicer-scaling resistance and water to portland cement ratio for several air-entrained concretes with and without fly ash. A scale rating of 0 is no scaling and 5 is severe scaling (after Whiting 1989).

CONCRETE MIXTURE PROPORTIONS

The optimum amounts of SCMs used with portland cement or blended cement are determined by testing, considering the exposure conditions and specified properties of the concrete, availability of the materials, and their relative costs.

Several test mixtures are required to determine the optimum amount of SCM to use in the concrete mixture. These test mixtures should cover a range of blends to establish the relationship between the property or properties being optimized and the SCM content. The water-to-cementitious materials ratio (w/cm) will also have a strong effect on many properties. These mixtures should be established according to ACI 211.1R-91 or ACI 211.2R-98, taking into account the relative densities and reactivity of the supplementary cementitious materials. The results of these tests will be a series of values for each mixture tested, often as a function of age. For example, a family of compressive strength curves might be developed to evaluate the effects of SCM content and age. In other cases, the permeability at 56 days might be the property of interest. The content of a cementitious material is usually stated as a mass percentage of all the cementitious materials in a concrete mixture.

Chapter 13 has more detail on concrete proportioning.

Multi-Cementitious Systems

Traditionally, fly ash, slag cement, silica fume, and other pozzolans have been used in concrete individually with portland cement or as a blended cement. However, there are often benefits in using more than one SCM in concrete in combination with portland cement to produce concrete mixtures (Taylor and others 2012). Ternary blends containing portland cement plus two SCMs or a blended cement and an SCM are often used where special properties are required. Quaternary blends consisting of portland cement plus three SCMs have also been used.

Some pre-blended combinations of SCMs are available in the US and may meet ASTM C1697, *Standard Specification for Blended Supplementary Cementitious Materials*, covers nomenclature, uniformity requirements, and physical requirements for blends of two or three SCMs that individually meet requirements of ASTM C989, ASTM C618, or ASTM C1240.

There can be a pronounced synergy in combining SCMs (Thomas and Wilson 2002, and Thomas 2013). Benefits include:

- fly ash (and, to a lesser extent, slag cement) offsets increased water demand of high fineness SCMs like silica fume and metakaolin;
- Class F fly ash (and, to a lesser extent, slag cement) compensates for high heat release from blended cement with silica fume;
- silica fume or metakaolin may compensate for lower early strength of concrete with Class F fly ash or slag cement;

- fly ash and slag cement increase long-term strength development of silica fume concrete;
- very high resistance to chloride and sulfate ion penetration can be obtained with ternary blends; and
- silica fume reduces the normally high levels of Class C fly ash or slag cement required for sulfate resistance and ASR prevention.

STORAGE

Although moisture will not affect the physical performance of some SCMs, in general these materials should be kept dry to avoid difficulties in handling and discharge. In particular, Class C fly ash, slag cement, and calcined shale must be kept dry as they will set and harden when exposed to moisture. Equipment for handling and storing these materials is similar to that required for portland cement. Some silica fumes are provided in liquid (slurry) form. Additional modifications may be required when using silica fume, because it does not have the same flow characteristics as other SCMs.

SCMs are usually kept in bulk storage facilities or silos, although some products are available in bags. Because the materials may resemble portland cement in color and fineness, the storage facilities should be clearly marked to avoid the possibility of misidentification and contamination with other materials at the batch plant. Good practice requires all valves and piping should also be clearly marked and properly sealed to avoid leakage and contamination.

AVAILABILITY

All SCMs may not be available in all areas. Consult local material suppliers on availability. Fly ash type and availability will vary by region. Slag cements are available in most regions. Silica fume is available in most locations because relatively small amounts are used. Natural pozzolans are available in select areas, as are ground-glass pozzolans.

REFERENCES

PCA's online catalog includes links to PDF versions of many of our research reports and other classic publications. Visit: cement.org/library/catalog.

ACI Committee 201, *Guide to Durable Concrete*, ACI 201.2R-16, American Concrete Institute, Farmington Hills, MI, 1998, 86 pages.

ACI Committee 211, *Standard Practice for Selecting Proportions for Normal, Heavyweight, and Mass Concrete*, ACI 211.1R-91, reapproved 2009, American Concrete Institute, Farmington Hills, MI, 1991, 38 pages.

ACI Committee 211, *Standard Practice for Selecting Proportions for Structural Lightweight Concrete*, ACI 211.2R-98, reapproved 2004, American Concrete Institute, Farmington Hills, MI, 1998, 14 pages.

ACI Committee 211, *Guide for Proportioning Concrete Mixtures with Ground Calcium Carbonate and Other Mineral Fillers*, ACI 211.7R-20, American Concrete Institute, Farmington Hills, Michigan, 2020, 21 pages.

ACI Committee 232, *Report of the Use of Fly Ash in Concrete*, ACI 232.2R-18, American Concrete Institute, Farmington Hills, MI, 2018, 60 pages.

ACI Committee 232, *Use of Raw or Processed Natural Pozzolans in Concrete*, ACI 232.1R-12, American Concrete Institute, Farmington Hills, MI, 2012, 33 pages.

ACI Committee 233, *Guide to the Use of Slag Cement in Concrete and Mortar*, ACI 233R-17, American Concrete Institute, Farmington Hills, MI, 2017, 36 pages.

ACI Committee 234, *Guide to the Use of Silica Fume in Concrete*, ACI 234R-06, American Concrete Institute, Farmington Hills, MI, 2006, 63 pages.

ACI Committee 301, *Specifications for Concrete Construction*, ACI 301-20, American Concrete Institute, Farmington Hills, MI, 2020.

ACI Committee 318, *Building Code Requirements for Structural Concrete (ACI 318-14) and Commentary*, ACI 318-14, American Concrete Institute, Farmington Hills, MI, 2014, 520 pages.

Anonymous, "An Investigation of the Pozzolanic Nature of Coal Ashes," *Engineering News*, 1914, pages 1334 to 1335.

Barger, G.S.; Lukkarila, M.R.; Martin, D.L.; Lane, S.B.; Hansen, E.R.; Ross, M.W.; and Thompson, J.L., "Evaluation of a Blended Cement and a Mineral Admixture Containing Calcined Clay Natural Pozzolan for High-Performance Concrete," *Proceedings of the Sixth International Purdue Conference on Concrete Pavement Design and Materials for High Performance*, Purdue University, West Lafayette, IN, 1997, 21 pages.

Bernhardt, C.J., "SiO₂-Stov som Cementtilsetning," ("SiO₂ Dust as an Admixture to Cement"), *Betongen Idag*, Vol. 17, No. 2, 1952, pages 29 to 53.

Bhatty, M.S.Y., "Mechanism of Pozzolanic Reactions and Control of Alkali-Aggregate Expansion," *Cement, Concrete, and Aggregates*, American Society for Testing and Materials, West Conshohocken, PA. Winter 1985, pages 69 to 77.

Bhatty, M.S.Y.; and Greening, N.R., "Some Long Time Studies of Blended Cements with Emphasis on Alkali-Aggregate Reaction," *7th International Conference on Alkali-Aggregate Reaction*, Ottawa, Canada, 1986, pages 85 to 92.

Brooks, J.J.; Wainwright, P.J.; and Boukendakji, M., "Influence of Slag Type and Replacement Level on Strength Elasticity, Shrinkage and Creep of Concrete," *Proceedings of the Fourth International Conference on Fly Ash, Silica Fume, Slag, and Natural Pozzolans in Concrete*, SP-132, American Concrete Institute, Farmington Hills, MI, 1992, pages 1325 to 1342.

Burg, R.G.; and Ost, B.W., *Engineering Properties of Commercially Available High-Strength Concretes (Including Three-Year Data)*, Research and Development Bulletin RD104, Portland Cement Association, Skokie, IL, 1994, 62 pages.

Campbell, D.H.; Sturm, R.D.; and Kosmatka, S.H., "Detecting Carbonation," *Concrete Technology Today*, PL911, Portland Cement Association, March 1991, pages 1 to 5.

CSA A3000, *Cementitious Materials Compendium*, CSA Group, Standards Council of Canada, Ottawa, Ontario, 2018, 263 pages.

Cochran, J.W.; and Boyd, T.J, "Beneficiation of Fly Ash by Carbon Burnout," *Proceedings of the Tenth International Ash Use Symposium*, EPRI TR 101774, Vol. 2, Paper 73, 1993, pages 73-1 to 73-9.

Cornelissen, H.A.W.; Hellewaard, R.E.; and Vissers, J.L.J., "Processed Fly Ash for High Performance Concrete," *Fly Ash, Silica Fume, Slag, and Natural Pozzolans in Concrete*, Proceedings of the Fifth International Conference, SP-153, V.M. Malhotra, ed., American Concrete Institute, Farmington Hills, MI, 1995, pages 67 to 80.

Davis, R.E.; Carlson, R.W.; Kelly, J.W.; and Davis, H.E.,"Properties of Cements and Concretes Containing Fly Ash," *Proceedings, American Concrete Institute*, Vol. 33, May-June 1937, pages 577 to 612.

Detwiler, R.J., "Controlling the Strength Gain of Fly Ash Concrete at Low Temperature," *Concrete Technology Today*, CT003, Portland Cement Association, Skokie, IL, 2000, pages 3 to 5.

Detwiler, R.J., *PCA's Guide Specification for Concrete Subject to Alkali-Silica Reactions: Mitigation Measures*, SN2407, Portland Cement Association, Skokie, IL, 2003, 13 pages.

Detwiler, R.J.; Bhatty, J.I.; and Bhattacharja, S., *Supplementary Cementing Materials for Use in Blended Cements*, Research and Development Bulletin RD112, Portland Cement Association, Skokie, IL, 1996, 108 pages.

Dhir, R.K., "Near – Surface Characteristics of Concrete: Prediction of Carbonation Resistance," *Magazine of Concrete Research*, Vol. 41, No. 148, 1989, pages 137 to 143.

Diaz-Loya, I.; Juenger, M.; Seraj, S.; and Minkara, R., "Extending Supplementary Cementitious Material Resources: Reclaimed and Remediated Fly Ash and Natural Pozzolans," *Cement and Concrete Composites*, Vol. 101, August 2019, pages 44 to 51.

Dodson, V., "Air-Entraining Admixtures," Chapter 6, *Concrete Admixtures*, Van Nostrand Reinhold, New York, NY, 1990, pages 129 to 158.

EPA, *Advancing Sustainable Materials Management: 2017 Fact Sheet*, EPA 530-F-19-007, US Environmental Protection Agency, Office of Land and Emergency Management, November 2019, 22 pages.

Farny, J.A.; and Kerkhoff, B., *Diagnosis and Control of Alkali-Aggregate Reactivity*, IS413, Portland Cement Association, Skokie, IL, 2007, 26 pages.

Fidjestøl, P.; and Frearson, J., "High-Performance Concrete Using Blended and Triple Blended Binders," *High-Performance Concrete – Proceedings, International Conference Singapore*, 1994, SP-149, American Concrete Institute, Farmington Hills, MI, 1994, pages 135 to 158.

Fiskaa, O.; Hansen, H.; and Moum, J., *Betong i Alunskifer, (Concrete in Alum Shale)*, Publication No. 86, Norwegian Geotechnical Institute, 1971, 32 pages. (in Norwegian with English summary)

Gajda, J.; and VanGeem, M., "Controlling Temperatures in Mass Concrete," *Concrete International*, Vol. 24, No. 1, January 2002, pages 58 to 62.

Gajda, J., *Mass Concrete for Buildings and Bridges*, EB547, Portland Cement Association, Skokie, IL, 2007, 44 pages.

Gebler, S.H.; and Klieger, P., *Effect of Fly Ash on Some of the Physical Properties of Concrete*, Research and Development Bulletin RD089, Portland Cement Association, 1986, 46 pages.

Gebler, S.H.; and Klieger, P., *Effect of Fly Ash on the Durability of Air-Entrained Concrete*, Research and Development Bulletin RD090, Portland Cement Association, 1986a, 44 pages.

Groppo, J., "The Recovery of High Quality Fuel From Ponded Utility Coal Combustion Ash," *ACAA Fourteenth International Symposium on Coal Combustion Products*, San Antonio, TX, Vol. 1, 2001, page 14-1.

Hansen, W.C., *Twenty-Year Report on the Long-Time Study of Cement Performance in Concrete*, Research Department Bulletin, RX175, Portland Cement Association, Skokie, IL, 1965, 44 pages.

Helmuth, R.A., *Fly Ash in Cement and Concrete*, SP040T, Portland Cement Association, Skokie, IL, 1987, 203 pages.

Hemmings, R.T.; and Berry, E.E. "On the Glass in Coal Fly Ashes: Recent Advances," *Fly Ash and Coal Conversion By-Products: Characterization, Utilization and Disposal IV*, G.J. McCarthy, F.P. Glasser, D.M. Roy and R. Hemmings, Eds., Materials Research Society Symposia Proceedings, Vol. 113, 1988, pages 3 to 38.

Ho, D.W.S.; and Lewis, R.K., "Carbonation of Concrete Incorporating Fly Ash or a Chemical Admixture," *Proceedings of the First International Conference on the Use of Fly Ash, Silica Fume, Slag, and Other By-Products in Concrete*, ACI SP-79, Vol. 1, American Concrete Institute, Farmington Hills, MI, 1983, pages 333 to 346.

Ho, D.W.S.; and Lewis, R.K., "Carbonation of Concrete and Its Prediction," *Cement and Concrete Research*, Vol. 17, No. 3, 1987, pages 489 to 504.

Hobbs, D.W., "Carbonation of Concrete Containing PFA," *Magazine Concrete Research*, 40, No. 143, 1988, pages 69 to 78.

Holland, T.C.; Krysa, A.; Luther, M.; and Liu, T., "Use of Silica-Fume Concrete to Repair Abrasion-Erosion Damage in the Kinzua Dam Stilling Basin," *Proceedings, CANMET/ACI Second International Conference on the Use of Fly Ash, Silica Fume, Slag, and Natural Pozzolans in Concrete*, Madrid, V.M. Malhotra, ed., SP-91, V. 2, American Concrete Institute, Detroit, MI, 1986, pages 841 to 864.

Holland, T. C., *Silica Fume User's Manual*, FHWA-IF-05-016, Federal Highway Administration, Washington, DC, 2005, 194 pages.

Hooton, R.D.; and Emery, J.J., "Sulfate Resistance of a Canadian Slag," *ACI Materials Journal*, Vol. 87, No. 6, November–December, 1990, pages 547 to 555.

Hooton, R.D.; and Emery, J.J., "Glass Content Determination and Strength Development Predictions for Vitrified Blast Furnace Slag," *Proceedings, First International Conference on the Use of Fly Ash, Silica Fume, Slag and Other Mineral By-Products in Concrete*, Montebello, Quebec, July 31-August 5, 1983, ACI SP79, Vol. 2, American Concrete Institute, Farmington Hill, MI, 1983, pages 943 to 962.

Kerkhoff, B., *Effect of Substances on Concrete and Guide to Protective Treatments*, IS001, Portland Cement Association, Skokie, IL, 2007, 36 pages.

Lane, D.S.; and Ozyilidirim, C., *Combinations of Pozzolans and Ground, Granulated, Blast-furnace Slag for Durable Hydraulic Cement Concrete*, VTRC 00-R1, Virginia Transportation Research Council, August 1999, 22 pages.

Lerch, W., *Studies of Some Methods of Avoiding the Expansion and Pattern Cracking Associated with the Alkali-Aggregate Reaction*, Research Department Bulletin RX031, Portland Cement Association, 1950, 28 pages.

Lewandowski, R., "Effect of Different Fly-Ash Qualities and Quantities on the Properties of Concrete," *Betonwerk and Fetigteil – Technik*, Nos. 1 to 3, 1983.

Lothenbach, B.; Scrivener, K.; and Hooton, R.D., "Supplementary Cementitious Materials," *Cement and Concrete Research*, Vol. 41, 2011, pages 1244 to 1256.

Malhotra, V.M.; Ramachandran, V.S.; Feldman, R.S.; and Aïtcin, P.C., *Condensed Silica Fume in Concrete*, CRC Press, Boca Raton, FL, 1987, 240 pages.

Mather, B., "Laboratory Tests of Portland Blast-Furnace Slag Cements," *ACI Journal, Proceedings*, Vol. 54, No. 3, September 1957, pages 205 to 232.

Matthews, J.D., "Carbonation of Ten-Year Old Concretes With and Without PFA," *Proceedings 2nd International Conference on Fly Ash Technology and Marketing*, London, Ash Marketing, CEGB, 398A, 1984.

Mehta, P.K., "Studies on Blended Portland Cements Containing Santorin Earth," *Cement and Concrete Research*, Vol. 11, No. 4, July 1981, pages 507 to 518.

Nagataki, S.; Ohga, H.; and Kim, E.K., "Effect of Curing Conditions on the Carbonation of Concrete with Fly Ash and the Corrosion of Reinforcement in Long Term Tests," *Fly Ash, Silica Fume, Slag and Natural Pozzolans in Concrete*, ACI SP-91, Vol. 1, American Concrete Institute, Farmington Hills, MI, 1986, pages 521 to 540.

Obla, K.H.; Hill, R.L.; Thomas, M.D.A.; Balaguru, P.N.; and Cook, J., "High Strength High Performance Concrete Containing Ultra Fine Fly Ash," *Seventh CANMET/ACI International Conference on Fly Ash, Silica Fume, Slag and Natural Pozzolans in Concrete*, July 22-27, Chennai, India, supplementary paper, 2001a, pages 227 to 239.

Obla, K.H.; Lobo, C.L.; and Kim, H., "The 2012 NRMCA Supplementary Cementitious Materials Use Survey," *Concrete InFocus*, National Ready-Mixed Concrete Association, Silver Spring, MD, Fall 2012, pages 16 to 18.

Omran, A.; Soliman, N.; Zidol, A.; and Tagnit-Hamou, A., "Performance of Ground-Glass Pozzolan as a Cementitious Material – A Review," *Advances in Civil Engineering Materials*, Vol. 7, No. 1, 2018, pages 237 to 270.

PCI, *Manual for Quality Control for Plants and Production of Architectural Precast Concrete Products*, 4th ed., MNL-117-13, Precast/Prestressed Concrete Institute, Chicago, IL, 2013, 334 pages.

Pinto, R.C.A.; and Hover, K.C., "Superplasticizer and Silica Fume Addition Effects on Heat of Hydration of Mortar Mixtures with Low Water-Cementitious Materials Ratio," *ACI Materials Journal*, Vol. 96, No. 5, September-October 1999, pages 600 to 605.

Shehata, M.H.; and Thomas, M.D.A., "The Effect of Fly Ash Composition on the Expansion of Concrete Due to Alkali Silica Reaction," *Cement and Concrete Research*, Vol. 30, No. 7, 2000, pages 1063 to 1072.

Shimada, Y.; Johansen, V.C.; Miller, F.M.; and Mason, T.O., *Chemical Path of Ettringite Formation in Heat-Cured Mortar and Its Relationship to Expansion: A Literature Review*, Research and Development Bulletin RD136, Portland Cement Association, Skokie, IL, 2005, 51 pages.

Soliman, N.A.; and Tagnit-Hamou, A., "Development of Ultra-High-Performance Concrete using Glass Powder – Towards Ecofriendly Concrete," *Construction and Building Materials*, Vol. 125, 2016, pages 600 to 612.

Stark, D., *Durability of Concrete in Sulfate-Rich Soils*, Research and Development Bulletin RD097, Portland Cement Association, Skokie, IL, 1989, 14 pages.

Stark, D.C.; Kosmatka, S.H.; Farny, J.A.; and Tennis, P.D., *Performance of Concrete Specimens in the PCA Outdoor Test Facility*, Research and Development Bulletin RD124, Portland Cement Association, Skokie, IL, 2002, 36 pages.

Taylor, H.F.W., "Delayed Ettringite Formation, Advances in Cement and Concrete," *Proceedings of an Engineering Foundation Conference*, American Society of Civil Engineers, July 24-29, 1994, pages 122 to 131.

Taylor, P.; Tikalsky, P.; Wang, K.; Fick, G.; and Wang, X., *Development of Performance Properties of Ternary Mixtures: Field Demonstrations and Project Summary*, DTFH61-06-H-00011, Federal Highway Administration, Washington, D.C., July 2012, 332 pages.

Thomas, A., "Metallurgical and Slag Cements, the Indispensable Energy Savers," *General Practices, IEEE Cement Industry 21 Technical Conference*, 1979, 108 pages.

Thomas, M.D.A.; and Matthews, J.D., "Carbonation of Fly Ash Concrete," *Magazine of Concrete Research*, Vol. 44, No. 158, 1992, pages 217 to 228.

Thomas, M.D.A.; and Matthews J.D., "Carbonation of Fly Ash Concrete," *Proceedings of the 4th ACI/CANMET International Conference on the Durability of Concrete*, ACI SP-192, Vol. 1, American Concrete Institute, Farmington Hills, MI, 2000, pages 539 to 556.

Thomas, M.D.A.; Fournier, B.; and Folliard, K., *Report on Determining the Reactivity of Concrete Aggregates and Selecting Appropriate Measures for Preventing Deleterious Expansion in New Concrete Construction*, HIF-09-001, Federal Highway Administration, Washington, D.C., April 2008, 28 pages.

Thomas, M., *Optimizing the Use of Fly Ash in Concrete*, IS548, Portland Cement Association, Skokie, IL, 2007, 24 pages.

Thomas, M., *Supplementary Cementing Materials in Concrete*, CRC Press, Boca Raton, FL, 2013, 210 pages.

Thomas, M.; and Wilson, M.L., *Supplementary Cementing Materials for Use in Concrete*, CD038, Portland Cement Association, Skokie, IL, 2002.

Tritsch, S.L.; Sutter, L.; and Diaz-Loya, I., *Use of Reclaimed Fly Ash in Highway Infrastructure*, Tech Brief, National Concrete Pavement Technology Center, Ames, IA, September 2020, 11 pages.

Tsukayama, R., "Long Term Experiments on the Neutralization of Concrete Mixed with Fly Ash and the Corrosion of Reinforcement," *Proceedings of the Seventh International Congressional on the Chemistry of Cement*, Paris, Vol. III, 1980, pages 30 to 35.

Whiting, D., *Strength and Durability of Residential Concretes Containing Fly Ash*, Research and Development Bulletin RD099, Portland Cement Association, Skokie, IL, 1989, 42 pages.

Whitlock, D.R., "Electrostatic Separation of Unburned Carbon from Fly Ash," EPRI TR 101774, *Proceedings of the Tenth International Ash Use Symposium*, Vol. 2, 1993, pages 70-1 to 70-12.

Zunino, F.; Bentz, D.P.; and Castro, J., "Reducing Setting Time of Blended Cement Paste Containing High-SO3 Fly Ash (HSFA) using Chemical/Physical Accelerators and by Fly Ash Pre-Washing," *Cement and Concrete Composites*, Vol. 90, July, 2018, pages 14 to 26.

MIXING WATER FOR CONCRETE

FIGURE 4-1. Water is a key ingredient in concrete.

Water is a key ingredient in concrete that, when mixed with cementitious materials, forms a paste that binds aggregates together. Water impacts the setting and hardening of concrete through a chemical reaction with portland cement, called hydration. Further details of hydration are discussed in Chapter 2.

The quality of water should not adversely impact the potential properties of concrete. Almost any water that is drinkable (potable water) and has no pronounced taste or odor, can be used as mixing water in concrete (Figure 4-1). Many sources of water that are not fit for drinking are still suitable for use as mixing water in concrete in support of environmental stewardship.

Requirements for mixing water used in concrete are stated in ASTM C1602, *Standard Specification for Mixing Water Used in the Production of Hydraulic Cement Concrete*. This specification is referenced in ACI 318, *Building Code Requirements for Structural Concrete* (2019) and ACI 301, *Specification for Concrete Construction* (2020). ASTM C1602 is incorporated by reference in ASTM C94, *Standard Specification for Ready-Mix Concrete*.

ASTM C1602 includes provisions for:

- *Potable water* – water fit for human consumption.
- *Non-potable water* – water sources not fit for human consumption, unrelated to process or storm water generated at concrete plants. This can include water from wells, streams, or lakes.
- *Water recovered from concrete production operations* – process (wash) water from mixers, storm water collected at concrete plants, or water containing various concrete ingredients.
- *Combined water* – a combination of one or more of the above defined sources. Water sources might be blended together before or during mixing of concrete.

Potable water used in municipal water supplies can be used in concrete without any additional testing or qualification. Water of questionable suitability, including non-potable water or water from concrete production operations, can be used in concrete if it complies with ASTM C1602. The primary requirements of ASTM C1602 are summarized in Table 4-1.

TABLE 4-1. Performance Requirements for Questionable Mixing Water Sources

	LIMITS
Compressive strength, minimum percentage of control at 7 days*	90
Time of setting, deviation from control, hr:min.*	From 1:00 earlier to 1:30 later

* Comparisons shall be based on fixed proportions for concrete or mortar mixtures. The control mixture shall be made with 100% potable or distilled water. The test mixture shall be made with the mixing water that is being evaluated (see ASTM C1602).

These criteria evaluate the impact of a questionable source of water on strength and setting time of concrete. Concrete produced with questionable water is compared to control batches produced using potable water. The 7-day strength of concrete cylinders or mortar cubes prepared from concrete using the test water must achieve at least 90% of the strength of the control batch. The setting time, as measured by ASTM C403, *Standard Test Method for Time of Setting of Concrete Mixtures by Penetration Resistance* (AASHTO T 197), of the test batch should not be accelerated by more than 1 hour or retarded by more than 1½ hours when compared to the control batch. NRMCA (2013) provides additional details of laboratory evaluation of water sources to comply with ASTM C1602.

The most critical water combination anticipated by a supplier should be tested and qualified. Less critical water combinations than those qualified may be used without further testing. For example, if the concrete supplier tests water containing 100,000 parts per million (ppm) of solids, or 10% by mass of the total mixing water, and the concrete or mortar made with this water meets the requirements for strength and setting time, then combinations of water with less restrictive conditions (for example, water containing 90,000 ppm of solids) are also permitted for use. Water with a specific gravity of 1.03 has a total solids content of approximately 50,000 ppm. This amounts to about 9 kg (15 pounds) of solids in 1 cubic meter (cubic yard) of a typical concrete mixture.

ASTM C1602 includes optional limits, as stated in Table 4-2, on chemistry and total solids content by mass in the combined mixing water. When a concrete producer uses non-potable water sources, information on the characteristics should be developed and recorded, and appropriate documentation maintained. Optional limits must be stated in project specifications or within purchase orders. The chemical composition of water is measured in accordance with methods described in ASTM C114, *Standard Test Methods for Chemical Analysis of Hydraulic Cement* (AASHTO T 105) while the solids content is measured in accordance with ASTM C1603, *Standard Test Method for Measurement of Solids in Water*. The density of water is measured during production of concrete to estimate the solids content using a pre-established correlation for the specific production facility as described in ASTM C1603.

The optional requirements set limits on those chemical species that can impact the durability of concrete including chlorides, alkali content, and sulfate content. The chemical limits are conservative considering the total concentration of these

TABLE 4-2. Optional Chemical Limits for Combined Mixing Water (ASTM C1602)

CHEMICAL	MAXIMUM CONCENTRATION, ppm*	TEST METHOD
Chloride, as Cl		
Prestressed concrete or concrete in bridge decks	500[†]	ASTM C114
Other reinforced concrete in moist environments or containing aluminum embedments or dissimilar metals or with stay-in-place galvanized metal forms	1000[†]	ASTM C114
Sulfate, as SO_4	3000	ASTM C114
Equivalent alkalies, as $(Na_2O + 0.658\,K_2O)$	600	ASTM C114
Total solids by mass	50,000	ASTM C1603

* ppm is the abbreviation for parts per million.

[†] The requirements for concrete in ACI 318-19 shall govern when the manufacturer can demonstrate that these limits for mixing water can be exceeded. For conditions allowing the use of calcium chloride ($CaCl_2$) accelerator as an admixture, the chloride limitation is permitted to be waived by the purchaser.

species in a concrete mixture. Observations of process water at concrete production facilities typically indicate the amounts of chemical species listed in the optional requirements are considerably less than the optional limits. No other chemical species or impurities are addressed in ASTM C1602. Any other impurities, such as organic materials or acids in the mixing water, will likely impact the strength and setting time of concrete and are evaluated by the performance-based requirements in Table 4-1.

The Appendix of ASTM C1602 also suggests minimum testing frequencies to qualify mixing water and demonstrate conformance with its requirements (listed in Table 4-1 and Table 4-2). More frequent testing is suggested when water contains a higher concentration of solids (higher density). Details of the evaluation of water, including the age and temperature of water at time of sampling, admixtures permitted, mixture proportions of the test and control batches, and sample temperatures are outlined in Annex of ASTM C1602.

AASHTO M 157, *Specification for Ready-Mixed Concrete*, is referenced by some transportation agencies and includes minor differences on water requirements as compared to ASTM C1602. The chemical limits of AASHTO M 157 are mandatory; the chloride limits stated for concrete are similar to those in ACI 318-19, however AASHTO M 157 references different test methods. The same procedures given in ASTM C1602 are used to analyze the chemistry of water. Also, it should be noted that the evaluation of water using AASHTO T 26, *Standard Method of Test for Quality of Water to be used in Concrete,* was found unsuitable for continuous evaluations of mixing water and was withdrawn as a standard method by AASHTO in 2014.

SOURCES OF MIXING WATER

It is important to account for all sources of water in the mixture when considering water quality in concrete production.

By far, the greatest volume of mixing water in concrete is the batchwater which may be supplied from a municipal water supply, a municipal reclaimed water supply, site-sourced water, or water from concrete production operations.

Municipal Water Supply

Municipal water supply systems obtain their water from a variety of sources including; aquifers, lakes and rivers, and desalinated seawater. In most cases, the water is purified and subjected to other treatments prior to use as drinking water. An atomic absorption spectrophotometer can be used to provide an analysis of water (Figure 4-2).

FIGURE 4-2. An atomic absorption spectrophotometer can be used to detect concentration of elements in the laboratory analysis of water.

Typical analyses from six municipal water supplies and also from seawater are shown in Table 4-3. These analyses are typical of domestic water supplies for most cities with populations over 20,000 in the United States and Canada. Water from any of these sources is suitable for use in concrete.

TABLE 4-3. Typical Analyses of City Water Supplies and Seawater

CHEMICAL SPECIES	SAMPLE NO. (ppm)						
	1	2	3	4	5	6	SEAWATER*
Silicate (SiO_2)	2.4	0.0	6.5	9.4	22.0	3.0	—
Iron (Fe)	0.1	0.0	0.0	0.2	0.1	0.0	—
Calcium (Ca)	5.8	15.3	29.5	96.0	3.0	1.3	50 to 480
Magnesium (Mg)	1.4	5.5	7.6	27.0	2.4	0.3	260 to 1410
Sodium (Na)	1.7	16.1	2.3	183.0	215.0	1.4	2190 to 12,200
Potassium (K)	0.7	0.0	1.6	18.0	9.8	0.2	70 to 550
Bicarbonate (HCO_3)	14.0	35.8	122.0	334.0	549.0	4.1	—
Sulfate (SO_4)	9.7	59.9	5.3	121.0	11.0	2.6	580 to 2810
Chloride (Cl)	2.0	3.0	1.4	280.0	22.0	1.0	3960 to 20,000
Nitrate (NO_3)	0.5	0.0	1.6	0.2	0.5	0.0	—
Total dissolved solids	31.0	250.0	125.0	983.0	564.0	19.0	35,000

* Chemical composition of seawater varies considerably depending on its source.

Municipal Reclaimed Water

Reclaimed water is wastewater treated for removal of solids and certain impurities. It is typically used for nonpotable applications such as irrigation, dust control, fire suppression, concrete production, construction, and other industrial uses. Reclaimed water use supports sustainable efforts to extend water supplies rather than discharge the treated wastewater to surface waters such as rivers and oceans (Abrams 1924). It also conserves supplies of potable water. Reclaimed water is particularly important in regions where water demand exceeds the available supply of municipal water sources. These conditions are prevalent during drought conditions or where the existing infrastructure to furnish potable water is inadequate to supply the needs of the local population.

Site-Sourced Water

Many large concrete paving projects and remote construction sites use site source water obtained either from wells, ponds or rivers. These natural sources of water are typically acceptable; however, when they contain significant amounts of suspended particles such as silt, organic impurities and algae, additional testing is warranted.

Recycled Water
(Water from Concrete Production)

Recycled water from concrete production processed from returned concrete is primarily a mixture of water, partially or completely hydrated cementitious materials, and aggregate fines. Recycled water can include mixer wash water and storm water collected at the concrete plant. The ready-mixed concrete industry returns on average about 5% of its estimated annual U.S. production of 300 million cubic meters (400 million cubic yards) as returned concrete (Lobo 2003). Managing returned concrete through reclaimers generates process water. In addition, about 80,000 truck mixers are washed out using on average 600 liters (150 gallons) each of water daily (Lobo 2003 and Mullings 2009). Most of this water is recirculated to keep equipment clean. The storage, conveyance, and final disposition of process water and storm water is required for permit compliance.

Given the strict regulations on discharge of water from concrete plants, the industry must look at recycling the process and storm water generated at ready mixed concrete plants.

ASTM C94 and AASHTO M 157 permit the use of water from concrete production operations as mixing water in concrete, provided it meets the limits in Table 4-1. The industry uses about 15% of the water for production as recycled water (NRMCA 2019). Environmental regulations and broader environmental practices of the concrete industry are moving the ready mixed industry towards zero-discharge production facilities, especially in larger metropolitan locations and environmentally sensitive areas.

The U.S. Environmental Protection Agency (EPA) and state environmental agencies prohibit the discharge of untreated process water from concrete operations into storm water drains that eventually funnel into the nation's waterways. This water recovered from processes of concrete production includes: wash water from mixers as part of concrete batching, water collected as a result of storm water runoff at a concrete production facility, or water that contains quantities of concrete ingredients from cleaning activities at concrete plants. In most situations, the recycled water is passed through settling ponds where the solids settle out, leaving clarified water (Figure 4-3). In some cases, the recycled water from a reclaimer unit, used to process returned fresh concrete, is continually agitated to maintain fine solids in suspension for reuse as a portion of the concrete batch water.

Solid content in recycled water typically varies from 2.5% to 10%. Solid contents exceeding 9 kg/m^3 (15 lb/yd^3) (represented by the 50,000 ppm limit in Table 4-2) may adversely impact the properties of concrete through increased water demand and accelerated setting. The increased water demand results in a higher w/cm, lower compressive strengths, and higher permeability (Lobo and Mullings 2003). The use of extended set-retarding admixtures (also referred to as hydration control or hydration stabilizing admixtures) has been shown to offset the effects of higher solids contents in mixing water (Table 4-4).

TABLE 4-4. Effect of Recycled Water on Concrete Properties *

RECYCLED WATER WITH	WATER DEMAND	SETTING TIME	COMPRESSIVE STRENGTH	PERMEABILITY	FREEZE-THAW RESISTANCE
Solid contents within ASTM C94 limits (≤ 8.9 kg/m^3 [≤ 15 lb/yd^3])	↔	↔	↔	↔	↔
High solid contents (> 8.9 kg/m^3 [> 15 lb/yd^3])	↑	↓	↓[†]	↑[†]	↔
High solid contents and treated with extended set retarding admixture	↔	↔	↔	no data	no data

Key: ↓ Decreased ↑ Increased ↔ No trend

* Compared to reference concrete produced with tap water (after Lobo and Mullings 2003).
[†] Strength and permeability effects were related to increased mixing water content.

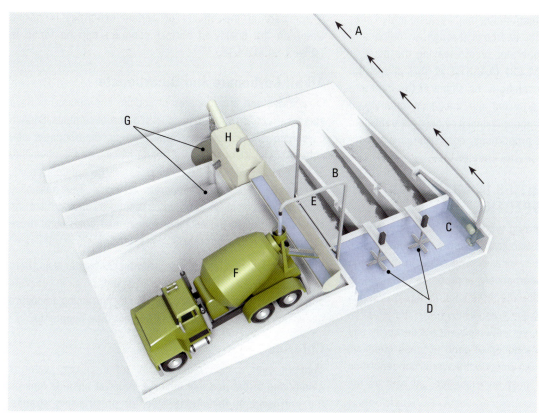

A - To water storage
B - Settling pits
C - Recycled water
D - Agitators
E - Washout station with fresh H_2O supply
F - Ready-mix truck with returned concrete
G - Reclaimed aggregates
H - Recycling screw station

FIGURE 4-3. A water reclamation system provides reuse of wash water in concrete production.

Seawater

Seawater containing up to 35,000 ppm of dissolved salts is generally suitable as mixing water for plain concrete not containing reinforcing steel. About 78% of the dissolved material in seawater is sodium chloride, and 15% is magnesium chloride and magnesium sulfate. Although concrete made with seawater may have higher early strength compared to normal concrete, strengths at later ages (after 28 days) may be lower in comparison to normal concrete. This strength reduction can be compensated by reducing the w/cm.

Seawater is not suitable for use in production of concrete with steel reinforcement and likewise, it should not be used in prestressed concrete due to the risk of corrosion of the reinforcement. If seawater is used in plain concrete (no reinforcing steel) for marine applications, moderate sulfate resistant cements, Types II, II(MH), MS, or blended cements with (MS) designations, should be used along with a low w/cm (see Chapter 2).

Sodium or potassium salts present in seawater used for mixing water can increase the alkali concentration in the concrete and increase the potential for deleterious expansions due to alkali-aggregate reactivity. Thus, seawater should not be used as mixing water for concrete with potentially reactive aggregates.

Seawater used for mixing water also tends to cause efflorescence and dampness on concrete surfaces exposed to air and water (Steinour 1960). Marine-dredged aggregates are discussed in Chapter 5.

Other Sources of Mixing Water

Other portions of the mixing water may come from additional materials selected for use in the concrete mixture including:

Free moisture on aggregates. Moisture adsorbed on the surface constitutes a substantial portion of the total mixing water. It is important that the free water on the aggregate is free from harmful materials. This might be the case when unwashed aggregate is extracted from marine sources (see Chapter 5).

Ice. During hot-weather concreting, ice might be used as part of the mixing water (see Chapter 18). The ice should be completely melted by the time mixing is completed.

Admixtures. Water contained in admixtures must be considered part of the mixing water if the water incorporated by the use of the admixture increases the water-to-cementitious materials ratio by 0.01 or more. Admixture products such as certain accelerators, corrosion inhibitors, shrinkage reducing admixtures, and viscosity modifiers are used at greater quantities in gallons per cubic yard. They should be included as part of the mixing water and for the calculation of w/cm of the concrete mixture.

Water addition after batching. Water might also be added after concrete is batched either in transit through an automated system or by the truck operator before leaving the plant or at the project site. ASTM C94 (AASHTO M 157) allows the addition of water after batching if the slump of the concrete is less than specified, provided the maximum quantity of water does not exceed the maximum amount established by the designed mixture proportions, or maximum water-to-cementitious materials ratio (see Chapter 14).

EFFECTS OF IMPURITIES IN MIXING WATER ON CONCRETE PROPERTIES

Excessive impurities in mixing water may affect setting time, concrete strength, and durability. Therefore, certain optional limits on chlorides, sulfates, alkalies, and solids in the mixing water may be invoked. Alternatively, appropriate tests can be performed to determine the effect the impurity has on various properties (Table 4-1). Some impurities may have little effect on strength and setting time, yet adversely impact durability and other properties. The contribution of each of these chemical species present in the mixing water to the overall concentration in the concrete mixture must be evaluated, as well as the impact on durability.

Water containing less than 2000 ppm of total dissolved solids is generally satisfactory for use in concrete. Water containing more than 2000 ppm of dissolved solids should be tested for its effect on strength and time of set (Table 4-1). Additional information on the effects of mixing water impurities can be found in Steinour (1960) and Abrams (1924). Over 100 different compounds and ions are discussed.

A summary of the effects of certain impurities in mixing water on the quality of normal concrete can be found in Tables 4-5A and 4-5B.

Alkali Carbonate and Bicarbonate

Carbonates and bicarbonates of sodium and potassium have varying effects on the setting times of different cements. Sodium carbonate can cause very rapid setting, bicarbonates can either accelerate or retard the set depending on the chemistry of the cement used in the concrete. In large concentrations, these salts can reduce concrete strength. The possibility of exacerbated alkali-aggregate reactions should also be considered due to the alkalies (sodium and potassium) these salts bring into the concrete.

Some chemical species impact the hardness of water. Using hard water can adversely impact the ability to entrain air in concrete or considerably increase the dosage of air entraining admixture necessary to obtain a required air content (See Chapter 6).

Chloride

Concern over a high chloride content in mixing water is chiefly due to the possible adverse effect of chloride ions on the corrosion of reinforcing steel or prestressing strands. Chloride ions attack the protective oxide film formed on the steel by the highly alkaline (pH often greater than 13.0) chemical environment present in concrete. The chloride ion concentration that initiates corrosion of steel reinforcement is about 0.2% to 0.4% by mass of cementitious material – referred to as the threshold concentration (Whiting 1997; Whiting and others 2002; and Taylor and others 2000).

TABLE 4-5A. Summary of Potential Adverse Effects of Water Impurities on Concrete Strength and Setting Time

STRENGTH			SETTING TIME	
HIGH IMPACT	**MODERATE IMPACT**	**LOW IMPACT**	**HIGH IMPACT**	**MODERATE IMPACT**
Sodium sulfide Salts of zinc, copper, and lead Sugar Algae	Alkali carbonate / bicarbonate Salts of manganese and tin Sodium iodate / phosphate / borate / arsenate Organic acids Industrial wastewater	Calcium / magnesium bicarbonate Magnesium sulfate / chloride Iron salts Inorganic acids Seawater Alkali hydroxides Oils	Sodium carbonate Salts of zinc, copper, and lead Sugar	Alkali bicarbonate Salts of manganese and tin Sodium iodate / phosphate / borate / arsenate Sodium sulfide Alkali hydroxides Industrial wastewater

TABLE 4-5B. Summary of Potential Adverse Effects of Water Impurities on Other Concrete Properties

WORKABILITY	AIR CONTENT	ALKALI-AGGREGATE REACTIVITY	CORROSION RESISTANCE	FREEZE-THAW RESISTANCE	SULFATE RESISTANCE
Silt Suspended particles	Hard water High salt content Seawater Oils Algae	Alkalies Seawater	Chlorides Seawater	Algae	Sulfates Seawater

Chlorides can be introduced internally into fresh concrete with any of the mixture ingredients – admixtures, aggregates, cementitious materials, or mixing water – or into hardened concrete through external exposure to deicing salts, seawater, or salt-laden air in coastal environments. Placing an acceptable limit on chloride content for any one ingredient, such as mixing water, is problematic considering the variety of chloride ion sources in concrete. An acceptable chloride limit in the concrete depends primarily upon the type of structure and the environment to which it is exposed during its service life.

A high dissolved solids content of natural water is sometimes due to a high content of sodium chloride or sodium sulfate. Both can be tolerated in rather large quantities. Concentrations of 20,000 ppm of sodium chloride are generally tolerable in plain (unreinforced) concrete that will be dry in service and has low potential for corrosive reactions.

Water used in prestressed concrete or in concrete designed with aluminum embedments should not contain significant amounts of chloride ions. The contribution of chlorides from ingredients other than water must also be considered. Calcium chloride admixtures, used to accelerate setting time in cold weather, should be avoided in steel reinforced concrete. See Chapter 11 for more information on chloride limits and resistance to corrosion.

Sulfate

Concern over a high sulfate content in mixing water is due to possible expansive reactions and deterioration caused by sulfate attack (see Chapter 11). Although mixing waters containing 10,000 ppm of sodium sulfate have been used satisfactorily, the limit in Table 4-2 should be considered unless special precautions in the composition of the concrete mixture are taken. The potential for sulfate attack increases with an increase in tricalcium aluminate (C_3A) content of portland cements and alumina (Al_2O_3) content of SCMs in the cementitious materials (See Chapters 2 and 3).

Calcium and Magnesium

Calcium carbonate and magnesium carbonate are not very soluble in water and are seldom found in sufficient concentration to affect the strength of concrete. Calcium bicarbonate and magnesium bicarbonate are present in some municipal waters. Total concentrations up to 400 ppm of bicarbonate in these forms are not considered harmful.

Magnesium sulfate and magnesium chloride can be present in high concentrations without harmful effects on concrete strength. Satisfactory strengths have been obtained using water with concentrations up to 40,000 ppm of magnesium chloride. Concentrations of magnesium sulfate should be less than 25,000 ppm.

Iron Salts

Natural ground waters seldom contain more than 20 ppm to 30 ppm of iron; however, acid mine waters may contain larger quantities. Iron salts in concentrations up to 40,000 ppm do not usually adversely affect concrete strengths. The potential for staining should be evaluated when aesthetics are important.

Miscellaneous Inorganic Salts

Salts of manganese, tin, zinc, copper, and lead in mixing water can cause a significant reduction in strength and large variations in setting time. Of these, salts of zinc, copper, and lead are the most active. Salts that are especially active as retarders include sodium iodate, sodium phosphate, sodium arsenate, and sodium borate. All can greatly retard both set and strength development when present in concentrations as low as a few tenths percent by mass of the cementitious materials. Generally, concentrations of these salts up to 500 ppm can be tolerated in mixing water.

Another salt that may be detrimental to concrete is sodium sulfide; even the presence of 100 ppm warrants testing.

Acid Waters

Acceptance of acid mixing water should be based on the concentration (in parts per million) of acids in the water. Generally, mixing waters containing hydrochloric, sulfuric, and other common inorganic acids in concentrations as high as 10,000 ppm have no adverse effect on strength.

Occasionally, acceptance is based on the measured pH, a log scale measure of the hydrogen-ion concentration. The pH value is an intensity index and is not the best measure of potential acid or base reactivity. The pH of neutral water is 7.0; values below 7.0 indicate acidity and those above 7.0 alkalinity (a base). Acid waters with pH values less than 3.0 may create handling problems and should be avoided if possible. Organic acids, such as tannic acid, can have a significant effect on strength at higher concentrations (Figure 4-4).

Alkaline Waters

Waters with sodium hydroxide concentrations of 0.5% by mass of cementitious materials do not greatly affect concrete strength provided quick set is not induced. Higher concentrations, however, may reduce concrete strength.

Potassium hydroxide in concentrations up to 1.2% by mass of cementitious materials has little effect on the concrete strength developed by some cements, but the same concentration level when used with other cementitious materials may substantially reduce the 28-day strength.

The possibility for increased expansion due to alkali-aggregate reactivity should be considered.

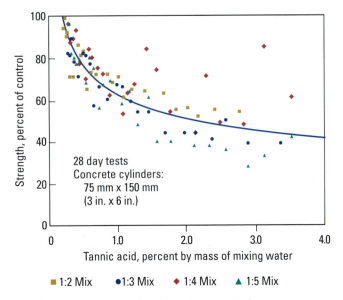

Strength, percent of control (y-axis)
Tannic acid, percent by mass of mixing water (x-axis)

28 day tests
Concrete cylinders:
75 mm x 150 mm
(3 in. x 6 in.)

■ 1:2 Mix   ● 1:3 Mix   ◆ 1:4 Mix   ▲ 1:5 Mix

FIGURE 4-4. Effect of tannic acid on the strength of concrete (Abrams 1920).

Industrial Wastewater

Most waters carrying industrial wastes have less than 4000 ppm of solids. When such water is used as mixing water in concrete, the reduction in compressive strength is generally not greater than 10% to 15%.

Wastewaters such as those from tanneries, paint factories, coke plants, and chemical and galvanizing plants may contain harmful impurities. It is best to test any wastewater that contains as little as a few hundred parts per million of unusual solids.

Silt or Suspended Particles

About 2000 ppm of suspended clay or fine rock particles can be tolerated in mixing water. Higher amounts might not affect strength but may influence other properties of some concrete mixtures. Before use, muddy or cloudy water should be passed through settling basins or otherwise clarified to reduce the amount of silt and clay added to the concrete mixture. When cementitious fines are returned to the concrete in reused wash water, 50,000 ppm can be tolerated when additional steps to offset the effects of higher solids content are taken – typically the use of extended set controlling admixtures.

ORGANIC IMPURITIES

The effect of organic substances on the setting time of portland cement or the ultimate strength of concrete is a problem of considerable complexity. Such substances, like surface loams, can be found in natural waters. Highly colored waters, waters with a noticeable odor, or those in which green or brown algae are visible should be regarded with suspicion and tested accordingly.

Waters Carrying Sanitary Sewage

A typical sewage may contain about 400 ppm of organic matter. After the sewage is diluted in a good disposal system, the concentration is reduced to about 20 ppm or less. This amount is usually considered too low to have any significant effect on concrete strength.

Sugar

Small amounts of sucrose (common sugar), as little as 0.03% to 0.15% by mass of cement, usually retards the setting of portland cement phases. The upper limit of this range varies with different cements. The 7-day compressive strength may be reduced while the 28-day strength may be improved. Sugar in quantities of 0.25% or more by mass of cement may cause rapid setting and a substantial reduction in 28-day strength. Each type of sugar can influence setting time and strength differently. Less than 500 ppm of sugar in mixing water generally has no adverse effect on strength, but if the concentration exceeds this amount, tests for setting time and strength should be conducted.

Oils

Various kinds of oil are occasionally present in mixing water. Mineral oil (petroleum) not mixed with animal or vegetable oils may have less effect on strength development than other oils. However, mineral oil in concentrations greater than 2.5% by mass of cementitious materials may reduce strength by more than 20%. Oils may also interfere with the action of air-entraining admixtures.

Algae

Water containing algae is unsuitable for concrete because the algae can cause an excessive reduction in strength. Algae in water leads to lower strengths either by influencing cement hydration or by causing a large amount of air to be entrained in the concrete. Algae may also be present on aggregates, in which case the bond between the aggregate and cement paste is reduced. A maximum limit on algae content of 1000 ppm is recommended.

INTERACTION WITH ADMIXTURES

When evaluating a water source for its effect on concrete properties, it is important to also test the water with chemical admixtures that will be used in the concrete mixture. Certain compounds in water can influence the performance and efficiency of certain admixtures. These interactions of cementitious materials, admixtures, and chemistry of water can become rather complicated. See Chapter 6 for more information on Chemical Admixtures. ASTM C1679, *Standard Practice for Measuring Hydration Kinetics of Hydraulic Cementitious Mixtures Using Isothermal Calorimetry*, and/or ASTM C1753, *Standard Practice for Evaluating Early Hydration of Hydraulic Cementitious Mixtures Using Thermal Measurements*, may be useful tools in evaluating some of the performance effects of these combinations.

REFERENCES

PCA's online catalog includes links to PDF versions of many research reports and other classic publications.
Visit: cement.org/library/catalog.

Abrams, D. A., *Effect of Tannic Acid on the Strength of Concrete*, LS007, Bulletin 7, Structural Materials Research Laboratory, Lewis Institute, Chicago, IL, 1920, 34 pages.

Abrams, D.A., *Tests of Impure Waters for Mixing Concrete*, LS012, Bulletin 12, Structural Materials Research Laboratory, Lewis Institute, Chicago, IL, 1924, 50 pages.

ACI Committee 301, *Specification for Concrete Construction*, ACI 301-20, American Concrete Institute, Farmington Hills, MI, 2020, 73 pages.

ACI Committee 318, *Building Code Requirements for Structural Concrete*, ACI 318-19, American Concrete Institute, Farmington Hills, MI, 2019, 629 pages.

Lobo, C., "Recycled Water in Concrete," *Concrete Technology Today*, Vol. 24, No. 3, CT033, Portland Cement Association, Skokie, IL, December 2003, pages 2 to 3.

Lobo, C.; and Mullings, G.M., "Recycled Water in Ready Mixed Concrete Operations," *Concrete in Focus*, 2003, 10 pages.

Meininger, R.C., *Recycling Mixer Wash Water*, National Ready Mixed Concrete Association, Silver Spring, MD, 2000.

Mullings, G.M., *Environmental Management in the Ready Mixed Concrete Industry*, 2PEMRM, 1st Edition, National Ready Mixed Concrete Association, Silver Spring, MD, 2009, 256 pages.

NRMCA, *Mixing Water Quality for Concrete*, Technology in Practice 10, National Ready Mixed Concrete Association, Silver Spring, MD, 2013, 8 pages.

NRMCA, *Ready Mixed Concrete Industry Data Report*, National Ready Mixed Concrete Association, Silver Spring, MD, 2019, 40 pages.

Steinour, H.H., *Concrete Mix Water—How Impure Can It Be?*, Research Department Bulletin RX119, Portland Cement Association, Chicago, IL, 1960, 20 pages.

Taylor, P. C.; Nagi, M.A.; and Whiting, D.A., *Threshold Chloride Content of Steel in Concrete: A Literature Review*, SN2169, Portland Cement Association, Skokie, IL, 1999, 32 pages.

Whiting, D.A., *Origins of Chloride Limits for Reinforced Concrete*, SN2153, Portland Cement Association, 1997, Skokie, IL, 16 pages.

Whiting, D.A.; Taylor, P.C.; and Nagi, M.A., *Chloride Limits in Reinforced Concrete*, SN2438, Portland Cement Association, Skokie, IL, 2002, 76 pages.

AGGREGATES FOR CONCRETE

FIGURE 5-1. Aggregate stockpiles at a ready-mix plant (courtesy of S. Parkes, Votorantim Cimentos).

Aggregates generally occupy 60% to 75% of concrete's volume (70% to 85% by mass). The proper selection of aggregates is critical to the long-term performance of any concrete mixture. Aggregates (Figure 5-1) strongly influence the concrete's fresh and hardened properties, proportions of other ingredients, sustainability, and economy. For optimum engineering performance, aggregates must be clean, hard, strong, and durable particles that are largely free of coatings of clay, and other fine materials in amounts that could affect hydration and/or bond with the cement paste.

A wide range of commonly occurring rocks, sand, and gravels are suitable aggregates for producing concrete. Some variation in the type, quality, cleanliness, grading, moisture content, and other properties of aggregates is expected and can be accommodated. Aggregates that are friable (capable of being easily broken or split) or containing appreciable amounts of soft and porous materials, including some varieties of siltstone, claystone, mudstone, shale, and shaley rocks, should be avoided. Certain types of chert have low resistance to weathering and can cause surface defects such as popouts in freezing and thawing and are sometimes reactive to alkalies in concrete.

Fine aggregates (Figure 5-2), or sand, generally consist of natural sand or manufactured sand from crushed stone with most particles smaller than 5 mm (0.2 in.). *Coarse aggregates* (Figure 5-3) typically consist of gravels, crushed stone, or a combination of both, with particles predominantly larger than 5 mm (0.2 in.) and generally range in between 9.5 mm and 37.5 mm (⅜ in. and 1½ in.).

More than 446,000,000 metric tons (492,000,000 tons) of natural sand and gravel was used in concrete applications (Willet 2020a), and more than 1,150,000,000 metric tons (1,268,000,000 tons) of crushed stone was used as construction aggregate (including non-concrete applications) (Willet 2020b) in the US in 2019.

FIGURE 5-2. Fine aggregate (sand) consists of particles smaller than 5 mm (0.2 in.).

FIGURE 5-3. Coarse aggregate generally range between 9.5 mm and 37.5 mm (⅜ in. and 1½ in.). Gravel (left) tends to be more rounded and crushed stone (right) tends to be more angular.

AGGREGATE GEOLOGY

Concrete aggregates are a mixture of rocks and minerals (see Table 5-1). A mineral is a naturally occurring solid substance with an orderly internal structure and a chemical composition that ranges within narrow limits. A rock is generally composed of several minerals. For example, granite contains quartz, feldspar, mica, and a few other minerals; and most limestones consist of calcite, dolomite, and minor amounts of quartz and clay. Rocks are classified as igneous, sedimentary, or metamorphic, depending on their geological origin.

AGGREGATE CLASSIFICATION

Aggregates are classified into three categories: normalweight, lightweight, and heavyweight. The approximate bulk densities of *aggregate* commonly used in concrete range from:

- normalweight concrete aggregate: about 1200 kg/m³ to 1750 kg/m³ (75 lb/ft³ to 110 lb/ft³),
- lightweight concrete aggregate: about 560 kg/m³ to 1120 kg/m³ (35 lb/ft³ to 70 lb/ft³), and
- heavyweight concrete aggregate: typically over 2100 kg/m³ (130 lb/ft³).

These lead to *concrete* with a typical densities (fresh concrete unit weight) ranging between:

- normalweight concrete: 2200 kg/m³ to 2400 kg/m³ (140 lb/ft³ to 150 lb/ft³),
- structural lightweight concrete: 1350 kg/m³ to 1850 kg/m³ (90 lb/ft³ to 120 lb/ft³),
- insulating lightweight concrete: 250 kg/m³ to 1450 kg/m³ (15 lb/ft³ to 90 lb/ft³), and
- heavyweight concrete: 2900 kg/m³ to 6100 kg/m³ (180 lb/ft³ to 380 lb/ft³).

Lightweight Aggregates

Structural lightweight aggregates are usually classified according to their production process because various processes produce aggregates with somewhat different properties. Processed lightweight aggregates for structural concrete applications should meet the requirements of ASTM C330, *Standard Specification for Lightweight Aggregates for Structural Concrete*, which includes:

- rotary kiln expanded clays, shales, and slates (Figure 5-4);
- pelletized or extruded fly ash; and
- expanded slags.

FIGURE 5-4. Expanded clay (left) and expanded shale (right).

Structural lightweight aggregates can also be produced by processing other types of material, such as naturally occurring pumice and scoria.

Structural lightweight aggregates may absorb 5% to 20% water by weight of dry material. To control the uniformity of structural lightweight concrete mixtures, the aggregates are prewetted prior to batching.

TABLE 5-1. Rock and Mineral Constituents in Aggregates

MINERALS	MINERALS (CONTINUED)	IGNEOUS ROCKS	SEDIMENTARY ROCKS	METAMORPHIC ROCKS
Silica	Carbonates	Granite	Conglomerate	Marble
Quartz	Calcite	Syenite	Sandstone	Metaquartzite
Opal	Dolomite	Diorite	Quartzite	Slate
Chalcedony	Sulfates	Gabbro	Graywacke	Phyllite
Tridymite	Gypsum	Peridotite	Subgraywacke	Schist
Cristobalite	Anhydrite	Pegmatite	Arkose	Amphibolite
Silicates	Iron sulfides	Volcanic glass	Claystone	Hornfels
Feldspars	Pyrite	Obsidian	Siltstone	Gneiss
Ferromagnesian	Marcasite	Pumice	Argillite	Serpentinite
Hornblende	Pyrrhotite	Tuff	Shale	
Augite	Iron oxide	Scoria	Carbonates	
Clay	Magnetite	Perlite	Limestone	
Illites	Hematite	Pitchstone	Dolomite	
Kaolins	Goethite	Felsite	Marl	
Chlorites	Ilmenite	Basalt	Chalk	
Montmorillonites	Limonite		Chert	
Mica				
Zeolite				

For brief descriptions, see ASTM C294, *Standard Descriptive Nomenclature for Constituents of Concrete Aggregates.*

High-Density Aggregates

High-density aggregates such as barite, ferrophosphorus, goethite, hematite (Figure 5-5), ilmenite, limonite, magnetite, and degreased steel punchings and steel shot are used to produce high-density concrete. Where high fixed-water content is desirable, serpentine (which is slightly heavier than normal-weight aggregate) or bauxite can be used (see ASTM C637, *Standard Specification for Aggregates for Radiation-Shielding Concrete*, and ASTM C638, *Standard Descriptive Nomenclature of Constituents of Aggregates for Radiation-Shielding Concrete*).

Table 5-2 gives typical bulk density, relative density (specific gravity), and percentage of fixed water for some of these materials. The values are a compilation of data from a wide variety of tests or projects reported in the literature. Steel punchings and shot are used where concrete with a density of more than 4800 kg/m³ (300 lb/ft³) is required. Higher fixed water contents of high-density aggregate improve radiation shielding.

In general, selection of an aggregate is determined by physical properties, availability, and cost. Heavyweight aggregates should be reasonably free of fine material, oil, and other

FIGURE 5-5. Iron ore, an example of a high-density aggregate (courtesy of S. Lee, CTLGroup).

foreign substances that may affect either the bond of paste to aggregate particle or the hydration of cement. For good workability, maximum density, and economy, aggregates should be roughly equidimensional in shape and free of excessive flat or elongated particles.

TABLE 5-2. Physical Properties of Typical High-Density Aggregates and Concrete

TYPE OF AGGREGATE	FIXED-WATER* % by weight	AGGREGATE RELATIVE DENSITY	AGGREGATE BULK DENSITY kg/m³ (lb/ft³)	CONCRETE DENSITY kg/m³ (lb/ft³)
Goethite	10–11	3.4–3.7	2080–2240 (130–140)	2880–3200 (180–200)
Limonite†	8–9	3.4–4.0	2080–2400 (130–150)	2880–3360 (180–210)
Barite	0	4.0–4.6	2320–2560 (145–160)	3360–3680 (210–230)
Ilmenite	†	4.3–4.8	2560–2700 (160–170)	3520–3850 (220–240)
Hematite	†	4.9–5.3	2880–3200 (180–200)	3850–4170 (240–260)
Magnetite	†	4.2–5.2	2400–3040 (150–190)	3360–4170 (210–260)
Ferrophosphorus	0	5.8–6.8	3200–4160 (200–260)	4080–5290 (255–330)
Steel punchings or shot	0	6.2–7.8	3680–4650 (230–290)	4650–6090 (290–380)

* Water retained or chemically bound in aggregates.

† Aggregates may be combined with limonite to produce fixed-water contents varying from about ½% to 5%.

Normalweight Aggregates

The remainder of this chapter focuses on normalweight aggregates. Normal-weight aggregates should meet the requirements of ASTM C33, *Standard Specification for Concrete Aggregates*, or AASHTO M 6, *Standard Specification for Fine Aggregate for Hydraulic Cement Concrete* or AASHTO M 80, *Standard Specification for Coarse Aggregate for Hydraulic Cement Concrete*; and for highway construction, ASTM D448, *Standard Classification for Sizes of Aggregate for Road and Bridge Construction* (AASHTO M 43).

Normalweight aggregates for use in concrete include natural aggregate, manufactured aggregate, recycled-concrete aggregate, and marine-dredged aggregate.

AGGREGATE SOURCES

Natural Aggregate

Natural gravel and sand are usually dug or dredged from a pit, river, lake, seabed or glacial deposits. Weathering and erosion of rocks produces particles of stone, gravel, sand, silt, and clay. Some natural aggregate deposits of gravel and sand can be readily used in concrete with minimal processing (Figure 5-6).

FIGURE 5-6. Natural gravel and sand being extracted from quarry (courtesy of S. Parkes, Votorantim Cimentos).

The quality of natural aggregate depends on the bedrock from which the particles were derived and the mechanism by which they were mined, processed, and transported. Sand and gravel derived from igneous and metamorphic rocks tend to be sound, while sand and gravel derived from rocks rich in shale and siltstone are more likely to be unsound. Natural aggregate deposited at higher elevations from glaciers tend to be superior to deposits in lower areas. Natural aggregates from higher elevations tend to consist of harder more sound rocks. Sand and gravel that have been smoothed to rounded or subrounded particle shape by prolonged agitation in water usually are considered higher quality because they are harder and have a more rounded shape than less abraded sand and gravel. However, the smooth surface of natural gravels have potential to reduce the bond strength with the cement paste and may result in lower overall concrete strength.

Manufactured Aggregate

Manufactured aggregate (including manufactured sand) is commonly called crushed stone and is produced by crushing sound parent rock (igneous, sedimentary, or metamorphic) at crushing plants. Crushed air-cooled blast-furnace slag may also be used as fine or coarse aggregate.

Manufactured aggregates differ from gravel and sand in their grading, shape, and surface texture. As a result of the crushing operation, manufactured aggregates often have a rough surface texture, are more angular in nature, tend to be cubical or elongated in shape (depending on the lithology and method of crushing), and more uniform in size (Wigum 2004). Producers can provide material meeting specified particle shapes and gradations (Owens 2009). In many cases, the particle elongation and layered flakes of manufactured sands can be reduced through appropriate crushing techniques.

FIGURE 5-7. Recycled-concrete aggregate.

FIGURE 5-8. A reinforced concrete structure is demolished and recycled.

Manufactured aggregates are less likely than gravel and sand to be contaminated by deleterious substances such as clay minerals or organic matter. Some specifications, such as ASTM C33, permit higher fines content in manufactured sands because of the expectation of less clay contamination (Owens 2009).

Recycled-Concrete Aggregate

Recycled-concrete aggregate (RCA) is existing concrete that has been crushed to produce aggregate (Figure 5-7). In the US in 2014, approximately 30 million metric tons of recycled concrete was used as concrete aggregate (Townsend and Anshassi 2017). The concept of recycling and reusing concrete pavements, buildings, and other structures as a source of aggregate is now relatively common practice, resulting in both material and energy savings (ECCO 1999).

Recycling concrete aggregates from existing structures involves demolishing and removing the existing concrete, removing any reinforcing steel and other embedded items, crushing the material in primary and secondary crushers (Figure 5-8), grading and washing, and stockpiling the resulting coarse and fine aggregate (ACI 555R 2001). Dirt, gypsum board, wood, and other foreign materials should be prevented from contaminating the final product. Recycled-concrete aggregate may also be produced by crushing returned concrete.

Recycled-concrete aggregate is primarily used in pavement reconstruction. It has been satisfactorily used as aggregate in granular subbases, lean-concrete subbases, soil-cement, and in new concrete as the primary source of aggregate or as a partial replacement of new aggregate. More than 300 million metric tons of RCA was recycled as road base and in other applications in 2014 in the US (Townsend and Anshassi 2017).

ASTM C33 includes recycled-concrete aggregates as part of its standard specification as does AASHTO M 80. ACI 301, *Specifications for Concrete Construction* (2020), and ACI 318, *Building Code Requirements for Structural Concrete* (2019), now permit use of RCA if documentation is provided as required by the contract documents and its use is determined suitable for a particular project.

Recycled-concrete aggregate generally has a higher absorption and a lower specific gravity than conventional aggregate (ACI 555R 2001), due to hardened cement paste within the recycled concrete aggregate having a higher porosity. Absorption values typically range from 3% to 10%, depending on the concrete being processed and the method being used to produce it. This absorption rate lies between those for natural and lightweight aggregate. Absorption rates increase as coarse particle size decreases (Figure 5-9). The high absorption of the recycled aggregate makes it necessary to add more water to achieve the same workability and slump as concrete with conventional aggregates. Dry recycled aggregate absorbs water during and after mixing. To avoid this increased absorption, RCA stockpiles should be prewetted or kept moist, as is the practice with lightweight aggregates.

The particle shape of recycled-concrete aggregate is similar to crushed rock. The relative density decreases progressively as particle size decreases. The chloride content of recycled-concrete aggregate should be determined where applicable.

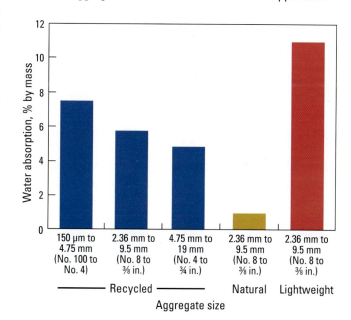

FIGURE 5-9. Comparison of water absorption of three different recycled aggregate particle sizes and one size of natural and lightweight coarse aggregate (Kerkhoff and Siebel 2001).

New concrete made from recycled-concrete aggregate generally has good durability. As with any new aggregate source, recycled-concrete aggregate should be tested for durability, gradation, and other properties. Carbonation, permeability, and freeze-thaw resistance have similar or improved performance relative to concrete produced with conventional aggregates.

Recycled concrete used as coarse aggregate in new concrete possesses some potential for alkali-silica reaction (ASR) if the old concrete contained alkali-reactive aggregate. Aggregates prone to alkali-silica reactivity are listed under the section on **Alkali-Aggregate Reactivity** in this chapter, and aggregate testing discussed in Chapter 20 can be used to evaluate RCA. The alkali content of the cement used in the old concrete has little effect on expansion due to alkali-silica reaction. For highly reactive aggregates made from recycled concrete, special measures discussed in Chapter 11 should be used to control ASR. Also, even if expansion from ASR did not develop in the original concrete, it cannot be assumed that it will not develop in the new concrete without appropriate control measures. Petrographic examination and expansion tests are recommended for careful evaluation (Stark 1996).

Using RCA from non-air entrained concrete could contribute to freeze-thaw issues. Concrete made with recycled coarse aggregates and conventional fine aggregate can generally obtain adequate compressive strength. The use of recycled fine aggregate can result in minor compressive strength reductions, unless mixture proportions are adjusted to compensate. However, drying shrinkage and creep of concrete made with recycled aggregates can be up to 100% greater than concrete made with conventional aggregate. This is due to the large amount of existing cement paste and mortar present, especially in the fine aggregate. Therefore, considerably lower values of drying shrinkage can be achieved using recycled coarse aggregate with natural sand, as compared to fine aggregate manufactured from recycled concrete (Kerkhoff and Siebel 2001).

Additional reductions in drying shrinkage and creep can be achieved using the Equivalent Mortar Volume (EMV) method of proportioning (Kim and Sadowski 2019). Using this method, the mortar fraction of the recycled concrete aggregate is treated as mortar for the new mixture, as opposed to the aggregate fraction. Values for shrinkage, creep, and other concrete properties for concrete made using this method are similar to those for concrete made with virgin aggregates (Fathifazl and others 2010). The EMV method requires additional testing of the RCA to accurately determine the amount of mortar that is present.

Concrete trial mixtures should be made with recycled-concrete aggregates to determine the proper mixture proportions. The variability in the properties of the old concrete may in turn affect the properties of the new concrete. This can partially be avoided by frequent monitoring of the properties of the old concrete that is being recycled. Adjustments in the mixture proportions may be needed.

Marine-Dredged Aggregate

Marine-dredged aggregate from tidal estuaries and sand and gravel from the seashore may be used with caution in concrete applications when other aggregate sources are not available. Marine dredged sands typically are mono-size and will not meet grading requirements. There are two primary concerns with aggregates obtained from seabeds: seashells and salt. Although seashells are a hard material that can produce good quality concrete, a higher paste content may be required due to the angularity of the shells. Aggregate containing complete shells (uncrushed) should be avoided as their presence may cause placement issues that result in voids in the concrete and may lower the compressive strength.

Marine-dredged aggregates often contain salt (chlorides) from the seawater. The highest salt content occurs in sands located just above the high-tide level, although the amount of salt on the aggregate is often not more than about 1% of the mass of the mixing water. See Chapter 11 for a discussion of chloride content limits for reinforced concrete.

The presence of chlorides may affect the concrete by: altering the time of set, increasing drying shrinkage, significantly increasing the risk of corrosion of steel reinforcement, and causing efflorescence. Generally, marine aggregates containing large amounts of chloride should not be used in reinforced concrete. To reduce the chloride content, marine-dredged aggregates can be washed with fresh water.

CHARACTERISTICS OF AGGREGATES

ASTM C33 (AASHTO M 6 or AASHTO M 80) specifications provide requirements for aggregate characteristics and limit the permissible amounts of deleterious substances. The important characteristics of aggregates for concrete are listed in Table 5-3. Compliance is determined using one or more of the several standard tests cited in the following sections and tables. More information on aggregate testing is found in Chapter 20.

Identification of the constituents of an aggregate cannot alone provide a basis for predicting the behavior of aggregates in service. Visual inspection will often disclose weaknesses in coarse aggregates. Service records are invaluable in evaluating aggregates. In the absence of a performance record, the aggregates should be tested before they are used in concrete.

Grading

Grading is the particle-size distribution of an aggregate as determined by a sieve analysis (ASTM C136/C117 or AASHTO T 27/T 11). The range of particle sizes in aggregate is illustrated in Figure 5-10. The aggregate particle size is determined using wire-mesh sieves with square openings.

TABLE 5-3. Characteristics and Tests of Aggregate

CHARACTERISTIC	SIGNIFICANCE	STANDARD DESIGNATION*		REQUIREMENT OR ITEM REPORTED
Definitions of constituents	Clear understanding and communication	ASTM C125 ASTM C294		
Resistance to abrasion and degradation	Index of aggregate quality; wear resistance of floors and pavements	ASTM C131 ASTM C535 ASTM C779	(AASHTO T 96)	Maximum percentage of weight loss. Depth of wear and time
Resistance to freezing and thawing	Surface scaling, roughness, loss of section, and aesthetics	ASTM C666 ASTM C672	(AASHTO T 161) (AASHTO T 103)	Maximum number of cycles or period of frost immunity; durability factor
Resistance to disintegration by sulfates	Soundness against weathering action D-cracking	ASTM C88	(AASHTO T 104)	Weight loss, particles exhibiting distress
Particle shape and surface texture	Workability of fresh concrete	ASTM C295 ASTM D3398 ASTM D4791		Maximum percentage of flat and elongated particles
Grading	Workability of fresh concrete; economy	ASTM C117 ASTM C136	(AASHTO T 11) (AASHTO T 27)	Minimum and maximum percentage passing standard sieves
Aggregate degradation	Index of aggregate quality; Resistance to degradation	ASTM D6928 ASTM D7428	AASHTO T 327	
Uncompacted void content of fine aggregate	Workability of fresh concrete	ASTM C1252	(AASHTO T 304)	Uncompacted voids and specific gravity values
Bulk density (unit weight)	Mixture proportioning classification	ASTM C29	(AASHTO T 19)	Compact weight and loose weight
Relative density (specific gravity)	Mixture Proportioning	ASTM C127 (fine aggregate) ASTM C128 (coarse aggregate)	(AASHTO T 85) (AASHTO T 84)	
Absorption and surface moisture	Control of concrete quality (water-cement ratio)	ASTM C70 ASTM C127 ASTM C128 ASTM C566	(AASHTO T 85) (AASHTO T 84) (AASHTO T 255)	
Compressive and flexural strength	Acceptability of fine aggregate failing other tests	ASTM C39 ASTM C78	(AASHTO T 22) (AASHTO T 97)	Strength to exceed 95% of strength achieved with purified sand
Aggregate constituents	Determine amount of deleterious and organic materials	ASTM C40 ASTM C87 ASTM C117 ASTM C123 ASTM C142 ASTM C295	(AASHTO T 21) (AASHTO T 71) (AASHTO T 11) (AASHTO T 113) (AASHTO T 112)	Maximum percentage allowed of individual constituents

* The majority of the tests and characteristics listed are referenced in ASTM C33 (AASHTO M 6/M 80). Table 5-9 lists several additional tests for potentially harmful materials.

FIGURE 5-10. Range of particle sizes found in coarse and fine aggregate for use in concrete.

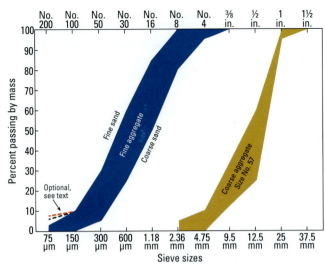

FIGURE 5-11. Curves indicate the limits specified in ASTM C33 for fine aggregate and for No. 57 coarse aggregate, a commonly used size (grading). Optional limits shown are discussed in footnotes to Table 5-4.

Results of sieve analyses are used to: determine whether or not the materials meet specifications; select the most suitable material if several aggregates are available; and detect variations in grading that are sufficient to warrant blending selected sizes or an adjustment of concrete mixture proportions.

The grading and grading limits are usually expressed as the percentage of material passing each sieve. Figure 5-11 shows these limits for fine aggregate and for one specific size of coarse aggregate.

In general, aggregates that do not have a large deficiency or excess of any one particle size and give a smooth grading curve will produce the most satisfactory results. ASTM C33 (AASHTO M 6/ AASHTO M 80) and ASTM D448 (AASHTO M 43) allow for the use of aggregates with gradings falling outside the specification limits provided it can be shown that good quality concrete can be produced from those aggregates.

There are several reasons for specifying grading limits and a nominal maximum aggregate size. Grading affects relative aggregate proportions as well as cement and water requirements, workability, pumpability, bleeding, economy, porosity, shrinkage, and durability of concrete. Variations in grading can seriously affect the uniformity of concrete from batch to batch. Very fine sands are often uneconomical while very coarse sands and coarse aggregate can produce harsh, unworkable mixtures. Coarse sands may be desirable with higher strength concrete containing higher quantities of cementitious materials.

Fine-Aggregate Grading. Requirements of ASTM C33 or AASHTO M 6/M 43 permit a relatively wide range in fine-aggregate grading, but specifications by other organizations are sometimes more restrictive. The most desirable fine-aggregate grading depends on the type of application, the paste content, and the maximum size of coarse aggregate. In leaner mixtures,

or when small-size coarse aggregates are used, a grading that approaches the maximum recommended percentage passing each sieve is desirable for workability.

Fine-aggregate grading within the limits of ASTM C33 (AASHTO M 6) is satisfactory for most concretes. The ASTM C33 (AASHTO M 6) limits with respect to sieve size are shown in Table 5-4. In general, if the water-to-cement ratio is kept constant and the ratio of fine-to-coarse aggregate is suitable, a wide range in grading can be used without measurable effect on strength. However, the most economical concrete mixture is typically achieved by adjusting the proportions to suit the grading of the locally available aggregates.

TABLE 5-4. Fine-Aggregate Grading Limits (ASTM C33/AASHTO M 6)

SIEVE SIZE		PERCENT PASSING BY MASS	
9.5 mm	(⅜ in.)	100	
4.75 mm	(No. 4)	95 to 100	
2.36 mm	(No. 8)	80 to 100	
1.18 mm	(No. 16)	50 to 85	
600 μm	(No. 30)	25 to 60	
300 μm	(No. 50)	5 to 30	(AASHTO M 6: 10 to 30)
150 μm	(No. 100)	0 to 10	(AASHTO M 6: 2 to 10)
75 μm	(No. 200)*	0 to 3.0[†,‡]	

* AASHTO M 6 does not include requirements for the 75-μm (No. 200) sieve in the standard grading requirements table.

[†] For concrete not subject to abrasion, the limit for material finer than the 75-μm (No. 200) sieve shall be 5.0% maximum.

[‡] For manufactured fine aggregate or recycled aggregate, if the material finer than the 75-μm (No. 200) consists of dust of fracture, essentially free of clay or shale, this limit shall be 5.0% maximum for concrete subject to abrasion, and 7.0% maximum for concrete not subject to abrasion.

ASTM C33 grading limits include a maximum percent passing the 75-μm (No. 200) sieve while AASHTO M 6 does not (controlled under deleterious material limits). Also, AASHTO M 6 grading limits for 300-μm (No. 50) and 150-μm (No. 100) sieve sizes vary slightly from the ASTM C33 limits. The AASHTO specifications permit the minimum percentages (by mass) of material passing the 300-μm (No. 50) and 150-μm (No. 100) sieves to be reduced to 5% and 0% respectively (matching ASTM C33 requirements), provided:

- The aggregate is used in air-entrained concrete containing more than 237 kg/m³ (400 lb/yd³) of cement and having an air content of more than 3%.

- The aggregate is used in concrete containing more than 297 kg/m³ (500 lb/yd³) of cement when the concrete is not air-entrained.

- An approved supplementary cementitious material is used to supply the deficiency in material passing these two sieves.

Other requirements of ASTM C33 (AASHTO M 6) for fine aggregate include:

- The fine aggregate must not have more than 45% retained between any two consecutive standard sieves.

- The fineness modulus (FM) must be not less than 2.3 or more than 3.1, and not vary more than 0.2 from the average value of the aggregate source being tested. If this value is outside the required 2.3 to 3.1 range, the fine aggregate should be rejected unless suitable adjustments are made in proportions of fine and coarse aggregate. If the FM varies by more than 0.2, adjustments may need to be made with regard to coarse and fine aggregate proportions as well as the water requirements for the concrete mixture. (See discussion under **Fineness Modulus** later in this chapter.)

- The amounts of fine aggregate passing the 300-μm (No. 50) and 150-μm (No. 100) sieves affect workability, surface texture, air content, and bleeding of concrete. Most specifications allow 5% to 30% to pass the 300-μm (No. 50) sieve. The lower limit may be sufficient for easy placing conditions or where concrete is mechanically finished, such as in pavements. However, for hand-finished concrete floors, or where a smooth surface texture is desired, fine aggregate with at least 15% passing the 300-μm (No. 50) sieve and 3% or more passing the 150-μm (No. 100) sieve should be considered.

- A small quantity of clean particles that pass a 150-μm (No. 100) sieve but are retained on a 75-μm (No. 200) sieve is desirable for workability. Most specifications permit up to 10% of this finely divided material in fine aggregate.

- Other specifications require an adjustment in mixture proportions if the amount retained on any two consecutive sieves changes by more than 10% by mass of the total fine aggregate sample.

Coarse Aggregate Grading. The requirements of ASTM C33 (AASHTO M 80) permit a wide range in grading and a variety of grading sizes for coarse aggregates (see Table 5-5). The particle size distribution, or grading, of an aggregate significantly affects concrete mixture proportioning and workability and are an important element in the assurance of concrete quality. Materials containing too much or too little of any one size should be avoided.

The grading for a given maximum-size coarse aggregate can be varied over a moderate range without appreciable effect on the cement and water requirement of a mixture if the proportion of fine aggregate to total aggregate produces concrete of good workability. Mixture proportions should be changed to produce workable concrete if wide variations occur in the coarse aggregate grading. Since variations are difficult to anticipate, it is often more economical to maintain uniformity in manufacturing and handling coarse aggregate than to reduce variations in grading.

The size of coarse aggregate used in concrete has a direct bearing on the economy of concrete. Usually more water and cement (paste volume) is required for small-size aggregates than for large sizes, due to an increase in total aggregate surface area. The water and cement required for a slump of approximately 75 mm (3 in.) is shown in Figure 5-12 for a wide range of coarse-aggregate sizes. For a given water-cement ratio, the amount of cement (and water) required decreases as the maximum size of coarse aggregate increases. Aggregates of different sizes may give slightly different concrete strengths for the same water-cement ratio. Typically, concrete with a smaller maximum-size aggregate will have a higher compressive strength. This is especially true for high-strength concrete (see Chapter 23). The optimum maximum size of coarse aggregate for higher strength depends on factors such as relative strength of the cement paste, paste-aggregate bond, and strength of the aggregate particles.

The terminology used to specify size of coarse aggregate must be chosen carefully. Particle size is determined by size of sieve and applies to the percentage of aggregate passing that sieve and retained on the next smaller sieve.

When speaking of an assortment of particle sizes, the size number (or grading size) of the grading is used. The size number applies to the collective amount of aggregate that passes through an assortment of sieves. As shown in Table 5-5, the weight of aggregate passing the respective sieves is given in percentages; it is called a sieve analysis.

Maximum Size vs. Nominal Maximum Size Aggregate. Often there is confusion surrounding the term "maximum size" of aggregate. ASTM C125, *Standard Terminology Relating to Concrete and Concrete Aggregates*, defines this term and distinguishes it from "nominal maximum size" of aggregate.

TABLE 5-5. Grading Requirements for Coarse Aggregates (ASTM C33 and AASHTO M 80)

SIZE NO.	NOMINAL SIEVE SIZE	AMOUNTS FINER THAN EACH LABORATORY SIEVE, mass percent passing														
		100 mm (4 in.)	90 mm (3½ in.)	75 mm (3 in.)	63 mm (2½ in.)	50 mm (2 in.)	37.5 mm (1½ in.)	25 mm (1 in.)	19 mm (¾ in.)	12.5 mm (½ in.)	9.5 mm (⅜ in.)	4.75 mm (No. 4)	2.36 mm (No. 8)	1.18 mm (No. 16)	300 µm (No. 50)	75 µm (No. 200)
1	90 mm to 37.5 mm (3½ in. to 1½ in.)	100	90 to 100	–	25 to 60	–	0 to 15	–	0 to 5	–	–	–	–	–	–	0 to 1.0*
2	63 mm to 37.5 mm (2½ in. to 1½ in.)	–	–	100	90 to 100	35 to 70	0 to 15	–	0 to 5	–	–	–	–	–	–	0 to 1.0*
3	50 mm to 25 mm (2 in. to 1 in.)	–	–	–	–	90 to 100	35 to 70	0 to 15	–	0 to 5	–	–	–	–	–	0 to 1.0*
357	50 mm to 4.75 mm (2 in. to No. 4)	–	–	–	100	95 to 100	–	35 to 70	–	10 to 30	–	0 to 5	–	–	–	0 to 1.0*
4	37.5 mm to 19.0 mm (1½ in. to ¾ in.)	–	–	–	–	100	90 to 100	20 to 55	0 to 15	–	0 to 5	–	–	–	–	0 to 1.0*
467	37.5 mm to 4.75 mm (1½ in. to No. 4)	–	–	–	–	100	95 to 100	–	35 to 70	–	10 to 30	0 to 5	–	–	–	0 to 1.0*
5	25.0 mm to 12.5 mm (1 in. to ½ in.)	–	–	–	–	–	100	90 to 100	20 to 55	0 to 10	0 to 5	–	–	–	–	0 to 1.0*
56	25.0 mm to 9.5 mm (1 in. to ⅜ in.)	–	–	–	–	–	100	90 to 100	40 to 85	10 to 40	0 to 15	0 to 5	–	–	–	0 to 1.0*
57	25.0 mm to 4.75 mm (1 in. to No. 4)	–	–	–	–	–	100	95 to 100	–	25 to 60	–	0 to 10	0 to 5	–	–	0 to 1.0*
6	19.0 mm to 9.75 mm (¾ in. to ⅜ in.)	–	–	–	–	–	–	100	90 to 100	20 to 55	0 to 15	0 to 5	–	–	–	0 to 1.0*
67	19.0 mm to 4.75 mm (¾ in. to No. 4)	–	–	–	–	–	–	100	90 to 100	–	20 to 55	0 to 10	0 to 5	–	–	0 to 1.0*
7	12.5 mm to 4.75 mm (½ in. to No. 4)	–	–	–	–	–	–	–	100	90 to 100	40 to 70	0 to 15	0 to 5	–	–	0 to 1.0*
8	9.5 mm to 2.36 mm (⅜ in. to No. 8)	–	–	–	–	–	–	–	–	100	85 to 100	10 to 30	0 to 10	0 to 5	–	0 to 1.0*
89	9.5 mm to 1.18 mm (⅜ in. to No. 16)	–	–	–	–	–	100	–	–	100	90 to 100	20 to 55	5 to 30	1 to 10	0 to 5	0 to 1.0*
9	4.75 mm to 1.18 mm (No. 4 to No. 16)	–	–	–	–	–	–	–	–	–	100	85 to 100	10 to 40	0 to 10	0 to 5	0 to 1.0*

* This maximum limit is permitted to be increased under either of the following conditions: (1) if the material is essentially free of clay or shale, to 1.5%; or (2) if the fine aggregate to be used contains less than the specified maximum amount passing the 75-µm (No. 200) sieve in Table 1 of ASTM C33, this maximum limit is based on a weighted formula limiting the total mass of material passing the 75-µm (No. 200) sieve in the concrete. See ASTM C33 for complete details.

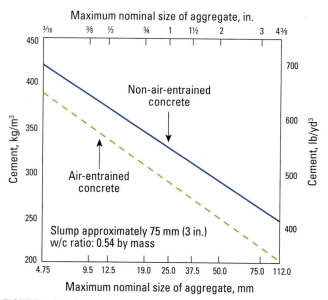

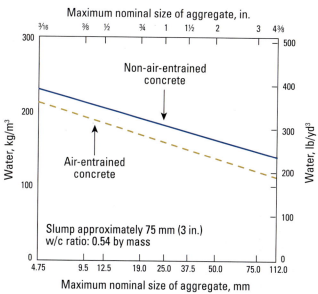

FIGURE 5-12. Cement and water contents in relation to maximum size of aggregate for air-entrained and non-air-entrained concrete. Less cement and water are required in mixtures having large coarse aggregate (Bureau of Reclamation 1981).

The maximum size of an aggregate is the smallest sieve that <u>all</u> of a particular aggregate will pass through (100% passing). The nominal maximum size of an aggregate is the smallest sieve size through which the <u>major portion</u> of the aggregate must pass (typically 85% to 95% passing). The nominal maximum-size sieve may retain 5% to 15% of the aggregate depending on the size number. For example, aggregate size number 67 has a maximum size of 25 mm (1 in.) and a nominal maximum size of 19 mm (¾ in.); 90% to 100% of this aggregate must pass the 19-mm (¾-in.) sieve and all of the particles must pass the 25-mm (1-in.) sieve.

The maximum size of aggregate that can be used generally depends on the size and shape of the concrete member and on the amount and distribution of reinforcing steel. Requirements for limits on nominal maximum size of aggregate particles are covered by ACI 318-19 and additional requirements for clear spacing between reinforcing bars and forms (cover) are found in CSA A23.1/A23.2, *Concrete Materials and Methods of Concrete Construction/Test Methods and Standard Practices for Concrete* (2019).

The nominal maximum size of aggregate (D_{max}) should not exceed:

• one-fifth the narrowest dimension of a vertical concrete member: $D_{max} \leq \frac{1}{5} B$,

• one-third the depth of slabs: $D_{max} \leq \frac{1}{3} T$,

• three-quarters the clear spacing between reinforcing bars, $D_{max} \leq \frac{3}{4} S$, and

• three-quarters the clear spacing between the reinforcing bars and forms: $D_{max} \leq \frac{3}{4} C$.

Variables are as defined in Figure 5-13.

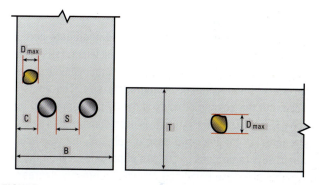

FIGURE 5-13. ACI 318-19 requirements for nominal maximum size of aggregates: D_{max}, based on concrete dimensions, B, and T; and reinforcement spacing, S. CSA A23.1 has additional requirements for cover depth, C.

These requirements may be waived if, in the judgment of the engineer, the mixture possesses sufficient workability that the concrete can be properly placed without honeycomb or voids. In highly congested sections or for difficult placements, it may be beneficial to cast a mockup to assure proper mixture placement is possible with aggregates larger than recommended above.

For concrete to be placed by pumping, the maximum size of aggregate should not exceed ⅓ of the hose diameter, nor ¼ of the diameter of a tremie hose. See Chapter 15 and ACI 304.2R, *Guide to Placing Concrete by Pumping Methods* (2017), for more information on these limits.

Combined Aggregate Grading. Aggregate is sometimes analyzed using the combined grading of fine and coarse aggregate in the proportions they would exist in concrete. This provides a more thorough analysis of how the aggregates will perform in concrete.

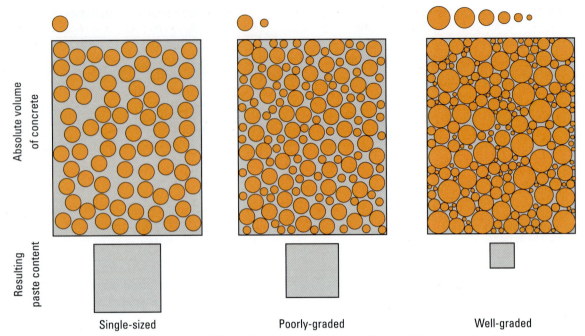

FIGURE 5-14. For equal absolute volumes when different sizes are combined, the void-content decreases, therefore the necessary paste content decreases.

The effect of a collection of various sized aggregate particles in reducing the total volume of voids is best illustrated in Figure 5-14. A single-sized aggregate results in higher volume of voids. This effect is independent of aggregate size. The voids are smaller, but the volume of voids is nearly the same (and high) when a single-size fine aggregate is used compared to a coarse aggregate. For the idealized case of spheres, the void volume is about 36% regardless of the size of particles. When the two aggregate sizes are combined, the void content is decreased. If several additional sizes are used, a further reduction in voids would occur. The paste requirement for concrete is dictated by the void content of the combined aggregates.

Voids in aggregates can be determined according to ASTM C29, *Standard Test Method for Bulk Density ("Unit Weight") and Voids in Aggregate* (AASHTO T 19). In reality, the amount of cement paste required in concrete is somewhat greater than the volume of voids between the aggregates to provide

sufficient lubricity between particles during placement and consolidation. This is illustrated in Figure 5-15. The illustration on the left represents coarse aggregates, with all particles in contact. The illustration on the right represents the dispersal of aggregates in a matrix of paste. The amount of paste necessary is greater than the void content in order to provide workability to the concrete. The actual amount is also influenced by the cohesiveness of the paste.

Figure 5-16 illustrates a theoretical ideal uniform gradation (well-graded aggregate). Well-graded aggregates contain particles on each sieve size, which maximizes the aggregate volume to the greatest extent. Well-graded aggregates enhance numerous characteristics and result in greater workability and durability. The more well-graded an aggregate is, the more it will pack together efficiently, thus reducing the volume between aggregate particles that must be filled by paste.

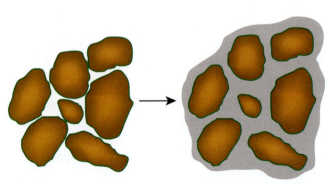

FIGURE 5-15. Illustration of the dispersion of aggregates in cohesive concrete mixtures.

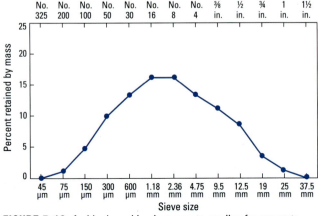

FIGURE 5-16. An ideal combined aggregate grading for concrete.

On the other hand, gap-graded aggregates – those having either a large quantity or a deficiency of one or more sieve sizes, typically mid-sized aggregate, around 9.5 mm (⅜ in.) – can result in reduced workability during mixing, pumping, placing, consolidation and finishing (see **Gap-graded Aggregates**). Durability can also suffer from using more fine aggregate and water to produce a more workable mixture. Finer aggregates require more paste because they have higher surface-to-volume ratios. The higher sand and paste requirements may cause higher water demand; resulting in higher shrinkage.

Concrete mixtures that are well-graded and made with a lower paste volume generally will have less shrinkage, have lower permeability, and be more economical. A combined gradation can be used to better control workability, pumpability, shrinkage, permeability, and other properties of concrete. Abrams (1918), Shilstone (1990), and Cook and others (2013) demonstrate the benefits of a combined aggregate analysis.

With constant cement content and consistency from batch to batch, there is an optimum for every combination of aggregates that will produce the most effective water to cement ratio and highest strength. The optimum mixture has the least particle interference and responds best to a high frequency, high amplitude vibrator.

Shilstone (1990) analyzed aggregate gradations by coarseness and workability factors to improve aggregate gradations. In Figure 5-17 the *workability factor* is considered the percentage of particles passing the 2.36 mm (No. 8) sieve. The *coarseness factor* is the percentage of particles retained on the 2.36-mm (No. 8) sieve that are also retained on the 9.5-mm (⅜ in.) sieve. The amount of fine aggregate in a mixture must be balanced against the needs of the larger, "inert" particles. A *W-Adjust factor* (Shilstone 1990) can also be determined which takes into consideration cementitious materials.

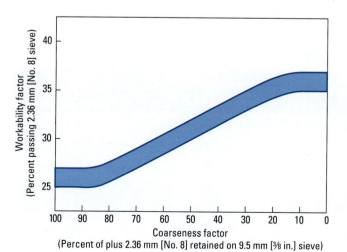

FIGURE 5-17. Shilstone workability factor chart for combined aggregate gradations. The optimum mixture should plot in or near the trend bar (Shilstone 1990).

Crouch and others (2000) found in their studies on air-entrained concrete, that the water-cement ratio could be reduced by over 8% using combined aggregate grading.

Cook and others (2013) found that desirable overall combinations that fall within the so-called "Tarantula Curve" (Figure 5-18) provide improved workability and resistance to segregation.

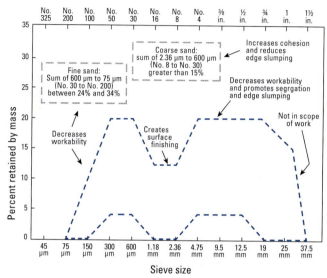

FIGURE 5-18. Tarantula Curve. A set of suggested limits was developed for pavements by comparing the workability and aggregate gradation of more than 500 different mixtures with 8 different aggregate sources (adapted from Cook and others 2013).

The workability performance was evaluated with the Box Test (see Chapter 20), a quick and inexpensive workability test designed for slip formed pavements (Cook and others 2014). In addition there are requirements for the amount of coarse sand needed for cohesion (the amount retained on the 2.36-mm [No. 8], 1.18-mm [No. 16], and 600-µm [No. 30] sieves shall be greater than 15%) and fine sand needed for workability (between 24% and 34% retained on the 600-µm to 75-µm [No. 30 to No. 200] sieves). Also, a limit for the flatness of the aggregate based on ASTM D4791, *Standard Test Method for Flat Particles, Elongated Particles, or Flat and Elongated Particles in Coarse Aggregate*, has been proposed (Cook and others 2013). More information can be found at Ley (2020).

However, aggregate optimizations cannot be used for all construction due to variations in placing and finishing needs and availability. Aggregates that are not necessarily well graded should not be prematurely dismissed. Studies by the NRMCA (Obla and others 2007) have indicated that some grading distributions that are not well-graded can outperform well-graded aggregate blends. Particle shape and texture can impact concrete performance. If problems develop due to a poor grading, alternative aggregates, blending, or special screening of existing aggregates should be considered.

Gap-Graded Aggregates. In gap-graded aggregates certain particle sizes are intentionally omitted. Typical gap-graded aggregates consist of only one size of coarse aggregate with all the particles of fine aggregate able to pass through the voids in the compacted coarse aggregate. Gap-graded mixtures are used in architectural concrete to obtain uniform textures in exposed-aggregate finishes. They are also used in pervious concrete mixtures to improve storm water management. They can also be used in normal structural concrete to improve other concrete properties and to permit the use of local aggregate gradations (Houston 1962 and Litvin and Pfeifer 1965).

For an aggregate of 19-mm (¾-in.) maximum size, the 4.75-mm to 9.5-mm (No. 4 to ⅜-in.) particles can be omitted without making the concrete unduly harsh or subject to segregation. In the case of 37.5-mm (1½-in.) aggregate, usually the 4.75-mm to 19-mm (No. 4 to ¾-in.) sizes are omitted.

Care must be taken in choosing the percentage of fine aggregate in a gap-graded mixture. A poor choice can result in concrete that is likely to segregate or honeycomb because of an excess of coarse aggregate. Also, concrete with an excess of fine aggregate may have a high water demand resulting in higher shrinkage. Fine aggregate is usually 25% to 35% by volume of the total aggregate. (Lower percentages are typically used with rounded aggregates and higher with crushed material.) For a smooth off-the-form finish, a somewhat higher percentage of fine aggregate to total aggregate may be used than for an exposed-aggregate finish, but both use a lower fine aggregate content than continuously graded mixtures. Fine aggregate content also depends upon cement content, type of aggregate, and workability.

Air entrainment can be used to improve workability since low-slump, gap-graded mixtures use a low fine aggregate percentage and produce harsh mixes without entrained air.

Segregation of gap-graded mixtures must be prevented by restricting the slump to the lowest value consistent with good consolidation. This may vary from zero to 75 mm (3 in.) depending on the thickness of the section, amount of reinforcement, and height of placement. However, even low slump concrete has been known to segregate when exposed to vibration, such as in a dump truck during delivery. Close control of grading and water content is also required because variations might cause segregation. If a stiff mixture is required, gap-graded aggregates may produce higher strengths than normal aggregates used with comparable cement contents. Because of their low fine-aggregate volumes and low water-cement ratios, gap-graded mixtures might be considered unworkable for some cast-in-place construction without the aid of chemical admixtures. When properly proportioned, however, these concretes are readily consolidated with vibration.

Fineness Modulus. The *fineness modulus* (FM) of either fine, coarse, or combined aggregate according to ASTM C125, is calculated by adding the cumulative percentages by mass retained on each of a specified series of sieves and dividing the sum by 100. The specified sieves for determining FM are: 150 µm (No. 100), 300 µm (No. 50), 600 µm (No. 30), 1.18 mm (No. 16), 2.36 mm (No. 8), 4.75 mm (No. 4), 9.5 mm (⅜ in.), 19.0 mm (¾ in.), 37.5 mm (1½ in.), 75 mm (3 in.), and 150 mm (6 in.).

Fineness modulus is an indicator of the fineness of an aggregate. In general, the higher the FM, the coarser the aggregate. However, different aggregate gradings may have the same FM. The FM of fine aggregate is useful in mixture proportioning estimating the proportions of fine and coarse aggregates. An example of how the FM of a fine aggregate is determined (for a specific sieve analysis) is shown in Table 5-6.

Aggregate Mineral Fillers. Ground calcium carbonate (GCC) and other aggregate mineral fillers (AMF) can improve

TABLE 5-6. Determination of Fineness Modulus of Fine Aggregates

SIEVE SIZE		PERCENTAGE OF INDIVIDUAL FRACTION RETAINED BY MASS	PERCENTAGE PASSING BY MASS	CUMULATIVE PERCENTAGE RETAINED BY MASS
9.5 mm	(⅜ in.)	0	100	0
4.75 mm	(No. 4)	2	98	2
2.36 mm	(No. 8)	13	85	15
1.18 mm	(No. 16)	20	65	35
600 µm	(No. 30)	20	45	55
300 µm	(No. 50)	24	21	79
150 µm	(No. 100)	18	3	97
	Pan	3	0	–
	Total	100		283

Fineness modulus = 283 / 100 = 2.8

TABLE 5-7. Gradation Requirements for Ground Calcium Carbonate and Aggregate Mineral Fillers, minimum % by mass passing

SIEVE SIZE	TYPE A	TYPE B	TYPE C
300 µm (No. 50)	100	100	100
150 µm (No. 100)	100	85	–
75 µm (No. 200)	95	70	65
45 µm (No. 325)	90	65	–

Adapted from ASTM C1797.

particle packing and rheological properties of fresh concrete. Requirements for mineral fillers are found in ASTM C1797, *Standard Specification for Ground Calcium Carbonate and Aggregate Mineral Fillers for use in Hydraulic Cement Concrete*, which defines three types of mineral fillers, Types A, B, and C. Types A and B are GCC, with 92% or more, or 70% or more calcium carbonate content respectively. Type C mineral fillers do not have a calcium carbonate content requirement and are typically derived from other types of stone. Table 5-7 shows grading requirements for mineral fillers.

Guidance on properties of these materials, effects on concrete properties and proportioning in concrete mixtures is provided in ACI 211.7R, *Guide for Proportioning Concrete Mixtures with Ground Limestone and Other Mineral Fillers* (2020).

Particle Shape and Surface Texture

The particle shape and surface texture of aggregate influence the fresh concrete properties more than the properties of hardened concrete. Rough-textured, angular, and elongated particles require more water to produce workable concrete than do smooth, rounded, compact aggregates. Aggregate particles that are angular also require more paste to maintain the same dispersion and workability (Figure 5-19). Angular or poorly graded aggregates may also be more difficult to pump.

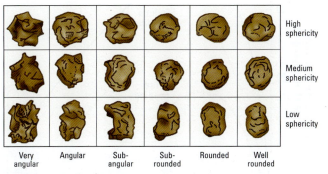

FIGURE 5-19. Particle shape influences the properties of concrete.

Void contents of uncompacted fine or coarse aggregate (see **Bulk Density (Unit Weight) and Voids**) can be used as an index of differences in the shape and texture of aggregates of the same grading. The mixing water and cement requirement tend to increase as aggregate void content increases. Voids between aggregate particles increase with aggregate angularity.

The bond between cement paste and a given aggregate generally increases as particles change from smooth and rounded to rough and angular. This increase in bond is a consideration in selecting aggregates for concrete where flexural strength is important or where high compressive strength is needed.

Flat and elongated aggregate particles should be avoided or at least limited to about 15% by mass of the total aggregate. A particle is called flat and elongated when the ratio of length to thickness exceeds a specified value, for example, a value of 3 has been used to define elongated particles (see USACE 1994).

In addition to several standard test methods (see Chapter 20), a number of automated test machines are available for rapid determination of the particle shape and size distribution of aggregate.

Densities

Several types of aggregate density value are commonly reported in the concrete industry: bulk density, relative density (specific gravity), and (particle) density. It is important to clearly communicate the type of density as the values are significantly different: The particle density is higher than the bulk density as the latter includes the volume of space between the particles. (See **Aggregate Classification** for more detail on bulk density.)

Bulk Density (Unit Weight) and Voids. The bulk density or unit weight of an aggregate is the mass or weight of the aggregate required to fill a container of a specified unit volume. The volume is occupied both by aggregates and the voids between aggregate particles.

The void content between particles affects paste requirements in concrete mix design (see preceding sections, **Grading** and **Particle Shape and Surface Texture**). Void contents range from about 30% to 45% for coarse aggregates to about 40% to 50% for fine aggregate. Angularity increases void content while larger sizes of well-graded aggregate and improved grading decreases void content.

Relative Density (Specific Gravity). The relative density (specific gravity) of an aggregate is the ratio of its mass to the mass of an equal volume of water. Most natural aggregates have relative densities between 2.40 and 2.90, which means that the aggregates are 2.40 to 2.90 times denser than water. Relative density is not generally used as a measure of aggregate quality, though some porous aggregates that exhibit accelerated freeze-thaw deterioration do have low specific gravities. It is commonly used in certain computations for mixture proportioning and control, such as the volume occupied by the aggregate in the absolute volume method of mix design (see Chapter 13).

Density. The density of aggregate particles used in mixture proportioning computations (not including voids between particles) is determined by multiplying the relative density (specific gravity) of the aggregate times the density of water. An approximate value of 1000 kg/m³ (62.4 lb/ft³) is often used for the density of water. Most natural aggregates have particle densities of between 2400 kg/m³ and 2900 kg/m³ (150 lb/ft³ and 181 lb/ft³).

Absorption and Surface Moisture

The absorption and surface moisture of aggregates should be determined (see **Absorbtion and Surface Moisture** in Chapter 20 for details on test methods), so that the total water content of the concrete can be controlled and corrected batch weights determined. The internal structure of an aggregate particle is made up of solid matter and voids that may or may not contain water.

The moisture conditions of aggregates are described in Figure 5-20. They are designated as:

• *Oven dry* – zero moisture content, fully absorbent

• *Air dry* – dry at the particle surface but containing some interior moisture, less than potential absorption

• *Saturated surface dry (SSD)* – neither absorbing water from nor contributing water to the concrete mixture, equal to potential absorption

• *Damp or wet* – containing an excess of moisture (free water) on the surface, greater than absorption

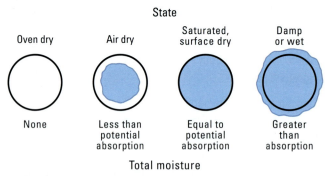

FIGURE 5-20. Moisture conditions of aggregate: oven dry, air dry, saturated surface dry, or wet.

The amount of water added at the concrete batch plant must be adjusted for the moisture conditions of the aggregates in order to accurately meet the water requirement of the mix design (see Chapter 13) and to minimize variability. If the water content of the concrete mixture is not kept constant, the water-cement ratio will vary from batch to batch causing other properties, such as

the compressive strength and workability to vary as well. Only the surface moisture, not the absorbed moisture, becomes part of the mixing water in concrete. As discussed in Chapter 13, the amount of water added to a batch is reduced by the amount of surface moisture. In addition, the batch weight of aggregates is increased by the percentage of total moisture present in each type of aggregate. If adjustments are not made during batching, the mass of surface water will be taken to be a portion of the aggregate mass and the batch will not have the correct w/cm or yield. The difference between the total oven-dry batch weights and the total adjusted batch weights represents the water to satisfy aggregate absorption.

Coarse and fine aggregate will generally have absorption levels (moisture contents at SSD) in the range of 0.2% to 4% and 0.2% to 2%, respectively. If wet (with moisture contents above SSD), free-water contents will usually range from 0.5% to 2% for coarse aggregate and 2% to 6% for fine aggregate. Further information on calculating moisture content of aggregates is covered in Chapter 20 and examples of adjusting mixtures for aggregate moisture corrections are given in Chapter 13.

Bulking. Bulking is the increase in total volume of moist fine aggregate over the same mass in a dry condition. Surface tension in the moisture holds the particles apart and keeps them from flowing naturally, causing an increase in volume. Bulking of a fine aggregate can occur when it is shoveled or otherwise moved in a damp condition, even though it may have been fully consolidated previously. Figure 5-21 illustrates how the amount of bulking of fine aggregate varies with moisture content and grading; fine gradings bulk more than coarse gradings for a given amount of moisture.

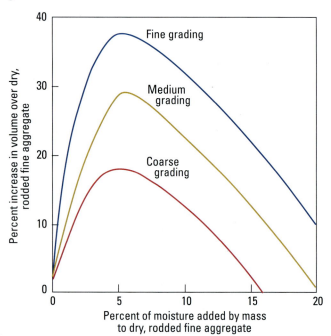

FIGURE 5-21. Surface moisture on fine aggregate can cause considerable bulking; the amount varies with the amount of moisture and the aggregate grading (PCA 1955 and PCA 1949).

Figure 5-22 shows similar information in terms of weight for a particular fine aggregate. Since most fine aggregates are delivered in a damp condition, wide variations can occur in batch quantities if batching is done by volume. For this reason, best practice has long favored weighing the aggregate and adjusting for moisture content when proportioning concrete.

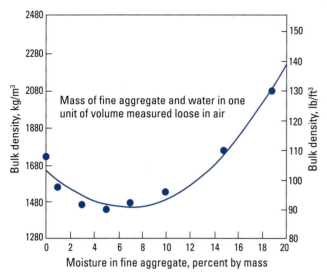

FIGURE 5-22. Bulk density is compared with the moisture content for a particular sand (PCA 1955).

Resistance to Freezing and Thawing

The frost resistance of an aggregate is related to its porosity, absorption, permeability, and pore structure. An aggregate particle with high absorption may not accommodate the expansion that occurs during the freezing of water if that particle becomes critically saturated. If enough unstable particles are present, the result can be expansion and cracking of the aggregate and possible cracking or disintegration of the concrete. If an individual low density particle is near the surface of the concrete, it can cause a popout. Popouts generally appear as conical fragments that break out of the concrete surface (see Chapter 16). The offending aggregate particle or a part of it is usually found at the bottom of the resulting void. Generally, it is coarse rather than fine aggregate particles with higher porosity values and medium-sized pores (0.1 μm to 5 μm) that are easily saturated and that cause concrete deterioration and popouts. Larger pores do not usually become saturated or cause concrete distress, and water in very fine pores may not freeze readily.

The critical size at which an aggregate will fail depends upon the rate of freezing and critical saturation, as well as and the porosity, permeability, and tensile strength of the particle (Stark 1976). For fine-grained aggregates with low permeability (cherts for example), the critical particle size may be within the range of normal aggregate sizes. It is larger for coarse-grained materials or those with capillary systems interrupted by numerous macropores (voids too large to hold moisture by capillary action). For these aggregates, the critical particle size may be sufficiently large to have no impact on freeze-thaw resistance in concrete even though the absorption may be high. Also, if potentially marginal aggregates are used in concrete subjected to periodic drying while in service, they may never become sufficiently saturated to cause failure.

The cracking of concrete pavements caused by freeze-thaw deterioration of the aggregate is called D-cracking. D-cracks are closely spaced crack formations oriented parallel to transverse and longitudinal joints that later multiply outward from the joints toward the center of the pavement panel (Figure 5-23). This type of cracking has been observed in pavements after three to 20 years of service. D-cracked concrete resembles frost-damaged concrete caused by freeze-thaw deterioration of the hardened paste. D-cracking is a function of the pore properties of certain types of aggregate particles and the environment in which the pavement is placed. If water accumulates under pavements in the base and subbase layers, the aggregate may eventually become critically saturated. With continued freezing and thawing cycles, cracking of the concrete starts in the saturated aggregate (Figure 5-24) at the bottom of the slab and progresses upward until it reaches the wearing surface. D-cracking can be reduced either by selecting aggregates that perform better in freeze-thaw cycles or, where marginal aggregates must be used, by reducing the maximum particle size.

FIGURE 5-23. D-cracking is a manifestation of freezing and thawing damage due to failure of the coarse aggregate in concrete pavements (courtesy of J. Grove).

FIGURE 5-24. Fractured aggregate particle as a source of distress in D-cracking.

Installation of permeable bases or effective drainage systems help carry free water out from under the pavement (Harrigan 2002). See Chapter 20 for information on testing aggregates for resistance to freezing and thawing.

Wetting and Drying Properties

Weathering due to wetting and drying can also affect the durability of aggregates. The expansion and contraction coefficients of aggregates vary with temperature and moisture content. If alternate wetting and drying occurs, severe strain can develop in some aggregates. With certain types of aggregate this can cause a permanent increase in the volume of the concrete and eventual concrete deterioration. Clay lumps and other friable particles can degrade rapidly with repeated wetting and drying. Popouts can also develop due to the moisture-swelling characteristics of certain aggregates, especially clay balls and shales. An experienced petrographer can assist in determining the potential for this distress.

Abrasion and Skid Resistance

The abrasion resistance of an aggregate is often used as a general index of its quality. Abrasion resistance is essential when the aggregate is to be used in concrete subject to strong mechanical stress or abrasion, as in heavy-duty floors or pavements. Hard aggregate such traprock, chert, and granite tend to result in concrete with higher abrasion resistance than softer aggregate, such as limestone. However, there is no clear correlation between results of aggregate abrasion tests and concrete abrasion tests, and concrete wear resistance should be evaluated by abrasion testing of the concrete mixture if this is a key property for a specific application.

Low abrasion resistance of an aggregate may increase the quantity of fines in the concrete during transporting, stockpiling, and mixing. Consequently, this may increase the water requirement and require an adjustment in the water-cement ratio.

For pavements, the wet-weather friction and micro-texture of the aggregate exposed at the concrete surface are important to the skid resistance. For concrete pavements, it is the fine aggregate that is exposed at the surface initially. With long-term wear and/or grinding and grooving of the pavement the coarse aggregate becomes exposed.

To provide good skid resistance on pavements, the siliceous particle content of the fine aggregate should be at least 25%. For specification purposes, the siliceous particle content is considered equal to the insoluble residue content after treatment in hydrochloric acid under standardized conditions (ASTM D3042, *Standard Test Method for Insoluble Residue in Carbonate Aggregates*). Certain carbonate manufactured sands from limestone and dolomite can produce slippery pavement surfaces and should be investigated for acceptance before use.

Strength and Shrinkage

The strength of an aggregate is rarely tested and generally does not influence the strength of conventional concrete as much as the strength of the paste and the paste-aggregate bond. However, aggregate strength does become important in high-strength concrete. Aggregate stress levels in concrete are often much higher than the average stress over the entire cross section of the concrete. Aggregate tensile strengths range from 2 MPa to 15 MPa (300 psi to 2300 psi) and compressive strengths from 65 MPa to 270 MPa (10,000 psi to 40,000 psi). Strength of bulk rock samples can be tested according to ASTM C170, *Standard Test Method for Compressive Strength of Dimension Stone*.

Different aggregate types have different compressibility, modulus of elasticity, and moisture-related shrinkage characteristics that influence the same properties in concrete as shown in Table 5-8. Aggregates with high absorption may have high shrinkage on drying. Quartz and feldspar aggregates, along with limestone, dolomite, and granite, are considered low shrinkage aggregates; while aggregates with sandstone, shale, slate, hornblende, and graywacke are often associated with high shrinkage in concrete (Figure 5-25).

TABLE 5-8. Relative Density and Shrinkage Characteristics of Selected Aggregates*

AGGREGATE	RELATIVE DENSITY (SPECIFIC GRAVITY)	ABSORPTION %	ONE-YEAR SHRINKAGE, 50% RH, millionths
Sandstone	2.47	5.0	1160
Slate	2.75	1.2	680
Granite	2.67	0.5	470
Limestone	2.74	0.2	410
Quartz	2.65	0.3	320

*Carlson 1938.

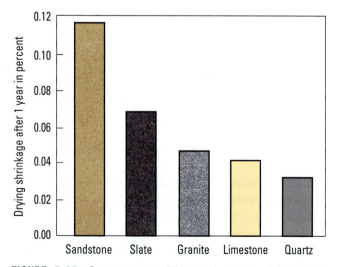

FIGURE 5-25. Concretes containing sandstone or slate tend to produce higher shrinkage concrete. Granite, limestone, and quartz are lower shrinkage-producing aggregates (Carlson 1938).

Resistance to Acid and Other Corrosive Substances

Portland cement concrete is durable in most natural environments; however, concrete in service is occasionally exposed to substances that will attack it.

Although acids generally attack and leach away the hydrated cement phases of the cement paste, they may not readily attack certain aggregates, such as siliceous aggregates. Calcareous aggregates often react readily with acids. However, the sacrificial effect of calcareous aggregates may be more beneficial than siliceous aggregates in mild acid exposures or in areas where water is not flowing. With calcareous aggregate, the acid attacks the entire exposed concrete surface uniformly, reducing the rate of attack on the paste and slowing the loss of aggregate particles at the surface. Calcareous aggregates also tend to neutralize the acid, especially in stagnant locations. Siliceous aggregate should be avoided when strong solutions of sodium hydroxide are present, as these solutions attack this type of aggregate.

Fire Resistance and Thermal Properties

The fire resistance and thermal properties of concrete (such as conductivity, diffusivity, and coefficient of thermal expansion) depend to some extent on the mineral constituents of the aggregates used. Manufactured lightweight and some naturally occurring lightweight aggregates are more fire resistant than normal-weight aggregates due to their insulating properties and high-temperature stability (see Figure 5-26).

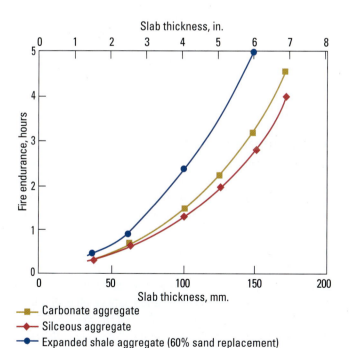

FIGURE 5-26. Fire endurance of concrete slabs of varying thickness and aggregate type (adapted from Abrams and Gustaferro 1968).

In general, concrete containing a calcareous coarse aggregate performs somewhat better under fire exposure than a concrete containing quartz or siliceous aggregate such as granite or quartzite. At about 590°C (1060°F), quartz expands 0.85% causing disruption to the concrete structure (document is withdrawn by ACI 221 1996). The coefficient of thermal expansion of aggregates ranges from 0.55×10^{-6} per °C to 12×10^{-6} per °C (1×10^{-6} per °F to 22×10^{-6} per °F). For more information refer to Chapter 10 for temperature-induced volume changes.

POTENTIALLY HARMFUL MATERIALS

Harmful substances that may be present in aggregates include organic impurities, silt, clay, shale, iron oxide, coal, lignite, and certain lightweight and soft particles (Table 5-9). In addition, rocks and minerals such as some cherts, strained quartz (Buck and Mather 1984), and certain dolomitic limestones are alkali reactive (see **Alkali-Aggregate Reactivity**). Gypsum and anhydrite may cause sulfate attack. Soluble chlorides in aggregates can decrease corrosion resistance for steel reinforcement. Certain aggregates, such as some shales, will cause popouts by swelling (simply by absorbing water) or by freezing of absorbed water. Most specifications limit the permissible amounts of these substances. The performance history of aggregates should be a determining factor in setting the limits for harmful substances. Test methods for detecting harmful substances qualitatively or quantitatively are listed in Table 5-9 and discussed further in Chapter 20.

Aggregates are potentially harmful if they contain compounds known to react chemically with portland cement paste and produce any of the following:

• significant volume changes of the paste, aggregates, or both;

• interference with the normal hydration of cement; and

• otherwise harmful reaction products.

Organic impurities may: delay setting and hardening of concrete, interfere with interaction of chemical admixtures, reduce strength gain, and in unusual cases, cause deterioration and localized staining. Organic impurities such as peat, humus, and organic loam may not be as detrimental but should be avoided.

Materials finer than the 75-μm (No. 200) sieve, especially silt and clay, may be present as loose dust and may form a coating on the aggregate particles. Even thin coatings of silt or clay on gravel particles can be harmful because they may weaken the bond between the cement paste and aggregate. If certain types of silt or clay are present in excessive amounts, water requirements may increase significantly. Aggregate particles containing or comprised of expansive clay minerals may cause localized disruption or deterioration of concrete upon saturation with water.

TABLE 5-9. Harmful Materials in Aggregates

SUBSTANCES	EFFECT ON CONCRETE	STANDARD DESIGNATION
Organic impurities	Affects setting and hardening, may cause deterioration	ASTM C40 (AASHTO T 21) ASTM C87 (AASHTO T 71)
Materials finer than the 75-μm (No. 200) sieve	Affects bond, increases water requirement	ASTM C117 (AASHTO T 11)
Coal, lignite, or other lightweight materials	Affects durability, may cause stains and popouts	ASTM C123 (AASHTO T 113)
Clay lumps and friable particles	Affects workability and and durability, may cause popouts	ASTM C142 (AASHTO T 112)
Chert of less than 2.40 relative density	Affects durability, may cause popouts	ASTM C123 (AASHTO T 113) ASTM C295
Alkali-reactive aggregates	Causes abnormal expansion, map cracking, and popouts	ASTM C295 ASTM C586 ASTM C1260 (AASHTO T 303) ASTM C1293 ASTM C1105 ASTM C 1778 (AASHTO R 80)
Chlorides	Can initiate corrosion in concrete reinforced with ferrous metals	ASTM C1524

Coal or lignite, or other low-density materials such as wood or fibrous materials, in excessive amounts will affect the durability of concrete. If these impurities occur at or near the surface, they might disintegrate, pop out, or cause stains. Potentially harmful chert in coarse aggregate can be identified using ASTM C123, *Standard Test Method for Lightweight Particles in Aggregate* (AASHTO T 113).

Soft particles and clay lumps in coarse aggregate are deleterious because they cause popouts and can affect the durability and wear resistance of concrete. If friable, they could break up during mixing and thereby increase the amount of water required. Local experience is often the best indication of the potential durability of concrete made with such aggregates.

Where abrasion resistance is critical, such as in heavy-duty industrial floors, testing may indicate further investigation or use of another aggregate source is needed.

Aggregates can occasionally contain particles of iron oxide and iron sulfide (for example, pyrite) that result in unsightly stains on exposed concrete surfaces (see Chapter 16). Tests described in Chapter 20 should be required when aggregates without a record of successful use are to be used in architectural concrete.

Pyrrhotite

Pyrite is a relatively common iron sulfide mineral, that generally does not present a serious distress issue for concrete, although it can cause surface staining or popouts that are unsightly. *Pyrrhotite* is another type of iron sulfide mineral which can cause deterioration in concrete. It is exceedingly rare in North American aggregates, and has only recently been identified as an issue for concrete in North America (CSA A23.1 2019). Detailed mechanisms and mitigation measures are being currently investigated. It is recommended that aggregates known to have pyrrhotite not be used as concrete ingredients. See Chapter 11 for more information on pyrrhotite-related distress.

ALKALI-AGGREGATE REACTIVITY

There are two primary types of potentially deleterious reactions between aggregates and the highly alkaline pore solution of concrete: alkali-silica reaction (ASR) and alkali-carbonate reaction (ACR). Table 5-10 includes a partial list of potentially susceptible materials.

Alkali-Silica Reaction

Aggregates containing certain constituents can react with alkali hydroxides in concrete. The reactivity is potentially harmful only when it produces significant expansion (Mather 1975). Field performance history is the best method of evaluating the susceptibility of an aggregate to ASR. Comparisons should be made between the existing and proposed concrete mixture proportions, ingredients, and service environments. This process should indicate whether special requirements are needed, or whether testing of the aggregate or job concrete is required. Guidance is provided in ASTM C1778, *Standard Guide for Reducing the Risk of Deleterious Alkali-Aggregate Reaction in Concrete* (AASHTO R 80), on evaluation and identification of potentially alkali-silica reactive aggregates and preventive measures. Chapters 11 and 20 provide more detail on identifying susceptible aggregate sources and mitigation techniques.

Alkali-Carbonate Reaction

Reactions observed with certain dolomitic rocks are associated with alkali-carbonate reaction (ACR). Reactive rocks usually contain large crystals of dolomite scattered in and surrounded by a fine-grained matrix of calcite and clay. ACR is relatively rare because aggregates susceptible to this reaction are usually unsuitable for use in concrete for other reasons, such as strength potential. Alkali reactivity of carbonate rocks is not usually dependent upon its clay mineral composition (Hadley 1961). Aggregates have potential for expansive ACR if the following lithological characteristics exist (Ozol 2006 and Swenson and Gillott 1967):

- clay content, or insoluble residue content, in the range of 5% to 25%;
- calcite-to-dolomite ratio of approximately 1:1; and
- small size of the discrete dolomite crystals (rhombs) suspended in a clay matrix.

See ASTM C1778, AASHTO R 80, Thomas and others (2008), Farny and Kerkhoff (2007), and Chapters 11 and 20 for more information on alkali-aggregate reactivity.

AGGREGATE BENEFICIATION

Aggregate processing consists of:

- basic processing – crushing, screening, and washing – to obtain proper gradation and cleanliness; and

- beneficiation – improving quality by processing methods such as heavy media separation, jigging, rising-current classification, and crushing.

In heavy media separation, aggregates are passed through a heavy liquid comprised of finely ground heavy minerals and water proportioned to have a relative density (specific gravity) less than that of the desirable aggregate particles but greater than that of the deleterious particles. The heavier particles sink to the bottom while the lighter particles float to the surface. This process can be used when acceptable and harmful particles have distinguishable relative densities.

Jigging separates particles with small differences in density by pulsating water current. Upward pulsations of water through a jig (a box with a perforated bottom) move the lighter material into a layer on top of the heavier material; the top layer is then removed.

Rising-current classification separates particles with large differences in density. Light materials, such as wood and lignite, are floated away in a rapidly upward moving stream of water.

Crushing is also used to remove soft and friable particles from coarse aggregates. This process is sometimes the only means of making material suitable for use. Unfortunately, with any process some acceptable material is always lost and removal of all harmful particles may not be possible.

TABLE 5-10. Some Potentially Harmful Reactive Minerals, Rock, and Synthetic Materials*

ALKALI-SILICA REACTIVE SUBSTANCES[†]		ALKALI-CARBONATE REACTIVE SUBSTANCES[‡]
Andesites	Opal	Calcitic dolomites
Argillites	Opaline shales	Dolomitic limestones
Certain siliceous limestones and dolomites	Phyllites	Fine-grained dolomites
Chalcedonic cherts	Quartzose	
Chalcedony	Rhyolites	
Cristobalite	Schists	
Dacites	Siliceous shales	
Glassy or cryptocrystalline volcanics	Strained quartz and certain other forms of quartz	
Granitic gneiss	Synthetic and natural siliceous glass	
Graywackes	Tridymite	
Metagraywackes		

* This list is not exhaustive.

[†] Several of the rocks listed (granitic gneiss and certain quartz formations for example) react very slowly and may not show evidence of any harmful degree of reactivity until the concrete is over 20 years old.

[‡] Only certain sources of these materials have shown reactivity.

HANDLING AND STORING AGGREGATES

Aggregates should be handled and stored in a way that minimizes segregation (separation of aggregates by size) and degradation and that prevents contamination by deleterious substances (Figure 5-27). Stockpiles should be built up in thin layers of uniform thickness to minimize segregation. An acceptable method of forming aggregate stockpiles is the truck-dump method, which discharges the loads in a way that keeps them tightly joined. The aggregate is then reclaimed with a front-end loader. The loader should remove slices from the edges of the pile from bottom to top so that every slice will contain a portion of each horizontal layer.

FIGURE 5-27. Proper storage and transportation of aggregates are important in minimizing segregation and contamination.

Crushed aggregates segregate less than rounded (gravel) aggregates and larger-size aggregates segregate more than smaller sizes. To avoid segregation of coarse aggregates, size fractions can be stockpiled and batched separately. Proper stockpiling procedures, however, should eliminate this requirement. Specifications provide a range in the amount of material permitted in any size fraction partly to accommodate some minor segregation in stockpiling and batching operations.

Washed aggregates should be stockpiled well before use so that they can drain to a uniform moisture content. Damp fine material has less tendency to segregate than dry material. When dry fine aggregate is dropped from buckets or conveyors, wind can blow away the fines; this should be avoided if possible.

Bulkheads or dividers should be used to avoid unintentional intermingling or contamination of aggregate stockpiles (Figure 5-28). Partitions between stockpiles should be high enough to prevent intermingling of materials. Storage bins, or hoppers, should be circular or nearly square. Their bottoms should slope not less than 50 degrees from the horizontal on all sides to a center outlet. When loading the bin, the material should fall vertically over the outlet into the bin. Chuting the material into a bin at an angle and against the bin sides will cause segregation. Baffle plates or dividers will help minimize segregation.

FIGURE 5-28. Bulkheads or dividers should be utilized and filled only to capacity to avoid intermixing of aggregate stockpiles.

Bins should be kept as full as possible since this reduces breakage of aggregate particles and their tendency to segregate. Recommended methods of handling and storing concrete aggregates are discussed at length in Matthews (1965 to 1967), NCHRP (1967), Bureau of Reclamation (1981), and NSSGA (2013).

REFERENCES

PCA's online catalog includes links to PDF versions of many of our research reports and other classic publications. Visit: cement.org/library/catalog.

Abrams, D.A., *Design of Concrete Mixtures*, LS001, Lewis Institute, Structural Materials Research Laboratory, Bulletin No. 1, Chicago, IL, 1918, 20 pages.

Abrams, M.S.; and Gustaferro, A.H., *Fire Endurance of Concrete Slabs as Influenced by Thickness, Aggregate Type, and Moisture*, RX223, Portland Cement Association, Skokie, IL, 1968, 22 pages.

ACI Committee 211, *Guide for Proportioning Concrete Mixtures with Ground Limestone and Other Mineral Fillers*, ACI 211.7R-15, American Concrete Institute, Farmington Hills, MI, 2015, 14 pages.

ACI Committee 221, *Guide for Use of Normal Weight and Heavy Weight Aggregates in Concrete*, ACI 221R-96, American Concrete Institute, Farmington Hills, MI, 1996 (Reapproved 2001), 28 pages.

ACI Committee 304, *Guide to Placing Concrete by Pumping Methods*, ACI 304.2R-17, American Concrete Institute, Farmington Hill, MI, 2017, 25 pages.

ACI Committee 318, *Building Code Requirements for Structural Concrete and Commentary*, ACI 318-19, American Concrete Institute, Farmington Hills, MI, 2019, 624 pages.

ACI Committee 555, *Removal and Reuse of Hardened Concrete*, ACI 555R-01, American Concrete Institute, Farmington Hills, MI, 2001, 26 pages.

Bentz, D.P.; and Weiss, W.J., *Internal Curing: A 2010 State-of-the-Art Review*, NISTIR 7765, US Department of Commerce, National Institute of Standards and Technology, Gaithersburg, MD, February 2011, 94 pages.

Bohan, R.P.; and Ries, J., *Structural Lightweight Concrete*, IS032, Portland Cement Association, Skokie, IL, 2008, 8 pages.

Buck, A.D.; and Mather, K., *Reactivity of Quartz at Normal Temperatures*, Technical Report SL-84-12, Structures Laboratory, Waterways Experiment Station, U.S. Army Corps of Engineers, Vicksburg, MS, July 1984.

Bureau of Reclamation, *Concrete Manual*, 8th ed., U.S. Bureau of Reclamation, Denver, CO, 1981.

Carlson, R.W., "Drying Shrinkage of Concrete as Affected by Many Factors," *ASTM Proceedings*, Vol. 38, Part II, 1938, pages 419 to 437.

Cook, D., *Aggregate Proportioning for Slipped Formed Pavements and Flowable Concrete*. Dissertation, Oklahoma State University, Stillwater, OK, 2015, 273 pages.

Cook, M.D.; Ghaeezadah, A.; and Ley, M.T., "A Workability Test for Slip Formed Concrete Pavements," *Construction and Building Materials*, Vol. 68, Elsevier, 2014, pages 376 to 383.

Cook, D.; Ghaeezadah, A.; Ley, T.; and Russell, B., *Investigation of Optimized Graded Concrete for Oklahoma – Phase 1*. Oklahoma Department of Transportation, 2013, 118 pages.

Crouch, L.K.; Sauter, H.J.; and Williams, J.A., "92-MPa Air-entrained HPC," *TRB-Record 1698*, Concrete 2000, page 24.

CSA Standard A23.1-19/A23.2-19, *Concrete Materials and Methods of Concrete Construction/Test Methods and Standard Practices for Concrete*, Canadian Standards Association, Toronto, Canada, 2019.

CSA A23.1:19, *Concrete Materials and Methods of Concrete Construction, Annex P (Informative), Impact of Sulphides in Concrete Aggregate on Concrete Behaviour*, Canadian Standards Association, Toronto, Canada, 2019.

ECCO, *Recycling Concrete and Masonry*, EV 22, Environmental Council of Concrete Organizations, Skokie, IL, 1999, 12 pages.

Farny, J.A.; and Kerkhoff, B., *Diagnosis and Control of Alkali-Aggregate Reactions*, IS413, Portland Cement Association, Skokie, IL, 2007, 26 pages.

Fathifazl, G.; Razaqpur, A.G.; Isgor, O.B.; Abbas, A.; Fournier, B.; and Foo, S., "Proportioning Concrete Mixtures with Recycled Concrete Aggregate," *Concrete International*, Vol. 32, No. 3, American Concrete Institute, Farmington Hills, MI, March 2010, pages 37 to 43.

Hadley, D.W., *Alkali Reactivity of Carbonate Rocks – Expansion and Dedolomitization*, Research Department Bulletin RX139, Portland Cement Association, Chicago, IL, 1961.

Harrigan, E., "Performance of Pavement Subsurface Drainage," *NCHRP Research Results Digest*, No. 268, Transportation Research Board, Washington, D.C., November 2002.

Houston, B.J., *Investigation of Gap-Grading of Concrete Aggregates; Review of Available Information*, Technical Report No. 6-593, Report 1, Waterways Experiment Station, U.S. Army Corps of Engineers, Vicksburg, MS, February 1962.

Kerkhoff, B.; and Siebel, E., "Properties of Concrete with Recycled Aggregates (Part 2)," *Beton,* 2/2001, Verlag Bau + Technik, 2001, pages 105 to 108.

Kim, J.; and Sadowski, L., "The Equivalent Mortar Volume Method in the Manufacturing of Recycled Aggregate Concrete," *Czasopismo Techniczne*, January 2019.

Lee, S.; Daugherty, A.; and Broton, D., "Assessing Aggregates for Radiation-Shielding Concrete," *Concrete International*, May 2013, pages 31 to 38.

Ley, T., *Tarantula Curve*, website, tarantulacurve.com, accessed August 2020.

Litvin, A.; and Pfeifer, D.W., *Gap-Graded Mixes for Cast-in-Place Exposed Aggregate Concrete*, DX090, Development Department Bulletin, Portland Cement Association, Chicago, IL, 1965, 31 pages.

Maerz, N.H.; and Lusher, M., "Measurement of flat and elongation of coarse aggregate using digital image processing," *80th Annual Meeting*, Transportation Research Board, Washington D.C., 2001, pages 2 to 14.

Mather, B., *New Concern Over Alkali-Aggregate Reaction*, Joint Technical Paper by National Aggregates Association and National Ready Mixed Concrete Association, NAA Circular No. 122 and NRMCA Publication No. 149, Silver Spring, MD, 1975.

Matthews, C.W., "Stockpiling of Materials," *Rock Products*, series of 21 articles, Maclean Hunter Publishing Company, Chicago, IL, August 1965 through August 1967.

NCHRP, *Effects of Different Methods of Stockpiling and Handling Aggregates*, NCHRP Report 46, National Cooperative Highway Research Program, Transportation Research Board, Washington, D.C., 1967.

NSSGA, *The Aggregates Handbook*, 2nd ed, National Stone, Sand and Gravel Association, Alexandria, VA, 2013, 908 pages.

Obla, K.; Kim, H.; and Lobo, C., *Effect of Continuous (Well-Graded) Combined Aggregate Grading on Concrete Performance, Phase B: Concrete Performance*, National Ready Mixed Concrete Association, Silver Spring, MD, May 2007, 45 pages.

Owens, G., *Fulton's Concrete Technology*, 9th edition, Cement and Concrete Institute, Midrand, South Africa, 2009, 465 pages.

Ozol, M.A., "Alkali-Carbonate Rock Reaction," Chapter 35 in *Significance of Tests and Properties of Concrete and Concrete-Making Materials*, ASTM STP 169D, ASTM International, West Conshohocken, PA, 2006, pages 410 to 424.

PCA, *Bulking of Sand Due to Moisture*, ST20, Portland Cement Association, Skokie, IL, 1949, 2 pages.

PCA, *Effect of Moisture on Volume of Sand* (1923), PCA Major Series 172, Portland Cement Association, Skokie, IL, 1955, 1 page.

Shilstone, J.M., Sr., "Concrete Mixture Optimization," *Concrete International*, American Concrete Institute, Farmington Hills, MI, June 1990, pages 33 to 39.

Stark, D., *Characteristics and Utilization of Coarse Aggregates Associated with D-Cracking*, RD047, Research and Development Bulletin, Portland Cement Association, Skokie, IL,1976.

Stark, D., *The Use of Recycled-Concrete Aggregate from Concrete Exhibiting Alkali-Silica Reactivity*, RD114, Research and Development Bulletin, Portland Cement Association, Skokie, IL, 1996.

Swenson, E.G.; and Gillott, J.E., "Alkali Reactivity of Dolomitic Limestone Aggregate," *Magazine of Concrete Research*, Vol. 19, No. 59, Cement and Concrete Association, London, June 1967, pages 95 to 104.

Townsend, T.G.; and Anshassi, M., *Benefits of Construction and Demolition Debris Recycling in the United States*, Construction Demolition Recycling Association, Chicago, IL, April 2017, 22 pages.

Thomas, M.D.A.; Fournier, B.; and Folliard, K.J., *Report on Determining the Reactivity of Concrete Aggregates and Selecting Appropriate Measures for Preventing Deleterious Expansion in New Concrete Construction*, FHWA-HIF-09-001, Federal Highway Administration, Washington, D.C., April 2008, 28 pages.

USACE, *Test Method for Flat and Elongated Particles in Fine Aggregate*, CRD-C 120-94, US Army Corps of Engineers, September 1994, 2 pages.

Vogler, R.H.; and Grove, G.H., "Freeze-Thaw Testing of Coarse Aggregate in Concrete: Procedures Used by MI Department of Transportation and Other Agencies," *Cement, Concrete, and Aggregates*, American Society for Testing and Materials, West Conshohocken, PA, Vol. 11, No. 1, Summer 1989, pages 57 to 66.

Weiss, W.J., *Internal Curing for Concrete Pavements*, FHWA-HIF-16-006, US Department of Transportation, Federal Highway Administration, July 2016, 7 pages.

Wigum, B.J., "Norwegian Petrographic Method – Development and Experiences During a Decade of Service," *Proceedings of the 12th International Conference on Alkali-Aggregate Reaction in Concrete*, Vol. I, International Academic Publishers – World Publishing Corporation, October 15-19, 2004, pages 444 to 452.

Willett, J.C., "Sand and Gravel (Construction)," *Mineral Commodity Summaries 2020*, US Geological Survey, US Department of the Interior, Reston, VA, January 2020, pages 140 to 141.

Willett, J.C., "Stone (Crushed)," *Mineral Commodity Summaries 2020*, US Geological Survey, US Department of the Interior, Reston, VA, January 2020, pages 154 to 155.

CHEMICAL ADMIXTURES FOR CONCRETE

FIGURE 6-1. Liquid admixtures, from left to right: anti-washout admixture, shrinkage reducer, water reducer, foaming agent, corrosion inhibitor, and air-entraining admixture.

Chemical admixtures are primarily water-soluble substances used to modify the properties of concrete, mortar, or grout in the plastic state, hardened state, or both (Figure 6-1).

Admixtures can be used to modify the fresh concrete properties in order to increase workability, reduce rate of slump loss, reduce segregation, modify setting time, modify the rate and capacity for bleeding, reduce the temperature rise due to heat of hydration, and improve pumpability, placeability, and finishability. Admixtures may also be used to modify hardened concrete properties including increasing strength (lower w/cm), improving impact and abrasion resistance, reducing shrinkage, reducing permeability, and improving durability. The aesthetics of concrete, including final color, can also be modified by chemical admixtures.

There are a variety of admixtures available for use in concrete mixtures to modify fresh and hardened concrete properties. Chemical admixtures can be classified by function as follows:

- Air entraining
- Water reducing
- Set retarding
- Set accelerating
- Strength enhancing
- Extended set control
- Workability retaining
- Viscosity modifying
- Corrosion inhibiting
- Shrinkage and crack reducing
- Permeability reducing
- Alkali silica reactivity inhibiting
- Clay mitigating
- Coloring admixtures
- Miscellaneous admixtures such as pumping aids, bonding, grouting, gas forming, air detraining, fungicidal, anti-washout, foaming, and admixtures for cold weather

TABLE 6-1. Concrete Admixtures by Classification

TYPE OF ADMIXTURE	DESIRED EFFECT	ACTIVE INGREDIENTS	TYPICAL DOSAGE RANGE*
Accelerators (ASTM C494, Type C)	Accelerate setting and early-strength development	Calcium chloride (ASTM D98 and AASHTO M 144) triethanolamine, sodium thiocyanate, calcium formate, calcium nitrite, calcium nitrate, triisopropanolamine, sodium nitrate, sodium formate, sodium lactate, and silicate and aluminate salts	Chloride-based: 520-3260 mL/100 kg (8-50 fl oz/100 lb) Nonchloride: 520-3260 mL/100 kg (8-50 fl oz/100 lb)
Admixture for Cold Weather, ASTM C1622	Allows concrete placement when concrete temperature reaches as low as 23°F (-5°C) prior to initial set.	Various calcium salts and organic set activators	Trial batching required
Air detrainers (ASTM C494, Type S)	Decrease air content	Tributyl phosphate, dibutyl phthalate, octyl alcohol, water-insoluble esters of carbonic and boric acid, silicones	15-325 mL/100 kg (0.2-5.0 fl oz/100 lb)
Air-entraining admixtures (ASTM C260 and AASHTO M 154)	Improve durability in freeze-thaw, deicer, sulfate, and alkali-reactive environments Improve workability	Salts of wood resins (vinsol resin), some synthetic detergents, salts of sulfonated lignin, salts of petroleum acids, salts of proteinaceous material, fatty and resinous acids and their salts, alkylbenzene sulfonates, salts of sulfonated hydrocarbons	Resin: 30-200 mL/100 kg (0.5-3 fl oz/100 lb) Synthetic: 7-400 mL/100 kg (0.1-6 fl oz/100 lb)
Alkali-aggregate reactivity inhibitors	Reduce expansion due to alkali-aggregate reactivity	Barium salts, lithium nitrate, lithium carbonate, lithium hydroxide	4.63 L (0.55 gal) per % equivalent alaklies in cement
Antiwashout admixtures (ASTM C494, Type S)	Cohesive concrete for underwater placements	Cellulose deritivites, acrylic polymer, biogums	325-975 mL/100 kg (5-15 fl oz/100 lb)
Bonding admixtures	Increase bond strength	Polyvinyl chloride, polyvinyl acetate, acrylics, butadiene-styrene copolymers	Equal parts admixture and water added to specified amount of cement and sand.
Coloring admixtures (ASTM C979)	Colored concrete	Modified carbon black, iron oxide, phthalocyanine, umber, chromium oxide, titanium oxide, cobalt blue	Trial batching suggested, up to 10% by mass of cement
Corrosion inhibitors (ASTM C1582)	Reduce steel corrosion activity in a chloride-laden environment	Amine carboxylates, aminoester organic emulsion, calcium nitrite, organic alkyidicarboxylic, chromates, phosphates	5-30 L/m³ (2-6 gal/yd³)
Extended set controlling, also referred to as hydration controlling admixtures	Suspend and reactivate cement hydration	Carboxylic acid salts, sugars, phosphorus-containing organic acid salts	130-650 mL/100 kg (2-10 fl oz/100 lb)
Foaming agents (ASTM C869)	Produce lightweight, foamed concrete with low density	Cationic and anionic surfactants, hydrolized proteins	Trial batching required
Fungicides, germicides, and insecticides	Inhibit or control bacterial and fungal growth	Polyhalogenated phenols, dieldrin emulsions, copper compounds	0.1-10% weight of cement; more than 3% affects strength
Gas formers	Cause expansion before setting	Aluminum powder	Trial batching required
Grouting admixtures	Adjust grout properties for specific applications	See Air-entraining admixtures, accelerators, retarders, and water reducers	Refer to manufacturer
Nano-admixtures	Multi-functional	Fine dispersion of nano-sized particles	Refer to manufacturer
Permeability-reducing admixture: non-hydrostatic conditions (PRAN)	Water-repellent surface, reduced water absorption	Long-chain fatty acid derivatives (stearic oleic, caprylic capric), soaps and oils, (tallows, soya-based), petroleum derivatives (mineral oil, paraffin, bitumen emulsions), silanes, and fine particle fillers (silicates, bentonite, talc)	190-390 mL/100 kg (3-6 fl oz/100 lb)

* Admixtures are typically dosed by 100 kg (100 lb) weight of cementitious materials content. Follow manufacturer's recommendations for all products on final dosage ranges.

TABLE 6-1. Concrete Admixtures by Classification (Continued)

TYPE OF ADMIXTURE	DESIRED EFFECT	ACTIVE INGREDIENTS	TYPICAL DOSAGE RANGE*
Permeability-reducing admixture: hydrostatic conditions (PRAH)	Reduced permeability, increased resistance to water penetration under pressure	Crystalline hydrophilic polymers (latex, water-soluble, or liquid polymer)	190-2340 mL/100 kg (3-36 fl oz/100 lb)
Pumping aids	Improve pumpability	Organic and synthetic polymers, organic flocculents organic emulsions of paraffin, coal tar, asphalt, acrylics, bentonite and pyrogenic silicas, hydrated lime (ASTM C141)	See Viscosity Modifier, Water Reducer
Retarding admixtures (ASTM C494 and AASHTO M 194, Type B)	Retard setting time	Lignin, borax, sugars, tartaric acid and salts corn syrup, sodium gluconate and glucoheptonate, sodium phosphate, zinc salts, and organophosphates and organophosphonates	130-325 mL/100 kg (2-5 fl oz/100 lb)
Rheology Controlling Admixture (ASTM C494, Type S)	Improves placement of low slump concrete (RCC, Pervious) extrusion rate of dry-cast concrete	polylvinyl alcohol	130-780 mL/100 kg (2-12 fl oz/100 lb)
Shrinkage and crack reducers (ASTM C494, Type S)	Reduce drying shrinkage and cracking potential	Polyoxyalkylene alkyl ether, propylene glycol, Hexylene glycol	455-2990 mL/100 kg (7-46 fl oz/100 lb)
Slump Retaining Admixtures	Maintain concrete workability with minimal set retardation	Special polycarboxylates	See Workability Retaining
Strength-enhancing admixture (ASTM C494, Type S)	Increase early-age strength development	C-S-H nanoparticles, alkanol amines and thiocyanates	260-980 mL/100 kg (4-15 fl oz/100 lb)
Viscosity modifying (ASTM C494, Type S)	Increase paste viscosity to enhance the stability of concrete.	Cellulose ether, biogum (welan, diutan), welan gum, diutan gum, polyethylene glycol	32-910 mL/100 kg (0.5-14 fl oz/100 lb)
Water reducer (ASTM C494, Type A), also referred to as plasticizers	Reduce water content at least 5%, or increase flowability	Lignosulfonates, hydroxylated carboxylic acids, carbohydrates (also tend to retard set so accelerator is often added)	Carboxylic Acids: 150-250 mL/100 kg (2-4 fl oz/100 lb) Lignosulfonates: 200-650 mL/100 kg (3-10 fl oz/100 lb)
Water reducer and accelerator (ASTM C494, Type E)	Reduce water content (minimum 5%) and accelerate set	See water reducer, Type A (accelerator is added)	650-2940 mL/100 kg (10-45 fl oz/100 lb)
Water reducer and retarder (ASTM C494, Type D)	Reduce water content (minimum 5%) and retard set	See water reducer, Type A (retarder is added)	195-390 mL/100 kg (3-6 fl oz/100 lb)
Water reducer – high range (ASTM C494, Type F), also referred to as Superplasticizers	Reduce water content (minimum 12%), or increase flowability	Sulfonated melamine formaldehyde condensates sulfonated naphthalene formaldehyde condensates lignosulfonates, polycarboxylates (including polyaryl ether)	Polycarboxylate: 130-980 mL/100 kg (2-15 fl oz/100 lb) Naphthalene: 390-1300 mL/100 kg (6-20 fl oz/100 lb)
Water reducer – high range – and retarder (ASTM C494, Type G)	Reduce water content (minimum 12%) and retard set	See superplasticizers and also water reducers	Polycarboxylate: 130-325 mL/100 kg (2-5 fl oz/100 lb) Naphthalene: 390-2080 mL/100 kg (6-32 fl oz/100 lb)
Water reducer – mid range	Reduce water content (between 6% and 12%) or increase flowability	Lignosulfonates, polycarboxylates	195-980 mL/100 kg (3-15 fl oz/100 lb)
Workability retaining (ASTM C494, Type S)	Maintain slump and workability of concrete and consistency between loads without retarding	Modified polycarboxylate	130-780 mL/100 kg (2-12 fl oz/100 lb)

* Admixtures are typically dosed by 100 kg (100 lb) weight of cementitious materials content. Follow manufacturer's recommendations for all products on final dosage ranges.

The major reasons for using chemical admixtures in concrete mixtures are to:

- achieve specific properties in fresh and hardened concrete more effectively than by other means;

- maintain consistency during mixing, transporting, placing, finishing, and curing (including adverse weather conditions and intricate placements);

- make the mixture more economical; and

- enable use of a wider selection of concrete materials (including marginal materials).

Despite these considerations, no admixture of any type or amount is a substitute for good concreting practice.

EFFECTIVENESS OF AN ADMIXTURE

The effectiveness of an admixture depends upon factors such as its composition, type, and dosage rate. The admixture will also be dependent on the concrete mixture composition and proportions, time of addition, sequencing, mixing time, slump, and temperature of the concrete. Classes of chemical admixtures and their general composition for use in concrete should meet applicable specifications presented in Table 6-1. Most admixtures are liquids, but some are available in powder form.

Proper use of chemical admixtures begins with reviewing the manufacturer's data sheet to confirm that the admixture under consideration is being used for the designed application. Trial mixtures should be made with the admixture and the other concrete ingredients at the temperature and relative humidity and simulated estimated time between mixing and discharge

anticipated during placement. Observations can then be made on the compatibility of the admixture with other ingredients, as well as its effects on the properties of the fresh and hardened concrete (see Compatibility of Admixtures and Cementitious Materials). A combination of admixtures may be needed to achieve desired performance. The type and amount of admixture used should be recommended by the manufacturer. The optimum dosage, combination of admixtures, and sequencing should be verified by laboratory and field testing.

AIR-ENTRAINING ADMIXTURES

Air entrainment provides purposefully sized and spaced microscopic air bubbles in concrete that dramatically improves the durability of concrete exposed to cycles of freezing and thawing and deicer chemicals (see Chapter 11). There are also other additional benefits of entrained air in both freshly mixed and hardened concrete (see Chapter 9).

Air entrainment for most concretes is achieved through the use of chemical air-entraining admixtures, rather than through the use of air-entraining cements. Air-entraining cement is a portland cement with an air-entraining addition interground with the cement clinker during manufacturing (see Chapter 2). An air-entraining admixture is added to concrete either immediately before or during mixing. Adequate control of material quality, mixing speed and time, placement, and consolidation is required to ensure the proper air content is maintained in the concrete.

Specifications and methods of testing air-entraining admixtures are given in ASTM C260, *Standard Specification for Air-Entraining Admixtures for Concrete* (AASHTO M 154, and

TABLE 6-2. Classification and Performance Characteristics of Common Air-Entraining Admixtures

CLASSIFICATION	CHEMICAL DESCRIPTION	NOTES AND PERFORMANCE CHARACTERISTICS
Wood derived acid salts	Alkali or alkanolamine salt of: A mixture of tricyclic acids, phenolics, and terpenes.	Quick air generation. Minor air gain with initial mixing. Air loss with prolonged mixing. Mid-sized air bubbles formed. Compatible with most other admixtures.
Wood rosin	Tricyclic acids-major component. Tricyclic acids-minor component.	Same as above.
Tall oil	Fatty acids-major component. Tricyclic acids-minor component.	Slower air generation. Air may increase with prolonged mixing. Smallest air bubbles of all agents. Compatible with most other admixtures.
Vegetable oil acids	Coconut fatty acids, alkanolamine salt.	Slower air generation than wood rosins. Moderate air loss with mixing. Coarser air bubbles relative to wood rosins. Compatible with most other admixtures.
Synthetic detergents	Alkyl-aryl sulfonates and sulfates (e.g., sodium dodecyl benzene sulfonate).	Quick air generation. Minor air loss with mixing. Coarser bubbles. May be incompatible with some HRWR. Also applicable to cellular concretes.
Synthetic workability aids	Alkyl-aryl ethoxylates.	Primarily used in masonry mortars.
Miscellaneous	Alkali-alkanolamine acid salts of lignosulfonate. Oxygenated petroleum residues. Proteinaceous materials. Animal tallows.	All these are rarely used as concrete air-entraining agents in current practice.

C233, *Standard Test Method for Air-Entraining Admixtures for Concrete* (AASHTO T 157). Air-entraining additions for use in the manufacture of air-entraining cements must meet requirements of ASTM C226, *Standard Specification for Air-Entraining Additions for Use in the Manufacture of Air-Entraining Hydraulic Cement*. Applicable requirements for air-entraining cements are given in ASTM C150, *Standard Specification for Portland Cement* (AASHTO M 85); ASTM C595, *Standard Specification for Blended Hydraulic Cements* (AASHTO M 240); and ASTM C1157, *Standard Performance Specification for Hydraulic Cement*. See Klieger (1966), and Whiting and Nagi (1998) for more information.

Composition of Air-Entraining Admixtures

The primary ingredients used in air-entraining admixtures are listed in Table 6-1. Numerous commercial air-entraining admixtures, manufactured from a variety of materials, are available. Most air-entraining admixtures consist of one or more of the following materials: wood resin, sulfonated hydrocarbons, fatty and resinous acids, and synthetic materials. Synthetic materials are generally considered to be surface active chemical compounds/polymers not derived from natural products. Chemical descriptions and performance characteristics of common air entrainers are shown in Table 6-2.

Mechanism of Air Entrainment

Air-entraining admixtures are surfactants (surface-active agents) which concentrate at the air-water interface and reduce the surface tension of the pore water, encouraging the formation of microscopic bubbles during the mixing process. The air-entraining admixture stabilizes those bubbles, produces bubbles of various sizes, impedes bubble coalescence, and anchors bubbles to cement and aggregate particles.

Air-entraining admixtures typically have a negatively charged 'head' that is hydrophilic and attracts water, and a hydrophobic 'tail' which repels water. As illustrated in Figure 6-2; the hydrophobic end is attracted to the air within bubbles generated during the mixing process. The polar end, that is hydrophilic, orients itself towards water (A). The air-entraining admixture forms a tough, water-repelling film, with sufficient strength and elasticity to contain and stabilize the air bubbles. The hydrophobic film also keeps water out of the bubbles while mixing is continued (B). The stirring and kneading action of mechanical mixing generates and disperses the air bubbles. The charge around each bubble leads to repulsive forces that prevent the coalescence of bubbles (C & D). The surface charge causes the air bubble to adhere to the charged surfaces of cement and aggregate particles. The fine aggregate particles also act as a three-dimensional grid to help stabilize the bubbles in the mixture (E). This improves the cohesion of the mixture and further stabilizes the air bubbles (F).

Entrained air bubbles are unlike entrapped air voids, which occur in all concretes as a result of mixing, handling, and placing. Entrapped air voids are largely a function of aggregate characteristics. Intentionally entrained air bubbles are extremely small in size, between 10 μm to 1000 μm in diameter, while entrapped voids are usually 1000 μm (1 mm) or larger. As shown in Figure 6-3, the bubbles are not normally interconnected. They are well dispersed and randomly distributed. See Chapter 11 or more information on spacing factor and specific size of air bubbles.

Non-air-entrained concrete with a 25-mm (1-in.) maximum-size aggregate typically has an air content of approximately 1% to 3%. This same mixture, air entrained for severe frost exposure,

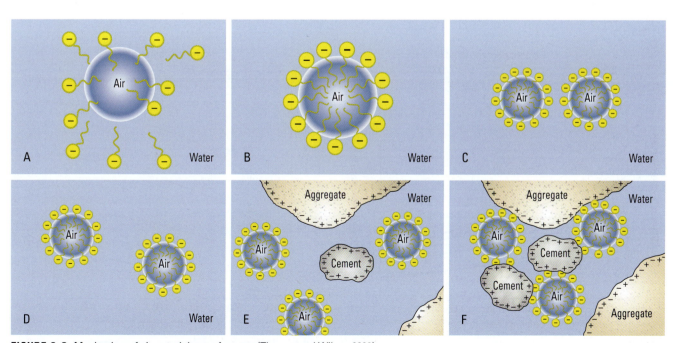

FIGURE 6-2. Mechanism of air-entraining surfactants (Thomas and Wilson 2002).

would require a total air content of about 6%, made up of both coarse entrapped air voids and fine entrained air voids in hardened concrete. However, it is the finely entrained air-void system that is most effective at providing frost resistance.

FIGURE 6-3. Polished concrete surface shows air voids clearly visible as dark circles.

Control of Air Content

The amount of air entrained in concrete for a given dose of air-entraining admixture will depend on concrete materials, mixture proportions, methods of transport, placing and finishing methods, and curing. Variations in air content can be expected with variations in aggregate proportions and gradation, mixing time and intensity, temperature, and slump. The order of batching and mixing concrete ingredients when using an air-entraining admixture can also have a significant influence on the amount of air entrained. The late addition of water and extended retempering along with the time of addition of other chemical admixtures, can cause clustering of air bubbles around aggregate resulting in potentially significant strength

reduction (Kozikowski and others 2005 and Camposagrado 2006). Therefore, consistency in batching is needed to maintain adequate control. For more information on the effects of constituent materials, mixture proportions, and placing and finishing operations on air content see Chapter 9.

An excessive amount of air content might have adverse effects on concrete properties (Whiting and Stark 1983). When entrained air is beyond that required for the application, it can be reduced by using one of the following defoaming (air-detraining) agents: tributyl phosphate, dibutyl phthalate, octyl alcohol, water-insoluble esters of carbonic acid and boric acid, and silicones. Only the smallest possible dosage of defoaming agent should be used to reduce the air content to the specified limit.

Impact of Air Content on Properties of Concrete

The presence of a finely distributed network of bubbles has a significant impact on the properties of plastic concrete. Entrained air improves the workability of concrete. It is particularly effective in lean (low cement content) mixtures that otherwise might be harsh and difficult to work (Cordon 1946). Workability of concrete mixtures with angular and poorly graded aggregates is similarly improved.

Because of improved workability with entrained air, water and sand contents can be reduced significantly as shown in Figure 6-4. A volume of air-entrained concrete requires less water than an equal volume of non-air-entrained concrete of the same consistency and maximum size aggregate. Water reductions in the range of 15 L/m³ to 25 L/m³ (25 lb/yd³ to 40 lb/yd³) may be achieved with adequate air entrainment (Figure 6-4). Freshly mixed concrete containing entrained air is cohesive, looks and feels "fatty" or workable, and can usually be handled with ease. The air bubbles reduce the tendency for segregation

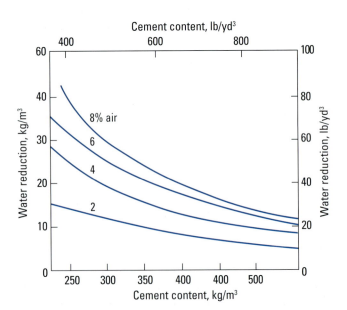

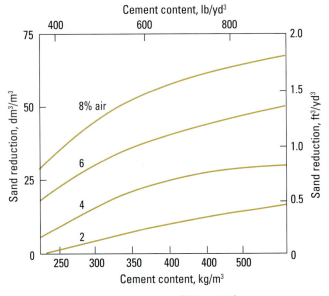

FIGURE 6-4. Reduction of water and sand content obtained at various levels of air and portland cement contents (Gilkey 1958).

and bleeding. On the other hand, high air contents can make a mixture sticky and more difficult to finish.

Improvements in the performance of hardened concrete with air entrainment obviously include improved resistance to freezing and thawing, and deicer-salt scaling. Furthermore, the incorporation of air also results in a reduced permeability and possibly improved resistance to sulfate attack and alkali-silica reactivity (see Chapter 11).

A detrimental effect of air entrainment is that an increase in air content results in a decrease in the strength of the concrete. When the air content is maintained constant, strength varies inversely with the water-cementitious materials ratio (see Figure 6-5). As a rule of thumb, each 1% increase in air content is accompanied by 5% to 6% reduction in later-age strength. More information on the impact of air content on fresh and hardened properties of concrete can be found in Chapter 9.

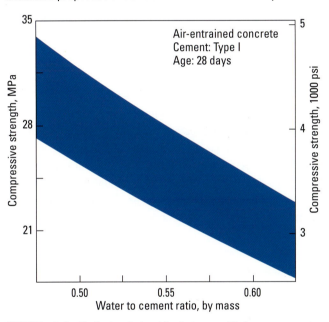

FIGURE 6-5. Typical relationship between 28-day compressive strength and water-cement ratio for a wide variety of air-entrained concretes using Type I cement.

WATER-REDUCING ADMIXTURES

A water-reducing admixture increases workability without increasing the water content of concrete, or permits a decrease in water content without decreasing the slump. They are referred to as plasticizers when used solely to increase slump and workability. High-range water reducers (HRWR) are also known as superplasticizers. Water reducers/plasticizers used in concrete must conform to ASTM C494, *Standard Specification for Chemical Admixtures for Concrete.* Previously, high-range water reducers were required to be tested for conformance with ASTM C1017, S*tandard Specification for Chemical Admixtures for Use in Producing Flowing Concrete.* However, this standard was withdrawn in 2022 and is no longer required by state agencies.

When used as a water-reducer, the water content is lowered while maintaining the slump; this reduces the water-to-cementitious materials ratio of the concrete and increases its strength and durability. In this situation, the admixture is used as a true water-reducer; allowing concrete to be produced with low w/cm while maintaining workability. When the same chemical admixture is used as a plasticizer, the workability is increased while the water content is kept constant. This can improve the placing characteristics of the concrete without adversely affecting the strength and durability. Advantage may be taken for both purposes of a water-reducing admixture when designing concrete mixtures, thereby producing concrete with improved fresh and hardened properties (Figure 6-6).

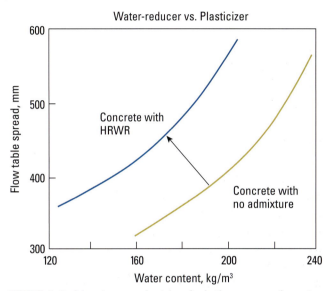

FIGURE 6-6. Advantage may be taken for both purposes of a water-reducing admixture (lowering w/cm and increasing flow) when designing concrete (Adapted from Neville 1995).

Normal (Conventional) Water Reducers

When used as a water reducer, normal range or conventional water reducers can reduce the water content by approximately 5% to 10% without exceeding the ASTM C494 time of set limit relative to a control mixture. Alternatively, they may be used as plasticizers to provide a moderate increase in workability. Normal water reducers are intended for concretes with slumps from 75 mm (3 in.) to 150 mm (6 in.), but also can be used in combination with mid-range and high-range water reducers for higher slump concrete mixtures.

Mid-Range Water Reducers

Mid-range water reducers were first introduced in 1984 to bridge the gap between normal range water reducers and high-range water reducers (superplasticizers). These admixtures provide water reduction typically (between 6% and 12%) for concretes with slumps of 125 mm to 200 mm (5 in. to 8 in.) without the retardation associated with high dosages of conventional (normal) water reducers. Mid-range water reducers can be used to reduce stickiness and improve finishability, pumpability, and placeability of concretes containing silica fume and other supplementary cementing materials. An ASTM specification for mid-range water reducing admixtures has yet to be established, however, these products meet ASTM C494 Type A requirements and some also meet the Type F requirements. For more information on mid-range water reducers, see Nmai and others (1998), and Schaefer (1995).

High-Range Water Reducers

High-range water reducers, ASTM C494 Types F (water reducing) and G (water reducing and retarding), commonly referred to as superplasticizers, can be used to impart properties induced by regular water reducers, only much more efficiently. They can greatly reduce water demand and cement contents and help produce low water-cement ratio, high-strength concrete with normal or enhanced workability and to generate slumps greater than 150 mm (6 in.). A water reduction of 12% to 40% can be obtained using these admixtures. The reduced water content and water-cement ratio can produce concretes with: ultimate compressive strengths in excess of 110 MPa (16,000 psi), increased early strength gain, reduced chloride-ion penetration, and other beneficial properties associated with low water-cement ratio concrete.

High-range water reducers are added to low-to-very low water-cement ratio concrete mixtures with a low-to-normal slump to make high-slump flowing concrete (Figure 6-7). Flowing concrete is by definition a cohesive concrete with a slump greater than 190 mm (7.5 in.), which includes self-consolidating concrete (SCC), that can be placed with little or no vibration or compaction while still remaining essentially free of excessive bleeding or segregation (See Chapter 15). Controlled flowing concrete may be readily produced with use of HRWR for moderate slump flows up to 635 mm (25 in.) (Burns and others, 2017). The cohesiveness provided by high-range water reducers may result in stickiness and lead to finishing difficulties. Applications where flowing concrete is used include: thin-section placements, areas of closely spaced and congested reinforcing steel, tremie pipe (underwater) placements, pumped concrete to reduce pump pressure (increasing lift and distance capacity), areas where conventional consolidation methods are impractical or cannot be used, and reducing handling costs. The addition of a high-range water reducer to a 75-mm (3-in.) slump concrete can easily produce a concrete with a 230-mm (9-in.) slump.

Composition of Water-Reducing Admixtures

The classifications and components of water reducers are listed in Table 6-1. The chemistry of water-reducing or plasticizing admixtures falls into broad categories: ligno-sulfonates, hydroxycarboxylic acid, hydroxylated polymers, salts of melamine formaldehyde sulfonates or naphthalene formaldehyde sulfonic acids, polycarboxylates, and polyaryl ethers. The use of organic materials to reduce the water content or increase the fluidity of concretes dates back to the 1930s. The most popular technology is the development of high-range water-reducers based on polycarboxylates and those based on polyaryl ether technology, a modified polycarboxylate (Daczko 2008).

Mechanisms of Water Reducers

Water reducers and plasticizers function as cement dispersants primarily through electrostatic and steric repulsive forces. Acidic groups within the polymer neutralize the surface charges on the cement particles (Ramachandran 1998, and Collepardi and Valente 2006). These groups bind to positive ions on the cement particle surfaces. These ions attach the polymer and give the cement a slight negative charge as well as create a layer on the surface. This negative charge and layer of adsorbed compounds create a combination of electrostatic and steric repulsion forces between individual cement particles, dispersing them, thus releasing the water tied up in agglomerations and reducing the viscosity and, to a smaller extent, the yield stress of the paste and concrete (Figures 6-8 and 6-9). A melamine-based, naphthalene-based, or lignin-based superplasticizer uses a molecule that has a size of about 1 μm to 2 μm. The effect of the water reducer depends on the dosage level, sequence of addition, and molecular weight. The water reducer will also contribute to dispersion by repelling negatively charged aggregate particles and air-entrained bubbles. The electrostatic repulsion for these materials is affected far more by dissolved ions (as compared to polycarboxylates) and rapidly diminishes as the hydrating cement releases more ions into the mixture.

Polycarboxylate Technology. Polycarboxylate high-range water reducers are polymers comprised of a main carbon chain with carboxylate groups and polyethylene oxide (PEO) side chains. The number of carboxylate groups and the number and length of PEO side chains can be adjusted to change the properties of the admixture. The PEO side chains extend out from the cement

FIGURE 6-7. High-range water reducers, also known as superplasticizers, produce flowing concrete with slumps greater than 190 mm (7.5 in.).

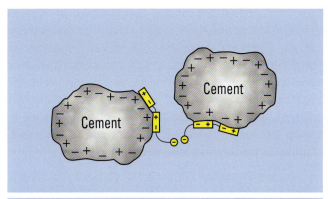

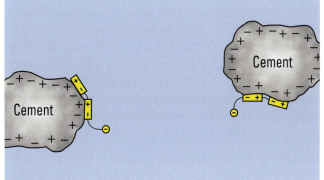

FIGURE 6-8. Mechanism of dispersive action of water-reducing admixtures (Thomas and Wilson 2002).

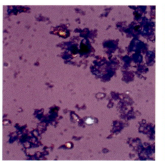 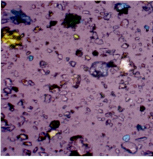

FIGURE 6-9. Dispersive action of water-reducing admixtures.

particles and add the mechanism of steric hindrance to the typical electrostatic repulsion (Jeknavorian and others 1997, Li and others 2005, and Nawa 2006).

The mechanism of steric hindrance by polycarboxylates is illustrated in Figure 6-10 (A-D). As with typical superplasticizers, the water reducer is dissolved in water, and the polar backbone of the polymer is absorbed at the solid-water interface (A). The long side chains physically help disperse and hold the

cement grains apart allowing water to totally surround the cement grains (steric hindrance) (B). Additionally, the polar chain imparts a slight negative charge causing the cement grains to repel one another (electrostatic repulsion) (C). As the electrostatic repulsion dispersing effect wears off due to cement hydration, the long side chains still physically keep the cement dispersed (D).

The PEO chains prevent particles from agglomerating through physical separations on the order of 10 nm (4×10^{-7} in) (Nawa 2006). This physical separation is still great enough to allow fluid to flow between the particles. This inhibition of agglomeration disperses the cement particles and allows the concrete to flow more easily. Because steric hindrance is a physical mechanism, it is not as sensitive to dissolved ions as an electrostatic repulsion mechanism. Concrete mixtures with polycarboxylate additions tend to retain fluidity for longer periods and they tend to require less water than concrete mixtures using other water reducers (Jeknavorian and others 1997). Polycarboxylate admixtures are commonly used in self-consolidating concrete (see Chapter 23 and Szecsy and Mohler [2009] for more information on self-consolidating concrete).

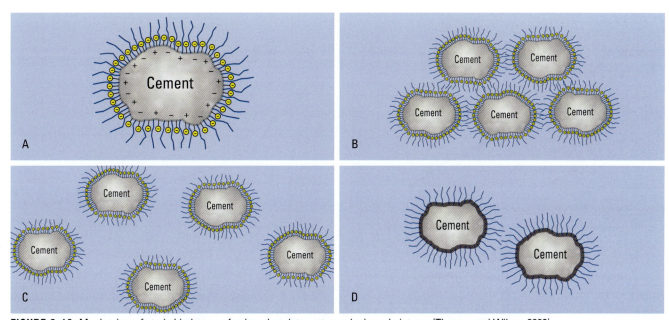

FIGURE 6-10. Mechanism of steric hindrance of polycarboxylate water-reducing admixtures (Thomas and Wilson 2002).

Polyaryl Ether Technology. A modified version of polycarboxylate technology, based on polyaryl ether, has been developed and is in use globally to produce low-viscosity concrete mixtures that significantly improve pumpability and placeability of concrete and help to minimize mix stickiness. Their action fundamentally differs from high-range water reducers based on naphthalene sulfonates and typical polycarboxylates and they can lower viscosity by about 30 percent compared to normal polycarboxylate-based high-range water reducers.

Polycarboxylate Admixtures for Clay-Bearing Aggregates. Many clay minerals expand in the presence of water and can intercalate and absorb the PEO component of polycarboxylate based admixtures, thereby reducing cement dispersion and water reduction capabilities. Polycarboxylate-based admixtures are available, which are formulated with sacrificial agents that are absorbed and intercalate with clays before the PCE component of the admixture. These admixtures allow for typical water reduction properties at normal dosage rates.

Impact of Water Reducers on Properties of Concrete

Adding a water-reducing admixture to concrete without also reducing the water content can produce a mixture with a higher slump. The rate of slump loss, however, is not reduced and in most cases is increased (Figures 6-11 and 6-12), with the exception of polycarboxylate technology. Rapid slump loss results in reduced workability and less time to place concrete.

High-range water reducers are generally more effective than regular water-reducing admixtures in producing highly workable concrete. The effect of earlier plasticizers in increasing workability or making flowing concrete is short-lived,

typically between 30 to 60 minutes. This period is often followed by a rapid loss in workability or slump loss. High temperatures can also aggravate slump loss. Due to their propensity for slump loss, these early generation admixtures were sometimes added to the concrete mixer at the jobsite. Extended-slump-life plasticizers added at the batch plant help reduce slump-loss problems. Additionally, some mixer trucks are equipped to monitor the slump and add HRWR to the concrete in transit, ensuring the proper slump upon arrival at the jobsite.

An increase in strength is generally obtained with water-reducing admixtures when the water-cement ratio is reduced. For concretes of equal cement content, air content, and slump, the 28-day strength of a water-reduced concrete containing a water reducer can be 10% to 25% greater than concrete without the admixture. Using a water reducer to reduce the cement and water content of a concrete mixture, while maintaining a constant water-cement ratio, can result in equal or reduced compressive strength, and can increase slump loss by a factor of two or more (Whiting and Dziedzic 1992).

The effect of water reducers on bleeding is dependent on the chemical composition of the admixture. A significant reduction in bleeding can result with large reductions of water content; possibly causing finishing difficulties on flat surfaces when rapid drying conditions are present (see Chapter 18). Tests have shown that some plasticized concretes bleed more than control concretes of equal water-cement ratio (Figure 6-13); but plasticized concretes bleed significantly less than control concretes of equally high slump and higher water content.

Despite a reduction in water content, certain water-reducing admixtures may cause increases in drying shrinkage. Usually the effect of the water reducer on drying shrinkage is small

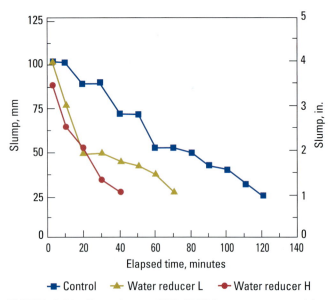

FIGURE 6-11. Slump loss at 23°C (73°F) in concretes containing conventional water reducers (ASTM C494 Type D) compared with a control mixture (Whiting and Dziedzic 1992).

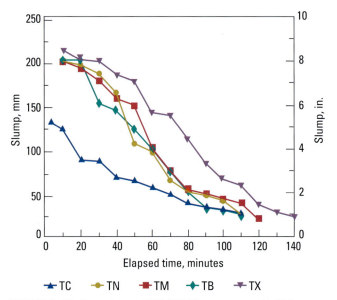

FIGURE 6-12. Slump loss at 32°C (90°F) in concretes with high-range water reducers (TN, TM, TB, and TX) compared with control mixture (TC) (Whiting and Dziedzic 1992).

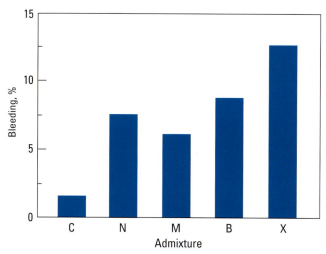

FIGURE 6-13. Bleeding of flowing concretes with plasticizers (N, M, B, and X) compared to control (C) (Whiting and Dziedzic 1992).

when compared to other more significant factors that affect shrinkage cracking in concrete. High-slump, low-water-content, plasticized concrete tends to develop less drying shrinkage than a high-slump, high-water-content conventional concrete. However, high slump plasticized concrete plasticized with napthalene and melamine HRWR has similar or higher drying shrinkage than conventional low-slump, low-water-content concrete (Whiting 1979, Gebler 1982, and Whiting and Dziedzic 1992).

Water reducers can delay time of set and early strength gain to varying degrees of retardation while others do not significantly affect the setting time. ASTM C494, Type A water reducers can have little effect on setting time at their typical dosages, while Type D admixtures provide water reduction with set retardation, and Type E admixtures provide water reduction with accelerated setting. Type D water-reducing admixtures usually retard the

setting time of concrete from one to four hours (Figure 6-14). Some water reducers meet the requirements of more than one category depending on the dosage rate. For example, a Type A water reducer may perform as a Type D water reducing and set retarding admixture as the dosage rate is increased. Setting time may be accelerated or retarded based on each admixture's chemistry, dosage rate, and interaction with other admixtures and cementing materials in the concrete mixture.

Some water-reducing admixtures may also entrain some air in concrete. Lignin-based admixtures can increase air contents by 1% to 2%. Polycarboxylate-based water reducers are normally formulated with defoamer(s) to minimize entrained air produced by the polymer. Concretes with water reducers generally have good air retention. Concretes with high-range water reducers can have larger entrained air voids and higher void-spacing factors than normal air-entrained concrete. Air loss can also be significant when compared to concretes without high range water reducers held at constant water-cement ratios (reduced cement and water contents) (Table 6-3). Some research has indicated poor frost- and deicer-scaling resistance for some flowing concretes when exposed to a continuously moist environment without the benefit of a drying period (Whiting and Dziedzic 1992). However, laboratory tests have shown that concretes with a moderate slump using high-range water reducers have good freeze-thaw durability, even with slightly higher void-spacing factors. This may be the result of the lower w/cm ratios often associated with these concretes.

The effectiveness of water reducers on concrete is a function of their chemical composition, concrete temperature, cement composition and fineness, cement content, and the presence of other admixtures. See Whiting and Dziedzic (1992) for more information on the effects of water reducers on concrete properties.

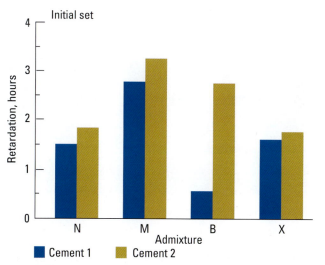

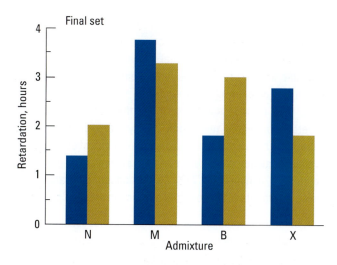

FIGURE 6-14. Retardation of set in cement-reduced mixtures relative to control mixture. Concretes N, M, B, and X contain high-range water reducer (Whiting and Dziedzic 1992).

TABLE 6-3. Loss of Air from Cement Reduced Concrete Mixtures

MIXTURE		INITIAL AIR CONTENT, %*	FINAL AIR CONTENT, %†	AIR RETAINED, %	RATE OF AIR LOSS, % / minute
Control	C	5.4	3.0	56	0.020
Water reducer	L	7.0	4.7	67	0.038
	H	6.2	4.6	74	0.040
High-range water reducer	N	6.8	4.8	71	0.040
	M	6.4	3.8	59	0.065
	B	6.8	5.6	82	0.048
	X	6.6	5.0	76	0.027

* Represents air content measured after addition of admixture.

† Represents air content taken at point where slump falls below 25 mm (1 in.) (Whiting and Dziedzic 1992).

ACCELERATING ADMIXTURES

An accelerating admixture is used to accelerate the rate of hydration (setting) and strength development of concrete at an early age. The strength development of concrete can also be accelerated by other methods: using Type III or Type HE high-early-strength cement, adding additional cement to the concrete, using a water reducer, or curing at higher temperatures. Accelerators are designated as Type C admixtures under ASTM C494.

Composition of Accelerating Admixtures

Calcium chloride ($CaCl_2$) is the most common material used as an accelerating admixture, especially for non-reinforced concrete. It should conform to the requirements of ASTM D98 (AASHTO M 144), *Standard Specification for Calcium Chloride*.

Calcium chloride is highly effective in this role; however, it is limited to a dosage of 2% or less (by mass of cementitious materials) and only for use in non-reinforced concrete. Specifications restrict its use in concrete; chlorides accelerate corrosion of the reinforcing steel. Calcium chloride is generally available in three forms – as flake, pellets, or in solution. The amount of calcium chloride added to concrete should be no more than is necessary to produce the desired results and never be permitted to exceed 2% by mass of cementitious material. When calculating the chloride content of commercially available calcium chloride, it can be assumed that regular flake contains a minimum of 77% $CaCl_2$; and concentrated flake, pellet, or granular forms contain a minimum of 94% $CaCl_2$.

Non-chloride containing accelerators are available for use with reinforced concrete (Rear and Chin 1990, and Jeknavorian and others 1994). These include organic compounds such as triethanolamine (TEA) and inorganic salts such as sodium and calcium salts of formate, nitrate, nitrite, thiocyanate, and lactate (see Table 6-1). These inorganic salts tend to be less effective than calcium chloride at accelerating setting times and are typically used at higher dosages. Triethanolamine is not used on its own as an accelerating admixture, but is often used in other chemical admixture formulations such as normal and mid-range water reducers. TEA offsets and compensates for possible set-retarding effects from other water reducing additives. Certain non-chloride set accelerators are specially formulated for use in cold weather applications with ambient temperatures down to -7°C (20°F). See Brook and others (1993), Jeknavorian and others (1994), and Korhonen and Jeknavorian (2005).

Mechanism of Accelerating Admixtures

The mechanisms responsible for acceleration are not well understood. Calcium chloride may play a role with deflocculation of C-S-H causing a different hydration rate (Juenger and others 2005) as illustrated in Figure 6-15.

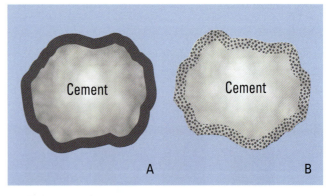

Unhydrated C_3S or cement grain	
Unflocculated C-S-H layer	
Flocculated C-S-H layer	

FIGURE 6-15. Mechanism of calcium chloride accelerating admixtures (after Juenger 2005).

In contrast to those inorganic salts, TEA acts on the C_3A component of the cement accelerating its reaction with gypsum and the production of ettringite and also promoting the subsequent conversion to monosulfate (see Chapter 2). At high dosage levels, TEA may cause flash set to occur. It can also retard, even permanently, the hydration of C_3S leading to reduced long-term strength. Calcium formate, another organic compound, works by accelerating the hydration of C_3S.

Effects of Accelerators on Concrete Properties

Set accelerators cause a reduction in the time to both the initial and final set, the effect generally increasing with admixture dosage. There will also typically be an increase in the early-age strength development dependent on the admixture dosage, composition, time of addition, and the chemical and physical properties of the cement. The increase in strength due to the addition of calcium chloride is particularly noticeable at low temperatures and early ages.

The widespread use of calcium chloride as an accelerating admixture has provided much data and experience on the effect of this chemical on the properties of concrete. Besides accelerating strength gain, calcium chloride causes an increase in potential reinforcement corrosion and may lead to rapid stiffening and discoloration (a darkening of concrete). The incorporation of calcium chloride can also affect other properties of concrete including shrinkage, long-term strength development, and resistance to freezing and thawing, sulfates, and alkali-silica reaction.

Applications where calcium chloride should be used with caution:

- concrete reinforced with steel;

- concrete subjected to steam curing;

- concrete containing embedded dissimilar metals, especially if in direct physical contact with steel reinforcement;

- concrete slabs supported on permanent galvanized-steel forms; and

- colored concrete.

Calcium chloride or admixtures containing soluble chlorides should not be used in the following:

- construction of parking garages;

- prestressed concrete because of possible steel corrosion hazards;

- concrete containing embedded aluminum (for example, conduit);

- concrete containing aggregates that, under standard test conditions, have potential to be deleteriously reactive;

- concrete exposed to sulfate from soil or water;

- floor slabs intended to receive dry-shake metallic finishes;

- hot weather concreting; and

- massive concrete placements.

The maximum chloride-ion content for corrosion protection of prestressed and reinforced concrete as recommended by the ACI 318-19 building code is provided in Chapter 11. Gaynor (1998) demonstrates how to calculate the chloride content of fresh concrete and compares it with recommended limits.

Calcium chloride is not an antifreeze agent. When used in allowable amounts, it will not reduce the freezing point of concrete by more than a few degrees. Instead, proven reliable precautions should be taken during cold weather (see Chapter 19).

Calcium chloride should be added to the concrete mixture in solution form as part of the mixing water. If added to the concrete in dry flake form, all of the dry particles may not completely dissolve during mixing. Undissolved lumps in the mixture can cause dark spots and popouts in the hardened concrete.

Strength-Enhancing Admixture

An emerging technology for accelerating early strength gain of concrete is based on nanoparticles which can act as seed crystals to enable rapid growth of calcium silicate hydrate (C-S-H) (Halford 2011). This process, known as "crystal seeding" significantly speeds up the hardening of concrete during the early stages of cement hydration. Reportedly, concrete treated with this nanoparticle crystal seeding admixture hardens and develops strength at 20°C (68°F) as rapidly as untreated concrete cured at 60°C (140°F), making it potentially useful in precast concrete production (BFT International 2010). The nanoparticle crystal seeding admixture is approved for use in concrete in Europe in lieu of conventional accelerating admixtures in accordance with DIN 1045-2 (2008). C-S-H nanoparticle-based strength-enhancing admixtures meeting the requirements of ASTM C494 Type S are also available in North America and are being used predominantly in precast concrete as well as some cast-in-place concrete applications.

SET-RETARDING ADMIXTURES

Set retarding admixtures are used to delay the rate of setting of concrete. Retarders may be used to: offset the accelerating effect of hot weather on the setting of concrete (see Chapter 18); delay the initial set of concrete when difficult or unusual conditions of placement occur, such as placing concrete in large piers and foundations, cementing oil wells, or pumping grout or concrete over considerable distances; delay the set for special finishing techniques, such as an exposed aggregate surface (Figure 6-16); or anticipating long transport time or delays between batching and placement.

FIGURE 6-16. Exposed aggregate finishes are commonly produced by applying a set retarder to the surface immediately after finishing and then removing the paste by brushing with a fine water spray to remove the surface paste and expose gap-graded aggregates.

A set retarder extends the period during which concrete remains plastic. This allows a large placement to be completed before setting occurs, which helps eliminate cold joints in large or complex pours and extends the time allowed for finishing and joint preparation. The reduced hydration rate is also helpful in reducing early temperature rises, which can induce internal stresses and cracking in concrete.

Composition of Set-Retarding Admixtures

The classifications and components of set retarders are listed in Table 6-1. Compounds used as set retarders fall into four general categories: lignosulfonates, hydroxycarboxylic acids, polysaccharides (corn syrup), organophosphates, organophosphonates, sugars and their derivatives, and selected inorganic salts.

Mechanism of Set Retarders

Set retarders function by slowing the normal cement hydration through complex processes. Theories on retarding mechanisms are based on adsorption, precipitation, and nucleation (Ramachandran 1995). There are two main processes involved: a blocking mechanism where the admixture adsorbs on the cement surface, slowing the formation of silicate hydrates (Hansen 1952); or chelation of calcium ions in solution, preventing the precipitation of calcium hydroxide (portlandite) (Suzuki and others 1969). One or both of these mechanisms may be involved, depending on the admixture type selected (Young 1976). The latter is probably the more common for most retarder types and prevents setting but not workability loss. Some retarding admixtures accelerate the initial formation of ettringite while others may delay it.

Effects of Set Retarders on Concrete Properties

The use of a set retarding admixture delays both the initial and final set of concrete. The extent of the delay is dependent on admixture composition, dosage, time of addition, and the temperature of the concrete. The effectiveness of a set retarder can also depend on when it is added to the concrete mixture. Set times will usually be longer when the set retarders

are added to the concrete after the cement and water have mixed (delayed addition).

Retarders do not decrease the initial temperature of concrete. When the retarding effects of the admixture wear off, the temperature rise due to heat of hydration will take effect. This is important to keep in mind when using a retarder to help delay setting time in hot weather concreting – thermal control measures will still need to be taken (see Chapter 18).

The amount of water reduction for an ASTM C494 Type B retarding admixture is normally less than that obtained with a Type A water reducer. Type D admixtures provide both water reduction and retardation.

In general, some reduction in strength at early ages (one day strength) accompanies the use of retarders. However, increased long-term strength may result from retarding the initial rate of hydration. Excessive addition rates of a retarding admixture may permanently inhibit the hydration of the cement.

The effects of these materials on the other properties of concrete, such as shrinkage, may be unpredictable. The incorporation of retarders can affect some of the other properties of concrete including workability, bleeding, and early-age strength development. Therefore, acceptance tests of retarders should be made with actual project materials under the anticipated job conditions.

EXTENDED-SET CONTROL ADMIXTURES

Extended-set control admixtures (also known as hydration controlling or hydration stabillizing admixtures) became available in the late 1980s and are essentially very potent retarders that can suspend the hydration of cementitious materials for controlled extended periods. The components of extended-set control admixtures most commonly include sugars, carboxylic acid salts, and phosphorus-containing organic acid salts. They are used at low dosages to provide normal degrees of retardation in lieu of conventional retarders or at higher dosages to maintain concrete in a stabilized non-hardened state during long hauls or when placements times are extended, such as in drilled shafts. Hydration-control admixtures were first developed to facilitate the reuse of concrete returned in a ready-mix operation by keeping the concrete plastic in the truck for reuse either the same day, overnight, or over a weekend, by suspending setting as the first part of a two-part chemical system (Kinney 1989). When used for the purposes of reusing returned concrete in accordance with ASTM C1798, *Specification for Returned Fresh Concrete for Use in a New Batch of Ready-Mixed Concrete*, an activator that reestablishes normal hydration and setting once added to the stabilized concrete may also be used as the second part of the two-part system. Extended-set control admixtures are classified as Type B (retarding) or Type D (water-reducing, retarding) admixtures under ASTM C494.

WORKABILITY-RETAINING ADMIXTURES

Workability retaining admixtures provide varying degrees of workability retention without significantly affecting the initial set of concrete or early-age strength development, as can be the case with retarding admixtures. These admixtures, normally based on a derivitized acrylate polymer with vinyl ether, can be used with mid-range or high-range water reducers to provide desired levels of workability retention in moderate to high slump concrete mixtures, including self-consolidating concrete (SCC) (Daczko 2008). Their main benefit is reducing the need for slump adjustments prior to concrete placement, thus helping to maintain consistency in concrete performance throughout a project. Workability-retaining admixtures should meet the requirements of ASTM C494 Type A or S. Workability-retaining admixtures can be used in combination with extended-set control admixtures if necessary.

VISCOSITY-MODIFYING ADMIXTURES

Another group of chemical admixtures important to improvements in rheology in concrete mixtures such as with self-consolidating concrete production is viscosity modifying admixtures (VMA) (Jeknavorian 2016). These are also commonly used as anti-washout admixtures for concrete placed underwater. Anti-washout admixtures increase the cohesiveness of concrete to a level that allows limited exposure to water with little loss of cement paste. This cohesiveness allows placement of concrete in water and under water without the use of tremies. These admixtures increase the viscosity of water resulting in a mixture with increased thixotropy and resistance to segregation. Viscosity modifying admixtures usually consist of water-soluble cellulose ethers, acrylic polymers, or high molecular weight biogums. Viscosity-modifying admixtures should meet the requirements of ASTM C494 Type S (see Chapter 9 for a summary of rheological properties of concrete.)

The two basic types of VMAs include thickening type and binding type VMAs. The thickening type increases the viscosity through molecular obstruction. The technology is based on an addition of a large polymer molecule into the paste. Thickened paste then translates to increased cohesion of the mortar system and the concrete as a whole. The binding type of VMA is much more effective than the thickening type (Bury and Buehler 2002). A binding type VMA chemically combines with water molecules, as opposed to just obstructing them. Binding VMAs are typically inorganic materials that produce a gel. The gel promotes thixotropic behavior that inhibits changes in viscosity from shearing action inherent in concrete transportation, handling, and placement. In both types, the increase in viscosity may also be accompanied by an increase in yield stress. Viscosity modifying admixtures dampen the changes in viscosity potentially caused by material and process variations. However, VMAs are not a substitute for good concrete quality control (EFNARC 2006).

Viscosity modifying admixtures can be used to replace fines or to supplement them. Trial batch evaluation, using the recommended dosage from the manufacturer as a starting point, is the best method to determine the appropriate use for each mixture. An increase in high-range water reducer dosage may be necessary when a VMA is used to counteract an increase in yield stress.

CORROSION INHIBITING ADMIXTURES

Corrosion inhibitors are chemical admixtures added to concrete to limit the corrosion of steel reinforcement. Commercially available corrosion inhibitors include: calcium nitrite, dimethyl ethanolamine, phosphates, and ester amines, as listed in Table 6-1. Corrosion inhibitors should conform to ASTM C1582, *Standard Specification for Admixtures to Inhibit Chloride-Induced Corrosion of Reinforcing Steel in Concrete.*

Corrosion inhibitors are used in concrete for parking structures, marine structures, and bridges where chloride salts are present. Chlorides can cause corrosion of steel reinforcement in concrete (Figure 6-17). Corrosion-inhibiting admixtures chemically arrest the corrosion reaction and reinstate the passive layer, which provides protection to the steel. Note that adequate cover depth and low permeability concrete to minimize chloride migration through the concrete is still necessary for protection (see Chapter 11).

FIGURE 6-17. The damage to this concrete parking structure resulted from chloride-induced corrosion of steel reinforcement.

When calcium nitrite is used as an admixture, nitrite anions are in solution with hydroxyl and chloride ions. The nitrite ions cause the ferric oxide passivation layer around the steel reinforcement to become more stable. This ferric oxide film is created by the high pH environment in concrete. In effect, the chloride ions are prevented from penetrating the passive film and making contact with the steel. A certain amount of nitrite can stop corrosion up to some level of chloride ion. Therefore, increased chloride levels require increased levels of nitrite to stop corrosion.

Organic inhibitors, based on a combination of amines and esters in a water medium, act in two ways. First, the esters provide some water repellency (see section on PRAN-type permeability reducing admixtures), thereby restricting the ingress of water-soluble chlorides. Second, the amines adsorb onto the steel surface forming a tight film which repels moisture and acts as a barrier to chemical attack. The result of this dual action is an increase in the time to corrosion initiation and a decrease in the rate of corrosion once it has started.

Cathodic inhibitors react with the steel surface to interfere with the reduction of oxygen. The reduction of oxygen is the principal cathodic reaction in alkaline environments (Berke and Weil 1994).

SHRINKAGE AND CRACK REDUCING ADMIXTURES

Because of their effectiveness in reducing drying shrinkage, SRAs and CRAs are also generally effective in reducing curling and cracking in slabs. They are also used in combination with synthetic macrofibers to reduce joints in slabs-on-ground. Both SRAs and CRAs should meet the requirements of ASTM C494 Type S. Shrinkage-reducing admixtures (SRAs) and crack-reducing admixtures (CRAs), are used in bridge decks, critical floor slabs, and buildings where cracks, curling, and warping must be minimized for durability or aesthetic reasons (Figure 6-18). As concrete dries, water is removed from the capillary pores and a meniscus is formed at the air-water interface due to surface tension. Surface tension forces also act on the solid phases and tend to draw the walls of the pore together. As the water meniscus recedes into smaller and smaller pores the surface tension forces increase, causing the concrete to shrink more.

FIGURE 6-18. Shrinkage cracks, such as shown on this bridge deck, can be reduced with the use of good concreting practices and shrinkage reducing admixtures.

Shrinkage-Reducing Admixtures

Shrinkage reducing admixtures reduce the surface tension of the liquid phase, which reduces the forces exerted on the pore walls thereby producing less drying shrinkage.

Propylene glycol and polyoxyalkylene alkyl ether have been used as shrinkage reducers. Drying shrinkage reductions between 25% and 50% have been demonstrated. These admixtures have negligible effects on workability, but can impact air content and may possibly require an increase in the dose of air-entraining admixture necessary to achieve a target air content. A delay in time of set and slower bleed rate may also result from the use of SRAs. They are generally compatible with other admixtures (Nmai and others 1998, and Shah and others 1998). The manufacturer's recommendations should be followed, particularly when used in air-entrained concrete. Due to their potential effects on bleeding and setting time, caution and proper planning are required when SRAs are used in slabs that receive a hard-troweled finish. Premature finishing of a SRA-treated concrete slab can trap bleedwater and, subsequently, lead to delamination of the concrete surface.

Crack-Reducing Admixtures

Crack-reducing admixtures (CRAs) are based on a specialty alcohol alkoxylate. CRAs perform the same as SRAs in reducing concrete drying shrinkage, but in the event of cracking, provide a smaller crack width. Compared with conventional SRAs, the CRA has been shown to provide internal stress relief in the ASTM C1581, *Standard Test Method for Determining Age at Cracking and Induced Tensile Stress Characteristics of Mortar and Concrete under Restrained Shrinkage*, ring test and, as a result, to change the mode of failure in the ring test from a sudden release of all the compressive strain in the inner ring to a gradual release of the compressive strain. Consequently, the CRA reportedly provides a greater delay in the time-to-cracking in the ring test and an initial crack width of about 0.1 mm (0.004 in.) compared to 1 mm (0.04 in.) in untreated concrete and SRA-treated concrete specimens (Nmai and others 2014).

PERMEABILITY-REDUCING ADMIXTURES

A class of materials referred to as permeability-reducing admixtures (PRAs) have been developed to control water and moisture movement (Roy and Northwood 1999), and reduce chloride ion ingress (Aldred 1988, and Munn and others 2003).

Water can enter concrete through two primary mechanisms; capillary absorption under non-hydrostatic conditions (often referred to as wicking) and the direct ingress of water under pressure. The term permeability refers only to concrete exposed to water under pressure. However, permeability is often used informally to describe any passage of water through concrete, whether by pressure driven ingress or by wicking.

Considering these two mechanisms of water ingress, permeability reducing admixtures (PRAs) can be divided into two categories; Permeability Reducing Admixture – Non-Hydrostatic (PRAN) and Permeability Reducing Admixtures – Hydrostatic (PRAH) (ACI 212.3R 2016).

Non-Hydrostatic (PRAN)

Permeability reducing admixtures that are non-hydrostatic (PRAN) have traditionally been referred to as "damp-proofers." Most PRANs are hydrophobic in nature. PRANs provide concrete water repellency and reduced absorption (wicking). Common materials include soaps such as stearates and other long chain fatty acids, or their derivatives, as well as petroleum products. These PRANs are sometimes used to reduce the transmission of moisture through concrete in contact with water or damp soil. However, hydrophobic admixtures are usually not effective when the concrete is in contact with water under pressure. Some PRANs contain finely divided solids such as bentonite or siliceous powders that restrict water absorption. Often referred to as "densifiers," finely divided solids may reduce permeability slightly, although the effect is relatively small. In practice, fine solid fillers similar to hydrophobic admixtures are usually used for non-hydrostatic applications.

Hydrostatic (PRAH)

Hydrostatic permeability reducing admixtures (PRAH) have often been referred to as "waterproofers," although permeability reducing admixture for hydrostatic conditions is a more technically correct term. PRAHs contain materials that act to block the pores and capillaries in concrete. These materials have been shown to be effective in reducing permeability to water under pressure. They have also been shown to reduce concrete corrosion in chemically aggressive environments. Products usually consist of hydrophilic crystalline materials that react in concrete to produce pore blocking deposits, or polymeric materials that coalesce in the concrete's pores. Reactive PRAHs have also been shown to increase the autogenous sealing of hairline cracks.

Additional Considerations

Permeability reducing admixtures (PRA) will not correct a poorly designed concrete mixture. Proper proportioning, placement, and curing are needed for effective performance from a PRA. Concrete joints should be treated with a suitable waterstop. Also, the selection of a PRA must take into account the expected service conditions as well as the features of the admixture (PRAN or PRAH). PRANs are usually evaluated using an absorption based test method such as ASTM C1585, *Standard Test Method for Measurement of Rate of Absorption of Water by Hydraulic-Cement Concretes*. PRAHs are best evaluated using a pressure driven penetration test such as the U.S. Army Corp of Engineers CRC C48-92, *Standard Test Method for Water Permeability of Concrete*, or the European standard BS EN 12390-8, *Testing Hardened Concrete – Depth of Penetration of Water under Pressure*. PRAs are discussed in detail in ACI 212.3R (2016).

ALKALI-AGGREGATE REACTIVITY INHIBITING ADMIXTURES

The use of lithium nitrate, lithium carbonate, lithium hydroxide, lithium aluminum silicate (decrepitated spodumene), and barium salts have shown reductions of alkali-silica reaction (ASR) in laboratory tests (Figure 6-19) (Thomas and Stokes 1999). Some of these materials may have potential for use as an additive to cement (Gajda 1996). AASHTO R 80, *Standard Practice for Determining the Reactivity of Concrete Aggregates and Selecting Appropriate Measures for Preventing Deleterious Expansion in New Concrete Construction*; and CSA A23.2-28a, *Standard Practice for Laboratory Testing to Demonstrate the Effectiveness of Supplementary Cementing Materials and Lithium-Based Admixtures to Prevent Alkali-Silica Reaction in Concrete* (2014), provide guidance and a test procedure for determining an appropriate dosage for lithium nitrate admixtures.

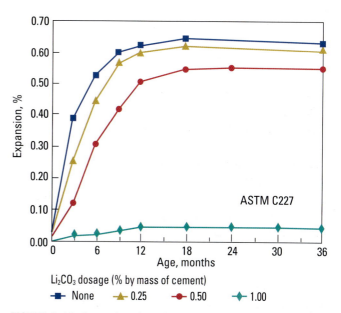

Li_2CO_3 dosage (% by mass of cement)
- None
- 0.25
- 0.50
- 1.00

FIGURE 6-19. Expansion of specimens made with lithium carbonate admixture (Stark 1992).

As discussed in Chapter 11, ASR gel has a great capacity to absorb moisture from within the concrete pores. This can cause a volumetric expansion of the gel, which in turn leads to the buildup of internal stresses and the eventual disruption of the cement paste surrounding the aggregate particle. If lithium nitrate is used as an admixture, the pore solution will contain lithium and nitrate ions in addition to sodium, potassium, and hydroxyl ions. Likewise, the reaction product (ASR gel) that forms will also contain appreciable quantities of lithium. The reduction in ASR-related expansion by lithium salts appears to result from an exchange of the lithium ion with sodium and potassium. The resulting lithium-bearing reaction product does not have the same propensity to absorb water and expand.

Where possible and for economy, lithium-nitrate based admixtures should be used in combination with mitigating supplementary cementitious materials, such as Class F fly ash, slag cement, or other pozzolans, to reduce the dosage of lithium admixture required.

Alkali-aggregate reactivity inhibitors should meet the requirements of ASTM C494 Type S.

COLORING ADMIXTURES (PIGMENTS)

Natural and synthetic materials are used to color concrete for aesthetic and safety reasons (Figure 6-20) and are available in either powder or liquid form. The general materials used for coloring concrete are given in Table 6-4. Pigments used in amounts less than 6% by mass generally do not affect concrete properties. Red, brown, and tan colors work especially well in concrete mixtures. Chromium oxide (green) and cobalt oxide (blue) usually cost significantly more than iron oxides. Untreated carbon black and lampblack should not be used because they are unstable, fade, or leach out of concrete, and reduce air content. Most carbon black used to color concrete contains an additive to offset this effect on air.

FIGURE 6-20. Pigments may be used to color concrete in a variety of shades (courtesy of Davis Colors).

Generally, the amount of pigments used in concrete should not exceed 10% by weight of the cement.

Before a coloring admixture is used on a project, it should be tested for color fastness in direct sunlight and autoclaving, chemical stability in cement, and effects on concrete properties. To avoid color distortions, calcium chloride should not be used in concrete containing pigments. Pigments should conform to ASTM C979, *Standard Specification for Pigments for Integrally Colored Concrete.*

TABLE 6-4. Guide to Mineral Pigments for Colored Concrete Finishes*

COLOR	PIGMENT
Black	Iron oxide Carbon black (indoors)
Blue	Cobalt blue Phthalocyanine blue (indoors)
Brown	Burnt umber Brown oxide of iron
Buff	Yellow ocher Yellow oxide of iron
Cream	Yellow oxide of iron
Gray	Normal portland cement
Green	Chromium oxide
Pink	Red oxide of iron
Red	Brown oxide of iron Red oxide of iron
White	Titanium dioxide with white portland cement, white sand
Yellow	Yellow oxide of iron

*Adapted from Kosmatka and Collins (2004), and ACI 212.3R (2016)

PUMPING AIDS

Pumping aids are added to concrete mixtures to improve pumpability. A partial list of materials used in pumping aids is given in Table 6-1. Some admixtures that serve other primary purposes but also improve pumpability are: air-entraining admixtures and some water-reducing and retarding admixtures, and viscosity-modifying admixtures.

Some pumping aids may increase water demand, reduce compressive strength, cause air entrainment, or retard setting time. These side effects can be corrected by adjusting the mixture proportions or by adding another admixture to offset the side effect.

Pumping aids are not a cure-all; they are best used to make marginally pumpable (harsh) concrete more pumpable. These admixtures increase viscosity or cohesion in concrete to reduce friction in the lubricating layer inside the pump line (and minimize dewatering of the paste) during pumping under pressure.

BONDING ADMIXTURES AND BONDING AGENTS

Bonding admixtures are usually water emulsions of organic materials including rubber, polyvinyl chloride, polyvinyl acetate, acrylics, styrene butadiene copolymers, and other polymers. They are added to portland cement mixtures to increase the bond strength between old and new concrete. Flexural strength and resistance to chloride-ion ingress can also be improved. They are added in proportions of about 5% to 20% by mass of the cementing materials, depending on job conditions and type of admixture used. Some bonding admixtures may increase the air content of mixtures. Non-reemulsifiable types are resistant to water and are better suited to exterior applications and in applications where moisture is present.

A bonding admixture is only as good as the surface to which the concrete is applied. The surface must be dry, clean, sound, free of dirt, dust, paint, and grease, and at the proper temperature. The International Concrete Repair Institute (ICRI) Guideline 310.2R, *Selecting and Specifying Concrete Surface Preparation for Sealers, Coatings, Polymer Overlays, and Concrete Repair* (2013) provides information needed to select and specify the methods for preparing concrete surfaces prior to the application of a protective system or repair material. Organic or polymer-modified concretes are acceptable for patching and thin-bonded overlays.

Bonding agents should not be confused with bonding admixtures. Admixtures are an ingredient in the concrete; bonding agents are applied to existing concrete surfaces before new concrete is placed. Timing of application is very important for these products to work properly. Bonding agents help "glue" the existing and the new materials together. Bonding agents are often used in restoration and repair work; they consist of portland cement or latex-modified portland cement grout or polymers such as epoxy resins (ASTM C881, *Standard Specification for Epoxy-Resin-Base Bonding Systems for Concrete*, or AASHTO M 235, *Standard Specification for Epoxy Resin Adhesives*) or latex (ASTM C1059, *Standard Specification for Latex Agents for Bonding Fresh to Hardened Concrete*).

GAS-FORMING ADMIXTURES

Aluminum powder and other gas-forming materials (zinc or magnesium, hydrogen peroxide, nitrogen, ammonium compounds, and certain forms of activated carbon or fluidized coke) are sometimes added to concrete and grout in very small quantities. These materials cause a slight expansion of the mixture prior to hardening. This may be of benefit where the complete grouting of a confined space is essential, such as under machine bases or in post-tensioning ducts of prestressed concrete. These materials are also used in larger quantities to produce autoclaved cellular concretes. The amount of expansion that occurs is dependent upon the amount of gas-forming material used, the temperature of the fresh mixture, the alkali loading and other variables. Where the amount of expansion is critical, careful testing through trial batching and jobsite control of mixtures and temperatures must be exercised. Gas-forming agents will not cause further expansion once hardening caused by drying or carbonation occurs.

AIR-DETRAINERS

Air-detraining admixtures, also known as defoamers, reduce the air content in concrete. They are used when the air content cannot be reduced by adjusting the mixture proportions or by changing the dosage of the air-entraining admixture and other admixtures. Air-detrainers do not typically remove all entrained air, their effectiveness and dosage rate should be established on trial mixtures. Materials used in air-detraining agents are listed in Table 6-1. The impact of air-detrainers on the quality of the air-void system of intentionally air-entrained has yet to be determined.

FUNGICIDAL, GERMICIDAL, AND INSECTICIDAL ADMIXTURES

Bacteria and fungal growth on concrete surfaces or in hardened concrete may be partially controlled through the use of fungicidal, germicidal, and insecticidal admixtures. The most effective materials are polyhalogenated phenols, dieldrin emulsions, and copper compounds. The effectiveness of these materials is generally temporary. In high dosages they may reduce the compressive strength of concrete.

SELF-SEALING ADMIXTURES

Admixtures can be used to provide autogenous sealing of hairline cracks in concrete. Some researchers have laced the concrete with spores filled with bacteria that secrete calcium carbonate to fill the cracks and pores (Chahal and Siddique 2013 and De Belie and others 2012), while others embedded glass capillaries with a healing agent (Achal and others 2013). When microscopic stress cracks begin to form in the concrete, the embedded capsules rupture and release the sealing agent into the adjacent areas. The sodium silicate reacts with the calcium hydroxide naturally present in the concrete to form calcium-silicate-hydrate that seals the cracks and blocks the pores in the concrete.

COMPATIBILITY OF ADMIXTURES AND CEMENTITIOUS MATERIALS

Fresh concrete problems of varying degrees of severity are encountered as a result of cement-admixture incompatibility and incompatibility between admixtures (Dodson and Hayden 1989, Cheung and others 2011, Hanehara and Yamada 1999, Sandberg and Roberts 2005, Roberts and Taylor 2007, Aïtcin and Mindess 2015, and Cost and Knight 2007). Incompatibility between supplementary cementing materials and admixtures or cements can also occur. Slump loss, air loss, early stiffening, and other factors affecting fresh concrete properties can result from material incompatibilities as well as the sequence of material addition. While these problems primarily affect the plastic-state performance of concrete, long-term hardened concrete performance may also be adversely affected. For example, early stiffening can cause difficulties with consolidation of concrete which may also compromise strength.

When incompatibility is encountered, it can often be solved by changing the admixture dosage rate or the sequence of addition to the mixture. However, some incompatibility issues may be solved by modifying the composition of the cement, particularly the C_3A content, alkali or sulfate content, or by modifying the composition of the admixture.

Taylor and others (2006) have developed protocols for testing the compatibility of various combinations of materials. This preconstruction testing can reduce the likelihood of performance problems in the field. However, adequately addressing all incompatibility issues due to variations in materials, mixing equipment, mixing time, and environmental factors may not be feasible. When incompatibility is discovered in the field, a common solution is to simply change admixtures or cementing materials. ASTM C1810, *Standard Guide for Comparing Performance of Concrete-Making Materials Using Mortar Mixtures* provides procedures that can be suitable for comparing the relative performance of combinations of concrete-making materials such as fine aggregate, chemical admixtures, supplementary cementitious materials (SCMs), water, and hydraulic cement. ASTM C1810 can be useful to identify unexpected performances due to combination of various materials. The relative trends in performance observed with the mortar method may suggest performance in concrete mixtures batched with the same materials and relative mixture proportions. For more information on incompatibility of cement and chemical admixtures refer to Taylor and others (2008), Helmuth and others (1995), Tagnit-Hamou and Aïtcin (1993), and Tang and Bhattacharja (1997).

Less-Than Expected Water Reduction (Plasticity)

If the water reduction achieved using an admixture is less than expected based on previous experience with the same admixture, this may be caused by: the composition of the cementitious materials, moisture condition of the aggregates, presence of other set-control admixtures, the time of addition, the temperature of the concrete, the presence of clay minerals or fines in the aggregates, and the dose of the admixture itself.

Slump loss. The rate of loss depends on: C_3A content and form (cubic vs orthorhombic), SO_3 content and form (gypsum, plaster, and anhydrite), alkali content, fineness of cementitious materials, and temperature. High-range water reducers are only effective for a limited period before they are overwhelmed by the build-up of hydration products (particularly ettringite) during the very early stages of hydration. Polycarboxylate-based high range water reducers generally have better slump retention compared to naphthaene and melamine sulfonate-based products (Jeknavorian 1997).

Slump loss can often be offset by delaying the time of addition of the admixture. For example, a high-range water reducer may be added on site rather than during batching. Concrete batched from a remote location, such as in municipal paving projects that employ a stationary ready-mix source is a prime example.

Incompatibility between some high-range water-reducers and cementing materials can result in very rapid losses in workability, shortly after mixing. While this can often be attributed to the temperature of the concrete, the reactivity of the cement and the continuous availability of admixture to disperse the hydrating cement grains is a key factor.

Certain minerals found in various aggregate sources, such as expansive clays, have been found to rapidly adsorb polycarboxylate-type superplasticizers, thus significantly reducing their effectiveness (Jeknavorian and others 2003).

Cement admixture compatibility, with regards to slump loss, can be examined in the laboratory by testing the flow properties of pastes. A suitable test is the mini-slump test. This test gives an indication of how long the plasticizing action of a high-range water reducer can be maintained (Tang and Bhattacharja 1997).

Another test used for this purpose is the Marsh cone method (ASTM D6910, *Standard Test Method for Marsh Funnel Viscosity of Construction Slurries,* and API 13B-1). This test can also be used to gauge the saturation point for a particular cement/admixture combination. The saturation point is the level at which further admixture addition will no longer produce any benefit.

ASTM C1679, *Standard Practice for Measuring Hydration Kinetics of Hydraulic Cementitious Mixtures Using Isothermal Calorimetry,* may include other useful approaches to evaluate the compatibility of cementitious mixtures containing chemical admixtures. Changes in the thermal power curve obtained from this practice may indicate changes in a material property.

Less-Than Expected Retardation

If the length of retardation is less than expected, this may be due to an increase in the C_3A content of the cement. Excessively retarded set may be caused by: a low C_3A content or low cement reactivity, excessive dosage of admixture with retarding properties, high levels of SCMs, or low temperature. Testing an admixture over a range of addition rates can often identify a critical dosage above or below which unacceptable set performance can result. Problems on site can be avoided by testing in accordance with ASTM C1679 to determine trial batch needs and then trial batching with job materials in environments (particularly temperatures) that are as close as possible to actual field conditions.

The performance of concrete produced with cements of low C_3A and SO_3 content should be carefully observed when water-reducing or set retarding admixtures are used, as the sulfate balance can be impacted by these admixtures.

Significant changes in the alkali content of the cement should alert the concrete producer to potential changes in admixture performance.

BATCHING AND SEQUENCING CHEMICAL ADMIXTURES

Admixtures can be batched by mass or by volumetric dispensers. The total quantity of chemical admixture is dosed based on the cementitious materials content of the concrete mixture. Liquid chemical admixtures are usually dispensed individually by volumetric means. Volumetric dispensers typically require a calibration tube or other means to visually verify the accuracy of dosage. Other more sophisticated volumetric dispensers include positive displacement flow meters. Both liquid and powdered admixtures can be measured by mass, however, powdered admixtures should not be measured by volume. Because of the low dosage rates typically used for admixtures in most concrete mixtures, it is often necessary to dilute admixture solutions with water to obtain a sufficient quantity for accuracy when measuring by mass. Any addition of water must be measured and accounted for in the total water content of the concrete mixture. Admixtures may also be added in transit or at the jobsite using tanks with pressurized dispensing systems. For more information on batching chemical admixtures see Chapter 14.

Chemical admixtures are commonly incorporated into the mixture through the water feed directly into the plant or truck mixer. Other admixtures, such as the air entrainment may be discharged onto the sand conveyor belt, or into the sand hopper. Care should be taken to keep certain admixtures separate before they are dispensed into the batch. Some combinations in sequencing may neutralize the effects desired. Consult the admixture manufacturer concerning compatible admixture combinations or perform laboratory tests to document performance.

When no other information or experience is provided, ACI 212.3R, *Report on Chemical Admixtures for Concrete* (2016), gives the following recommendations on order of sequencing of admixtures:

- air entrainers,
- water-reducers,
- set retarders,
- accelerators,
- high-range water reducers,
- permeability reducers,
- shrinkage-reducing admixtures,
- corrosion-inhibiting admixtures, and
- viscosity-modifying admixtures.

If after trial batching, it is necessary to increase slump, air, or both–consider increasing the dosages of high-range water-reducers and air-entraining admixtures.

STORING AND DISPENSING CHEMICAL ADMIXTURES

Liquid admixtures can be stored in barrels or bulk tankers (Figure 6-21). Powdered admixtures can be placed in special storage bins and some are available in premeasured plastic bags. Admixtures added to a truck mixer at the jobsite are often stored in plastic jugs or bags. Some truck mixers are equipped with calibrated equipment and tanks to dispense admixtures into the load in transit. Powdered admixtures, or an admixture drum or barrel may be stored at the project site.

FIGURE 6-21. Liquid admixture tank at a ready-mix plant (courtesy of S. Parkes).

Dispenser tanks at concrete plants should be properly labeled for specific admixtures to avoid contamination and to avoid dosing the wrong admixture. Most liquid chemical admixtures should not be allowed to freeze; therefore, they should be stored in heated environments. Consult the admixture manufacturer for proper storage temperatures. Powdered admixtures are usually less sensitive to temperature restrictions, but may be sensitive to moisture. Care should be taken to assure that all the materials used for storage and the dispenser system are compatible with the admixture.

For more information on chemical admixtures for use in concrete see ACI 212.3R (2016), Thomas and Wilson (2002), Hewlett (1998), Ramachandran (1995), Dodson (1990), Rixom (1999), and Spiratos and others (2006).

REFERENCES

PCA's online catalog includes links to PDF versions of many research reports and other classic publications.
Visit: cement.org/library/catalog.

Abrams, D.A., *Calcium Chloride as an Admixture in Concrete*, Structural Materials Research Laboratory, Lewis Institute, Bulletin No. 13, PCA LS013, Chicago, IL, 1924, 64 pages.

Achal, V.; Mukerjee, A.; and Sudhakara, R., "Biogenic Treatment Improves the Durability and Remediates the Cracks of Concrete Structures," *Construction and Building Materials*, Vol. 48, 2013, pages 1 to 5.

ACI Committee 212, *Report on Chemical Admixtures for Concrete*, ACI 212.3R-16, American Concrete Institute, Farmington Hills, MI, 2016, 80 pages.

ACI Committee 318, *Building Code Requirements for structural Concrete*, ACI 318-19, American Concrete Institute, Farmington Hills, MI, 2019, 624 pages.

Aïtcin, P-C.; and Mindess, S., "Back to the Future," *Concrete International*, May 2015, pages 37 to 40.

Aldred, J., "HPI Concrete," *Concrete International*, November 1988, pages 52 to 57.

Berke, N.S.; and Weil, T.G., "World Wide Review of Corrosion Inhibitors in Concrete," *Advances in Concrete Technology*, CANMET, Ottawa, 1994, pages 891 to 914.

BFT International, "Crystal Seeding to Master the Current Challenges of the Precast Industry," *Concrete Plant + Precast Technology*, Issue 01/2010, Gütersloh.

Brook, J.W.; Factor, D.F.; Kinney, F.D.; and Sarkar, A.K., "Cold Weather Admixture," *Concrete International*, October 1988, pages 44 to 49.

Burns, E.; Rieder, K.A.; Curto, J.; and Tregger, N., "When It Comes to Concrete, Just Go with the Flow, *Concrete Construction*, December 6, 2017.

Bury, M.A.; and Buehler, E., "Methods and Techniques for Placing Self-Consolidating Concrete – An Overview of Field Experiences in North American Applications," *Conference Proceedings: First North American Conference on the Design and Use of Self-Consolidating Concrete*, Advanced Cement-Based Materials Center, Evanston, IL, 2002, pages 281 to 286.

Camposagrado, G., *An Investigation on the Cause and Effect of Air-Void Coalescence in Air-Entrained Concrete Mixes*, SN2624, Portland Cement Association, Skokie, IL, 2006, 136 pages.

Chahal, N.; and Siddique, R., "Permeation Properties of Concrete Made With Fly Ash and Silica Fume: Influence of Ureolytic Bacteria," *Construction and Building Materials*, Vol. 49, 2013, pages 161 to 174.

Cheung, J.; Jeknavorian, A.; Roberts, L.; and Silva, D., "Impact of Admixtures on the Hydration Kinetics of Portland Cement," *Cement and Concrete*, Vol. 41, 2011, pages 1289 to 1309.

Collepardi, M.; and Valente, M., "Recent Developments in Superplasticizers," *Eighth CANMET/ACI International Conference on Superplasticizers and Other Chemical Admixtures in Concrete*, SP-239, American Concrete Institute, Farmington Hills, MI, 2006, pages 1 to 14.

Cordon, W.A., *Entrained Air—A Factor in the Design of Concrete Mixes*, Materials Laboratories Report No. C-310, Research and Geology Division, Bureau of Reclamation, Denver, CO, March 15, 1946.

Cost, V.T.; and Knight, G., "Use of Thermal Measurements to Detect Potential Incompatibilities of Common Concrete Materials," *Concrete Heat Development: Monitoring, Prediction, and Management*, ACI SP-241-4, Atlanta, GA, April 2007, pages 39-58.

Daczko, J., "The New Variety of Polycarboxylate Dispersants," *Concrete Construction*, January 2008, (Accessed June 2020).

De Belie, N.K.; Van Tittelboom, K.; Snoeck, D.; and Wang, J., *Smart Additives for Self-Sealing and Self-Healing Concrete*, 21st International Materials Research Congress, IMRC 2012, August 12, 2012 – August 17, 2012, Cancun, Mexico, Materials Research Society.

Dodson, V., *Concrete Admixtures*, Van Mostrand Reinhold, New York, NY, 1990, 211 pages.

Dodson, V.; and Hayden, T., "Another Look at the Portland Cement/Chemical Admixture Incompatibility Problem," *Cement, Concrete and Aggregates*, Vol. 11, 1989, pages 52 to 59.

DIN 1045-2 (2008), *Concrete, Reinforced and Prestressed Concrete Structures* – Part 2: Concrete – Specification, properties, production and conformity – Application rules for DIN EN 206-1 (Foreign Standard).

European Federation of Producers and Contractors of Specialist Products for Structures (EFNARC), *Guidelines for Viscosity Modifying Admixtures For Concrete*, September 2006, (Accessed November 2008).

Gajda, J., *Development of a Cement to Inhibit Alkali-Silica Reactivity*, Research and Development Bulletin RD115, Portland Cement Association, 1996, 58 pages.

Gaynor, R.D., "Calculating Chloride Percentages," *Concrete Technology Today*, PL983, Portland Cement Association, 1998, pages 4 to 5.

Gebler, S.H., *The Effects of High-Range Water Reducers on the Properties of Freshly Mixed and Hardened Flowing Concrete*, Research and Development Bulletin RD081, Portland Cement Association, 1982, 15 pages.

Gilkey, H.J., "Re-Proportioning of Concrete Mixtures for Air Entrainment," *Journal of the American Concrete Institute*, Proceedings, Vol. 29, No. 8, Farmington Hills, MI, February 1958, pages 633 to 645.

Halford, B., "Building Small," *Chemical and Engineering News*, June 13, 2011, pages 12 to 17.

Hanehara, S.; and Yamada, K., "Interaction Between Cement and Chemical Admixture From the Point of Cement Hydration, Absorption Behaviour of Admixture, and Paste Rheology," *Cement and Concrete Research*, Vol. 29, 1999, pages 1159 to 1165.

Hansen, W.C., *Actions of Calcium Sulfate Admixtures in Portland Cement Paste*, ASTM, Tech. Pub. 266, 1960, pages 3 to 25.

Helmuth, R.; Hills, L.M.; Whiting, D.A.; and Bhattacharja, S., *Abnormal Concrete Performance in the Presence of Admixtures*, RP333, Portland Cement Association, 1995, 92 pages.

Hewlett, P.C., *Lea's Chemistry of Cement and Concrete*, 1998, 4th Edition, Arnold, London, 1998, 1092 pages.

Jeknavorian, A.A., "Viscosity Modifying Admixtures," *Concrete Construction*, Hanley Wood, June 28, 2016.

Jeknavorian, A.A.; Jardine, L.; Ou, C.-C.; Koyata, H.; and Folliard, K., "Interaction of Superplasticizers with Clay-Bearing Aggregates," *Seventh CANMET/ACI International Conference on Superplasticizers and Other Chemical Admixtures in Concrete*, SP-217, W.R. Grace and Co. Conn., Cambridge, MA. American Concrete Institute, 2003, pages 143 to 159.

Jeknavorian, A.A.; Berke, N.S.; and Shen, D.F., "Performance Evaluation of Set Accelerators for Concrete," *Fourth CANMET/ACI International Conference on Superplasticizers and Chemical Admixtures in Concrete*, SP-148, American Concrete Institute, 1994, pages 385 to 405.

Jeknavorian, A.A.; Roberts, L.R.; Jardine, L.; Koyata, H.; Darwin, D.C., "Condensed Polyacrylic Acid-Aminated Polyether Polymers as Superplasticizers for Concrete," *Fifth CANMET/ACI International Conference on Superplasticizers and Other Chemical Admixtures in Concrete*, SP-173, American Concrete Institute, Farmington Hills, MI, 1997, pages 55 to 81.

Juenger, M.C.G.; Monteiro, P.; Gartner, E.; and Denbeaux, G. "A Soft X-Ray Microscope Investigation into the Effects of Calcium Chloride on Tricalcium Silicate Hydration," *Cement and Concrete Research*, Vol. 35, No. 1, January 2005, pages 19 to 25.

Kinney, F.D., "Reuse of Returned Concrete by Hydration Control: Characterization of a New Concept," *Superplasticizers and Other Chemical Admixtures in Concrete*, SP-119, American Concrete Institute, Farmington Hills, MI, 1989, pages 19 to 40.

Klieger, P., *Air-Entraining Admixtures*, Research Department Bulletin RX199, Portland Cement Association, 1966, 12 pages.

Korhonen, C.J.; and Jeknavorian, A.A, "Breaking the Freeze Barrier," *Concrete International*, November 2005, pages 38 to 43.

Kosmatka, S.H.; and Collins, T.C., *Finishing Concrete with Color and Texture*, PA124, Portland Cement Association, Skokie, IL, 2004, 63 pages.

Kozikowski, Jr., R.L.; Vollmer, D.B.; Taylor, P.C.; and Gebler, S.H., *Factor(s) Affecting the Origin of Air-Void Clustering*, SN2789, Portland Cement Association, 2005, 22 pages.

Lackey, H.B., "Factors Affecting Use of Calcium Chloride in Concrete," *Cement, Concrete and Aggregates*, Vol. 14, No. 2, Winter 1992, pages 97 to 100.

Li, C.Z.; Feng, N.Q.; Li, Y.D.; and Chen, R.J., "Effects of Polyethlene Oxide Chains on the Performance of Polycarboxylate-Type Water-Reducers," *Cement and Concrete Research*, Vol. 35, May 2005, pages 867 to 873.

Munn, R.L.; Kao, G.; and Chang, Z-T., "Performance and Compatability of Permeability Reducing and Other Chemical Admixtures in Australian Concretes," *Seventh CANMET/ACI International Conference on Superplasticizers and Other Chemical Admixtures in Concrete*, SP-217, American Concrete Institute.

Nawa, T., "Effect of Chemical Structure on Steric Stabilization of Polycarboxylate-based Superplasticizer," *Journal of Advanced Concrete Technology*, Vol. 4, No. 2, June 2006, pages 225 to 232.

Neville, A.M., *Properties of Concrete*, Pearson Education Limited, Essex, England, 1995.

Nmai, C.K.; Schlagbaum, T.; and Violetta, B., "A History of Mid-Range Water-Reducing Admixtures," *Concrete International*, American Concrete Institute, Farmington Hills, MI, April 1998, pages 45 to 50.

Nmai, C.K.; Tomita, R.; Hondo, F.; and Buffenbarger, J., "Shrinkage-Reducing Admixtures," *Concrete International*, American Concrete Institute, Farmington Hills, MI, April 1998, pages 31 to 37.

Nmai, C.K.; Vojtko, D.; Schaef, S.; Attiogbe, E.K.; and Bury, M.A., "Crack-Reducing Admixture," *Concrete International*, American Concrete Institute, Farmington Hills, MI, January 2014, pages 53 to 57.

Ramachandran, V.S., *Superplasticizers: Properties and Applications in Concrete*, Canada Centre for Mineral and Energy Technology, Ottawa, Ontario, 1997, 404 pages.

Ramachandran, V.S., *Concrete Admixtures Handbook*, Noyes Publications, Park Ridge, NJ, 1995, 1184 pages.

Rear, K.; and Chin, D., "Non-Chloride Accelerating Admixtures for Early Compressive Strength," *Concrete International*, October 1990, pages 55 to 58.

Rixom, R.; and Mailvaganam, N., *Chemical Admixtures for Concrete*, 3rd. ed., Routledge, New York, NY, 1999.

Roberts, L.R.; and Taylor, P.C., "Understanding Cement-SCM-Admixture Interaction Issues," *Concrete International*, Vol. 29, No. 1, January 2007, pages 33 to 41.

Roy, S.K.; and Northwood, D.O., "Admixtures to Reduce the Permeability of Concrete," *Fourth CANMET/ACI International Conference on Durability of Concrete*, SP-170, American Concrete Institute, Farmington Hills, MI, 1997, pages 267 to 284.

Sandberg, P.; and Roberts, L.R., "Cement-Admixture Interactions Related to Aluminate Control," *Journal of ASTM International*, Vol. 2, No. 6, 2005, pages 1 to 14.

Schaefer, G., "How Mid-range Water Reducers Enhance Concrete Performance," *Concrete Construction*, 1995, 3 pages.

Shah, S.P.; Weiss, W.J.; and Yang, W., "Shrinkage Cracking—Can it be Prevented?," *Concrete International*, American Concrete Institute, Farmington Hills, MI, April 1998, pages 51 to 55.

Spiratos, N.; Page, M.; Mailvaganam, N.P.; Malhotra, V.M.; and Jolicoeur, C., *Superplasticizers for Concrete – Fundamentals, Technology, and Practice*, Marquis, Quebec, Canada, 2006, 322 pages.

Stark, D.C., *Lithium Salt Admixtures – An Alternative Method to Prevent Expansive Alkali-Silica Reactivity*, RP307, Portland Cement Association, 1992, 10 pages.

Suzuki, S.; Watanabe, Y.; and Nishi, S., "Influence of Saccharides and Other Organic Compounds on the Hydration of Portland Cement," *Journal of Research, Onoda Cement Co.*, Vol. 11, 1969, pages 184 to 196.

Szecsy, R.; and Mohler, N., *Self-Consolidating Concrete*, IS546, Portland Cement Association, 2009, 24 pages.

Tagnit-Hamou, A.; and Aïtcin, P.C., "Cement and Superplasticizer Compatibility," *World Cement*, Palladian Publications Limited, Farnham, Surrey, England, August 1993, pages 38 to 42.

Tang, F.J.; and Bhattacharja, S., *Development of an Early Stiffening Test*, RP346, Portland Cement Association, 1997, 36 pages.

Taylor, P.C.; Johansen, V.C.; Graf, L.A.; Kozikowski, R.L.; Zemajtis, J.Z.; and Ferraris, C.F., *Identifying Incompatible Combinations of Concrete Materials: Volume I-Final Report*, FHWA HRT-06-079, Federal Highway Administration, Washington D.C., 2006, 162 pages.

Taylor, P.C.; Johansen, V.C.; Graf, L.A.; Kozikowski, R.L.; Zemajtis, J.Z.; and Ferraris, C.F., *Identifying Incompatible Combinations of Concrete Materials: Volume II-Test Protocol*, FHWA HRT-06-080, Federal Highway Administration and Portland Cement Association, Washington D.C., 2006a, 86 pages.

Taylor, P.; VanDam, T.; Sutter, L.; and Fick, C., *Integrated Materials and Construction Practices for Concrete Pavement: A State-of-the-Practice Manual*, FHWA Intrans Project 13-482, National Concrete Paving Technology Center, May 2019, 338 pages.

Thomas, M.D.A.; Fournier, B.; Folliard, K.J.; Ideker, J.H.; and Resendez, Y., *The Use of Lithium To Prevent or Mitigate Alkali-Silica Reaction in Concrete Pavements and Structures*, FHWA-HRT-06-133, Federal Highway Administration, Turner-Fairbank Highway Research Center, Maclean, VA, March 2007, 50 pages.

Thomas, M.D.A.; and Stokes, D.B., "Use of a Lithium-Bearing Admixture to Suppress Expansion in Concrete Due to Alkali-Silica Reaction," *Transportation Research Record*, Vol. 1668, Transportation Research Board, Washington, D.C., 1999, pages 54 to 59.

Thomas, M.D.A.; and Wilson, M.L., *Admixtures Use in Concrete*, CD039, Portland Cement Association, Skokie, IL, 2002.

Whiting, D., *Effects of High-Range Water Reducers on Some Properties of Fresh and Hardened Concretes*, Research and Development Bulletin RD061, Portland Cement Association, 1979, 17 pages.

Whiting, D.; and Dziedzic, W., *Effects of Conventional and High-Range Water Reducers on Concrete Properties*, Research and Development Bulletin RD107, Portland Cement Association, 1992, 28 pages.

Whiting, D.A.; and Nagi, M.A., *Manual on the Control of Air Content in Concrete*, EB116, National Ready Mixed Concrete Association and Portland Cement Association, 1998, 52 pages.

Whiting, D.; and Stark, D., *Control of Air Content in Concrete*, NCHRP Report No. 258 and Addendum, Transportation Research Board and National Research Council, Washington, D.C., May 1983, 84 pages.

Young, J.F., *Reaction Mechanisms of Organic Admixtures with Hydrating Cement Compounds*, Research Record 564, Transportation Research Board and National Research Council, Washington, D.C., 1976, pages 1 to 9.

FIBERS

FIGURE 7-1. Steel, glass, synthetic, and natural fibers with different lengths and shapes can be used in concrete applications (courtesy of Fiber Reinforced Concrete Association).

Fibers have been used in construction materials for centuries. Fibers made from steel, plastic, glass, natural materials (such as wood cellulose) are available in a variety of shapes, sizes, and thicknesses (Figure 7-1) and are commonly used in many applications including ready-mixed concrete, precast concrete, and shotcrete applications. They may be round, flat, crimped, and deformed with typical lengths of 6 mm to 65 mm (0.25 in. to 2½ in.) and thicknesses ranging from 0.005 mm to 0.75 mm (0.0002 in. to 0.03 in.).

Fibers are generally described as micro or macro, depending on the equivalent diameter of the fiber. A microfiber is defined as a fiber for use in concrete with an equivalent diameter less than 0.3 mm (0.012 in.) and a macrofiber has an equivalent diameter greater than or equal to 0.3 mm (0.012 in.). There are also fibers on the commercial market that approach an equivalent diameter of 1.0 mm (0.04 in).

ASTM C1116, *Standard Specification for Fiber Reinforced Concrete*, classifies fiber-reinforced concrete into four different categories based on the type of the fiber material used:

- *Type I* – Steel Fiber-Reinforced Concrete – contains stainless steel, alloy steel, or carbon steel fibers.
- *Type II* – Glass Fiber-Reinforced Concrete – contains alkali-resistant (AR) glass fibers.
- *Type III* – Synthetic Fiber-Reinforced Concrete – contains synthetic fibers including polyolefin, nylon, and carbon fibers.
- *Type IV* – Natural Fiber-Reinforced Concrete – contains natural fibers including cellulose fibers.

Fibers are added to concrete either during batching or at the project site and, typically, in relatively low dosages (often less than 1% by volume).

Fibers are uniformly dispersed into the concrete during mixing and become integral to the concrete matrix at the time of placement. A uniform dispersion of fibers is intended to provide multi-directional orientation within the concrete matrix, resulting in potential benefits in fresh and hardened concrete properties. The main factors that control the performance of the composite material are:

- physical properties of fibers and matrix,
- strength of bond between fibers and matrix, and
- fiber dosage.

Although some of the basic governing principles are the same, there are characteristic differences between conventional and fiber reinforcement:

- Fibers are generally randomly distributed and oriented throughout a given cross section, whereas reinforcing bars or wires are spaced to be located where required by design.
- Most fibers are relatively short and closely spaced as compared with continuous reinforcing bars or wires.

ADVANTAGES AND DISADVANTAGES OF USING FIBERS

Fresh Properties

In fresh concrete, a uniform distribution of fibers in the concrete matrix disrupts the movement of bleed water to the concrete surface and minimizes the development of large and continuous bleed water channels.

Consequently, fibers, especially microfibers, help reduce bleeding, plastic settlement, and cracking associated with plastic settlement. Fibers also arrest plastic cracks that may form under adverse ambient conditions, minimizing the potential for plastic shrinkage cracking (see Chapter 16). The effectiveness of a fiber in reducing plastic shrinkage cracking can be evaluated using ASTM C1579, *Standard Test Method for Evaluating Plastic Shrinkage Cracking of Restrained Fiber Reinforced Concrete (Using a Steel Form Insert)*. Microfibers typically do not significantly alter the free drying shrinkage after hardening.

Fiber reinforced concrete may lower workability in concrete mixtures, especially at higher dosages. The use of fibers can sometimes result in a lowered slump measurement, which can be offset by the use chemical admixtures. It is not recommended to add water as this may change other properties of the concrete, such as the compressive strength and permeability. Proper proportioning and mixing are needed to uniformly distribute fibers throughout the mixture to avoid clumping of fibers.

Hardened Properties

In hardened concrete, the primary benefits of fibers, particularly macrofibers, come from the post-crack load-carrying capacity, ductility, and flexural toughness (a measure of energy absorption capacity) that they impart to the concrete. As a result of increased toughness, fibers improve the shatter resistance and impact strength of concrete as well as fatigue resistance. The improvement in post-crack load-carrying capacity provided by fiber-reinforced concrete has led to the use of fibers in lieu of conventional temperature and shrinkage reinforcement. Applications include: slabs-on-ground, concrete parking lots, pavements and overlays, composite metal deck slabs, mining and tunneling, shotcrete, and other excavation support work. Increasingly, fibers are also being used to help mitigate cracking in bridge decks and in precast concrete applications such as septic tanks and utility vaults.

Fibers may be relatively inefficient when compared to traditional reinforcement for structural use due to their random orientation. At high enough dosages, fibers can increase the resistance to cracking and decrease crack width (Shah and others 1998) and, consequently, can be used to resist applied loads in some applications such as precast concrete pipes (Park and others 2014). Standard specifications that govern the use of steel fibers and rigid synthetic fibers in concrete pipe are, respectively, ASTM C1765, *Standard Specification for Steel Fiber Reinforced Concrete Culvert, Storm Drain, and Sewer Pipe*, and ASTM C1818, *Standard Specification for Rigid Synthetic Fiber Reinforced Concrete Culvert, Storm Drain and Sewer Pipe*.

There are three ASTM standard test methods that can be used to assess the post-crack flexural performance of fiber-reinforced concrete: namely, ASTM C1399, *Standard Test Method for Obtaining Average Residual-Strength of Fiber-Reinforced Concrete* – a beam test that yields an average residual strength value; ASTM C1609, *Standard Test Method for Flexural Performance of Fiber-Reinforced Concrete (Using Beam With Third-Point Loading)* – a beam test, and the preferred method, that yields several parameters for use in design equations including residual strengths, toughness, an equivalent flexural strength, and an equivalent flexural strength ratio; and, ASTM C1550, *Standard Test Method for Flexural Toughness of Fiber Reinforced Concrete (Using Centrally Loaded Round Panel)* – a round panel test typically required for shotcrete and tunneling applications that yields a flexural toughness expressed as energy absorbed in joules (J).

Some fibers may adversely affect durability: where steel fibers are exposed to the environment, they may be susceptible to corrosion, some glass fibers react with the alkaline environment of concrete, and natural fibers can deteriorate over time.

TYPES AND PROPERTIES OF FIBERS AND THEIR EFFECT ON CONCRETE

Steel Fibers

Steel fibers are short, discrete lengths of steel with an aspect ratio (ratio of length to diameter) from about 30 to 100, and with a variety of cross sections and profiles (Figure 7-2). Some steel fibers have hooked ends to improve resistance to pullout from a cement-based matrix. ASTM A820, *Standard Specification for Steel Fibers for Fiber-Reinforced Concrete*, classifies five different types based on their manufacture:

- *Type I* – Cold-drawn wire fibers are the most commercially available, manufactured from drawn steel wire.

- *Type II* – Cut sheet fibers are manufactured as the name implies: steel fibers are laterally sheared off steel sheets.

- *Type III* – Melt-extracted fibers are manufactured with a relatively complicated technique where a rotating wheel is used to lift liquid metal from a molten metal surface by capillary action. The extracted molten metal is then rapidly frozen into fibers and spun off the wheel. The resulting fibers have a crescent-shaped cross section.

- *Type IV* – Mill cut fibers are manufactured by a high-speed rotor equipped with special cutting plates which comminute the sample material through special milling techniques. This results in fibers with a triangular longitudinal twist, rough surface, hooked end and a soft longitudinal torsion.

- *Type V* – Modified cold-drawn wire fibers are made by cutting cold drawn wires into short fibers according to specified length with a cutting knife, punch or rotary tool. Common methods to improve smooth surface for bond include pressing edges and forming waves.

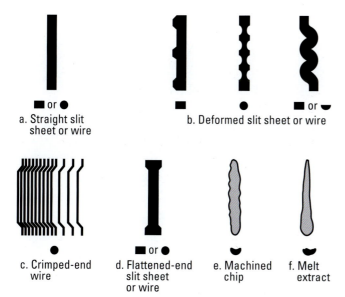

a. Straight slit sheet or wire

b. Deformed slit sheet or wire

c. Crimped-end wire

d. Flattened-end slit sheet or wire

e. Machined chip

f. Melt extract

FIGURE 7-2. Various steel fiber geometries, top view and typical cross section(s).

Steel-fiber volumes used in concrete typically range from 0.25% to 2%. Volumes of more than 2% generally reduce workability and fiber dispersion and require special design of concrete mixture or placement techniques.

The compressive strength of concrete is only slightly affected by the presence of fibers. The addition of 1.5% by volume of steel fibers can increase the direct tensile strength by up to 40% and the residual flexural strength up to 150% in comparison to concrete without steel fibers.

Steel fibers do not affect free shrinkage. Steel fibers delay the fracture of restrained concrete during shrinkage without influencing the stress at failure (Altoubat and Lange 2001).

The durability of steel-fiber concrete is contingent on the same factors as conventional concrete. Freeze-thaw durability is not diminished by the addition of steel fibers provided the mixture proportions are adjusted to accommodate the fibers. The concrete must also be air-entrained and have a low water-cementitious materials ratio (Chapter 11). Steel fibers have a relatively high modulus of elasticity (Table 7-1). Their bond to the cement matrix can be enhanced by mechanical anchorage or surface roughness. When fully encapsulated, they are protected from corrosion by the alkaline environment in the concrete matrix (ACI 544.3R 2008).

Steel fibers are most commonly used in industrial floors and they have been used to reduce contraction joints and extend joint spacing at a volume fraction of 0.5%, a dosage of about 40 kg/m³ (66 lb/yd³). They have also been used in bridge decks, airport runway/taxi overlays, and highway pavements. Structures containing steel fibers and exposed to high-velocity water flow have lasted about three times longer than conventional concrete alternatives. Steel fibers are also used in many precast concrete applications because of the improved impact resistance or toughness imparted by the fibers. In utility boxes and septic tanks, steel fibers have been used to replace conventional reinforcement. Use of steel fibers for shear reinforcement in beams is addressed in ACI 318, *Building Code Requirements for Structural Concrete* (2019). High-strength steel fibers with 0.2 mm (0.008 in.) diameter and 13 mm (½ in.) or 20 mm (¾ in.) in length are being used to produce ultra-high-performance concrete (UHPC), a class of concrete that has a minimum specified compressive strength of 150 MPa (22,000 psi) with specified durability, tensile ductility, and toughness requirements (ACI 239R 2018). Additional information on the unique applications of UHPC can be found in ACI 239R, *Ultra-High-Performance Concrete: An Emerging Technology Report* (2018) and in Chapter 23.

Steel fibers are also widely used with shotcrete in thin-layer applications, especially rock-slope stabilization and tunnel linings. Steel-fiber shotcrete has been successfully applied

TABLE 7-1. Properties of Selected Fibers

FIBER TYPE	RELATIVE DENSITY (SPECIFIC GRAVITY)	DIAMETER, µm (0.001 in.)	TENSILE STRENGTH, MPa (ksi)	MODULUS OF ELASTICITY, MPa (ksi)	STRAIN AT FAILURE, %
Steel					
	7.80	100 – 1000 (4 – 40)	500 – 2600 (70 – 380)	210,000 (30,000)	0.50 – 3.5
Glass					
E	2.54	8 – 15 (0.3 – 0.6)	2000 – 4000 (290 – 580)	72,000 (10,400)	3.0 – 4.8
AR	2.70	12 – 20 (0.5 – 0.8)	1500 – 3700 (220 – 540)	80,000 (11,600)	2.5 – 3.6
Synthetic					
Acrylic	1.18	5 – 17 (0.2 – 0.7)	200 – 1000 (30 – 145)	17,000 – 19,000 (2500 – 2800)	28 – 50
Aramid	1.44	10 – 12 (0.4 – 0.47)	2000 – 3100 (300 – 450)	62,000 – 120,000 (9000 – 17,000)	2 – 3.5
Carbon	1.90	8 – 9 (0.3 – 0.35)	1800 – 2600 (260 – 380)	230,000 – 380,000 (33,400 – 55,100)	0.5 – 1.5
Nylon	1.14	23 (0.9)	1000 (140)	5200 (750)	20
Polyester	1.38	10 – 80 (0.4 – 3.0)	280 – 1200 (40 – 170)	10,000 – 18,000 (1500 – 2500)	10 – 50
Polyethylene	0.96	25 – 1000 (1 – 40)	80 – 600 (11 – 85)	500 (725)	12 – 100
Polypropylene	0.90	20 – 200 (1 – 40)	450 – 700 (11 – 85)	3500 – 5200 (725)	12 – 100
Polyvinyl alcohol (PVA)	1.3	27 – 660 (1.0 – 26)	900 – 1,600 (130 – 232)	23,000 – 40,000 (3300 – 5800)	7 – 8
Natural					
Wood cellulose	150	25 – 125 (1 – 5)	350 – 2000 (51 – 290)	10,000 – 40,000 (1500 – 5800)	
Sisal			280 – 600 (40 – 85)	13,000 – 25,000 (1900 – 3800)	
Coconut coir	1.12 – 1.15	100 – 400 (4 – 16)	120 – 200 (17 – 29)	19,000 – 25,000 (2800 – 3800)	10 – 25
Bamboo	1.50	50 – 400 (2 – 16)	350 – 500 (51 – 73)	33,000 – 40,000 (4800 – 5800)	
Jute	1.02 – 1.04	100 – 200 (4 – 8)	250 – 350 (36 – 51)	25,000 – 32,000 (3800 – 4600)	1.5 – 1.9
Elephant grass			180 (26)	4900 (710)	3.6

Adapted from PCA (1991) and ACI 544.1R-96.

with fiber volumes up to 2%. For further details about properties of steel fibers and its use in steel-fiber reinforced concrete, see ACI 544.1R, *State-of-the-Art Report on Fiber Reinforced Concrete* (1996), and ACI 544.4R, *Guide to Design with Fiber-Reinforced Concrete* (2018).

Glass Fibers

Initial research on glass fibers in the early 1960s used conventional borosilicate glass (E-glass) (Table 7-1) and soda-lime-silica glass fibers (A-glass). The test results showed that alkali reactivity between the E-glass fibers and the cement paste reduced the strength of the concrete. Continued research resulted in alkali-resistant glass fibers (AR-glass) (Table 7-1), that improved long-term durability, but sources of other strength-loss trends were observed. AR-glass fibers were subject to embrittlement, caused by infiltration of calcium hydroxide particles and by-products of cement hydration into fiber bundles (Bentur 1985). Alkali reactivity and cement hydration are the basis for the following two widely held theories explaining strength and ductility loss, particularly in exterior glass fiber concrete:

- Alkali attack on glass-fiber surfaces reduces fiber tensile strength and, subsequently, lowers compressive strength.
- Ongoing cement hydration causes calcium hydroxide penetration of fiber bundles, thereby increasing fiber-to-matrix bond strength and embrittlement; the latter lowers tensile strength by inhibiting fiber pullout.

Fiber modifications to improve long-term durability involve specially formulated chemical coatings to help combat hydration-induced embrittlement, and employment of a dispersed silica fume slurry to adequately fill fiber voids, thereby reducing potential for calcium hydroxide infiltration.

A low-alkaline cement developed in Japan produced no calcium hydroxide during hydration and accelerated tests with the cement in alkali-resistant-glass fiber-reinforced concrete samples showed greater long-term durability than previously achieved (PCA 1991).

Metakaolin can be used in glass-fiber-reinforced concrete without significantly affecting flexural strength, strain, modulus of elasticity, and toughness (Marikunte and others 1997).

Alkali-resistant glass fibers have allowed the development of thin concrete, typically 13 mm (½ in.) that has made possible many glass fiber-reinforced concrete (GFRC) products with the sustainability and durability offered by concrete, but with much lighter weight. The single largest application of glass-fiber-reinforced concrete has been the manufacture of exterior building façade panels (Figure 7-3). Other applications are listed in PCA SP039, *Fiber Reinforced Concrete* (1991). The standard for AR-glass fibers is ASTM C1666, S*tandard Specification for Alkali Resistant (AR) Glass Fiber for GFRC and Fiber-Reinforced Concrete and Cement*.

Synthetic Fibers

Synthetic fibers are man-made fibers which were developed during research and development in the petrochemical and textile industries. Typical synthetic fiber types used in portland cement concrete are: acrylic, aramid, carbon, nylon, polyolefins such as polyethylene and polypropylene, and polyvinyl alcohol (PVA). Table 7-1 summarizes the range of physical properties of these fibers. The standard for polyolefin fibers is ASTM D7508, *Standard Specification for Polyolefin Chopped Strands for Use in Concrete*.

Synthetic microfibers are typically either monofilaments or fibrillated in nature. During the mixing process each fibrillated fiber turns into a unit with several fibrils at its end. The fibrils provide better mechanical bonding than conventional monofilaments. The high number of fine fibrils also reduces plastic shrinkage cracking and may increase the ductility and toughness of the concrete (Trottier and Mahoney 2001). Synthetic macrofibers are provided in varying materials, lengths, forms, and configurations. These fibers can be rigid or more pliable. Some fibers are also embossed to improve mechanical anchorage. Bond to concrete is typically mechanical and is achieved primarily through friction, but a synthetic polyolefin-based macrofiber that also provides chemical bonding in concrete has been developed (Attiogbe and others 2014).

Synthetic fibers are also used in stucco and mortar. For this use the fibers are shorter than synthetic fibers used in concrete. Usually small amounts of 13 mm (½ in.) long alkali-resistant fibers are added to base coat plaster mixtures. They can be used in small diameter stucco and mortar pumps and spray guns. They should be added to the mixture in accordance with manufacturer's recommendation.

FIGURE 7-3. (left) Glass-fiber-reinforced concrete panels are light and strong enough to reduce this building's structural requirements. (right) Spray-up fabrication made it easy to create their contoured profiles.

Problems associated with synthetic fibers include: change in workability if measures are not taken, use of inconclusive performance testing for low fiber-volume usage with polypropylene, polyethylene, polyester, and nylon; and a hairy surface finish. ACI 544.3R *Guide for Specifying, Proportioning, and Production of Fiber-Reinforced Concrete* (2008), provides guidance on proper use of synthetic fibers.

Polypropylene fibers. The most popular of the synthetics, polypropylene fibers, are lightweight, hydrophobic, and typically inert (with exception of polyolefin-based macrofiber technology). They are produced as continuous cylindrical monofilaments that can be cut to specified lengths or cut as films and tapes and formed into fine fibrils of rectangular cross section (Figures 7-4).

FIGURE 7-4. Polypropylene fibers.

Used at a rate of up to 0.1% by volume of concrete (0.9 kg/m^3 [1.5 lb/yd^3]), polypropylene microfibers reduce plastic shrinkage cracking and subsidence cracking over steel reinforcement (Suprenant and Malisch 1999). The presence of polypropylene fibers in concrete may reduce settlement of aggregate particles, thus reducing capillary bleed channels. Polypropylene fibers can help reduce explosive spalling of high-strength, low-permeability concrete exposed to fire in a moist condition. Typically, a minimum of about 2 kg/m^3 (3.4 lb/yd^3) of short (6 mm [¼ in.]) monofilament, polypropylene microfibers will be effective in reducing explosive spalling. Fire testing of concrete panels containing the fiber should be performed to determine the actual amount of fiber required.

Polypropylene macrofibers are typically added at volume fractions of 0.2% to 1% of concrete for drying shrinkage crack control, which equates to fiber dosages of 1.8 kg/m^3 to 8.9 kg/m^3 (3.0 lb/yd^3 to 15 lb/yd^3). Adjustments in concrete mixture proportions will typically be required to provide an adequate paste content for the dosage of synthetic macrofiber being used, thereby ensuring adequate workability and, in slab applications, the desired finishability. Due to their efficiency in reducing crack width, synthetic macrofibers have also been used to extend joint spacings beyond the recommendations provided in Chapter 15 and from ACI 360R, *Guide to Design of Slabs-on-Ground* (2010). Polypropylene macrofiber dosages of 3 kg/m^3 to 4.5 kg/m^3 (5 lb/yd^3 to 7.5 lb/yd^3) have been used to either extend joints, without any intermediate sawcut contraction joints, or to produce jointless slabs-on-ground as shown in Figure 7-5 (Walker 2007, Holland and others 2008, and Nmai and others 2014). Generally, limiting the number of joints or increasing the joint spacing can be accomplished without increasing the number of random cracks. A key benefit of extended joint spacing is reduced maintenance.

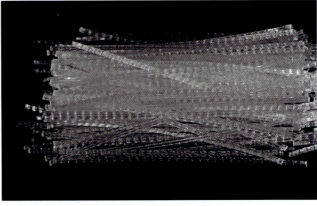

FIGURE 7-5. Jointless slab (top) 27.4 m × 18.3 m × 190 mm (90 ft × 60 ft × 7.5 in.), cast using concrete dosed with a rigid synthetic polyolefin-based macrofibers (bottom) (courtesy of Master Builders Solutions).

Acrylic fibers. Acrylic fibers are the most suitable replacement for asbestos fibers that were commonly used in cement board and roof-shingle production. Fiber volumes of up to 3% can produce a composite with mechanical properties similar to that of an asbestos-cement composite. Acrylic-fiber concrete composites exhibit high postcracking toughness and ductility. Although lower than that of asbestos-cement composites, acrylic-fiber-reinforced concrete's flexural strength is ample for many building applications.

Aramid fibers. Aramid fibers have high tensile strength and a high tensile modulus. Aramid fibers are two and a half times as strong as E-glass fibers and five times as strong as steel fibers. In addition to excellent strength characteristics, aramid fibers have excellent strength retention up to 160°C (320°F), dimensional stability up to 200°C (392°F), static and dynamic fatigue resistance, and creep resistance. Aramid strand is available in a wide range of diameters.

Carbon fibers. Carbon fibers were developed primarily for their high strength and elastic modulus and stiffness properties for applications within the aerospace industry. Carbon fibers have high tensile strength and modulus of elasticity (Table 7-1). They are also inert to most chemicals. Carbon fiber is typically produced in strands that may contain up to 12,000 individual filaments. The strands are commonly spread prior to incorporation in concrete to facilitate cement matrix penetration and to maximize fiber effectiveness.

Nylon fibers. Nylon fibers are commonly produced for use in apparel, home furnishing, industrial, and textile applications. Nylon fibers are spun from nylon polymer and transformed through extrusion, stretching, and heating to form an oriented, crystalline, fiber structure. For concrete applications, high tenacity (high tensile strength), heat and light stable yarn is spun and subsequently cut into shorter lengths. Nylon fibers exhibit good toughness, and elastic recovery. Nylon can be used at comparatively small dosages and has potentially greater reinforcing value than low volumes of polypropylene or polyester fiber. Nylon is relatively inert and resistant to a wide variety of organic and inorganic materials, including strong alkalies.

Nylon is hydrophilic, with moisture retention of 4.5%, and may increase the water demand of concrete. While this does not affect concrete hydration or workability at low prescribed contents ranging from 0.1% to 0.2% by volume, the impact of increased water demand should be considered at higher fiber volume contents.

Polyvinyl alcohol (PVA). Polyvinyl alcohol is a water-soluble polymer that is used in a variety of applications that include: papermaking, geo-textiles, rope, nets, coatings, and fibers. Polyvinyl alcohol fibers have high tenacity, high modulus of elasticity, low elongation, low creep, and a good affinity to water, which makes them unique in their ability to bond to a cementitious matrix (Noushini and others 2013, and Klemenc 2009). They can be used as a substitute for asbestos and glass fibers and, as a result, they are widely used in fiber cement boards and tiles. Polyvinyl alcohol fibers with modified surface characteristics, and at a volume fraction of about 2%, have made it possible to produce an engineered cementitious composite (ECC), also called bendable concrete, that has extremely high ductility and can deform up to 3% to 5% in tension before it fails (see Chapter 24, Wang and Li 2005, and Li 2012).

Interground fiber cement. The technology of interground fiber cement takes advantage of the fact that some synthetic fibers are not destroyed or pulverized in the cement finishing mill. The fibers are mixed with dry cement during grinding where they are uniformly distributed. The surface of the fibers is roughened during grinding, which offers a better mechanical bond to the cement paste (Vondran 1995).

For further details about chemical and physical properties of synthetic fibers and design of synthetic fiber concrete, see ACI 544.1R-96, and ACI 544.4R-18.

Natural Fibers

Natural fibers were used as reinforcement long before the advent of conventional reinforced concrete. Mud bricks reinforced with straw and mortars reinforced with horsehair are examples of how natural fibers were used historically as a form of reinforcement. Many natural reinforcing materials can be obtained at low cost and energy. Such fibers are used in the manufacture of low-fiber-content concrete and occasionally have been used in thin sheet concrete with high-fiber content. For typical properties of natural fibers see Table 7-1.

Unprocessed Natural Fibers. Research on the engineering properties of natural fibers, and concrete made with these fibers began in the 1960s. The results indicated that these fibers can be used successfully to make thin sheets for walls and roofs. Products were made using portland cement and unprocessed natural fibers such as coconut coir, sisal, bamboo, jute, wood, and vegetable fibers. Although the concretes made with unprocessed natural fibers show good mechanical properties, they have some deficiencies in durability. Many of the natural fibers are highly susceptible to volume changes due to variations in fiber moisture content. Fiber volumetric changes that accompany variations in fiber moisture content can drastically affect the bond strength between the fiber and cement matrix.

Wood Fibers (Processed Natural Fibers). The properties of wood cellulose fibers are greatly influenced by the method by which the fibers are extracted and the refining processes involved. The process by which wood is reduced to a fibrous mass is called pulping. The kraft process is most commonly used for producing wood cellulose fibers. This process involves cooking wood chips in a solution of sodium hydroxide, sodium carbonate, and sodium sulfide. Wood cellulose fibers have relatively good mechanical properties compared to many manmade fibers such as polypropylene, polyethylene, polyester, and acrylic. Delignified cellulose fibers (lignin removed) can be produced with a tensile strength of up to approximately 2000 MPa (290 ksi) for selected grades of wood and pulping processes. Fiber tensile strengths of approximately 500 MPa (73 ksi) can be routinely achieved using a chemical pulping process with the more common, less expensive grades of wood. Cellulose-based fibers are typically used at a dosage of 0.9 kg/m³ (1.5 lb/yd³) to minimize plastic shrinkage cracking and to replace lighter-gage welded-wire reinforcement in slabs.

MULTIPLE FIBER SYSTEMS (HYBRID FIBERS)

For a multiple fiber system, two or more fibers are blended into one system to form a hybrid fiber. Hybrid fibers currently available in the marketplace are either blends of steel macro- and microfibers, blends of synthetic macro- and microfibers, or blends of steel macrofibers and synthetic microfibers. Hybrid fibers provide fibers of varying length that result in a closer fiber-to-fiber spacing, which reduces microcracking and increases tensile strength. In addition, hybrid fibers combine the toughness and increased post-crack flexural toughness provided by macrofibers with the reduced plastic cracking benefit obtained with microfibers. Applications for hybrid fibers include thin repairs and patching (Banthia and Bindiganavile 2001), slab-on-ground, bridge decks, and other types of flatwork. As is typical with fibers, concrete with blended fibers typically has a lower slump compared to plain concrete, but has enhanced elastic and post-elastic strength.

Work by Blunt and others (2015) has shown that a hybrid fiber reinforced concrete composite (HyFRC) had significantly improved flexural performance and damage resistance over plain concrete and reinforced concrete with the same steel reinforcement ratio (Figures 7-6 and 7-7). The HyFRC includes coarse aggregate and three fibers with a total fiber volume fraction of 1.5% (0.2% PVA microfibers 8 mm [0.3 in.] in length, 0.5% steel macrofibers 30 mm [1.2 in.] in length, and 0.8% steel macrofibers 60 mm [2.4 in.] in length). Further research has been performed on HyFRC in applications such as precast, self-consolidating, and post-tensioned concrete construction.

 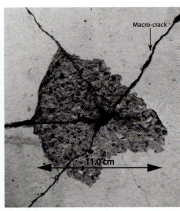

FIGURE 7-7. (left) Damage resistance of HyFRC panel after being subjected to a steel projectile with velocity of 167 m/s (374 mph). (right) Damage of a plain concrete panel after being subjected to a steel projectile with velocity of 127 m/s (284 mph) (courtesy of C. Ostertag).

Current research is being done to investigate the effects of hybrid fiber reinforcement in ultra-high performance concrete. Turker and others (2019) tested the flexural capacity of UHPC beams with 1.5% steel fibers, reinforced with either only microfibers or 1% microfibers and 0.5% macrofibers; the hybrid fiber reinforcement was found to be more efficient in increasing flexural capacity. Smarzewski and Barnat-Hunek (2017) found improved durability properties when comparing steel and polypropylene fibers in UHPC, compared to none or one type of fiber.

REFERENCES

PCA's online catalog includes links to PDF versions of many research reports and other classic publications.
Visit: cement.org/library/catalog.

ACI Committee 239, *Ultra-High-Performance Concrete: An Emerging Technology Report*, ACI 239R-18, American Concrete Institute, Farmington Hills, MI, 2018, 29 pages.

ACI Committee 318, *Building Code Requirements for Structural Concrete*, ACI 318-19, American Concrete Institute, Farmington Hills, MI, 2019, 629 pages.

ACI Committee 360, *Guide to Design of Slabs-on-Ground*, ACI 360R-10, American Concrete Institute, Farmington Hills, MI, 2010, 77 pages.

ACI Committee 544, *Guide to Design with Fiber-Reinforced Concrete*, ACI 544.4R-18, American Concrete Institute, Farmington Hills, MI, 2018, 45 pages.

ACI Committee 544, *Guide for Specifying, Proportioning, and Production of Fiber-Reinforced Concrete*, ACI 544.3R-08, American Concrete Institute, Farmington Hills, MI, 2008, 14 pages.

ACI Committee 544, *State-of-the-Art Report on Fiber Reinforced Concrete*, ACI 544.1R-96, reapproved 2009, American Concrete Institute, Farmington Hills, MI, 1997, 66 pages.

Altoubat, S.A.; and Lange, D.A., "Creep, Shrinkage, and Cracking of Restrained Concrete at Early Age," *ACI Materials Journal*, American Concrete Institute, Farmington Hills, MI, July-August 2001, pages 323 to 331.

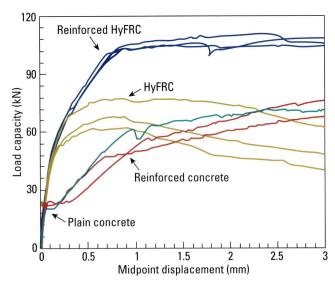

FIGURE 7-6. Flexural performance of hybrid fiber reinforced concrete composite (HyFRC), reinforced HyFRC (reinforced with conventional steel reinforcing bar), compared to plain concrete and reinforced concrete with the same steel reinforcement ratio (Blunt and others 2015).

Attiogbe, E.K.; Schaef, S.; Kerobo, C.O.; Vojtko, D.; and Nmai, C.K., "A New Fiber for Enhanced Crack Control," *Concrete International*, American Concrete Institute, Farmington Hills, MI, December 2014, pages 35 to 39.

Banthia, N.; and Bindiganavile, V., "Repairing with Hybrid-Fiber-Reinforced Concrete," *Concrete International*, American Concrete Institute, Farmington Hills, MI, June 2001, pages 29 to 32.

Bentur, A., "Mechanisms of Potential Embrittlement and Strength Loss of Glass Fiber Reinforced Cement Composites," *Proceedings-Durability of Glass Fiber Reinforced Concrete Symposium*, Prestressed Concrete Institute, Chicago, IL, 1985, pages 109 to 123.

Blunt, J.; Jen, G.; and Ostertag, C.P., "Enhancing Corrosion Resistance of Reinforced Concrete Structures with Hybrid Fiber Reinforced Concrete," *Corrosion Science*, 92, 2015, pages 182 to 191.

FHWA *Development of Non-Proprietary Ultra-High Performance Concrete for Use in the Highway Bridge Sector*, FHWA-HRT-13-100, Federal Highway Administration, Turner-Fairbank Highway Research Center, McLean, VA, October 2013, 8 pages.

Holland, J.A.; Simonelli, R.M.; and Walker, W.W., "Macro Polymeric Fibers for Slabs on Ground," *Concrete Construction*, June 2008.

Klemenc, S.E., "An Overview of PVA Fibers and Concrete," *Building with Concrete, Concrete Countertops*, Vol. 9, No. 7, Nov/Dec 2009.

Li, V., *Bendable Concrete, with a Design Inspired by Seashells, Can Make US infrastructure Safer and More Durable*, The MI Engineer News Center, May 30, 2018.

Marikunte, S.; Aldea, C.; and Shah, S., "Durability of Glass Fiber Reinforced Cement Composites: Effect of Silica Fume and Metakaolin," *Advanced Cement Based Materials*, Volume 5, Numbers 3/4, April/May 1997, pages 100 to 108.

Nmai, C.K.; Vojtko, D.; Schaef, S.; Attiogbe, E.K.; and Bury, M.A., "Crack-Reducing Admixture," *Concrete International*, American Concrete Institute, January 2014, pages 53 to 57.

Noushini, A.; Samali, B.; and Vessalas, K., "Influence of Polyvinyl Alcohol Fibre Addition on Fresh and Hardened Properties of Concrete," Conference Proceedings, *Thirteenth East Asia-Pacific Conference on Structural Engineering and Construction* (EASEC-13), August 2013.

Park, Y.; Abolmaali, A.; Attiogbe, E., and Lee, S-H, "Time-Dependent Behavior of Synthetic Fiber–Reinforced Concrete Pipes Under Long-Term Sustained Loading," *Transportation Research Record: Journal of the Transportation Research Board*, No. 2407, Transportation Research Board of the National Academies, Washington, DC, 2014, pages 71 to 79.

PCA, *Fiber Reinforced Concrete*, SP039, Portland Cement Association, Slokie, IL, 1991, 54 pages.

Shah, S.P.; Weiss, W.J.; and Yang, W., "Shrinkage Cracking – Can it be Prevented?," *Concrete International*, American Concrete Institute, Farmington Hills, MI, April 1998, pages 51 to 55.

Smarzewski, P.; and Barnat-Hunek, D., "Effect of Fiber Hybridization on Durability Related Properties of Ultra-High Performance Concrete," *International Journal of Concrete Structures and Materials*, Vol. 11, No. 2, June 2017, pages 315 to 325.

Suprenant, B.A.; and Malisch, W.R., "The Fiber Factor," *Concrete Construction*, Addison, IL, October 1999, pages 43 to 46.

Trottier, J.F.; and Mahoney, M., "Innovative Synthetic Fibers," *Concrete International*, American Concrete Institute, Farmington Hills, MI, June 2001, pages 23 to 28.

Turker, K.; Hasgul, U; Birol, T.; Yavas, A.; and Yazici, H., "Hybrid Fiber Use on Flexural Behavior of Ultra High Performance Fiber Reinforced Concrete Beams," *Composite Structures*, Vol. 229, 2019.

Vondran, G.L., "Interground Fiber Cement in the Year 2000," *Emerging Technologies Symposium on Cements for the 21st Century*, SP206, Portland Cement Association, March 1995, pages 116 to 134.

Walker, W., "Reinforcement for Slabs on Ground: What It Will and Will Not Do," *Concrete Construction, Annual Floors Issue 2007*, November, 2007.

Wang, S.; and Li, V., "Polyvinyl Alcohol Fiber Reinforced Engineered Cementitious Composites: Material Design and Performances," *Proceedings of International RILEM Workshop on HPFRCC in Structural Applications*, RILEM Publications SARL, 2006, pages 65 to 73.

REINFORCEMENT

Reinforcement is used in most types of concrete structures and components: elevated floor slabs, pavements, walls, beams, columns, mats, frames, and more (Figure 8-1). Reinforcement typically consists of four distinct types: reinforcing bars, welded wire reinforcement, prestressing steel, and fibers. See Chapter 7 for more information on fibers.

WHY USE REINFORCEMENT IN CONCRETE

Concrete is very strong in compression, but relatively weak in tension. The tensile strength of concrete is only about one tenth of its compressive strength. Reinforcing steel is strong in tension and can be utilized in reinforced concrete to offset the low tensile strength of concrete.

FIGURE 8-1. Reinforcement is used for structural concrete including foundation carriers (left) and elevated slabs and beams (right) (courtesy of J. Turner and J. Hausfeld).

In addition to resisting tensile forces in structural members, reinforcement is also used in concrete construction for the following reasons:

- To resist a portion of compression loading. For example; increasing a column's axial load carrying capacity, or reducing the element size. Compression in steel also reduces long term creep deflections.

- To resist diagonal tension from shear in beams, walls, and columns. This type of reinforcement is commonly available in the form of stirrups, ties, hoops, or spirals.

- To resist internal pressures in circular structures such as tanks, pipes, bins, and silos.

- To reduce the size of cracks in concrete by distributing stresses resulting in numerous smaller cracks in place of a few large cracks.

- To limit crack widths and control spacing of cracks caused by stresses induced by temperature and moisture changes.

- To provide confinement of the concrete to increase ductility.

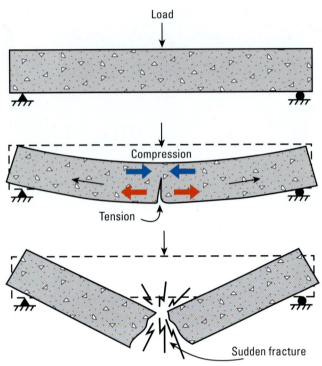

FIGURE 8-2. Unreinforced concrete can have a sudden failure under loading.

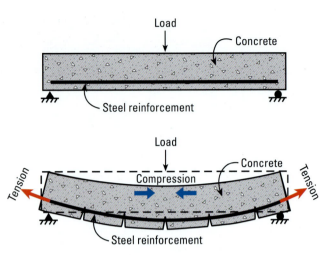

FIGURE 8-3. Reinforcement provides ductility and crack control in concrete.

Consider the beam shown in Figure 8-2. Load applied to the beam produces compression in the top of the beam, and tension in the bottom at mid-span. Unreinforced, plain concrete beams can fail by sudden fracture upon cracking, with little warning. The addition of steel reinforcement increases the load capacity and ductility as shown in Figure 8-3. Placing reinforcement in the tension zone of concrete also provides crack control.

REINFORCING SUPPORT

All reinforcement, including bars and wire, must be supported in location prior to concrete placement. The various types of supports used are covered in CRSI RB4.1, *Supports for Reinforcement Used in Concrete* (2016). One of the most important parameters when designing a structural member is the structural depth, known as the distance from the compression face to the reinforcement. Proper support of reinforcement in forms and above grade during construction assures that the location of reinforcement and structural depth is maintained. ACI 117, *Specifications for Tolerances for Concrete Construction and Materials* (2010), addresses the permissible deviation of reinforcement location from its design location.

STEEL REINFORCING BARS

Steel reinforcing bars (also referred to as rebar) are remarkably well-suited for concrete reinforcement because they exhibit high tensile strength and thermal strain compatibility with concrete. Because of steel's high tensile strength, relatively small amounts are required. For example, steel reinforcing bars are generally about 15 times stronger than conventional

concrete (420 MPa [60,000 psi] versus 28 MPa [4000 psi]) and has about 100 times the tensile strength. A flexural beam typically requires only a small amount of reinforcement to provide tensile strength to the concrete. Reinforcement ratios (area of steel divided by the gross area of concrete) typically vary from 0.0018 to 0.04 in most structures.

The bond between concrete and steel allows for an effective transfer of stress or load between the steel and concrete so that both materials act together in composite action. Both materials expand and contract to about the same degree with temperature changes, about 12 millionths per °C (6.5×10^{-6} per °F) for steel vs. an average of 10 millionths per °C (5.5×10^{-6} per °F) for concrete. Because of this unique compatibility, steel is the most common material used to reinforce concrete. For more information on volume change of hardened concrete due to temperature see Chapter 10.

Reinforcing bars have improved over the last 100 years from the low strength "twisted" bars of the early 1900s to today's high strength/coated deformed bars. Reinforcing bars are specified by ASTM A615, *Standard Specification for Deformed and Plain Carbon-Steel Bars for Concrete Reinforcement* (AASHTO M 31), ASTM A706, *Standard Specification for Deformed and Plain Low-Alloy Steel Bars for Concrete Reinforcement*, ASTM A996, *Standard Specification for Rail-Steel and Axle-Steel Deformed Bars for Concrete Reinforcement*, and ASTM A1035, *Standard Specification for Deformed and Plain, Low-Carbon, Chromium, Steel Bars for Concrete Reinforcement*. ASTM specifications define the material properties as well as ductility requirements.

Grades

Reinforcing bars are available in different grades and sizes for use in concrete. Table 8-1 lists the minimum yield and tensile strengths for the most common ASTM types of steel reinforcement.

TABLE 8-1. Summary of ASTM Strength Requirements for Reinforcement.

REINFORCEMENT TYPE	ASTM SPECIFICATION	DESIGNATION	MINIMUM YIELD STRENGTH MPa (ksi)	MINIMUM TENSILE STRENGTH MPa (ksi)
Reinforcing bars	A615	Grade 40	280 (40)	420 (60)
		Grade 60	420 (60)	550 (80)
		Grade 80	550 (80)	690 (100)
		Grade 100	690 (100)	790 (115)
	A706	Grade 60	420 (60), 540 (78) max	550 (80), not less than 1.25 actual yield
		Grade 80	550 (80), 675 (98) max	690 (100), not less than 1.25 actual yield
	A996	Grade 40	280 (40)	500 (70)
		Grade 50	350 (50)	550 (80)
		Grade 60	420 (60)	620 (90)
	A1035	Grade 100	690 (100)	1030 (150)
		Grade 120	830 (120)	1030 (150)
Wire – plain	A1064	Grade 70	485 (70)	550 (80)
Wire – deformed	A1064	Grade 75	515 (75)	585 (85)
Welded wire reinforcement				
Plain W1.2 and larger	A1064	Grade 65	450 (65)	515 (75)
Deformed	A1064	Grade 70	485 (70)	550 (80)
Prestressing tendons				
Seven-wire strand	A416	Grade 250 (stress-relieved)	85% of tensile	1725 (250)
		Grade 250 (low-relaxation)	90% of tensile	1725 (250)
		Grade 270 (stress-relieved)	85% of tensile	1860 (270)
		Grade 270 (low-relaxation)	90% of tensile	1860 (270)
Wire	A421	Type BA (stress-relieved)	85% of tensile	Varies by size from 1655 (240) to 1620 (235)
		Type BA (low-relaxation)	90% of tensile	Varies by size from 1655 (240) to 1620 (235)
		Type WA (stress-relieved)	85% of tensile	Varies by size from 1725 (250) to 1620 (235)
		Type WA (low-relaxation)	90% of tensile	Varies by size from 1725 (250) to 1620 (235)
Bars	A722	Type I (plain)	85% of tensile	1035 (150)
		Type II (deformed)	80% of tensile	1035 (150)

The most common grade of reinforcement for use in concrete is Grade 60, which has a minimum yield strength of 420 MPa (60,000 psi) and a minimum tensile strength of 550 MPa (80,000 psi). When designing concrete using ACI 318, *Building Code Requirements for Structural Concrete and Commentary* (2019), ASTM A706 Grade 60, Grade 80, and Grade 100 rebar are permitted to resist moments, axial forces, and shear forces in special structural walls, and Grade 60 and Grade 80 are permitted for special moment frames. However, ASTM A615 Grade 80 and Grade 100 are not permitted to resist moments and axial forces in special seismic systems because of concern associated with low-cycle fatigue behavior (Slavin and Ghannoum 2015).

Welding of reinforcing bars must be performed with caution, due to possible changes in metallurgy that could be detrimental to the strength and ductility of the bar. If weldability is important for construction purposes, ASTM A706, Grade 60, includes restrictions on chemical composition for rebar that allows for reliable welding. All welding of reinforcement bars must be in accordance with AWS D1.4, *Structural Welding Code – Reinforcing Steel* (2018).

If using ASTM A996 rail steel as concrete reinforcement, ACI 318-19 requires Type R bars as they meet more restrictive provisions for bend tests than other types of rail steel.

Stress-Strain Curves

Reinforcing bars have distinct stress-strain curves with a definable yield plateau for the lower strength low-carbon steels (Helgason and others 1976). Typical stress-strain curves for U.S. reinforcing steels are shown in Figure 8-4. The complete stress-strain curves are shown in the figure on the left; the right figure gives the initial portions of the curves magnified 10 times. The strain hardening of reinforcing bars is not considered in the design of structural reinforced concrete. The slope of the initial elastic portion of the steel stress-strain curve is the modulus of elasticity (E_s) and is given as 200 GPa (29,000,000 psi) (ACI 318 2019).

Some reinforcing bars, particularly those steels with a grade of 75 and higher, do not always exhibit a well-defined yield point (Fintel 1985). In such instances, the yield strength of the reinforcement is determined using the "0.2 percent offset" method (ACI 318 2019). This method determines yield strength by measuring the stress at an elongation equivalent to a 0.2% deviation from the straight-line approximation of the modulus of elasticity (Figure 8-4 right). This does not change the method of design or the expected performance of the structure. All steels exhibit strain hardening and the length of the yield plateau will vary due to many factors.

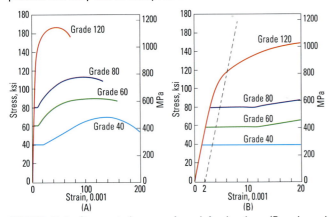

FIGURE 8-4. Stress-strain curve for reinforcing bars (Darwin and others 2015).

Bar Marks

To uniquely identify the reinforcement, bars are marked during production with a letter and number signifying the type of steel (Figure 8-5). The grade of steel is designated with a grade mark or grade line as listed in Table 8-2. The bar mark also contains a letter or symbol identifying the mill in which the bar was produced.

Bars are also marked with their bar size, such as the number '4' indicating a No. 4 bar. The inch-pound bar size number is approximately the bar nominal diameter in eighths of an inch (that is, a No. 4 bar has approximately a ⅘ in., or ½ in., diameter). Rebar manufactured outside of the U.S. may be produced and

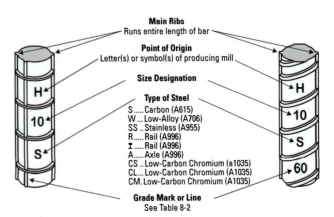

FIGURE 8-5. ASTM bar markings (courtesy of CRSI).

TABLE 8-2. ASTM Bar Markings.

MINIMUM YIELD DESIGNATION		
GRADE OF STEEL	**GRADE MARK***	**GRADE LINE†**
40	blank	no lines
50	blank	no lines
60	60	1 line
75	75	2 lines
80	80	3 lines
100	100 or C (A615)	4 lines (A615)
100	100 (A1035)	3 lines (A1035)
120	120	4 lines

* For stainless-steel (A955) reinforcing bars: Gr 60 = "•", for Gr 75 = "••"
† Grade lines must be at least 5 deformation spaces long.
(Adapted from CRSI 2020.)

marked to true metric sizes. Table 8-3 lists typical bar sizes for standard ASTM reinforcing bars.

Deformed Reinforcing Bars

Reinforced concrete design is predicated on the fact that once the plastic concrete has hardened, the concrete and reinforcing bars undergo the same strains (that is, they behave as a composite material). In order to enhance bond, reinforcing bars are produced with deformations, as shown in Figure 8-5, to improve mechanical bond between the bars and the concrete. Minimum requirements for these deformations have been developed based on an experimental research (Zuo and Darwin 2000). Different bar producers use different patterns, all of which satisfy these requirements.

Non-deformed bars, referred to as plain bars, are available. However, ACI 318-19 permits their usage only as spiral reinforcement. Plain bars may also be used as unbonded dowels to maintain alignment of concrete members, such as along movement joints in walls and slabs.

TABLE 8-3. ASTM Standard Reinforcing Bars.

BAR SIZE DESIGNATION NO.	ASTM STANDARD METRIC (INCH-POUND) REINFORCING BARS		
	NOMINAL DIMENSIONS		
	CROSS-SECTIONAL AREA mm² (in.²)	MASS kg/m (lb/ft)	DIAMETER mm (in.)
10 (3)	71 (0.11)	0.560 (0.376)	9.5 (0.375)
13 (4)	129 (0.20)	0.994 (0.668)	12.7 (0.500)
16 (5)	199 (0.31)	1.552 (1.043)	15.9 (0.625)
19 (6)	284 (0.44)	2.235 (1.502)	19.1 (0.750)
22 (7)	387 (0.60)	3.042 (2.044)	22.2 (0.875)
25 (8)	510 (0.79)	3.973 (2.670)	25.4 (1.000)
29 (9)	645 (1.00)	5.060 (3.400)	28.7 (1.128)
32 (10)	819 (1.27)	6.404 (4.303)	32.2 (1.270)
36 (11)	1,006 (1.56)	7.907 (5.313)	35.8 (1.410)
43 (14)	1,452 (2.25)	11.38 (7.65)	43.0 (1.693)
57 (18)	2,581 (4.00)	20.24 (13.60)	57.3 (2.257)
64 (20)	3,167 (4.91)	24.84 (16.69)	63.5 (2.500)

ASTM A615 Grade 40 is only available in bar sizes No. 10 (3) through No. 19 (6). ASTM A706 and ASTM A996 are not available in bar size No. 64 (20). ASTM A996 is not available in bar sizes No. 43 (14) through No. 64 (20).

COATED REINFORCING BARS AND CORROSION PROTECTION

In exterior environments exposed steel will eventually corrode. Fortunately, when concrete surrounds the steel, the high pH of the concrete cover provides excellent corrosion protection. The corrosion resistance may be reduced if the concrete paste is subjected to carbonation that reduces alkalinity of the concrete or the concrete is cracked, permitting direct pathway of chlorides to the level of the reinforcement. Corrosion can be initiated by certain levels of chloride ions from seawater or deicing salts reaching the reinforcing steel.

Corrosion of steel reinforcement causes byproducts (rust) to expand in volume. The expansion generates large internal stresses that may result in cracking and spalling of the concrete, exposing the steel reinforcement (Figure 8-6). Corrosion can result in a reduction in the strength of the element.

To improve the corrosion resistance of reinforcing bars, bars may be epoxy-coated, zinc-coated (galvanized), both (dual-coated), or under special circumstances, stainless steel bars or fiber reinforced polymer reinforcement may be used. Coated and uncoated reinforcement should not be placed in the same member.

Corrosion inhibiting admixtures, surface treatments, concrete overlays, and cathodic protection may also be employed for increasing resistance to corrosion or reducing the rate of corrosion as discussed in Chapter 11.

FIGURE 8-6. Reinforcing steel corrosion and concrete spalling due to reduced cover.

Epoxy-Coated Reinforcement. Epoxy coating provides a waterproof, dielectric surface to enhance corrosion resistance. The epoxy coating prevents chloride ions and other corrosive chemicals, moisture, and oxygen from reaching the steel. Fusion-bonded epoxy-coated reinforcing steel is popular for the construction of marine structures, pavements, bridge decks, and parking garages exposed to deicer chemicals. As shown in Figures 8-7 and 8-8 epoxy-coated reinforcement is generally recognized in the field by a green coating. ASTM A775, *Standard Specification for Epoxy-Coated Steel Reinforcing Bars*, and ASTM A934, *Standard Specification for Epoxy-Coated Prefabricated Steel Reinforcing Bars*, cover the requirements for bars coated before or after fabrication (respectively) including surface preparation.

FIGURE 8-7. Epoxy-coated reinforcing bars.

FIGURE 8-8. Shipping of epoxy-coated reinforcing bars.

Care must be taken to minimize damage to the epoxy-coating on both epoxy and dual coated reinforcement during shipping and placement (Figure 8-7). Damage to the epoxy-coating may result in localized corrosion and reduced performance. If the epoxy-coating contains pinholes or is damaged during construction, its protection ability is effectively reduced. Repairs to all visible damage to the coating should be made prior to concrete placement. Epoxy repair kits are available and should be used to recoat the damaged portion of the bar prior to concrete placement. Epoxy coated rebar should also conform to ASTM D3963, *Specification for Fabrication and Jobsite Handling of Epoxy-Coated Steel Reinforcing Bars*.

As the coating on epoxy-coated reinforcing bars interferes with the surface bond between concrete and the reinforcing steel, longer lap splice and development lengths are required for epoxy-coated reinforcement compared with uncoated reinforcement. Changes in development lengths are calculated within the design codes and are typically 20% longer than for uncoated bars.

Zinc-Coated (Galvanized) Reinforcement. Corrosion of the zinc coating on galvanized bars provides a sacrificial physical barrier and cathodic protection to uncoated areas, such as scratches in the coating. As the zinc coating corrodes, it may also create a barrier along the reinforcement, protecting it against further corrosion.

Galvanized bars are covered by ASTM A767, *Standard Specification for Zinc-Coated (Galvanized) Steel Bars for Concrete Reinforcement*, and ASTM A1094, *Standard Specification for Continuous Hot-Dip Galvanized Steel Bars for Concrete Reinforcement*. Bars are typically galvanized after cutting and bending to the project requirements. Zinc and epoxy dual coated bar requirements are covered in ASTM A1055, *Standard Specification for Zinc and Epoxy Dual-Coated Steel Bars*.

Chloride-ions can cause corrosion of galvanized steel in concrete and may lead to severe cracking and spalling of the surrounding concrete. The use of chloride admixtures should be avoided in concrete containing reinforcing including galvanized steel exposed to corrosive or wet environments. Stark (1989) illustrates the effect of humidity and chloride content on corrosion of bare (untreated) and galvanized steel bars.

Additional production time may be required to fabricate and galvanize reinforcement. Design using galvanized bars is the same as that for uncoated bars, except that bend diameters may be greater than those for the uncoated bars.

Stainless-Steel Reinforcement. Use of stainless steel reinforcement in zones exposed to high chloride concentrations can ensure a long service life in that part of the structure, provided the concrete itself is made sufficiently resistant to avoid other types of deterioration. Stainless steel can be coupled with carbon steel reinforcement without causing galvanic corrosion (Gjørv 2009). Some stainless steels may be used in concretes containing chloride if test data support their performance.

Stainless steel reinforcement should comply with ASTM A955, *Standard Specification for Deformed and Plain Stainless Steel Bars for Concrete Reinforcement*, and the designation or alloy of stainless steel must be specified.

Fiber-Reinforced Polymer (FRP) Reinforcement. Fiber-reinforced polymers are composite materials that typically consist of strong fibers embedded in a resin matrix (Figure 8-9), with the fibers providing strength and stiffness and carrying most of the loads, while the matrix acts to bond and protect the fibers and to provide shear resistance within the material (ACI 440R 2007). FRP can occasionally be used to replace part or all of the steel reinforcement in portland cement concrete exposed to extremely corrosive chemicals.

Most often, FRP products are constructed using thermoset resins such as polyester, vinylester, epoxy, and phenolics, but they can also be made using thermoplastic resins such as polypropylene, nylon, polycarbonate, and others.

FIGURE 8-9. FRP can be manufactured in shapes that mimic steel rebar.

Generally, thermoplastic polymers can be remelted and recycled and thermoset resins cannot.

The lightweight, nonmagnetic, nonconductive, high strength (tensile strength greater than 690 MPa [100,000 psi]) bars are chemically resistant to many acids, salts, and gases and are unaffected by electrochemical attack. Commercially available FRP reinforcement is made of continuous aramid, carbon, or glass fibers embedded in a resin matrix. The resin allows the composite action of the fibers to work as a single element.

FRP reinforcing bars are available in most conventional bar sizes. The reinforcing fiber can be discontinuous (chopped) or continuous (long strand). Fiber-reinforced polymers shapes include bars, tendons, sheets, dowels, and connectors. All forms have good corrosion resistance and a high strength-to-weight ratio, with a strength exceeding that of steel. Bars are used similarly to steel rebar in concrete. Tendons are used for prestressed concrete. Sheets are bonded to flat or curved surfaces to strengthen existing concrete or masonry structures, often to provide enhanced seismic resistance to older structures that were designed by past codes. In the cases of sheet materials, ensuring good bond with the substrate is key to performance. Tie connectors and grid shear connectors are used in concrete sandwich walls, which can significantly improve energy efficiency by removing the thermal bridges (ACI 440R 2007).

FRPs lack ductility, so failure modes are brittle and differ from the behavior of concrete beams under-reinforced with steel. Flexural failure of concrete members reinforced with currently available FRP materials is governed by either concrete crushing or FRP tensile rupture. To compensate, a higher reserve of strength is required, which means lower strength-reduction factors for FRPs. Shear capacity of concrete beams reinforced with FRPs is also reduced. See ACI 440R, *Report on Fiber-Reinforced Polymer (FRP) Reinforcement for Concrete Structures* (2007), for special design considerations.

WELDED WIRE REINFORCEMENT

For reinforcing slabs and other structural elements, an alternative to reinforcing bars is welded wire reinforcement (WWR), also known as welded wire mesh or fabric. Welded wire reinforcement is comprised of orthogonal longitudinal and transverse cold-drawn steel wires that are assembled by welding at every wire intersection (Figure 8-10). The size and spacing of the wires can vary in each direction, based on the requirements of the project.

FIGURE 8-10. Welded wire reinforcement (courtesy of Master Builders Solutions USA, LLC).

Welded wire reinforcement is typically designated by the spacing of the wires in each direction along with the wire diameter, where the spacing is in millimeters (inches) and the wire size is the cross-sectional area in square millimeters (hundredths of a square inch). Plain (smooth) wires are designated with an "MW" ("W") and deformed wires are designated with an "MD" ("D"). As per Table 8-4, a 152 × 152-MW 26 × MW 26 (6 × 6-W4 × W4) would provide 170 mm^2 per meter (0.08 $in.^2$ per foot) of width of welded wire reinforcement (WRI 2016). Typically, the yield strength of WWR is taken as 420 MPa (60,000 psi) but is available with higher yield strengths, including 551 MPa to 689 MPa (80,000 psi to 100,000 psi).

Plain and deformed WWR is covered in a combined standard, ASTM A1064, *Standard Specification for Carbon-Steel Wire and Welded Wire Reinforcement, Plain and Deformed, for Concrete*. Plain steel wires are available in sizes ranging from MW9 to MW290 (W1.4 to W45) and deformed steel wires are available in sizes ranging from MD26 to MD290 (D4 to D45). Stainless steel wires are specified according to ASTM A1022, *Standard Specification for Deformed and Plain Stainless Steel Wire and Welded Wire for Concrete Reinforcement*. Epoxy-coated WWR is available in accordance with ASTM A884, S*tandard Specification for Epoxy-Coated Steel Wire and Welded Wire Reinforcement.* Galvanized WWR is available in accordance with ASTM A1060, *Standard Specification for Zinc-Coated (Galvanized) Steel Welded Wire Reinforced, Plain and Deformed, for Concrete.*

TABLE 8-4. Welded Wire Reinforcement.

W & D WIRE SIZE PLAIN*	W & D METRIC WIRE SIZE (CONVERSION) PLAIN†	CUSTOMARY UNITS			METRIC UNITS (CONVERSION)		
		AREA, in.²	DIAMETER, in.	NOMINAL WEIGHT, lb/ft	NOMINAL AREA, mm²	NOMINAL DIAMETER, mm	NOMINAL MASS, kg/m
W45	MW 290	0.45	0.757	1.530	290	19.23	2.28
W31	MW 200	0.31	0.628	1.054	200	15.96	1.57
W20	MW 130	0.200	0.505	0.680	129	12.8	1.01
	MW 122	0.189	0.490	0.643	122	12.4	0.96
W18	MW 116	0.180	0.479	0.612	116	12.2	0.91
	MW 108	0.168	0.462	0.571	108	11.7	0.85
W16	MW 103	0.160	0.451	0.544	103	11.5	0.81
	MW 94	0.146	0.431	0.495	94	10.9	0.74
W14	MW 90	0.140	0.422	0.476	90	10.7	0.71
	MW 79	0.122	0.394	0.414	79	10.0	0.62
W12	MW 77	0.120	0.391	0.408	77	9.9	0.61
W11	MW 71	0.110	0.374	0.374	71	9.5	0.56
W10.5	MW 68	0.105	0.366	0.357	68	9.3	0.53
	MW 67	0.103	0.363	0.351	67	9.2	0.52
W10	MW 65	0.100	0.357	0.340	65	9.1	0.51
W9.5	MW 61	0.095	0.348	0.323	61	8.8	0.48
W9	MW 58	0.090	0.338	0.306	58	8.6	0.45
	MW 56	0.086	0.331	0.292	55.5	8.4	0.43
W8.5	MW 55	0.085	0.329	0.289	54.9	8.4	0.43
W8	MW 52	0.080	0.319	0.272	52	8.1	0.40
W7.5	MW 48	0.075	0.309	0.255	48.4	7.8	0.38
W7	MW 45	0.070	0.299	0.238	45	7.6	0.35
W6.5	MW 42	0.065	0.288	0.221	42	7.3	0.33
	MW 41	0.063	0.283	0.214	41	7.2	0.32
W6	MW 39	0.060	0.276	0.204	39	7.0	0.30
W5.5	MW 36	0.055	0.265	0.187	35.5	6.7	0.28
	MW 35	0.054	0.263	0.184	34.8	6.7	0.27
W5	MW 32	0.050	0.252	0.170	32	6.4	0.25
	MW 30	0.047	0.244	0.158	30	6.2	0.24
	MW 29	0.045	0.239	0.153	29	6.1	0.23
W4	MW 26	0.040	0.226	0.136	26	5.7	0.20
W3.5	MW 23	0.035	0.211	0.119	23	5.4	0.18
W2.9	MW 19	0.029	0.192	0.098	19	4.9	0.15
W2.0	MW 13	0.020	0.160	0.068	13	4.1	0.10
W1.4	MW 9	0.014	0.135	0.048	9	3.4	0.07

(Courtesy of Wire Reinforcement Institute.)
*U.S. customary sizes can be specified in 0.001 in.² increments.
† Metric wire sizes can be specified in 1 mm² increments.

PRESTRESSING STEEL

Prestressing a concrete member takes maximum advantage of the beneficial aspects of both the concrete and steel. In prestressed members, compressive stresses are purposefully introduced into the concrete to reduce tensile stresses resulting from applied loads, including the self weight of the member (dead load). Prestressing steel comes in three standard types: wires, tendons composed of several strands of wires, and high strength alloy steel bars.

Pretensioning is a method of prestressing in which the reinforcement is tensioned before the concrete is placed. The prestressing force is primarily transferred to the concrete through bond. Post-tensioning is a method of prestressing in which the reinforcement is tensioned after the concrete has hardened. The prestressing force is primarily transferred to the concrete through the end anchorages. In both methods, prestressing is transferred to the concrete at an early age. In pretensioned applications, steam curing helps achieve

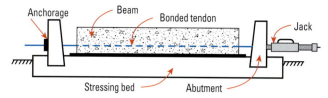

FIGURE 8-11. Pretensioning prestressed concrete.

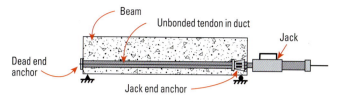

FIGURE 8-12. Post-tensioning prestressed concrete.

28 MPa to 55 MPa (4000 psi to 8000 psi) in about 16 hours. In post-tensioned construction, concrete strengths of 21 MPa to 28 MPa (3000 psi to 4000 psi) are reached at early age- in one to three days for stressing time, and design strengths can be much higher, depending on the application.

Pretensioning consists of tensioning steel in the forms prior to placing the concrete and cutting the stressed wires after the concrete develops the necessary strength. Prestressing adds a precompression to the concrete to offset tension stresses induced later during loading (Figure 8-11) (PCI 2017).

Post-tensioning consists of casting the concrete around tendons placed in ducts, or an extruded, greased sheathing. After the concrete hardens, the tendons are stressed (Figures 8-12 and 8-13). After stressing, the tendons, the post-tensioning may be grouted, bonded or unbonded, and encapsulated (PTI 2006).

FIGURE 8-13. Post-tensioning tendon layout shown with green sheathing (courtesy of Neal Khosa).

Tendons must be properly placed and stressed only after concrete has developed enough strength to withstand the load. Tensioning operations are dangerous and appropriate safety precautions must be taken including training and certification of personnel conducting the installation (PTI 2019).

The requirements for prestressing steels are covered by ASTM A421, *Standard Specification for Stress-Relieved Steel Wire for Prestressed Concrete* (AASHTO M 204), ASTM A416, *Standard Specification for Low-Relaxation, Seven-Wire Steel Strand for Prestressed Concrete* (AASHTO M 203), and ASTM A722, *Standard Specification for High-Strength Steel Bars for Prestressed Concrete*. The most common form of tendon consists of low-relaxation seven-wire strand ASTM A416 Grade 270 (f_{pu} = 1750 MPa [250,000 psi]) (Table 8-5). Prestressing wires are cold-drawn, high-strength, high-carbon steel with diameters ranging from 4.9 mm to 7.0 mm (0.192 in. to 0.276 in.).

Prestressing wires, strands, and alloy bars do not exhibit the same elastic plastic stress-strain curves of conventional reinforcing steel. The yielding of the prestressing steel occurs gradually. The ratio of the fracture strain to yield strain for a prestressing wire is significantly less than that for Grade 60 reinforcing bar, therefore, prestressing tendons tend to be less ductile than conventional reinforcement. Prestressing tendons also exhibit some relaxation under service loads which needs to be accounted for in design.

Since a prestressed structure counteracts the loading induced tensions with a compressive force, prestressed structures can be designed for no cracking. The induced tension also results in less deflection under load. For extremely aggressive environments, epoxy-coated strand is available and is covered under ASTM A882, *Standard Specification for Filled Epoxy-Coated Seven-Wire Steel Prestressing Strand*. For building use, to resist corrosion, encapsulated tendon systems are utilized in post-tensioned concrete per ACI 423.7, *Specifications for Unbonded Single Strand Tendon Materials* (2014).

For more information on prestressed and post-tensioned concrete see the Precast/Prestressed Concrete Institute (2017) and Post-Tensioning Institute (2006).

For more information on reinforcing for use in concrete refer to ACI 439.4R, *Report on Steel Reinforcement-Material Properties and US Availability* (2009) and the CRSI *Design Guide on the ACI 318 Building Code Requirements for Structural Concrete* (2020).

TABLE 8-5A. Standard Prestressing Tendons (SI).

TYPE*	NOMINAL DIAMETER, mm	NOMINAL AREA, mm²	NOMINAL MASS, kg/m
Seven-wire strand (Grade 1725)	6.4	23.2	0.182
	7.9	37.4	0.294
	9.5	51.6	0.405
	11.1	69.7	0.548
	12.7	92.9	0.730
	15.2	139	1.090
Seven-wire strand (Grade 1860)	9.53	55.0	0.430
	11.1	74.2	0.580
	12.70	98.7	0.780
	13.2	108	0.840
	14.3	124	0.970
	15.2	140	1.100
	15.7	150	1.200
	15.8	190	1.506
Prestressing wire	4.88	18.7	0.146
	4.98	19.5	0.149
	6.35	31.7	0.253
	7.01	38.6	0.298
Prestressing bars (plain)	19	284	2.23
	22	387	3.04
	25	503	3.97
	29	639	5.03
	32	794	6.21
	35	955	7.52
Prestressing bars (deformed)	15	181	1.46
	20	271	2.22
	26	548	4.48
	32	806	6.54
	36	1,019	8.28
	46	1,664	13.54
	65	3,331	27.10
	75	4,419	35.85

*Availability of some tendon sizes should be investigated in advance.

TABLE 8-5B. Standard Prestressing Tendons (inch-pound).

TYPE*	NOMINAL DIAMETER, in.	NOMINAL AREA, in.²	NOMINAL MASS, lb/ft
Seven-wire strand (Grade 250)	¼ (0.250)	0.036	0.122
	⅝ (0.313)	0.058	0.197
	⅜ (0.375)	0.080	0.272
	⁷⁄₁₆ (0.438)	0.108	0.367
	½ (0.500)	0.144	0.490
	0.600	0.216	0.737
Seven-wire strand (Grade 270)	⅜ (0.375)	0.085	0.290
	⁷⁄₁₆ (0.438)	0.115	0.390
	½ (0.500)	0.153	0.520
	0.520	0.167	0.570
	0.563	0.192	0.650
	0.600	0.217	0.740
	0.620	0.231	0.780
	0.700	0.294	1.000
Prestressing wire	0.192	0.029	0.098
	0.196	0.030	0.100
	0.250	0.049	0.170
	0.276	0.060	0.200
Prestressing bars (plain)	¾	0.44	1.50
	⅞	0.60	2.04
	1	0.78	2.67
	1⅛	0.99	3.38
	1¼	1.23	4.17
	1⅜	1.48	5.05
Prestressing bars (deformed)	⅝	0.28	0.98
	¾	0.42	1.49
	1	0.85	3.01
	1¼	1.25	4.39
	1⅜	1.58	5.56
	1¾	2.58	9.10
	2½	5.16	18.20
	3.0	6.85	24.09

*Availability of some tendon sizes should be investigated in advance.

REFERENCES

PCA's online catalog includes links to PDF versions of many of our research reports and other classic publications.
Visit: cement.org/library/catalog.

ACI Committee 117, *Specification for Tolerances for Concrete Construction and Materials* and Commentary, ACI 117-10, American Concrete Institute, Farmington Hills, MI, 2010, reapproved 2015, 76 pages.

ACI Committee 318, *Building Code Requirements for Structural Concrete and Commentary*, ACI 318-19, American Concrete Institute, Farmington Hills, MI, 2019, 520 pages.

ACI Committee 423, *Specification for Unbonded Single Strand Tendon Materials*, ACI 423.7-14, American Concrete Institute, Farmington Hills, MI, 2014, 8 pages.

ACI Committee 439, *Report on Steel Reinforcement-Material Properties and US Availability*, ACI 439.4R-09, American Concrete Institute, Farmington Hills, MI, 2009, reapproved 2017, 25 pages.

ACI Committee 440, *Report on Fiber-Reinforced Polymer (FRP) Reinforcement for Concrete Structures*, ACI 440R-07, American Concrete Institute, Farmington Hills, MI, 2007, 104 pages.

AWS, *Structural Welding Code – Reinforcing Steel*, AWS D1.4/D1.4M: 2018, American Welding Society, Miami, FL, 2018.

CRSI, *Design Guide on the ACI 318 Building Code Requirements for Structural Concrete*, Concrete Reinforcing Steel Institute, Schaumburg, IL, 2020, 996 pages.

CRSI, *Placing Reinforcing Bars*, 10th Edition, Concrete Reinforcing Steel Institute, Schaumburg, IL, 2019, 296 pages.

CRSI, *Supports for Reinforcement Used in Concrete*, RB-4.1, Concrete Reinforcing Steel Institute, Schaumburg, IL, 2016, 25 pages.

Darwin, D.; Dolan, C.; and Nilson, A., *Design of Concrete Structures*, 15th Edition, McGraw Hill Education, January 2015, 800 pages.

Fintel, M., *Handbook of Concrete Engineering*, 2nd edition, Van Nostrand Reinhold Company Inc., New York, NY, 1985, 892 pages.

Helgason, T.; Hanson, J.M.; Somes, N.F.; Corley, W.G.; and Hognestad, E., *Fatigue Strength of High-Yield Reinforcing Bars*, RD045, Portland Cement Association, Skokie, IL, 1976, 34 pages.

PCI, *PCI Design Handbook, Precast and Prestressed Concrete*, MNL-120-17, 8th Edition, Precast/Prestressed Concrete Institute, Chicago, IL, 2017.

PTI, *Post-Tensioning Manual*, Sixth Edition, Post-Tensioning Institute, 2006, 354 pages.

PTI/ASBI M50.3 – *Specification for Multistrand and Grouted Post-Tensioning*, Post-Tensioning Institute, 2019.

PTI M55.1 – *Specification for Grouting of Post-Tensioned Structures*, Post-Tensioning Institute, 2019.

Slavin, C.M.; and Ghannoum, W.M., *Defining Structurally Acceptable Properties of High-Strength Steel Bars through Material and Column Testing, Part I: Material Testing Report*, RGA #05-14, Charles Pankow Foundation, McLean, VA, August 2015, 135 pages.

WRI, *Structural Welded Wire Reinforcement Manual of Standard Practice*, Wire Reinforcement Institute, Hartford, CT, 9th Edition, December 2016, 52 pages.

Zuo, J.; and Darwin, D., "Splice Strength of Conventional and High Relative Rib Area Bars in Normal and High Strength Concrete," *ACI Structural Journal*, Vol. 97, No. 4, July-August 2000, 12 pages.

PROPERTIES OF CONCRETE

Quality concrete possesses well defined and accepted principal characteristics. For freshly mixed concrete, those include:

- *Workability* – ease of placing, consolidating, and finishing.
- *Consistency* – the ability to flow.
- *Uniformity* – a homogeneous mixture, with evenly dispersed constituents.
- *Stability* – resistance to segregation.
- *Finishability* – ease of performing finishing operations to achieve specified surface characteristics.

For hardened concrete, they include:

- *Strength* – resists strain or rupture induced by external forces (compressive, flexural, tensile, torsion, and shear).
- *Durability* – resists weathering, chemical attack, abrasion, and other service conditions.
- *Appearance* – meets the desired aesthetic characteristics.

For different applications, concrete properties need to be controlled within certain ranges. By understanding the nature and basic characteristics of concrete, fresh and hardened properties can be more readily designed to meet and perform within a given budget.

FRESHLY MIXED CONCRETE

Freshly mixed concrete should be plastic or semifluid and generally capable of being molded by hand (Figure 9-1). A flowable concrete mixture can be molded in the sense that it can be cast in a mold, but this is not within the definition of *plastic* – that which is pliable and capable of being molded or shaped.

FIGURE 9-1. Concrete of plastic consistency does not crumble, but flows cohesively without segregation, maintaining stability.

In a plastic concrete mixture, all grains of sand and particles of gravel or stone are encased and held in suspension. The ingredients are not apt to segregate during transport. During placing, concrete of proper consistency does not crumble, but flows cohesively without segregation, maintaining stability.

A plastic mixture is required for strength and homogeneity during handling and placement. A slump ranging from 75 mm to 150 mm (3 in. to 6 in.) is suitable for most concrete work; however, plasticizing admixtures may be used to make concrete more flowable for use in thin or heavily reinforced concrete members.

In general, fresh concrete must be capable of satisfying the following requirements:

• be easily mixed and transported,
• be uniform throughout a given batch (and consistent between batches),
• have flow properties that allow it to completely fill formwork,
• be compacted (consolidated) without requiring an excessive amount of energy,
• not segregate during transportation, placing, and consolidation, and
• be capable of being finished to the required surface treatment.

Workability

The ease of placing, consolidating, and finishing freshly-mixed concrete and the degree to which it resists segregation is called *workability*. Concrete should be workable, but the ingredients should not separate during transport and handling. Concrete properties related to workability include the *consistency* (flow) and *stability* (segregation resistance). Further defined flow characteristics (used in high-performance mixtures such as self-consolidating concrete) include: *unconfined flowability*, called the filling ability; and *confined flowability*, called the passing ability. *Dynamic stability* is the ability to resist separation during transport and placement. *Static stability* is the ability to maintain a uniform distribution of all mixture components after the fluid concrete has stopped moving.

The degree of workability required for proper placement of concrete is controlled by the placement method, type of consolidation, and type of concrete. For example, self-consolidating concrete has the unique properties of high workability without loss of stability and allows for complex shapes and rigorous construction schedules (Szecsy and Mohler 2009). Different types of placements require different levels of workability (see Chapter 15).

Factors that influence the workability of concrete are:

• method and duration of transportation;
• quantity and characteristics of cementitious materials;

• concrete consistency;
• grading, shape, and surface texture of aggregates;
• air content;
• water content;
• concrete and ambient air temperatures; and
• selection of admixtures.

A uniform distribution of aggregate particles and the presence of entrained air significantly help control segregation and improve workability. Figure 9-2 illustrates the effect of casting temperature on the consistency, or slump, and relative workability of concrete mixtures.

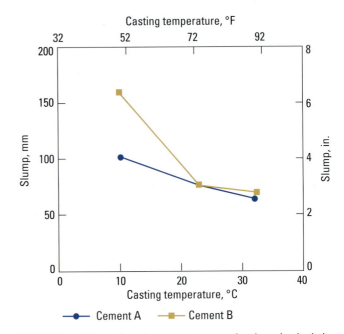

FIGURE 9-2. Effect of casting temperature on the slump (and relative workability) of two concretes made with different cements (Burg 1996).

Consistency. Consistency is considered a close indication of workability. The slump test, ASTM C143, *Standard Test Method for Slump of Hydraulic-Cement Concrete* (AASHTO T 119), is the most generally accepted method used to measure the consistency of concrete (Figure 9-3). The primary benefit of the slump test is that it measures the consistency from one batch of concrete to the next. However, it does not characterize the rheology or workability of a concrete mixture completely. Other test methods are available for measuring consistency of concrete (see Chapter 20) more technically. A low-slump concrete has a stiff consistency. If the consistency is too dry and harsh, the concrete will be difficult to place and compact. The larger aggregate particles may separate from the mixture. However, it should not be assumed that a wetter, more fluid mixture is necessarily more workable. Segregation, honeycombing and reduced hardened properties can occur if the mixture is too wet.

FIGURE 9-3. Concrete should have a consistency suitable for placement applications.

Excessive water used to produce high slump is a primary cause of poor concrete performance – it leads to higher bleeding, segregation, and increased drying shrinkage. If a finished concrete surface is to be level, uniform in appearance, and wear resistant, all batches placed in the floor must have nearly the same slump and meet specification criteria. Concrete should be placed at the desired consistency with the lowest practical water content for placement and consolidation equipment. For more information on consistency, see Powers (1932) and Daniel (2006).

Rheology. The science of rheology has been widely used to address the concepts of flowability and stability of concrete mixtures, particularly for self-consolidating concrete. Rheology is the study of material deformation and flow. Rheology allows researchers, practitioners, mixture developers, and others a more scientific approach to determine the flow and workability of concrete.

The Bingham model (Bingham 1916) describes two properties of the material, the yield stress and the plastic viscosity (Figure 9-4). In fresh concrete, *yield stress* defines the threshold between static and fluid behavior. Consider trying to pull a come-along through a pile of freshly mixed concrete. As soon as one exerts enough force on the pile of concrete, it will begin to move, indicating that the force has overcome the yield stress of the concrete. As soon as the stress applied to the concrete no longer exceeds the yield stress, the concrete stops moving. The *viscosity* of the concrete determines how fast it moves (rate of deformation). In order for concrete to flow without assistance, the yield stress of the freshly mixed concrete must be low enough that it can move by the effects of gravity. While concrete does not completely follow Bingham model behavior, it is sufficiently close at low shear rates to be useful in understanding the rheological behavior of concrete.

The Bingham model is a means to characterize concrete by two measurable parameters determined with established engineering principles. From a practical standpoint, those

two parameters give contractors and producers a means to explicitly specify the fresh property performance parameters best suited for the placement application.

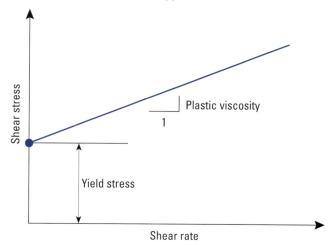

FIGURE 9-4. Basic Bingham behavior curve.

Bleeding and Settlement

Bleeding is the development of a layer of water at the top or surface of freshly placed concrete. It is caused by sedimentation (settlement) of solid particles (cement and aggregate) and the simultaneous upward migration of water (Figure 9-5). Some bleeding is normal and it should not diminish the quality of properly placed, finished, and cured concrete.

FIGURE 9-5. Bleed water on the surface of a freshly placed concrete slab.

Excessive bleeding increases the water-cement ratio near the top surface which creates a weak top layer with poor durability, especially if finishing operations take place while bleed water is present. A water pocket or void can develop under a prematurely finished surface (which can cause a future surface delamination). Bleed water can accumulate under and alongside coarse aggregate particles (Figure 9-6). This is especially likely when differential settlement occurs between the aggregate and paste, or between the paste and reinforcement. Once the aggregate can no longer settle, the paste continues to settle allowing bleed water to rise and collect under the aggregate.

FIGURE 9-6. Trapped bleed water pocket left void under a coarse aggregate particle.

Bleed-water channels also tend to migrate along the sides of coarse aggregate. This leads to a reduction in paste-aggregate bond, which reduces concrete strengths.

The bleeding properties of fresh concrete can be determined by two methods described in ASTM C232, *Standard Test Methods for Bleeding of Concrete* (AASHTO T 158). Because most concrete ingredients today provide concrete with a normal and acceptable level of bleeding, bleeding is usually not a concern and bleeding tests are rarely performed. However, there are situations in which bleeding properties of concrete should be reviewed prior to construction. In some instances, lean concretes placed in very deep forms have accumulated large amounts of bleed water at the surface. In contrast, lack of bleed water on concrete flatwork may lead to plastic shrinkage cracking or a dry surface that is difficult to finish (Chapter 16).

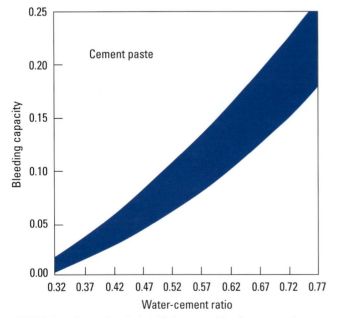

FIGURE 9-7. Range in relationship between bleeding rate and water-cement ratio of pastes made with normal portland cement and water. The range is attributed to different cements having different chemical composition and fineness (Steinour 1945).

The bleeding rate and bleeding capacity (total settlement per unit of original concrete height) increases with initial water content and concrete height and pressure (Figure 9-7). The accumulation of water at the surface of a concrete mixture can occur slowly by uniform seepage over the entire surface or at localized channels carrying water to the surface.

Uniform seepage is referred to as normal bleeding. Water rising through the concrete in discrete paths, sometimes carrying fine particles with it, is termed *channel bleeding*. This can occur in concrete mixtures with very low cement contents, high water contents, or concretes with very high bleeding properties (Figure 9-8).

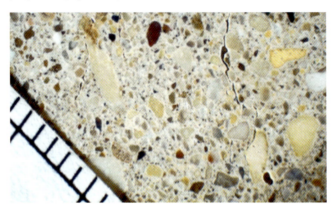

FIGURE 9-8. Large bleed channels may form from concrete mixtures with very low cement contents, high water contents, or other mixtures with high bleeding characteristics.

As bleeding proceeds, the water layer at the surface maintains the original height of the concrete sample in a vessel, assuming that there is no pronounced temperature change or evaporation. The surface subsides as the solids settle through the liquid. After evaporation of all bleed water, the hardened surface will be slightly lower than the freshly placed surface. This decrease in height from time of placement to initial set is called settlement shrinkage.

Reduced bleeding may be required for a variety of reasons including facilitating finishing operations, minimizing the formation of weak concrete at the top of lifts, reducing sand streaking in wall forms, or to stabilize the hardened volume with respect to the plastic volume and settlement of the concrete.

The most effective means of reducing bleeding in concrete include:

• reducing the water content, water-cementitious material ratio, and slump;

• increasing the amount of cement resulting in a reduced water-cement ratio;

• using finer cementitious materials (Figure 9-9);

• using blended hydraulic cements or supplementary cementing materials;

• increasing the amount of supplementary cementing materials;

- increasing the amount of fines in the sand;
- using chemical admixtures that permit reduced water-to-cementitious materials ratios or provide other means capable of reducing the bleeding of concrete; and
- Using air-entrained concrete.

For more information on bleeding, see Powers (1939), Steinour (1945), and Kosmatka (2006).

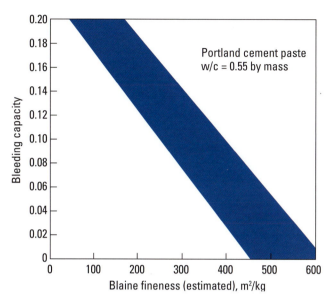

FIGURE 9-9. Effect of portland cement fineness on bleeding capacity of paste (after Steinour 1945, with the original Wagner fineness data converted to approximate Blaine values).

Air Content

Air entrainment is recommended for nearly all exterior concretes, principally to improve freeze-thaw resistance when exposed to freezing water and deicing chemicals (see Chapter 11). Entrained air may also be useful for concrete mixtures that do not require freeze-thaw protection because it reduces bleeding and increases plasticity, among other important benefits of entrained air.

Air-entrained concrete is produced by adding an air-entraining admixture during batching, or less commonly, using an air-entraining cement. The air-entraining admixture stabilizes bubbles formed during the mixing process, enhances the incorporation of bubbles of various sizes by lowering the surface tension of the mixing water, impedes bubble coalescence, and adheres bubbles to cement and aggregate particles (see Chapter 6).

While minimum air contents are well established for durability (see Chapters 11 and 13), there is also a reason to consider setting a maximum air content to control strength and potential surface delaminations. Entrained air lowers the compressive strength of concrete (a general rule is 5% to 6% strength reduction for every percent of entrained air). A maximum total air content of 3% has also been established for interior floor

slabs receiving troweled finishes to reduce the possibility of blistering (ACI 302.1R 2015). This occurs because steel trowels can seal the surface and trap air pockets beneath it.

The total air content developed in hardened concrete is impacted by constituent materials, mixture proportions, production and handling, delivery, placing, finishing methods, and the environment as discussed in the following sections.

Concrete Constituents. As summarized in Table 9-1, the constituent materials may have a significant effect on air content.

As cementitious materials contents increase, the air content decreases for a fixed dosage of air-entraining admixture per unit of cementitious material within the normal range of cement contents (Figure 9-10). For example, in going from 240 kg/m³ to 360 kg/m³ of cement (400 lb/yd³ to 600 lb/yd³), the dosage rate may have to be doubled to maintain a constant air content. However, studies indicate that when the admixture dosage is increased, the air-void spacing factor generally decreases and for a given air content the specific surface increases, improving durability.

An increase in cement fineness will result in a decrease in the amount of air entrained. Type III cement, a very finely ground material, may require twice as much air-entraining admixture as a Type I cement of normal fineness.

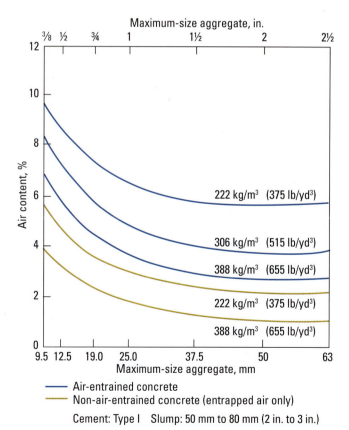

FIGURE 9-10. Relationship between aggregate size, portland cement content, and air content of concrete. The air-entraining admixture dosage per unit of cement was constant for air-entrained concrete (PCA 1948).

TABLE 9-1. Effect of Concrete Constituents on Control of Air Content in Concrete*

MATERIALS / CHARACTERISTICS		EFFECTS	GUIDANCE
Portland cement	Alkali content	Air content increases with increase in cement alkali level. Less air-entraining admixture dosage needed for higher alkali cements. Air-void system may be more unstable with some combinations of alkali level and air-entraining admixture used.	Changes in alkali content or cement source require that air-entraining admixture dosage be adjusted. Decrease dosage as much as 40% for high-alkali cements.
	Fineness	Decrease in air content with increased fineness of cement.	Use up to 100% more air-entraining admixture for very fine (Type III) cements. Adjust admixture if cement source or fineness changes.
	Cement content in mixture	Decrease in air content with increase in cement content. Smaller and greater number of voids with increased cement content.	Increase air-entraining admixture dosage rate as cement content increases.
Fly ash		Air content decreases with increase in loss on ignition (carbon content). Air-void system may be more unstable with some combinations of fly ash/cement/air-entraining admixture.	Changes in LOI or fly ash source will likely require that air-entraining admixture dosage be adjusted. Perform "foam index" test to estimate increase in dosage. Prepare trial mixes and evaluate air-void systems.
Slag cement		Decrease in air content with increased fineness of slag cement.	Use up to 100% more air-entraining admixture for finely ground slags.
Silica fume		Decrease in air content with increase in silica fume content.	Increase air-entraining admixture dosage up to 100% for fume contents up to 10%.
Metakaolin		No apparent effect.	Adjust air-entraining admixture dosage if needed.
Water reducers		Air content increases with increases in dosage of lignin-based materials. Spacing factors may increase when water-reducers used.	Reduce dosage of air-entraining admixture. Select formulations containing air-detraining admixtures. Prepare trial mixes and evaluate air-void systems.
Retarders		Effects similar to water-reducers.	Adjust air-entraining admixture dosage.
Accelerators		Minor effects on air content.	No adjustments normally needed.
High-range water reducers (Plasticizers)		Moderate increase in air content when formulated with lignosulfonate. Spacing factors increase.	Only slight adjustments needed. No significant effect on durability.
Aggregate	Maximum size	Air content requirement decreases with increase in maximum size. Little increase over 37.5 mm (1½ in.) maximum size aggregate.	Decrease air content.
	Sand-to-total aggregate ratio	Air content increases with increased sand content.	Decrease air-entraining admixture dosage for mixtures having higher sand contents.
	Sand grading	Middle fractions of sand promote air-entrainment.	Monitor gradation and adjust air-entraining admixture dosage accordingly.

*Whiting and Nagi 1998.

The effect of fly ash on the required dosage of air-entraining admixtures can range from no effect to an increase in required dosage of up to five times the normal amount (Gebler and Klieger 1986). Class C fly ash typically requires less air-entraining admixture than Class F fly ash and tends to lose less air during mixing (Thomas 2007). The carbon content of fly ash, can play a significant role: air-entraining admixtures are adsorbed on the surface of carbon particles, making them ineffective in stabilizing entrained air. Slag cements have variable effects on the required dosage rate of air-entraining admixtures. Silica fume has a marked influence on the air-entraining admixture requirement. In most cases, AEA dosage rapidly increases with an increase in the amount of silica fume used in the concrete. Large quantities of slag and silica fume can double the dosage of air-entraining admixtures required (Whiting and Nagi 1998). The inclusion of both fly ash and silica fume in non-air-entrained concrete will generally reduce the amount of entrapped air.

Concretes with high alkali-loadings may entrain more air than those with low alkali-loading with the same amount of air-entraining material. A concrete with a low alkali-loading may require 20% to 40% (occasionally up to 70%) more air-entraining admixture than a concrete with a high alkali-loading to achieve an equivalent air content. Precautions are necessary when changing sources of cementitious materials, or when using more than one source of cementitious materials in a batch plant (Greening 1967).

Water-reducing and set-retarding admixtures generally increase the efficiency of air-entraining admixtures by 50% to 100%. Therefore, less air-entraining admixture will usually give the desired air content. Also, the time of addition of these admixtures into the mixture affects the amount of entrained air. Delayed additions generally increase air content.

Set retarders may increase the air-void spacing in concrete. Some water-reducing or set-retarding admixtures are not compatible with some air-entraining admixtures. If they are added together to the mixing water before being dispensed into the mixer, a precipitate may form. This will settle out and result in large reductions in entrained air. The fact that some individual admixtures interact in this manner does not mean that they will not be fully effective if dispensed separately into a batch of concrete.

Superplasticizers (high-range water reducers) may increase or decrease the air content of a concrete mixture. The effect is based on the admixture's chemical formulation and the slump of the concrete. Naphthalene-based superplasticizers tend to increase the air content while melamine-based materials may decrease or have little effect on air content. The normal air loss in flowing concrete during mixing and transport is about 2% to 4% (Whiting and Dziedzic 1992). Polycarboxylate admixtures may increase air content.

Superplasticizers can also affect the air-void system of hardened concrete by increasing the general size of the entrained air voids. This results in a higher-than-normal spacing factor, occasionally higher than what may be considered desirable for freeze-thaw durability. However, tests on superplasticized concrete with slightly higher spacing factors have indicated that superplasticized concretes can demonstrate good freeze-thaw durability. This may be caused by the reduced water-cementious materials ratio often associated with superplasticized concretes.

A small quantity of calcium chloride is sometimes used in cold weather to accelerate the hardening of concrete. It can be used successfully with air-entraining admixtures if it is added separately in solution form to the mix water. However, calcium chloride is not intended for use with reinforced concrete as it will lead to corrosion of the reinforcing steel. Calcium chloride will slightly increase air content. However, if calcium chloride comes in direct contact with some air-entraining admixtures, a chemical reaction can take place that makes the air-entraining admixture less effective. Non-chloride accelerators may increase or decrease air content, depending upon the chemistry of the specific admixture. Generally they have little effect on air content.

Coloring agents such as carbon black usually decrease the amount of air entrained for a given amount of admixture. This is especially true for coloring materials with increasing percentages of carbon (Taylor 1948).

The size of coarse aggregate has a pronounced effect on the air content of both air-entrained and non-air-entrained concrete, as shown in Figure 9-11. There is little change in air content when the size of aggregate is increased above 37.5 mm (1½ in.). The fine-aggregate content of a mixture affects the percentage of entrained air. As shown in Figure 9-10, increasing the amount of fine aggregate causes more air to be entrained for a given amount of air-entraining cement or admixture (more air is also entrapped in non-air-entrained concrete). Fine-aggregate particles passing the 600-μm to 150-μm (No. 30 to No. 100) sieves entrap more air than either very fine or coarser particles. Appreciable amounts of material passing the 150-μm (No. 100) sieve will result in a significant reduction of entrained air, unless the AEA dosage is adjusted. Fine aggregates from different sources may entrap different amounts of air even though they have identical gradations. This may be due to differences in shape and surface texture or as a result of contamination by organic materials.

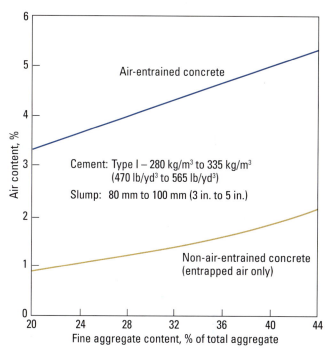

FIGURE 9-11. Relationship between percentage of fine aggregate and air content of concrete (PCA 1948).

Mixing water may also affect air content. Algae-contaminated water increases air content. Highly alkaline wash water from truck mixers can affect air contents, increasing the air content compared to neutral water. The effect of water hardness in most municipal water supplies is generally insignificant. Very hard water from wells used in many communities may decrease the air content in concrete.

Mixture Design. The concrete mixture design's effect on the air content is summarized in Table 9-2.

An increase in the mixing water makes more water available for the generation of air bubbles, thereby increasing the air content as slumps increase up to about 150 mm or 175 mm (6 in. or 7 in.). Air contents also increase with an increase in the water-cementitious materials ratio. A portion of the air increase is due to the relationship between slump and air content.

TABLE 9-2. Effect of Concrete Mixture Design on Control of Air Content in Concrete*

CHARACTERISTIC	EFFECTS	GUIDANCE
Water chemistry	Very hard water reduces air content. Batching of admixture into concrete wash water decreases air. Algae may increase air.	Increase air entrainer dosage. Avoid batching into wash water.
Water-to-cement ratio	Air content increases with increased water to cement ratio.	Decrease air-entraining admixture dosage as water to cement ratio increases.
Slump	Air increases with slumps up to about 150 mm (6 in.). Air decreases with very high slumps. Difficult to entrain air in low-slump concretes.	Adjust air-entraining admixture dosages for slump. Avoid addition of water to achieve high-slump concrete. Use additional air-entraining admixture; up to ten times normal dosage.

*Whiting and Nagi 1998

Air content increases with slump even when the w/cm is held constant. The spacing factor, $\bar{L}$, of the air-void system also increases. That is, the voids become coarser at higher water-cementitious materials ratios, thereby reducing concrete freeze-thaw durability (Stark 1986).

The addition of 5 kg/m³ (8.4 lb/yd³) of water to concrete can increase the slump by about 25 mm (1 in.). A 25-mm (1-in.) increase in slump increases the air content by approximately 0.5% to 1% for concretes with a low to moderate slump and constant air-entraining admixture dosage.

This approximation is greatly affected by concrete temperature, slump, and the type and amount of cement and admixtures present in the concrete. A low slump concrete with a high dosage of water-reducing and air-entraining admixtures can undergo large increases in slump and air content with a small addition of water. Alternatively, a very fluid concrete mixture with a 200-mm to 250-mm (8-in. to 10-in.) slump may lose air with the addition of water.

Production Procedures. The effect of the production procedures on the air content is summarized in Table 9-3. Mixing action is one of the most important factors in the production of entrained air in concrete. Uniform distribution of entrained air voids is essential to produce freeze-thaw resistant and scale-resistant concrete. Non-uniformity might result from inadequate dispersion of the entrained air during short mixing periods. In production of ready mixed concrete, it is especially important that adequate and consistent mixing be maintained at all times.

The amount of entrained air varies with the type and condition of the mixer, the amount of concrete being mixed, and the rate and duration of mixing. The amount of air entrained in a given mixture will decrease appreciably as the mixer blades become worn, or if hardened concrete is allowed to accumulate in the drum or on the blades. Because of differences in mixing action and time, concretes made in a stationary mixer and those made in a transit mixer may differ significantly in amounts of air entrained. The air content may increase or decrease when the size of the batch departs significantly from the rated capacity

of the mixer. Little air is entrained in very small batches in a large mixer. However, the air content increases as the mixer capacity is approached.

Figure 9-12 shows the effect of mixing speed and duration of mixing on the air content of freshly mixed concretes made in a transit mixer. Generally, more air is entrained as the speed of mixing is increased up to about 20 rpm, beyond which air entrainment decreases. In the tests from which the data in Figure 9-12 were derived, the air content reached an upper limit during mixing and a gradual decrease in air content occurred with prolonged mixing. Mixing time and speed will have different effects on the air content of different mixtures. Significant amounts of air can be lost during mixing with certain types of air-entrainment, mixing equipment, and mixture proportions.

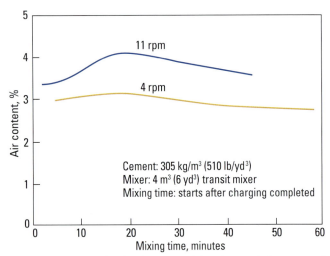

FIGURE 9-12. Relationship between mixing time and air content of concrete (PCA 1948).

Figure 9-13 shows the effect of continued mixer agitation on air content. The changes in air content with prolonged agitation can be explained by the relationship between slump and air content. For high-slump concretes, the air content increases with continued agitation as the slump decreases to about 150 mm or 175 mm (6 in. or 7 in.). Prolonged agitation will decrease

TABLE 9-3. Effect of Production Procedures on Control of Air Content in Concrete*

PROCEDURE/VARIABLE	EFFECTS	GUIDANCE
Batching sequence	Simultaneous batching lowers air content. Cement-first raises air content.	Add air-entraining admixture with initial water or on sand.
Mixer capacity	Air increases as capacity is approached.	Run mixer close to full capacity. Avoid overloading.
Mixing time	Central mixers: air content increases up to 90 seconds of mixing. Truck mixers: air content increases with mixing. Short mixing periods (30 seconds) reduce air content and adversely affect air-void system.	Establish optimum mixing time for particular mixer. Avoid overmixing. Establish optimum mixing time (about 60 seconds).
Mixing speed	Air content gradually increases up to approximately 20 rpm. Air may decrease at higher mixing speeds.	Follow truck mixer manufacturer recommendations. Maintain blades and clean truck mixer.
Admixture metering	Accuracy and reliability of metering system will affect uniformity of air content.	Avoid manual-dispensing or gravity-feed systems and timers. Positive-displacement pumps interlocked with batching system are preferred.

*Whiting and Nagi 1998

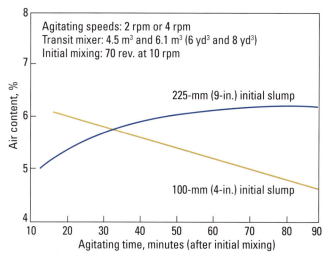

FIGURE 9-13. Relationship between agitating time, air content, and slump of concrete (PCA 1948).

slump further and will also decrease air content. For initial slumps lower than 150 mm (6 in.), both the air content and slump decrease with continued agitation. When concrete is retempered (the addition of water and remixing to restore original slump), the air content is increased.

However, after about 4 hours, retempering is ineffective in increasing air content and may cause clustering of air bubbles (Kozikowski and others 2005). Prolonged mixing or agitation of concrete is accompanied by a progressive reduction in slump.

Transportation and Delivery. The effect of transportation and delivery of concrete on the air content is summarized in Table 9-4.

Generally, some air, approximately 1% to 2%, is lost during transportation of concrete from the mixer to the jobsite. The stability of the air content during transport is influenced by several variables including concrete ingredients, haul time, amount of agitation or vibration during transport, temperature, slump, and amount of retempering.

Placement and Consolidation. The effect of placing techniques and internal vibration on air content is summarized in Table 9-5.

Once at the jobsite, the concrete air content remains essentially constant during handling by chute discharge, wheelbarrow, power buggy, and shovel. However, concrete pumping, crane and bucket, and conveyor-belt handling can cause some loss of air, especially with high-air-content mixtures. Pumping concrete can cause a loss of up to 3% of air (Whiting and Nagi 1998).

TABLE 9-4. Effect of Transportation and Delivery on Control of Air Content in Concrete*

PROCEDURE/VARIABLE	EFFECTS	GUIDANCE
Transport and delivery	Some air (1% to 2%) normally lost during transport. Loss of air in nonagitating equipment is slightly higher.	Normal retempering with water to restore slump will restore air. If necessary, retemper with air-entraining admixture to restore air. Dramatic loss in air may be due to factors other than transport.
Haul time and agitation Retempering	Long hauls, even without agitation, reduce air, especially in hot weather. Regains some of the lost air. Does not usually affect the air-void system. Retempering with air-entraining admixtures restores the air-void system. May cause clustering of air bubbles.	Optimize delivery schedules. Maintain concrete temperature in recommended range. Retemper only enough to restore workability. Avoid addition of excess water. Higher admixture dosage is needed for jobsite admixture additions.

*Whiting and Nagi 1998

TABLE 9-5. Effect of Placement Techniques and Internal Vibration on Control of Air Content in Concrete*

PROCEDURE/VARIABLE	EFFECTS	GUIDANCE
Belt conveyors	Reduces air content by an average of 1%.	Avoid long conveyed distance if possible. Reduce the free-falling effect at the end of conveyor.
Pumping	Reduction in air content ranges from 2% to 3%. Does not significantly affect air-void system. Minimum effect on freeze-thaw resistance.	Use of proper mix design provides a stable air-void system. Avoid high-slump, high-air-content concrete. Keep pumping pressure as low as possible. Use loop in descending pump line.
Shotcrete	Generally reduces air content in wet-process shotcrete.	Air content of mixture should be at high end of target zone.

*Whiting and Nagi 1998

The effect of slump and vibration on the air content of concrete is shown in Figure 9-14. For a constant amount of air-entraining admixture, air content increases as slump increases up to about 150 mm or 175 mm (6 in. or 7 in.). Beyond that, air content begins to decrease with further increases in slump. At all slumps, however, even 15 seconds of vibration (ACI 309R 2005) will cause a considerable reduction in air content. Prolonged vibration of concrete should be avoided.

The greater the slump, air content, and vibration time, the larger the percentage of reduction in air content during vibration (Figure 9-14). However, if vibration is properly applied, little of the intentionally entrained air is lost. The air lost during handling and moderate vibration consists mostly of the larger bubbles, or entrapped air. Reduction or removal of large entrapped air voids will increase strengths. While the average size of the air voids is reduced, the air-void spacing factor remains relatively constant.

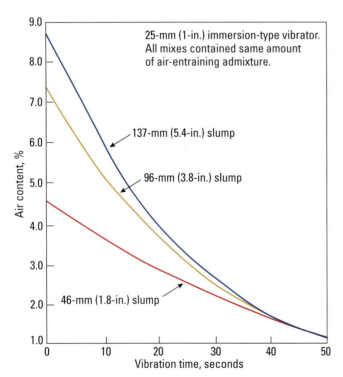

FIGURE 9-14. Relationship between slump, duration of vibration, and air content of concrete (Brewster 1949).

Internal vibrators reduce air content more than external vibrators. The air loss due to vibration increases as the volume of concrete is reduced or as the vibration frequency is significantly increased. Lower vibration frequencies (8000 vpm) have less effect on spacing factors and air contents than high vibration frequencies (14,000 vpm). High frequencies can significantly increase spacing factors and decrease air contents after 20 seconds of vibration (Brewster 1949 and Stark 1986).

Specified air contents and uniform air void distributions can be achieved in pavement construction by operating paving machine speeds at 1.2 m/min to 1.9 m/min (4 ft/min to 6 ft/min) and by using vibrator frequencies of 5000 vibrations/min to 8000 vibrations/min. One research study found that the most uniform distribution of air voids throughout the depth of concrete, both in and out of the vibrator trails, was obtained with the combination of a vibrator frequency of approximately 5000 vibrations/minute and a slipform paving machine forward track speeds of 1.2 m/min (4 ft/min). Higher frequency speeds, singularly or in combination, can result in discontinuities and lack of required air content in the upper portion of the concrete pavement. This in turn provides a greater opportunity for water and salt to enter the pavement and reduce the durability and life of the pavement (Cable and others 2000).

Finishing and Environment. The effect of finishing and environment on air content is summarized in Table 9-6.

Proper screeding, floating, and general finishing practices should not affect the air content. McNeal and Gay (1996) and Falconi (1996) demonstrated that the sequence and timing of finishing and curing operations are critical to surface durability. Overfinishing (excessive finishing) may reduce the amount of entrained air in the surface region of slabs, making the concrete surface vulnerable to scaling. However, as shown in Figure 9-15, early finishing does not necessarily affect scale resistance unless bleed water is present (Pinto and Hover 2001). Concrete to be exposed to deicers, requiring entrained air content should not be steel troweled.

Temperature of the concrete affects air content, as shown in Figure 9-16. Less air is entrained as the temperature of the

TABLE 9-6. Effect of Finishing and Environment on Control of Air Content in Concrete*

PROCEDURE/VARIABLE	EFFECTS	GUIDANCE
Internal vibration	Air content decreases under prolonged vibration or at high frequencies. Proper vibration does not influence the air-void system.	Do not overvibrate. Avoid high-frequency vibrators (greater than 10,000 vpm). Avoid multiple passes of vibratory screeds. Closely spaced vibrator insertion is recommended for better consolidation.
Finishing	Air content reduced in surface layer by excessive finishing.	Avoid finishing with bleed water still on surface. Avoid overfinishing. Do not sprinkle water on surface prior to finishing. It is not recommended to hard trowel concrete with air contents greater than 3%.
Temperature	Air content decreases with increase in temperature. Changes in temperature do not significantly affect spacing factors.	Increase air-entraining admixture dosage as temperature increases.

*Whiting and Nagi 1998

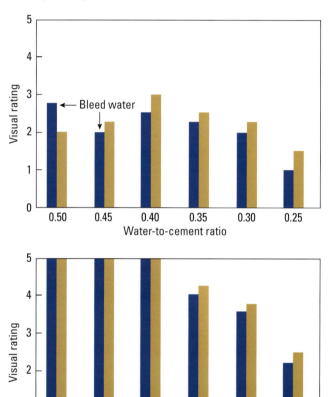

FIGURE 9-15. Effect of early finishing on scale resistance – magnesium floating 20 minutes after casting – for (top) 6% air entrained concrete and (bottom) non-air-entrained concrete (Pinto and Hover 2001).

the temperature of the concrete ingredients have equalized. Although increased concrete temperature during mixing generally reduces air volume, the spacing factor and specific surface are only slightly affected.

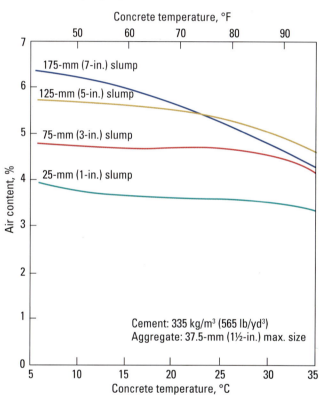

FIGURE 9-16. Relationship between temperature, slump, and air content of portland cement concrete (PCA 1948 and Lerch 1960).

Uniformity

Uniformity is a measure of the homogeneity of the concrete. This measurement includes within-batch uniformity and uniformity between batches of concrete mixtures. Samples of fresh concrete collected at the point of discharge are tested on site to determine properties such as: slump, temperature, air content, unit weight, and yield. Sampling of fresh concrete is performed in accordance with ASTM C172, *Standard Practice for Sampling Freshly Mixed Concrete* (AASHTO M 241). Test specimens may be cast for later testing of strength and other hardened concrete

concrete increases, particularly as slump is increased. This effect is especially important during hot-weather concreting when the concrete might be quite warm. A decrease in air content can be offset, when necessary, by increasing the quantity of air-entraining admixture. During cold-weather concreting, the air-entraining admixture may lose some of its effectiveness if hot mix water is used during batching. To offset this loss, these admixtures should be added to the batch after properties. The results of these tests are used to determine that the concrete meets specification requirements. These results also provide a measure of the uniformity of the concrete both between batches and, when necessary, within a batch.

ASTM C94, *Standard Specification for Ready-Mixed Concrete*, covers criteria for determining the within-batch uniformity of concrete. If the within-batch uniformity is low, this can be indicative of inadequate or inefficient mixing. Samples of concrete are taken at two locations within the batch to represent the first and last portions on discharge. The samples are tested separately for density, air, slump, and strength; the difference between the test results must be less than the requirements of the specification. For example, if the average of the slump for the two samples is 90 mm (3.5 in.), the difference between the measured individual values cannot vary more than 25 mm (1 in.).

Consolidation. Uniformity of concrete is typically achieved by consolidation. Vibration sets into motion the particles in freshly mixed concrete, reducing friction between them and giving the mixture the mobile qualities of a thick fluid. The vibratory action permits use of a stiffer mixture containing a larger proportion of coarse aggregate and a smaller proportion of fine aggregate. The larger the maximum size aggregate in concrete with a well-graded aggregate, the less volume there is to fill with paste and the less aggregate surface area there is to coat with paste. Therefore less water and cementitious materials are needed. Concrete with an optimally graded aggregate will be easier to consolidate and place. Consolidation of coarser as well as stiffer mixtures results in improved quality and economy. On the other hand, poor consolidation results in porous, weak concrete (Figure 9-17) with poor durability (Figure 9-18). Mechanical vibration has many advantages. Vibrators make it possible to economically place mixtures that are impractical to consolidate by hand. For more information on consolidation, see Chapter 15.

FIGURE 9-18. Poor consolidation can result in early corrosion of reinforcing steel and low compressive strength.

Hydration, Setting, and Hardening

The binding quality of portland cement paste is due to the chemical reaction between the cement and water, called hydration. As discussed in Chapter 2, portland cement is not a simple chemical compound, rather, it is a mixture of many compounds. The two calcium silicates, which constitute about 75% of the weight of portland cement, react with water to form new compounds: calcium hydroxide and calcium silicate hydrate (C-S-H). The latter is by far the most important cementing component in concrete. The engineering properties of concrete – setting and hardening, strength, and dimensional stability – depend primarily on calcium silicate hydrate. It is the heart of concrete.

The composition of calcium silicate hydrate is somewhat variable, but it is chemically represented by lime (CaO) and silicate (SiO_2) in a ratio on the order of 3 to 2. The surface area of calcium silicate hydrate is around 300 m^2/g. In hardened cement paste, the calcium silicate hydrate forms dense, bonded aggregations between the other crystalline phases and the remaining unhydrated cement grains. These aggregations also adhere to grains of sand and to pieces of coarse aggregate, cementing all materials together (Copeland and Schulz 1962).

The less porous the cement paste, the stronger the concrete. When mixing concrete, therefore, no more water than is absolutely necessary to make the concrete plastic and workable should be used. Even then, the water used is usually more than is required for complete hydration of the cement. About 0.4 times as much water (by mass) as cement is needed to completely hydrate cement (Powers 1948 and 1949). However, complete hydration is rare in field concrete placements due to a lack of moisture and the long time (decades) required to achieve complete hydration. Knowledge of the rate of reaction between cement and water is important because it determines the rate of hardening. The initial reaction must be slow enough to allow time for the concrete to be transported and placed. However, rapid hardening is typically desirable once the concrete has been placed and finished. Gypsum, added at the cement mill

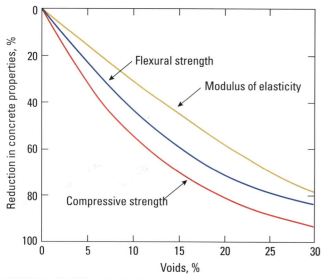

FIGURE 9-17. Effect of voids in concrete due to a lack of consolidation on modulus of elasticity, compressive strength, and flexural strength.

when clinker is ground, acts as a regulator of the initial rate of setting of portland cement. Other factors that influence the rate of hydration include portland cement fineness, chemical admixtures, type and amount of supplementary cementitious materials, amount of water added, and temperature of the materials at the time of mixing. Figure 9-19 illustrates the setting properties of a concrete mixture at different temperatures.

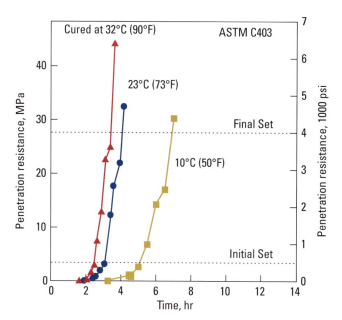

FIGURE 9-19. Initial and final set times for a concrete mixture at different temperatures (Burg 1996).

The setting and hardening of portland cement can be explained using a simple model showing unhydrated cement grains dispersed in water (Figure 9-20). In this example, time starts when the water is first added to the cement. Upon the addition of water, a chemical reaction occurs between the water and cement – the reaction is called hydration. The solid products resulting from hydration occupy a greater volume than the original cementitious materials and consequently some of the space between the cement grains is filled in. Eventually the hydration products will connect adjacent grains and a continuous solid network is formed. This is referred to as initial set.

Between the addition of water and just before initial set occurs the paste has little rigidity. If the rigidity of the paste was plotted against time – there would be only a small increase in stiffness during this period. This is referred to as the dormant period. During the dormant period the paste is still plastic and the concrete can still be handled and placed. As the cement continues to hydrate, more hydration products are formed and the solid matrix becomes more dense and rigid. This period is called the setting, or transition period, as it represents the period during which the paste transforms from a fluid to a solid. Eventually, the paste can be considered a rigid and solid material with mechanical properties such as strength and stiffness.

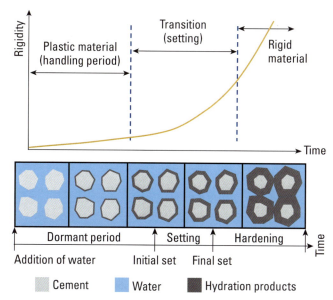

FIGURE 9-20. Setting of concrete (adapted from Young and others 1998).

Setting times of concrete are determined in accordance with ASTM C403, *Standard Test Method for Time of Setting of Concrete Mixtures by Penetration Resistance* (AASHTO T 197). The setting time is determined by measuring the resistance of a sieved mortar sample to the penetration of a standard needle. Initial and final set are defined as a penetration resistance of 3.5 MPa and 27.5 MPa (500 psi and 4000 psi), respectively. However, laboratory settings cannot adequately demonstrate all of the variables involved in setting of concrete and timing of finishing operations in the field (see Chapters 15).

The setting time of concrete is affected by the type and amount of cement used, the type and level of supplementary cementing materials, the presence of set modifying admixtures, the water-to-cementitious materials ratio and the temperature of the concrete (Bullard and others 2006).

Knowledge of the amount of heat released as cementitious materials hydrate can be useful in planning construction. In winter, containing the heat of hydration will help protect the concrete against damage from freezing temperatures. The heat may be harmful, however, in mass concrete structures because it may produce undesirable temperature rise and differences in temperature between the interior and surface of the concrete (see Chapter 18).

Figure 9-21 shows that initially all the mixtures demonstrate a similar rate of heat release irrespective of the w/cm. After the first 24 hours a difference begins to be noticed between the different w/c mixtures. Mixtures with a higher water to cement ratio show a greater heat release. This suggests that lower w/cm mixtures have a reduction in their hydration due to either a lack of water to participate in the hydration (or relative humidity) or a lack of space for the particles to hydrate.

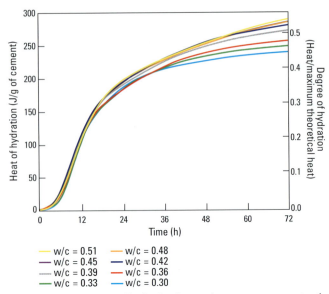

w/c = 0.51	w/c = 0.48
w/c = 0.45	w/c = 0.42
w/c = 0.39	w/c = 0.36
w/c = 0.33	w/c = 0.30

FIGURE 9-21. Degree of hydration for varying water to cement ratio mixtures for plain mortars (Lura and others 2007).

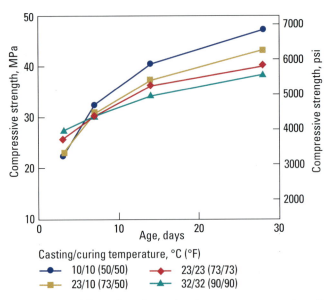

Casting/curing temperature, °C (°F)

10/10 (50/50)	23/23 (73/73)
23/10 (73/50)	32/32 (90/90)

FIGURE 9-22. Effect of casting and curing temperature on strength development. Note that cooler temperatures result in lower early strength and higher later strength (Burg 1996).

HARDENED CONCRETE

When the concrete hardens, it becomes a homogeneous mixture of all the components. The following sections will discuss the properties of hardened concrete including: curing, drying rate, strength, density, permeability and watertightness, volume stability and crack control, durability, and aesthetics in more detail.

Curing

Increase in strength with age continues provided:

- unhydrated cement is still present,
- the concrete remains moist or has an internal relative humidity above approximately 80%,
- the concrete temperature remains favorable, and
- sufficient space is available for hydration products to form.

When the relative humidity within the concrete drops to about 80% or the temperature of the concrete drops below -10°C (14°F), hydration and strength gain virtually stop (Powers 1948). Figure 9-22 illustrates the relationship between strength gain and curing temperature.

If concrete is resaturated after a drying period, hydration resumes and strength will again increase. However, it is best to wet-cure concrete continuously from the time concrete is placed until it has attained the desired quality. Figure 9-23 illustrates the long-term strength gain of concrete in an outdoor exposure. Outdoor exposures often continue to provide moisture through ground contact and rainfall. Indoor concretes often dry out after curing and do not continue to gain strength. See Chapter 17 for more information on curing concrete.

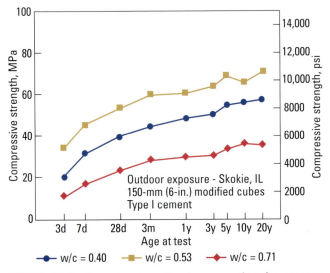

w/c = 0.40	w/c = 0.53	w/c = 0.71

FIGURE 9-23. Concrete strength gain versus time for concrete exposed to outdoor conditions. Concrete continues to gain strength for many years when moisture is provided by rainfall and other environmental sources (Wood 1992).

Drying Rate of Concrete

The drying rate of concrete is helpful in understanding the properties or physical condition of concrete. Concrete must continue to hold enough moisture throughout the curing period for the cement to hydrate to the extent that desired properties are achieved. Freshly cast concrete usually has an abundance of water, but as drying progresses from the surface inward, strength gain will continue at each depth only as long as the relative humidity at that point remains above about 80%.

Concrete will lose or absorb water until it comes into equilibrium with the relative humidity of the surrounding air. The relationship between the equilibrium moisture content of the concrete and the RH of the air is described by sorption isotherms. Figure 9-24 is a schematic of the moisture sorption isotherms of a porous material, such as concrete. These isotherms summarize what happens as the material is cycled from an oven-dry to a saturated condition and then back to the oven-dry condition. It is seen that there are two curves depending on whether it is the wetting cycle or the drying cycle. So, a concrete with an internal humidity of 60% means that the concrete is in moisture

equilibrium with surrounding air at 60% RH. The value of the equilibrium moisture content, however, depends on whether the concrete is undergoing drying or wetting.

During the first stage of drying in concrete, liquid water is present at the surface and evaporates into the air over the concrete (Figure 9-25A). The rate of evaporation at the concrete surface depends upon temperature, relative humidity, and air flow over the surface. Warm, dry, rapidly moving air will cause faster evaporation than cool, stagnant air. As the liquid water evaporates, it is replenished with water from within the body of the concrete. As liquid water moves from within the body of the concrete to replace water that has evaporated at the surface, the concrete must shrink to make up the volume of water that has left. If the rate of evaporation is very high, the concrete may shrink excessively before the cement paste has developed much strength. This is the cause of plastic shrinkage cracking that may occur within the first few hours after the concrete is placed.

When concrete dries, it shrinks as it loses water. Drying shrinkage is a primary cause of cracking, and the width of cracks is a function of the degree of drying and buildup of stress. While the surface of a concrete element will dry quite rapidly, it takes a much longer time for concrete in the interior to dry.

When the concrete can no longer shrink to accommodate the volume lost due to water evaporation, the second stage of drying begins (Figure 9-25B). Liquid water recedes from the exposed surface of the concrete into the pores. Within each pore, water clings to the sidewalls and forms a curved surface called a meniscus. At the surface of the concrete, water evaporates from the meniscus in each pore into the air over the concrete. Therefore, the rate of evaporation still depends mostly on the temperature, relative humidity, and air flow over the concrete surface.

FIGURE 9-24. Schematic of the sorption isotherms of a porous material indicating the relationship between the equilibrium moisture content as a function of the relative humidity of the surrounding environment (after Powers and Brownyard 1948).

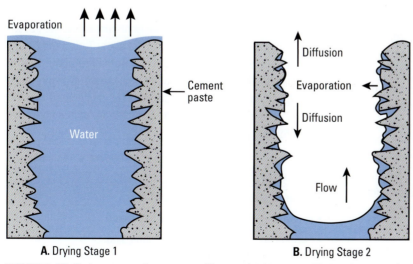

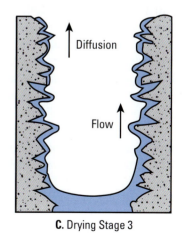

A. Drying Stage 1   **B.** Drying Stage 2   **C.** Drying Stage 3

FIGURE 9-25. Drying stages of concrete as illustrated using a magnified surface pore (adapted from Scherer 1990).

At this point, water still fills the bulk of the pore structure of the concrete. There are continuous paths for liquid water to flow from within the body of concrete to the partially filled pores at the surface where the water can evaporate. The surface may appear to be dry, but the concrete is just beginning to dry in a very thin layer. The rate of drying during this period steadily decreases.

The third stage of drying begins when enough water has evaporated from just below the surface that the pores are no longer continuously filled with liquid (Figure 9-25C). Pockets of liquid water exist but moisture must now move by vapor diffusion within the body of the concrete before arriving at the surface where it can evaporate. This stage is called the second falling rate period because the rate of drying continuously decreases over time and is slower than the previous stage of drying (Figure 9-26).

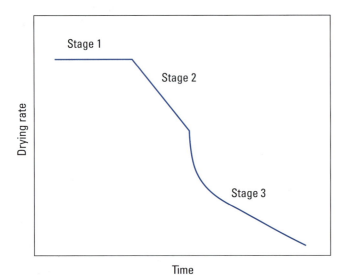

FIGURE 9-26. Stages of drying for concrete. Stage 1 has a constant rate and depends on air movement and relative humidity over the slab while Stages 2 and 3 depend more on the properties of the cement paste (adapted from Hughes 1966).

The rate of drying depends less on temperature, relative humidity, and air flow above the concrete surface because moisture must evaporate and diffuse within the body of the concrete before arriving at the surface. The rate of drying during this stage is determined by the quality of the cement paste: pastes with low water-cementitious materials ratios offer more resistance to vapor diffusion than high water-cementitious materials ratios. Concrete made with a water-cement ratio, greater than approximately 0.65, will have a continuously connected capillary pore system. In these concretes moisture vapor moves with much less resistance than concretes made with lower water cement ratios.

The moisture content of concrete depends on the concrete's constituents, original water content, drying conditions, and the size of the concrete element (Hedenblad 1996 and 1997). Size and shape of a concrete member have an important bearing on the rate of drying. Concrete elements with large surface area in relation to volume (such as floor slabs) dry faster than large volume concrete members with relatively small surface areas (such as bridge piers).

Concrete to receive flooring material must be dry enough to permit the adhesive to bond properly and to prevent damage to the flooring. For coatings, the concrete must be sufficiently dry to develop and maintain adequate bond and to allow the coating to chemically cure. A concrete surface may appear dry, but the slab can still contain sufficient moisture to cause problems after it is covered (Kanare 2008).

Theoretically, it is possible to estimate the drying time for a given concrete (Hall 1997). This calculation requires the absorption characteristics, diffusion coefficients for water and water vapor, porosity and pore size distribution, and degree of hydration. Since this information usually is not available, current practice relies on experimental data combined with measurements of the actual moisture condition of the concrete slab in the field (Kanare 2008).

Many other properties of hardened concrete are also affected by its moisture content: elasticity, creep, insulating value, fire resistance, abrasion resistance, electrical conductivity, frost resistance, scaling resistance, corrosion resistance, and resistance to alkali-aggregate reactivity.

Also see the sections on **Density** and **Volume Stability** in this chapter and see Chapter 10 for additional information on drying effects, mass loss, and volume change.

Strength

Compressive strength is the measured maximum resistance of a concrete specimen to axial loading. It is generally expressed in megapascals (MPa) or pounds per square inch (psi) at an age of 28 days. Other test ages are also used. However, it is important to realize the relationship between the 28-day strength and other test ages. For ordinary portland cement concretes, the seven-day strengths are often estimated to be about 75% of the 28-day strength while 56-day and 90-day strengths are about 10% to 15% greater than 28-day strengths, as shown in Figure 9-27. The specified compressive strength is designated by the symbol f'_c, and ideally is exceeded by the actual compressive strength, designated by f'_{cr}.

The compressive strength that a concrete achieves is governed by the following:

• water-cementitious materials ratio,

• extent to which hydration has progressed,

• curing and environmental conditions, and

• age of concrete.

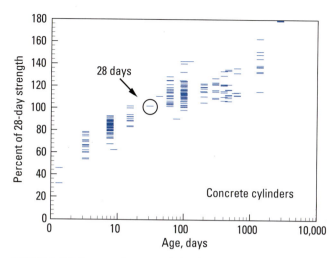

FIGURE 9-27. Compressive strength development of various concretes illustrated as a percentage of the 28-day strength (Lange 1994).

The relationship between strength and water-cement ratio has been studied since the late 1800s and early 1900s (for example, Feret 1897 and Abrams 1918). Figure 9-28 shows compressive strengths for a wide range of concrete mixtures and water-cement ratios at an age of 28 days. Note that strengths increase as the water-cement ratios decrease. These factors also affect flexural and tensile strengths and bond of concrete to steel.

The water-cement ratio compressive strength relationships in Figure 9-28 are for typical non-air-entrained concretes. When more precise values for concrete are required, graphs should be developed for the specific materials and mix proportions to be used on the job, due to the wide range of concrete ingredients and their characteristics that can impact strength.

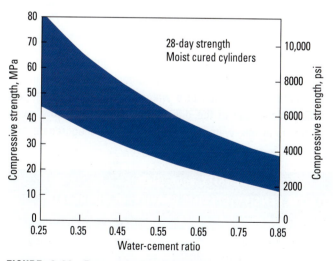

FIGURE 9-28. Range of typical strength to water-cement ratio relationships of portland cement concrete based on over 100 different concrete mixtures cast between 1985 and 1999.

For a given workability and a given amount and type of cementitious materials, air-entrained concrete requires less mixing water than non-air-entrained concrete. The lower water-cementitious materials ratios possible for air-entrained

concretes tends to offset the somewhat lower strengths of air-entrained concretes, particularly in lean- to medium-cementitious materials content mixtures.

To determine compressive strength, tests are made on specimens of mortar or concrete. In the United States, unless otherwise specified, compression tests of mortar are made on 50-mm (2-in.) cubes, while compression tests of concrete are made on cylinders 150 mm (6 in.) in diameter and 300 mm (12 in.) high or smaller cylinders sized at 100 mm × 200 mm (4 in. × 8 in.) (Figure 9-29).

FIGURE 9-29. Testing a 150-mm × 300-mm (6-in. × 12-in.) concrete cylinder in compression. The load on the test cylinder is registered on the display.

The compressive strength of concrete is a fundamental physical property frequently used in design calculations for bridges, buildings, and other structures. Most general-use concrete has a compressive strength between 20 MPa and 40 MPa (3000 psi and 6000 psi). Compressive strengths of 70 MPa to 140 MPa (10,000 psi to 20,000 psi) or more have been used in special structures such as bridge and high-rise building applications (see Chapter 23).

The flexural strength, or modulus of rupture, of concrete is used to design pavements and other slabs on ground. Compressive strength, which is easier to measure than flexural strength, can be used as an index of flexural strength, once the empirical relationship between them has been established for the materials and the size of the element involved. The flexural strength of normal-weight concrete is often approximated as 0.7 to 0.8 times the square root of the compressive strength in MPa (7.5 to 10 times the square root of the compressive strength in psi). Wood (1992) illustrates the relationship between flexural strength and compressive strength for concretes exposed to moist curing, air curing, and outdoor exposure.

The direct tensile strength of concrete is about 8% to 12% of the compressive strength and is often estimated as 0.4 to 0.7 times the square root of the compressive strength in MPa

(5 to 7.5 times the square root of the compressive strength in psi). Splitting tensile strength is 8% to 14% of the compressive strength (Hanson 1968). Splitting tensile strength versus time is presented by Lange (1994).

The torsional strength for concrete is related to the modulus of rupture and the dimensions of the concrete element. Hsu (1968) presents torsional strength correlations.

Shear strength to compressive strength relationships are discussed in ACI 318, *Building Code Requirements for Structural Concrete* (2019). The correlation between compressive strength and flexural, tensile, torsional, and shear strength varies with concrete ingredients and environment.

Modulus of elasticity, denoted by the symbol *E*, may be defined as the ratio of normal stress to corresponding strain for tensile or compressive stresses below the proportional limit of a material. For normal-weight concrete, *E* ranges from 1 GPa to 4 GPa (2 million psi to 6 million psi) and can be approximated as 5,000 times the square root of the compressive strength in MPa (57,000 times the square root of the compressive strength in psi). Like other strength relationships, the modulus of elasticity to compressive strength relationship is mixture specific and should be verified in a laboratory (Wood 1992).

Density

Conventional concrete, normally used in pavements, buildings, and other structures, has a density (unit weight) in the range of 2200 kg/m³ to 2400 kg/m³ (137 lb/ft³ to 150 lb/ft³). The density of concrete varies depending on the amount and density of the aggregate, the amount of air that is entrapped or purposely entrained, and the water and cement contents, which in turn are influenced by the maximum size of the aggregate. Reducing the cement paste content (increasing aggregate volume) increases density for normal-weight aggregates. Values of the density of fresh concrete are given in Table 9-7. For the design of reinforced concrete structures, the combination of conventional concrete and reinforcing steel is commonly assumed to weigh 2400 kg/m³ (150 lb/ft³).

The weight of dry concrete equals the weight of the freshly mixed concrete ingredients less the weight of mix water that evaporates during drying. Some of the mix water combines chemically with the cement during the hydration process, converting the cement phases into hydrates. Also, some of the water remains tightly held in pores and capillaries and does not evaporate under normal conditions. The amount of mix water that will evaporate from concrete exposed to ambient air at 50% relative humidity is about 0.5% to 3% of the concrete weight. The actual amount depends on initial water content of the concrete, absorption characteristics of the aggregates, and size and shape of the concrete element and duration of drying.

There is a wide spectrum of special concretes to meet various needs. Their densities range from lightweight insulating concretes with a density of as little as 240 kg/m³ (15 lb/ft³) to heavyweight concrete with a density of up to 6000 kg/m³ (375 lb/ft³) used for counterweights or radiation shielding.

Low-density (lightweight) concrete. Lightweight concrete has a lower density made with lightweight aggregates or with a combination of lightweight and normal-weight aggregates. Material proportions vary significantly for different materials and strength requirements. Structural lightweight aggregate concrete is used primarily to reduce the dead-load weight in concrete members, such as floors in high-rise buildings.

According to ACI 213R, *Guide for Structural Lightweight-Aggregate Concrete* (2014), structural lightweight concrete has an equilibrium density in the range of 1120 kg/m³ to 1920 kg/m³ (70 lb/ft³ to 120 lb/ft³) and a minimum 28-day compressive strength of 17 MPa (2500 psi). The density of structural lightweight aggregate concrete can be measured by following ASTM C567, *Standard Test Method for Determining Density of Structural Lightweight Concrete*.

The compressive strength of structural lightweight concrete can be related to the cementitious content at a given slump and air content, rather than to a water-to-cementitious materials ratio. This is due to the difficulty in determining how much of the total mix water is absorbed into the aggregate and

TABLE 9-7. Observed Average Density of Fresh Concrete*

MAXIMUM SIZE OF AGGREGATE, mm (in.)	AIR CONTENT, %	WATER, kg/m³ (lb/yd³)	CEMENT, kg/m³ (lb/yd³)	DENSITY, kg/m³ (lb/ft³)				
				RELATIVE DENSITY (SPECIFIC GRAVITY) OF AGGREGATE‡				
				2.55	2.60	2.65	2.70	2.75
19 (¾)	6.0	168 (283)	336 (566)	2194 (137)	2227 (139)	2259 (141)	2291 (143)	2323 (145)
37.5 (1½)	4.5	145 (245)	291 (490)	2259 (141)	2291 (143)	2339 (146)	2371 (148)	2403 (150)
75 (3)	3.5	121 (204)	242 (408)	2307 (144)	2355 (147)	2387 (149)	2435 (152)	2467 (154)

*Adapted from Bureau of Reclamation 1981.
†Air-entrained concrete with indicated air content.
‡On saturated surface-dry basis. Multiply relative density by 1000 (62.4) to obtain density of aggregate particles in kg/m³ (lb/ft³).

unavailable for reaction with the cement. Typical compressive strengths range from 20 MPa to 35 MPa (3000 psi to 5000 psi). High-strength concrete can also be made with structural lightweight aggregates.

Due to lower aggregate density, structural lightweight aggregate concrete does not slump as much as normal-weight concrete with the same workability. A slump of 50 mm to 100 mm (2 in. to 4 in.) produces the best results for finishing. Greater slumps may cause segregation, delay finishing operations, and result in rough, uneven surfaces. With higher slumps, the large aggregate particles tend to float to the surface, causing segregation and making finishing difficult.

If pumped concrete is being considered, the specifier, suppliers, and contractor should all be consulted for field trials using the pump and mixture planned for the project. Adjustments to the mixture may be necessary; pumping pressure causes the aggregate to absorb more water, thus reducing the slump and increasing the density of the concrete. The relationship between the slump at the point of delivery (truck) and the point of placement (pump) should be correlated.

Some lightweight aggregates, particularly lightweight sands, can provide additional water for internal curing, which can be important for high-performance concrete (HPC) (See Bentz and Weiss 2010, Weiss 2016, and Chapter 17 for more information on internal curing).

Lightweight aggregate concrete will have a longer period of drying time than normalweight concrete (see **Drying Rate of Concrete**). If the concrete is to receive a moisture sensitive floor covering, this must be taken into consideration (see Kanare 2008). See Bohan and Ries (2008) for more information on lightweight aggregates.

High-density (heavyweight) concrete. Heavyweight concrete has a density of up to about 6400 kg/m^3 (400 lb/ft^3 and is used principally for radiation shielding but is also used for counterweights and other applications where high-density is important.

As a shielding material, heavyweight concrete protects against the harmful effects of x-rays, gamma rays, and neutron radiation. Selection of concrete for radiation shielding is based on space requirements and on the type and intensity of radiation. Type and intensity of radiation usually determine the requirements for density and water content of shielding concrete. Effectiveness of a concrete shield against gamma rays is approximately proportional to the density of the concrete; the denser the concrete, the more effective the shield.

Boron additions such as colemanite (hydrated calcium borate hydroxide), boron frits, and borocalcite are sometimes used to improve the neutron shielding properties of concrete. However, they may adversely affect setting and early strength of concrete; therefore, trial mixtures should be made with the addition under field conditions to determine suitability. Admixtures such as pressure-hydrated lime can be used with coarse-sand sizes to minimize any retarding effect.

ACI 304.3R, *Heavyweight Concrete: Measuring, Mixing, Transporting, and Placing* (2020), provides recommendations for mixing and placing heavyweight concrete. Conventional methods of mixing and placing are often used, but care must be taken to avoid overloading the mixer, especially with very heavy aggregates such as steel punchings. Batch sizes should be reduced to about 50% of the rated mixer capacity. Because some heavy aggregates are quite friable, excessive mixing should be avoided to prevent aggregate breakup and its resultant detrimental effects on workability and bleeding.

Preplaced aggregate methods can be used for placing normal and high-density concrete in confined areas and around embedded items. This practice will minimize segregation of coarse aggregate, especially steel punchings or steel shot. The method also reduces drying shrinkage and produces concrete of uniform density and composition. With this method, the coarse aggregate is preplaced in the forms and grout made of cement, sand, and water is then pumped through pipes to fill the voids between aggregate particles. Puddling heavyweight concrete is a method whereby a 50-mm (2-in.) layer or more of mortar is placed in the forms and then covered with a layer of coarse aggregate that is rodded or internally vibrated into the mortar. Care must be taken to ensure uniform distribution of aggregate throughout the concrete. Pumping of heavyweight concrete through pipelines may be advantageous in locations where space is limited. Heavyweight concretes cannot be pumped as far as normal-weight concretes because of their higher densities.

Permeability and Watertightness

Concrete used in water-retaining structures or exposed to weather or other severe exposure conditions must be of low permeability or watertight. Watertightness is the ability of concrete to hold back or retain water without visible leakage. Permeability refers to the amount of water migration through concrete when the water is under pressure or to the ability of concrete to resist penetration by water or other substances (liquid, gas, or ions). Generally, the same properties of concrete that make it less permeable also make it more watertight.

The overall permeability of concrete to water migration is a function of:

• permeability of the paste,
• permeability and gradation of the aggregate,
• quality of the paste and aggregate transition zone, and
• relative proportion of paste to aggregate.

Decreased permeability improves concrete's resistance to freezing and thawing, resaturation, sulfate attack, chloride-ion penetration, and other chemical attack (see Chapter 11).

The permeability of the paste is particularly important because the paste envelops all constituents in the concrete. Paste permeability is related to type and amount of cementitious materials, water-cementitious materials ratio, degree of cement hydration, and length of moist curing. A low-permeability concrete requires a low water-cement ratio and an adequate moist-curing period. Air entrainment aids watertightness but has little effect on permeability. Permeability increases with drying.

The permeability of mature hardened portland cement paste kept continuously moist ranges from 0.1×10^{-14} m/s to 1.2×10^{-12} m/s for water-cement ratios ranging from 0.3 to 0.7 (Powers and others 1954). The permeability of rock commonly used as concrete aggregate varies from approximately 1.7×10^{-11} m/s to 3.5×10^{-15} m/s. The permeability of mature, good-quality concrete is approximately 1×10^{-12} m/s.

Test results obtained by subjecting 25-mm (1-in.) thick non-air-entrained mortar disks to 140-kPa (20-psi) water pressure are given in Figure 9-30.

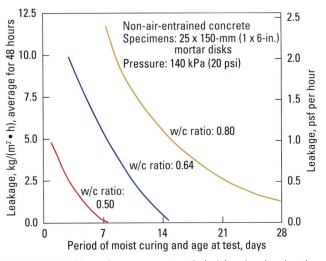

FIGURE 9-30. Effect of water-cement ratio (w/c) and curing duration on permeability of mortar. Note that leakage is reduced as the water-cement ratio is decreased and the curing period increased (McMillan and Lyse 1929 and PCA 1929).

Mortar disks that had a water-cement ratio of 0.50 by weight or less and were moist-cured for seven days showed no water leakage. Where leakage occurred, it was greater in mortar disks made with high water-cement ratios. Also, for each water-cement ratio, leakage was less as the length of the moist-curing period increased. In disks with a water-cement ratio of 0.80, the mortar still permitted leakage after moist curing for one month. These results clearly show that a low water-cement ratio and a reasonable period of moist curing significantly reduce permeability.

Figure 9-31 illustrates the effect of different water cement ratios on concrete's resistance to chloride ion penetration as indicated by electrical conductance. The total charge in coulombs was

significantly reduced with a low water-cement ratio. Also, the results showed that a lower charge passed when the concrete contained a higher air content.

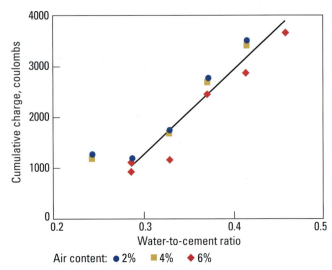

FIGURE 9-31. Total charge at the end of the ASTM C1202 (rapid chloride permeability test) as a function of water to cement ratio (Pinto and Hover 2001).

A low water-cement ratio also reduces segregation and bleeding, further contributing to watertightness. Watertight concrete must also be free from cracks, honeycomb, or other large visible voids.

The impact of cementing materials on the permeability of concrete is shown in Table 9-8. Silica fume and metakaolin are especially effective in this regard (Figure 9-32).

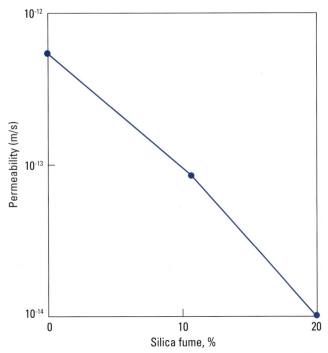

FIGURE 9-32. Effect of silica fume on concrete permeability (Hooton 1986).

TABLE 9-8. Permeability and Absorption of Concretes Moist Cured 7 Days and Tested After 90 Days

MIX NO.	CEMENT, kg/m³ (lb/yd³)	W/CM	COMPRESSIVE STRENGTH AT 90 DAYS, MPa (psi)	PERMEABILITY				POROSITY, %[†]	VOLUME OF PERMEABLE VOIDS, %	ABSORPTION AFTER IMMERSION, %	ABSORPTION AFTER IMMERSION AND BOILING, %
				RCPT, coulombs	90 DAYS PONDING, % Cl	WATER, m/s[†]	AIR, m/s[†]				
			ASTM C39 (AASHTO T 22)	ASTM C1202 (AASHTO T 277)	(AASHTO T 259)	API RP 27	API RP 27		ASTM C642	ASTM C642	ASTM C642
1	445 (750)	0.26*	104.1 (15,100)	65	0.013	—	2.81×10^{-10}	7.5	6.2	2.43	2.56
2	445 (750)	0.29*	76.7 (11,130)	852	0.022	—	3.19×10^{-10}	8.8	8.0	3.13	3.27
3	381 (642)	0.40*	46.1 (6690)	3242	0.058	2.61×10^{-13}	1.16×10^{-9}	11.3	12.2	4.96	5.19
4	327 (550)	0.50	38.2 (5540)	4315	0.076	1.94×10^{-12}	1.65×10^{-9}	12.5	12.7	5.45	5.56
5	297 (500)	0.60	39.0 (5660)	4526	0.077	2.23×10^{-12}	1.45×10^{-9}	12.7	12.5	5.37	5.49
6	245 (413)	0.75	28.4 (4120)	5915	0.085	8.32×10^{-12}	1.45×10^{-9}	13.0	13.3	5.81	5.90

Adapted from Whiting (1988).
* Note on ingredients: Mix 1: 59.4 kg/m³ (100 lb/yd³) silica fume and 25.4 mL/kg of cement (30 fl. oz/cwt) of HRWR;
Mix 2: 13.0 mL/kg (20 fl. oz/cwt) HRWR;
Mix 3: 2.2 mL/kg (3.4 fl. oz/cwt) WR.
[†] To convert from m/s to m², multiply by 1.02×10^{-7}.
[‡] Measured with helium porosimetry.

Tests show that the permeability of concrete decreases as the quantity of hydrated cementitious materials increases and is age dependent. With appropriate dosages, fly ash, slag cement, and natural pozzolans can greatly reduce the permeability and absorption of concrete. See Chapter 3 for more information on SCMs.

Volume Stability and Crack Control

Hardened concrete changes volume due to changes in temperature, moisture content, and stress. These volume or length changes may range from about 0.01% to 0.08%. Two basic causes of cracks in concrete are: stress due to applied loads; and stress due to volume changes from drying shrinkage, temperature changes, durability related distress, and restraint.

Concrete under stress deforms elastically. Sustained stress results in an additional deformation called creep. The rate of creep (deformation per unit of time) decreases with time.

Concrete kept continually moist will expand slightly. When permitted to dry, concrete will shrink. The primary factor influencing the amount of drying shrinkage is the water content of the freshly mixed concrete. Drying shrinkage increases directly with increases in water content of a concrete mixture. The amount of shrinkage also depends upon several other factors, such as: the amount of aggregate used; properties of the aggregate; size and shape of the concrete element; relative humidity and temperature of the ambient air; method of curing; degree of hydration; and time.

Drying shrinkage is an inherent, unavoidable property of concrete. However, properly positioned reinforcing steel is used to reduce crack widths, or joints are used to predetermine and control the location of cracks. Thermal stress due to fluctuations in ambient temperature also can cause cracking, particularly at an early age.

Concrete shrinkage cracks are often the result of restraint. As drying shrinkage occurs, if there is no restraint, the concrete will not crack. Restraint comes from several sources. Drying shrinkage is typically greater near the surface of concrete; the moist inner portions restrain the concrete near the surface, which can cause cracking. Other sources of restraint are reinforcing steel embedded in concrete, the interconnected parts of a concrete structure, and the friction of the subgrade on which concrete is placed.

Joints. Joints are the most effective method of controlling unsightly cracking. If a sizable expanse of concrete (a wall, slab, or pavement) is not provided with properly spaced joints to accommodate drying shrinkage and temperature contraction, the concrete will crack in a random manner.

Contraction (shrinkage control) joints are grooved, formed, or sawed into sidewalks, driveways, pavements, floors, and walls so that cracking will occur in these joints rather than in a random manner. Contraction joints permit movement in the plane of a slab or wall. They extend to a depth of approximately one-quarter the concrete thickness.

Isolation joints separate a concrete placement from other parts of a structure and permit horizontal and vertical movements. They should be used at the junction of floors with walls, columns, footings, and other points where restraint can occur. They extend the full depth of slabs and include a premolded joint filler.

Construction joints occur where concrete work is concluded for the day. They separate areas of concrete placed at different times. In slabs-on-ground, construction joints usually align with, and function as, control or isolation joints. They may also require dowels for load transfer.

For more information on volume changes in concrete, see Chapter 10.

Durability

The durability of concrete may be defined as the ability of concrete to resist weathering action, chemical attack, and abrasion while maintaining its desired engineering properties. Concrete is exposed to a greater variety of potentially harmful exposure conditions than any other construction material.

There are many causes of concrete deterioration and most of these involve either the movement of moisture or the movement of chemical species, such as chlorides and sulfates, dissolved in the water. Generally, the greater the resistance of the concrete to the movement of water, the lower its permeability and the greater its resistance to deterioration. The following sections discuss several deterioration mechanisms. For more information on each of these topics and preventive measures, see Chapter 11.

Freeze-thaw and deicer salts. Deterioration due to freezing and thawing is a result of the expansive forces that are generated when the water in saturated concrete freezes. Water expands by approximately 9% upon freezing. If the concrete is not designed to resist freeze-thaw cycles, cracking will occur. As the number of freeze-thaw cycles increase, the cracking will become more advanced and eventually severe deterioration may occur. Damage due to freezing and thawing is exacerbated in the presence of deicing salts (see Chapter 11). When concrete is exposed to these conditions particular attention must be paid to ensure that the mixture proportions are appropriate and that the concrete is finished and cured properly. D-cracking or durability cracking occurs when frost-susceptible aggregates are used in concrete exposed to freezing and thawing. In such cases, it is the expansion of water in the aggregate that leads to cracking of the concrete.

Corrosion. Concrete may be exposed to chloride ions during service. Common sources of chlorides include deicing salt, seawater, or chloride-contaminated groundwater. This is generally not an issue for non-reinforced concrete; however, over time, the chlorides will penetrate through the concrete cover and eventually reach the embedded steel reinforcement. The chlorides break down the passive layer, allowing corrosion of the steel to begin. Because the products of corrosion, rust, occupy more volume than the metallic steel, expansive forces develop which can lead to cracking in the concrete. Eventually corrosion can lead to spalling and delamination of the concrete cover. Corrosion of embedded steel reinforcement is the most prevalent form of deterioration of reinforced concrete structures.

Carbonation. Another form of concrete deterioration due to corrosion is caused by carbonation. Carbonation of concrete occurs when carbon dioxide from the atmosphere penetrates concrete and reacts with the products of cement hydration and reduces the alkalinity of the concrete. If the carbonation depth reaches the steel, the steel passivation layer can become unstable and the steel can start to corrode. Carbonation is typically a very slow process.

Alkali-silica reactivity. Alkali-silica reaction (ASR) is the reaction between the alkalies (sodium and potassium) in concrete and certain siliceous rocks or minerals, such as opaline chert, strained quartz, and acidic volcanic glass, present in some aggregates. In some cases, alkalies may migrate from sources outside the concrete.

The reaction product is an alkali-silica gel which has the capacity to adsorb water and swell. Under certain conditions the products of the reaction may cause abnormal expansion and cracking of concrete in service.

Abrasion. Floors, pavements, and hydraulic structures are subjected to abrasion. In these applications concrete must have a high abrasion resistance. Test results indicate that abrasion resistance is closely related to the compressive strength of concrete. Strong concrete has more resistance to abrasion than weak concrete. The type of aggregate and surface finish or treatment used also have a strong influence on abrasion resistance: Hard aggregate is more wear resistant than soft aggregate and a steel-troweled surface resists abrasion better than an untroweled surface.

Sulfate attack. Excessive amounts of sulfates in soil or water can attack and destroy a concrete that is not properly designed. Sulfates (for example, calcium sulfate, sodium sulfate, and magnesium sulfate) can attack concrete by reacting with hydrated compounds in the hardened cement paste. These reactions can induce sufficient pressure to disrupt the cement paste, resulting in disintegration of the concrete (loss of paste cohesion and strength).

Other forms of concrete deterioration. There are other forms of concrete deterioration less common or which occur only in special conditions:

• delayed ettringite formation (DEF) – which only occurs in concrete exposed to excessive temperatures at early ages,

• thaumasite form of sulfate attack – which differs from classical sulfate salt attack,

• alkali-carbonate reaction (ACR) – which involves the attack by alkalis on carbonate phases of the rock and is much less widespread than alkali-silica reaction,

• salt crystallization, and

• attack by chemicals other than sulfates.

Different concretes require different degrees of durability depending on the exposure environment and the desired properties. The concrete ingredients, proportioning of those ingredients, interactions between the ingredients, and placing and curing practices determine the ultimate durability and service life of the concrete. For more information and an extensive discussion on the durability of concrete, see Chapter 11.

Aesthetics

Attractive decorative finishes can be built into concrete during construction. Variations in the color and texture of concrete surfaces are limited only by the imagination of the designer and the skill of the concrete craftsman.

Color may be added to the concrete through the use of white cement and pigments, exposure of colorful aggregates, or addition of score lines to create borders for the application of penetrating or chemically reactive stains. Desired textured finishes can be varied, from a smooth polish to the roughness of gravel. Geometric patterns can be scored, stamped, rolled, or inlaid into the concrete to resemble stone, brick, or tile paving (Figure 9-33). Other interesting patterns are obtained

using divider strips (commonly redwood) to form panels of various sizes and shapes – rectangular, square, circular, or diamond. Special techniques are also available to make concrete slip-resistant and to impart various sheens and other surface characteristics.

Decorative concrete can be used in exterior and interior applications in homes or commercial buildings. Colored and imprinted concrete is an excellent flooring material combining the economy, durability, decorative qualities, and strength of concrete and the thermal mass needed for passive solar buildings. Special concrete finishes (interior or exterior) enhance the aesthetic appeal and value of any property. For more information on decorative concrete, see ACI 303R, *Guide to Cast-in-Place Architectural Concrete Practice* (2012); ACI 310R, *Guide to Decorative Concrete* (2019); and Kosmatka and Collins (2004).

REFERENCES

PCA's online catalog includes links to PDF versions of many of our research reports and other classic publications. Visit: cement.org/library/catalog.

Abrams, D.A., *Design of Concrete Mixtures*, LS001, Lewis Institute, Structural Materials Research Laboratory, Bulletin No. 1, Chicago, IL, 1918, 20 pages.

ACI Committee 213, *Guide for Structural Lightweight-Aggregate Concrete*, ACI 213R-14, American Concrete Institute, Farmington Hills, MI, 2014, 53 pages.

ACI Committee 302, *Guide for Concrete Floor and Slab Construction*, ACI 302.1R-15, American Concrete Institute, Farmington Hills, MI, 2015, 76 pages.

ACI Committee 303, *Guide to Cast-in-Place Architectural Concrete Practice*, ACI 303R-12, American Concrete Institute, Farmington Hills, MI, 2012, 32 pages.

ACI Committee 304, *Heavyweight Concrete: Measuring, Mixing, Transporting, and Placing*, ACI 304.3R-20, American Concrete Institute, Farmington Hills, MI, 2020, 8 pages.

ACI Committee 309, *Guide for Consolidation of Concrete*, ACI 309R-05, American Concrete Institute, Farmington Hills, MI, 2005, 36 pages.

ACI Committee 310, *Guide to Decorative Concrete*, ACI 310R-19, American Concrete Institute, Farmington Hills, MI, 2019, 48 pages.

ACI Committee 318, *Building Code Requirements for Structural Concrete*, ACI 318-19, American Concrete Institute, Farmington Hills, MI, 2019, 624 pages.

Bingham, E.C., "An Investigation of the Laws of Plastic Flow," *Bulletin of the Bureau of Standards*, Vol. 13, No. 2, August 1916, pages 309 to 353.

Brewster, R.S., *Effect of Vibration Time upon Loss of Entrained Air from Concrete Mixes*, Materials Laboratories Report No. C-461, Research and Geology Division, Bureau of Reclamation, Denver, CO, November 25, 1949.

FIGURE 9-33. Pattern-stamped finish and colored surfaces are popular for decorative concretes.

Bullard, J.W.; D'Ambrosia, M.; Grasley, Z.; Hansen, W.; Kidner, N.; Lange, D.; Lura, P.; Mason, T.O.; Moon, J.; Rajabipour, F.; Sant, G.; Shah, S.; Voigt, T.; Wansom, N.; Weiss, W.J.; and Woo L., "A Comparison of Test Methods for Early-Age Behavior of Cementitious Materials," *Advances in Concrete through Science and Engineering*, RILEM, Quebec, 2006, 20 pages.

Bureau of Reclamation, *Concrete Manual*, 8th Edition, Bureau of Reclamation, Denver, CO, 1981, 661 pages.

Burg, R.G., *The Influence of Casting and Curing Temperature on the Properties of Fresh and Hardened Concrete*, Research and Development Bulletin RD113, Portland Cement Association, Skokie, IL, 1996, 20 pages.

Cable, J.K.; McDaniel, L.; Schlorholtz, S.; Redmond, D.; and Rabe, K., E*valuation of Vibrator Performance vs. Concrete Consolidation and Air Void System*, SN2398, Portland Cement Association, Skokie, IL, 2000, 60 pages.

Copeland, L.E.; and Schulz, E.G., *Electron Optical Investigation of the Hydration Products of Calcium Silicates and Portland Cement*, Research Department Bulletin RX135, Portland Cement Association, Skokie, IL, 1962, 12 pages.

Daniel, D.G., "Factors Influencing Concrete Workability," *Significance of Tests and Properties of Concrete and Concrete-Making Materials*, ASTM STP 169D, ASTM International, West Conshohocken, PA, 2006, pages 59 to 72.

Falconi, M.I., *Durability of Slag Cement Concretes Exposed to Freezing and Thawing in the Presence of Deicers*, Master of Science Thesis, Cornell University, Ithaca, NY, 1996, 306 pages.

Feret, R., "Etudes Sur la Constitution Intime Des Mortiers Hydrauliques" ("Studies on the Intimate Constitution of Hydraulic Mortars"), *Bulletin de la Societe d'Encouragement Pour Industrie Nationale*, 5th Series, Vol. 2, Paris, 1897, pages 1591 to 1625.

Gebler, S.H.; and Klieger, P., *Effect of Fly Ash on Durability of Air-Entrained Concrete*, Research and Development Bulletin RD090, Portland Cement Association, Skokie, IL, 1986, 40 pages.

Greening, N.R., *Some Causes for Variation in Required Amount of Air-Entraining Agent in Portland Cement Mortars*, Research Department Bulletin RX213, Portland Cement Association, Skokie, IL, 1967, 20 pages.

Hall, C., "Barrier Performance of Concrete: A Review of Fluid Transport Theory," *Penetration and Permeability of Concrete: Barriers to Organic and Contaminating Liquids*, RILEM Report 16, E&FN Spon, London, 1997, pages 7 to 40.

Hanson, J.A., *Effects of Curing and Drying Environments on Splitting Tensile Strength of Concrete*, Development Department Bulletin DX141, Portland Cement Association, Skokie, IL, 1968, 29 pages.

Hedenblad, G., *Drying of Construction Water in Concrete-Drying Times and Moisture Measurement*, Stockholm, Byggforskningsrådet, 1997, 54 pages.

Hedenblad, G., *Data for Calculation of Moisture Migration*, Swedish Council for Building Research, S-11387, Stockholm, Sweden, 1996.

Hooton, R.D., "Permeability and Pore Structure of Cement Pastes Containing Fly-Ash, Slag, and Silica Fume," *Blended Cements*, ASTM STP 897, G. Frohnsdorff, ed., American Society of Testing and Materials, Philadelphia, PA, 1986, pages 128 to 143.

Hsu, T.T.C., *Torsion of Structural Concrete – Plain Concrete Rectangular Sections*, Development Department Bulletin DX134, Portland Cement Association, Skokie, IL, 1968, 42 pages.

Hughes, B.P.; Lowe, I.R.G.; and Walker, J., "The Diffusion of Water in Concrete at Temperatures Between 50 and 95°C," *British Journal of Applied Physics*, Vol. 17, No. 12, December 1966, pages 1545 to 1552.

Kanare, H.M., *Concrete Floors and Moisture*, EB119, Portland Cement Association, Skokie, IL, and National Ready Mixed Concrete Association, Silver Spring, MD, 2008, 176 pages.

Kosmatka, S.H., "Bleed Water," *Significance of Tests and Properties of Concrete and Concrete-Making Materials*, RP328, ASTM STP 169D, ASTM International, West Conshohocken, PA, 2006, pages 99 to 122.

Kosmatka, S.H.; and Collins, T.C., *Finishing Concrete with Color and Texture*, PA124, Portland Cement Association, Skokie, IL, 2004, 76 pages.

Kozikowski, Jr.; R.L.; Vollmer, D.B.; Taylor, P.C.; and Gebler, S.H., *Factor(s) Affecting the Origin of Air-Void Clustering*, SN2789, Portland Cement Association, Skokie, IL, 2005, 22 pages.

Lange, D.A., *Long-Term Strength Development of Concrete*, RP326, Portland Cement Association, Skokie, IL, 1994, 33 pages.

Lerch, W., *Basic Principles of Air-Entrained Concrete*, T-101, Portland Cement Association, Chicago, IL, 1960, 40 pages.

Lura, P.; Pease, B.; Mazzotta, G.; Rajabipour, F.; and Weiss, W.J., "Influence of Shrinkage-Reducing Admixtures on Development of Plastic Shrinkage Cracks," *ACI Materials Journal*, Vol. 104, No. 2, American Concrete Institute, Farmington Hills, MI, 2007, pages 187 to 194.

McMillan, F.R.; and Lyse, I., "Some Permeability Studies of Concrete," *Journal of the American Concrete Institute, Proceedings*, Vol. 26, Farmington Hills, MI, December 1929, pages 101 to 142.

McNeal, F.; and Gay, F., "Solutions to Scaling Concrete," *Concrete Construction*, Addison, IL, March 1996, pages 250 to 255.

PCA, "Permeability Studies of Concrete," *Proceedings of the Portland Cement Association*, MJ227, Portland Cement Association, Chicago, IL, Vol. 28, November 1929, pages 79 to 84.

PCA, *Study of the Air Content of Mortar and Concrete*, Major Series 336, Portland Cement Association, Chicago, IL, 1948.

Pinto, R.C.A.; and Hover, K.C., *Frost and Scaling Resistance of High-Strength Concrete*, Research and Development Bulletin RD122, Portland Cement Association, Skokie, IL, 2001, 70 pages.

Powers, T.C., "Studies of Workability of Concrete," *Journal of the American Concrete Institute*, Vol. 28, American Concrete Institute, Farmington Hills, MI, February 1932, page 419.

Powers, T.C., *A Discussion of Cement Hydration in Relation to the Curing of Concrete*, Research Department Bulletin RX025, Portland Cement Association, Chicago, IL, 1948, 14 pages.

Powers, T.C., *The Bleeding of Portland Cement Paste, Mortar, and Concrete*, Research Department Bulletin RX002, Portland Cement Association, Chicago, IL, 1939, 182 pages.

Powers, T.C., *The Nonevaporable Water Content of Hardened Portland Cement Paste – Its Significance for Concrete Research and Its Method of Determination*, Research Department Bulletin RX029, Portland Cement Association, Chicago, IL, 1949, 20 pages.

Powers, T.C.; and Brownyard, T.L., *Studies of the Physical Properties of Hardened Portland Cement Paste*, Research Department Bulletin RX022, Portland Cement Association, Chicago, IL, 1948, 356 pages.

Powers, T.C.; Copeland, L.E.; Hayes, J.C.; and Mann, H.M., *Permeability of Portland Cement Pastes*, Research Department Bulletin RX053, Portland Cement Association, Chicago, IL, 1955, 19 pages.

Scherer, G.W., "Theory of Drying," *Journal of the American Ceramic Society*, Vol 73, No. 1, 1990, pages 3 to 14.

Stark, D.C., *Effect of Vibration on the Air-System and Freeze-Thaw Durability of Concrete*, Research and Development Bulletin RD092, Portland Cement Association, Skokie, IL, 1986, 13 pages.

Steinour, H.H., *Further Studies of the Bleeding of Portland Cement Paste*, Research Department Bulletin RX004, Portland Cement Association, Chicago, IL, 1945, 108 pages.

Steinour, H.H., *The Setting of Portland Cement*, Research Department Bulletin RX098, Portland Cement Association, Chicago, IL, 1958, 124 pages.

Szecsy, R.; and Mohler, N., *Self-Consolidating Concrete*, IS546D, Portland Cement Association, Skokie, IL, 2009, 24 pages.

Taylor, T.G., *Effect of Carbon Black and Black Iron Oxide on Air Content and Durability of Concrete*, Research Department Bulletin RX023, Portland Cement Association, Chicago, IL, 1948, 15 pages.

Thomas, M., *Optimizing the Use of Fly Ash in Concrete*, IS548, Portland Cement Association, Skokie, IL, 2007, 24 pages.

Whiting, D.A.; and Nagi, M.A., *Manual on Control of Air Content in Concrete*, EB116, National Ready Mixed Concrete Association, Silver Spring, MD, and Portland Cement Association, Skokie, IL, 1998, 42 pages.

Whiting, D.; and Dziedzic, D., *Effects of Conventional and High-Range Water Reducers on Concrete Properties*, Research and Development Bulletin RD107, Portland Cement Association, Skokie, IL, 1992, 25 pages.

Whiting, D., "Permeability of Selected Concretes," *Permeability of Concrete*, SP-108, American Concrete Institute, Farmington Hills, MI, 1988, pages 195 to 222.

Wood, S.L., *Evaluation of the Long-Term Properties of Concrete*, Research and Development Bulletin RD102, Portland Cement Association, Skokie, IL, 1992, 99 pages.

Young, J.F.; Mindess, S.; Gray, R.J.; and Bentur, A., *The Science and Technology of Civil Engineering Materials*, Prentice Hall, Upper Saddle River, NJ, 1998, 398 pages.

VOLUME CHANGES OF CONCRETE

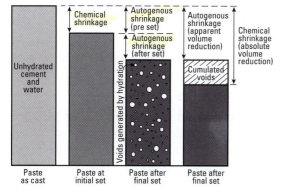

FIGURE 10-1. Chemical shrinkage and autogenous shrinkage volume changes of fresh and hardened paste. Not to scale.

Concrete undergoes slight changes in volume due to a number of factors; understanding the nature of volume change is useful in planning or analyzing concrete work. If concrete is free of restraints, normal volume changes would be unimpeded and cracking would not result. However, concrete in service is commonly restrained by foundations, subgrades, reinforcement, and connecting members; and significant stresses can develop under deformation. This is particularly true of tensile stresses.

While quite strong in compression, cracks develop because concrete is relatively weak in tension (see Chapter 9). Controlling the variables that affect volume change can minimize high stresses and cracking. Joint location and tolerable crack widths should be considered in the structural design.

Characteristic volume changes consist of linear expansion and contraction due to hydration reactions, temperature swings, and moisture variation. Chemical effects such as carbonation, sulfate attack, and the disruptive expansion of alkali-aggregate reactions also cause volume changes. In addition, creep, or deformation caused by sustained stress, is possible. Equally important is the elastic or inelastic change in dimensions or shape that occurs instantaneously under applied load.

Changes in length are often expressed as a coefficient of length in parts per million, or simply as millionths. It is applicable to any length unit (for example, m/m or ft/ft); one millionth is 0.000001 m/m (0.000001 in./in.) and 600 millionths is 0.000600 m/m (0.000600 in./in.). Change of length can also be expressed as a percentage; for example: 0.06% is the same as 0.000600, which is also the same as 6 mm per 10 m (¾ in. per 100 ft). The sum of length changes from all causes that ordinarily occur in concrete are generally small, ranging in length from approximately 10 millionths up to about 1000 millionths.

EARLY AGE VOLUME CHANGES

Volume change in concrete begins immediately after water comes into contact with cement. Early volume changes, within 24 hours, can influence the long-term volume changes and potential uncontrolled crack formation in hardened concrete. Following are discussions on various forms of early volume change; for more information, see ACI 231R, *Report on Early-Age Cracking: Causes, Measurement, and Mitigation* (2010).

Chemical Shrinkage

Chemical shrinkage refers to the reduction in absolute volume of solids and liquids in paste resulting from cement hydration. The absolute volume of the hydrated products is less than the absolute volume of cement and water before hydration. This change in volume of cement paste during the plastic state is illustrated by the first two bars in Figure 10-1. This does not include air voids generated during mixing.

Chemical shrinkage continues to occur as long as cement hydrates. After initial set, the paste cannot deform as much as it can in a plastic state. Further hydration and chemical shrinkage is compensated by the consumption of water from capillary pores. This process is called self-desiccation. Most of this volume change is internal and does not significantly change the visible external dimensions of a concrete element until the capillary pores are drained to a diameter of about 50 nm. The small pore size generates internal capillary pressure which drives external volume change. This process is called autogenous shrinkage (Tazawa 1999).

The amount of volume change due to chemical shrinkage can be estimated from the hydrated cement phases and their densities, or it can be determined by physical testing as illustrated in Figure 10-2 in accordance with ASTM C1608, *Standard Test Method for Chemical Shrinkage of Hydraulic Cement Paste.* An

example of long-term chemical shrinkage of portland cement paste is shown in Figure 10-3. Early researchers sometimes referred to chemical shrinkage as the absorption of water during hydration (Powers 1935). Le Chatelier (1900) was the first to study chemical shrinkage of cement pastes. It is important to recognize that the shrinkage potential of a concrete mixture is affected by the water content and the quantity, gradation, and volume stability of the aggregates.

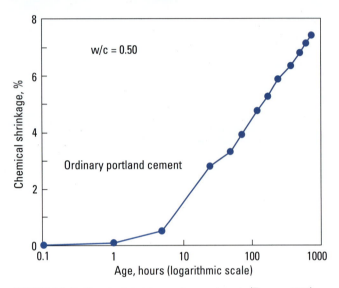

FIGURE 10-3. Chemical shrinkage of cement paste (Tazawa 1999).

Autogenous Shrinkage

Autogenous shrinkage is the macroscopic volume reduction (visible dimensional change) of cement paste, mortar, or concrete caused by cement hydration. Autogenous shrinkage is measured in accordance with ASTM C1698, *Standard Test Method for Autogenous Strain of Cement Paste and Mortar.* The macroscopic volume reduction of autogenous shrinkage is much less than the absolute volume reduction of chemical shrinkage because of the rigidity of the hardened paste structure. Chemical shrinkage is the driving force behind autogenous shrinkage. The relationship between autogenous shrinkage and chemical shrinkage is illustrated in Figures 10-1, 10-4, and 10-5. Some researchers consider that autogenous shrinkage starts at initial set while others evaluate autogenous shrinkage from time of placement (Bentur 2003).

When external water is available, autogenous shrinkage cannot occur. If external water is not available (due to a lack of moist curing, low permeability concrete, or thick sections that limit the transport of curing water), cement hydration consumes pore water resulting in self desiccation of the paste and a uniform reduction of volume (Copeland and Bragg 1955 and Radlinska and others 2008). Autogenous shrinkage increases with a decrease in water to cement ratio and with an increase in the fineness and amount of cementitious material. Autogenous shrinkage is most prominent in concrete with a water to

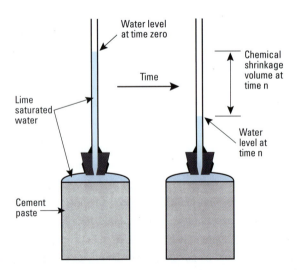

FIGURE 10-2. Test for chemical shrinkage of cement paste showing flask for cement paste and pipet for absorbed water measurement (ASTM C1608).

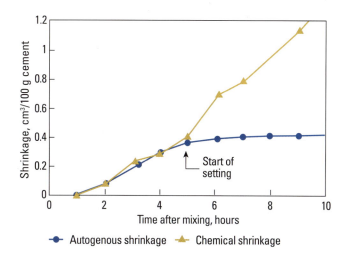

FIGURE 10-4. Relationship between autogenous shrinkage and chemical shrinkage of cement paste at early ages (Hammer 1999).

cement ratio under 0.42 (Holt 2001). High-strength, low water to cement ratio (0.30) concrete can experience 200 millionths to 400 millionths of autogenous shrinkage. Autogenous shrinkage can be half that of drying shrinkage for concretes with a water to cement ratio of 0.30 (Holt 2001 and Aïtcin 1999).

Use of high performance, lower w/cm concretes in bridges and other structures has renewed interest in designing for autogenous shrinkage to control crack development. Concretes susceptible to large amounts of autogenous shrinkage should be cured with external water for at least 7 days to minimize crack development, and fogging should be provided as soon as the concrete is cast. The hydration of supplementary cementing materials also contributes to autogenous shrinkage, although at different levels than with portland cement. In addition to adjusting paste content and water to cement ratios, autogenous shrinkage can be reduced by using shrinkage reducing or shrinkage compensating admixtures or internal curing techniques. Some cementitious systems may experience autogenous expansion. Tazawa (1999), Holt (2001), Shah and others (1998), and Radlinska and others (2008) review techniques to control autogenous shrinkage. Test methods for autogenous shrinkage and expansion of cement paste, mortar, and concrete include direct testing methods that measure volumetric or linear dimensional changes. Indirect methods establish correlations between porosity measurements and changes in relative humidity (Tazawa 1999).

Subsidence

Subsidence refers to the vertical shrinkage of fresh cementitious materials before initial set. It is caused by a combination of bleeding, air voids rising to the surface, and chemical shrinkage. Subsidence is also called settlement shrinkage. Subsidence of well-consolidated concrete with minimal bleed water is insignificant to concrete's total volume

change. The relationship between subsidence and other shrinkage mechanisms is illustrated in Figure 10-5. Excessive subsidence is often caused by a lack of consolidation of fresh concrete. Excessive subsidence over embedded items, such as steel reinforcement, can result in cracking located over embedments. Concretes made with air entrainment, sufficient fine materials, and low water contents can help minimize subsidence cracking. Also, fibers have been reported to reduce subsidence cracking (Suprenant and Malisch 1999).

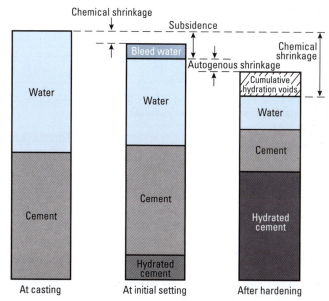

FIGURE 10-5. Volumetric relationship between subsidence, bleed water, chemical shrinkage, and autogenous shrinkage. Only autogenous shrinkage after initial set is shown. Chemical shrinkage occurs both at initial set and after hardening. Not to scale.

Plastic Shrinkage

Plastic shrinkage refers to volume change occurring while the concrete is still fresh, before hardening. It is usually observed in the form of plastic shrinkage cracks occurring before or during finishing. The cracks often resemble tears in the surface. Plastic shrinkage results from a combination of chemical and autogenous shrinkage and rapid evaporation of moisture from the surface that exceeds the bleeding rate. Plastic shrinkage cracking can be controlled by minimizing surface evaporation through use or combination of fogging, wind breaks, shading, plastic sheet covers, wet burlap, evaporation retarders, and fibers (see Chapters 16 and 17 for more information).

Swelling

Concrete, mortar, and cement paste swell in the presence of external water. External water can come from wet curing or submersion. Swelling occurs due to a combination of crystal growth, absorption of water, and osmotic pressure. When water drained from capillaries by chemical shrinkage is replaced by external water, the volume of the concrete mass initially increases. The swelling is not large, only about 50 millionths at

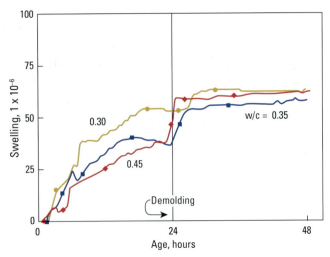

FIGURE 10-6. Early age swelling of 100 × 100 × 375-mm (4 × 4 × 15-in.) concrete specimens cured under water (Aïtcin 1999).

early ages (Figure 10-6). As there is no self-desiccation, there is no autogenous shrinkage. When the external water source is removed, autogenous shrinkage and drying shrinkage reverse the volume increase.

Expansion due to Exothermic Reaction

As cement hydrates, exothermic reactions provide a significant amount of heat. In large elements, the heat is retained, rather than dissipated as with thin elements. This temperature rise, occurring over the first few hours and days, can induce a small amount of expansion that can partially counteract autogenous and chemical shrinkage (Holt 2001). For more on temperature rise due to heat of hydration, see Chapters 2 and 9.

Moisture Changes (Drying Shrinkage) of Hardened Concrete

Hardened concrete expands slightly with a gain in moisture and contracts with a moisture loss. The effects of these cycles are illustrated in Figure 10-7. Specimen A represents concrete stored continuously in water from time of casting. Specimen B represents the same concrete exposed first to air drying and then to alternate cycles of wetting and drying. For comparative purposes, it should be noted that the swelling that occurs during continuous wet storage over a period of several years is usually less than 150 millionths; this is about one-fourth of the typical shrinkage of air-dried concrete for the same period. Figure 10-8 shows swelling of concretes wet cured for 7 days followed by shrinkage when sealed or exposed to air drying. Autogenous shrinkage reduces the volume of the sealed concretes to a level about equal to the amount of swelling at 7 days. Note that the concretes wet cured for 7 days had less shrinkage due to drying and autogenous effects than the concrete that had no water curing. This illustrates the importance of early, wet curing to minimize shrinkage (Aïtcin 1999).

Drying shrinkage can be evaluated in accordance with ASTM C157, *Standard Test Method for Length Change of Hardened*

Hydraulic-Cement Mortar and Concrete (AASHTO T 160) (Figure 10-9 left) and ASTM C1581 *Standard Test Method for Determining Age at Cracking and Induced Tensile Stress Characteristics of Mortar and Concrete under Restrained Shrinkage* (ring test) (Figure 10-9 right). Tests indicate that the typical ultimate drying shrinkage of small, plain concrete specimens (without reinforcement) ranges from about 400 millionths to 800 millionths when exposed to air at 50% relative humidity. Concrete with a drying shrinkage of 550 millionths shortens about the same amount as the thermal contraction caused by a decrease in temperature of 55°C (100°F). Preplaced aggregate concrete has a drying shrinkage of 200 millionths to 400 millionths; this is considerably less than normal concrete due to point-to-point contact of aggregate particles in preplaced aggregate concrete. The drying shrinkage of structural lightweight concrete ranges from slightly less than that of normal-density concrete, to 30% more depending on the type of aggregate used, aggregate conditioning, and method of curing.

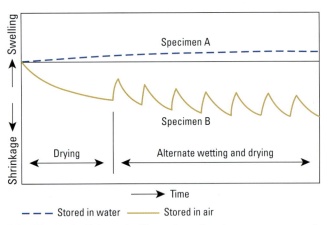

FIGURE 10-7. Schematic illustration of moisture movements in concrete. If concrete is kept continuously wet, a slight expansion occurs. However, drying usually takes place in air, causing shrinkage. Further wetting and drying causes alternate cycles of swelling and shrinkage (Roper 1960).

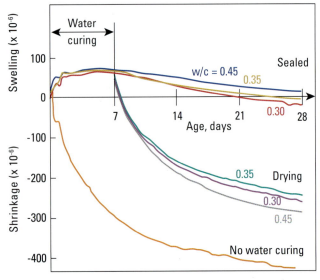

FIGURE 10-8. Length change of concrete samples exposed to different curing regimes (Aïtcin 1999).

FIGURE 10-9. (left) Standard test method for length change (drying shrinkage) in concrete (ASTM C157). (right) Restrained shrinkage testing (ASTM C1581) (courtesy of M. D'Ambrosia).

The drying shrinkage of reinforced concrete is less than that for plain concrete, the difference is dependent on the amount of reinforcement. Steel reinforcement restricts but does not prevent drying shrinkage. In reinforced concrete structures with normal amounts of reinforcement, drying shrinkage is assumed to be 200 to 300 millionths. Similar values are found for slabs on ground restrained by subgrade. Tarr (2012) and ACI 209R, *Prediction of Creep, Shrinkage, and Temperature Effects in Concrete Structures* (1992), provide approaches to predicting and modeling shrinkage in reinforced structures.

Some specifiers require a 28-day shrinkage which is only a portion of the ultimate shrinkage. Common specifications, including ACI 301, *Specification for Concrete Construction* (2020), requires 0.040% maximum shrinkage at 28 days for concrete with a maximum aggregate size of 25 mm (1 in.) and smaller and 0.035% maximum shrinkage at 28 days for concrete with a maximum aggregate size above 25 mm (1 in.).

The amount of moisture in concrete is affected by the relative humidity of the ambient air. The free moisture content of concrete elements after drying in air at relative humidity ranges of 50% to 90% for several months is about 1% to 2% by weight of the concrete. The actual amount depends on the concrete's constituents, original water content, drying conditions, and the size and shape of the concrete element.

After concrete has dried to a constant moisture content at a particular relative humidity condition, a decrease in humidity causes moisture loss; likewise, an increase in humidity causes a gain in moisture. The concrete shrinks or swells with each fluctuation in moisture content due primarily to responses of the cement paste. Most aggregates show little response to changes in moisture content, with the exception of a few types of aggregates that may swell or shrink.

As drying takes place, concrete shrinks to its center of mass. Where there is no restraint, movement occurs freely and no stresses or cracks develop (Figure 10-10A top).

If the tensile stress that results from restrained drying shrinkage exceeds the tensile strength of the concrete, cracks may develop (Figure 10-10A bottom). The magnitude of the tensile stress increases with an increased drying rate (Grasley 2003).

Random cracks may develop if joints are not properly provided in a timely manner and the concrete element is restrained from shortening. Contraction joints for slabs on ground (Figure 10-10B) should be spaced at distances of 24 to 30 times the slab thickness to control random cracks (Tarr and Farny 2008). Joints in walls are equally important for crack control (Figure 10-10C).

Shrinkage and cracking

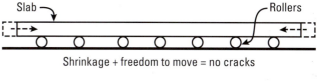

Shrinkage + freedom to move = no cracks

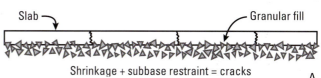

Shrinkage + subbase restraint = cracks

A

B

C

FIGURE 10-10. (A) Illustration showing no crack development in concrete that is free to shrink (slab on rollers); however, in reality a slab on ground is restrained by the subbase (or other elements) creating tensile stresses and cracks. (B) A properly functioning contraction joint controls the location of shrinkage cracking. (C) Contraction joints in the slabs and walls shown will minimize the formation of random cracks.

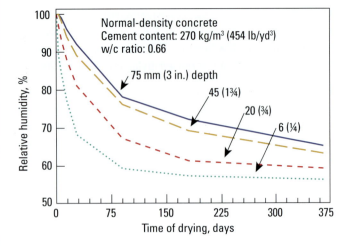

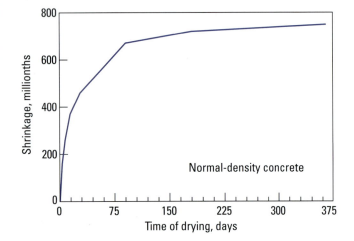

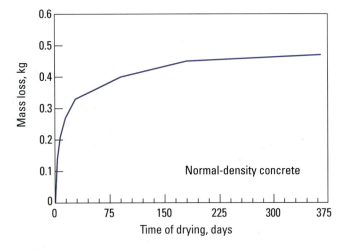

FIGURE 10-11. Relative humidity distribution at various depths, drying shrinkage, and mass loss of 150-mm × 300-mm (6-in. × 12-in.) cylinders moist-cured for 7 days followed by drying in laboratory air at 23°C (73°F) and 50% RH (Hanson 1968).

Figure 10-11 illustrates the relationship between drying rate at different depths, drying shrinkage, and mass loss for normal-density concrete (Hanson 1968). Yang and others (2000) and Weiss and others (2000) discuss shrinkage stress in slabs and the effects of geometry on shrinkage. See PCA EB086, *Building Movements and Joints* (1982) and Chapter 15 for more information on jointing concrete.

Shrinkage may continue for a number of years, depending on the size and shape of the concrete and ambient conditions. The rate and ultimate amount of shrinkage are usually smaller for larger masses of concrete; however, shrinkage continues longer for large masses. Also, higher volume-to-surface ratios experience lower shrinkage as shown in Figure 10-12.

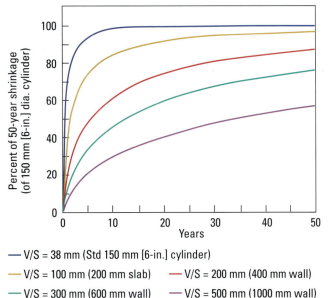

— V/S = 38 mm (Std 150 mm [6-in.] cylinder)

— V/S = 100 mm (200 mm slab) — V/S = 200 mm (400 mm wall)

— V/S = 300 mm (600 mm wall) — V/S = 500 mm (1000 mm wall)

FIGURE 10-12. Drying shrinkage of various volume/surface (V/S) ratios based on the 50-year shrinkage of a standard 150-mm × 300-mm (6-in. × 12-in.) cylinder (Baker and others 2007).

The general uniformity of shrinkage of concretes with different types of cement at different ages is illustrated in Figure 10-13. Specimens were initially moist-cured for 14 days at 21°C (70°F), then stored for 38 months in air at the same temperature and 50% relative humidity. Shrinkage recorded at the age of 38 months ranged from 600 to 790 millionths. An average of 34% of this shrinkage occurred within the first month. At the end of 11 months an average of 90% of the 38-month shrinkage had taken place. This study showed a tendency for similar patterns of shrinkage, however it should be noted that not all cements or cementing materials will have similar results.

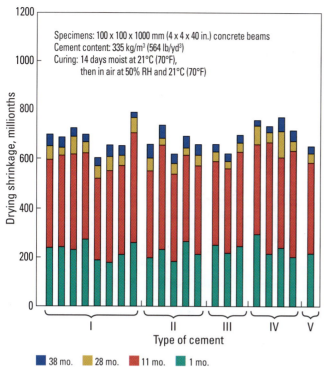

FIGURE 10-13. Results of long-term drying shrinkage tests by the U.S. Bureau of Reclamation. Shrinkage ranged from 600 to 790 millionths after 38 months of drying. The shrinkage of concretes made with air-entraining cements was similar to that for non-air-entrained concretes in this study (Bureau of Reclamation 1947 and Jackson 1955).

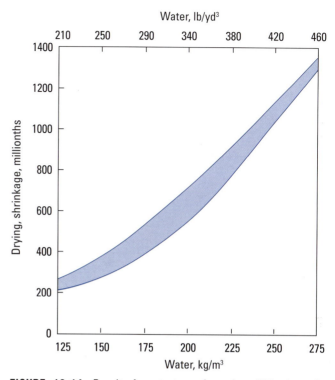

FIGURE 10-14. Results from tests performed at PCA show the relationship between total water content and drying shrinkage. A large number of mixtures with various proportions is represented within the shaded area of the curves. Drying shrinkage increases with increasing water contents.

Effect of Concrete Ingredients on Drying Shrinkage

The most important controllable factor affecting drying shrinkage is the amount of water per unit volume of concrete. Typically, about half the water in concrete mixtures is consumed by the chemical reaction of hydration, and the other half is necessary to achieve workability and finishability during placement. Some of this excess water is expelled from the concrete during the bleeding phase. Water that does not come to the surface as bleedwater during placement remains in the hardened concrete to eventually evaporate and contribute to drying shrinkage. The evaporation process at the concrete surface is limited by diffusion of capillary water through the pore microstructure and driven by the difference in relative humidity (drying gradient). The relationship between total water content and drying shrinkage is shown in Figure 10-14.

As shown in Figure 10-14, shrinkage can be minimized by keeping the water content of concrete as low as possible. Lowering the w/cm ratio also slows the drying process by limiting diffusion. Using an optimal blend of aggregates to maximize the packing density and increase the total coarse aggregate content of the concrete (minimizing paste content) also minimizes the total shrinkage potential. Therefore, use of placing methods that minimize water requirements are major factors in controlling concrete shrinkage, although slump alone is a poor predictor of shrinkage potential. Any practice that increases the water requirement of the cement paste; such as the use of high slumps (without superplasticizers), excessively high freshly-mixed concrete temperatures, higher fine-aggregate contents, or use of smaller-sized coarse aggregate (increasing paste content), will increase shrinkage.

Also, anything that decreases the bleed rate and capacity can possibly result in greater drying shrinkage. The effect of accelerators is generally offset by low temperatures which increase the bleed period. It has been found that a small amount of water can be added to ready mixed concrete at the jobsite without affecting drying shrinkage properties provided the additions are within specifications (Suprenant and Malisch 2000). Radlinska and Weiss (2012) provide guidance on performance specifications to limit concrete shrinkage.

Cements containing limestone may exhibit similar or less drying shrinkage depending on the fineness (Bucher and others 2008 and Bentz and others 2009). Fly ash and slag have little effect on shrinkage at normal dosages. Figure 10-15 shows that concretes with normal dosages of selected fly ashes performed similar to the control concrete made with only portland cement. However, very fine SCMs, such as silica fume or metakaolin that increase water demand can also increase shrinkage properties in concrete mixtures if not accounted for in the mixture design.

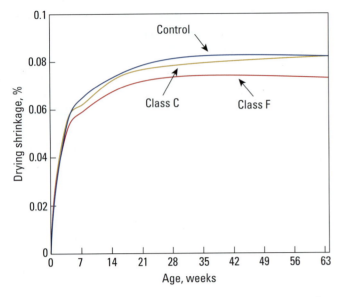

FIGURE 10-15. Drying shrinkage of fly ash concretes compared to a control mixture. The graphs represent the average of four Class C ashes and six Class F ashes, with the range in drying shrinkage rarely exceeding 0.01 percentage points. Fly ash dosage was 25% of the cementing material (Gebler and Klieger 1986).

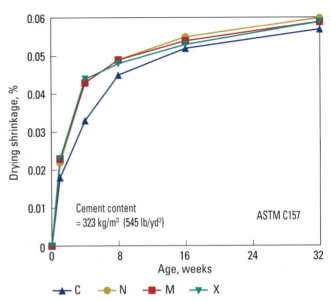

FIGURE 10-16. Drying shrinkage of concretes made with selected high-range water reducers (N, M, and X) compared to a control mixture (C) (Whiting and Dziedzic 1992).

Aggregates in concrete, especially coarse aggregate, physically restrain the shrinkage of hydrating and drying cement paste. That is why paste content affects the drying shrinkage of mortar more than that of concrete. Drying shrinkage is also highly dependent on the type of aggregate used in the concrete mixture. Except for total water content, the type of aggregate has the greatest influence on the drying shrinkage potential. Studies have shown that the type of aggregate can increase the drying shrinkage by as much as 120% to 150% (Powers 1959, Meininger 1966, and Tremper and Spellmen 1963). Hard, rigid aggregates are difficult to compress and provide more restraint to shrinkage than softer, less rigid aggregates. For example, heavyweight or shielding concrete made with steel punchings has significantly less shrinkage than conventional concrete; in this instance shrinkage would be reduced 30% or more. Drying shrinkage can also be reduced by avoiding aggregates that have high drying shrinkage properties such as some sandstone and graywacke, and aggregates containing excessive amounts of clay. Quartz, granite, feldspar, limestone, and dolomite aggregates generally produce concretes with lower drying shrinkages (ACI 224R 2001).

Most chemical admixtures have little effect on shrinkage. However, the use of accelerators such as calcium chloride will increase drying shrinkage of concrete. Despite reductions in water content, some water-reducing admixtures can increase drying shrinkage, particularly those that contain an accelerator to counteract the retarding effect of the admixture. Air entrainment has little or no effect on drying shrinkage. High-range water reducers usually have little effect on drying shrinkage (Figure 10-16).

Effect of Curing on Drying Shrinkage

The amount and type of curing can affect the rate and ultimate amount of drying shrinkage. Generally, if the initiation of drying is delayed, the aging of concrete due to continued hydration will reduce the rate of shrinkage and may reduce the total potential shrinkage. Curing compounds, sealers, and coatings can trap free moisture in the concrete for long periods, resulting in delayed or reduced shrinkage. Wet curing methods, such as fogging or wet burlap, hold off shrinkage until curing is terminated, after which the concrete dries and shrinks at a normal rate. Cooler initial curing temperatures can reduce shrinkage if protected from moisture loss during curing (Figure 10-17). Steam curing will also reduce drying shrinkage. Computer software, such as HIPERPAV III is available to predict the effect of curing and environmental conditions on shrinkage and cracking (FHWA and Transtec 2015). Hedenblad (1997) provides tools to predict the drying of concrete as affected by different curing methods and type of construction. See Chapter 17 for more information on methods of curing concrete.

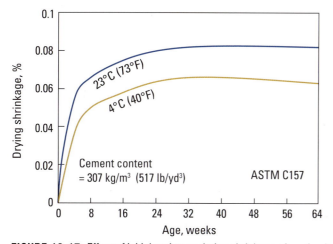

FIGURE 10-17. Effect of initial curing on drying shrinkage of portland cement concrete prisms. Concrete with an initial 7-day moist cure at 4°C (40°F) had less shrinkage than concrete with an initial 7-day moist cure at 23°C (73°F). Similar results were found with concretes containing 25% fly ash as part of the cementing material (Gebler and Klieger 1986).

TEMPERATURE CHANGES OF HARDENED CONCRETE

Concrete expands slightly as temperature rises and contracts as temperature falls, although concrete may also expand as free water in the concrete freezes. Temperature changes may be caused by environmental conditions or by cement hydration. An average value for the coefficient of thermal expansion (CTE) of concrete is about 10 millionths per °C (5.5 millionths per °F), although values ranging from 6 to 13 millionths per °C (3.2 to 7.0 millionths per °F) have been observed. This amounts to a length change of about 5 mm for 10 m of concrete (about ⅔ in. for 100 ft of concrete) subjected to a rise or fall of 50°C (100°F). The coefficient of thermal expansion for structural low-density (lightweight) concrete varies from 7 to 11 millionths per °C (3.6 to 6.1 millionths per °F). The coefficient of thermal

expansion of concrete can be determined by AASHTO T 336, *Standard Method of Test for Coefficient of Thermal Expansion of Hydraulic Cement Concrete.*

Thermal expansion and contraction of concrete varies with factors such as aggregate type, cement content, water-cement ratio, temperature range, concrete age, and relative humidity. Of these, aggregate type has the greatest influence because it constitutes the largest percentage of both volume and mass of a concrete mixture.

Table 10-1 shows experimental values of the thermal coefficient of expansion of concretes made with aggregates of various types. These data were obtained from tests conducted on over 2000 concrete pavements following AASHTO T 336.

The thermal coefficient of expansion for steel is about 12 millionths per °C (6.5 millionths per °F), which is comparable to that for concrete. The coefficient for reinforced concrete can be assumed as 11 millionths per °C (6 millionths per °F), the average for concrete and steel.

Temperature changes that result in shortening can crack concrete members that are highly restrained by another part of the structure or by ground friction. Consider a long restrained concrete member cast without joints that, after moist curing, has a drop in temperature. As the temperature drops, the concrete will shorten because it is restrained longitudinally. The resulting tensile stresses cause the concrete to crack. Tensile strength and modulus of elasticity of concrete both may be assumed proportional to the square root of concrete compressive strength. Calculations show that a large enough temperature drop will crack concrete regardless of its age or strength, provided the coefficient of expansion does not vary with temperature and the concrete is fully restrained (FHWA and Transtec 2015 and PCA 1982).

TABLE 10-1. Effect of Aggregate Type on Coefficient of Thermal Expansion of Concrete

PRIMARY AGGREGATE CLASS	AVERAGE CTE, millionths per °F	STANDARD DEVIATION, millionths per °F	AVERAGE CTE, millionths per °C	STANDARD DEVIATION, millionths per °C	SAMPLE COUNT
Andesite	4.32	0.42	7.78	0.75	52
Basalt	4.33	0.43	7.80	0.77	141
Chert	6.01	0.42	10.83	0.75	106
Diabase	4.64	0.52	8.35	0.94	91
Dolomite	4.95	0.40	8.92	0.73	433
Granite	4.72	0.40	8.50	0.71	331
Limestone	4.34	0.52	7.80	0.94	813
Quartzite	5.19	0.50	9.34	0.90	131
Sandstone	5.32	0.52	9.58	0.94	84
Schist	4.43	0.39	7.98	0.70	30

Adapted from Hall and Tayabji (2011).

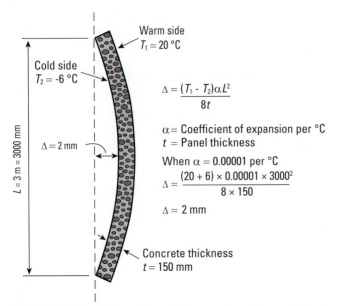

FIGURE 10-18. Curling of a plain concrete wall panel due to temperature that varies uniformly from inside to outside. Not to scale.

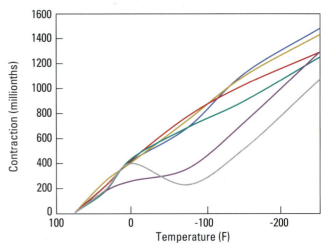

— NWC, cement content of 390 kg/m³ (658 lb/yd³)
— NWC, cement content of 307 kg/m³ (517 lb/yd³), Moist cured
— NWC, cement content of 307 kg/m³ (517 lb/yd³), Cured at 50% RH
— NWC, cement content of 307 kg/m³ (517 lb/yd³), Dry cured
— NWC, cement content of 223 kg/m³ (376 lb/yd³)
---- LWC, cement content of 334 kg/m³ (564 lb/yd³)

FIGURE 10-19. Contraction of concretes at low temperatures (Monfore and Lentz 1962). Note that normal weight concrete (NWC) materials are sand and gravel; lightweight concrete (LWC) materials are expanded shale.

Precast wall panels and slabs on ground are susceptible to bending and curling caused by temperature gradients that develop when concrete is cool on one side and warm on the other. The calculated amount of curling in a wall panel is illustrated in Figure 10-18.

Low Temperatures

For outdoor applications in winter, the volume changes due to increase in moisture content and the decrease in average temperature tend to offset each other.

Concrete continues to contract as the temperature is reduced below freezing. The amount of volume change at subfreezing temperatures is greatly influenced by the moisture content, behavior of the water (physical state – ice or liquid), and type of aggregate in the concrete. In a study by Monfore and Lentz (1962), the coefficient of thermal expansion for a temperature range of 24°C to -157°C (75°F to -250°F) varied from 6×10^{-6} per °C (3.3×10^{-6} per °F) for a low density (lightweight) aggregate concrete to 8.2×10^{-6} per °C (4.5×10^{-6} per °F) for a sand and gravel mixture (Figure 10-19).

Subfreezing temperatures can significantly increase the compressive and tensile strength and modulus of elasticity of moist concrete. Dry concrete properties are not as affected by low temperatures. In the same study, moist concrete with an original compressive strength of 35 MPa at 24°C (5,000 psi at 75°F) achieved over 117 MPa at -101°C (17,000 psi at -150°F). The same concrete tested oven dry or at a 50% internal relative humidity had strength increases of only about 20%. The modulus of elasticity for sand and gravel concrete with 50% relative humidity was only 8% higher at -157°C than at 24°C (-250°F than at 75°F), whereas the moist concrete had a 50% increase

in modulus of elasticity. Cooling from 24°C to -157°C (75°F to -250°F), the thermal conductivity of normal-weight concrete also increased, especially for moist concrete. The thermal conductivity of lightweight aggregate concrete is little affected (Monfore and Lentz 1962 and Lentz and Monfore 1966).

High Temperatures

Temperatures greater than 95°C (200°F) that are sustained for several months or even several hours can have significant effects on concrete. The total volume change of concrete is the sum of the individual volume changes of the cement paste and aggregate. At high temperatures, the paste shrinks due to dehydration while the aggregate expands. For normal-weight aggregate concrete, the expansion of the aggregate exceeds the paste shrinkage resulting in an overall expansion of the concrete. Some aggregates such as expanded shale, andesite, or pumice with low coefficients of thermal expansion can produce a very volume-stable concrete in high-temperature environments (Figure 10-20). However, other aggregates undergo extensive and abrupt volume changes at a particular temperature, causing disruption in the concrete. For example, in one study a dolomitic limestone aggregate containing an iron sulfide impurity caused severe expansion, cracking, and disintegration in concrete exposed to a temperature of 150°C (302°F) for four months; yet at temperatures above and below 150°C (302°F) there was no detrimental expansion (Carette and others 1982). Concrete's coefficient of thermal expansion tends to increase with temperature rise.

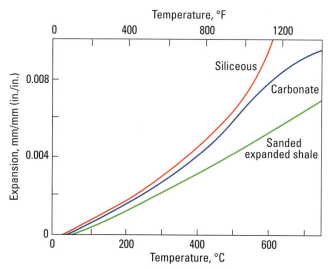

FIGURE 10-20. Thermal expansion of concretes containing various types of aggregate (Abrams 1977).

Besides volume change, sustained high temperatures can also have other, usually irreversible, effects such as a reduction in strength, modulus of elasticity, and thermal conductivity. Creep increases with temperature. Above 100°C (212°F), the paste begins to dehydrate (lose chemically combined water of hydration) resulting in significant strength losses. Strength decreases with increases in temperature until the concrete loses essentially all its strength. The effect of high-temperature exposure on compressive strength of concretes made with various types of aggregate is illustrated in Figure 10-21. Several factors including concrete moisture content, aggregate type and stability, cement content, exposure time, rate of temperature rise, age of concrete, restraint, and existing stress all influence the behavior of concrete at high temperatures. Figure 10-22 shows the difference in response between normal-strength concrete and high-strength concrete (HSC) (see Chapter 23).

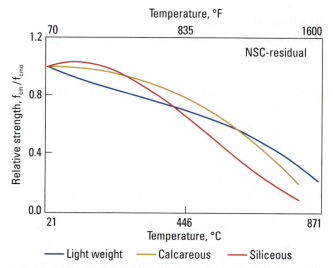

FIGURE 10-21. Relationships between high temperatures and the residual compressive strength of normal strength concretes (NSC) containing various types of aggregate (Knaack and others 2009).

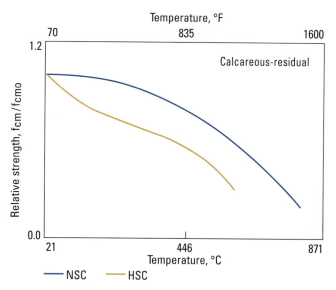

FIGURE 10-22. Relationships between high temperatures and the residual compressive strength of normal-strength concrete (NSC) and high-strength concretes (HSC) containing calcareous aggregate (Knaack and others 2009).

If stable aggregates are used and strength reduction and the effects on other properties are accounted for in the mixture design, high quality concrete can be exposed to temperatures of 90°C to 200°C (200°F to 400°F) for long periods. Some concrete elements have been exposed to temperatures up to 250°C (500°F) without distress; however, special materials (such as heat-resistant calcium aluminate cement) should be considered for exposure temperatures greater than 200°C (400°F). Before any structural concrete is exposed to high temperatures (greater than 90°C or 200°F), laboratory testing should be performed to determine the thermal properties. This will avoid any unexpected distress.

Curling and Warping

Curling and warping caused by differences in moisture content and temperature between the top and bottoms of slabs can be problematic (Figure 10-23). Curling and warping are closely related to the shrinkage potential of the concrete mixture. While the terms have been used interchangeably, there is a difference between "curling" and "warping" (Tarr 2004). Curling is the deformation of the concrete slab due to a difference in temperature between the top surface and the bottom of the slab. Like most materials, concrete expands and contracts with change in temperature. If the slab surface is cooler than the slab bottom, the surface contracts causing the slab edges to curl upward. Warping is the deformation of the slab surface profile due to a difference in moisture between the surface and bottom of the slab. As with a sponge, if the slab surface is allowed to dry and the bottom is kept moist, the edges will tend to warp upward. Exterior pavement slabs typically have a permanent upward edge warp and experience downward curling on a daily basis due to surface warming and cooling cycles related to exposure to the sun. In general, the edges of interior concrete

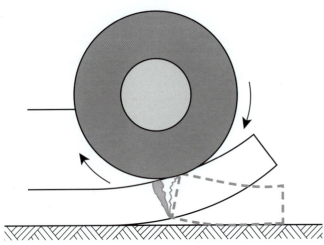

FIGURE 10-23. Illustration of curling of a concrete slab on ground or pavement. The edge of the concrete at a joint or free end lifts off the subbase creating a cantilevered section of concrete that can break off under heavy loading. Not to scale.

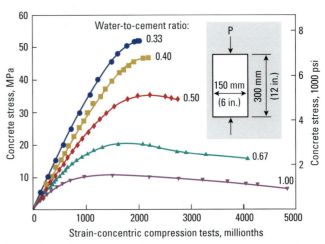

FIGURE 10-24. Stress-strain curves for compression tests on 150 × 300-mm (6 × 12-in.) concrete cylinders at an age of 28 days (Hognestad and others 1955).

floor slab panels warp upward due to a moisture difference between the top and bottom of the slab (Tarr and Farny 2008).

A slab will assume a reverse curl when the surface is wetter or warmer than the bottom. However, enclosed slabs, such as floors on ground, only curl upward. When the edges of an industrial floor slab are curled upward they lose support from the subbase and act as a cantilever. Lift-truck traffic passing over joints causes a repetitive vertical deflection that creates a great potential for fatigue cracking in the slab. The amount of vertical upward curl (curling) is small for a short, thick slab.

If the top surface of slab is covered or coated, such as in many commercial floors, warped edges can relax back downward as the moisture gradient through the slab equalizes. As the deformation in the slab reduces, the warping stresses diminish as well. However, as joints and cracks narrow, floor coverings and coatings can be put into compression. If the amount of warping relaxation is great enough, the floor covering can buckle (Tarr and others 2006).

Curling can be reduced or eliminated using design and construction techniques that minimize shrinkage potential and by using techniques described earlier to reduce temperature and moisture-related volume changes. Thickened edges, shorter joint spacings, permanent vapor-impermeable sealers, and heavy reinforcing steel placed 50 mm (2 in.) below the surface all help reduce curling (Ytterberg 1987). Design options such as post-tensioning and shrinkage-compensating concrete can also be used to control curling/warping (Tarr and Farny 2008).

ELASTIC AND INELASTIC DEFORMATION

Compression Strain

The series of curves in Figure 10-24 illustrate the amount of compressive stress and strain that results instantaneously due to loading of unreinforced concrete. With water-cement ratios

of 0.50 or less and strains up to 1500 millionths, the upper three curves show that strain is closely proportional to stress; and the concrete is almost elastic. The upper portions of the curves and beyond show that the concrete is inelastic. The curves for high-strength concrete have sharp peaks, whereas those for lower-strength concretes have long and relatively flat peaks. Figure 10-24 also shows the sudden failure characteristics of higher strength, low water to cement ratio, concrete cylinders.

When load is removed from concrete in the inelastic zone, the recovery line usually is not parallel to the original line for the first load application. Therefore, the amount of permanent set may differ from the amount of inelastic deformation.

The term "elastic" is not favored for general discussion of concrete behavior because frequently the strain may be in the inelastic range. For this reason, the term "instantaneous strain" is often used.

Modulus of Elasticity

The ratio of stress to strain in the elastic range of a stress-strain curve for concrete defines the modulus of elasticity (E) of that concrete (Figure 10-25). Normal-density concrete has a modulus of elasticity of 14,000 MPa to 41,000 MPa (2000,000 psi to 6,000,000 psi), depending on factors such as compressive strength and aggregate type. For normal-density concrete with compressive strengths (f'_c) between 20 MPa and 35 MPa (3000 psi and 5000 psi), the modulus of elasticity can be estimated as 5000 times the square root of f'_c (57,000 times the square root of f'_c in psi). The modulus of elasticity for structural lightweight concrete is between 7000 MPa and 17,000 MPa (1,000,000 psi and 2,500,000 psi). The modulus of elasticity for any particular concrete can be determined in accordance with ASTM C469, *Standard Test Method for Static Modulus of Elasticity and Poisson's Ratio of Concrete in Compression*. See Puri and Weiss (2006) and Weiss (2006) for stress-strain relationships.

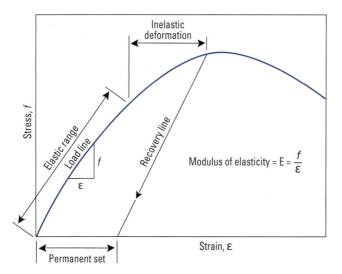

FIGURE 10-25. Generalized stress-strain curve for concrete.

Deflection

Deflection of concrete beams and slabs is one of the more common and obvious building movements. The deflections are the result of flexural strains that develop under dead and live loads. They may result in cracking in the tensile zone of concrete members. Reinforced concrete structural design anticipates these tension cracks (see Chapter 8). Concrete members are often cambered, that is, built with an upward bow, to compensate for the expected later deflection.

Poisson's Ratio

When a block of concrete is loaded in uniaxial compression, as in Figure 10-26, it will compress and simultaneously develop a lateral strain or bulging. The ratio of lateral to axial strain is called Poisson's ratio, μ. A common value used for μ in concrete is 0.20 to 0.21, but the value may vary from 0.15 to 0.25 depending upon the aggregate, moisture content, concrete age, and compressive strength. Poisson's ratio (ASTM C469) is generally

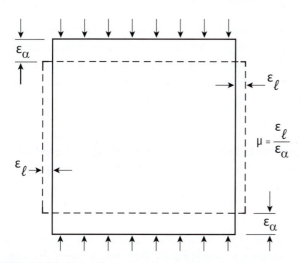

FIGURE 10-26. Ratio of lateral to axial strain is Poisson's ratio, μ.

not a concern to the structural designer. It is used in advanced structural analysis of flat-plate floors, shell roofs, arch dams, and mat foundations.

Shear Strain

Concrete, like other materials, deforms under shear forces. The shear strain produced is important in determining the load paths or distribution of forces in indeterminate structures—for example where shear-walls and columns both participate in resisting horizontal forces in a concrete building frame. The amount of movement, while not large, is significant in short members. In larger members, flexural strains are much greater and more significant to designers. Calculation of the shear modulus (modulus of rigidity), G, is shown in Figure 10-27; G varies with the strength and temperature of the concrete.

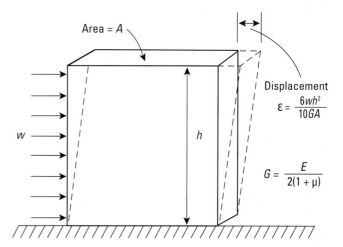

FIGURE 10-27. Strain that results from shear forces on a body. G = shear modulus. μ = Poisson's ratio. Strain resulting from flexure is not shown.

Torsional Strain

Plain rectangular concrete members can also fail in torsion, that is, a twisting action caused by bending about an axis parallel to the wider face and inclined at an angle of about 45 degrees to the longitudinal axis of a member. Microcracks develop at low torque. However, concrete behaves reasonably elastic up to the maximum limit of the elastic torque (Hsu 1968).

CREEP

When concrete is loaded, the deformation caused by the load can be divided into two parts: a deformation that occurs immediately (elastic strain) and a time-dependent deformation that begins immediately but continues at a decreasing rate as long as the concrete is loaded. This latter deformation is called creep.

An early study by Pickett (1947) found that creep of concrete does not respond linearly to stress. The amount of creep is dependent upon: the magnitude of the applied stress, the age and strength of the concrete when stress is applied, and the length of time the concrete is stressed.

Creep is also affected by other factors related to the quality of the concrete and conditions of exposure, such as: type, amount, and maximum size of aggregate; type of cementing materials; amount of cement paste; size and shape of the concrete element; volume to surface ratio of the concrete element; amount of steel reinforcement; prior curing conditions; and the ambient temperature and humidity. See ACI 209.1R, *Report on Factors Affecting Shrinkage and Creep of Hardened Concrete* (2005) for more on the factors that affect creep, and ACI 209.2R, *Guide for Modeling and Calculating Shrinkage and Creep in Hardened Concrete* (2008).

Within normal stress ranges, creep is proportional to stress. In relatively young concrete, the change in volume or length due to creep is largely unrecoverable; in older or drier concrete it is largely recoverable.

The creep curves shown in Figure 10-28 are based on tests conducted under laboratory conditions in accordance with ASTM C512, *Standard Test Method for Creep of Concrete in Compression* (Figure 10-29). Cylinders were loaded to almost 40% of their compressive strength. Companion cylinders not subject to load were used to measure drying shrinkage; this was then deducted from the total deformation of the loaded specimens to determine creep. Cylinders were allowed to dry while under load except for those marked "sealed." The two 28-day curves for each concrete strength in Figure 10-28 show that creep of concrete loaded under drying conditions is greater than creep of concrete sealed against drying. Concrete

FIGURE 10-29. ASTM C512 creep test frames loaded with both sealed and unsealed specimens (courtesy of M. D'Ambrosia).

specimens loaded at a later age will creep less than those loaded at an early age. It can be seen that as concrete strength decreases, creep increases. Figure 10-30 illustrates recovery from the elastic and creep strains after load removal.

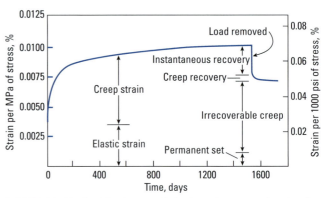

FIGURE 10-30. Combined curve of elastic and creep strains showing amount of recovery. Specimens (cylinders) were loaded at 8 days immediately after removal from fog curing room and then stored at 21°C (70°F) and 50% RH. The applied stress was 25% of the compressive strength at 8 days (Hansen and Mattock 1966).

A combination of strains occurring in a reinforced column is illustrated in Figure 10-31. The curves represent deformations and volume changes in a 14th-story column of a 76-story reinforced concrete building while under construction (Russell and Corley 1977). The 400-mm × 1200-mm (16-in. × 48-in.) column contained 2.08% vertical reinforcement and was designed for 60-MPa (9000-psi) concrete. The method of curing prior to loading has a marked effect on the amount of creep in concrete. The effects on creep of three different methods of curing are shown in Figure 10-32. Note that very little creep occurs in concrete that is cured by high-pressure steam (autoclaving). Note also that atmospheric steam-cured concrete has considerably less creep than 7-day moist-cured concrete. The two methods of steam curing shown in Figure 10-32 reduce drying shrinkage of concrete about half as much as they reduce creep.

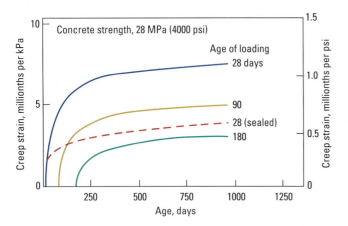

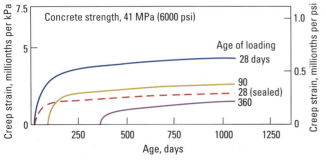

FIGURE 10-28. Relationship of time and age of loading to creep of two different strength concretes. Specimens were allowed to dry during loading, except for those labeled as sealed (Russell and Corley 1977).

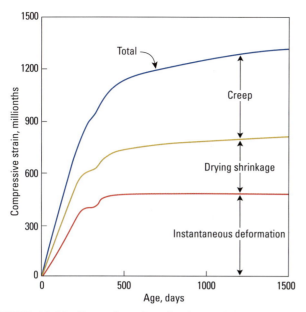

FIGURE 10-31. Summation of strains in a reinforced concrete column during construction of a tall building (Russell and Corley 1977).

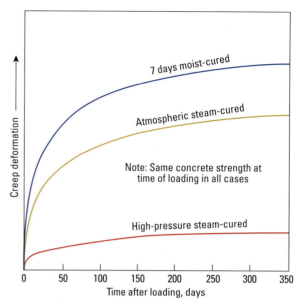

FIGURE 10-32. Effect of curing method on magnitude of creep for typical normal-density concrete (Hanson 1964).

CHEMICAL CHANGES AND EFFECTS

Some volume changes of concrete result from chemical reactions; these may take place shortly after placing and finishing or later due to reactions within the hardened concrete in the presence of water or moisture.

Carbonation

Hardened concrete containing some moisture reacts with carbon dioxide present in air, a reaction that results in a slight shrinkage of the surface paste of the concrete. The effect, known as carbonation, is not destructive, and increases the chemical stability and strength of the concrete. The depth of carbonation is very shallow in dense, high-quality concrete, but can penetrate deeply in porous, poor-quality concrete. Because so little of a concrete element carbonates, carbonation shrinkage of cast-in-place concrete is insignificant and does not usually affect engineering properties.

Carbonation of paste proceeds slowly and produces little direct shrinkage at relative humidities of 100% and 25%. Maximum carbonation and carbonation shrinkage occurs at about 50% relative humidity. Irreversible shrinkage and weight gain occur during carbonation. The carbonated product may show improved volume stability and reduced permeability to subsequent moisture change (Verbeck 1958).

However, carbonation also reduces the pH of concrete. If steel is present in the carbonated area, steel corrosion can occur due to the absence of the protective oxide film provided by concrete's high pH. Rust, a product of corrosion of reinforcing steel, is an expansive material and results in cracking and spalling of the concrete. For more information on carbonation and corrosion (see Chapter 11).

Sulfate Attack

Sulfate attack of concrete can occur where soil and groundwater have a high sulfate content and measures to reduce sulfate attack, such as use of a low water to cementitious materials ratio and sulfate resistant cementitious materials, have not been taken. The attack is more aggressive in concrete that is exposed to wetting and drying, such as foundation walls and posts that are in contact with the ground and air. Sulfate attack usually results in an expansion of the concrete because of the formation of solids from the chemical action or salt crystallization. The amount of expansion in severe circumstances has been significantly higher than 0.1%, and the disruptive effect within the concrete can result in extensive cracking and disintegration. For more information on sulfate attack, see Chapter 11.

Alkali-Aggregate Reactions

Certain aggregates can react with alkali hydroxides in concrete, causing expansion and cracking over a period of years. The reaction is greater in those parts of a structure exposed to moisture. A knowledge of the characteristics of local aggregates is essential. There are two types of alkali-aggregate reactions, alkali-silica reactivity (ASR) and alkali-carbonate reactivity (ACR). Expansion due to alkali-aggregate reaction expansion may exceed 0.5% in concrete and can cause the concrete to fracture and break apart.

In concrete that includes aggregate susceptible to ASR, special measures must be taken to prevent the occurrence of deleterious expansion (see Chapter 11).

Design and Control of Concrete Mixtures

REFERENCES

PCA's online catalog includes links to PDF versions of many of our research reports and other classic publications. Visit: cement.org/library/catalog.

Abrams, M.S., *Performance of Concrete Structures Exposed to Fire*, Research and Development Bulletin RD060, Portland Cement Association, Skokie, IL, 1977.

ACI Committee 209, *Prediction of Creep, Shrinkage, and Temperature Effects in Concrete Structures*, ACI 209R-92, reapproved 2008, American Concrete Institute, Farmington Hills, MI, 1992, 47 pages.

ACI Committee 209, *Report on Factors Affecting Shrinkage and Creep of Hardened Concrete*, ACI 209.1R-05, American Concrete Institute, Farmington Hills, MI, 2005, 12 pages.

ACI Committee 209, *Guide for Modeling and Calculating Shrinkage and Creep in Hardened Concrete*, ACI 209.2R-08, 2008, 45 pages.

ACI Committee 224, *Control of Cracking in Concrete Structures*, ACI 224R-01, reapproved 2008, American Concrete Institute, Farmington Hills, MI, 2001, 46 pages.

ACI Committee 231, *Report on Early-Age Cracking: Causes, Measurement, and Mitigation*, ACI 231R-10, reapproved 2020, American Concrete Institute, Farmington Hills, MI, 2010, 46 pages.

ACI Committee 301, *Specifications for Concrete Construction*, ACI 301-20, American Concrete Institute, Farmington Hills, MI, 2020.

Aïtcin, P.-C., "Does Concrete Shrink or Does it Swell?," *Concrete International*, American Concrete Institute, Farmington Hills, MI, December 1999, pages 77 to 80.

Baker, W.F.; Korista, D.S.; Novak, L.C.; Pawlikowski, J.; and Young, B., "Creep and Shrinkage and the Design of Supertall Buildings – A Case Study: The Burj Dubai Tower," *Structural Implications of Shrinkage and Creep of Concrete*, SP-246CD, American Concrete Institute, Farmington Hills, MI, 2007, pages 133 to 148.

Bentur, A., ed., *Early Age Cracking in Cementitious Systems*, RILEM Report 25, RILEM Publications s.a.r.l., Bagneux, France, 2003, 351 pages.

Bentz, D. P.; Irassar, E. F.; Bucher, B. E.; and Weiss, W. J., "Limestone Fillers Conserve Cement–Part 2: Durability Issues and the Effects of Limestone Fineness on Mixtures," *Concrete International*, Vol. 31, No. 12, December 2009, pages 35 to 39.

Bucher, B.; Radlinska, A.; and Weiss, J., *Preliminary Comments on Shrinkage and Shrinkage Cracking Behavior of Cement Systems that Contain Limestone*, Concrete Technology Forum 2008, National Ready Mixed Concrete Association, Silver Spring, MD, 2008, 8 pages.

Bureau of Reclamation, "Long-Time Study of Cement Performance in Concrete—Tests of 28 Cements Used in the Parapet Wall of Green Mountain Dam," *Materials Laboratories Report No. C-345*, U.S. Department of the Interior, Bureau of Reclamation, Denver, CO, 1947.

Carette, G.G.; Painter, K.E.; and Malhotra, V.M., "Sustained High Temperature Effect on Concretes Made with Normal Portland Cement, Normal Portland Cement and Slag, or Normal Portland Cement and Fly Ash," *Concrete International*, American Concrete Institute, Farmington Hills, MI, July 1982.

Copeland, L.E.; and Bragg, R.H., *Self Desiccation in Portland Cement Pastes*, RX052, Portland Cement Association, Chicago, IL, 1955, 17 pages.

Davis, R.E., "A Summary of the Results of Investigations Having to Do with Volumetric Changes in Cements, Mortars, and Concretes Due to Causes Other Than Stress," *Proceedings of the American Concrete Institute*, American Concrete Institute, Farmington Hills, MI, Vol. 26, 1930, pages 407 to 443.

FHWA and Transtec, HIPERPAV III [Computer software], Washington, DC, 2015.

Gebler, S. H.; and Paul, K., *Effect of Fly Ash on Some of the Physical Properties of Concrete*, Research and Development Bulletin RD089, Portland Cement Association, Skokie, IL, 1986, 46 pages.

Grasley, Z., *Internal Relative Humidity, Drying Stress Gradients, and Hygrothermal Dilation of Concrete*, SN2625, Master's Thesis, University of Illinois at Urbana-Champaign, 2003, 81 pages.

Hall, K.; and Tayabji, S., "Coefficient of Thermal Expansion in Concrete Pavement Design," *Techbrief*, FHWA-HIF-09-015, Advanced Concrete Pavement Technology Products Program, FHWA, Lanham, MD, 2011, 6 pages.

Hammer, T.A., "Test Methods for Linear Measurement of Autogenous Shrinkage Before Setting," *Autogenous Shrinkage of Concrete* edited by E. Tazawa, E&FN Spon and Routledge, New York, NY, 1999, pages 143 to 154.

Hansen, T.C.; and Mattock, A.H., *Influence of Size and Shape of Member on the Shrinkage and Creep of Concrete*, DX103, Portland Cement Association, Chicago, IL, 1966, 41 pages.

Hanson, J.A., *Prestress Loss As Affected by Type of Curing*, DX075, Portland Cement Association, Skokie, IL, 1964, 32 pages.

Hanson, J.A., *Effects of Curing and Drying Environments on Splitting Tensile Strength of Concrete*, DX141, Portland Cement Association, Skokie, IL, 1968, 29 pages.

Hedenblad, G., *Drying of Construction Water in Concrete*, Byggforskningsrådet, The Swedish Council for Building Research, Stockholm, 1997, 54 pages.

Hognestad, E.; Hanson, N.W.; and McHenry, D., *Concrete Stress Distribution in Ultimate Strength Design*, DX006, Portland Cement Association, Chicago, IL, 1955.

Holt, E.E., *Early Age Autogenous Shrinkage of Concrete*, Espoo. Technical Research Centre of Finland, VTT Publications 446, 2001, pages 184-193.

Hsu, T.T.C., *Torsion of Structural Concrete – Plain Concrete Rectangular Sections*, DX134, Portland Cement Association, Skokie, IL, 1968, 42 pages.

Jackson, F.H., *Long-Time Study of Cement Performance in Concrete – Chapter 9. Correlation of the Results of Laboratory Tests with Field Performance Under Natural Freezing and Thawing Conditions*, RX060, Portland Cement Association, Chicago, IL, 1955, 41 pages.

Knaack, A.; Kurama, Y.; and Kirkner, D., *Stress-Strain Properties of Concrete at Elevated Temperatures*, SN3111, Structural Engineering Research Report #NDSE-09-01, University of Notre Dame, April 2009, 153 pages.

Le Chatelier, H., "Sur les Changements de Volume qui Accompagent le durcissement des Ciments," *Bulletin Societe de l'Encouragement pour l'Industrie Nationale*, Paris, 1900.

Lentz, A.E.; and Monfore, G.E., *Thermal Conductivities of Portland Cement Paste, Aggregate, and Concrete Down to Very Low Temperatures*, RX207, Portland Cement Association, Chicago, IL, 1966, 10 pages.

Meininger, R.C., "Drying Shrinkage of Concrete," *Engineering Report*, No. RD3, National Ready Mixed Concrete Association, Silver Spring, MD, June 1966.

Monfore, G.E.; and Lentz, A.E., *Physical Properties of Concrete at Very Low Temperatures*, RX145, Portland Cement Association, Chicago, IL, 1962, 13 pages.

PCA, *Building Movements and Joints*, EB086, Portland Cement Association, Skokie, IL, 1982, 72 pages.

Pickett, G., *The Effect of Change in Moisture Content on the Creep of Concrete Under a Sustained Load*, RX020, Portland Cement Association, Chicago, IL, 1947, 26 pages.

Powers, T.C., "Absorption of Water by Portland Cement Paste during the Hardening Process," *Industrial and Engineering Chemistry*, Vol. 27, No. 7, July 1935, pages 790 to 794.

Powers, T. C., "Causes and Control of Volume Change," Volume 1, Number 1, Journal of the *PCA Research and Development Laboratories*, Portland Cement Association, Chicago, IL, January 1959.

Puri, S.; and Weiss, W.J., "Assessment of Localized Damage in Concrete Under Compression Using Acoustic Emission," *ASCE Journal of Civil Engineering Materials*, Vol. 18, No. 3, 2006, pages 325 to 333.

Radlinska, A; and Weiss, W.J., "Toward the Development of a Performance Related Specification for Concrete Shrinkage," *ASCE Journal of Civil Engineering Materials*, Vol. 23, 2012, pages 64-71.

Radlinska, A.; Rajabipour, F.; Bucher, B.; Henkensiefken, R.; Sant, G.; and Weiss, J., "Shrinkage Mitigation Strategies in Cementitious Systems: a Closer Look at Differences in Sealed and Unsealed Behavior," *Transportation Research Record*, Vol. 2070, 2008, pages 59 to 67.

Roper, H., *Volume Changes of Concrete Affected by Aggregate Type*, RX123, Portland Cement Association, Chicago, IL, 1960, 10 pages.

Russell, H.G.; and Corley, W.G., *Time-Dependent Behavior of Columns in Water Tower Place*, Research and Development Bulletin RD052, Portland Cement Association, Skokie, IL, 1977, 13 pages.

Shah, S.P.; Weiss, W. J.; and Yang, W., "Shrinkage Cracking – Can It Be Prevented?" *Concrete International*, Vol. 20, No. 4, 1998, pages 51-55.

Suprenant, B. A.; and Malisch, W. R., "The Fiber Factor," *Concrete Construction*, Addison, IL, October 1999, pages 43 to 46.

Suprenant, B. A.; and Malisch, W. R., "A New Look at Water, Slump, and Shrinkage," *Concrete Construction*, Addison, IL, April 2000, pages 48 to 53.

Tarr, S.M., "Concrete Mixture Shrinkage Potential," *Concrete Construction*, September 2012.

Tarr, S.M., "Interior Cement Floors Don't Curl: But Concrete Floors Do Warp and Joints Suffer!" *L&M Concrete News*, L&M Construction Chemicals, Omaha, NE, Vol. 5, No. 2, Fall 2004, page 5.

Tarr, S. M., Craig, P., and Kanare, H., "Concrete Slab Repair: Getting Flat is One Thing, Staying Flat is Another!", *Concrete Repair*, January/February, 2006, 4 pages.

Tarr, S. M.; and Farny, J. A., *Concrete Floors on Ground*, EB075, Fourth Edition, Portland Cement Association, Skokie, IL, 2008, 252 pages.

Tazawa, E., *Autogenous Shrinkage of Concrete*, E&FN Spon and Routledge, New York, NY, 1999, 428 pages.

Tremper, B.; and Spellman, D.L., "Shrinkage of Concrete – Comparison of Laboratory and Field Performance," *Highway Research Record Number 3*, Properties of Concrete, Transportation Research Board, National Research Council, Washington, D.C., 1963, 32 pages.

Verbeck, G.J., *Carbonation of Hydrated Portland Cement*, RX087, Portland Cement Association, Chicago, IL, 1958, 21 pages.

Weiss, W.J., "Chapter 19 – Elastic Properties, Creep, and Relaxation," ASTM 169D, *Significance of Tests and Properties of Concrete and Concrete Making Materials*, ASTM International, West Conshohocken, PA, 2006, pages 194 to 206.

Weiss, W. J.; Yang, W.; and Shah, S.P., "Influence of Specimen Size and Geometry on Shrinkage Cracking," *Journal of Engineering Mechanics Division*, American Society of Civil Engineering, Vol. 126, No. 1, 2000, pages 93-101.

Whiting, D.; and Dziedzic, W., *Effects of Conventional and High-Range Water Reducers on Concrete Properties*, Research and Development Bulletin RD107, Portland Cement Association, Skokie, IL, 1992, 28 pages.

Yang, W.; Weiss, W. J.; and Shah, S.P., "Prediction of Shrinkage Stress and Displacement Fields in a Concrete Slab Restrained by an Elastic Subgrade." *Journal of Engineering Mechanics Division*, American Society of Civil Engineering, Vol. 126, No. 1, 2000, pages 35-42.

Ytterberg, R.F., "Shrinkage and Curling of Slabs on Grade, Part I – Drying Shrinkage, Part II – Warping and Curling, and Part III – Additional Suggestions," *Concrete International*, American Concrete Institute, Farmington Hills, MI, April 1987, pages 22 to 31; May 1987, pages 54 to 61; and June 1987, pages 72 to 81.

DURABILITY

FIGURE 11-1. The Confederation Bridge was designed for a 100-year service life in a severe environment (see Chapter 23 for more information).

Well-designed and constructed concrete is an inherently durable material, capable of providing decades of service life. Concrete's durability improves the sustainability of infrastructure, pavements, and other structures by conserving resources and reducing environmental impacts related to repair and replacement.

Concrete durability is defined as the ability of concrete to resist weathering action, chemical attack, and abrasion while maintaining its desired engineering properties for the expected service life. To ensure durability, care in concrete materials selection, design, and construction is important, with increased care and design considerations necessary in more severe environments (Figure 11-1).

Although a durable structure is expected to serve without deterioration or major repair before the end of its design life, concrete's durability does not eliminate the need for maintenance. Even high-performance concrete structures (see Chapter 23) that are designed and constructed to a higher durability standard require regular inspection and routine maintenance (Figure 11-2).

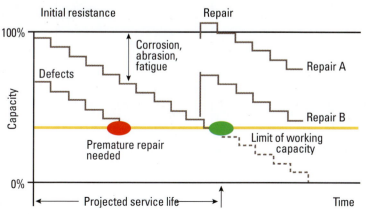

FIGURE 11-2. Service life of a durable structure.

FACTORS AFFECTING DURABILITY

Permeability and Diffusion

Concrete is resistant to most natural environments; however, it is sometimes exposed to substances that can cause deterioration (Kerkhoff 2007). Most concrete durability deterioration mechanisms are controlled by the availability of moisture to enter concrete or for aggressive chemicals to move through the concrete. The resistance of concrete to the ingress and movement of fluids or harmful ions (that is, low permeability) is fundamental to durability.

Deterioration processes typically involve two stages. Initially, aggressive fluids (water or solutions with dissolved solids or gases) penetrate or are transported through the capillary pore structure of the concrete to reaction sites (such as chlorides penetrating to steel reinforcement, or sulfates penetrating to aluminates). In a second step, these aggressive fluids trigger chemical or physical deterioration mechanisms.

Concrete subjected to severe exposure conditions should be designed and constructed to have a low permeability. *Permeability* refers to the ease of fluid migration through concrete when the fluid is under pressure or to the ability of concrete to resist penetration by water or other substances (liquid, gas, or ions) as discussed in Chapter 9. *Diffusivity* refers to the ease with which dissolved ions move through water-filled pores in concrete. Decreased permeability and diffusivity improves concrete's resistance to freezing-thawing cycles, resaturation, sulfates, and chloride-ion penetration, and other forms of chemical attack.

The size of the molecules or ions that are transported through the concrete, the viscosity of the fluid, and the valence of the ions, along with the interaction between the penetrating species and the hydrated cementitious phases can all affect the transport properties. Therefore, permeability and diffusivity must be expressed in terms of the substance that is migrating through the concrete.

Permeability and diffusivity are influenced by porosity but are distinct from porosity. Porosity is the volume of voids as a percent (or fraction) of the total volume. Permeability and diffusivity are affected by the size and connectivity of the pores. Figure 11-3 shows two hypothetical porous materials with approximately the same porosity. The image on the left shows pores that are discontinuous (as with entrained air bubbles), while the image on the right shows interconnected (continuous) pores. Discrete pores have almost no effect on permeability, while interconnected pores increase permeability.

Figure 11-4 shows the relative sizes of various pores and solids found in concrete. Capillary pores are primarily responsible for the transport properties. As a rough guide, Powers (1958) plotted the permeability versus capillary porosity for portland cement paste, as shown in Figure 11-5. As the porosity of the paste increases above about 30% (as with high water-cementitious materials ratios), the permeability increases dramatically.

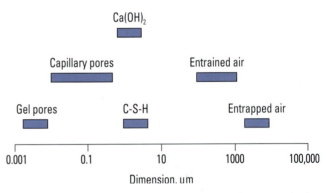

FIGURE 11-4. Relative sizes of different types of pores and other microstructural features.

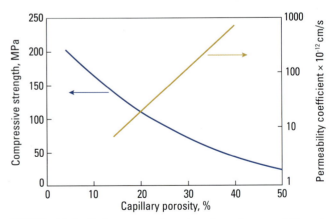

FIGURE 11-5. Both compressive strength and permeability are related to the capillary porosity of the cement paste (adapted from Powers 1958).

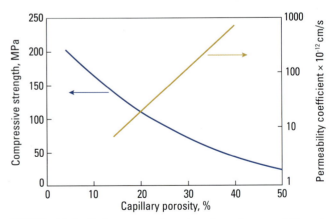

FIGURE 11-3. Porosity and permeability are related but distinct. These two hypothetical materials have about the same porosity (volume of pores), but very different permeabilities.

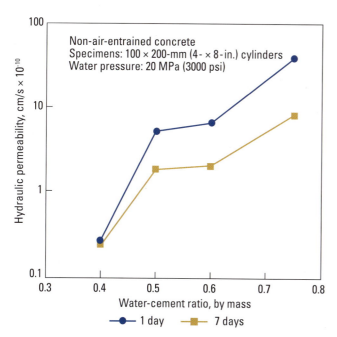

FIGURE 11-6. Relationship between hydraulic (water) permeability, water-cement ratio, and initial curing on concrete (Whiting 1988).

TABLE 11-1. Approximate Age Required to Produce Maturity at Which Time Capillary Pores Become Discontinuous for Portland Cement Concrete Continuously Moist-Cured (Powers and others 1959)

WATER-CEMENT RATIO	TIME REQUIRED
0.40	3 days
0.45	7 days
0.50	14 days
0.60	6 months
0.70	1 year
Over 0.70	Impossible

Decreasing the porosity of the paste below 30% reduces the permeability. The pore system of portland cement paste becomes discontinuous at about 30% porosity and below. Figure 11-6 and Table 11-1 illustrate the relationship between water-cement ratio, permeability, and curing. Powers and others (1959) estimated the time required for capillary pores to become discontinuous with increasing hydration of portland cement. As shown in Table 11-1, mixtures with a w/c ratio greater than 0.70 will always have connected capillary pores.

Low water-cementitious materials ratios result in less distance between hydrating particles, and less space to fill in as hydration progresses (see Chapters 1 and 9). The hydrates can then become more densely packed and the resulting concrete can be more impermeable. Typically, the water-to-cementitious materials ratio is limited to a maximum of 0.40 to 0.50 when concrete is being designed for durability, depending on the exposure category (ACI 318-19). Favorable curing conditions are also necessary to ensure durability, as the hydrating cementitious materials stop filling in pore space if water or temperature conditions are unfavorable (see Chapter 17).

The use of SCMs can reduce the permeability and diffusivity of concrete. These materials may not greatly reduce the total porosity, but instead, act to refine and subdivide the capillary pores so that they become discontinuous. Some other benefits are obtained by the improvements in fresh concrete properties when using some SCMs. For instance, the use of fly ash can reduce the water demand of the concrete, allowing the water-to-cementitious materials ratio to be reduced with equivalent workability.

Fly ash and other pozzolans benefit concrete from the standpoint of durability from the pozzolanic reaction – calcium hydroxide reacts with silica and water to form calcium silicate hydrate. Because calcium silicate hydrate has a greater volume than the calcium hydroxide and pozzolan from which it originates, the pozzolanic reaction results in a finer and less interconnected system of capillary pores. Slag cement contributes its own calcium silicate hydrate during hydration, also leading to a less connected capillary pore system.

Excessive quantities of SCMs are not necessarily beneficial. Because the pozzolanic reaction takes time, care must be taken to ensure that the properties of the concrete at early ages are satisfactory. Extended moist curing may be necessary to achieve required results. For more information on the use of SCMs in concrete mixtures and their impact on concrete properties, see Chapter 3.

Test Methods to Determine Permeability. Various methods (both direct and indirect) are available to determine permeability of concrete as shown in Table 11-2. Refer to Chapter 20 for more information on test methods.

Cracking and Durability

No matter how good the quality of concrete, if the concrete cracks extensively, moisture and aggressive chemicals like chlorides and sulfates can enter the concrete and adversely affect durability. Cracks in concrete are caused from stresses that exceed capacity due to applied loads and volume changes when concrete is restrained. Random cracking can be avoided with proper joint spacing to predetermine the location of the cracks, and by properly sized and positioned reinforcement to reduce crack widths. In addition, there are many measures for reducing the total volume change of concrete reducing the risk of cracking (see Chapter 10).

TABLE 11-2. Test Methods Used to Determine Various Permeability-Related Properties

TEST METHOD	TITLE	COMMENTS
AASHTO T 259	Standard Test Method for Determining the Penetration of Chloride Ion into Concrete by Ponding	Used by highways agencies. The test requires between 6 and 12 months to complete. No clear interpretation of results.
AASHTO T 357	Standard Method of Test for Predicting Chloride Penetration of Hydraulic Cement Concrete by the Rapid Migration Procedure	Less variable than ASTM C1202. Results are not influenced by corrosion inhibiting admixtures.
AASHTO T 358	Standard Method of Test for Surface Resistivity Indication of Concrete's Ability to Resist Chloride Ion Penetration	Relatively rapid assessment of concrete permeability.
AASHTO TP 119	Standard Method of Test for Electrical Resistivity of a Concrete Cylinder Tested in Uniaxial Resistance Test	Provisional test method. Similar to C1876.
ASTM C1202 (AASHTO T 277)	Standard Test Method for Electrical Indication of Concrete's Ability to Resist Chloride Ion Penetration	Widely used in specifications for reinforced concrete exposed to chlorides.
ASTM C1556	Standard Method of Test for Determining the Apparent Chloride Diffusion Coefficient of Cementitious Mixtures by Bulk Diffusion	Considered a useful method for prequalification of concrete mixtures but takes about three months to complete.
ASTM C1585	Standard Test Method for Measurement of Rate of Absorption of Water by Hydraulic-Cement Concretes	New test method. Initial ingress of aggressive ions by absorption into unsaturated concrete is much faster than by diffusion or permeability.
ASTM C1876	Standard Test Method for Bulk Electrical Resistivity or Bulk Conductivity of Concrete	Relatively rapid assessment of concrete permeability.
CRD-C 48	Standard Test Method for Water Permeability of Concrete	Only suitable for concretes with low-cement contents, such as mass concrete in hydraulic structures.
CRD-C 163	Test Method for Water Permeability of Concrete Using Triaxial Cell	Best for evaluating concretes with a w/cm between 0.40 and 0.70.

Protective Treatments

The first line of defense against chemical and mechanical attack is the use of good-quality concrete. In addition to using concrete with a low permeability, surface treatments can be used to keep aggressive substances from coming into direct contact with a concrete surface or reach its interior. PCA IS001, *Effects of Substances on Concrete and Guide to Protective Treatments* (Kerkhoff 2007), discusses the effects of hundreds of chemicals on concrete and provides a list of treatments to help control chemical attack.

Exposure Categories

Assessing the environment to which the concrete will be exposed is a fundamental part of designing durable concrete. This assessment will influence both the design of concrete mixtures as well as troubleshooting and repair of distressed concrete. Table 11-3 shows the four main durability-related exposure categories covered in ACI 318, *Building Code Requirements for Structural Concrete and Commentary* (2019), and Figure 11-7 provides examples of how those categories apply to concrete structures. The designer selects the relevant exposures for each component of the concrete structure and determines the critical requirements in terms of the maximum water-to-cementitious materials ratio (w/cm), the minimum concrete strength (f_c'), and other characteristics. The minimum concrete strength requirement for durability governs the design if higher than that needed for structural loading. Additional requirements may apply for some exposures.

ACI 318-19 addresses exposures to freezing and thawing, soluble sulfates in soil or water, chlorides, and concrete that needs low permeability when in contact with water. Additional durability concerns specific to a project (for example abrasion) need to be addressed separately by a project specification.

TABLE 11-3. Exposure Categories for Durable Concrete (adapted from ACI 318-19*)

EXPOSURE CLASS		EXPOSURE CATEGORIES	MAXIMUM w/cm	MINIMUM DESIGN COMPRESSIVE STRENGTH, MPa (psi)	ADDITIONAL REQUIREMENTS*
F	Freezing and thawing	**F0** (Not applicable) – for concrete not exposed to cycles of freezing and thawing	—	17 (2500)	None.
		F1 (Moderate) – Concrete exposed to freezing and thawing cycles and occasional exposure to moisture	0.55	24 (3500)	Minimum air content requirements apply.
		F2 (Severe) – Concrete exposed to freezing and thawing cycles and in frequent exposure with moisture	0.45	31 (4500)	Minimum air content requirements apply.
		F3 (Very Severe) – Concrete exposed to freezing and thawing cycles that will be in frequent exposure with moisture and exposure to deicing chemicals or seawater[†]	0.40 [0.45 for plain (nonreinforced) concrete]	35 (5000) [31 (4500) for plain (nonreinforced) concrete]	Minimum air content requirements and maximum supplementary cementitious materials contents apply.
S	Sulfate[†]	**S0** (Not applicable) – Soil: SO_4 <0.10% by mass – Water: SO_4 < 150 ppm	—	17 (2500)	None.
		S1 (Moderate) – Soil: 0.10% ≤ SO_4 < 0.20% by mass – Water: 150 ppm ≤ SO_4 < 1500 ppm – Seawater	0.50	28 (4000)	Limited to moderately (or highly) sulfate resistant cementitious materials (see Table 11-9). For seawater, portland cements with up to 10% C_3A content are permitted with a maximum water:cement ratio of 0.40.
		S2 (Severe) – Soil: 0.20% ≤ SO_4 < 2.0% by mass – Water: 1500 ppm ≤ SO_4 < 10,000 ppm	0.45	31 (4500)	Limited to highly sulfate resistant cementitious materials (see Table 11-9). Calcium chloride admixtures are not allowed.
		S3 (Very severe) – Soil: SO_4 > 2.0% by mass – Water: SO_4 > 10,000 ppm — Option 1	0.45	31 (4500)	Limited to highly sulfate resistant cementitious materials (see Table 11-9) with additional pozzolans or slag cement determined by service record or testing to improve sulfate resistance. Calcium chloride admixtures are not allowed.
		Option 2	0.40	35 (5000)	Limited to highly sulfate resistant cementitious materials (see Table 11-9). Calcium chloride admixtures are not allowed.
C	Corrosion protection for reinforcement	**C0** (Not applicable) – Concrete that will be dry or protected from moisture in service	—	17 (2500)	Water-soluble chloride ion limits apply.
		C1 (Moderate) – Concrete exposed to moisture but not to an external source of chlorides in service	—	17 (2500)	Water-soluble chloride ion limits apply.
		C2 (Severe) – Concrete exposed to moisture and an external source of chlorides in service (including seawater, brackish water, deicing chemicals, salt, or spray from these sources)	0.40	35 (5000)	Water-soluble chloride ion limits and minimum cover requirements apply (see ACI 318-19 and Table 11-8).
W	In contact with water	**W0** (Not applicable) – Concrete that is dry in service	—	17 (2500)	None.
		W1 Concrete in contact with water but not required to have low permeability to water	—	17 (2500)	Aggregates are not ACR-susceptible. Aggregates are not ASR-susceptible, or adequate mitigation measures are taken.
		W2 Concrete in contact with water where low permeability is required	0.50	28 (4000)	Aggregates are not ACR-susceptible. Aggregates are not ASR-susceptible, or adequate mitigation measures are taken.

*This table is only a summary; See ACI 318-19 for complete details.

[†]ASTM C1580 method is recommended to determine sulfate content of soil and ASTM D516 for testing water.

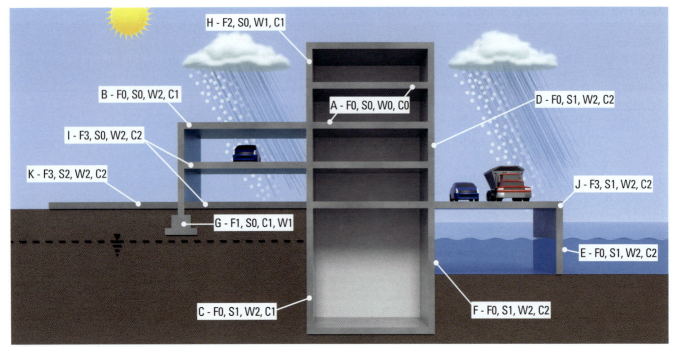

FIGURE 11-7. Example exposure classes as defined in ACI 318-19 (see Table 11-3 for details): (A) Interior concrete, apartment building; (B) Exterior concrete, reinforced parking structure, non-freeze-thaw region; (C) Underground, below the water table, office building, non-freeze-thaw region, groundwater sulfate concentration determined as 250 ppm; (D) Exterior, beachfront hotel, non-freeze-thaw region; (E) Concrete pier in seawater, non-freeze-thaw region; (F) Wall under seawater, non-freeze-thaw region; (G) Underground, above the water table, soil sulfate content determined as 0.05% by mass, freeze-thaw region; (H) Above ground, exterior concrete, freeze-thaw region; (I) Reinforced parking deck, freeze-thaw region; (J) Concrete bridge supports, exposed to salt spray, freeze-thaw region; (K) Roadway pavement in a freeze-thaw region, soil sulfate content determined as 1.1% by mass.

DURABILITY TESTING

Durability testing of concrete continually evolves as our knowledge of distress mechanisms increases, leading to improvements in existing test methods or development and standardization of new tests. Table 11-4 provides a list of common tests related to durability. Chapter 20 provides detail on a wide range of concrete testing, including test methods related to durability. No standard test methods have been developed for salt crystallization or physical salt attack, delayed ettringite formation, pyrrhotite, thaumasite, or acid attack (see ACI 201.2R 2016).

DETERIORATION MECHANISMS AND MITIGATION

Abrasion and Erosion

Concrete surfaces that are exposed to strong mechanical stress require high abrasion and erosion resistance. Abrasion damage in concrete is defined as *"wearing away of a surface by rubbing and friction"* and erosion is defined as *"progressive loss of material from a solid surface due to a mechanical interaction between that surface and a fluid, a multi-component fluid, or solid particles carried with the fluid"* (ACI CT 2020).

Wear on concrete surfaces can occur in the following situations:

• floors and slabs due to pedestrian and wheeled traffic,

• pavements and slabs subject to vehicular traffic (in particular, studded tire use and/or use of chains during winter weather events),

• sliding bulk material (for example in silos),

• impact stress of heavy objects (for example factory floors or loading ramps),

• erosion of hydraulic structures from the impact of objects transported by the fluid,

• cavitation in hydraulic structures, and

• abrasion by ice in marine structures.

Mechanism of Abrasion and Erosion. As listed above, abrasion and erosion can be due to sliding, revolving, and/or impact stresses or from cavitation in hydraulic structures. When concrete is exposed to sliding stress, fine material can be dislodged, depending on friction and roughness of the contact surfaces. This leads to abrasion of the surface. Revolving stress involves wear caused by rubber tires (soft) or plastic tires (hard). Hard tires in particular create an abrasive attack on the concrete surface. This leads to loss of paste around individual aggregate pieces and, ultimately, breaking away. Figure 11-8 shows aggregate exposed after decades of service for a concrete pavement. An impact stress is caused by an object hitting the concrete surface.

TABLE 11-4. Standard Test Methods Related to Concrete Durability*

STANDARD	TITLE	COMMENTS
ABRASION AND EROSION		
ASTM C131 (AASHTO T 96)	*Standard Test Method for Resistance to Degradation of Small-Size Coarse Aggregate by Abrasion and Impact in the Los Angeles Machine*	Abrasion resistance of coarse aggregates smaller than 37.5 mm (1½ in.) nominal size. Also called the "rattler method."
ASTM C418	*Standard Test Method for Abrasion Resistance of Concrete by Sandblasting*	Depth of wear under sandblasting used to determine abrasion resistance
ASTM C535	*Standard Test Method for Resistance to Degradation of Large-Size Coarse Aggregate by Abrasion and Impact in the Los Angeles Machine*	Abrasion resistance of coarse aggregates larger than 75 mm (¾ in.) nominal size.
ASTM C779	*Standard Test Method for Abrasion Resistance of Horizontal Concrete Surfaces*	Includes methods offering different degrees of abrasive force: revolving-disc with grit, dressing-wheel, and ball-bearing machines.
ASTM C944	*Standard Test Method for Abrasion Resistance of Concrete or Mortar Surfaces by the Rotating-Cutter Method*	Method uses rotating cutters (more useful for smaller samples).
ASTM C1138	*Standard Test Method for Abrasion Resistance of Concrete (Underwater Method)*	Simulates the effects of swirling water or cavitation.
FREEZING AND THAWING		
ASTM C173 (AASHTO T 196)	*Test Method for Air Content of Freshly Mixed Concrete by the Volumetric Method*	Determines volume of total air in fresh concrete.
ASTM C231 (AASHTO T 152)	*Test Method for Air Content of Freshly Mixed Concrete by the Pressure Method*	Determines volume of total air in fresh concrete.
ASTM C457	*Standard Test Method for Microscopical Determination of Parameters of the Air-Void System in Hardened Concrete*	Determines volume and spacing factor of entrained air for hardened concrete.
ASTM C666 (AASHTO T 161)	*Standard Test Method for Resistance of Concrete to Rapid Freezing and Thawing*	Used to determine the relative durability factor based on change in dynamic modulus of elasticity, as measured by resonant frequency measurements, after up to 300 cycles of freezing.
ASTM C672-12	*Standard Test Method for Scaling Resistance of Concrete Surfaces Exposed to Deicing Chemicals*	(Withdrawn 2021.) Uses a visual rating of 1 (good) to 5 (poor). Mass of scaled material after a specific number of freeze-thaw cycles may be also be determined.
ALKALI-AGGREGATE REACTIVITY		
See Table 11-7.		ASTM C1778 and AASHTO R 80 provide detailed guidance.
CARBONATION		
Phenolphthalein pH indicator test		Pink color denotes pH above 9.5, indicating no significant carbonation has occurred. May be included in a petrographic evaluation.
CORROSION		
ASTM C876	*Test Method for Corrosion Potentials of Uncoated Reinforcing Steel in Concrete*	Method is suitable for both field and laboratory determination of corrosion activity of reinforcing steel.
ASTM C1152	*Test Method for Acid-Soluble Chloride in Mortar and Concrete.*	Determines the acid-soluble chloride ion content.
ASTM C1218	*Test Method for Water-Soluble Chloride in Mortar and Concrete*	Determines water-soluble chloride ion content.
ASTM C1524	*Test Method for Water-Extractable Chloride in Aggregate (Soxhlet Method)*	Used to evaluate aggregates that contain a high amount of naturally occurring chloride.
ASTM G109	*Standard Test Method for Determining Effects of Chemical Admixtures on Corrosion of Embedded Steel Reinforcement in Concrete Exposed to Chloride Environments*	Used to determine the effects of admixtures on the corrosion of embedded steel reinforcement in concrete exposed to chloride environments.
SULFATE ATTACK		
ASTM C114	*Standard Test Methods for Chemical Analysis of Hydraulic Cement*	Sulfate resistance of Type II, Type II(MH), and Type V portland cements is based on C_3A content determined by Bogue calculation (see Chapter 2), which is based on chemical analyses.
ASTM C452	*Standard Test Method for Potential Expansion of Portland-Cement Mortars Exposed to Sulfate*	Optional test for sulfate resistance of Type V cements.
ASTM C1012	*Standard Test Method for Length Change of Hydraulic-Cement Mortars Exposed to a Sulfate Solution*	Test for blended cements and combinations of portland cements and SCMs to verify sulfate resistance based on expansion results at 6 months or 1 year.
ASTM C1580	*Standard Test Method for Water-Soluble Sulfate in Soil*	Used to test soils in contact with concrete to define sulfate exposure (see Table 11-3).
ASTM D516	*Standard Test Method for Sulfate Ion in Water*	Used to test water in contact with concrete to define sulfate exposure (see Table 11-3).
MICROBIALLY-INDUCED CORROSION		
ASTM C1904	*Standard Test Methods for Determination of the Effects of Biogenic Acidification on Concrete Antimicrobial Additives and/or Concrete Products*	Can be used to evaluate antimicrobial additives.
SEAWATER EXPOSURE		
See Table 11-2 for permeability tests.		

* Chapter 20 includes additional detail on many of these test methods.

FIGURE 11-8. Picture of aggregate exposed on a concrete pavement after decades of use (courtesy of R.D. Hooton).

FIGURE 11-9. An example of erosion due to cavitation in the stilling basin of Kinzua Dam, Pennsylvania.

The softer cement paste is attacked, gradually uncovering the coarser aggregates, and eventually removing them from their embedment. Cavitation (Figure 11-9) is the result of bubbles collapsing in a fast-moving stream of water. Vapor bubbles are formed as the water moves over surface irregularities and later collapse explosively, causing damage to the concrete surface.

Due to the variety of mechanisms causing abrasion, there are a number of test methods available for assessing the ability of a concrete sample to resist abrasion. The form of loading is somewhat different in each case, making it impossible to employ a single standardized test that covers all possible damage mechanisms for abrasion and erosion.

Materials and Methods for Abrasion and Erosion Resistance. Higher compressive strength concrete has increased resistance to abrasion. A low water-cement ratio and adequate curing are essential for abrasion resistance. The type of aggregate and the surface finish also have a strong influence on final abrasion resistance. Hard dense aggregate is more wear resistant than soft, porous aggregate. Figure 11-10 shows results of abrasion tests on concretes of different compressive strengths and different aggregate types. The total aggregate content in the concrete mixture should be reasonably high so that the thickness of the paste layer at the surface is kept to a minimum without compromising the surface finish.

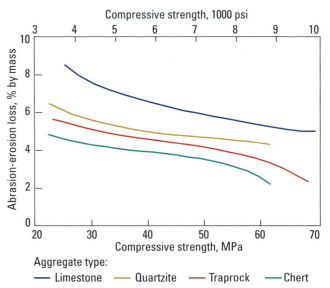

FIGURE 11-10. Effect of compressive strength and aggregate type on the abrasion resistance of concrete (ASTM C1138). High strength concrete made with a hard aggregate is highly resistant to abrasion (Liu 1981).

It is critical that the surface of the concrete be as durable as possible, which requires careful selection of finishing techniques for interior and exterior applications. The mixture should be designed to minimize bleeding. A steel-troweled surface is recommended for abrasion resistance. Figure 11-11 illustrates the effect surface treatments, such as metallic or mineral aggregate surface hardeners, have on abrasion resistance of hard steel troweled surfaces. ACI 201.2R, *Guide to Durable Concrete* (2016), recommends that surfaces that are likely to be heavily abraded should be treated with a dry-shake surface hardener to provide additional protection.

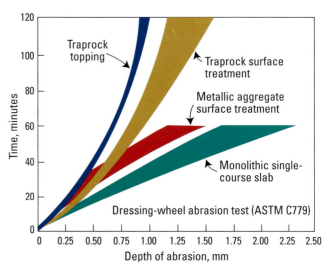

FIGURE 11-11. Effect of surface treatments on the abrasion resistance of concrete (ASTM C779). Base slab compressive strength was 40 MPa (6000 psi) at 28 days. All slabs were steel troweled (Brinkerhoff 1970).

Freezing and Thawing

Concrete elements exposed to weathering in colder climates need sufficient resistance to freezing and thawing cycles and deicer chemicals. Freeze-thaw exposures range from moderate (Category F1), such as a vertical facade element exposed to occasional moisture; to very severe (Category F3), including pavement or bridge decks in continual contact with moisture and exposed to deicer chemicals or freezing seawater. Freeze-thaw damage in concrete can also be a consequence of using aggregates that are not resistant to freeze-thaw exposures (see Chapter 5). Concrete generally exhibits lower resistance to the combined effects of freezing and thawing cycles in conjunction with deicing chemicals as compared to freezing and thawing cycles alone.

Mechanism of Freeze-Thaw Damage. As water in concrete freezes, it produces osmotic and hydraulic pressures in the capillary and other pores of the cement paste and aggregate. Hydraulic pressures are caused by the 9% expansion of water upon freezing; in this process, growing ice crystals displace unfrozen water. If the concrete is above critical saturation (78% to 91.7% of the capillary pores filled with water), hydraulic pressures result as freezing progresses (ACI 201.3T 2019 and Todak and others 2015). If the pressure exceeds the tensile strength of the surrounding paste or aggregate, some of the pores will dilate and rupture. The accumulative effect of successive freeze-thaw cycles is the disruption of paste and aggregate eventually causing significant deterioration in the form of cracking, scaling, and disintegration (Figures 11-12 and 11-13).

FIGURE 11-12. Severe scaling of a concrete pavement.

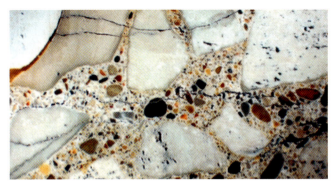

FIGURE 11-13. Sample of sawn and polished concrete damaged by freeze-thaw cycles.

The mechanisms of freeze-thaw damage and salt scaling in concrete are quite complex. Powers (1962) and Pigeon and Pleau (1995) conducted extensive reviews of the mechanisms of freeze-thaw action. Concrete damage due to freezing and thawing cycles is the result of complex microscopic and macroscopic interactions closely related to the freezing behavior of the pore solution (Weiss 2015, and Valenza and Scherer 2007a, and 2007b).

Ice in capillary pores (or any ice in large voids or cracks) draws water from surrounding pores to advance its growth. Since most pores in cement paste and some aggregates are too small for ice crystals to form, water attempts to migrate to locations where it can freeze.

The pore solution of concrete contains a high quantity of dissolved ions which lower the freezing point. Use of deicers will increase the ion concentration in the pore solution and therefore impact the freezing behavior. In addition, the freezing point of the pore solution is dependent on pore size; the smaller the pore, the lower the freezing point (Table 11-5). Pore volume and pore size distribution are therefore important variables for freeze-thaw resistance. As concrete ages, the pore size distribution changes due to continuing hydration and carbonation at the concrete surface.

The moisture content of the pore system is a crucial factor of the extent of freeze-thaw damage. Fagerlund (1975) introduced the term critical saturation. Critical saturation is influenced by pore volume and pore size distribution. It is reached when the

TABLE 11-5. Pore Size Distribution (Adapted from Setzer 1997)

TYPE	SIZE	FILLING OF PORES	TEMPERATURE AT FREEZING
Macro pores	≥1 mm	Empty	
Capillary pores	<1 mm	Suction, immediately filling	
Meso capillaries	<30 µm	Suction, filling in minutes to weeks	Water freezes between 0°C (32°F) and -20°C (4°F)
Micro capillaries	<1 µm	Complete filling through capillary suction not possible	
Meso gel pores	<30 nm	Filling through condensation at relative humidity of 50% to 98%	Water freezes between -20°C (-4°F) and -40°C (-40°F)
Micro gel pores	<1 nm	Filling through sorption at relative humidity of <50%	Water freezes at about -90°C (-130°F)

moisture content is high enough to damage concrete during one or a few freeze-thaw cycle(s). Setzer (1999) found that pastes subjected to freeze-thaw cycles have the potential to acquire more water with each cycle due to a micropump effect. It is therefore likely to find concrete to be durable during initial freeze-thaw cycles, but then observe increasingly higher vulnerability with an increased number of cycles. As well, additional water from the surrounding environment can enter the pores of concrete during thaw periods and raise the level of saturation as winter progresses (Fagerlund 1975). Cracking due to freeze-thaw (or other distresses) increases the accessibility of water, allowing the concrete to reach critical saturation more readily.

Osmotic pressures develop from differential concentrations of alkali solutions in the paste (Powers 1965a). As pure water freezes, the alkali concentration increases in the adjacent unfrozen water. A high-alkali solution, through the mechanism of osmosis, draws water from lower-alkali solutions in the pores. This osmotic transfer of water continues until equilibrium in the fluids' alkali concentration is achieved. Osmotic pressure is considered a minor factor, if present at all, in aggregate frost action, whereas it may be dominant in the cement paste fraction of concrete. Osmotic pressures are considered a major factor in salt scaling, where ionic concentrations in the pore solution are increased due to the penetration of deicing salts.

Monteiro and others (2006) showed the presence of a transition layer surrounding the air void. This layer can have a significant effect on the ice formation inside the air voids. Stresses arise from differences in thermal expansion of ice and concrete and place the ice in tension as the temperature drops. Higher air contents and more dense concretes (lower water cementitious materials ratios and adequate curing) make concrete more difficult to critically saturate (Figure 11-14). At lower water contents (less than critical saturation levels), no hydraulic pressure should exist.

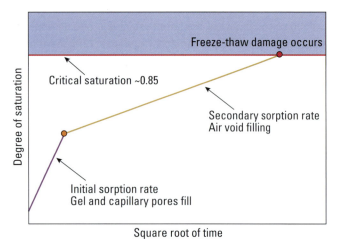

FIGURE 11-14. Impacts of air content and concrete density on ease of saturation of concrete. Once concrete becomes critically saturated, freeze-thaw damage can result (after Todak and others 2015).

Exposure to Deicers and Anti-icers

Deicers are solid or liquid chemicals that are applied on concrete to melt ice or snow. Anti-icers are liquids applied before a precipitation event and work to keep water from freezing or refreezing. Deicing and anti-icing chemicals used for snow and ice removal can cause and exacerbate surface scaling and joint deterioration (Figure 11-15). Deicer scaling and cracking of inadequately air-entrained or non-air-entrained concrete during freezing is believed to be caused by a buildup of osmotic and hydraulic pressures in excess of the normal hydraulic pressures produced when water in concrete freezes. These pressures become critical and result in scaling unless entrained air voids are present at both the surface and throughout the paste fraction of the concrete to relieve the pressure. The hygroscopic (moisture absorbing) properties of deicing salts also attract water and keep the concrete more saturated, increasing the potential for freeze-thaw deterioration.

FIGURE 11-15. V-shaped joints are a common sign of the effects of freeze-thaw damage in concrete pavements. Some joints exhibit an inverted V-shaped deterioration (courtesy of D. Harrington).

Valenza and Scherer (2006) found cracking of the brine ice layer on the concrete surface to be the origin of salt scaling. Considering the mechanical and viscoelastic properties of ice, the researchers found that the differences will cause moderately concentrated brine solutions to crack. The stresses induced from crack formations in the brine ice will then result in cracks penetrating into the concrete surface, resulting in surface damage.

Studies have also shown that, in the absence of freezing, the formation of salt crystals in concrete (from external sources of chloride, sulfate, and other salts) may contribute to concrete scaling and deterioration similar to the disintegration of rocks by salt weathering. The entrained air voids in concrete allow space for salt crystals to grow. This mechanism relieves internal stress similar to the way the voids relieve stress from freezing water in concrete (ASCE 1982 and Sayward 1984). See **Salt Crystallization or Physical Salt Attack** in this chapter.

The extent of scaling depends upon the amount of deicer used and the frequency of application. Relatively low concentrations of deicer (on the order of 2% to 4% by mass) produce more surface scaling than higher concentrations or the absence of deicer (Verbeck and Klieger 1956). However, some highly concentrated anti-icer solutions can potentially include other surface damage due to chemical attack.

Deicers can reach concrete surfaces in ways other than direct application, such as splashing by vehicles and dripping from the undersides of vehicles. Scaling is more severe in poorly drained areas because more of the deicer solution remains on the concrete surface during freezing and thawing. Air entrainment is effective in preventing surface scaling and is recommended for all concretes that may come in contact with deicing or anti-icing chemicals (Figure 11-16).

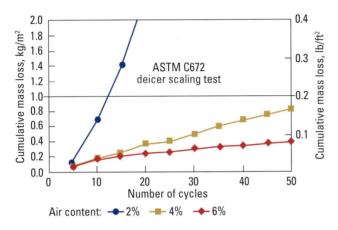

FIGURE 11-16. Cumulative mass loss for mixtures with a water-to-cement ratio of 0.45 and on-time finishing (Pinto and Hover 2001).

Table 11-6 summarizes the most commonly used snow and ice control materials (NCHRP 2007). Chloride-based salts containing sodium, calcium, magnesium, and potassium (NaCl, CaCl$_2$, MgCl$_2$, and KCl) comprise the majority of deicers used to melt snow and ice. These chemicals work well because they lower the freezing point of the precipitation that falls on concrete pavements.

In recent years, there has been considerable interest in enhancing the deicing effectiveness at low temperatures by substituting calcium chloride or magnesium chloride for a portion of the sodium chloride brine. Blending salt (sodium chloride) brine with these and other liquid ice control chemicals is considered an economical method to maintain low temperature surface performance properties. The most common chemical blended with salt brine is calcium chloride, but magnesium chloride is also used along with a myriad of other products that are mixtures of mostly calcium chloride and a number of carbohydrate chemicals.

Deterioration of concrete pavement joints has been observed due to a potential chemical reaction between the deicer salt and the cement matrix that forms calcium oxychloride (Villani and others 2015, Farnam and others 2015, and ACI 201.3T 2019).

Deicers can have many effects on concrete and the immediate environment. All deicers can aggravate scaling of concrete that is not properly air entrained. Sodium chloride (ASTM D632 or AASHTO M 143), and calcium chloride (ASTM D98 or AASHTO M 144) are the most frequently used deicers. In the absence of freezing, sodium chloride has little to no chemical effect on concrete but can damage plants and corrode metals. Calcium chloride in weak solutions generally has little chemical effect on concrete and vegetation but does corrode metal including embedded steel reinforcement if the chlorides penetrate the concrete cover.

Studies have shown that concentrated calcium chloride solutions can chemically attack concrete (Brown and Cady 1975 and Sutter and others 2008). There is disagreement about the effect of magnesium chloride on concrete deterioration resulting from laboratory studies (Leiser and Dombrowski 1967; Cody and others 1996; Lee and others 2000; Kozikowski and others 2007; and Sutter and others 2008). However, no damage of field concrete attributed to magnesium chloride use has been observed (NCHRP 2007). Nonchloride deicers are used to minimize corrosion of reinforcing steel and minimize groundwater chloride contamination. The use of deicers containing ammonium nitrate and ammonium sulfate should be strictly prohibited as they rapidly attack and disintegrate concrete (Stark and others 2002).

The chemicals used for aircraft and airfield deicing are distinctly different from those commonly used for pavement deicing. For pavements at airports, only non-chloride deicing agents are used. These include: urea, potassium acetate, sodium acetate, sodium formate, calcium magnesium acetate, propylene glycol, and ethylene glycols (Mericas and Wagoner 2003). The latter two glycol deicers although not used for deicing pavement are commonly used for aircraft deicing. Propylene glycol and ethylene glycols make up 30% to 70% of the as-applied solution, with an increased use of propylene glycol because of the toxicity concerns related to ethylene glycol (Ritter 2001). Urea does not chemically damage concrete, vegetation, or metal.

As an indication of the degree of freeze-thaw resistance, Cordon (1966) suggested that concrete with good frost resistance should have durability factor greater than 80% (Figure 11-17), when tested in accordance with ASTM C666, *Standard Test Method for Resistance of Concrete to Rapid Freezing and Thawing* (AASHTO T 161). Natural freezing of concrete in service normally occurs at lower cooling rates, at later ages, and after some period of drying (Lin and Walker 1975, Pigeon and others 1985, and Vanderhost and Jansen 1990).

TABLE 11-6. Snow and Ice Control Materials (adapted from NCHRP 2007)*

MATERIAL TYPE*		FORMS USED	PRACTICAL TEMPERATURE LIMIT	COMMENT
Chlorides	Sodium, NaCl	Solid and liquid brine	-10°C (15°F)	The most common ice melting salt. Has little chemical effect on concrete, but can damage lawns and shrubs and contribute to disintegration of low-quality, non-air-entrained concrete. Promotes corrosion of metal.
	Magnesium, $MgCl_2$	Liquid brine, some solid flake	-15°C (5°F)	Releases about 40% less chlorides into the environment than either rock salt or calcium chloride. Promotes corrosion of metal. Studies evaluating the effect of magnesium chloride deicing salts on concrete show conflicting results. (Leiser and Dombrowski 1967; Cody and others 1996; Lee and others 2000; Kozikowski and others 2007; Sutter and others 2008; and NCHRP 2007).
	Calcium, $CaCl_2$	Liquid brine, some solid flake	-32°C (-25°F)	At low concentrations has little chemical effect on concrete, lawns, and shrubs, but does promote corrosion of metals and can contribute to damage of low-quality concrete. Can absorb moisture from the air, causing it to clump, harden, or even liquefy during storage. Calcium chloride can be hazardous to human health, can leave a slippery residue that is difficult to clean, and tends to refreeze quickly, which may require frequent reapplication (Peeples 1998). At high concentrations, it has been reported to damage concrete through the formation of calcium oxychlorides (Sutter and others 2008).
	Potassium, KCl	Solid	-4°C (25° F)	Common plant nutrient. It is not a skin irritant and is perceived to be less damaging to vegetation. Impractical as a deicer unless used in conjunction with other ingredients. For example, a natural product containing a blend of sodium, magnesium, and potassium chloride is used in some western states (NCHRP 2007).
Organics	Potassium acetate	Liquid	-32°C (-25°F)	Biodegradable deicers primarily used for airports. Environmentally friendly, slightly corrosive, often mixed with a corrosion inhibitor (Peeples 1998).
	Potassium formate		-32°C (-25°F)	
	Sodium acetate	Solid	-18°C (0°F)	
	Sodium formate		-18°C (0°F)	
	Calcium magnesium acetate, CMA	Mostly liquid, some solid	-4°C (20°F)	Made from limestone and acetic acid. Biodegradable, no toxic effects on terrestrial or aquatic animals or soils or vegetation (Wyatt and Fritzsche 1989). Can effectively prevent the formation of ice-surface bonds when applied prior to precipitation. The high cost has limited its practical use. CMA is not effective at very low temperatures.
	Agricultural by-products	Liquid	varies	For example alpha methyl glucoside, a corn by-product that is most effective when combined with other ingredients.
	Manufactured organic materials (glycols, methanol)	Liquid	varies	
Ammonium	Ammonium nitrate			Should be strictly prohibited as they rapidly attack and disintegrate concrete.
	Ammonium sulfate			
Nitrogen products	Urea	Solid	-10°C (15°F)	A common fertilizer nutrient. In its pure form not corrosive. However, most of the commercially available products are not suitable for use in corrosion sensitive environments (Peeples 1998). Might damage low-quality concrete. Does not damage lawns, shrubs, or properly made air-entrained concrete.

*Sand and other abrasives are also used for snow and ice control.

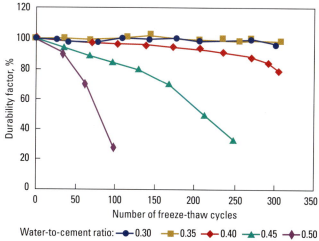

Water-to-cement ratio: ●—0.30 ■—0.35 ◆—0.40 ▲—0.45 ◆—0.50

FIGURE 11-17. ASTM C666 durability factors vs. number of freeze-thaw cycles for selected non-air-entrained concretes (Pinto and Hover 2001).

Materials and Methods to Control Freeze-Thaw and Deicer Damage. The resistance of hardened concrete to freezing and thawing and deicers in a moist condition is significantly improved by the use of intentionally entrained air as shown in Figures 11-18 and 11-19.

Entrained air voids act as empty chambers in the paste where freezing (expanding) and migrating water can enter, thus relieving the pressure and preventing damage to the concrete (Powers 1955, Lerch 1960, and Powers 1965). Upon thawing, most of the water returns to the capillaries as a result of capillary action and pressure from air compressed within the bubbles.

The spacing and size of air voids are important factors contributing to the effectiveness of air entrainment in concrete. ASTM C457, *Standard Test Method for Microscopical Determination of Parameters of the Air-Void System in Hardened Concrete*, describes a method of evaluating the air-void system in hardened concrete. The pressure developed by water as it expands during freezing depends largely upon the distance the water must travel to the nearest air void for relief. Therefore, the voids must be spaced closely enough to reduce the pressure below that which would exceed the tensile strength of the

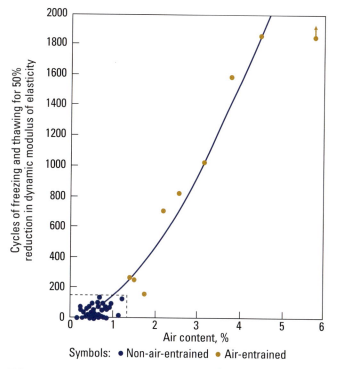

Symbols: ● Non-air-entrained ● Air-entrained

FIGURE 11-18. Effect of entrained air on the resistance of concrete to freezing and thawing in laboratory tests. Concretes were made with cements of different fineness and composition and with various w/c and cement contents (Bates and others 1952, and Lerch 1960).

concrete. The amount of hydraulic pressure is also related to the rate of freezing and the permeability of the paste.

The following air-void characteristics are representative of a system with adequate freeze-thaw resistance (Powers 1949, Powers 1954, Klieger 1952, Klieger 1956, Mielenz and others 1958, Powers 1965, Klieger 1966, Whiting and Nagi 1998, and Pinto and Hover 2001):

• calculated spacing factor, $\bar{L}$, (an index related to the distance between bubbles in three dimensions, but not the average spacing in the system) – less than 0.200 mm (0.008 in.); and

• specific surface, (surface area of the air voids) – 24 mm²/mm³ (600 in.²/in.³) of air-void volume, or greater.

FIGURE 11-19. Effect of weathering on boxes and slabs on ground at the Long-Time Study outdoor test plot, PCA, Skokie, Illinois (Stark and others 2002). Specimens A and B are air-entrained, specimens C and D exhibiting severe crumbling and scaling are non-air-entrained. All concretes were made with 335 kg/m³ (564 lb/yd³) of Type I portland cement. Periodically, calcium chloride deicer was applied to the slabs. Specimens were 40 years old when photographed (see Klieger 1963 for concrete mixture information).

Figure 11-20 illustrates the relationship between spacing factor and total air content. Measurement of air volume alone does not permit full evaluation of the important characteristics of the air-void system; however, air-entrainment is generally considered effective for freeze-thaw resistance when the total volume of air in the mortar fraction of the concrete – material passing the 4.75-mm (No. 4) sieve – is about 9 ± 1% (Klieger 1952) or about 18% by paste volume. For equal admixture dosage rates per unit of cement, the air content of ASTM C185, *Standard Test Method for Air Content of Hydraulic Cement Mortar* (AASHTO T 137) mortar is about 19% due to the standard aggregate's properties. The air content of concrete with 19-mm (¾-in.) maximum-size aggregate is approximately 6% for effective freeze-thaw resistance.

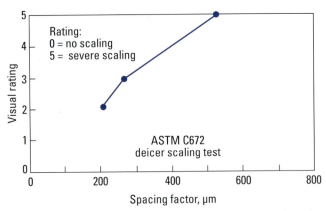

FIGURE 11-21. Visual rating, after 50 cycles as a function of spacing factor, for a concrete mixture with a water-to-cement ratio of 0.45 (Pinto and Hover 2001).

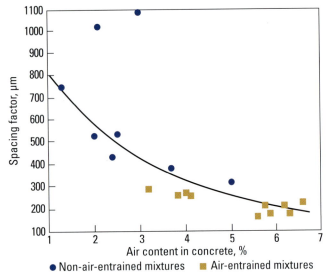

FIGURE 11-20. Spacing factor as a function of total air content in concrete (Pinto and Hover 2001).

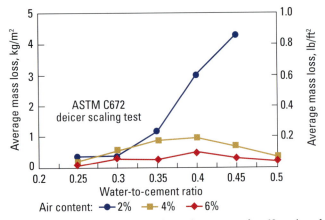

FIGURE 11-22. Measured mass loss of concrete after 40 cycles of deicer and frost exposure at various water to cement ratios (Pinto and Hover 2001).

Pinto and Hover (2001) address paste air content versus frost resistance. The total required concrete air content for durability increases as the maximum-size aggregate is reduced (due to greater paste volume) and as the exposure conditions become more severe (see Chapter 9).

A good air-void system with a low spacing factor (maximum of 200 μm or 0.008 in.) is perhaps more important to deicer environments than saturated frost environments without deicers. Properly designed and placed air-entrained concrete will withstand the majority of deicers for decades.

The relationship between spacing factor and deicer scaling is illustrated in Figure 11-21. A low water to cement ratio helps minimize scaling, but alone is not sufficient to control scaling at normal water-cement ratios. Figure 11-22 illustrates the overriding impact of air content over water-cement ratio in controlling scaling.

Current field quality control practice for concrete placements in the US in accordance with ACI 301, *Specifications for Concrete Construction* (2020) involves the measurement of total air volume in freshly mixed concrete and does not distinguish air-void size or spacing factor in most requirements. C173, *Test Method for Air Content of Freshly Mixed Concrete by the Volumetric Method* (AASHTO T 196), and C231, *Test Method for Air Content of Freshly Mixed Concrete by the Pressure Method* (AASHTO T 152), are each used to determine the volume of air in fresh concrete mixtures.

Canadian practice, CSA A23.1 (2019), requires the average spacing factor to be less than 230 μm (0.009 in.) with no single test value to exceed 260 μm (0.010 in.). For w/cm less than or equal to 0.36 these values are 260 μm (0.010 in.) and 300 μm (0.012 in.), respectively. In addition, CSA also recommends targeting a value of 170 μm (0.007 in.) to account for the variability of ASTM C457.

Field service and extensive laboratory testing have shown that when the proper spacing and volume of air voids are present, air-entrained concrete will have excellent resistance to freeze-thaw cycles and surface scaling due to freezing and thawing and the application of deicer chemicals, provided the concrete is properly proportioned, placed, finished, and cured. As shown in Table 11-3, a maximum w/cm for reinforced concrete of 0.40 (and for plain concrete of 0.45) and minimum compressive strength of 35 MPa (5000 psi) (31 MPa [4500 psi] for plain concrete) is required for scale resistance of concrete in continuous contact with deicers per ACI 318-19. There are also limits on dosages of SCMs for resistance to scaling unless otherwise demonstrated by field performance. See Chapter 16 for additional guidance on scaling resistance of concrete.

Air Drying. The resistance of air-entrained concrete to freeze-thaw cycles and deicers is greatly increased by an air drying period after initial moist curing. Air drying removes excess moisture from the concrete which in turn reduces the internal stress caused by freeze-thaw conditions and deicers. Water-saturated concrete will deteriorate faster than an air-dried concrete when exposed to moist freeze-thaw cycling and deicers. See Chapter 17 for more information on curing and drying.

Treatment of Scaled Surfaces. If surface scaling (an indication of an inadequate air-void system, insufficient curing, or poor finishing practices) develops during the first frost season, a breathable surface treatment can be applied to the dry concrete to help protect it against further damage. Treatment often consists of a penetrating sealer made with boiled linseed oil (ACPA 1996), breathable methacrylate, silanes, siloxanes, or other materials. Nonbreathable formulations should be avoided as they can trap water below the surface and cause delamination. See Chapter 16 for more information on treating scaled surfaces.

Alkali-Aggregate Reactivity

Alkali-aggregate reactivity (AAR) has two forms – alkali-silica reaction (ASR) and alkali-carbonate reaction (ACR). Both forms of AAR can lead to potentially deleterious expansive reactions between alkalies in the concrete pore solution and reactive minerals in some aggregates. Types of potentially harmful reactive minerals, rock, and synthetic materials are discussed in Chapter 5. The occurrence of aggregates containing reactive silica minerals contributing to ASR is more widespread than argillaceous dolomitic rocks that are associated with ACR. It is possible to prevent deleterious expansion due to ASR. The use of ACR-susceptible aggregates should be avoided because successful mitigation options have not been identified. Guidance is provided in ASTM C1778, *Standard Guide for Reducing the Risk of Deleterious Alkali-Aggregate Reaction in Concrete* (AASHTO R 80), on evaluation and identification of potentially reactive aggregates and preventive measures for concrete.

Alkali-Silica Reaction. Alkali-silica reactivity has been recognized as a potential source of distress in concrete since the late 1930s (Stanton 1940). Even though potentially reactive aggregates exist throughout North America, ASR distress is not widespread in concrete. There are a number of reasons for this:

- most aggregates are chemically stable in hydraulic cement concrete,
- aggregates with good service records are available in many areas,
- concrete dry in service can inhibit ASR,
- use of certain pozzolans or slag cement in appropriate amounts can control ASR,
- the alkali loading of the concrete is low enough in many mixtures to control harmful ASR, and
- certain forms of ASR do not produce significant deleterious expansion.

The reduction of ASR-induced expansion requires understanding the ASR mechanism; properly using tests to identify potentially reactive aggregates and the degree of their reactivity; and, if needed, taking steps to minimize the potential for expansion and related damage.

Typical indicators of ASR might be any of the following: a network of cracks (Figures 11-23 and 11-24); cracks with straining or exuding gel; closed or spalled joints; relative displacements of different parts of a structure; or fragments breaking out of the surface of the concrete (popouts). Because ASR deterioration is slow, the risk of catastrophic failure is low. However, ASR can cause serviceability problems and can exacerbate other deterioration mechanisms such as those that occur in frost, chloride, or sulfate exposures.

FIGURE 11-23. Cracking of concrete from alkali-silica reactivity.

FIGURE 11-24. Indication of potential alkali-silica reaction is provided by the classic map cracks on the surface of the concrete. Since this cracking may be also be caused by other mechanisms, ASR should be verified by petrographic examination.

To pinpoint ASR as the cause of damage in concrete, the presence of deleterious ASR gel must be verified. A site of expansive reaction can be defined as an aggregate particle that is recognizably reactive or potentially reactive and is at least partially replaced by gel. Gel can be present in cracks and voids and may also be present in a ring surrounding an aggregate particle at its edges. A network of internal cracks connecting reacted aggregate particles is an almost certain indication that ASR is responsible for cracking. A petrographic examination (ASTM C856, *Standard Practice for Petrographic Examination of Hardened Concrete*) is the most conclusive method available for identifying ASR gel in concrete (Powers 1999). Petrography, when used to study a damaged concrete specimen, can confirm the presence of reaction products and verify ASR as an underlying cause of deterioration (Figure 11-25).

FIGURE 11-25. Polished section view of an alkali reactive aggregate in concrete. Observe the alkali-silica reaction rim around the reactive aggregate and the crack formation.

Mechanism of ASR. Alkali-silica reaction forms gel, some that swells as it draws water from the surrounding cement paste. Reaction products from ASR have a great affinity for moisture. In absorbing water, these gels can expand, inducing pressure, expansion, and cracking of the aggregate and

surrounding paste (Figure 11-26). The reaction can be visualized as a two-step process:

1. Alkali hydroxide + reactive silica gel → alkali-silica gel
2. Alkali-silica gel + moisture → expansion

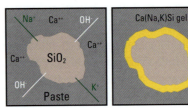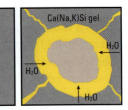

FIGURE 11-26. Mechanism of ASR (adapted from Thomas and Wilson 2002).

Factors Affecting ASR. For alkali-silica reaction to occur, the following three conditions must be present:

- Reactive forms of silica in the aggregate,
- High-alkali (pH) pore solution, and
- Sufficient moisture.

If one of these conditions is absent, ASR cannot occur.

Low-Alkali Portland Cement and Concrete Alkali Loading

Low-alkali portland cement was historically defined in ASTM C150, *Standard Specification for Portland Cement* (AASHTO M 85), as a cement with a total equivalent alkali content (Na_2Oeq = % Na_2O + 0.658 % K_2O) of not more than 0.60%. A low-alkali cement used in concrete with a moderate to high cement content can have a relatively high alkali loading and a sufficiently reactive aggregate may suffer from deleterious expansion due to ASR, if no other mitigation strategies are used. To encourage best practices to be followed, reference to low-alkali cement was removed from the national portland cement specifications (ACI 225.1T 2020). Equivalent alkali contents of portland cements are reported on mill test reports to permit concrete alkali loading to be calculated:

$$C = P \times N / 100$$

where:

C = concrete alkali loading in kg/m³ (lb/yd³),

P = portland cement content of the concrete in kg/m³ (lb/yd³), and

N = equivalent alkali content (Na_2Oeq) of the portland cement (or of the portland cement portion of a blended cement in % by mass).

For example, in a concrete with 355 kg/m³ (600 lb/yd³) of portland cement, and an Na_2Oeq of 0.50%, the concrete alkali loading would be:

355 × 0.50 /100 = 1.8 kg/m³ (600 × 0.50/100 = 3.0 lb/yd³)

The amount of gel formed in the concrete depends on the amount, type and reactivity of silica, the alkali hydroxide concentration, and the availability of moisture. The presence of gel does not always coincide with distress, and gel presence does not necessarily indicate destructive ASR. Conversely, certain aggregates produce relatively little gel, yet can lead to significant and deleterious expansion.

Table 11-7 summarizes different test methods used to evaluate the potential for alkali-silica reactivity and prevention and mitigation measures. Chapter 20 and ASTM C1778 (AASHTO R 80) provide detailed guidance on interpretation.

ASR tests should not be used to disqualify use of potentially reactive aggregates, but should be used to determine how reactive an aggregate is. Often, reactive aggregates can be safely used with the careful selection of cementitious materials to prevent harmful expansion.

Materials and Methods to Prevent ASR. The best way to avoid deleterious ASR is to take appropriate precautions in the design of the concrete mixture. Modifications to the mixture design should be carefully tailored to avoid limiting the concrete producer's options. This permits careful analysis of available cementitious materials and aggregates and choosing a control strategy that optimizes effective and economic selection of locally-available materials. If an aggregate is not reactive based on historical use or testing, no special requirements are needed.

ASTM C1778 and AASHTO R 80 are similar in approach and include similar prescriptive and performance approaches to controlling ASR. Broadly, the prescriptive approach can be described in 4 steps:

- Classify aggregate reactivity, based on ASTM C1293 or ASTM C1260 testing; higher levels of expansion in these tests indicate more reactive aggregates. Reactivity ranges from R0 (unreactive) to R3 (very highly reactive).

- Decide on the level of ASR risk, based on the reactivity of the aggregate and factors such as the size of structure and exposure to moisture and alkalies in service. Risk levels range from Level 1 (low risk) to Level 6 (high risk).

- Assign a structure classification from S1 (low) to S4 (high), which is based on the consequences of deleterious ASR: Structures with more serious life safety, economic, or environmental concerns that would result from ASR distress are classified with a higher structure classification.

- The level of prevention needed is then decided based on the structure classification and the level of ASR risk. Prevention levels range from V (no prevention needed), through increasingly more stringent levels of W, X, Y, Z, and ZZ (high). (There is one level beyond ZZ as well.).

Stepping through the decisions and assignments are made more straightforward by following guidance given in ASTM C1778

and AASHTO R 80. Once the level of prevention is assigned, minimum dosages of some types of SCMs (either as concrete ingredients or blended cements) can be read from a table, as can guidance on maximum concrete alkali loading, or, for the most stringent levels of prevention, both.

Alkali loading limits of 3.0 kg/m³ (5.0 lb/yd³), 2.4 kg/m³ (4.0 lb/yd³) or 1.8 kg/m³ (3.0 lb/yd³) are referenced in ASTM C1778 and AASHTO R 80, depending on the reactivity of the aggregate and the prevention level chosen.

For supplementary cementitious materials not listed in the prescriptive tables in ASTM C1778 and AASHTO R 80 of minimum SCM contents, a performance approach is also included to qualify combinations of cementitious materials and reactive aggregates. When SCMs are used to control ASR, their effectiveness must be determined by tests such as ASTM C1567 or ASTM C1293. ASTM C1567 with a 14-day expansion limit is the most common method used to evaluate effectiveness of control measures, although the 2-year ASTM C1293 is typically regarded as the best available test. When possible, a range of amounts of pozzolan or slag cement contents should be tested to determine the optimum dosage. Expansion usually decreases as the dosage of the SCM increases (Figure 11-27).

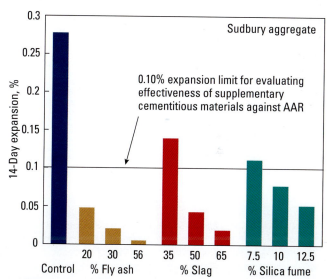

FIGURE 11-27. Influence of different amounts of fly ash, slag, and silica fume by mass of cementing material on mortar bar expansion (modified ASTM C1260/C1567) after 14 days when using reactive aggregate (Fournier 1997).

Lithium-based admixtures are available to control ASR in fresh concrete (Stark 1992, Thomas and others 2007). AASHTO R 80 includes methods of evaluating the appropriate dosage of lithium nitrate admixture to prevent deleterious expansion due to ASR.

See ASTM C1778 (AASHTO R 80) for complete guidance on designing concrete mixtures to minimize the risk of ASR-related expansion in concrete. Additional information can be found in Farny and Kerkhoff (2007) and Thomas and others (2008a).

TABLE 11-7. Test Methods for Alkali-Silica Reactivity

TEST NAME	PURPOSE	TYPE OF TEST	TYPE OF SAMPLE
ASTM C295, *Guide for Petrographic Examination of Aggregates for Concrete*	To evaluate possible aggregate reactivity through petrographic examination	Visual and microscopic examination of prepared samples – sieve analysis, microscopy, scratch or acid tests	Varies with knowledge of quarry: cores 53 mm to 100 mm in diameter (2⅛ in. to 4 in.), 45 kg (100 lb) or 300 pieces, or 2 kg (4 lb)
ASTM C856, *Petrographic Examination of Hardened Concrete*	To outline petrographic examination procedures of hardened concrete – useful in determining condition or performance	Visual and microscopic examination of prepared samples	At least one core 150 mm diameter by 300 mm long (6 in. diameter by 12 in. long)
ASTM C856 Annex, Uranyl-Acetate Treatment Procedure, (AASHTO T 299)	To identify products of ASR in hardened concrete	Staining of a freshly-exposed concrete surface and viewing under UV light	Varies: core with lapped surface, core with broken surface
Los Alamos staining method (Powers 1999)	To identify products of ASR in hardened concrete	Staining of a freshly-exposed concrete surface with two different reagents	Varies: core with lapped surface, core with broken surface
ASTM C1260, *Potential Alkali-Reactivity of Aggregates (Mortar-Bar Method),* (AASHTO T 303)	To determine the potential ASR susceptibility of aggregate	Immersion of 25-mm × 25-mm × 285-mm (1-in. × 1-in. × 11¼-in.) mortar bars in alkaline solution at 80°C (176°F)	At least 3 mortar bars for each aggregate being evaluated
ASTM C1293, *Determination of Length Change of Concrete Due to Alkali-Silica Reaction (Concrete Prism Test)*	To determine the potential ASR susceptibility of aggregates and potential for deleterious ASR of cementitious materials and aggregate combinations	75-mm × 75-mm × 285-mm (3-in. × 3-in. × 11¼-in.) concrete prisms stored over water at 38°C (100°F)	3 concrete prisms per cement-aggregate combination. Alkalies added to mix water to bring alkali loading of the concrete to 5.25 kg/m³ (8.85 lb/yd³) for portland cement only mixtures; or to proportionally lower alkali loading for mixtures with SCMs.
ASTM C1567, *Potential Alkali-Silica Reactivity of Combinations of Cementitious Materials and Aggregates (Accelerated Mortar-Bar Method)*	To determine the potential for deleterious ASR of cementitious materials and aggregate combinations	Test specimens and conditions similar to ASTM C1260.	At least 3 mortar bars for each cementitious materials and aggregate combination
AASHTO T 380, *Potential Alkali Reactivity of Aggregates and Effectiveness of ASR Mitigation Measures (Miniature Concrete Prism Test, MCPT)*	To determine the potential ASR susceptibility of aggregates and potential for deleterious ASR of cementitious materials and aggregate combinations	51-mm × 51-mm × 285-mm (2-in. × 2-in. × 11¼-in.) concrete prisms stored 1 N NaOH solution at 60°C (140°F)	3 concrete prisms per cement-aggregate combination. Alkalies added to mix water to bring alkali loading of the concrete to 5.25 kg/m³ (8.85 lb/yd³) for portland cement only mixtures; or to proportionally lower alkali loading for mixtures with SCMs.

DURATION OF TEST	MEASUREMENT	CRITERIA	COMMENTS
Short duration – visual examination does not involve long test periods	Particle characteristics, like shape, size, texture, color, mineral composition, physical condition	Identification of potentially deleterious constituents	Usually includes optical microscopy; may include other analytical tests
Short duration – includes preparation of samples and visual and microscopic examination	Orientation and geometry of cracks. Presence of ASR gel.	See measurement – this examination determines if ASR reactions have taken place and their effects upon the concrete. Used in conjunction with other tests	Specimens can be examined with stereo microscopes, polarizing microscopes, metallographic microscopes, and SEM
Immediate results	Intensity of fluorescence	Lack of fluorescence	Identifies small amounts of ASR gel whether they cause expansion or not. Opal, a natural aggregate, and carbonated paste can indicate – interpret results accordingly. Test must be supplemented by petrographic examination and physical tests for determining concrete expansion
Immediate results	Color of stain	Dark pink stain corresponds to ASR gel and indicates an advanced state of degradation	
16 days (14 days immersion after 2 days of curing)	Length change	See ASTM C1778: If bars expand more than 0.10% at 14 days, indicative of potentially reactive aggregate.	Rapid method used to assess aggregate reactivity. Test is quite aggressive and many aggregates will fail this test but will not be classified by ASTM C1293 as potentially reactive or perform well in the field (Thomas and others 2006). There are also reported cases of coarse aggregates that pass this test but classify as reactive in ASTM C1293 or exhibit expansion in field concrete (Folliard and others 2006).
1 year to evaluate aggregates; 2 years to evaluate mitigation strategies	Length change	Per ASTM C1778, aggregate is potentially deleteriously reactive if expansion equals or exceeds 0.04% at 1 year or 0.04% at 2 years if mitigation using SCMs is being evaluated. AASHTO R 80 includes provisions for determining minimum dosages of lithium-based admixtures to prevent deleterious expansion.	Preferred method of assessment. Best represents the field. Long test durations required, but may not be practical. Use as a supplement to ASTM C295, ASTM C1260, and ASTM C1567. Similar to CSA A23.2-14A
		Per ASTM C1778: minimum SCM content resulting in less than 0.10% expansion at 14 days	Rapid alternative to ASTM C1293, although with less certainty. Allows for evaluation of SCMs and determination of minimum dosage to reduce the potential for deleterious expansion due to ASR.
Varies: 56 days for most aggregates; 84 days for some slowly reacting aggregates. SCM mitigation 56 days.	Length change	See AASHTO T 380 for details. Expansion at 56 days of 0.030% or less is generally considered non-reactive if 2-week average expansion from 8 to 12 weeks is 0.10% or less. Expansion of less than 0.020% at 56 days indicates effective SCM content for mitigation.	New method attempting to accelerate the ASTM C1293 test procedure.

Mitigation of ASR in Hardened Concrete. Structures and pavements that carry the potential for ASR damage should be monitored on a regular basis for signs of ASR. Typically, this process will begin with a visual condition survey (see ACI 201.1R 2008) and progress forward if any symptoms are found. If ASR-related expansion has been identifi in concrete, the options for remediation are limited to treating the symptoms, rather than the causes. Efforts to treat ASR-damaged concrete to mitigate further damage have been unsuccessful, because the alkalies and reactive aggregate components are found throughout the concrete; attempts to inject lithium solutions, for example, have not been successful due to the low permeability of concrete. One example of treating the symptoms is sawing movement joints in concrete to create room for expansion, which may provide some stress relief. See Fournier and others (2010) for a detailed explanation of diagnosing and mitigating ASR in hardened concrete. Landgren and Hadley (2002) provide techniques to control popouts caused by ASR.

Alkali-Carbonate Reaction. Reactions observed with certain argillaceous dolomitic rocks are associated with alkali-carbonate reaction (ACR). Reactive rocks usually contain crystals of dolomite, typically between 10 μm and 50 μm, scattered in, and surrounded by, a finer-grained matrix of calcite and clay. Calcite is one of the mineral forms of calcium carbonate; dolomite is the common name for calcium-magnesium carbonate. Argillaceous dolomitic limestone contains calcite and dolomite with appreciable amounts of clay and can also contain small amounts of reactive silica. ACR is relatively rare because aggregates susceptible to this reaction are typically deemed unsuitable for use in concrete for reasons other than ACR including strength potential.

Mechanism of ACR. Although there is debate about the mechanisms of ACR, some researchers attribute the expansion to dedolomitization, or the breaking down of dolomite (Hadley 1961). Concrete that contains dolomite and has expanded also contains brucite [magnesium hydroxide, $Mg(OH)_2$], which is formed by dedolomitization. Dedolomitization proceeds according to Equation 11-1 (Ozol 2006).

The dedolomitization reaction and subsequent crystallization of brucite may cause considerable expansion (Figure 11-28). Whether dedolomitization causes expansion directly or indirectly, it is usually a precursor to other expansive processes (Tang and others 1994).

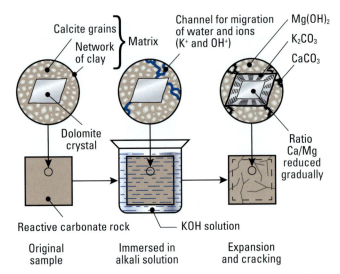

FIGURE 11-28. Schematic diagram of the mechanism of ACR. A dolomite crystal combines with alkalies in solution to form brucite, and potassium and calcium carbonates (after Tang and others 1987).

Test Methods for Identifying the Potential for ACR Distress. The four test methods commonly used to identify potentially alkali-carbonate reactive aggregate are:

- Petrographic examination (ASTM C295, *Standard Guide for Petrographic Examination of Aggregates for Concrete*);

- Rock cylinder method [ASTM C586, *Standard Test Method for Potential Alkali Reactivity of Carbonate Rocks as Concrete Aggregates* (Rock-Cylinder Method)];

- Concrete prism test (ASTM C1105, *Standard Test Method for Length Change of Concrete Due to Alkali-Carbonate Rock Reaction*) or C1293; and

- Chemical composition (CSA A23.2-26A, *Determination of Potential Alkali-Carbonate Reactivity of Quarried Carbonate Rocks by Chemical Composition*).

See ASTM C1778 and AASHTO R 80 for detailed information.

Materials and Methods to Control ACR. The best preventative measure has been to avoid the use of ACR-susceptible aggregates. Alkali reactivity of carbonate rocks is not usually dependent upon its clay mineral composition (Hadley 1961). ACR-susceptible aggregate has a specific texture that is readily identified by petrography.

Concrete with low-alkali loadings, use of SCMs, and lithium solutions are generally not found to be effective in controlling

$CaMgCO_3$	+	(Na or K)OH	→	$Mg(OH)_2$	+	$CaCO_3$	+	K_2CO_3	+	(Na or K)OH
dolomite		alkali hydroxide in solution		brucite		calcium carbonate		potassium carbonate		alkali hydroxide in solution

EQUATION 11-1.

expansive ACR (Swenson and Gillot 1964, Rogers and Hooton 1992, Thomas and Innis 1998, Shehata and others 2009, and Wang and others 1994).

Carbonation

Carbonation is a process by which carbon dioxide, typically from the atmosphere, penetrates concrete and reacts with the various hydration products, such as calcium hydroxide, to form calcium carbonate (Verbeck 1958).

Carbonation may also occur in freshly placed, unhardened concrete. This carbonation can result in a soft, chalky surface called dusting. Dusting usually takes place during cold-weather concreting when there is an unusual amount of carbon dioxide in the air due to unvented heaters or gasoline-powered equipment operating in an enclosure (see Chapter 16).

Mechanism of Carbonation. Concrete may undergo carbonation from CO_2 in the atmosphere or in carbonated water. The CO_2 reacts with calcium hydroxide and other sources of calcium in the concrete to form calcium carbonate, reducing the pH. A high pH protects the reinforcement from corrosion by maintaining a passive layer on the surface of the steel. In addition, the ability of hydrated cement paste to bind chloride ions increases with pH (Page and Vennesland 1983). See discussion of carbonation test methods in Chapter 20.

The progression of carbonation depends on the following:

- concrete composition (type and amount of cement and SCMs),
- permeability (water-cement ratio, curing),
- exposure conditions during carbonation (relative humidity, moisture content of the concrete element, CO_2 content of the air), and
- duration of exposure.

The rate of carbonation ingress into concrete is most rapid at intermediate relative humidities, around 50%. Since the reaction involves the dissolution of CO_2 in water, some moisture must be present. However, in very wet concrete the transport of CO_2 is actually slower because its solubility in water is limited (Herholdt and others 1979); its rate of diffusion in water is four orders of magnitude slower than in air (Neville 1996). Therefore, the carbonation rate of water-saturated concrete is negligible.

The time before the carbonation front reaches the reinforcement for a given concrete cover impacts the service life of concrete, if the moisture and oxygen in the carbonated area are sufficient to trigger reinforcement corrosion. The lower pH begins to disrupt the passivation layer around the reinforcement, allowing corrosion to begin. However corrosion is typically slow as oxygen and possibly moisture still has to diffuse through the cover concrete.

The depth of carbonation, d_c, is approximately linear to the square root of the time of carbonation, t_c, as shown in Equation 11-2:

$$d_c = d_0 + \alpha\sqrt{t_c}$$

EQUATION 11-2.

where:

d_0 = a parameter that depends on curing and early exposure. It becomes smaller with later start of carbonation t_c.

α = a factor that contains parameters resulting from concrete composition, curing, and exposure conditions.

The equation is conservative. Carbonation is typically slower than predicted by Equation 11-2 if the concrete element is at least occasionally exposed to moisture (Neville 1996).

Materials and Methods to Resist Carbonation. The amount of carbonation is significantly increased in concrete with a high w/c or w/cm ratio, low cementitious materials content, short curing period, low strength, and highly porous and permeable paste. Ensuring that the concrete exhibits sufficiently low permeability best reduces the rate of carbonation. To reduce early carbonation, concrete needs to be protected from drying during the curing period. The concentration of CO_2 in the atmosphere averages about 0.04% (400 ppm) (Blunden and Arndt 2017); local concentrations can be higher in enclosed structures such as parking garages and tunnels.

Carbonation of portland cement concrete converts the calcium hydroxide into calcium carbonate, which tends to reduce permeability and slightly increase strength (Taylor 1997). For concrete containing SCMs, this effect is reversed, and the permeability of the surface layer increases somewhat after carbonation. Concrete containing moderate to high levels of fly ash or slag may carbonate more rapidly than portland cement concrete of the same w/cm, especially if the concrete is not properly cured (Osborne 1989 and Thomas and others 2000). Because concretes made with high volumes of SCMs typically gain strength more slowly, longer curing may be needed for sufficient resistance to carbonation.

The depth of carbonation in good-quality, well-cured concrete is generally of little practical significance provided the embedded steel has adequate concrete cover. Finished surfaces tend to experience less carbonation than formed surfaces. Carbonation of finished surfaces is often observed to a depth of 1 mm to 10 mm (0.04 in. to 0.4 in.) and for formed surfaces, between 2 mm and 20 mm (0.1 in. and 0.9 in.) after several years of exposure, depending on the concrete properties, ingredients, age, duration of curing, and environmental exposure (Campbell and others 1991). ACI 201.2R-16 has more information on atmospheric and water carbonation and ACI 318-19 provides reinforcing steel cover requirements for different exposures.

Corrosion

Concrete protects embedded steel from corrosion through its highly alkaline nature. The high pH environment in concrete (usually greater than 12.5) causes a passive and non-corroding protective oxide film to form around the steel. However, over time, carbonation can lower the concrete pH disrupting the film and the chloride ions from deicers or seawater can destroy or penetrate this film allowing corrosion to begin. Corrosion of steel is an expansive chemical process – the products of corrosion, various forms of rust, result in significant volume increases over time, leading to internal stresses and eventual spalling of the concrete over reinforcing steel (Figure 11-29).

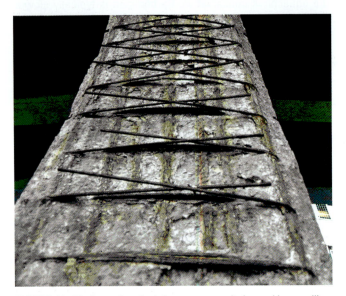

FIGURE 11-29. Corrosion of reinforcement results in cracking, spalling, and discoloration of the concrete (courtesy of R.D. Hooton).

Mechanism of Corrosion. The corrosion of steel reinforcement is an electrochemical process. For corrosion to take place, all elements of a corrosion cell must be present: oxygen, an anode, a cathode, an electrolyte, and an electrical connection. Once the chloride corrosion threshold of concrete, about 0.15% water-soluble chloride by mass of cement (Whiting 1997), is reached at the surface of the steel, an electric cell is formed along the steel or between steel bars and the electrochemical process of corrosion begins (Figure 11-30).

Some steel areas along the bar act as the anode, discharging current in the electric cell; from there the iron goes into solution. Steel areas that receive current are the cathodes where hydroxide ions are formed. The iron and hydroxide ions form iron hydroxide. The iron hydroxide further oxidizes to form rust or other iron (ferrous or ferric) oxides and hydroxides as illustrated in Figure 11-31. The volume of the final product may be more than six times the volume of the original iron, which can generate stresses that result in cracking and spalling of the concrete. The cross-sectional area of the steel can also be significantly reduced.

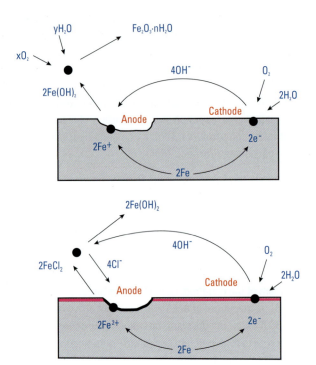

FIGURE 11-30. Dissolution of the iron takes place at the anode. (top) The ferrous ions combine with hydroxyl ions, oxygen, and water to form various corrosion products. (bottom) Chloride ions act as catalysts to the corrosion of iron. The localized corrosion that takes place is known as pitting corrosion (Detwiler and Taylor 2005).

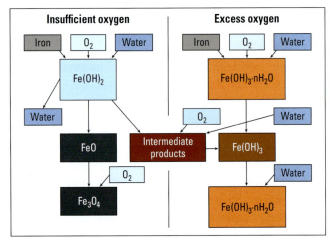

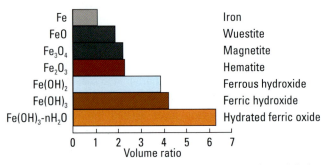

FIGURE 11-31. (top) The specific corrosion products formed during corrosion of iron depend on the availability of oxygen. (bottom) The volume of iron corrosion products may be more than six times that of the original steel (adapted from Herholdt and others 1979).

Once corrosion starts, the rate of steel corrosion is influenced by the concrete's electrical resistivity, moisture content, and the rate at which oxygen migrates through the concrete to the steel. Chloride ions alone can also penetrate the passive film on the reinforcement. They combine with iron ions to form a soluble iron chloride complex that carries the iron into the concrete for later oxidation (rust) (Whiting 1997, Taylor and others 2000, and Whiting and others 2002).

Various factors affect the rate of corrosion of steel. These include:

- Water – Water both participates in the cathodic reaction and transports the chloride ions.

- Oxygen – Corrosion is two orders of magnitude slower when oxygen is not present.

- The pH of the concrete – Illustrated in Figure 11-32, the pH affects the rate of corrosion. Below a pH of about 11, the passive layer of concrete begins to break down. Below a pH of about 4, the protective film on the steel dissolves.

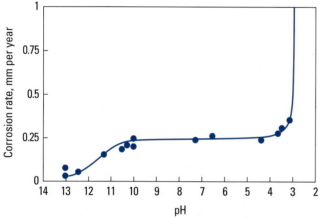

FIGURE 11-32. The pH of the surrounding medium affects the rate of corrosion. Below a pH of about 4, the passive oxide layer dissolves. The pH of concrete pore solution is normally greater than 12.5 (adapted from Uhlig and Revie 1985).

- Chlorides – Chloride ions act as catalysts to the corrosion reaction by breaking down the passive layer. The localized microcell corrosion that takes place under these circumstances is called pitting.

- Temperature – The rate of corrosion for a given concentration of oxygen approximately doubles for every 30°C (54°F) increase in temperature (Uhlig and Revie 1985).

- Electrical resistivity of the concrete – In some installations, stray currents may induce corrosion in the steel, thereby artificially driving the reactions. The presence of dissimilar metals embedded in concrete can also form a corrosion cell. Concrete with a high electrical resistivity will reduce the rate of corrosion by reducing the rate of ion transport.

- Permeability/Diffusivity of the concrete – Oxygen and chloride ions must reach the anode through the concrete cover. Thus, the diffusivity of the concrete to these ions affects

the rate of corrosion. In addition, the availability of oxygen at the cathode, and the availability of oxygen and water at the anode (which will affect the degree of expansion of the corrosion product) are a function of the permeability of the concrete.

- Cathode-to-anode area – When the cathode is much larger than the anode, localized corrosion proceeds rapidly, resulting in loss of area of the steel cross section. When the areas of the cathode and anode are approximately equal, corrosion proceeds much more slowly and is distributed more evenly.

Nonferrous Metals. Nonferrous metals are also frequently used for construction in contact with portland cement concrete. Metals such as zinc, aluminum, and lead – and alloys containing these metals – may be subject to corrosion when embedded in, or in surface contact with, concrete. Galvanic corrosion will occur if aluminum and steel or other metals of dissimilar composition are both embedded in concrete and in contact with each other. See PCA (2002), Kerkhoff (2007), and ACI 201.2R-16 for more information on dissimilar metal corrosion and nonferrous metals in concrete.

Corrosion Testing. Deductions concerning corrosion activity of embedded steel can be made using the information obtained from ASTM C876, *Test Method for Corrosion Potentials of Uncoated Reinforcing Steel in Concrete*.

Acid-soluble chloride content of concrete is measured in accordance with ASTM C1152, *Test Method for Acid-Soluble Chloride in Mortar and Concrete*. Testing to determine water-soluble chloride ion content should be performed in accordance with ASTM C1218, *Test Method for Water-Soluble Chloride in Mortar and Concrete*. ASTM C1524, *Test Method for Water-Extractable Chloride in Aggregate (Soxhlet Method)*, can be used to evaluate aggregates that contain a high amount of naturally occurring chloride. ASTM G109, *Standard Test Method for Determining Effects of Chemical Admixtures on Corrosion of Embedded Steel Reinforcement in Concrete Exposed to Chloride Environments*, can be used to determine the effects of admixtures on the corrosion of embedded steel reinforcement in concrete exposed to chloride environments.

Where stray currents are expected, the electrical conductivity of the concrete should be explicitly addressed by limiting the allowable charge passed as measured by ASTM C1202, *Standard Test Method for Electrical Indication of Concrete's Ability to Resist Chloride Ion Penetration* (AASHTO T 277).

Materials and Methods to Control Corrosion. Concrete materials and mixture proportions should be designed to lower corrosion activity and to optimize protection of embedded steel. To maximize chloride (corrosion) resistance, reduce permeability by specifying a maximum w/cm of 0.40 or less (Stark 1989) and at least seven days of moist curing. Judicious use of one or more SCM, combined with extended moist curing, can effectively reduce the permeability, diffusivity, and electrical conductivity of concrete. Cements with high C_3A contents, slag cement, or alumina-bearing pozzolans are also frequently used due to their effectiveness in binding chloride ions so that they do not migrate further into the concrete. This binding effect is particularly important in applications, such as parking garages, bridge decks, and marine construction, where the reinforcing steel is vulnerable to chloride-induced corrosion.

The threshold level at which corrosion starts in normal concrete with no inhibiting admixture is about 0.15% water-soluble chloride ion (0.20% acid-soluble) by weight of cement. Admixtures, aggregate, and mixing water containing chlorides should be avoided. In any case, the total acid-soluble chloride content of the concrete should be limited to a maximum of 0.08% and 0.20% by mass of cement for prestressed and reinforced concrete, according to ACI 201.2R-16 and ACI 222R, *Guide to Protection of Metals in Concrete Against Corrosion* (2019). ACI 318-19 bases the chloride limit on water-soluble chlorides, with maximum limits of 0.06% for prestressed concrete and maximum limits for reinforced concrete that vary with the exposure class: 1.00% for C0, 0.30% for C1, and 0.15% for C2.

Increasing the concrete cover depth extends the time required for chlorides to reach the depth of reinforcing steel. Requirements for concrete cover based on exposure and application are shown in Table 11-8. Sufficient concrete cover must be provided for reinforcement where the surface is to be exposed to corrosive substances. It is good practice to increase the concrete cover over the reinforcing steel above the amount specified in ACI 318-19. Additional cover extends the time that corrosive chemicals, such as chlorides, need to reach the reinforcing steel. ACI 201.2R-16 recommends a minimum cover of 40 mm (1½ in.) and preferably at least 50 mm (2 in.) for concrete in moderate-to-severe corrosion environments. Oesterle (1997) and Stark (1989) confirm the need for 65 mm to 75 mm (2½ in. to 3 in.) of cover over reinforcement to provide corrosion protection. Some engineers specify 90 mm (3½ in.) or more of concrete cover over steel in concrete exposed to chlorides or other corrosive solutions. However, large depths of cover on the tension side of concrete members can lead to excessive crack widths because they lose the benefit of the top mat of reinforcement in terms of controlling thermal and drying shrinkage.

Several types of reinforcement are resistant to corrosion, including epoxy-coated reinforcing steel, stainless steel, nickel-plated steel, galvanized steel, and fiber-reinforced plastic (such as GFRP). For more information on types of reinforcement see Chapter 8.

Corrosion inhibitors. Corrosion inhibitors such as calcium nitrite are used as an admixture to reduce corrosion. Organic-based corrosion inhibitors, based on amine and amine/fatty ester derivatives, are also available (Nmai and others 1992 and Berke and others 2003). See Chapter 6 for more on corrosion inhibitors.

Cathodic protection. Cathodic protection reverses the natural electron flow through concrete and reinforcing steel by inserting an anode in the concrete. This forces the steel to act as the cathode when electrically charging the system, as with an impressed current. Since corrosion occurs where electrons leave the steel, the reinforcement cannot corrode, as the steel is constantly receiving electrons.

The methods noted above can be combined with other corrosion protection methods. Some additional protective strategies stop or reduce chloride-ion or chemical penetration at the concrete surface and include concrete surface sealers, water repellents, surfacings, and overlays. Materials commonly used include silanes, siloxanes, methyl methacrylates, epoxies, and other compounds. Latex-modified and polymer concrete are often used in overlays to reduce chloride-ion or chemical penetration. Impermeable interlayer membranes (primarily used on bridge decks), prestressing for crack control, or polymer impregnation are also available to help protect reinforcement.

Using more than one protection method simultaneously can result in significant savings in maintenance costs. For example, using epoxy-coated rebar and an additional protection system, such as a corrosion-inhibitor or supplementary cementitious materials in the concrete mixture, can further protect the steel. Protection systems delay the time to corrosion, and likely reduce corrosion rates, once initiated.

Sulfate Attack

High amounts of sulfates in soil or water (for example calcium sulfate, sodium sulfate, and magnesium sulfate) can attack concrete that is not properly designed for sulfate resistance by reacting with hydrated compounds in the hardened cement paste. These reactions can induce sufficient pressure to disrupt the cement paste, resulting in disintegration of the concrete (loss of paste cohesion and strength).

Figure 11-33 illustrates this point, showing 7-year-old beams with multiple concrete mixtures partially buried in a sulfate-rich soil and ponded with a highly concentrated sulfate solution, while subjected to wetting and drying. The beams that are in better condition have low water-cementitious materials ratios and used sulfate resistant cements. The inset shows two of the beams tipped on their side to expose deterioration at decreasing levels with increased depth and moisture level. Deterioration of the concrete exposed to sulfates in this study was due to a combination of chemical sulfate attack and physical salt crystallization (Stark 2002). See also **Salt Crystallization or Physical Salt Attack** for the related role of sulfates on physical salt attack.

TABLE 11-8. Summary of Concrete Cover Requirements*

CONCRETE EXPOSURE	TYPE OF MEMBER	REINFORCEMENT TYPE / BAR SIZES metric (inch-pound)	CAST-IN-PLACE		PRECAST, MANUFACTURED IN A PLANT
			NONPRESTRESSED	PRESTRESSED	PRESTRESSED OR NONPRESTRESSED
			MINIMUM SPECIFIED COVER DEPTH, mm (in.)		
Cast against and permanently in contact with earth	All	All	75 (3)	75 (3)	–
Exposed to weather or in contact with ground	Slabs, Joists, and Walls	All	–	25 (1)	–
	Walls	No. 43 and No. 57 (No. 14 and No. 18) bars; tendons larger than 40 mm (1½ in.) diameter	–	–	40 (1½)
		No. 36 (No. 11) bars and smaller; MW200 and MD200 (W31 and D31) wire and smaller; tendons and strands 40 mm (1½ in.) diameter and smaller	–	–	20 (¾)
	All other	No. 19 through No. 57 (No. 6 through No. 18) bars	50 (2)	–	–
		No. 16 (No. 5) bar, MW200 or MD200 (W31 or D31) wire, and smaller; tendons and strands 16 mm (⁵⁄₈ in.) diameter and smaller	40 (1½)	–	30 (1¼)
		No. 43 and No. 57 (No. 14 and No. 18) bars; tendons larger than 40 mm (1½ in.) diameter	–	–	50 (2)
		No. 19 through No. 36 (No. 6 through No. 11) bars; tendons and strands larger than 16 mm (⁵⁄₈ in.) diameter through 40 mm (1½ in.) diameter	–	–	40 (1½)
Not exposed to weather and not in contact with ground	Slabs, Joists and Walls	No. 43 and No. 57 (No. 14 and No. 18) bars	40 (1½)	–	30 (1¼)
		Tendons larger than 40 mm (1½ in.) diameter	–	–	
		Tendons and strands 40 mm (1½ in.) diameter and smaller			20 (¾)
		No. 36 bar (No. 11), MW200 or MD200 (W31 or D31) wire, and smaller	20 (¾)	–	16 (⁵⁄₈)
		All	–	20 (¾)	–
	Beams, columns, pedestals, and tension ties	Primary reinforcement	40 (1½)	40 (1½)	Greater of d_b† and 16 mm (⁵⁄₈) and need not exceed 40 mm in (1½ in.)
		Stirrups, ties, spirals and hoops		25 (1)	10 (³⁄₈)

*Adapted from ACI 318-19. Does not include deep foundation requirements; see ACI 318-19 for complete requirements.
†d_b is the nominal diameter of bar, wire or prestressing strand, mm (in.)

FIGURE 11-33. Concrete beams after seven years of exposure to sulfate-rich wet soil in a Sacramento, California, test plot. The inset shows two of the beams tipped on their side to expose deterioration at decreasing levels with increased depth and moisture level (Stark 2002).

Sulfate ions may be found in soil and groundwater in North America, most notably in the Western United States and the Canadian Prairie provinces, and may also be present in other locations throughout the world. Sulfates are also found in seawater, in some industrial environments, and in sewers.

Sulfate attack can lead to loss of strength, expansion, spalling of surface layers, and ultimately the disintegration of concrete. Flowing water is more aggressive than stagnant water, since new sulfate ions are constantly being transported to the concrete for chemical reaction.

Mechanisms of Sulfate Attack. After infiltrating the concrete, sulfate ions react chemically with calcium hydroxide and the hydration products of C_3A, forming gypsum and ettringite in expansive reactions. Equations 11-3 and 11-4 provide example chemical equations for these reactions:

Magnesium sulfate attacks in a manner similar to sodium sulfate but also forms brucite (magnesium hydroxide). Brucite forms primarily on the concrete surface; it consumes calcium hydroxide, lowers the pH of the pore solution, and then decomposes the calcium silicate hydrates (Santhanam and others 2001). Depending on the cation(s) involved, sulfate attack can be more or less aggressive; with magnesium sulfate attack being the most aggressive and calcium sulfate the least aggressive.

Materials and Methods to Control Sulfate Attack. The first step to designing a sulfate-resistant concrete is to determine the exposure class for sulfates and use the lowest water-cementing materials ratio required (or below). ACI 318-19 defines four exposure classes, S0 to S3, for concrete, depending on the sulfate content in the soil or groundwater (S3 is the most severe) shown in Table 11-3.

Cements specially formulated to improve sulfate resistance (Chapter 2), such as ASTM C150 (AASHTO M 85) Types II, II(MH), or V cements, ASTM C595 (AASHTO M 240) cements designated with (MS) or (HS), or ASTM C1157 Types MS or HS, should be used in sulfate exposures (see Table 11-9). ACI 318-19 requires at least moderately sulfate resistant cements for S1 exposures and highly sulfate resistant cements for S2 exposures. For S3 exposures ACI 318-19 has two options:

- maximum w/cm of 0.45, using highly sulfate resistance cements with at least the minimum amounts of pozzolans or slag cements that have been shown to improve sulfate resistance by ASTM C1012 testing or service record; or

- maximum w/cm of 0.40, using a highly sulfate resistant cement. The optional ASTM C452 requirement in ASTM C150 (AASHTO M 85) for Type V cements applies.

ACI 318-19 also permits alternative combinations of cementitious materials to be qualified for use in S1, S2, or S3 exposures using ASTM C1012, with comparable limits to those in ASTM C595 (AASHTO M 240), except that S3 exposures using Option (1) have a 0.10% expansion limit at 18 months.

Figure 11-34 shows average values for concretes containing a wide range of cementitious materials, including cement Types I, II, V, blended cements, pozzolans, and slags. See Figure 11-35 for rating illustration and a description of the concrete beams (Stark 2002).

Sulfate-resistant cements alone are not adequate to resist most forms of sulfate attack. It is essential to limit the ability of the sulfates to enter the concrete. This is accomplished by reducing the permeability of the concrete by minimizing the w/cm (Stark 2002). Even Type V cement concrete cannot withstand a severe

$$4\,CaO \cdot Al_2O_3 \cdot SO_3 \cdot 12H_2O \quad + \quad 2\,(CaO \cdot SO_3 \cdot 2\,H_2O) \quad + \quad 16\,H_2O \quad \rightarrow \quad 6\,CaO \cdot Al_2O_3 \cdot 3\,SO_3 \cdot 32\,H_2O$$

calcium monosulfoaluminate   sulfate (shown as gypsum)   water   ettringite

EQUATION 11-3.

$$CaO \cdot H_2O \quad + \quad Na_2O \cdot SO_3 \quad \rightarrow \quad CaSO_4 \cdot 2H_2O \quad + \quad Na_2O \cdot H_2O \quad + \quad H_2O$$

calcium hydroxide   sulfate (shown as sodium sulfate)   gypsum   sodium hydroxide   water

EQUATION 11-4.

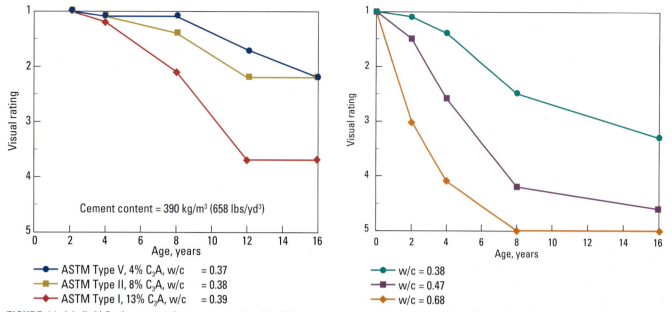

FIGURE 11-34. (left) Performance of concretes made with different cements in sulfate soil. Type II and Type V cements have lower C₃A contents that improve sulfate resistance. (right) Improved sulfate resistance results from low water-to-cementitious materials ratios as demonstrated over time for concrete beams made with a variety of cementitious materials and exposed to sulfate soils in a wetting and drying environment (Stark 2002).

FIGURE 11-35. Specimens used in the outdoor sulfate test plot in Sacramento, California, are 150-mm × 150-mm × 760-mm (6-in. × 6-in. × 30-in.) beams. A comparison of ratings is illustrated: (left) a rating of 5 for 12-year-old concretes made with Type V cement and a water-to-cement ratio of 0.65; and (right) a rating of 2 for 16-year-old concretes made with Type V cement and a water-to-cement ratio of 0.37 (Stark 2002).

sulfate exposure if the concrete has a high water-cementitious materials ratio. As noted in Table 11-9, the use of SCMs (as ingredients in blended cements or as concrete ingredients) is often beneficial to sulfate resistance of concrete, however some Class C fly ashes may not be beneficial at normal fly ash contents (Thomas 2007). Silica fume, Class F fly ashes, slag cements, metakaolin and many natural pozzolans have proven beneficial to sulfate resistance by reducing the permeability of concrete (with proper curing), increasing the strength of concrete, and in reducing calcium hydroxide contents. Testing by ASTM C1012 is recommended to verify appropriate SCM types and contents are used. See Chapter 3 for more information on SCMs and sulfate resistance.

Type V cement, like other portland cements, is not resistant to acids and other highly corrosive substances. In very severe sulfate exposures, a waterproof barrier protecting the concrete

may be beneficial (Stark 2002 and Kanare 2008). It is also important to provide required curing measures during concrete placement (see Chapter 17).

Sulfate attack can also occur when the sulfate is internally supplied to the concrete. ACI 221R, *Guide for Use of Normal Weight and Heavyweight Aggregates in Concrete* (1996) reports that the presence of sulfates occurs in a variety of aggregate types, either as an original component of the aggregate, or due to the oxidation of sulfides originally present. The most common form of sulfate in aggregates is gypsum ($CaSO_4 \cdot 2H_2O$). Gypsum may occur as a coating on sand and gravel, as a component of some sedimentary rock, or in weathered slags. Aggregates made from recycled building materials may contain sulfates in the form of contamination from plaster or gypsum wallboard. It is difficult to eliminate the gypsum present. On sieving, these particles break apart, becoming part of the sand fraction.

TABLE 11-9. Types of Cementitious Materials Required for Concrete Exposed to Sulfate in Soil or Water*

EXPOSURE CLASS			CEMENTITIOUS MATERIALS REQUIREMENTS[†]			MAXIMUM W/CM
			ASTM C150	ASTM 595[§]	ASTM C1157	
S0	Negligible		NSR	NSR	NSR	None
S1	Moderate, Seawater		II or II(MH)	Types with MS designation	MS	0.50
S2	Severe		V	Types with HS designation	HS	0.45
S3[‖]	Very severe	Option 1	V with pozzolan or slag cement[¶]	Types with HS designation with pozzolan or slag cement[¶]	HS with pozzolan or slag cement[¶]	0.45
		Option 2	V[‡]	Types with HS designation	HS	0.40

* Adapted from USBR (1988), ACI 201.2R-16, ACI 318-19, and PCA recommendations. Exposure classes are defined in Table 11-3. "NSR" indicates no special requirements for sulfate resistance.

† Cementitious materials suitable for use in higher exposure classes may be used in lower exposure classes. For example, a Type V cement may be used in S1 or S0 exposures. Alternative combinations of cementitious materials can be qualified using ASTM C1012: Moderate exposure (S1) – 0.10% at 6 months; Severe exposure (S2) – 0.05% at 6 months or 0.10% at 12 months; Very Severe exposure (S3) – Option 1: 0.10% at 18 months or (Option 2) 0.05% at 6 months or 0.10% at 12 months. For example, a Type I cement combined with SCMs can be used in S2 or S3 exposures if ASTM C1012 expansion limits are met.

‡ When Type V cement is used as the sole cementitious material in Exposure Class S3, the optional sulfate resistance limit in ASTM C150 based on ASTM C452 shall be specified.

§ When qualified by C1012 testing, with limits specified in ASTM C595, Type IL, IP, IS(<70) and IT(S<70) cements are suitable for use in sulfate exposures are identified with MS and HS designations. Type IS(≥70) cement is not used for structural concrete applications.

‖ ACI 318-19 requires SCMs (tested to verify improved sulfate resistance) with Type V, IL(HS), IP(HS), IS(<70)(HS), and HS cements for Option 1 exposure Class S3.

¶ Pozzolans and slag cement that have been determined by testing according to ASTM C1012 or by service record to improve sulfate resistance may also be used in concrete.

In some quarries, it is necessary to reject the entire fraction of aggregate smaller than 2.36 mm (No. 8) in size because it contains 1% to 2% SO_3 by total mass of aggregate.

Salt Crystallization or Physical Salt Attack

Similar to natural rock formations like limestone, concretes are susceptible to distress caused by salt crystallization. These salts may or may not contain sulfates or react with the hydrated compounds in concrete. Examples of salts known to cause weathering of exposed concrete include sodium carbonate and sodium sulfate. Laboratory studies have also related saturated solutions of calcium chloride and other salts to concrete deterioration. The greatest damage occurs with drying of saturated solutions of these salts, often in an environment with specific cyclical changes in relative humidity and temperature that alter mineralogical phases. Both aggregate particles and cement paste can be attacked by salts, and similar distress can occur with porous rocks, brick, and masonry.

Mechanism of Salt Attack. Sakr and others (2020), Haynes and others (1996), and Folliard and Sandberg (1994), describe the mechanisms of physical attack on concrete by salts present in groundwater. Groundwater with deleterious ions enters the concrete pores by capillary action and diffusion. When pore water evaporates from above-ground concrete surfaces, those ions concentrate at an evaporative front within the concrete until salts crystals precipitate, sometimes generating pressures large enough to cause cracking or flaking of the surface.

Changes in ambient temperature and relative humidity cause some salts to undergo cycles of dissolution and crystallization, or hydration-dehydration (Figure 11-36). When either of these processes is accompanied by volumetric expansion, repeated cycles can cause deterioration of concrete similar to that caused by cycles of freezing and thawing.

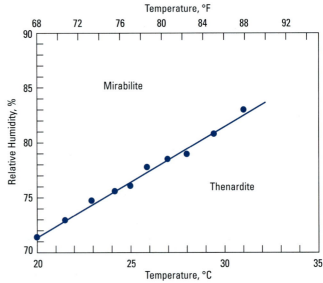

FIGURE 11-36. One form of physical salt attack involves a transition between two forms of sodium sulfate: thenardite and mirabilite. Cycles of relative humidity and temperature across the transition line lead to phase transformations and development of expansive stresses (after Sperling and Cooke 1985).

Physical attack by sulfate salts can be distinguished from conventional, chemical sulfate attack, for example, by evaluating the sulfate phases and microstructure of the concrete. Chemical sulfate attack can be evidenced by significant amounts of ettringite or gypsum, as well as the characteristic decalcification of the paste and cracking due to expansion. In physical sulfate attack, damage in the form of scaling is usually limited to the exterior surface of the concrete just above the soil line (Figure 11-37); the concrete is typically not affected below the surface. ACI 201.2R (2016) has more detail on physical salt attack, and note that it is also referred to as salt weathering, salt damp, salt hydration distress and salt crystallization.

Materials and Methods to Control Physical Salt Attack. The ideal way to prevent damage by salt crystallization is to prevent the salt-laden water from entering the concrete. Concrete structures exposed to salt solutions should have a low water-cementing materials ratio (0.40 to 0.45 maximum depending on severity) to reduce permeability. Where conditions conducive to salt crystallization exist – that is, where the climate is arid but the local groundwater table is near the surface – a barrier on the surface in contact with external sulfates is the most effective means to prevent damage.

Coating of the concrete surface, or creating a barrier (such as plastic sheeting) or break between concrete and groundwater (such as an open granular base, Figure 11-38) under the concrete, helps keep the salt-bearing solution out (Kanare 2008). It should be noted that over-watering and over fertilization of lawns and plants located near the concrete can locally raise the elevation of the water table or can introduce salts into the concrete via splashing or incidental contact.

Appropriate landscaping practice (use of native plants, locating lawns and plants away from the concrete) mitigates this problem. Any measure that reduces the permeability of the concrete –

low water-cementitious materials ratio, use of certain SCMs (Sakr and others 2020), and good curing – also reduces the vulnerability of the concrete to salt crystallization damage.

Thaumasite Sulfate Attack

Although extremely rare, thaumasite sulfate attack (TSA) may occur in moist conditions at temperatures usually between 0°C and 10°C (32°F to 50°F) as a result of a reaction between calcium silicate hydrate, sulfate, carbonate, and water (TEG 1999). A small amount of thaumasite can be accommodated in concrete without causing distress; however deleterious thaumasite sulfate attack results from decomposition of the calcium silicate hydrate by the formation of thaumasite. In concretes where deterioration is associated with excess thaumasite formation, cracks can be filled with thaumasite and haloes of white thaumasite are present around aggregate particles. At the concrete/soil interface the surface concrete layer can be soft with complete replacement of the cement paste by thaumasite (Hobbs 2001).

Recommendations for mitigating the potential for TSA generally follow those for other forms of sulfate attack: require concrete to have low permeability and appropriate levels of supplementary cementitious materials.

Delayed Ettringite Formation

Primary and Secondary Ettringite. Ettringite, a form of calcium sulfoaluminate (in oxide notation: $CaO·Al_2O_3·SO_3·32H_2O$), is found in all portland cement paste as a product of hydration. Calcium sulfate sources, such as gypsum, are added to portland cement during final grinding at the cement mill to prevent rapid setting and improve strength development. Sulfate can be also be present in SCMs and some admixtures. Gypsum and other sulfate compounds react with tricalcium aluminate to form ettringite within the first few hours after mixing cement with

FIGURE 11-37. Salt crystallization attack is often observed near the soil line. Here concrete posts have been attacked by sulfate salts near the soil line.

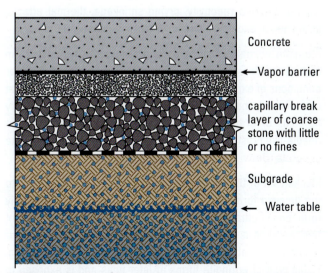

FIGURE 11-38. A capillary break and vapor barrier can reduce the potential for physical salt attack.

water. Most of the sulfate in cement is normally consumed to form ettringite or calcium monosulfoaluminate within 24 hours (Klemm and Miller 1997). At this stage ettringite is uniformly and discretely dispersed throughout the cement paste at a submicroscopic level (less than a micrometer in cross-section). This ettringite is often called primary ettringite.

If concrete is exposed to moisture for long periods (typically many years) ettringite can slowly dissolve and re-form in less confined locations. Upon microscopic examination, innocuous white, acicular (needle-like) crystals of ettringite can be observed lining air voids or cracks. This re-formed ettringite is usually called secondary ettringite (Figure 11-39).

FIGURE 11-39. White secondary ettringite deposits in void. Field width 64 μm.

Concrete deterioration accelerates the rate at which ettringite leaves its original location in the paste to go into solution and then recrystallize in larger spaces such as air voids or cracks. Both water and sufficient space must be present for the crystals to form. Cracks can form due to damage caused by frost action, alkali-aggregate reactivity, drying shrinkage, thermal effects, strain due to excessive stress, or other mechanisms.

Ettringite crystals in air voids and cracks are typically 2 μm to 4 μm in cross section and 20 μm to 30 μm in length. Under conditions of extreme deterioration or long exposure to a moist environment, the ettringite crystals can appear to completely fill voids or cracks. However, secondary ettringite, as large needle-like crystals, should not be interpreted as harmful to the concrete (Detwiler and Powers-Couche 1997).

Mechanism of DEF. Delayed ettringite formation (DEF) refers to the delayed formation of ettringite, in which the normal early formation of ettringite that occurs in concrete cured at ambient temperatures is interrupted as a result of exposure to high-temperatures, above 70°C (158°F), during placement or curing. If this type of ettringite forms at later ages and is exposed to moisture in service, it can lead to paste expansion and damage to the hardened concrete.

Since ettringite is not stable at elevated temperatures, monosulfoaluminate forms instead, even when sufficient sulfate is present. Later, after the concrete cools, ettringite forms from monosulfoaluminate, aluminates, and sulfate initially absorbed on hydration products (C-S-H). These conditions create an expansive reaction in the paste (Famy and Taylor 2001, Scrivener and others 1999, Taylor and others 2001, and Ramlochan and others 2004). This process may take many years for the monosulfoaluminate to convert to ettringite. As a result of an increase in paste volume, separation of the paste from the aggregates is usually observed with DEF. It is characterized by the development of rims around the aggregates (Figure 11-40), sometimes filled with ettringite. An abundance of water (saturation or near saturation of the concrete) is necessary for the formation of ettringite as each mole of ettringite contains 32 moles of water.

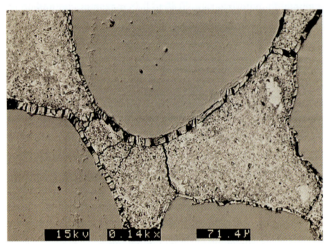

FIGURE 11-40. Delayed ettringite formation (DEF) creates gaps at the paste-aggregate interface. The gap can subsequently be filled with secondary ettringite crystals as shown here (courtesy of Z. Zhang and J. Olek).

The mechanism causing expansion in the paste due to DEF is not fully understood and is still under investigation. Shimada (2005) proposed that DEF depends on the concrete curing temperature as well as on both physical and chemical properties of the system. Several competing factors are at work, including: the amount of ettringite that decomposes, the rate at which ettringite re-forms, the rate at which stress can be relieved through reprecipitation in larger spaces (Ostwald ripening), and the strength of the system to resist expansive stresses. In addition, the onset of DEF-related expansion is associated with reductions in alkalinity (Famy and others 2001, Ramlochan and others 2004) and is therefore sometimes linked with alkali-silica reaction. Graf (2007) found that DEF-related expansion is influenced by the level of moisture exposure. A threshold level of relative humidity for expansion to occur seems to be slightly above 90%. Expansion of a system due to DEF is controlled by the curing temperature, the environment in which it exists, and the chemistry of the system.

DEF Testing. There are no commonly accepted test methods for DEF. The phenomenon is rarely observed in the field. The primary means of controlling DEF is to prevent the concrete from reaching excessive temperatures from heat of hydration or from thermal curing treatments (as used in some precast operations).

Materials and Methods to Control DEF. Because of the risk of delayed ettringite formation, as well as other deleterious effects of elevated temperature on durability, concrete temperatures above 70°C (158°F) should be avoided (Shimada 2005 and Detwiler and Taylor 2005). The heat of hydration itself might raise the internal temperature of concrete above 70°C (158°F), even if no additional energy is applied to the concrete. This is especially the case with mass concrete, particularly if a high cement content is used. Gajda (2007) discusses precautions that can be taken to prevent unacceptably high temperatures occurring during placement and curing of mass concrete. Chapter 18, *Hot Weather Concreting*, also discusses reducing concrete temperatures.

There is some evidence that the use of certain SCMs will help reduce the risk of DEF when concrete may reach excessive temperatures (Ramlochan and others 2003, Ramlochan and others 2004, Thomas and others 2008b). Excerpts from ACI 201.2R (2016) recommendations are in Table 11-10. Refer to PCI (1999 and 2013), NCPA (2020), and ACI 201.2R (2016) for more information on DEF control measures.

TABLE 11-10. Measures for Reducing Potential for DEF in Concrete Exposed to Elevated Temperatures at Early Ages

MAXIMUM CONCRETE TEMPERATURE	PREVENTION MEASURE
≤ 70°C (158°F)	No prevention required.
> 70°C (158°F) and ≤ 85°C (185°F)	Use one of the following: 1. Sulfate resistant portland cement with equivalent alkali contents less than 0.60% by mass and Blaine fineness ≤ 430 m²/kg 2. Portland cement with a 1-day ASTM C109 strength less than or equal to 20 MPa (2850 psi) 3. Any portland cement in combination with the following: • ≥ 25% Class F fly ash • ≥ 35% Class C fly ash • ≥ 35% slag cement • ≥ 5% silica fume in combination with ≥ 25% slag cement • ≥ 5% silica fume in combination with ≥ 20% Class F fly ash • ≥ 10% metakaolin 4. A blended hydraulic cement with the same pozzolan or slag cement content as listed in 3.
> 85°C (185°F)	Not permitted.

Source: Adapted from ACI 201.2R (2016) which cites Ghorab 1980, Thomas 2001, Ramlochan and others 2003, and Thomas and others 2008b.

Pyrrhotite

Pyrrhotite is an iron sulfide-based mineral found in some aggregate sources. Another iron sulfide-based mineral is pyrite. While pyrite is relatively common, it generally does not present a serious distress issue for concrete, although it can cause surface popouts that are unsightly. Pyrrhotite has led to serious deterioration; however, rarely in North America. Distress related to pyrrhotite has been associated with just two aggregate sources: a quarry in Connecticut and one in Quebec.

Mechanisms of Pyrrhotite Distress. Pyrrhotite is an iron sulfide mineral (chemical formula: $Fe_{(1-x)}S$, with x ranging from 0 to 0.125) that appears to slowly oxidize, an expansive reaction, which releases sulfate and sulfuric acid within the interior of concrete (CSA A23 2018). Thus, this distress is due to a combined effects of the expansive reaction, sulfate attack, and acid attack. These simultaneous distresses can lead to severe cracking of the concrete.

Methods to Control Pyrrhotite Distress. Generally accepted guidance to prevent the occurrence of pyrrhotite is limited, and research is ongoing. In Europe aggregates are limited to a maximum of 1% of sulfur by mass or a maximum of 0.1% of sulfur by mass if pyrrhotite is present (EN 12620 *Aggregates for Concrete*).

Acid Attack

Most acidic solutions will disintegrate portland cement concrete. The rate of disintegration will be dependent on the type, volume, and concentration of acid. Certain weak acids, such as oxalic acid, are relatively harmless, particularly if contact is of short duration and small amounts. Strong acids in sufficient quantity or exposed to concrete over a long time, can dissolve concrete.

Mechanism of Acid Attack. Acids attack concrete by dissolving both hydrated and unhydrated cement phases (compounds) as well as calcareous aggregate. Siliceous aggregates are resistant to most acids and other chemicals and are sometimes specified to improve the chemical resistance of concrete, especially with the use of chemically resistant cement. Siliceous aggregate should be avoided when a strongly *basic* solution, like sodium hydroxide, is present as it attacks siliceous aggregate.

Materials and Methods to Control Acid Attack. For concretes that will be often or continuously exposed to acids, especially strong acids, a protective barrier coating may be required. Kerhoff (2007) provides a discussion of acid attack and protective measures for concrete, including coating.

In certain acidic solutions, it may be impossible to apply an adequate protective treatment to the concrete. The use of a sacrificial calcareous aggregate should be considered, particularly in those locations where the acidic solution may pond. Replacement of siliceous aggregate by limestone or dolomite having a minimum calcium oxide concentration of 50% will aid in neutralizing the acid. The acid will attack the entire exposed surface more uniformly, reducing the rate of attack on the paste and preventing loss of aggregate particles at the surface. Langelier Saturation Index (LSI) values for a water solution and calcium absorption test data on a soil sample can be used to test for this condition (Hime and others 1986 and Steinour 1975). Negative LSI values indicate a lime deficiency in the solution, which will accelerate leaching of lime bearing phases from the concrete.

The use of calcareous aggregate may also retard expansion resulting from sulfate attack caused by some acid solutions. Within practical limits, the paste content of the concrete should be minimized – primarily by reducing water content and using a well-graded aggregate – to reduce the area of paste exposed to attack and the concrete permeability. High cement contents are not necessary for acid resistance. Concrete deterioration increases as the pH of the acid decreases below about 6.5 (Kong and Orbison 1987 and Fattuhi and Hughes 1988).

Properly cured concrete with reduced calcium hydroxide contents, as occur with concretes using SCMs, may experience a slightly slower rate of attack from acids. This is because acid resistance is linked to the total quantity of calcium-containing phases, not just the calcium hydroxide content (Matthews 1992). Resistance to acid attack is primarily dependent on the concrete's permeability, cementitious material content, and water cementitious materials ratio. Acid rain (often with a pH of 4 to 4.5) can slightly etch concrete surfaces, usually without affecting the performance of exposed concrete structures. Extreme acid rain or strong acids may warrant special concrete designs or precautions, especially in submerged areas.

The American Concrete Pressure Pipe Association (ACPPA) provides guidelines for granular soils with a pH below 5 and the total acidity of the soil exceeding 25 meq/100 g and requires one of the following precautions to be used (ACPPA 2000):

- backfill in the pipe zone with consolidated clay material or calcareous material;
- acid resistant membrane on or around the pipe; or
- 8% to 10% silica fume in the mortar coating.

Where soil pH is below 4, the pipe should be installed with an acid resistant membrane or in an envelope of non-aggressive consolidated clay (ACPPA 2000). Natural waters usually have a pH of more than 7 and seldom less than 6. Waters with a pH greater than 6.5 may be aggressive if they contain bicarbonates. Water that contains bicarbonates also contains dissolved free carbon dioxide (CO_2) and carbonic acid (H_2CO_3) which can dissolve calcium carbonate unless it is saturated. This aggressive carbon dioxide acts by acid reaction and can attack concrete products regardless of whether they are carbonated. Methods are presented in Steinour (1975) for estimating

the amount of aggressive carbon dioxide from an ordinary water analysis when the pH is between 4.5 and 8.6, and the temperature is between 0° C (32°F) and 65°C (145°F).

Other anions in aggressive water may react with the components of cement paste to form insoluble calcium salts. Such reactions are not harmful provided the products are neither expansive nor removed by erosion.

The German Institute of Standardization Specification DIN 4030-2, *Assessment of Water, Soil and Gases for Their Aggressiveness to Concrete Part 2: Sampling and Analysis of Water and Soil Samples* (2008), includes criteria and a test method for assessing the potential for damage from carbonic acid-bearing water.

Microbially-Induced Corrosion

Microbially-induced corrosion (MIC) is a specialized form of acid attack. In some environments, microbes produce acids that can result in acid attack on concrete, as well as sulfate attack. As an example, the oxidation of hydrogen sulfide (H_2S), present in sanitary sewers, forms sulfuric acid (H_2SO_4), a strong acid that dissolves the calcium, magnesium, and aluminum phases in concrete. Certain species of bacteria feed on any form of sulfide in wastewater, converting it to sulfuric acid. Odor complaints, which are often the first indication of the presence of H_2S, must be investigated so that the concrete does not undergo prolonged acid attack.

Mechanisms of MIC. ASTM C1894, *Standard Guide for Microbially Induced Corrosion of Concrete Products* (2019), provides additional information on MIC, noting that it is believed to be a 3-stage process (Figure 11-41):

- In the first step, the pH at the surface of the concrete is lowered to below about 10, possibly through carbonation or exposure to low pH liquids.
- Biofilms can then form at the surface in a second stage, leading to further pH reduction to the range of 4 to 6.
- In the final stage, acids are produced through biological action, further deteriorating the concrete with pH less than 4.

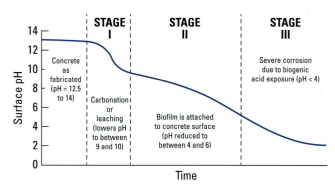

FIGURE 11-41. Microbially-induced corrosion (MIC) can be explained as a 3-stage process adapted from (adapted from ASTM C1894).

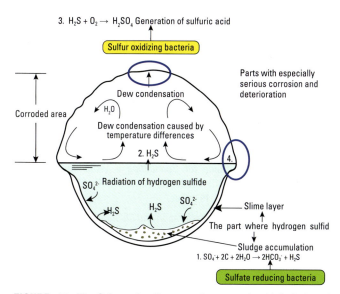

FIGURE 11-42 chemical labels:

3. $H_2S + O_2 \rightarrow H_2SO_4$ Generation of sulfuric acid

Sulfur oxidizing bacteria

Dew condensation

Parts with especially serious corrosion and deterioration

Corroded area

H_2O

Dew condensation caused by temperature differences

2. H_2S

4.

SO_4^{2-} Radiation of hydrogen sulfide

H_2S H_2S SO_4^{2-}

Slime layer

The part where hydrogen sulfid

Sludge accumulation

1. $SO_4^- + 2C + 2H_2O \rightarrow 2HCO_3^- + H_2S$

Sulfate reducing bacteria

FIGURE 11-42. Schematic diagram of some microbially-induced corrosion mechanisms (Wu and others 2019).

Figure 11-42 illustrates several of the common mechanisms of concrete attack due to MIC. Microbes are responsible for both reducing conditions producing hydrogen sulfide gas, and oxidizing to produce sulfuric acid.

Materials and Methods to Mitigate MIC. Guidance on minimizing or preventing damage due to MIC is being developed for the United States (ACI 201.1R 2016). In addition to steps for prevention of acid attack, such as low permeability concrete, additional concrete thickness, and protective barriers; some anti-microbial admixtures may be useful. ASTM C1904, *Standard Test Methods for Determination of the Effects of Biogenic Acidification on Concrete Antimicrobial Additives and/or Concrete Products*, includes a method for testing antimicrobial admixtures in a simulated wastewater solution and a method for evaluating such admixtures' abilities to delay or prevent damage in the second stage (biofilm) of MIC.

Seawater Exposure

Concrete can serve in seawater exposures for decades with excellent performance (Figure 11-43). However, special care

FIGURE 11-43. Concrete exposed to a marine environment.

FIGURE 11-44. Concrete specimens at the US Army Corps of Engineers test site at Treat Island, Maine. Specimens exposed to daily wetting/drying cycles along with multiple annual freeze-thaw cycles in seawater.

in mixture design and material selection is necessary for these severe environments. A structure exposed to brackish water, seawater or seawater spray is most vulnerable in the tidal or splash zone where there are repeated cycles of wetting and drying, with the additional threat of freezing and thawing in some areas (Figure 11-44).

All of the necessary ingredients are present for at least three different deterioration mechanisms:

• corrosion of reinforcement (chlorides, oxygen, water),

• sulfate attack (sulfates, water), and

• magnesium ion substitution (magnesium, water).

For reinforced concrete, corrosion is typically the most significant concern, although several additional deterioration mechanisms may also occur or be exacerbated:

• alkali-aggregate reactions (sodium and water); only if reactive aggregates are used,

• abrasion /erosion (due to wave action, sand, gravel, icebergs, and ice flow),

• freeze-thaw damage (due to possible saturation),

• carbonation, and

• salt crystallization.

Mechanisms of Seawater Attack. When several deterioration mechanisms take place in the same concrete element, they generally interact in a way such that the net effect is greater than the sum of individual effects. For example, cracking due to corrosion will allow additional chloride, carbon dioxide, sulfate, and other ions to penetrate more readily into the concrete to attack the interior.

One notable exception is the effect of chloride ions on sulfate attack. Despite the high sulfate content of seawater, marine concrete does not generally suffer from classical sulfate attack. Instead, deterioration often takes the form of erosion or loss of the solid constituents. Expansion due to ettringite formation from monosulfoaluminate is suppressed in environments where OH^- ions are essentially replaced by Cl^- ions. Mindess and others (2002) reported that gypsum and ettringite are both more soluble in chloride ion solutions, and Verbeck (1975) noted that aluminate phases react to form (nonexpansive) chloroaluminates, so that deleterious expansions are prevented. Therefore, seawater is considered a moderate sulfate exposure.

In concrete submerged in water, the rate of corrosion is limited by the supply of oxygen. The deeper the concrete, the slower the deterioration. The splash zone is the most severe exposure for several reasons: both oxygen and seawater are in abundant supply; cycles of wetting and drying facilitate adsorption of sulfate, chloride, sodium, and magnesium ions and concentrate them within the concrete; cycles of freezing and thawing (if any) take place when the concrete is fully saturated; and the action of waves, floating objects, and sand create severe conditions of abrasion and erosion. Floating ice may adhere to the concrete surface and later be pulled away by currents, removing some of the concrete surface with it.

Materials and Methods for a Marine Environment. The use of SCMs should be considered for a marine environment (whether in or near the ocean). Cement used in concrete for a marine environment must balance the benefit of higher C_3A content to bind chlorides with the need for sulfate resistance (that is, low C_3A content). A cement with an intermediate C_3A content that is resistant to moderate sulfate exposure is desirable. Portland cements with tricalcium aluminate (C_3A) contents that range from 4% to 10% have been found to provide satisfactory protection against seawater sulfate attack, as well as protection against reinforcement corrosion by chlorides. Slag cement and alumina-bearing pozzolans are effective in binding chlorides as well as providing sulfate resistance. For this reason, specially formulated marine cements generally contain high volumes (65% or more by mass) of slag cement. Other supplementary cementing materials can also be used to advantage in this environment. And, as always, for the best durability the concrete should be of high quality, that is, well consolidated, well cured, and with low permeability. Water-cementitious material ratios should not exceed 0.40 for marine exposures.

In northern climates, the concrete must be properly air entrained with at least 6% air. High-strength concrete should be considered where large ice formations abrade the structure. Proper concrete cover over reinforcing steel must be provided (see ACI 318-19) and Table 11-8. See Stark (1995 and 2001), and Kerkhoff (2007) for examples of durability studies in marine environments.

REFERENCES

PCA's online catalog includes links to PDF versions of many of our research reports and other classic publications.
Visit: cement.org/library/catalog.

ACI, *ACI Concrete Terminology*, CT-20, American Concrete Institute, Farmington Hills, MI, March 2020, 81 pages.

ACI Committee 201, *Guide to Durable Concrete*, ACI 201.2R-16, American Concrete Institute, Farmington Hills, MI, 2016, 86 pages.

ACI Committee 201, *Guide for Conducting a Visual Inspection of Concrete in Service*, ACI 201.1R-08, American Concrete Institute, Farmington Hills, MI, 2008, 19 pages.

ACI Committee 201, *Joint Deterioration and Chloride-Based Chemicals*, ACI 201.3T-19, Technical Note, American Concrete Institute, Farmington Hills, MI, 2019, 6 pages.

ACI Committee 221, *Guide for Use of Normal Weight and Heavyweight Aggregates in Concrete*, ACI 221R-96, American Concrete Institute, Farmington Hills, MI, 1996

ACI Committee 222, *Guide to Protection of Metals in Concrete Against Corrosion*, ACI 222R-19, American Concrete Institute, Farmington Hills, MI, 2019, 60 pages.

ACI Committee 225, *Removal of the Optional Alkali Limit in Standard Specifications for Portland Cement*, ACI 225.1T-20, Technical Note, American Concrete Institute, Farmington Hills, MI, 2020, 4 pages.

ACI Committee 301, *Specification for Concrete Construction*, ACI 301-20, American Concrete Institute, Farmington Hills, MI, 2020, 73 pages.

ACI Committee 318, *Building Code Requirements for Structural Concrete and Commentary*, ACI 318-19, American Concrete Institute, Farmington Hills, MI, 2019, 624 pages.

ACPA, *Scale-Resistant Concrete Pavements*, IS117, American Concrete Pavement Association, Skokie, IL, 1996.

ACPPA, *External Protection of Concrete Cylinder Pipe*, American Concrete Pressure Pipe Association, Reston, VA, 2000, 12 pages.

ASCE, "Entrained Air Voids in Concrete Help Prevent Salt Damage," *Civil Engineering*, American Society of Civil Engineers, New York, NY, May 1982.

Berke, N.S.; Aldykiewicz, Jr., A.J.; and Li, L., "What's New in Corrosion Inhibitors," *Structure*, July/August 2003, pages 10 to 12.

Brinkerhoff, C.H., *Report to ASTM C-9 Subcommittee III-m (Testing Concrete for Abrasion) Cooperative Abrasion Test Program*, University of California and Portland Cement Association, 1970. Reprinted as PCA SN2913, Portland Cement Association, Skokie, IL, 66 pages.

Bates, A.A.; Woods, H.; Tyler, I.L.; Verbeck, G.; and Powers, T.C., "Rigid-Type Pavement," Association of Highway Officials of the North Atlantic States, *28th Annual Convention Proceedings*, March 1952, pages 164 to 200.

Blunden, J.; and D.S. Arndt, "Abstract, State of the Climate in 2016," *Bulletin of the American Meteorological Society*, Vol. 98, No. 8, August 2017, 298 pages.

Brown, F.P.; and Cady, P.D., "Deicer Scaling Mechanisms in Concrete," *Durability of Concrete*, ACI SP-47, American Concrete Institute, Farmington Hills, MI, 1975, pages 101 to 119.

Campbell, D.H.; Sturm, R.D.; and Kosmatka, S.H., "Detecting Carbonation," *Concrete Technology Today*, PL911, Portland Cement Association, Skokie, IL, March 1991, pages 1 to 5.

Cody, R.D.; Cody, A.M.; Spry, P.G.; and Gan, G.L., "Concrete Deterioration by Deicing Salts: An Experimental Study," *Semisequicentennial Transportation Conference Proceedings*, Center for Transportation Research and Education, Ames, IA, 1996.

Cordon, W.A., "Freezing and Thawing of Concrete – Mechanisms and Control," *ACI Monograph No. 3*, American Concrete Institute, Farmington Hills, MI, 1966, 99 pages.

CSA Committee A23, *Concrete Materials and Methods of Concrete Construction/Test Methods and Standard Practices for Concrete*, CSA A23.1-19/A23.2-19, Canadian Standards Association, Toronto, Canada, 2019.

CSA Committee A23, *Standard Practice for Laboratory Testing to Demonstrate the Effectiveness of Supplementary Cementing Materials and Lithium-Based Admixtures to Prevent Alkali-Silica Reaction in Concrete*, CSA A23.2-28A, Canadian Standards Association, Toronto, Canada, 2009.

CSA Committee A23, *Concrete Materials and Methods of Concrete Construction/Test Methods and Standard Practices for Concrete – Annex P Impact of Sulphides in Concrete Aggregate on Concrete Behaviour*, CSA A23.1-14, CSA Group, June 2018, pages 240 to 250.

Detwiler, R.J.; and Powers-Couche, L., "Effect of Ettringite on Frost Resistance," *Concrete Technology Today*, PL973, Portland Cement Association, Skokie, IL, 1997, pages 1 to 4.

Detwiler, R.J.; and Taylor, P.C., *Specifier's Guide to Durable Concrete*, EB221, Portland Cement Association, Skokie, IL, 2005, 68 pages.

DIN, *Assessment of Water, Soil and Gases for Their Aggressiveness to Concrete - Part 2: Sampling and Analysis of Water and Soil Samples*, DIN 4030-2, Deutsches Institut Für Normung E.V. (German National Standard), January 2008, 20 pages.

EN 12620:2003, *Aggregates for Concrete*, European Committee for Standardization, 59 pages.

Fagerlund, G., "The Significance of Critical Degrees of Saturation at Freezing of Porous and Brittle Materials," *ACI Special Publication*, SP– 47, American Concrete Institute, Farmington Hills, MI, 1975, pages 13 to 65.

Famy, C.; and Taylor, H.F.W., "Ettringite in Hydration of Portland Cement Concrete and its Occurrence in Mature Concretes," *ACI Materials Journal*, Vol. 98, No. 4, American Concrete Institute, Farmington Hills, MI, July-August 2001, pages 350 to 356.

Famy, C.; Scrivener, K.L.; Atkinson, A.; and Brough, A.R., "Influence of the Storage Conditions on the Dimensional Changes of Heat-cured Mortars," *Cement and Concrete Research*, Vol. 31, No. 5, May 2001, pages 795 to 803.

Farnam, Y.; Washington, T.; and Weiss, W.J., "The Influence of Calcium Chloride Salt Solution on the Transport Properties of Cementitious Materials," *Advances in Civil Engineering*, June 2015, 13 pages.

Farny, J.A.; and Kerkhoff, B., *Diagnosis and Control of Alkali-Aggregate Reactions*, IS413, Portland Cement Association, Skokie, IL, 2007, 26 pages.

Fattuhi, N.I.; and Hughes, B.P., "Ordinary Portland Cement Mixes with Selected Admixtures Subjected to Sulfuric Acid Attack," *ACI Materials Journal*, American Concrete Institute, Farmington Hills, MI, November-December 1988, pages 512 to 518.

Folliard, J.J.; and Sandberg, P., "Mechanisms of Concrete Deterioration by Sodium Sulfate Crystallization," ACI SP-145, *Third International ACI/CANMET Conference on Concrete Durability*, Nice, France, 1994, pages 993 to 946.

Folliard, K.J.; Barborak, R.; Drimalas, T.; Du, L.; Garber, S.; Ideker, J.; Ley, T.; Williams, S.; Juenger, M.; Thomas, M.D.A.; and Fournier, B., *Preventing ASR/DEF in New Concrete: Final Report*, CTR 4085-5, The University of Texas at Austin, Center for Transportation Research, Austin, TX, 2006.

Fournier, B., *CANMET/Industry Joint Research Program on Alkali-Aggregate Reaction – Fourth Progress Report*, Canada Centre for Mineral and Energy Technology, Ottawa, 1997.

Fournier, B.; Berube, M.A.; Folliard, K.J.; and Thomas, M., *Report on the Diagnosis, Prognosis, and Mitigation of Alkali-Silica Reaction (ASR) in Transportation Structures*, FHWA-HIF-09-004, Federal Highway Administration, Washington, D.C., January 2010, 154 pages.

Gajda, J., *Mass Concrete for Buildings and Bridges*, EB547, Portland Cement Association, Skokie, IL, 2007, 44 pages.

Ghorab, H.Y.; Heinz, D.; Ludwig, U.; Meskendahl, T.; and Wolter, A., 1980, "On the Stability of Calcium Aluminate Sulphate Hydrates in Pure Systems and in Cements," *Proceedings of the 7th International Congress on the Chemistry of Cement*, Editions Septima, Paris, Vol. 4, 1980, pages 496 to 503.

Graf, L.A., *Effect of Relative Humidity on Expansion and Microstructure of Heat-Cured Mortars*, Research and Development Bulletin RD139, Portland Cement Association, Skokie, IL, 2007, 50 pages.

Hadley, D.W., *Alkali Reactivity of Carbonate Rocks – Expansion and Dedolomitization*, Research Department Bulletin RX139, Portland Cement Association, Skokie, IL, 1961, 18 pages.

Haynes, H.; O'Neill, R.; and Mehta, P.K., "Concrete Deterioration from Physical Attack by Salts," *Concrete International*, Vol. 18, No. 1, American Concrete Institute, Farmington Hills, MI, January 1996, pages 63 to 68.

Herholdt, A.G.; Justesen, C.F.P.; Nepper-Christensen, P.; and Nielsen, A., editors, *Beton-Bogen*, Aalborg Portland, Aalborg, Denmark, 1979, 719 pages.

Hime, W.G.; Erlin, B.; and McOrmond, R.R., "Concrete Deterioration Through Leaching with Soil-Purified Water," *Cement, Concrete, and Aggregates*, American Society for Testing and Materials, PA, Summer 1986, pages 50 to 51.

Hobbs, D.W., "Concrete Deterioration: Causes, Diagnosis and Minimizing Risk," *International Materials Review*, February 2001, pages 117 to 144.

Kanare, H.M., *Concrete Floors and Moisture*, EB119, Portland Cement Association, Skokie, IL, and National Ready Mixed Concrete Association, Silver Spring, MD, 2008, 176 pages.

Kerkhoff, B., *Effects of Substances on Concrete and Guide to Protective Treatments*, IS001, Portland Cement Association, Skokie, IL, 2007, 36 pages.

Klemm, W.A.; and Miller, F.M., "Plausibility of Delayed Ettringite Formation as a Distress Mechanism – Considerations at Ambient and Elevated Temperatures," Paper 4iv059, *Proceedings of the 10th International Congress on the Chemistry of Cement*, Gothenburg, Sweden, June 1997, 10 pages.

Klieger, P., *Air-Entraining Admixtures*, Research Department Bulletin RX199, Portland Cement Association, Skokie, IL, 1966, 23 pages.

Klieger, P., *Extensions to the Long-Time Study of Cement Performance in Concrete*, Research Department Bulletin RX157, Portland Cement Association, Skokie, IL, 1963, 20 pages.

Klieger, P., *Further Studies on the Effect of Entrained Air on Strength and Durability of Concrete with Various Sizes of Aggregate*, Research Department Bulletin RX077, Portland Cement Association, Chicago, IL, 1956, 21 pages.

Klieger, P., *Studies of the Effect of Entrained Air on the Strength and Durability of Concretes Made with Various Maximum Sizes of Aggregates*, Research Department Bulletin RX040, Portland Cement Association, Chicago, IL, 1952, 30 pages.

Kong, H.; and Orbison, J.G., "Concrete Deterioration Due to Acid Precipitation," *ACI Materials Journal*, American Concrete Institute, Farmington Hills, MI, March-April 1987, pages 110 to 116.

Kozikowski, R.; Taylor, P.; and Pyc, W.A., *Evaluation of Potential Concrete Deterioration Related to Magnesium Chloride (MgCl₂) Deicing Salts*, SN2770, Portland Cement Association, Skokie, IL, 2007, 30 pages.

Landgren, R.; and Hadley, D.W., *Surface Popouts Caused by Alkali Aggregate Reaction*, Research and Development Bulletin RD121, Portland Cement Association, Skokie, IL, 2002, 20 pages.

Lee, H.; Cody, R.D.; Cody, A.M.; and Spry, P.G., "Effects of Various Deicing Chemicals on Pavement Concrete Deterioration," *Proceedings of the Mid-Continent Transportation Symposium*, Center for Transportation Research and Education, Iowa State University, Ames, IA, 2000.

Leiser, K.; and Dombrowski, G., "Research Work on Magnesium Chloride Solution Used in Winter Service on Roads," *Strasse*, Vol. 7, No. 5, Berlin, Germany, May 1967.

Lerch, W., *Basic Principles of Air-Entrained Concrete*, T-101, Portland Cement Association, Skokie, IL, 1960.

Lin, C.; and Walker, R.D., "Effects of Cooling Rates on the Durability of Concrete," *Transportation Research Record 539*, Transportation Research Board, Washington, D.C., 1975, pages 8 to 19.

Liu, T.C., "Abrasion Resistance of Concrete," *Journal of the American Concrete Institute*, Farmington Hills, MI, September-October 1981, pages 341 to 350.

Matthews, J.D., "The Resistance of PFA Concrete to Acid Groundwaters," *Proceedings of the 9th International Congress on the Chemistry of Cement*, New Delhi, India, Vol. V, 1992, Building Research Establishment, Watford, England, page 355.

Mercias, D.; and Wagoner, B., "Runway Deicers: A Varied Menu," *Airport Magazine*, Vol. 15, 2003.

Mielenz, R.C.; Wokodoff, V.E.; Blackstrom, J.E.; and Flack, H.L., "Origin, Evolution and Effects of the Air-Void System in Concrete. Part 1–Entrained Air in Unhardened Concrete," July 1958, "Part 2–Influence of Type and Amount of Air-Entraining Agent," August 1958, "Part 3–Influence of Water-Cement Ratio and Compaction," September 1958, and "Part 4–The Air-Void System in Job Concrete," October 1958, *Journal of the American Concrete Institute*, Farmington Hills, MI, 1958.

Mindess, S.; Young, J.F.; and Darwin, D., *Concrete*, 2nd ed., Pearson Education, Inc., Upper Saddle River, NJ, 2002, 644 pages.

Monteiro, P.J.M.; Coussy, O.; and Silva, D.A., "Effect of Cryo-Suction and Air Void Transition Layer on the Hydraulic Pressure Developed During Freezing of Concrete," *ACI Materials Journal*, Vol. 103, No. 2, 2006, pages 136 to 140.

NCHRP, *Guidelines for the Selection of Snow and Ice Control Materials to Mitigate Environmental Impacts*, Report 577, National Cooperative Highway Research Program, Transportation Research Board, Washington D.C., 2007, 211 pages.

Neville, A.M., *Properties of Concrete*, 4th edition, John Wiley & Sons, New York, NY, 1996, 844 pages.

Nmai, C.K, "Freezing and Thawing," *Significance of Tests and Properties of Concrete and Concrete-Making Materials*, ASTM STP 169D, edited by Lamond, J.F. and Pielert, J.H., ASTM International, West Conshohocken, PA, 2006, pages 154 to 163.

Nmai, C.K.; Farrington, S.; and Bobrowski, G.S., "Organic-Based Corrosion-Inhibiting Admixture for Reinforced Concrete," *Concrete International*, Vol. 14, No. 4, April 1992, pages 45 to 51.

NPCA, *NPCA Quality Control Manual for Precast Concrete Plants*, 14th edition, National Precast Concrete Association, QCM-001, January 2020, 195 pages.

Oesterle, R.G., *The Role of Concrete Cover in Crack Control Criteria and Corrosion Protection*, SN2054, Portland Cement Association, Skokie, IL, 1997, 88 pages.

Osborne, G.J., "Carbonation and Permeability of Blast Furnace Slag Cement Concretes from Field Structures," *Proceedings of the 3rd International Congress on the Use of Fly Ash, Silica Fume, Slag and Natural Pozzolans in Concrete*, ACI SP-114, Vol. 2, American Concrete Institute, Detroit, MI, 1989, pages 1209 to 1237.

Ozol, M.A., "Alkali-Carbonate Rock Reaction," *Significance of Tests and Properties of Concrete and Concrete-Making Materials*, ASTM STP 169D, edited by Lamond, J.F. and Pielert, J.H., ASTM International, West Conshohocken, PA, 2006, pages 410 to 424.

Page, C.L.; and Vennesland, Ø., "Pore Solution Composition and Chloride Binding Capacity of Silica-Fume Cement Pastes," *Materials and Structures*, No. 91, January-February 1983, pages 19 to 25.

PCA, *Types and Causes of Concrete Deterioration*, IS536, Portland Cement Association, Skokie, Illinois, 2002, 16 pages.

PCI, *Manual for Quality Control for Plants and Production of Structural Precast Concrete Products*, MNL 116, 4th Edition, Precast/Prestressed Concrete Institute, Chicago, IL, 1999, 328 pages.

PCI, *Manual for Quality Control for Plants and Production of Architectural Precast Concrete Products*, MNL 117, 4th Edition,, Precast/Prestressed Concrete Institute, Chicago, IL, 2013, 334 pages.

Peeples, B., "Re: Using Salt to Melt Ice," *MadSci Network*, madsci.org, 1998.

Pigeon, M.; and Pleau, R., *Durability of Concrete in Cold Climates*, E&FN Spon Publishing House, London, UK, 1995, 244 pages.

Pigeon, M.; Prevost, J.; and Simard, J.M., "Freeze-Thaw Durability versus Freezing Rate," *ACI Journal*, American Concrete Institute, Farmington Hills, MI, September-October 1985, pages 684 to 692.

Pinto, R.C.A.; and Hover, K.C., *Frost and Scaling Resistance of High-Strength Concrete*, Research and Development Bulletin RD122, Portland Cement Association, Skokie, IL, 2001, 70 pages.

Powers, L.J., "Developments in Alkali-Silica Gel Detection," *Concrete Technology Today*, PL991, Portland Cement Association, Skokie, IL, April 1999.

Powers, T.C.; Copeland, H.E.; and Mann, H.M., *Capillary Continuity or Discontinuity in Cement Pastes*, Research Department Bulletin RX110, Reprinted from the Journal of the PCA Research and Development Laboratories, Vol. 1, No. 2, Portland Cement Association, Skokie, IL, May 1959, pages 38 to 48.

Powers, T.C., "The Mechanism of Frost Action in Concrete," *Stanton Walker Lecture Series on the Materials Sciences*, Lecture No. 3, National Sand and Gravel Association and National Ready Mixed Concrete Association, Silver Spring, MD, 1965a.

Powers, T.C., *Topics in Concrete Technology: (1) Geometric Properties of Particles and Aggregates, (2) Analyis of Plastic Concrete Mixtures, (3) Mixtures Containing Intentionally Entrained Air, and (4) Characteristics of Air-Void Systems*, Research Department Bulletin RX174, Portland Cement Association, Skokie, IL, 1965b, 83 pages.

Powers, T.C., *Prevention of Frost Damage to Green Concrete*, Research Department Bulletin RX148, Portland Cement Association, Skokie, IL, 1962, 18 pages.

Powers, T.C., "Structure and Physical Properties of Hardened Portland Cement Paste," *Journal of the American Ceramic Society*, Vol. 41, No. 1, The American Ceramic Society, Westerville, OH, January 1, 1958, pages 1 to 6. Reprinted as PCA Research Department Bulletin RX094, Portland Cement Association, Skokie, IL.

Powers, T.C., *Basic Considerations Pertaining to Freezing and Thawing Tests*, Research Department Bulletin RX058, Portland Cement Association, Skokie, IL, 1955, 30 pages.

Powers, T.C., *Void Spacing as a Basis for Producing Air-Entrained Concrete*, Research Department Bulletin RX049, Portland Cement Association, Skokie, IL, 1954, 25 pages.

Powers, T.C., *The Air Requirements of Frost-Resistant Concrete*, Research Department Bulletin RX033, Portland Cement Association, Skokie, IL, 1949, 33 pages.

Ramlochan, T.; Zacarias, P.; Thomas, M.D.A.; and Hooton, R.D., "The Effect of Pozzolans and Slag on the Expansion of Mortars Cured at Elevated Temperature, Part I: Expansive Behaviour," *Cement and Concrete Research*, Vol. 33, No. 6, 2003, pages 807 to 814.

Ramlochan, T.; Hooton, R.D.; and Thomas, M.D.A., "The Effect of Pozzolans and Slag on the Expansion of Mortars Cured at Elevated Temperature, Part II: Microstructural and Microchemical Investigations," *Cement and Concrete Research*, Vol. 34, No. 8, 2004, pages 1341 to 1356.

Ritter, S., "Aircraft Deicers," *Chemical and Engineering News*, Volume 79, No. 1, January 2001, page 30.

Rogers, C.A.; and Hooton, R.D., "Comparison between Laboratory and Field Expansion of Alkali-Carbonate Reactive Concrete," *Proceedings of the 9th International Conference on Alkali-Aggregate Reaction in Concrete*, 1992, pages 877 to 884.

Santhanam, M.; Cohen, M.D.; and Olek, J., "Sulfate Attack Research – Whither Now?," *Cement and Concrete Research*, 2001, pages 845 to 851.

Sakr, M.R.; Bassuoni, M.T.; Hooton, R.D.; Drimalas, T.; Haynes, H.; and K.J. Folliard, "Physical Salt Attack on Concrete: Mechanisms, Influential Factors and Protection," *ACI Materials Journal*, Vol. 117, No. 6, 2020, pages 253 to 268.

Sayward, J.M., *Salt Action on Concrete*, Special Report 84-25, U.S. Army Cold Regions Research and Engineering Laboratory, Hanover, NH, August 1984, 75 pages.

Scrivener, K.L.; Damidot, D.; and Famy, C., "Possible Mechanisms of Expansion of Concrete Exposed to Elevated Temperatures During Curing (Also Known as DEF) and Implications for Avoidance of Field Problems," *Cement, Concrete, and Aggregates*, CCAGDP, Vol. 21, No. 1, American Society for Testing and Materials, West Conshohocken, PA, June 1999, pages 93 to 101.

Setzer, M.J., "Basis of Testing the Freeze-Thaw Resistance: Surface and Internal Deterioration," *Frost Resistance of Concrete: Proceedings of the International RILEM Workshop on Resistance of Concrete to Freezing and Thawing With or Without De-icing Chemicals*, Setzer, M.J.; Auberg, R. (Editors), (RILEM Proceedings 4, E & F.N. Spon, 1997, pages 157 to 173.

Setzer, M.J., "Micro Ice Lens Formation and Frost Damage," *Frost Damage in Concrete, Proceedings of the International RILEM Workshop*, Janssen, D.J.; Setzer, M.J.; and Snyder, M.B. (Editors), RILEM Publication PRO 25, Minneapolis, MN, June 28 to 30, 1999, pages 1 to 15.

Shehata, M.H.; Jagdat, S.; Lachemi, M.; and Rogers, C., "Do Supplementary Cementing Materials Control Alkali-Carbonate Reaction?," *Proceedings, 17th Annual Symposium*, Ed. Fowler, D.; and Allen, J., International Centre for Aggregate Research, University of Texas, Austin, TX, 2009.

Shimada, Y.E., *Chemical Path of Ettringite Formation in Heat Cured Mortar and Its Relationship to Expansion*, Ph.D. Thesis, Northwestern University, Evanston, IL, 2005, 507 pages.

Sperling, C.H.B.; and Cooke, R.U., "Laboratory Simulation of Rock Weathering by Salt Crystallizations and Hydration Processes in Hot, Arid Environments," *Earth Surface Processes and Landforms*, Vol. 10, 1985, pages 541 to 555.

Stanton, T.E., "Expansion of Concrete through Reaction between Cement and Aggregate," *Proceedings, American Society of Civil Engineers*, Vol. 66, New York, NY, 1940, pages 1781 to 1811.

Stark, D., *Durability of Concrete in Sulfate-Rich Soils*, Research and Development Bulletin RD097, Portland Cement Association, Skokie, IL, 1989.

Stark, D., "Lithium Salt Admixture – An Alternative Method to Prevent Expansive Alkali-Silica Reactivity," *Proceedings of the 9th International Conference on Alkali-Aggregate Reaction in Concrete*, RP307, The Concrete Society, London, July 1992, 10 pages.

Stark, D., *Long-Time Performance of Concrete in a Seawater Exposure*, RP337, Portland Cement Association, Skokie, IL, 1995, 58 pages.

Stark, D., *Long-Term Performance of Plain and Reinforced Concrete in Seawater Environments*, Research and Development Bulletin RD119, Portland Cement Association, Skokie, IL, 2001, 14 pages.

Stark, D., *Performance of Concrete in Sulfate Environments*, Research and Development Bulletin RD129, Portland Cement Association, Skokie, IL, 2002, 28 pages.

Stark, D.; Kosmatka, S.H.; Farny, J.A.; and Tennis, P.D., *Performance of Concrete Specimens in the PCA Outdoor Test Facility*, Research and Development Bulletin RD124, Portland Cement Association, Skokie, IL, 2002, 36 pages.

Steinour, H.H., *Estimation of Aggressive CO_2 and Comparison with Langelier Saturation Index*, SN1905, Portland Cement Association, Skokie, IL, 1975, 52 pages.

Sutter, L.; Petersen, K.; Julio-Betancourt, G.; Hooton, R.D.; VanDam, T.; and Smith, K., *The Deleterious Chemical Effects of Concentrated Deicing Solutions on Portland Cement Concrete*, Final Report, South Dakota Department of Transportation, Research Report 2002-01-F, April 30, 2008, 198 pages.

Swenson, E.G.; and Gillott, J.E., "Alkali-Carbonate Rock Reaction," *Symposium on Alkali-Carbonate Rock Reactions*, Highway Research Record No. 45, Highway Research Board, Washington, D.C., 1964, pages 21 to 40.

Tang, M.; Deng, M.; Lon, X.; and Han, S., "Studies on Alkali-Carbonate Reaction," *ACI Materials Journal*, American Concrete Institute, Farmington Hills, MI, January-February 1994, pages 26 to 29.

Tang, M.; Liu, Z.; and Han, S., "Mechanism of Alkali-Carbonate Reaction," in *Concrete Alkali-Aggregate Reactions, Proceedings of the 7th International Conference*, P.E. Grattan-Bellew, ed., Noyes Publications, Park Ridge, NJ, 1987, pages 275 to 279.

Taylor, H.F.W., *Cement Chemistry*, 2nd edition, Thomas Telford, London, UK, 1997, 459 pages.

Taylor, H.F.W.; Famy, C.; and Scrivener, K.L., "Delayed Ettringite Formation," *Cement and Concrete Research*, Vol. 31, 2001, pages 683 to 693.

Taylor, P.C.; Whiting, D.A.; and Nagi, M.A., *Threshold Chloride Content for Corrosion of Steel in Concrete: A Literature Review*, SN2169, Portland Cement Association, Skokie, IL, 2000, 34 pages.

TEG, *The Thaumasite Form of Sulfate Attack: Risks, Diagnosis, Remedial Works and Guidance on New Construction*, Report of the Thaumasite Expert Working Group, Thaumasite Expert Group, Department of Environment, Transport, and the Regions, London, January 1999, 180 pages.

Thomas, M., *Optimizing the Use of Fly Ash in Concrete*, IS548, Portland Cement Association, Skokie, IL, 2007, 24 pages.

Thomas, M.D.A., "Delayed Ettringite Formation in Concrete: Recent Developments and Future Directions," *Materials Science of Concrete VI*, American Ceramics Society, Westerville, OH, 2001, pages 435 to 482.

Thomas, M.D.A.; Fournier, B.; and Folliard, K.J., *Report on Determining the Reactivity of Concrete Aggregates and Selecting Appropriate Measures for Preventing Deleterious Expansion in New Concrete Construction*, FHWA-HIF-09-001, Federal Highway Administration, U.S. Department of Transportation, Washington, D.C., April 2008a, 28 pages.

Thomas, M.D.A.; Folliard, K.; Drimalas, T.; and Ramlochan, T., "Diagnosing Delayed Ettringite Formation in Concrete Structures," *Cement and Concrete Research*, Vol. 38, No. 6, 2008b, pages 841 to 847.

Thomas, M.D.A.; Fournier, B.; Folliard, K.J.; Ideker, J.H.; and Resendez, Y., *The Use of Lithium to Prevent or Mitigate Alkali-Silica Reaction in Concrete Pavements and Structures*, FHWA-HRT-06-133, Federal Highway Administration, U.S. Department of Transportation, Washington, D.C., March 2007, 62 pages.

Thomas, M.D.A.; Fournier, B.; Folliard, K.J.; Ideker, J.H.; and Shehata, M., "Test Methods for Evaluating Preventive Measures for Controlling Expansion Due to Alkali-Silica Reaction in Concrete," *Cement and Concrete Research*, Vol. 36, No. 10, October 2006, pages 1842 to 1856.

Thomas, M.D.A.; and Innis, F.A., "Effect of Slag on Expansion Due to Alkali-Aggregate Reaction in Concrete," *ACI Materials Journal*, Vol. 95, No. 6, 1998, pages 716 to 724.

Thomas, M.D.A.; Matthews, J.D.; and Haynes, C.A., "Carbonation of Fly Ash Concrete," *Proceedings of the 4th ACI/CANMET International Conference on the Durability of Concrete*, (Ed. V.M. Malhotra), ACI SP-192, Vol. 1, 2000, pages 539 to 556.

Thomas, M.D.A.; and Wilson, M.L., *Supplementary Cementing Materials for Use in Concrete: Fly Ash, Slag, Silica Fume, Natural Pozzolans*, CD038, Portland Cement Association, Skokie, IL, 2002.

Todak, H.; Lucero, C.; and Weiss, W.J., "Why is the Air There? Thinking about Freeze-Thaw in Terms of Saturation," *Concrete inFocus*, January 2015, pages 3 to 7.

Uhlig, H.H.; and Revie, R.W., *Corrosion and Corrosion Control*, 3rd edition, John Wiley & Sons, New York, NY, 1985, 458 pages.

USBR, *Concrete Manual*, 8th ed., revised, U.S. Department of the Interior, Bureau of Reclamation, Denver, CO, 661 pages.

Valenza II, J.J.; and Scherer, G.W., "Mechanism for Salt Scaling," *Journal of the American Ceramic Society*, Vol. 89, No. 4, 2006, pages 1161 to 1179.

Valenza II, J.J.; and Scherer, G.W., "A Review of Salt Scaling: I. Phenomenology," *Cement and Concrete Research*, Vol.37, Iss. 7, July 2007a, pages 1007 to 1021.

Valenza II, J.J.; and Scherer, G.W., "A Review of Salt Scaling: II. Mechanisms," *Cement and Concrete Research*, Vol. 37, Iss. 7, 2007b, pages 1022 to 1034.

Vanderhost, N.M.; and Jansen, D.J., "The Freezing and Thawing Environment: What is Severe?," *Paul Klieger Symposium on Performance of Concrete*, ACI SP–122, David Whiting, editor, American Concrete Institute, Farmington Hills, MI, 1990, pages 181 to 200.

Verbeck, G.; and Klieger, P., *Studies of "Salt" Scaling of Concrete*, Research Department Bulletin RX083, Portland Cement Association, Skokie, IL, 1956, 14 pages.

Verbeck, G.J., *Carbonation of Hydrated Portland Cement*, Research Department Bulletin RX087, Portland Cement Association, Skokie, IL, 1958, 21 pages.

Verbeck, G.J., "Mechanisms of Corrosion of Steel in Concrete," *Corrosion of Metals in Concrete*, SP-49, American Concrete Institute, Farmington Hills, MI, 1975, pages 21 to 38.

Villani, C.; Farnam, Y.; Washington, T; Jian, J.; and Weiss, W.J., "Conventional Portland Cement and Carbonated Calcium Silicate-Based Cement Systems: Performance During Freezing and Thawing in the Presence of Calcium Chloride Deicing Salts," *Transportation Research Record: Journal of The Transportation Research Board*, Vol. 2508, Iss. 1, January 2015, pages 1 to 12.

Wang, H.; Tysl, S.; and Gillott, J.E., "Practical Implications of Lithium Based Chemicals and Admixtures in Controlling Alkali-Aggregate Reactions," ACI SP-148-20, *American Concrete Institute Special Publication 148*, American Concrete Institute, Farmington Hills, MI, 1994, pages 353 to 366.

Weiss, W.J., "Concrete Pavement Joint Durability: Absorption- Based Model for Saturation, the Role of Distributed Cracking, and Calcium Oxychloride Formation," *Proceedings of the 10th International Conference on Mechanics and Physics of Creep, Shrinkage, and Durability of Concrete and Concrete Structures* (CONCREEP 10), Eds. C. Hellmich; B. Pichler; and J. Kollegger, Vienna, Austria, September 21-23, 2015, pages 211 to 218.

Whiting, D.A., *Origins of Chloride Limits for Reinforced Concrete*, SN2153, Portland Cement Association, Skokie, IL, 1997, 18 pages.

Whiting, D.A., "Permeability of Selected Concretes," *Permeability of Concrete*, SP108-11, American Concrete Institute, Detroit, MI, 1988, pages 195 to 225.

Whiting, D.A.; and Nagi, M.A., *Manual on Control of Air Content in Concrete*, EB116, National Ready Mixed Concrete Association, Silver Spring, MD, and Portland Cement Association, Skokie, IL, 1998, 42 pages.

Whiting, D.A.; Taylor, P.C.; and Nagi, M.A., *Chloride Limits in Reinforced Concrete*, SN2438, Portland Cement Association, Skokie, IL, 2002, 76 pages.

Wu, M.; Wang, T.; Wu, K.; and Kan, L., "Microbially Induced Corrosion of Concrete in Sewer Structures: A Review of the Mechanisms and Phenomena," *Construction and Building Materials*, Vol. 239, April 2020.

Wyatt, J.; and Fritzsche, C., "The Snow Battle: Salt vs. Chemicals," *American City and County*, April 1989, pages 30 to 36.

SUSTAINABILITY

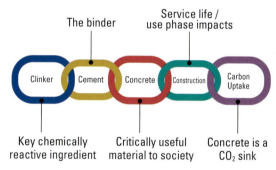

The binder

Service life /
use phase impacts

Clinker — Cement — Concrete — Construction — Carbon Uptake

Key chemically
reactive ingredient

Critically useful
material to society

Concrete is a
CO_2 sink

FIGURE 12-1. Each link of the value chain provides a specific contribution to sustainability. The combined contributions of the entire value chain are far greater.

Concrete is a sustainable building material – providing energy efficiency, long life-cycle, lower life-cycle costs, and resilience following natural and man-made disasters. Although users have typically selected concrete for its strength, durability, and other attributes, the use of cement and concrete can be placed into a much larger value chain that is foundational to modern civilization. All parts of our daily life – from residential homes and commercial structures to transportation systems, drinking water, and sanitary systems – benefit from using concrete.

VALUE CHAIN

As defined by PCA's Roadmap to Carbon Neutrality, cement and concrete are an integral part of an even larger value chain (PCA 2021). While each link of the value chain (clinker, cement, concrete, construction, and carbon uptake: known as the "5 Cs") provides a specific contribution to sustainability, the combined contributions of the entire chain are far greater. None of the contributions should be viewed in isolation because each link is dependent on both the previous link and the next link in the value chain (Figure 12-1).

The value chain starts with the production of clinker at a cement plant. Clinker production begins with quarry operations and ends with clinker nodules cooling in a clinker cooler. From the moment that the clinker is interground with gypsum and limestone to make portland cement, and possibly other supplementary cementing materials when making blended cements, its value is increased even further through the transformation of clinker into cement. Once cement is mixed with water, aggregates, other cementitious materials, and chemical admixtures, the concrete gains even more value. However, concrete is not the final step in the value chain. The final applications and end uses (paving and structures) are more valuable than the material used in construction.

The cement and concrete value chain is a microcosm of the circular economy. At its most efficient, this circular value chain is a zero-waste operation. Using this concept of value chain is a necessary step to understanding sustainability and the impact that clinker, cement, concrete, construction, and carbon uptake have on the environment, the economy, and our society.

The value chain also provides examples of specific societal contributions by producing a durable built environment. This includes providing resilient structures to fill society's needs; economic contributions through the value of put-in-place construction; and environmental contributions through concrete's contribution as a CO_2 sink. Just as the combined contributions of clinker, cement, concrete, construction, and carbon uptake are greater in total, the combined benefits they provide for society, the economy, and the environment are also far greater in total. It is this leverage that magnifies the contribution of cement and concrete and their associated value chain's role in sustainability.

THE TRIPLE BOTTOM LINE

As defined by John Elkington (1998), *sustainability* is typically referred to as: "The triple bottom line" representing the social, environmental, and economic impacts of development decision-making (Figure 12-2). More formally, sustainability means creating places that are environmentally responsible, healthful, equitable, and profitable. The U.S. Environmental Protection Agency (EPA) notes that sustainable construction expands and complements the classical building design concerns of economy, utility, durability, and comfort (EPA 2008a).

FIGURE 12-2. The triple bottom line considers social, environmental, and economic impacts in decision making.

Societal Impacts

Concrete in the built environment provides safe, sanitary, and sustainable housing and infrastructure, and directly impacts the public health and welfare of occupants and users of those structures. Concrete supplies optimal solutions for conveyance and storage of clean drinking water and environmentally-friendly wastewater treatment and storage.

Concrete buildings afford longer column to column spans, shorter building story heights, and increased load capacity, all of which optimize land use in built up areas. Concrete's acoustic control characteristics and thermal mass have direct impacts on the quality of life in service and indirectly also address further safeguarding the health of building occupants.

Critical infrastructure facilities requiring uninterrupted building occupancy and use (including police, fire services, and hospitals), are better served by concrete structures due to protection and rapid recovery from earthquakes, floods, fires, hurricanes, and tornadoes.

In developing communities, concrete provides safe, simple, and solid building solutions accessible to the local workforce through basic health and safety training coupled with workforce development training on concrete construction techniques.

Economic Impacts

The social benefits of concrete construction to health and welfare also include the economic impacts of the cement and concrete industry. Construction spending in the United States alone frequently amounts to more than 5% of the gross domestic product, directly employing nearly 8 million workers and indirectly employing an additional 226 workers for every 100 construction workers (USBEA 2020 and Bivens 2019).

Concrete's longevity is important to the US economy; recent construction put in place in the US alone has been valued at almost $1.4 trillion dollars annually (Holland 2019). Concrete structures last decades and this permanence reduces the disruption that demolition brings, and service delays from reconstruction and disposal of buildings, pavements, and other structures that are built with less durable and less resilient construction materials. Concrete also provides designers unique opportunities to repeatedly repurpose structures with minimal impact to the environment.

Environmental Impacts

The environmental impact of cement and concrete production is directly linked to the availability of its raw materials and selection of fuels and energy efficiency. The CO_2 generated from the manufacture of portland cement and production of concrete must be weighed against the benefits that concrete structures provide to modern society.

Over time, process efficiencies have reduced the CO_2 generated by combustion. Alternative fuels or alternative raw materials can reduce the CO_2 generated from the extraction and processing of virgin fuels and raw materials, beneficially utilizing materials that might otherwise have been landfilled. Other means of reducing CO_2 are by either lowering clinker contents of cement, or through replacing a portion of clinker with other suitable materials, or a combination of these two approaches. Blended cements like portland-limestone cements are excellent examples of lowering the clinker factor in cement.

The CO_2 generated by cement and concrete are offset by the use phase impacts of concrete, including improved fuel economy for automobiles, reduced energy use for buildings, albedo effects, and adsorption of CO_2 by concrete in service.

QUANTIFYING THE IMPACTS OF CEMENT AND CONCRETE

The impact actions have on society and the environment can be measured and quantified using widely accepted procedures adopted by a variety of standards developing organizations. In the case of the built environment, the basis for these procedures is the life cycle. Life cycle considers the production, construction, use, and end-of-life stages termed the 'cradle to grave' approach. Cement production includes all impacts from quarry operations to final shipment of product. Concrete production includes the upstream impact of cement, aggregate, water, and other ingredients along with the batching, mixing, transportation, and delivery of concrete. For concrete, in particular, a full life cycle assessment is critical to accurately quantify environmental impact because it is the use phase (and even the end-of-life stage) where many benefits of concrete accrue. See **Concrete in Use** for more information.

Life Cycle

Measurements using the life cycle approach include life cycle assessment, life cycle cost analysis, life cycle inventory, and life cycle impact assessment.

- *Life Cycle Assessment (LCA)* – estimates potential cumulative environmental impacts of production from cradle to grave

- *Life Cycle Cost Assessment (LCCA)* – estimates the cost impacts of a structure or pavement

- *Life Cycle Inventory (LCI)* – identifies and quantifies the inputs and outputs necessary to manufacture the product

- *Life Cycle Impact Assessment (LCIA)* – addresses the magnitude and significance of the potential environmental impacts of those inputs and outputs.

Product Category Rules and Environmental Product Declarations

The most widely accepted approach for quantifying the potential impact of cement and concrete is to use Product Category Rules

(PCRs) to develop Environmental Product Declarations (EPDs).

- *Product Category Rules* – A PCR is a set of rules, requirements, and guidelines for developing an EPD for one or more product categories.

- *Environmental Product Declaration* – An EPD is an independently verified and registered summary report of environmental impacts of a material's production based on a life cycle inventory (LCI).

A PCR sets the boundaries for an EPD by defining the life cycle stages of the product covered and the data quality required. A product category is a group of products that can fulfill equivalent functions. A goal of PCRs is to ensure that EPDs within a similar product type are developed from comparable data and analysis methods referencing appropriate standards. To ensure credibility, the rules are prepared with input from a wide range of stakeholders to achieve consensus. A panel of independent experts reviews and verifies that the draft PCR offers a substantive characterization of the environmental impacts for the product.

For construction products, EPDs and PCRs are prepared in accordance with two International Organization for Standardization (ISO) standards:

ISO 21930: 2007 *Sustainability in building construction – Environmental declaration of building products.*

ISO 14025: 2006 *Environmental labeling and declarations – Type III environmental declarations – Principles and procedures.*

The Portland Cement Association has developed version 2.0 of the PCR for Portland, Blended, Masonry, Mortar, and Plastic (Stucco) Cements, which is available from NSF International (NSF 2020). The Cement PCR is considered a "cradle-to-gate" PCR, as it only looks at the manufacturing (or production stage) of life cycle.

EPDs provide a clear, consistent, and transparent basis for reporting broad environmental performance for similar types of materials or products (often compared to the nutritional label found on food products). EPDs cannot tell the full story about the impact of construction unless they are used as data to inform the broader context of a life-cycle assessment (LCA) because LCAs allow for the use phase impacts of concrete to be included. An EPD is not a specification: within an EPD there is no evaluation of the environmental information since no predetermined environmental performance levels are set.

There are many types of EPDs depending on their scope:

- Industry-wide. These include results from many manufacturers of similar products based on the same PCR.

- Regional EPDs. For a particular country or region of a country.

- Product-specific Type III EPDs. These can be plant-specific or country-specific.

EPDs are valid for five years after the date of publication. In early 2021, four new industry-wide EPDs based on the 2020 Cement PCR (NSF 2020) were published to cover portland cement, portland-limestone cement, blended cements, and masonry cement produced in North America (PCA 2023a, 2023b, 2023c, and 2023d). Individual cement producers may also develop their own EPDs for their specific products.

Concrete EPDs for North America are also available from NSF International, produced by the National Ready Mixed Concrete Association (2022), based on a current PCR (NSF 2021). This industry-wide EPD covers 72 concrete products, which represent a range of concrete mixtures. Some concrete producers also have developed product-specific EPDs for their specific concrete mixtures and ingredients. Table 12-1 lists the range of characteristics reported in EPDs for concrete (NSF 2021) and provides typical values for 1 m³ of an example concrete mixture. Since concrete has a wide range of strength levels and mix designs appropriate for different exposure conditions and applications, there is a wide range of potential values for EPD characteristics. For concrete structures and pavements, values such as those in Table 12-1, but appropriate for the mixtures used, can be multiplied with the total volume of concrete necessary for a project to estimate the environmental footprint of the concrete used on a project. Environmental product declarations for other products used in the final structure or pavement can similarly be evaluated and summed to determine the total embodied environmental footprint of the project.

EMBODIED CO$_2$ IN CONCRETE

Cement and CO$_2$

The impacts of global warming have generated intense public awareness (Calvo Buendia and others 2019). There are four primary greenhouse gases: carbon dioxide, methane, nitrous oxide, and fluorinated gases. Each of these gases has a different global warming potential. Carbon dioxide (CO$_2$) has a global warming potential of one and is the baseline while methane, for example, has a global warming potential of 21 and nitrous oxide (N$_2$O) has a global warming potential of 300. The CO$_2$e or CO$_2$ equivalent is a summation of the contribution of each of the greenhouse gases and represents the number of metric tons of CO$_2$ emissions with the same global warming potential as one metric ton of another greenhouse gas. U.S. federal law (40 CFR Part 98) requires regulated industries, including the cement industry, to report their emissions of greenhouse gases. In the U.S., CO$_2$ constitutes more than 80% of all greenhouse gases emitted (EPA 2021a). Although cement industries contribute a small part of U.S. and international CO$_2$ emissions (Figure 12-3), each step in the value chain has a role to play. In 2021, the U.S. cement industry accounted for about 1.1% of all U.S. CO$_2$e emissions, which is about 2.5% of U.S. CO$_2$e from industrial sources (EPA 2020a and EPA 2023).

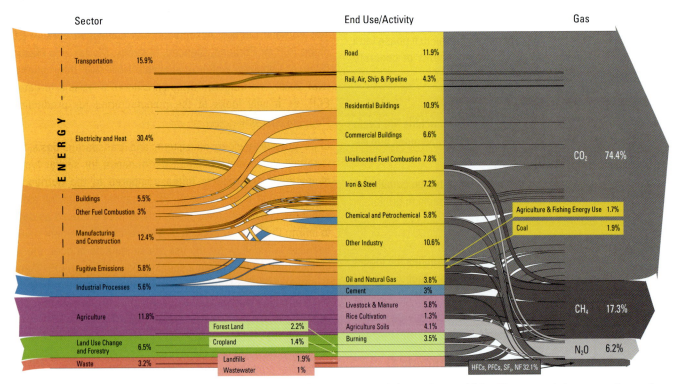

FIGURE 12-3. Global greenhouse gas emissions in 2016 (Ge 2020). The total is 49.4 Gt (54.5 billion tons) CO$_2$e. The cement industry contributes about 3% to the total (in blue) through calcination emissions.

TABLE 12-1. EPD Characteristics for an Example Ready-Mixed Concrete Mixture (Specified Compressive Strength of 27.6 MPa to 35 MPa [4001 to 5000 psi])*

CHARACTERISTIC	ABBREVIATION	UNIT‡	MAXIMUM VALUE per cubic meter of concrete	MINIMUM VALUE per cubic meter of concrete
Core Mandatory Impact Indicators				
Global warming potential	GWP	kg CO_2e	468.89	288.10
Depletion potential of the stratospheric ozone layer	ODP	kg CFC_{11}e	1.21E-05	8.13E-06
Acidification potential of soil and water sources	AP	kg SO_2e	1.37	1.02
Eutrophication potential	EP	kg Ne	0.50	0.38
Photochemical smog creation potential	POCP	kg O_3e	29.20	22.35
Abiotic depletion potential (ADPfossil)†	ADPf	MJ, NCV	729.22	640.71
Abiotic depletion potential (ADPelements)†	ADPe	kg Sbe	1.9E-04	1.55E-04
Use of Primary Resources				
Renewable primary energy resources as energy (fuel)	RPRE	MJ, NCV	28.78	15.82
Renewable primary resources as material	RPRM	MJ, NCV	0.00	0.00
Nonrenewable primary resources as energy (fuel)	NRPRE	MJ, NCV	2587.20	1925.85
Nonrenewable primary resources as material	NRPRM	MJ, NCV	0.00	0.00
Secondary Material, Secondary Fuel, and Recovered Energy				
Secondary materials	SM	kg	0.00	0.00
Renewable secondary fuels	RSF	MJ, NCV	0.00	0.00
Nonrenewable secondary fuels	NRSF	MJ, NCV	238.70	119.35
Recovered energy	RE	MJ, NCV	0.00	0.00
Mandatory Inventory Parameters				
Consumption of freshwater resources	FW	m³	0.83	0.64
Calcination and carbonation emissions	CCE	kg CO_2e	219.08	109.54
Indicators Describing Waste				
Hazardous waste disposed†	HWD	kg	0.13	0.13
Non-hazardous waste disposed†	NHWD	kg	9.64	9.64
High-level radioactive waste†	HLRW	m³	7.63E-08	2.07E-08
Intermediate- and low-level radioactive waste†	LLRW	m³	2.73E-07	1.98E-07
Components for re-use†	CRU	kg	0.00	0.00
Materials for recycling†	MR	kg	0.00	0.00
Materials for energy recovery†	MER	kg	0.00	0.00
Recovered energy exported from the product system†	EE	MJ, NCV	0.00	0.00

*See NRMCA (2022) for representative values for a wide range of concrete mixtures.

† Data in these categories have a high level of uncertainty as LCA impact categories and inventories protocols are still under development.

‡ Abbreviations: CO_2e = CO_2 equivalent; CFC_{11}e = chlorofluorocarbon-11 equivalent; SO_2e = sulfur dioxide equivalent; Ne = nitrogen equivalent; O_3e = ozone equivalent; Sbe = antimony equivalent; and NCV = net calorific value.

Calcination. Cement manufacture produces approximately 0.92 metric tons of CO_2 per metric ton of clinker. Approximately 60% of this CO_2 is produced as a result of calcination of calcium carbonate (limestone, the primary raw ingredient in the manufacture of clinker) and about 40% of this CO_2 is produced through fuel for combustion and electrical energy consumption. The calcination of calcium carbonate to calcium oxide is a key chemical reaction in the manufacture of cement (see Chapter 2). Calcination commences at approximately 800°C (around 1470°F) and is substantially complete at a temperature of 950°C (about 1740°F). Calcination requires sufficient energy to maintain temperatures at the intense level needed for combustion reactions to complete.

Energy efficiency. The manufacture of cement is one of the most energy efficient production processes of any man-made material. Average cement kilns operate at between 70% and 80% of theoretical efficiency (ECRA 2016). The production of CO_2 in the cement manufacturing process continues to decline as a result of energy efficiency improvements (Figure 12-4). Between 1970 and 2010, primary physical energy intensity for cement production dropped an average of 1.2% per year from 8.5 GJ/tonne (7.3 MBtu/ton) to 5.2 GJ/tonne (4.5 MBtu/ton) (Figure 12-4). Carbon dioxide intensity due to fuel consumption and raw material calcination dropped 24%, from 1.14 tonne CO_2/tonne of cement (1.14 ton CO_2/ton) to 0.84 tonne CO_2/tonne of cement (0.84 ton CO_2/ton) (Worrell and others 2013).

The amount of energy required to produce one ton of cement in the US averaged 5.06 GJ/tonne (4.35 MBtu/ton) in 2019. The long-term trend continues to move downward as more plants invest in energy efficient technologies. Since 1974, U.S. cement manufacturers have, on average, reduced their energy intensity by about 40% per tonne (ton) of cement.

Further improvements in cement manufacturing energy efficiency as a result of technology may be limited to marginal declines. The thermal process efficiencies in modern plants are now typically above 80% of the theoretical maximum efficiency for the manufacture of cement, which is high compared to other industrial processes (Schneider 2015).

Alternative Materials and Fuels. Cement manufacturing continues to incorporate increased use of alternative raw materials and alternative fuels as a sustainable practice. Industrial wastes, by-products, and other marginal materials are

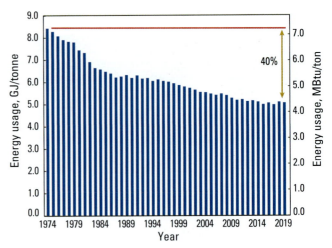

FIGURE 12-4. The average amount of energy required to produce a ton of portland cement in the US has dropped significantly in the last several decades, through process improvements; however, as the limits of available technology are approached, additional energy savings are more difficult to achieve (Sullivan and others 2020).

increasingly used as primary raw materials in the manufacture of cement. A summary of commonly used wastes and by-products in the cement industry is provided in Table 12-2. There is significant potential for the trend in alternative material use to continue and increase as raw materials become increasingly costly and scarce. However, their use ultimately depends on the economics of replacement and the compatibility of the alternate materials with the materials they are to replace (Bhatty and others 2011).

Alternative fuels range from engineered fuel and refuse-derived fuel to agriwaste, ashes, biofuel, biomass, carpet, charcoal, cherry pits, coal pond fines, coke breeze, filter fluff, flexicoke, glycerin, landfill gas, nylon fluff, pecan shells, plastics, rice hulls, sawdust, shingles, spent activated carbon, solvents, spent pot liner, textile waste, tire derived fuel, waste oil, wind, wood, and other materials. These materials might normally be discarded as waste (Table 12-3). Alternative fuel in U.S. cement plants was virtually non-existent in 1980 and now accounts for more than 13% of the energy consumed on average (Sullivan and others 2020). Nearly three quarters of U.S. plants incorporate at least one form of alternative fuel into their energy strategy. Alternative fuels used in 2019 include tire-derived fuel, waste oil, solvents, spent solids, and renewable energy including biofuels and biomass (Sullivan and others 2020).

TABLE 12-2. Waste Materials and By-Products Used in Cement Manufacture

LIME RICH	SILICA RICH	SILICA/ALUMINA RICH	IRON RICH	GYPSUM RICH
• Marginal limestone • Waste carbonate • Paper sludge • Wastewater lime • Sugar sludge • Fertilizer sludge • Metal slag	• Foundry sand • Sand washings • Catalytic fines • Rice husk ash	• Fly ash • Ponded ash • Bottom ash • Ore tailings • Basalt rocks • Bauxite waste	• Red mud • Mill scale • Laterite waste	• Desulfurization sludge

TABLE 12-3. Waste Materials Used as Alternative Fuels in Cement Kilns

Gaseous waste	Landfill gas
Liquid waste	Cleansing solvents
	Paint sludges
	Solvent contaminated waters
	"Slope" – residual washing liquid from oil and oil products storage tanks
	Used cutting and machining oils
	Waste solvents from chemical industry
Solid or pasty waste	Farming residues (rice husk, peanut husk, etc.)
	Municipal waste
	Plastic shavings
	Residual sludge from pulp and paper production
	Rubber shavings
	Sawdust and wood chips
	Sewage treatment plant sludge
	Tannery waste
	Tars and bitumens
	Used catalyst
	Used tires

An estimated 44 million scrap tires were diverted from landfills for energy recovery in cement kilns from the approximately 300 million scrap tires generated each year (RMA 2014). The EPA recognizes tire derived fuel (TDF) as an environmental best practice and encourages industries to use this resource to recover energy and conserve fossil fuel resources otherwise destined for landfills or unregulated disposal locations (Figure 12-5).

FIGURE 12-5. Whole rubber tires introduced directly into the kiln by conveyor as an alternative fuel.

Cement Kiln Dust. Cement production requires grinding of raw materials and clinker, along with other ingredients (see

Chapter 2) to fine powders. Grinding and transportation of these powders, as well as high gas flow through the kiln can result in a significant amount of dust. This dust is referred to as cement kiln dust (CKD), and is captured in bag houses. The drop in energy use discussed above has been accompanied by a drop in the amount of CKD produced at cement plants. Historically, CKD was often landfilled or in some cases sold as an agricultural amendment. It is now common practice for cement plants to either recycle CKD produced within the clinker manufacturing process or in the finish milling process.

Clinker versus Cement. The CO_2 per metric ton resulting from the manufacture of *cement* is less than that for *clinker* as clinker is interground with gypsum, limestone, and inorganic processing additions to produce portland cements and other materials to make blended cements. There are approximately 0.905 metric tons of clinker per metric ton of portland cement. Blended cements include supplementary cementitious materials and/or limestone that further reduce the clinker content. Cement makes up roughly 7% to 15% of concrete by volume (see Chapter 13). Therefore, the amount of CO_2 accounted for in the initial CO_2 footprint of concrete is substantially lower.

Portland-limestone cements. Portland-limestone cements (PLC) have demonstrated performance since their first use in Germany around 1965. In the U.S., these cements comprise 5% to 15% limestone and are specified using ASTM C595, *Standard Specification for Blended Hydraulic Cements* (AASHTO M 240), Type IL or Type IT (Thomas and Hooton 2010, Tennis and others 2011, and greenercement.com). Each replacement percentage of limestone eliminates an equal percentage of clinker, thereby reducing the amount of CO_2 that would normally be generated through the calcination process. On average, this amounts to about 10% lower CO_2 emissions for PLC compared to portland cement (Bushi and Meil 2014).

Other Blended Cements. ASTM C595 (AASHTO M 240) blended cements also include Type IP, Type IS, and other Type IT cements.

- Type IP typically replaces 15% to 25% of clinker with pozzolan,
- Type IS typically replaces 30% to 50% clinker with slag cement, and
- Type IT typically replaces 15% to 30% clinker with slag cement, and 10% to 20% clinker with a pozzolan, or up to 15% clinker with limestone.

As with portland-limestone cements, each percentage replacement of limestone, slag cement, pozzolan, or combination thereof, reduces approximately that same amount of CO_2 that would normally be generated through the calcination process in producing an ASTM C150 (AASHTO M 85) portland cement.

Carbon Capture, Utilization, and Storage. International Energy Agency (IEA 2009) has estimated that by 2050, cement manufacturing will require additional technologies to reduce its greenhouse gas emissions to target levels. These technologies are actively being researched and are referred to collectively as carbon capture utilization and storage (CCUS). In short CCUS technologies use various means (chemical absorption, physical adsorption, membrane technologies, mineralization, and calcium looping [Schneider 2015]) to capture CO_2 from calcination and fuel combustion for permanent storage or reuse in other products. For more information see DOE (2020) and ECRA (2017).

Concrete and CO_2

Depending upon the amount of cement used in a concrete mixture, the amount of CO_2 generated by concrete production is equivalent to approximately 5% to 13% of the weight of concrete (Marceau and VanGeem 2007). While the cement content of a concrete mixture has the greatest impact on concrete's overall CO_2 footprint, concrete itself remains a sustainable building material. The embodied CO_2 of reinforced concrete is approximately 0.2 kg CO_2/kg (0.2 lb CO_2/lb) of concrete while the embodied energy of concrete ranges from approximately 0.6 MJ/kg (about 250 Btu/lb) to 2 MJ/kg (about 850 Btu/lb) of concrete (Hammond and Jones 2011). This compares very favorably with a wide range of other construction materials (Figure 12-6).

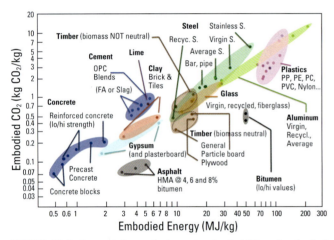

FIGURE 12-6. Concrete is one the lowest embodied CO_2 and embodied energy construction materials on a mass basis (Barcelo and others 2014, after data published by Hammond and Jones 2011). Data plotted as a log-log scale. To convert embodied energy from MJ/kg to Btu/lb, multiply by 429.85.

Common estimates suggest that cement production accounts for 5% to 7% of all manmade CO_2. This footprint is the result of the sheer amount of concrete made each year – potentially 20 billion tons to 35 billion tons produced world-wide annually (Barcelo and others 2014). Since concrete totals about half of all manmade materials produced in a given year (Figure 12-7), accounting for 5% to 7% of manmade CO_2 emissions is rather efficient.

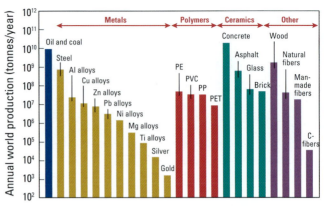

FIGURE 12-7. Concrete is the most widely used construction material, by more than a factor of ten. Concrete totals about half of all manmade materials in a given year (after Ashby 2009). Vertical axis is a log scale. To convert from tonnes to tons, multiply by 1.102.

Concrete presents several options for reducing its embodied CO_2. As demonstrated in Table 12-1 and the Concrete EPD (NSF 2019), a range of mixture designs are often available to provide concrete with a desired specified strength. Further improvements can be achieved, for example, by using plant-specific data to aid proportioning calculations and improved quality control on concrete batching operations (NRMCA 2021). Plant-specific strength data allows more accurate initial mixture proportioning estimates with locally available materials. Better quality control can reduce variability in properties, and thereby reduce overdesign requirements for strength. Both of these approaches permit better optimization of concrete mixtures while accounting for embodied CO_2.

When attempting to optimize concrete for low embodied CO_2, it is important remove unnecessary restrictions on the composition of the concrete in project specifications. Preference should be given to identifying and specifying the actual performance required of the fresh and hardened concrete. This approach also allows for optimizing paste contents of concrete through careful consideration of types and quantities of cementitious materials, aggregate characteristics like nominal maximum size and gradation, and admixture use, to provide strength, durability, and other fresh and hardened properties necessary in a CO_2-efficient manner.

Uniform comparisons of concrete to other building materials should consider the resource efficiency of concrete and concrete-making materials. Concrete aggregates, water, cementitious materials, and other ingredients are often locally available or locally produced and do not require bulk shipment from distant locations.

All of these embodied impacts and improvements are only a first step in a holistic approach provided by life-cycle assessments, which include use phase impacts.

CONCRETE IN USE

The use phase of concrete is typically measured in terms of decades and uniform comparisons of concrete to other building materials highlight the fact that concrete structures have service life expectancies in excess of 50 years. In some cases, designers are now specifying concrete structures having 100-year service life expectancy (ACI 365.1R 2017).

The built environment is an investment. Prematurely replacing that investment is a societal cost that disrupts users and an economic cost due to higher maintenance and repair expenditures. Building for extended service life reduces cost and provides users and owners a secure investment. Life cycle cost analyses (LCCAs) demonstrate concrete's economic advantage and environmental competitiveness even with potentially higher initial costs or possibly higher initial embodied carbon content. The benefits of concrete over its service life can often more than compensate for those differences.

Thermal Mass Effects for Buildings

Thermal mass is the property that enables structures to absorb, store, and later release significant amounts of heat. Concrete has thermal mass which moderates temperature swings between the interior and exterior of a structure (Figure 12-8). This moderation reduces heating and cooling energy for buildings and can be used to more efficiently size HVAC units and thereby lower heating and cooling costs (VanGeem 2013). This is critical as HVAC represents 35% of total building energy consumption (DOE 2015).

Thermal mass effects become critically important when realizing that 88% to 98% of the life cycle global warming potential (GWP) occurs in the use phase of a building (Ochsendorf and others 2011). Designers should consider all phases of building life cycle assessment (LCA), rather than solely focusing on the embodied

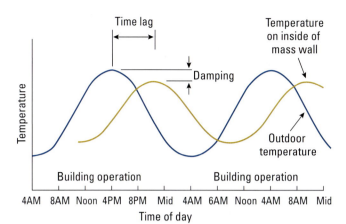

FIGURE 12-8. Damping and lag effects of thermal mass (CAC 2007).

energy resulting from the construction phase. For example, concrete buildings can have 3% to 10% lower *use-phase* GWP impacts and 5% to 8% lower *life-cycle* GWP impacts than comparative designs using other building materials (Figure 12-9; Ochsendorf and others 2011).

For a more detailed description of thermal mass performance and benefits, see de Saulles (2019).

Fuel Efficiency of Pavements

Concrete is a relatively stiff material and its use in pavement lowers both fuel consumption and related emissions of vehicles travelling over the pavement. Annual savings of 500 million gallons in fuel consumption resulting in annual saving of more than $2 billion and reduction of 5 million metric tons of CO_2 could be realized if the State of Florida's highway system were constructed using rigid pavement (Bienvenu and Jiao 2013). Concrete pavement in service provides at least 10% reductions in greenhouse gas emissions from vehicular traffic

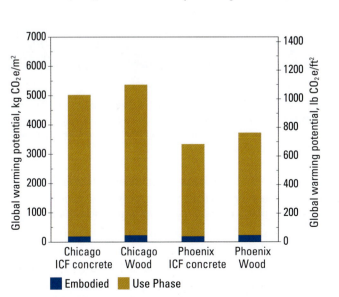

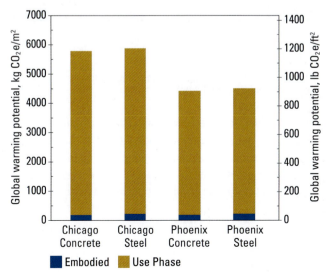

FIGURE 12-9. Global warming potential estimates per unit of floor surface area for (left) single-family homes with average air tightness and (right) commercial buildings for a 75-year lifespan (after Ochsendorf and others 2011).

when compared to unimproved baseline design scenarios (Santero 2013). Research also demonstrates that pavement maintenance decisions focused on road sections demonstrating high excess CO_2 emissions can reduce that excess fuel consumption, resulting in reductions of 10% to 20% of excess CO_2 emissions (Louhghalam 2017). These results demonstrate the significance of pavement-vehicle interaction on concrete pavements (Akbarian 2012).

Key factors in assessing pavement-vehicle interaction (PVI) are the roughness of a pavement, the texture of the pavement, and the deflection of a pavement under load (Figure 12-10). Pavement roughness might range from a severely worn pavement experiencing failure and evidenced by shoving, rutting, and potholes, to a newly placed pavement with no surface defects. Pavement texture relates to traction and skid resistance. Deflection in a pavement is a function of the load and the material properties. Concrete pavements are stiffer with far less pavement deflection than other paving materials. The energy required to overcome pavement deflection becomes important when considering the use phase of pavement. A pavement that deflects requires that vehicles exert more effort to overcome that deflection: it is like continually driving uphill, slightly, even on level ground. The PVI variables of stiffness, roughness, and texture together contribute to up to 4% improvement in fuel economy (MIT 2018) for heavy trucks traveling on a stiffer, smoother pavement, like concrete; and smaller amounts for passenger cars. This is significant when considering the daily and annual travel of hundreds of millions of vehicles in the US alone.

PVI-Roughness   PVI-Deflection

FIGURE 12-10. Concrete's stiffer surface results in less deflection, and its durability ensures less roughness for longer than competing materials. Both impacts improve fuel consumption and lower fuel emission for vehicles (MIT 2016).

Focusing solely on the production phase of pavement CO_2 footprint can miss the greater opportunities for sustainability. Pavement-vehicle interaction can effectively quantify potential pavement improvements by reducing fuel consumption and lowering fuel emissions of vehicles. Research using PVI models found that over a 5-year period, 1 billion gallons of excess fuel was used in California (MIT 2018), which amounts to about 9 million metric tons of CO_2 (EPA 2020a), due to roughness and deflection of pavements. In Virginia only 1.3% of the interstate network accounts for 10% of total greenhouse gas emissions from driving (MIT 2018). Replacing those pavements, and similar

roadways nationwide, with stiffer, smoother ones would result in significant CO_2 savings.

Resiliency

Disaster resilience refers to the ability to prepare for anticipated hazards, adapt to changing conditions, and withstand and recover rapidly from disruptions (GAO 2019). Performance in an earthquake, flood, fire, hurricane, or tornado is an extreme condition for any structure. Concrete structures have the ability to withstand these events while providing continuous use before, during, and after the event.

From an economic perspective, resiliency also includes understanding the potential costs and potential savings of mitigation strategies and valuing the protective functions and services of infrastructure and ecosystems. On a community level and on an individual building level, additional costs in the short-term can reduce much greater, longer-term losses. For example, the Multihazard Mitigation Council found that every dollar spent on pre-event mitigation related to earthquakes, wind, and flooding saved about $4 in post-event damages (NRC 2012).

Concrete's performance in natural disasters has showcased concrete's superior resilience. Figure 12-11 shows a concrete house that withstood hurricane-force winds while buildings surrounding it were destroyed. Rapid repair of minor damage to concrete homes and buildings after these events allows communities to return to normalcy much faster.

FIGURE 12-11. An elevated concrete house in Mexico Beach, Florida, came through Hurricane Michael almost unscathed (courtesy of J. Milano).

There are no current standards related to the definition, measurement, or evaluation of building resilience. There are however, efforts that address performance-based design and resilience-based design. The REDi™ Rating System (Almufti and Willford 2013) addresses resilient construction with respect to earthquake hazards while the U.S. Resiliency Council's *Building Rating System for Earthquake Hazards* relies upon methodology based on ASCE 41, *Seismic Rehabilitation of Existing Buildings*, as applied by the Structural Engineers

Association of Northern California *Earthquake Performance Rating System* (SEAONC 2015) and FEMA P-58, *Development of Next Generation Performance-Based Seismic Design Procedures for New and Existing Buildings*. Both examples address earthquake and earthquake-related hazards only (USRC 2013 and USRC 2019).

A recent metric developed to evaluate resilience is the break-even mitigation percent (BEMP), defined as the maximum investment in hazard mitigation while still reaching a break-even point on hazard repair savings considered over the lifetime of a building (Noori and others 2018). For example, this cost-benefit analysis shows a BEMP of 17.3% for buildings in Miami Dade County: by accounting for hurricane risk and costs of reconstruction, it was found that spending up to 17.3% more to enhance the resiliency of a structure during initial construction is cost effective.

Albedo

Heat islands are the result of decreased open land and vegetation caused by urbanization (EPA 2008b and 2012). Studies have shown that urban environments' annual mean air temperatures can be 1°C to 3°C (1.8°F to 5.4°F) warmer than surrounding areas and in extreme cases on a clear, calm night, temperature differences as much as 12°C (22°F) may be observed.

Strategies to reduce the amount of dark horizontal surfaces include white colored roofing, shade trees in parking lots, and lighter-colored surfaces for paving, parking lots, and sidewalks. Light-colored concrete materials have higher solar reflectance (also known as albedo) further reducing the heat-island effect (Figure 12-12). Surface finishing techniques and curing time also affect solar reflectance. Regardless of mixture constituents and placement techniques, concrete in the US reduces heat islands and qualifies for points in the LEED *Green Building Rating System* (USGBC 2020).

FIGURE 12-12. Thermal image of a pavement in Mesa, Arizona. Note the temperature difference between the concrete pavement (foreground) and the asphalt pavement (background) (courtesy of ACPA).

Solar reflectance (sometimes called albedo) is the ratio of solar energy that falls on a surface to the amount reflected. It is measured with a solar spectrum reflectometer on a scale of 0 (not reflective) to 1 (100% reflective). Generally, materials that appear light-colored have high solar reflectance and those that appear dark-colored have low solar reflectance. Ordinary portland cement concrete generally has a solar reflectance of approximately 0.35 to 0.45 and a Solar Reflective Index (SRI) between 38 to 52 (Table 12-4), although values can vary over time. A study of 45 concrete mix designs revealed that all 45 concretes tested according to ASTM C1549, *Standard Test Method for Determination of Solar Reflectance Near Ambient Temperature Using a Portable Solar Reflectometer*, had a solar reflectance of at least 0.30 (an SRI of at least 29), and meet or exceed LEED requirements (Marceau and VanGeem 2007).

TABLE 12-4. Solar Reflectance, Emittance, and Solar Reflective Index (SRI) of Concrete Material Surfaces

MATERIAL SURFACE	SOLAR REFLECTANCE	EMITTANCE	SRI
New concrete (ordinary)	0.35 to 0.45	0.9	38 to 52
New white portland cement concrete	0.7 to 0.8	0.9	86 to 100

Adapted from: Berdahl and Bretz 1994; Pomerantz and others 2000; Levinson and Akbari 2001; and Pomerantz and others 2002.

RECYCLING AT END OF SERVICE LIFE

The Construction and Demolition Recycling Association estimates that more than 275 million metric tons (303 million tons) of concrete are recycled in the U.S. annually (Townsend and Anshassi 2017). Concrete can be recycled as a variety of materials and reusing it at its original site has numerous cost saving benefits, including disposal cost savings and cost avoidance from road damage due to transportation of old concrete and new concrete, and less use of virgin aggregate. Some estimates note potential savings of 50% to 60% from the use of recycled aggregates in comparison to new aggregates (Verian 2013).

Larger pieces of concrete that are fractured or sawn may be used in specialized applications such as artificial reefs or breakwaters (ACPA 2009).

Typically, demolished concrete that is crushed and pulverized can be either recycled as aggregate or base material for engineered fill or pavement base, embankment fill, gravel roads, roof ballast, and railroad ballast. The increase in surface area generated through the comminution process greatly increases the rate of carbon uptake thereby adsorbing additional CO_2. Research is on-going to quantify the amount of CO_2 adsorbed by demolished concrete.

CARBON UPTAKE

Figure 12-13 depicts the cement and concrete carbonation cycle. Although CO_2 is emitted during the production of cement from calcination and energy requirements, exposed concrete surfaces slowly absorbs CO_2 from the air around them. Cement hydration products [primarily calcium hydroxide and calcium silicate hydrate (C-S-H), see Chapter 2] will slowly react with CO_2 to form calcium carbonate within concrete permanently sequestering that CO_2. Carbonation can make plain concretes more dense and less porous, adding to their durability. However, carbonation of reinforced concretes can be problematic as carbonation reduces the pH of concrete which may reduce the reinforcement passivity and increase the potential for corrosion and expansive rust formation for steel reinforcement (see Chapter 11). This potential is accounted for in design.

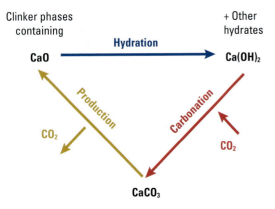

FIGURE 12-13. The cement/concrete carbonation cycle. Limestone calcination in cement production releases CO_2, which is partially reabsorbed and sequestered by concrete over its service life, and after demolition (Scrivener and others 2018).

A relatively new technology has been developed which injects carbon dioxide into fresh concrete mixtures which binds some CO_2 in concrete. Injection processes can be adapted for ready-mixed, precast, and other cement-based products (Monkman and other 2016).

The uptake of CO_2 through the carbonation of concrete can be beneficial provided best concrete practices are used. Current research includes three approaches to quantification: Tier 1, a simplified approach; Tier 2, a more sophisticated approach; and Tier 3, an advanced user model (Andersson and others 2019, Stripple and others 2018). These three approaches are intended to integrate with the United Nations Intergovernmental Panel on Climate Change (Eggleston and others 2006) *Guidelines*. For the cement industry, CO_2 emitted due to the calcination of limestone, is reported annually through a national inventory report (NIR) (see EPA 2018). For these reasons, the CO_2 uptake is determined on the basis of limestone calcination emissions.

Tier 1. The simplest approach for CO_2 uptake references data from a range of literature sources to estimate the CO_2 uptake by concrete over its service life. That approach (Andersson and others 2019) indicates that about 20% of the CO_2 emitted due to calcination of limestone when clinker is produced is adsorbed by concrete produced from cement made with that clinker, over that concrete's service life. This is believed to be a conservative assumption.

If mortar, plaster, and stucco make up more than about 10% of the cement usage within a country, the amount of carbonation is increased somewhat due to the higher surface area/volume of those applications (Stripple and others 2018).

An additional 2% calcination CO_2 uptake is assumed based on the demolition of the concrete at the end of its service life (as demolition frees up additional concrete surface area previously not exposed directly to atmospheric CO_2), and 1% for CO_2 uptake during subsequent secondary use of recycled concrete. These are again anticipated to be conservative estimates, which will likely increase once additional research is available.

Tier 1 estimates for the US indicate that concrete produced between 1990 and 2018 will adsorb more than 300 million metric tons (more than 330 million tons) of CO_2 over the service life of that concrete. This is equivalent (EPA 2020b) to CO_2 emissions from electricity use of more than 54 million homes in the US for one year.

> ### National Estimates of Concrete CO_2 Uptake
>
> The Tier 1 approach outlined in Stripple and others (2018) can be used as an estimate of the CO_2 uptake by concrete within a country or region over its service life. A factor of 0.23 is multiplied by the quantity of clinker used in a country each year (clinker produced + clinker imported − clinker exported) with adjustments for clinker imported or exported as cement. Another adjustment is included if a significant fraction of cement is used in thin applications like mortar.

Tier 2. A more advanced approach uses data on the distribution of cement to various concrete applications, allowing more accurate assumptions related to surface area/volume for concretes in a range of applications, and uses various factors for carbonation rate related to concrete strength class and cementitious materials use. EN 16757:2017, *Sustainability of Construction Works – Environmental Product Declarations – Product Category Rules for Concrete and Concrete Elements,* provides a basis for this approach. However, the more detailed data required for this method may not be readily available in all countries.

Tier 3. More sophisticated modeling efforts are in preliminary development.

GREEN RATING SYSTEMS

One effort to quantify sustainability in the built environment is through the LEED rating system established by the U.S. Green Building Council. The first version of LEED (LEED Version 1.0) was pilot tested among 19 projects in 1998. LEED currently operates on LEED Version 4, and USGBC has recently published LEED Version 4.1 for use in projects (usgbc.org).

LEED v 4.1 provides separate rating systems for the following:

• *Building Design and Construction (BD+C)*, new construction and major renovations, core and shell development, schools, retail, data centers, warehouses and distribution centers, hospitality, and health care.

• *Interior Design and Construction (ID+C)*, interior spaces that are a complete interior fit-out including commercial, retail, and hospitality.

• *Operations and Maintenance (O+M)*, existing building improvements with little or no construction involved.

• *Residential*, single family homes, multifamily homes, and multifamily homes core and shell,

• *Cities and Communities* – measures performance of existing and planned political jurisdictions or places defined by their municipal public sector (cities), and neighborhoods within cities such as sub-city locations (districts) and meta-city regions (counties).

• *Recertification*, applies to maintaining and improving buildings that have already achieved certification.

In addition to the rating systems listed above, the LEED Zero program is a complement to LEED that verifies projects with net carbon zero goals. LEED programs require a variety of prerequisites as well as assignment of points available in various categories. The BD+C program, for example, includes point categories for integrative process, location and transportation, sustainable sites, water efficiency, energy and atmosphere, materials and resources, indoor environmental quality, innovation, and regional priority.

Another effort to quantify sustainability is the Green Globes® program developed using the ANSI consensus development process. Green Globes® emphasizes use of life cycle assessments with multiple attribute evaluations.

Other rating systems include: BOMA BEST® Sustainable Buildings, version 3.0; Building Research Establishment's Environmental Assessment Method (BREEAM®) In-Use USA, version 2016; and ILFI's Living Building Challenge (LBC™), version 4.0.

INDUSTRY ACTIONS

PCA is committed to achieving carbon neutrality across the cement and concrete value chain by 2050 (PCA 2021).

The PCA Roadmap involves the entire value chain starting at the cement plant and extending through the entire life cycle of the built environment to incorporate the circular economy. This approach to carbon neutrality leverages relationships at each step of the value chain. The cement and concrete industry cannot do this alone. Government agencies, non-governmental organizations, and academic institutions all have a role. Likewise, Cembureau (2020), the Global Cement and Concrete Association (GCCA 2020), and the German Cement Association (VDZ 2020) have all made commitments to reach carbon neutrality by 2050.

The five links in the value chain include the production of clinker, the manufacture and shipment of cement, the manufacture of concrete, the construction of the built environment, and the capture of carbon dioxide using concrete as a carbon sink. Each link identifies a range of levers to be implemented in tandem to reach the goal of carbon neutrality:

• Increase the use of decarbonated raw materials;

• Decrease the use of traditional fossil fuels by 5X;

• Increase the use of alternative fuels;

• Push efficiency and decrease energy intensity for one metric ton of clinker;

• Utilize carbon capture to avoid the release of CO_2 emissions;

• Reduce clinker production emissions;

• Increase supplementary cementitious materials;

• Leverage new cement blends;

• Lower concrete manufacturing emissions to zero at the plant;

• Transition to zero emission fleets;

• Optimize concrete mixes; and

• Reduce overdesign.

The cement and concrete industries began addressing climate issues in the early 1970s, reducing average energy consumption by more than 40%. Investing in new innovative technologies, encouraging the use of life-cycle analyses, developing more complete analytical tools and procedures, and building awareness of sustainable construction options among architects, specifiers, developers, and the general public, will be key strategies for future improvements.

REFERENCES

PCA's online catalog includes links to PDF versions of many of our research reports and other classic publications.
Visit: cement.org/library/catalog.

ACI Committee 365, *Report on Service Life Prediction*, ACI 365.1R-17, American Concrete Institute, Farmington Hills, MI, 2017, 61 pages.

ACPA, *Recycling Concrete Pavements*, EB043P, American Concrete Pavement Association, Skokie, IL, 2009, 102 pages.

Akbarian, M.A., *Model Based Pavement-Vehicle Interaction Simulation for Life Cycle Assessment of Pavements*, M.S. Thesis, Massachusetts Institute of Technology, 2012, 121 pages.

Almufti, I.; and Willford, M., *REDiTM Rating System: Resilience-based Earthquake Design Initiative for the Next Generation of Buildings*, Version 1.0, Arup, October 2013, 134 pages.

Ashby, M.F., *Materials and the Environment: Eco-informed Material Choice*, Butterworth-Heinemann, Burlington, MA, 2009, 400 pages.

Andersson, R.; Stripple, H.; Gustafsson, T.; and Ljungkrantz, C., "Carbonation as a Method to Improve Climate Performance for Cement Based Material," *Cement and Concrete Research*, Vol. 124, October 2019, 10 pages.

ASCE/SEI, *Seismic Evaluation and Retrofit of Existing Buildings*, ASCE/SEI 41-17, American Society of Civil Engineers Structural Engineering Institute, Reston, VA, 2017, 550 pages.

Barcelo, L.; Kline, J.; Walenta, G.; and Gartner, E., "Cement and Carbon Emissions," *Materials and Structures*, Vol. 47, 2014, pages 1055 to 1065.

Berdahl, P.; and Bretz, S., "Spectral Solar Reflectance of Various Roof Materials," *Cool Building and Paving Materials Workshop*, Gaithersburg, MD, July 1994, 14 pages.

Bhatty, J.I.; Miller, F.M.; Kosmatka, S.H.; Bohan, R.P., editors, *Innovations in Portland Cement Manufacturing*, SP400, Portland Cement Association, Skokie, IL, 2011, 1734 pages.

Bienvenu, M.; and Jiao, X., *Comparison of Fuel Consumption on Rigid Versus Flexible Pavements Along I-95 In Florida*, Civil and Environmental Engineering, Florida International University, July 27, 2013, 15 pages.

Bivens, J., *Updated Employment Multipliers for the U.S. Economy*, Report 160282, Economic Policy Institute, Washington, DC, January 2019, 29 pages.

BOMA, *Application Guide: BOMA BEST Sustainable Buildings 3.0*, BOMA Canada, Toronto, Ontario, Canada, April 2018, 112 pages.

BRE, *BREEM USA In-Use 2016: Technical Standard Summary*, BRE Global, San Francisco, CA, April 2019, 6 pages.

Bushi, L.; and Meil, J., *An Environmental Life Cycle Assessment of Portland-Limestone and Ordinary Portland Cements in Concrete*, Technical Brief, Athena Sustainable Materials Institute, Ottawa, Ontario, Canada, January 2014, 10 pages

CAC, *Guide to Sustainable Design with Concrete*, Version 2.0, Cement Association of Canada, Ottawa, Ontario, Canada, 2007, 137 pages.

Calvo Buendia, E.; Tanabe, K.; Kranjc, A.; Baasansuren, J.; Fukuda, M.; Ngarize, S.; Osako, A.; Pyrozhenko, Y.; Shermanau, P.; and Federici, S. (eds), *2019 Refinement to the 2006 IPCC Guidelines for National Greenhouse Gas Inventories*, 5 vols., United Nations Intergovernmental Panel on Climate Change, Switzerland, 2019.

Cembureau, *Cementing the European Green Deal*, Cembureau, The European Cement Association, Brussels, Belgium, December 5, 2020, 38 pages.

Code of Federal Regulations, *Title 40-Protection of the Environment*, Chapter I, Subchapter C, Part 98-Mandatory Greenhouse Gas Reporting, Subpart H-Cement Production, ecfr.gov, November 2020.

de Saulles, T., *Thermal Mass Explained*, Mineral Products Association, The Concrete Centre, London, UK, February 2019, 20 pages.

DOE, *Quadrennial Technology Review: An Assessment of Energy Technologies and Research Opportunities*, U.S. Department of Energy, Washington, D.C., September 2015, 504 pages.

DOE, *2020 Compendium of Carbon Capture Technology*, National Energy Technology Laboratory, US Department of Energy, Albany, OR, May 2020, 818 pages.

ECRA, *Development of State of the Art-Techniques in Cement Manufacturing: Trying to Look Ahead*, CSI/ECRA-Technology Papers 2017, Cement Sustainability Initiative - World Business Council for Sustainable Development, Geneva, Switzerland, and European Cement Research Academy, Duesseldorf, Germany, 2017, 190 pages.

ECRA, *Evaluation of Energy Performance in Cement Kilns in the Context of Co-Processing*, Technical Report A-2016/1039, European Cement Research Academy, Duesseldorf, Germany, 2016, 53 pages.

Eggleston, H.S.; Buendia, L.; Miwa, K.; Ngara, T.; and Tanabe, K., eds., *2006 IPCC Guidelines for National Greenhouse Gas Inventories*, 5 vols, IPCC Task Force on National Greenhouse Gas Inventories, Institute for Global Environmental Strategies, Hayama, Kanagaawa, Japan, 2006.

Elkington, J., *Cannibals with Forks: The Triple Bottom Line of 21st Century Business*, New Society Publishers, Stony Creek, CT, 1998, 424 pages.

EN 16757:2017, *Sustainability of construction works - Environmental product declarations – Product Category Rules for concrete and concrete elements*, CEN/TC 350 Sustainability of Construction Works, European Committee for Standardization, Brussels, Belgium, June 2017, 58 pages.

EPA, *EPA Green Building Strategy*, EPA-100-F-08-073, U.S. Environmental Protection Agency, Washington, DC, November 2008a, 2 pages.

EPA, "Urban Heat Island Basics," *Reducing Urban Heat Islands: Compendium of Strategies*. Draft. epa.gov/heat-islands/ heat-island-compendium, US Environmental Protection Agency, Washington, D.C., 2008b, 22 pages.

EPA, *Facility Level Information on Greenhouse Gases Tool (FLIGHT)*, ghgdata.epa.gov, U.S. Environmental Protection Agency, Washington, DC, 2021.

EPA, *Greenhouse Gas Equivalencies Calculator*, epa.gov/energy/greenhouse-gas-equivalencies-calculator, U.S. Environmental Protection Agency, Washington, D.C., December 2020a.

EPA, *Fast Facts 1990-2018 National Level U.S. Greenhouse Gas Inventory*, EPA 430-F-20-002, U.S. Environmental Protection Agency, Washington, DC, April 2020b, 2 pages.

EPA, Inventory of U.S. Greenhouse Gas Emissions and Sinks: 1990-2021, EPA 430-R-23-002, U.S. Environmental Protection Agency, Washington D.C., 2023, 881 pages.

GAO, *Disaster Resilience Framework: Principles for Analyzing Federal Efforts to Facilitate and Promote Resilience to Natural Disasters*, GAO-20-100SP, U.S. Government Accounting Office, Washington, DC, October 2019, 24 pages.

GBI, *About Green Globes*, Green Building Initiative, Inc. thegbi.org, McAdam, NB, Canada, 2020.

GCCA, *GCCA Climate Ambition Statement: Toward Carbon Neutral Concrete*, Global Cement and Concrete Association, London, UK, September 2020, 19 pages.

Ge, M., *World Greenhouse Gas Emissions: 2016, Data Visualization*, wri.org/resources/data-visualizations/world-greenhouse-gas-emissions-2016, World Resources Institute, Washington, DC, February 2020.

Hammond, G.P.; and Jones, C.I., *Embodied Carbon: The Inventory of Carbon and Energy (ICE)*, BSRIA, Bracknell, Berkshire, UK, 2011, 136 pages.

Holland, L.S. "Construction Spending Survey, Annual Value of Construction Put in Place 2008 – 2019," *Annual 2008-2019*, US Census Bureau, Construction Expenditures Branch, August 29, 2019.

IEA, *Cement Technology Roadmap 2009 – Carbon Emissions Reductions up to 2050*, World Business Council for Sustainable Development, Geneva, Switzerland, and International Energy Agency, Paris, France, 2009, 36 pages.

ILFI, *Living Building Challenge 4.0 SM*, International Living Future Institute, June 2019, 84 pages.

ISO 14025:2006, *Environmental Labels and Declarations – Type III Environmental Declarations – Principles and Procedures*, ISO/TC 207/SC 3 Environmental Labelling, International Standards Organization, Geneva, Switzerland, 2006, 25 pages.

ISO 14044:2006, *Environmental Management – Life Cycle Assessment – Requirements and Guidelines*, ISO/TC 207/SC 5 Life Cycle Assessment, International Standards Organization, Geneva, Switzerland, 2006, 46 pages.

ISO 21930:2017, *Sustainability in Buildings and Civil Engineering Works – Core Rules for Environmental Product Declarations of Construction Products and Services*, ISO/TC 59/SC 17 Sustainability in Buildings and Civil Engineering Works, International Standards Organization, Geneva, Switzerland, 2017, 80 pages.

Levinson, R.; and Akbari, H., *Effects of Composition and Exposure on the Solar Reflectance of Portland Cement Concrete*, Publication No. LBNL- 48334, Lawrence Berkeley National Laboratory, Berkeley, CA, 2001, 39 pages.

Louhghalam, A.; Akbarian, M.; and Ulm, F.J., "Carbon Management of Infrastructure Performance: Integrated Big Data Analysis and Pavement-Vehicle-Interactions," *Journal of Cleaner Production*, Vol. 142, Part 2, 2017, pages 956 to 964.

Marceau, M.L.; and VanGeem, M.G., *Solar Reflectance of Concretes for LEED Sustainable Site Credit: Heat Island Effect*, SN2982, Portland Cement Association, Skokie, IL, 2007, 94 pages.

MIT, *Pavement Vehicle Interaction: Lowering Vehicle Fuel Consumption and Emissions Through Better Pavement Design and Maintenance*, Research Brief, Concrete Sustainability Hub, Massachusetts Institute of Technology, Cambridge, MA, October 2016, 3 pages.

MIT, *Pavement Vehicle Interaction: Lowering Vehicle Fuel Consumption and Emissions Through Better Pavement Design and Maintenance*, Research Brief, Concrete Sustainability Hub, Massachusetts Institute of Technology, Cambridge, MA, July 2018, 2 pages.

Monkman, S.; MacDonald, M.; and Hooton, R.D., "Using CO2 to Reduce the Carbon Footprint of Concrete," *1st International Conference on Grand Challenges in Construction Materials*, Los Angeles, CA, March 17-18, 2016, 8 pages.

Noori, M.; Miller, R.; Kirchain, R.; and Gregory, J., "How Much should be Invested in Hazard Mitigation? Development of a Streamlined Hazard Mitigation Cost Assessment Framework," *International Journal of Disaster Risk Reduction*, Vol. 28, June 2018, pages 578 to 584.

NRC, *Disaster Resilience: A National Imperative*, National Research Council, The National Academies Press, Washington, DC, 2012.

NRMCA, *Environmental Product Declaration: NRMCA Member Industry Average EPD for Ready Mixed Concrete*, National Ready Mixed Concrete Association, Silver Spring, MD, January 3, 2022, 33 pages.

NRMCA, "Reducing Embodied Carbon in Concrete Mixtures," *Technology in Practice: What, Why, and How?*, TIP 22, National Ready Mixed Concrete Association, Alexandria, VA, 2021, 8 pages.

NSF International, *Product Category Rule for Environmental Product Declarations: PCR for Concrete*, Version 2.1, Ann Arbor, Michigan, August 2021, 40 pages.

NSF International, *Product Category Rule for Environmental Product Declarations: PCR for Portland, Blended, Masonry, Mortar, and Plastic (Stucco) Cements*, Version 3.1, Ann Arbor, MI, September 2020, 33 pages.

Ochsendorf, J.; Norford, L.K.; Brown, D.; Durschlag, H.; Hsu, S.L.; Love, A.; Santero, N.; Swei, O.; Webb, A.; and Wildnauer, M., *Methods, Impacts, and Opportunities in the Concrete Building Life Cycle*, Research Report R11-01, Concrete Sustainability Hub, Massachusetts Institute of Technology, Cambridge, MA, August 2011, 119 pages.

Pomerantz, M.; Akbari, H.; Chang, S.C.; Levinson, R.; and Pon, B., *Examples of Cooler Reflective Streets for Urban Heat-Island Mitigation: Portland Cement Concrete and Chip Seals*, Lawrence Berkeley National Laboratory, Publication No. LBNL- 49283, 2002, 24 pages.

Pomerantz, M.; Pon, B.; and Akbari, H., *The Effect of Pavements' Temperatures on Air Temperatures in Large Cities*, Publication No. LBNL- 43442, Lawrence Berkeley National Laboratory, Berkeley, CA, 2000, 20 pages.

PCA, *Environmental Product Declaration for Portland Cements*, Portland Cement Association, Washington, D.C., and ASTM International, West Conshohocken, PA, Revised October 2023a, 13 pages.

PCA, *Environmental Product Declaration: Portland-Limestone Cements*, Portland Cement Association, Washington D.C., and ASTM International, West Conshohocken, PA, Revised October 2023b, 13 pages.

PCA, *Environmental Product Declaration for Blended Cements*, Portland Cement Association, Washington D.C., and ASTM International, West Conshohocken, PA, Revised October 2023c, 13 pages.

PCA, *Environmental Product Declaration for Masonry Cements*, Portland Cement Association, Washington D.C., and ASTM International, West Conshohocken, PA, Revised October 2023d, 13 pages.

PCA Roadmap to Carbon Neutrality, Portland Cement Association, Washington D.C., October 2021, 73 pages.

RMA, *2013 Scrap Tire Management Summary*, Rubber Manufacturers Association, Washington, D.C., November 2014.

Santero, N.; Loijos, A.; and Ochsendorf, J., "Greenhouse Gas Emissions Reduction Opportunities for Concrete Pavements," *Journal of Industrial Ecology*, September 2013, 10 pages.

Schneider, M., "Process Technology for Efficient and Sustainable Cement Production," *Cement and Concrete Research*, Vol. 78, December 2015, pages 14 to 23.

Scrivener, K.L.; John, V.M.; and Gartner, E.M., "Eco-efficient Cements: Potential Economically Viable Solutions for a Low-CO$_2$ Cement-Based Materials Industry," *Cement and Concrete Research*, Vol 114, 2018, pages 2 to 26.

SEAONC, *Earthquake Performance Rating System User's Guide*, Building Ratings Committee of the Structural Engineers Association of Northern California, Sacramento, CA, February 2, 2015, 34 pages.

Stripple, H.; Ljungkrantz, C.; Gustafsson, T.; and Andersson, R., CO$_2$ *Uptake in Cement Containing Products: Background and Calculation Models for IPCC Implementation*, B 2309, IVL Swedish Environmental Research Institute, October 2018, 66 pages.

Sullivan, E.J.; Bohan, R.P.; and Storck, T., *U.S. Labor-Energy Input Survey 2019*, Portland Cement Association, November 2020, 37 pages.

Tennis, P. D.; Thomas, M.D.A.; and Weiss, W. J., *State-of-the-Art Report on Use of Limestone in Cements at Levels of up to 15%*, SN3148, Portland Cement Association, Skokie, IL, 2011, 78 pages.

Thomas, M.D.A.; and Hooton, R.D., *The Durability of Concrete Produced with Portland-Limestone Cement: Canadian Studies*, SN3142, Portland Cement Association, Skokie, IL, 2010, 28 pages.

Townsend, T.G.; and Anshassi, M., *Benefits of Construction and Demolition Debris Recycling in the United States*, Construction Demolition Recycling Association, Chicago, IL, April 2017, 22 pages.

USBEA, *Value Added by Private Industries: Construction as a Percentage of GDP [VAPGDPC]*, US Bureau of Economic Analysis, retrieved from FRED, Federal Reserve Bank of St. Louis; fred.stlouisfed.org/series/VAPGDPC, December 22, 2020.

USGBC, *LEED v4.1 Building Design and Construction: Getting Started Guide for Beta Participants*, US Green Building Council, Washington, D.C., July 2020, 259 pages.

USRC, *Implementation Manual USRC Building Rating System for Earthquake Hazards*, U.S. Resiliency Council, Atherton, CA, January 2019, 54 pages.

USRC, *REDi™ Rating System Resilience-based Earthquake Design Initiative for the Next Generation of Buildings*, Version 1.0, U.S. Resiliency Council, Atherton, CA, October 2013, 68 pages.

VanGeem, M.G. "Optimal Thermal Mass and R-Value in Concrete," *First International Conference on Concrete Sustainability*, Conference Proceedings, Tokyo, May 2013.

Verian, K.P.; Whiting, N.M.; Olek, J.; Jain, J.; and Snyder, M.B., *Using Recycled Concrete as Aggregate in Concrete Pavements to Reduce Materials Cost*, FHWA/IN/JTRP-2013/18, Joint Transportation Research Program, Indiana Department of Transportation and Purdue University, West Lafayette, IN, 2013.

VDZ, *Executive Summary Decarbonising Cement and Concrete: A CO$_2$ Roadmap for German Cement Industry*, Verein Deutscher Zementwerke, Duesseldorf, Germany, November 2020, 10 pages.

Worrell, E.; Kermeli, K.; and Galitsky, C., *Energy Efficiency Improvement and Cost Saving Opportunities for Cement Making: An ENERGY STAR® Guide for Energy and Plant Managers*, U.S. Environmental Protection Agency, Washington, DC, August 2013, 141 pages.

SPECIFYING, DESIGNING, AND PROPORTIONING CONCRETE MIXTURES

FIGURE 13-1. The design and proportioning of concrete mixtures involves reviewing particular conditions of use, establishing specific concrete characteristics, and selecting proportions of available materials to produce concrete of required properties, with the greatest economy.

The objective of designing and proportioning concrete mixtures is to determine the most economical and practical combination of readily available materials to produce a concrete that will satisfy the constructability and project performance requirements under particular conditions of use (Figure 13-1). The mixture requirements are often established within standards and specifications which may be prescriptive, performance-based, or a combination of both.

The terms mixture design (mix design) and mixture proportioning are often incorrectly used interchangeably (Taylor and others 2019). As far back as 1918, this distinction was recognized:

> *The term "design" is used in the title of this article as distinguished from "proportioning" since it is the intention to imply that each element of the problem is approached with a deliberate purpose in view which is guided by a rational method of accomplishment. (Abrams 1918)*

Designing concrete mixtures is the process by which the concrete mixture performance characteristics are defined. It is in the design process that parameters such as air content, workability, and required strength and durability are established. These parameters are driven by factors such as the concrete's service environment, construction method, and structural requirements. Once the mixture's design parameters are established, the materials characteristics and placement methods are identified and determined. Then, the concrete mixture proportions can be developed using relationships established through research or from past experience.

The output from the design process becomes part of the input for proportioning concrete mixtures (Figure 13-2). In turn, the output from the proportioning process becomes the actual batch quantities of the various ingredients in the concrete mixture.

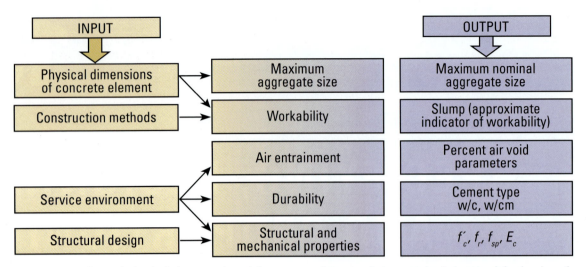

FIGURE 13-2. Inputs to mixture design include constraints of the concrete element and clear spacing between reinforcing, how the concrete will be batched and placed, the service environment with its associated durability requirements, and the required mechanical properties for the concrete. Output is the set of required characteristics for the mixture and each ingredient (courtesy of K. Hover).

Figure 13-3 outlines the steps in the process of design and proportioning concrete mixtures. Knowledge of the local materials and the relationships among selected ingredients combine to produce a set of mixture proportions. These proportions are designed to meet the requirements for both fresh and hardened concrete provided that the concrete is properly batched, mixed, transported, placed, and cured. For a given concrete application, there may be many different mixture proportions that satisfy the design requirements.

Performance records from the lab or field are required to determine whether any given set of materials and proportions can reliably meet the design requirements. Concrete materials are themselves variable, and the relationships among proportions are mixture- and materials-specific. Therefore, any set of proposed mixture proportions is referred to as a trial mixture, until satisfactory performance has been demonstrated through laboratory or field testing.

SPECIFYING CONCRETE MIXTURES

Contract documents contain performance requirements for fresh and hardened concrete, typically along with criteria for acceptable concrete ingredients including the cementitious materials, water, aggregates, and admixtures. Requirements may differ in the level of detail concerning proportions or ratios of ingredients, varying from a fully prescriptive specification that defines all ingredients and batch weights to a more performance-oriented specification that lists end-result concrete properties with no limitations on ingredients or proportions. Contract documents may either fully state material and mixture requirements or they may incorporate such requirements by referencing a standard specification such as ACI 301, *Specifications for Concrete Construction* (2020); or a variety of other public-agency standards and specifications including ASTM International (ASTM), American Association of State Highway and Transportation Officials (AASHTO), and other American Concrete Institute (ACI) standards.

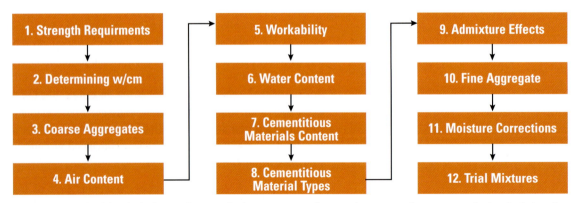

FIGURE 13-3. Steps involved in designing and proportioning concrete mixtures. Inputs to mixture proportioning include mixture design requirements, characteristics of the specific materials to be used, and requirements based on production technology, which then produce a set of mixture proportions expressed as batch weights.

Prescriptive and Performance Based Specifications

A prescriptive specification sets limits on the amounts of ingredients (such as cement content and air content). The limits are based on experience and an accepted general relationship between a value, such as w/cm, and an expected outcome (such as freeze-thaw resistance). Proportioning the concrete ingredients is a relatively easy task.

A performance approach defines the properties needed by the concrete for the application and uses tests to demonstrate compliance. There are no pre-established ratios or required amounts of ingredients. A performance approach lends opportunity to make better use of locally available materials; providing potential for improved constructability, engineering properties, durability, sustainability, and economy.

There are several challenges with a pure performance approach. The engineer or specifier must fully understand what concrete properties are needed for the project. These properties need to be properly conveyed within the specification in order to find the best proportions to meet the criteria. Tests can take months or years to perform. Also, some performance properties do not have reliable or fast corresponding test methods. There are potential risks if the specifier does not align the performance and prescriptive aspects of the design and project schedule.

Most concrete mixtures are designed and proportioned using a combination of prescriptive and performance criteria relative to experience with local materials. Several references are available on the topic of changing from a prescriptive approach to a performance approach. For more information, see ACI 329R, *Report on Performance-Based Requirements for Concrete* (2014), PCA EB233, *Guide Specification for High-Performance Concrete for Bridges* (Caldarone and others 2005), and Hover and others (2008).

DESIGNING CONCRETE MIXTURES

Before a concrete mixture can be proportioned, mixture characteristics are determined from the intended use of the concrete, the exposure conditions, the size and shape of building elements, and the physical properties of the concrete required for the structure.

Factors to be considered when designing concrete mixtures include:
• workability,
• placement conditions,
• strength requirements,
• durability,
• appearance, and
• economy.

Proportions in a concrete mixture will vary widely depending on the properties of the constituent materials and the final performance required from the concrete. This is the design process. The performance requirements considered in the design process should include the properties of the fresh concrete. A properly proportioned concrete mixture should possess sufficient workability to allow completion of placement and finishing operations, taking into consideration how the concrete is to be placed and consolidated. Concrete must be designed to meet required hardened properties. The concrete must have adequate strength to be able to carry the intended loads. Although strength is an important characteristic, other properties such as volume stability, durability, permeability, and wear resistance may be just as, if not more, important, especially when considering designing structures for a target service life. The mixture characteristics should reflect the needs of the application; for example, resistance to chloride ions should be verifiable and the appropriate test methods specified. Often, the aesthetics of the finished concrete are a major consideration, and materials and mixtures may need to be prequalified for architectural purposes. When aesthetics are an acceptance criteria, the project may require a mock-up. To meet construction budgets, concrete must be designed so the performance is met in the most economical fashion.

Understanding the basic principles of designing concrete mixtures is as important as the calculations used to establish mixture proportions. Only with proper selection of materials and mixture characteristics can the above qualities be obtained in concrete construction (Abrams 1918, Hover 1998, and Shilstone 1990).

Concrete mixtures should be kept as simple as reasonably possible. An excessive number of ingredients often makes a concrete mixture difficult to reliably produce. However, opportunities provided by modern concrete technology should not be overlooked.

The following breaks down design considerations for each of the twelve steps illustrated in Figure 13-3.

1. Strength Requirements

Strength (compressive or flexural) is the most universally used performance measure for concrete quality. The specified strength of a concrete mixture is selected considering both the structural and durability requirements of the concrete. The strength required to resist the loads applied to the structure is part of the original structural design (see Chapter 9).

Compressive strength is most commonly used for the design of concrete structures. Flexural strength is often used for the design of concrete pavements (see Chapter 21). However, flexural testing is generally avoided due to greater variability in the test method. Strength tests of standard cured compressive strength cylinders can be correlated to the flexural strength of concrete (Kosmatka 1985). Because of the robust empirical relationship between these two strength methods and the economics of testing cylinders instead of beams, most state departments of transportation are now using compression tests to monitor concrete quality for their pavement and bridge projects.

Strength requirements for various exposure conditions are given in ACI 318, *Building Code Requirements for Structural Concrete* (2019). ACI 318-19 specifies minimum compressive strengths for concrete depending on the exposure conditions (Table 13-1). For example, as shown in Table 13-1, structural concrete exposed to freezing and thawing and deicers is required to have a minimum specified strength of 35 MPa (5000 psi). For concrete mixtures that have more than one exposure classifications (for example sulfates and corrosion), the critical exposure class will be the one with highest strength requirement.

Within the normal range of strengths used in concrete construction, the compressive strength is inversely related to the water-cementitious materials ratio. For fully compacted concrete made with clean, sound aggregates, the strength and other desirable properties of concrete under given job conditions are governed by the quantity of mixing water used per unit of cement or cementing materials (Abrams 1918).

The strength of the cementitious paste binder in concrete depends on the quality and quantity of the reacting paste components and on the degree to which the hydration reaction has progressed. Concrete becomes stronger with time as long as there is available moisture (above 80% relative humidity) and a favorable temperature. Therefore, the strength at any particular age is both a function of the original water cementitious material ratio and the degree to which the cementitious materials have hydrated. The importance of prompt and thorough curing is easily recognized (see Chapter 17).

Differences in concrete strength for a given water cementing materials ratio may result from: changes in the aggregate size, grading, surface texture, shape, cleanliness, strength, and stiffness; differences in types and sources of cementing materials; entrained-air content; the presence, and type of admixtures; and the length of curing time and temperature.

Specified Compressive Strength and Acceptance Criteria. The specified compressive strength, f'_c, at 28 days, is the design strength. ACI 318-19 requires for f'_c to be at least 17.5 MPa (2500 psi). The strength of concrete is deemed satisfactory provided the acceptance criteria in ACI 318-19 and ACI 301-20 is met. The strength is expected to be equal to or exceeded by the average of any set of three consecutive strength tests. No individual test (average of two cylinders) can be more than 3.5 MPa (500 psi) below the specified strength. Specimens must be cured under laboratory conditions for an individual class of concrete (ACI 318 2019). Some specifications have alternative requirements.

Average Strength. The average strength, f'_{cr}, is the strength required in designing the mixture. For concrete mixtures, f'_{cr} should equal the specified strength plus an allowance to account for variations in materials; variations in methods of

TABLE 13-1. Maximum Water-Cementitious Material Ratios and Minimum Design Strengths for Various Exposure Conditions

EXPOSURE CATEGORY	EXPOSURE CONDITION	MAXIMUM WATER-CEMENTITIOUS MATERIAL RATIO	MINIMUM DESIGN COMPRESSIVE STRENGTH f'_c, MPa (psi)
F0, S0, W0, C0	Concrete protected from exposure to freezing and thawing, application of deicing chemicals, or aggressive substances	Select water-cementitious ratio on basis of strength, workability, and finishing needs	Select strength based on structural requirements
W1, S1	Concrete intended to have low permeability when exposed to water (W1) or moderate sulfates (S1)	0.50	28 (4000)
F1	Concrete exposed to freezing and thawing with limited exposure to moisture	0.55	25 (3500)
F2, S2	Concrete exposed to freezing and thawing with exposure to moisture (F2) or severe sulfates (S2)	0.45	31 (4500)
F3* S3	Concrete exposed to freezing-and-thawing cycles with frequent exposure to water and exposure to deicing chemicals (F3) or very severe sulfates (S3)	0.40	35 (5000)
C2	For corrosion protection for reinforced concrete exposed to chlorides from deicing salts, salt water, brackish water, seawater, or spray from these sources	0.40	35 (5000)

Adapted from ACI 318-19.
The following four exposure categories determine durability requirements for concrete (see Chapter 11):
F – Freezing and Thawing; S – Sulfates; W – Water; and C – Corrosion.
Increasing numerical values represent increasingly severe exposure conditions.
* For plain (unreinforced) concrete, the maximum w/cm shall be 0.45 and the minimum design strength shall be 31 MPa (4500 psi).

mixing, transporting, and placing the concrete; and variations in making, curing, and testing concrete cylinder specimens.

Proportioning from field data. Existing or previously used concrete mixture proportion can be used for a new project provided strength-test data and standard deviations show that the mixture is acceptable. Durability requirements from Table 13-1 must also be met. The statistical data should essentially represent the same materials, similar proportions, and concreting conditions to be used in the new project. The data used for proportioning should also be from concrete with an f'_c that is within 7 MPa (1000 psi) of the strength required for the proposed work. Also, the data should represent at least 30 consecutive tests or two groups of consecutive tests totaling at least 30 tests (one test is the average strength of two cylinders from the same sample). If only 15 to 29 consecutive tests are available, an adjusted standard deviation can be obtained by multiplying the standard deviation (S) for the 15 to 29 tests and a modification factor from Table 13-2. The data must represent at least 45 days of tests.

TABLE 13-2. Modification Factor for Standard Deviation When Less Than 30 Tests Are Available

NUMBER OF TESTS*	MODIFICATION FACTOR FOR STANDARD DEVIATION†
Less than 15	Use Table 13-3
15	1.16
20	1.08
25	1.03
30 or more	1.00

*Interpolate for intermediate numbers of tests.

† Modified standard deviation to be used to determine required average strength, f'_{cr}.

Adapted from ACI 301-20 and 214R-11.

The standard or modified deviation is then used in Equations 13-1 to 13-3. The average compressive strength from the test record must equal or exceed the required average compressive strength, f'_{cr}, in order for the concrete proportions to be acceptable. The f'_{cr} for the selected mixture proportions is equal to the larger of Equations 13-1 and 13-2 (for $f'_c \leq 35$ MPa [5000 psi]), or the larger of Equations 13-1 and 13-3 (for $f'_c > 35$ MPa [5000 psi]).

$$f'_{cr} = f'_c + 1.34S$$

EQUATION 13-1.

$$f'_{cr} = f'_c + 2.33S - 3.45 \text{ (MPa)} \quad \text{(SI)}$$
$$f'_{cr} = f'_c + 2.33S - 500 \text{ (psi)} \quad \text{(Inch-Pound)}$$

EQUATION 13-2.

$$f'_{cr} = 0.90 \, f'_c + 2.33S$$

EQUATION 13-3.

where:

f'_{cr} = required average compressive strength of concrete used as the basis for selection of concrete proportions, MPa (psi)

f'_c = specified compressive strength of concrete, MPa (psi)

S = standard deviation, MPa (psi)

When field strength test records do not meet the previously discussed requirements, f'_{cr} can be obtained from Table 13-3. A field strength record, several strength test records, or tests from trial mixtures must be used for documentation showing that the average strength of the mixture is equal to or greater than f'_{cr}.

TABLE 13-3A (SI). Required Average Compressive Strength When Data Are Not Available to Establish a Standard Deviation

SPECIFIED COMPRESSIVE STRENGTH, f'_c, MPa	REQUIRED AVERAGE COMPRESSIVE STRENGTH, f'_{cr}, MPa
Less than 21	$f'_c + 7.0$
21 to 35	$f'_c + 8.5$
Over 35	$1.10 \, f'_c + 5.0$

TABLE 13-3B (Inch-Pound). Required Average Compressive Strength When Data Are Not Available to Establish a Standard Deviation

SPECIFIED COMPRESSIVE STRENGTH, f'_c, psi	REQUIRED AVERAGE COMPRESSIVE STRENGTH, f'_{cr}, psi
Less than 3000	$f'_c + 1000$
3000 to 5000	$f'_c + 1200$
Over 5000	$1.10 \, f'_c + 700$

If less than 30, but not less than 10 tests are available, the tests may be used for average strength documentation if the tests were obtained in a single period of 45 days. Mixture proportions may also be established by interpolating between two or more test records if each meets the above requirements and the project requirements. If a significant difference exists between the mixtures that are used in the interpolation, a trial mixture should be considered to check strength gain. If the test records meet the above requirements, the proportions for the mixture may then be considered acceptable for the proposed work.

If the average strength of the mixtures with the statistical data is less than f'_{cr}, or statistical data or test records are insufficient or not available, the mixture should be proportioned by the trial-mixture method. The approved mixture must have a compressive strength that meets or exceeds f'_{cr}.

Trial mixtures using at least three different water-cementing materials ratios or three different cementing materials contents should be tested. A water-to-cementing-materials ratio vs. strength curve (similar to Figure 13-4) can then be plotted and the proportions interpolated from the data.

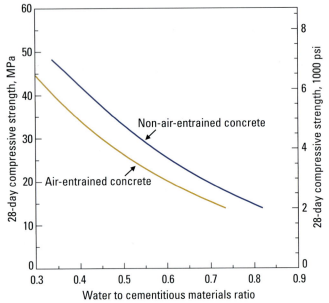

FIGURE 13-4. Approximate relationship between compressive strength and water-to-cementing materials ratio for concrete using 19-mm to 25-mm (¾-in. to 1-in.) nominal maximum size coarse aggregate. Strength is based on cylinders moist cured 28 days per ASTM C31 (AASHTO T 23). Adapted from ACI 211.1-91 and Hover 1995.

ACI 214 (2011) provides statistical analysis methods for monitoring the strength of concrete in the field to ensure that the mix properly meets or exceeds the design strength, f'_c.

Trial Mixtures. When field test records are not available or are insufficient for proportioning by field experience methods, the concrete proportions selected to meet design strength and other mixture requirements should be based on trial mixtures. The trial mixtures should use the same materials proposed for the work. Mixtures with three different water-cementing materials ratios or cementing materials contents should be made to produce a range of strengths that encompass f'_{cr}. The trial mixtures should have a slump within ±20 mm (±0.75 in.) and air content within 0.5% of the maximum permitted. Three cylinders for each water-cementing materials ratio should be made and cured according to ASTM C192, *Standard Practice for Making and Curing Concrete Test Specimens in the Laboratory* (AASHTO R 39). At 28 days, or the designated test age, the compressive strength of the concrete should be determined by testing representative cylinders. The test results should be plotted to produce a strength versus water-cementing materials ratio curve (similar to Figure 13-4) that is used to proportion a mixture.

2. Determining w/cm

The importance of the water-cementitious materials ratio (w/cm) and its impact on the strength and durability of concrete has been well established (Abrams 1918) and is discussed throughout this book (see Chapters 1, 9, 11). The w/cm selected for designing a concrete mixture must be the lowest value required to meet strength and anticipated exposure conditions. Table 13-1 shows w/cm requirements for various exposure conditions. When durability is not critical, the water-cementitious materials ratio should be selected on the basis of concrete compressive strength.

In designing concrete mixtures, the water-to-cementitious materials ratio, w/cm, is often used synonymously with water-to-cement ratio (w/c); however, some specifications differentiate between the two ratios. The w/c is simply the mass of water divided by the mass of portland cement. Likewise, the w/cm is the mass of water divided by the mass of all cementitious materials (portland cement, blended cement, fly ash, slag cement, silica fume, and natural pozzolans).

The water-cementitious materials ratio and mixture proportions for the required strength should be based on adequate field data or trial mixtures made using the same materials planned for the project. When no other data are available for a trial mixture, Table 13-4 or Figure 13-4 may be used as a starting point for trial mixtures to estimate a water-cementitious materials ratio with respect to the required average strength, f'_{cr}. Trial batching is required to establish the relationship between the w/cm and strength as discussed in Step 1.

3. Coarse Aggregates

Aggregates for use in concrete should meet the requirements of ASTM C33, *Standard Specification for Concrete Aggregates*, or AASHTO M 80, *Standard Specification for Coarse Aggregate for Hydraulic Cement Concrete*, as discussed in Chapter 5. The grading characteristics and nature of aggregate particles (shape, porosity, and surface texture) have an important influence on concrete mixtures because they affect the workability of the fresh concrete, as well as other concrete properties such as shrinkage, flexural strength, and compressive strength. The characteristics of the aggregate should be qualified for use in the concrete mixture. This includes testing the aggregates for deleterious substances and determining if any mitigation measures are needed such as for alkali silica reactivity (see Chapter 5 and Chapter 11).

Grading is important for attaining an economical mixture. It affects the amount of concrete that can be made with a given amount of cementitious materials and water. Coarse aggregates should be graded up to the largest size practical under job conditions.

TABLE 13-4A (SI). Relationship Between Water-to-Cementitious Materials Ratio and Compressive Strength of Concrete

COMPRESSIVE STRENGTH AT 28 DAYS, MPa	WATER-CEMENTITIOUS MATERIALS RATIO BY MASS	
	NON-AIR-ENTRAINED CONCRETE	AIR-ENTRAINED CONCRETE
45	0.38	0.30
40	0.42	0.34
35	0.47	0.39
30	0.54	0.45
25	0.61	0.52
20	0.69	0.60
15	0.79	0.70

Strength is based on cylinders moist-cured 28 days in accordance with ASTM C31 (AASHTO T 23). Relationship assumes nominal maximum size aggregate of about 19 mm to 25 mm. Adapted from ACI 211.1-91.

TABLE 13-4B (Inch-Pound). Relationship Between Water-to-Cementitious Materials Ratio and Compressive Strength of Concrete

COMPRESSIVE STRENGTH AT 28 DAYS, psi	WATER-CEMENTITIOUS MATERIALS RATIO BY MASS	
	NON-AIR-ENTRAINED CONCRETE	AIR-ENTRAINED CONCRETE
7000	0.33	–
6000	0.41	0.32
5000	0.48	0.40
4000	0.57	0.48
3000	0.68	0.59
2000	0.82	0.74

Strength is based on cylinders moist-cured 28 days in accordance with ASTM C31 (AASHTO T 23). Relationship assumes nominal maximum size aggregate of about ¾ in. to 1 in. Adapted from ACI 211.1-91.

The amount of mixing water required to produce a unit volume of concrete of a given slump is dependent on the shape and the maximum size and amount of coarse aggregate. Larger sizes minimize the water requirement and thus potentially allow the cement content to be reduced. Also, rounded aggregate requires less mixing water than crushed aggregate in concretes of equal slump (see **Water Content**).

The maximum size of coarse aggregate that will produce concrete of maximum strength for a given cement content depends upon the aggregate source as well as its shape and grading. For high compressive-strength concrete (greater than 70 MPa or 10,000 psi), the maximum size is about 19 mm (¾ in.). Higher strengths can also sometimes be achieved using crushed stone aggregate rather than rounded-gravel aggregate.

The maximum size that can be used depends on factors such as the size and shape of the concrete member to be cast, the amount and distribution of reinforcing steel in the member, and the thickness of slabs. Grading also influences the workability and placeability of the concrete. Sometimes mid-sized aggregate, around the 9.5 mm (⅜ in.) size, is lacking in an aggregate supply. This can result in a concrete with high shrinkage properties, high water demand, poor workability, and poor placeability. Durability may also be affected. Various options are available for obtaining optimal grading of aggregate (see Chapter 5, Abrams (1918), Shilstone (1990), and Cook and others (2013).

Requirements for Nominal Maximum Size Aggregate. Limits on nominal maximum size of aggregate particles reduce the risk of voids in a structure. The nominal maximum size of aggregate should not exceed (see Chapter 5):

• ⅕ the narrowest dimension of a vertical concrete member;

• ¾ the clear spacing between reinforcing bars and between the reinforcing bars and forms; and

• ⅓ the depth of slabs.

These requirements may be waived if, in the judgment of the designer, the mixture possesses sufficient workability that the concrete can be properly placed without honeycomb or voids. Smaller sizes can be used when availability or economic consideration require them.

Fineness Modulus. Fineness modulus (FM), is an indicator of the fineness of an aggregate which is calculated from the particle size distribution of sand. In general, the higher the FM, the coarser the aggregates. However different gradings can have the same FM. Values of the fineness modulus for concrete mixtures should be between 2.3 and 3.1 and within 0.2 for the same concrete mixture. Chapter 5 provides example to determine the FM. The most desirable fine-aggregate grading will depend upon the type of work, the paste content of the mixture, and the size of the coarse aggregate. For leaner mixtures, a fine grading (lower fineness modulus) is desirable for workability. For richer mixtures, a coarse grading (higher fineness modulus) is used for greater economy.

Bulk Volume of Coarse Aggregate. The bulk volume of coarse aggregate can be determined from Table 13-5 or Figure 13-5. These bulk volumes are based on aggregates in a dry-rodded condition as described in ASTM C29, *Standard Test Method for Bulk Density ("Unit Weight") and Voids in Aggregate* (AASHTO T 19). The volume includes the void space between the particles of stone. They are selected from empirical relationships between nominal maximum size of aggregates and the fineness modulus (FM) of the fine aggregates. The bulk volumes shown in Table 13-5 should produce concrete with a degree of workability suitable for general reinforced concrete construction. For less workable concrete, such as required for concrete pavement construction, bulk volume of coarse aggregate may be increased about 10%. For more workable concrete, such as may be required when placement by pump, bulk volume of coarse aggregate may be reduced up to 10%.

TABLE 13-5. Bulk Volume of Coarse Aggregate Per Unit Volume of Concrete

NOMINAL MAXIMUM SIZE OF AGGREGATE, mm (in.)	FINENESS MODULI OF FINE AGGREGATE*			
	2.40	2.60	2.80	3.0
9.5 (⅜)	0.50	0.48	0.46	0.44
12.5 (½)	0.59	0.57	0.55	0.53
19 (¾)	0.66	0.64	0.62	0.60
25 (1)	0.71	0.69	0.67	0.65
37.5 (1½)	0.75	0.73	0.71	0.69
50 (2)	0.78	0.76	0.74	0.72
75 (3)	0.82	0.80	0.78	0.76
150 (6)	0.87	0.85	0.83	0.81

*Bulk volumes are based on aggregates in a dry-rodded condition as described in ASTM C29 (AASHTO T 19).

Adapted from ACI 211.1 (1991).

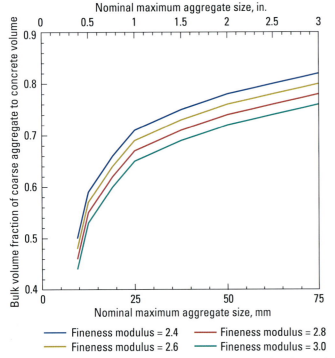

FIGURE 13-5. Bulk volume of coarse aggregate per unit volume of concrete. Bulk volumes are based on aggregates in a dry-rodded condition as described in ASTM C29 (AASHTO T 19). For more workable concrete, such as may be required when placement is by pump, bulk volume may be reduced up to 10%. Adapted from ACI 211.1 (1991) and Hover (1995 and 1998).

4. Air Content

Entrained air should be used in all concrete that will be exposed to freezing and thawing and deicing chemicals and meet the requirements of ASTM C260, *Standard Specification for Air-Entraining Admixtures for Concrete* (AASHTO M 154), as discussed in Chapter 6. Entrained air can also be used to improve workability in concrete mixtures even where not required. Air entrainment is accomplished using an air-entraining portland cement or by adding an air-entraining admixture at the mixer.

The amount of admixture should be adjusted to meet variations in concrete ingredients and job conditions. The amount recommended by the admixture manufacturer will, in most cases, produce the desired air content (see Chapters 6 and 9).

Recommended target air contents for air-entrained concrete are shown in Table 13-6, and typical air contents generated in non-air entrained concretes are shown in Table 13-7. They are also presented in Figure 13-6. Note that the amount of air required to provide adequate freeze-thaw resistance is dependent upon the nominal maximum size of aggregate and the level of exposure. In properly proportioned mixtures, the mortar content decreases as maximum aggregate size increases; decreasing the required concrete air content. This is evident in Figure 13-6 and Tables 13-6 and 13-7.

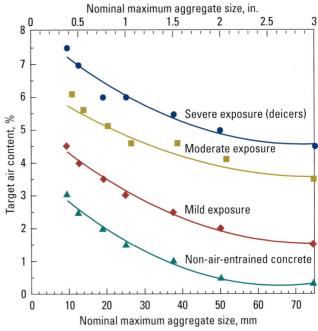

FIGURE 13-6. Target total air content requirements for concretes using different sizes of aggregate. The air content in job specifications should be specified to be delivered within ± 1.5% of the target value for moderate and severe exposures. Adapted from ACI 211.1 (1991), and Hover (1995 and 1998).

The levels of exposure for freeze-thaw are defined by ACI 318-19 as follows:

Mild Exposure, F0. Exposure Class F0 is assigned to concrete that will not be exposed to cycles of freezing and thawing or deicing agents. When air entrainment is desired for a beneficial effect other than durability, such as to improve workability or cohesion or in low cement content concrete to improve strength, entrained air contents lower than those needed for durability can be used. There are applications when air-entrainment should be restricted such as surfaces specified to be finished with steel trowels (see Chapter 16). Note, even in non-air entrained concrete, some air will be generated into the concrete mixture during the mixing process. This volume should be accounted for as shown in Table 13-7.

TABLE 13-6. Target Air* Content Requirements for Concrete Exposed to Cycles of Freezing and Thawing

NOMINAL MAXIMUM AGGREGATE SIZE, mm (in.)	TARGET AIR CONTENT, %	
	F1	F2 AND F3
9.5 (⅜)	6.0	7.5
12.5 (½)	5.5	7.0
19 (¾)	5.0	6.0
25 (1)	4.5	6.0
37.5 (1 ½)	4.5	5.5
50 (2)	4.0	5.0
75 (3)	3.5	4.5
150 (6)	3.0	4.0

*The air content in job specifications should be specified to be delivered within ± 1.5% of the table target value for moderate and severe exposures.

Adapted from ACI 211.1 (1991) and ACI 318-19. Hover (1995) presents this information in graphical form.

Moderate Exposure, F1. Exposure Class F1 is assigned to concrete exposed to cycles of freezing and thawing and that will be occasionally exposed to moisture before freezing and will not be exposed to deicing or other aggressive chemicals. Examples include exterior beams, columns, walls, girders, or slabs that are not in contact with wet soil and are located where they will not receive direct applications of deicing chemicals.

Severe Exposure, F2. Exposure Class F2 is assigned to concrete exposed to cycles of freezing and thawing that is in continuous contact with moisture before freezing where exposure to deicing salt is not anticipated. An example is an exterior water tank or vertical members in contact with soil.

Very Severe Exposure, F3. Exposure Class F3 is assigned to concrete exposed to cycles of freezing and thawing, in continuous contact with moisture, and where exposure to deicing chemicals is anticipated. Examples include pavements, bridge decks, curbs, gutters, sidewalks, canal linings, or exterior water tanks or sumps.

A specific air content may not be readily or repeatedly achieved because of the many variables affecting air content. Therefore, providing a total air content within a permissible range is acceptable. The range of ± 1.5% of the Figure 13-6 or Table 13-6 values is often used in project specifications, (ACI 301-20 and ASTM C94). For example, for a target value of 6% air, the specified range for the concrete delivered to the jobsite is 4.5% to 7.5%.

5. Workability

Concrete must always be made with a workability, consistency, and plasticity suitable for job conditions. Workability is a measure of how easy or difficult it is to place, consolidate, and finish concrete. Consistency is the ability of freshly mixed concrete to flow. Plasticity determines concrete's ease of molding. If more aggregate is used in a concrete mixture, or if less water is added, the mixture becomes stiff (less plastic, and less workable) and difficult to mold. Neither very dry, crumbly mixtures nor very watery, fluid mixtures have sufficient plasticity.

The slump test is used to measure concrete consistency. For a given proportion of cement and aggregate without admixtures, the higher the slump, and the more readily it will flow. Slump is indicative of workability when assessing similar mixtures. However, slump should not be used to compare mixtures of different proportions or make assumptions of hardened concrete properties. When used with different batches of the same mix design, a change in slump indicates a change in consistency and in the characteristics of materials, mixture proportions, water content, mixing, time of test, or the testing itself.

Slump is usually indicated in the job specifications as a range, or as a maximum value not to be exceeded. Current technology allows for a wide range of slump values. ASTM C94, *Standard Specification for Ready-Mixed Concrete*, addresses slump tolerances in detail. For minor batch adjustments, the slump can be changed by about 10 mm by adding or subtracting 2 kg of water per cubic meter of concrete (or changed by 1 in. by adding or subtracting 10 lb [or approximately 1 gal] of water per cubic yard of concrete). Slumps may also be modified with the use of a water-reducing admixture, plasticizer, or viscosity modifier.

Slump should be specified based on the method of placement. Different slumps are needed for various types of concrete construction. For example: for pavements placed by mechanical paver, slump is typically 25 mm to 75 mm (1 in. to 3 in.); for floors placed by chute or pump, slump is typically 75 mm to 125 mm (3 in. to 5 in.); for walls and foundations placed by chute or

TABLE 13-7. Amount* of Entrapped Air Contents in Non-Air-Entrained Concrete

	NOMINAL MAXIMUM AGGREGATE SIZE, mm (in)							
Non-Air-Entrained Concrete	9.5 (⅜)	12.5 (½)	19 (¾)	25 (1)	37.5 (1½)	50 (2)	75 (3)	150 (6)
Entrapped Air, %	3.0	2.5	2.0	1.5	1.0	0.5	0.3	0.2

*The amount of entrapped air is an approximation.
Adapted from ACI 211.1 (1991). Hover (1995) presents this information in graphical form.

pump, slumps typically range from 100 mm to 200 mm (4 in. to 8 in.). Once a slump is selected for the intended application, the mixture should be designed for a target slump range or maximum slump and monitored for only small changes in slump from batch to batch for consistency.

6. Water Content

Mixing water for use in concrete should meet the requirements of ASTM C1602, *Standard Specification for Mixing Water Used in the Production of Hydraulic Cement Concrete*, as discussed in Chapter 4. The required water content of concrete is influenced by aggregate size, aggregate shape, aggregate texture, slump, water-to-cementing materials ratio, air content, cementing materials type and content, admixtures, and environmental conditions. An increase in air content and aggregate size, a reduction in water-cementing materials ratio and slump, and the use of rounded aggregates, water-reducing admixtures, or certain SCMs such as fly ash can reduce water demand. On the other hand, increased temperatures, cement contents, slump, water-cement ratio, aggregate angularity, and a decrease in the proportion of coarse aggregate to fine aggregate will increase water demand.

The approximate water contents in Table 13-8 and Figure 13-7, used in proportioning, are for angular coarse aggregates (crushed stone). For some concretes and aggregates, the water estimates in Table 13-8 and Figure 13-7 can be reduced by approximately 10 kg (20 lb) for subangular aggregate, 20 kg (35 lb) for gravel with some crushed particles, and 25 kg (45 lb) for a rounded gravel.

Supplementary cementitious materials have varied effects on water demand and air contents (see Chapter 3). This illustrates the need for trial batch testing of local materials, as each aggregate source is different and can influence concrete properties differently.

When mixing water is held constant, the entrainment of air will increase slump. When cement content and slump are held constant, the entrainment of air results in the need for less mixing water, particularly in leaner concrete mixtures. In batch adjustments, to maintain a constant slump while changing the air content, the water should be decreased by about 3 kg/m³ (5 lb/yd³ [or approx ½ gal/yd³]) for each percentage point increase in air content or increased by about 3 kg/m³ (5 lb/yd³ [or approx ½ gal/yd³]) for each percentage point decrease. This is accounted for by separate rows on water contents for non-air entrained concrete and air-entrained concrete in Table 13-8.

7. Cementitious Materials Content

The cementing materials content is commonly determined from the selected water-cementing materials ratio and water content. It is generally desirable to specify the minimum amount of mixing water and cementitious materials necessary to achieve the desired consistency while achieving other required fresh and hardened properties. For severe freeze-thaw, deicer, and sulfate exposures, this must be done without exceeding the maximum water-cementitious materials ratios shown in Table 13-1.

The amount of cementitious material necessary for an optimum concrete mixture is dependent on the other mixture characteristics, particularly the aggregates. The quantity of cementitious paste needed should be enough cover each individual aggregate particle and provide workability. Paste requirements will change based on the placement and finishing methods.

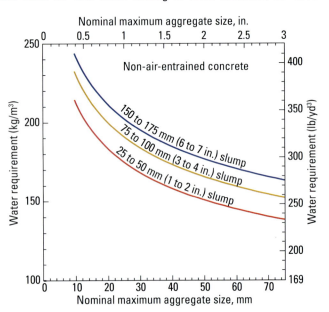

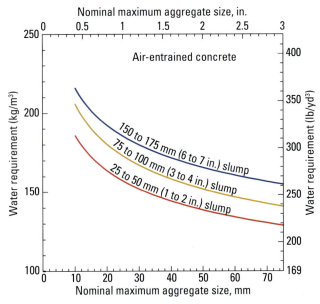

FIGURE 13-7. Approximate water requirement for various slumps and crushed aggregate sizes for (left) non-air-entrained concrete and (right) air-entrained concrete. Adapted from ACI 211.1 (1991), and Hover (1995 and 1998).

TABLE 13-8A (SI). Approximate Mixing Water and Target Air Content Requirements for Different Slumps and Nominal Maximum Sizes of Aggregate

SLUMP, mm	WATER, kg/m³ OF CONCRETE, FOR INDICATED SIZES OF AGGREGATE*							
	9.5 mm	12.5 mm	19 mm	25 mm	37.5 mm	50 mm†	75 mm†	150 mm†
Non-air-entrained concrete								
25 to 50	207	199	190	179	166	154	130	113
75 to 100	228	216	205	193	181	169	145	124
150 to 175	243	228	216	202	190	178	160	–
Air-entrained concrete								
25 to 50	181	175	168	160	150	142	122	107
75 to 100	202	193	184	175	165	157	133	119
150 to 175	216	205	197	184	174	166	154	–

*These quantities of mixing water are for use in computing cementitious material contents for trial batches. They are maximums for reasonably well-shaped angular coarse aggregates graded within limits of accepted specifications.

† The slump values for concrete containing aggregates larger than 37.5 mm are based on slump tests made after removal of particles larger than 37.5 mm by wet screening.

Adapted from ACI 211.1-91. Hover (1995) presents this information in graphical form.

TABLE 13-8B (Inch-Pound). Approximate Mixing Water and Target Air Content Requirements for Different Slumps and Nominal Maximum Sizes of Aggregate

SLUMP, in.	WATER, lb/yd³ OF CONCRETE, FOR INDICATED SIZES OF AGGREGATE*							
	³⁄₈ in.	½ in.	¾ in.	1 in.	1½ in.	2 in.†	3 in.†	6 in.†
Non-air-entrained concrete								
1 to 2	350	335	315	300	275	260	220	190
3 to 4	385	365	340	325	300	285	245	210
6 to 7	410	385	360	340	315	300	270	–
Air-entrained concrete								
1 to 2	305	295	280	270	250	240	205	180
3 to 4	340	325	305	295	275	265	225	200
6 to 7	365	345	325	310	290	280	260	–

*These quantities of mixing water are for use in computing cement factors for trial batches. They are maximums for reasonably well-shaped angular coarse aggregates graded within limits of accepted specifications.

† The slump values for concrete containing aggregates larger than 1½ in. are based on slump tests made after removal of particles larger than 1½ in. by wet screening.

Adapted from ACI 211.1 (1991). Hover (1995) presents this information in graphical form.

Minimum or maximum cementitious materials contents may be included in specifications in addition to a maximum water-cementing materials ratio. However, current trends are to limit prescriptive requirements on the paste proportions in lieu of performance requirements for the mixture. Excessively large amounts of cementing materials should be avoided to maintain economy in the mixture and to avoid adversely affecting workability, shrinkage and internal temperature rise.

For concrete placed underwater, it is recommended that not less than 390 kg of cementing materials per cubic meter of concrete (650 lb/yd³) should be used with a maximum w/cm of 0.45.

For workability, finishability, and abrasion resistance in flatwork, the quantity of cementing materials to be used should generally not be less than shown in Table 13-9. To obtain economy, proportioning should minimize the amount of cement required without sacrificing concrete quality.

Table 13-10 shows ACI 318-19 limits on the maximum amount of supplementary cementing materials in concrete exposed to deicers. Local practices should be consulted, as dosages different than those shown in Table 13-10 may be used without jeopardizing scale-resistance, depending on the exposure severity.

TABLE 13-9. Minimum Requirements of Cementing Materials for Concrete Used in Flatwork

NOMINAL MAXIMUM SIZE OF AGGREGATE, mm (in.)	CEMENTING MATERIALS, kg/m³ (lb/yd³)*
37.5 (1½)	280 (470)
25 (1)	310 (520)
19 (¾)	320 (540)
12.5 (½)	350 (590)
9.5 (⅜)	360 (610)

*Cementing materials quantities may need to be greater for severe exposure.
Adapted from ACI 302.1R-15

8. Cementitious Material Types

A typical specification may call for portland cements meeting ASTM C150, *Standard Specification for Portland Cement* (AASHTO M 85), blended cements meeting ASTM C595, *Standard Specification for Blended Hydraulic Cements* (AASHTO M 240), or for performance-based hydraulic cements meeting ASTM C1157, *Standard Performance Specification for Hydraulic Cement*. Table 13-11 provides a matrix of commonly used cements and their typical applications in concrete construction (see Chapter 2 for more information). The use of SCMs should also be considered, whenever practical.

When specifying cementitious materials for a project, the availability of cement types and SCMs should be verified. Specifications should allow flexibility in cementing materials selection. Cements with special properties and specific SCMs should not be required unless specific characteristics are necessary.

Concrete that will be exposed to sulfate conditions should be made with the cement types and SCMs shown in Table 13-12 depending on exposure class (exposure classes and other requirements for resistance to sulfate exposures are defined in Chapter 11). In addition, certain concrete applications

TABLE 13-10. Cementitious Materials Requirements for Concrete Exposed to Deicing Chemicals (Exposure Class F3)

CEMENTITIOUS MATERIALS*	MAXIMUM PERCENT OF TOTAL CEMENTITIOUS MATERIALS BY MASS†
Fly ash and natural pozzolans	25
Slag	50
Silica fume	10
Total of fly ash, slag, silica fume, and natural pozzolans	50§
Total of natural pozzolans and silica fume	35§

*Includes supplementary cementing materials in blended cements.
† Total cementitious materials include the summation of portland cements, blended cements, fly ash, slag, silica fume, and other pozzolans.
§ Silica fume should not constitute more than 10% of total cementitious materials and fly ash or other pozzolans shall not constitute more than 25% of cementitious materials.
Adapted from ACI 318-19.

such as those requiring control of temperature rise, chloride resistance, or resistance to ASR may require careful selection of cement and use of SCMs.

Modifications to the mixture design are usually necessitated when SCMs are incorporated in the mixture including their impacts on strength and durability, air content, and workability. Chapter 3 discusses the impact of SCMs on the properties of concrete.

SCMs have lower specific gravities than portland cement – this results in an increase in the volume of cementitious material (if the mass is kept constant). This is usually accounted for by reducing the sand content of the mixture. The use of SCMs is also likely to change the relationship between the water-to-cementitious-material ratio and the strength. Trial batching is necessary to ensure desired properties are achieved.

TABLE 13-11. Applications for Hydraulic Cements Used in Concrete Construction*

CEMENT SPECIFICATION	GENERAL PURPOSE	MODERATE HEAT OF HYDRATION	HIGH EARLY STRENGTH	LOW HEAT OF HYDRATION	MODERATE SULFATE RESISTANCE	HIGH SULFATE RESISTANCE	AIR-ENTRAINING
ASTM C150 portland cements	I	II(MH)	III	IV	II, II(MH)	V	IA, IIA, II(MH)A, IIIA
ASTM C595 blended hydraulic cements	IL, IP, IS(<70), IT(S<70)	IL(MH), IP(MH), IS(<70)(MH), IT(S<70)(MH)	–	IL(LH), IP(LH), IS(<70)(LH), IT(S<70)(LH)	IL(MS), IP(MS), IS(<70)(MS), IT(S<70)(MS)	IL(HS), IP(HS), IS(<70)(HS), IT(S<70)(HS)	Option A†
ASTM C1157 hydraulic cements	GU	MH	HE	LH	MS	HS	Option A†

*Check the local availability of specific cement types as some cements are not available everywhere.
† Any cement in the columns to the left can be used to make air-entraining cement.

TABLE 13-12. Types of Cementitious Materials Required for Concrete Exposed to Sulfate in Soil or Water*

EXPOSURE CLASS			CEMENTITIOUS MATERIALS REQUIREMENTS**		
			ASTM C150	ASTM 595	ASTM C1157
S0	Negligible		NSR	NSR	NSR
S1	Moderate, Seawater		II or II(MH)	Types with MS designation	MS
S2	Severe		V	Types with HS designation	HS
S3[II]	Very severe	Option 1	V with pozzolan or slag cement[†]	Types with HS designation with pozzolan or slag cement[†]	HS with pozzolan or slag cement[†]
		Option 2	V[‡]	Types with HS designation	HS

* Adapted from USBR (1988), ACI 201.2R-16 and ACI 318-19, and PCA recommendations. Exposure classes and other requirements for sulfate resistance are defined in Chapter 11. "NSR" indicates no special requirements for sulfate resistance.

[†] Pozzolans and slag cement that have been determined by testing according to ASTM C1012 or by service record to improve sulfate resistance may also be used in concrete. Maximum expansions when using ASTM C1012: Moderate exposure – 0.10% at 6 months; Severe exposure – 0.05% at 6 months or 0.10% at 12 months; Very Severe exposure – 0.10% at 18 months. Cementitious materials suitable for use in higher exposure classes may be used in lower exposure classes. For example, a Type V cement may be used in S1 or S0 exposures.

[‡] When Type V cement is used as the sole cementitious material in Exposure Class S3, the optional sulfate resistance limit in ASTM C150 based on ASTM C452 shall be specified.

[§] When qualified by C1012 testing, with limits specified in ASTM C595, Type IL, IP, IS(<70) and IT(S<70) cements are suitable for use in MS and HS environments. Type IS(≥70) cement is not used alone for structural concrete applications.

[II] ACI 318-19 requires SCMs (tested to verify improved sulfate resistance) with Type V, IL(HS), IP(HS), IS(<70)(HS), and HS cements for exposure Class S3.

**See Table 13-1 for additional requirements on water:cementitious material contents for sulfate exposures.

9. Admixture Effects

Chemical admixtures should be considered in concrete mixtures to aid in improvement of fresh and hardened properties as discussed in Chapter 6. The incorporation of certain chemical admixtures will result in changes to the water requirement or the air content of the concrete.

Water-reducing admixtures are added to concrete to reduce the water-cementing materials ratio, reduce cementing materials content, reduce water content, reduce paste content, or to improve the workability of a concrete without changing the water-cementing materials ratio. Water reducers will usually decrease water contents by 5% to 10% and some will also increase air contents by 0.5% to 1%. Retarders may also increase the air content.

High-range water reducers (plasticizers) reduce water contents between 12% and 30% and some can simultaneously increase the air content up to 1 percentage point; others can reduce or have no affect the air content.

Calcium chloride-based admixtures reduce water contents by about 3% and increase the air content by about 0.5%. When using a chloride-based admixture, the risks of reinforcing steel corrosion should be considered. Table 13-13 provides recommended limits on the water-soluble chloride-ion content in reinforced and prestressed concrete for various conditions.

When using more than one admixture in concrete, the compatibility of intermixing admixtures should be assured by

the admixture manufacturer or the combination of admixtures should be tested in trial batches. The water contained in admixtures should be considered part of the mixing water if the admixture's water content is sufficient to affect the water-cementing materials ratio by 0.01 or more. Consult the manufacturer's data sheet for information concerning % solids and total water content in admixtures.

Excessive use of multiple admixtures should be minimized to allow better control of the concrete mixture in production and to reduce the risk of admixture incompatibility. Sequencing of admixtures also impacts their compatibility (see Chapter 6).

TABLE 13-13. Maximum Water-Soluble Chloride-Ion Content (Cl⁻) in Concrete for Corrosion Protection

TYPE OF MEMBER	% BY MASS OF CEMENTITIOUS MATERIALS*
Prestressed concrete (Classes C0, C1, C2)	0.06
Reinforced concrete exposed to chloride in service (Class C2)	0.15
Reinforced concrete that will be dry or protected from moisture in service (Class C0)	1.00
Other reinforced concrete construction (Class C1)	0.30

*ASTM C1218, *Standard Test Method for Water-Soluble Chloride in Mortar and Concrete*.
Adapted from ACI 318-19.

10. Fine Aggregate

Fine aggregate should be evaluated for its gradation and deleterious materials as with coarse aggregates covered in Step 3. The final amount of fine aggregate used in a concrete mixture may be increased or decreased dependent on paste and workability requirements, and selection of bulk volume of coarse aggregate as discussed in Step 3.

When determining the amount of fine aggregate, the required quantities of coarse aggregate, air, water, and cement are already known from the previous steps in designing and proportioning concrete mixtures. These volumes are then subtracted from a unit volume (1 m³ or 1 yd³) to give the required volume of sand. The volume of sand is then converted to a mass proportion using its specific gravity.

11. Moisture Corrections

Prior to batching, corrections are needed to compensate for moisture in the aggregates. Mixture proportions are calculated assuming that the aggregates, coarse and fine, contain just enough moisture to fill all the internal voids with no excess moisture on the surface. This is called the saturated, surface dry condition (SSD). In a laboratory setting, the aggregates are often stored in an SSD condition until they are used.

In reality, the aggregates on the stockpile may have less than or more than the amount of moisture required for SSD condition. If the aggregate is below SSD moisture condition – it will absorb some of the water in the mixture required for workability. If the aggregate is above SSD – and has excess moisture on the surface – it will increase the water content and result in a higher water-to-cementitious materials ratio.

The dry-batch weights of aggregates, therefore, have to be increased to compensate for the moisture that is absorbed in and contained on the surface of each particle and between particles. The mixing water added to the batch must be adjusted to compensate for the absorption or free moisture contributed by the aggregates.

The final amount of aggregate required when correcting for moisture is calculated by multiplying the SSD mass determined from designing the mixture by a correction factor that includes the moisture content of the aggregate at the time of batching and its absorption value as shown in Equation 13-4.

$$M_{batch} = M_{SSD} \times \frac{1 + mc}{1 + abs}$$

EQUATION 13-4.

where:

M_{batch} = Mass of aggregate to be batched

M_{SSD} = Mass required in mix design (in SSD)

mc = Moisture content of aggregate

abs = Aggregate absorption

The moisture content of the aggregates can be determined by ASTM C566, *Standard Test Method for Total Evaporable Moisture Content of Aggregate by Drying* (AASHTO T 255). The absorption value is essentially the amount of water required to bring the aggregate to the SSD condition from oven-dry condition and is a fixed property of the aggregate determined by testing in accordance with ASTM C127, *Standard Test Method for Relative Density (Specific Gravity) and Absorption of Coarse Aggregate* (AASHTO T 85), for coarse aggregate; and ASTM C128, *Standard Test Method for Relative Density (Specific Gravity) and Absorption of Fine Aggregate* (AASHTO T 84), for fine aggregate. If the aggregate used for batching is oven dry, as may be the case in laboratories – then the moisture content of the aggregate is zero.

The moisture content of the aggregates should be determined immediately prior to batching, and the batch weights corrected accordingly. A correction also needs to be made to the batch water content to account for the water absorbed by aggregate that is less than SSD or water contributed by aggregate that is above SSD as shown in Equation 13-5. The magnitude of the water correction should be equal to the correction made to the aggregate. The overall mass of material in the unit volume remains unchanged.

$$W_{corr} = M_{SSD} \times \frac{abs - mc}{1 + abs}$$

EQUATION 13-5.

12. Trial Mixtures

Trial batching determines if the fresh and hardened properties of the concrete mixture are satisfactory. Appropriate adjustments should be made, when necessary, and subsequent trial batches produced. Trial batches should be produced in the mixer that will be used for the project. All relevant concrete properties (such as workability, density, air content, and strength) should be measured to ensure that the mixture performs as desired. Changing the amount of any single ingredient in a concrete mixture normally affects the proportions of other ingredients. The properties of the mixture are also altered. This iterative process continues until a satisfactory combination of the constituent materials is identified.

Figure 13-8 illustrates the change in mix proportions relative to a w/cm for various types of concrete mixtures using a particular aggregate source. Information for concrete mixtures using particular ingredients can be plotted in several ways to illustrate the relationship between ingredients and properties.

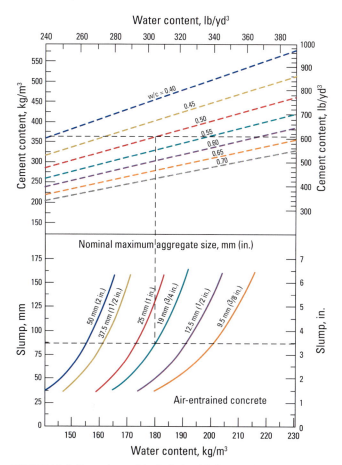

FIGURE 13-8. Example graphical relationship for a particular aggregate source demonstrating the relationship between slump, aggregate size, water-to-cementitious materials ratio, and cementitious materials content (Hover 1995).

This is especially useful when optimizing concrete mixtures for best economy or to adjust to specification or material changes.

The size of the trial batch is dependent on the equipment available and on the number and size of test specimens to be made. The trial mixtures can be relatively small batches made with laboratory precision or batches made to the quantity that will be yielded during the course of normal concrete production. Larger batches will produce more accurate data. Machine mixing is recommended since it more nearly represents job conditions. This is mandatory if the concrete is to contain entrained air. The mixing procedures outlined in ASTM C192 (AASHTO R 39) should be used.

PROPORTIONING CONCRETE MIXTURES

Once the design characteristics of the concrete are selected, the mixture can be proportioned from field or laboratory data. Proportioning methods have evolved from the arbitrary volumetric method (1:2:3 – cement: sand: coarse aggregate) of the early 1900s (Abrams 1918) to the present-day weight and absolute-volume methods described in ACI 211.1, *Standard Practice for Selecting Proportions for Normal, Heavyweight and Mass Concrete* (1991).

Proportioning methods are fairly simple and quick for estimating mixture proportions using an assumed or known weight of concrete per unit volume. A number of different methods of proportioning concrete ingredients have been used including:

- arbitrary assignment (1:2:3),
- volumetric,
- void ratio,
- fineness modulus,
- aggregate gradation, and
- cement content.

Any of these methods can produce approximately the same final mixture after adjustments are made in the field. The best approach, however, is to select proportions based on past experience and reliable test data with an established relationship between strength and water-to-cementing materials ratio for the materials to be used in the concrete. A concrete mixture also can be proportioned from field experience (statistical data) or from trial mixtures. A combination of these two approaches is often necessary to reach a satisfactory mixture for a given project.

A more accurate method, *absolute volume*, involves use of relative density (specific gravity) values for all the ingredients to calculate the absolute volume occupied by each other in a specific unit volume of concrete. The absolute volume method will be highlighted.

Other documents to help proportion concrete mixtures include ACI 213R, *Guide for Structural Lightweight-Aggregate Concrete* (2014); ACI 211.4R, *Guide for Selecting Proportions for High-Strength Concrete with Portland Cement and Fly Ash* (2008); ACI 211.5R, *Guide for Submittal of Concrete Proportions* (2014); and ACI 211.7, *Guide for Proportioning Concrete Mixtures with Ground Calcium Carbonate and Other Mineral Fillers* (2020).

Internet web sites and computer models also provide assistance with designing and proportioning concrete mixtures (Bentz 2001; NIST 2010; and Transtec Group, Inc. 2010). Use caution when using computer models to design concrete mixtures, as the locally available materials may not perform exactly as modeled by the program. Therefore, trial batching and field verification are still the best method to determine the performance of a concrete mixture.

Proportioning by the Absolute Volume Method

Concrete mixture proportions are usually expressed on the basis of the mass of ingredients per unit volume. The absolute volume of a granular material (such as cement and aggregates) is the volume of the solid matter in the particles; it does not include the volume of air spaces between particles.

The unit of volume used is either a cubic meter or a cubic yard of concrete as shown in Figure 13-9. For example, a cubic yard might contain approximately 43% by volume of coarse aggregate (stone), 25% fine aggregate (sand), 11% cementitious materials, 15% water, and 6% air. If the precise proportions are known, they should add up to equal 1 cubic meter (cubic yard).

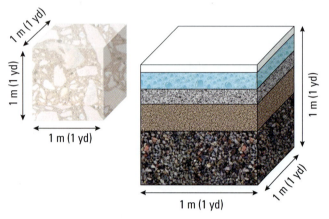

FIGURE 13-9. A unit of volume expressed as a cubic meter or cubic yard of volume.

Density (Unit Weight) and Yield. The density (unit weight) of freshly mixed concrete is expressed in kilograms per cubic meter (pounds per cubic foot). The yield is the volume of fresh concrete produced in a batch, usually expressed in cubic meters (cubic feet). The yield is calculated by dividing the total mass of the materials batched by the density of the freshly mixed concrete as shown in Equation 13-6. Density and yield are determined in accordance with ASTM C138, *Standard Test Method for Density (Unit Weight), Yield, and Air Content (Gravimetric) of Concrete* (AASHTO T 121).

The density of the material is calculated by multiplying the specific gravity by the density of water, 1000 kg/m³ (62.4 lb/ft³) as shown in Equation 13-7. (For inch-pound units, multiply the value in lb/ft³ by 27 ft³/yd³ to obtain a density in pounds per cubic yard.)

$$V = M / D$$

EQUATION 13-6.

$$D = SG \times 1000 \text{ kg/m}^3 \text{ (metric units)}$$
$$D = SG \times 62.4 \text{ lb/ft}^3 \text{ (inch-pound units)}$$

EQUATION 13-7.

where:

V = absolute volume,

M = mass of loose material,

D = specific gravity of a material × density of water, and

SG = specific gravity (relative density).

The volume of concrete in a batch can be determined by either of two methods: if the relative densities of the aggregates and cementing materials are known, these can be used to calculate concrete volume; or if relative densities are unknown, or they vary, the volume can be computed by dividing the total mass of materials in the mixer by the density of concrete. In some cases, both determinations are made, one serving as a check on the other.

The absolute volume of air in concrete, expressed as cubic meters per cubic meter (cubic feet per cubic yard), is equal to the total air content in percent divided by 100 (for example, 6% ÷ 100) and then multiplied by the volume of the concrete batch.

Relative Density (Specific Gravity). To determine the mass of each constituent – the specific gravity of the materials must be known. Portland cement has a relative density (specific gravity) generally ranging from 3.10 to 3.25, averaging 3.15 (see Chapter 2). Blended cements have relative densities ranging from 2.90 to 3.15. The relative density of fly ash varies from 1.9 to 2.8, slag cement from 2.85 to 2.95, and silica fume from 2.20 to 2.25. The relative density of water is 1.0 and the density of water is 1000 kg/m³ (62.4 lb/ft³) at 4°C (39°F), which is accurate enough for mix calculations at room temperature. More accurate water density values are given in Table 13-14. Relative density of normalweight aggregate usually ranges between 2.4 and 2.9.

The relative density of aggregate as used in mix-design calculations is the relative density of either SSD material or oven dry material. Relative densities of admixtures, such as water reducers, can also be considered if needed. Absolute volume is usually expressed in cubic meters (cubic feet).

TABLE 13-14. Density off Water Versus Temperature

TEMPERATURE, °C	DENSITY, kg/m³	TEMPERATURE, °F	DENSITY, lb/ft³
16	998.93	60	62.368
18	998.58	65	62.337
20	998.19	70	62.302
22	997.75	75	62.261
24	997.27	80	62.216
26	999.75	85	62.166
28	996.20		
30	995.61		

EXAMPLES OF MIXTURE PROPORTIONING

Example 1. Absolute Volume Method (Metric)

Conditions and Specifications. Concrete is required for a pavement that will be exposed to moisture in a severe freeze-thaw environment. A specified compressive strength, f'_c, of 35 MPa is required at 28 days. Air entrainment is required. Slump should be between 25 mm and 75 mm. A nominal maximum size aggregate of 25 mm is required. No statistical data on previous mixtures are available. The materials available are as follows:

- Cement: Type IL with a relative density of 3.0.

- Coarse aggregate: well-graded, 25-mm nominal maximum-size rounded gravel with an oven dry relative density of 2.68, absorption of 0.5% (SSD), and oven dry rodded bulk density (unit weight) of 1600 kg/m³. The laboratory sample for trial batching has a moisture content of 2%.

- Fine aggregate: natural sand with an oven dry relative density of 2.64 and absorption of 0.7%. The laboratory sample moisture content is 6%. The fineness modulus is 2.80.

- Air-entraining admixture: wood-resin type.

- Water reducer: this particular admixture is known to reduce water demand by 10% when used at a dosage rate of 3 g/kg (or 3 mL/kg) of cement. Assume that the chemical admixtures have a density close to that of water, meaning that 1 mL of admixture has a mass of 1 g.

From this information, the task is to proportion a trial mixture that will meet the above conditions and specifications.

Step 1

Strength Requirements. The design strength of 35 MPa is greater than the 31 MPa required in Table 13-1 for the exposure condition, F2. Since no statistical data are available, f'_{cr} (required compressive strength for proportioning) from Table 13-3 is equal to $f'_c + 8.5$ MPa. Therefore, $f'_{cr} = 35 + 8.5 = 43.5$ MPa.

Step 2

Determining w/cm. For an environment with moist freezing and thawing, F2 exposure, the maximum w/cm should be 0.45. The recommended water-cementitious material ratio for an f'_{cr} of 43.5 MPa is 0.31 from Figure 13-4 or, as interpolated from Table 13-4A:

$$w/cm = [(45 - 43.5)(0.34 - 0.30)/(45 - 40)] + 0.30 = 0.31$$

Since the lower w/cm governs, the mixture must be designed for 0.31. If a plot from trial batches or field tests had been available, the w/cm could have been extrapolated from that data.

Step 3

Coarse-Aggregate. The quantity of 25-mm nominal maximum-size coarse aggregate can be estimated from Figure 13-5 or Table 13-5. The bulk volume of coarse aggregate recommended when using sand with a fineness modulus of 2.80 is 0.67.

Since it has a bulk density of 1600 kg/m³, the oven dry weight of coarse aggregate (CA_{OD}) for a cubic meter of concrete is:

$$CA_{OD} = 1600 \text{ kg/m}^3 \times 0.67 \text{ m}^3 = 1072 \text{ kg}$$

and the mass of coarse aggregate at SSD (CA_{SSD}) is:

$$CA_{SSD} = CA_{OD} \times (1 + abs/100) =$$
$$CA_{SSD} = 1072 \text{ kg} \times (1 + [0.5/100]) = 1077 \text{ kg}$$

Step 4

Air Content. For a severe freeze-thaw exposure, Table 13-6 recommends a target air content of 6.0% for a 25-mm aggregate. Therefore, design the mixture for 5% to 7% air and use 7% (or the maximum allowable) for batch proportions. The trial-batch air content must be within ±0.5% of the maximum allowable air content.

Step 5

Workability. A slump was specified for the pavement between 25 mm and 75 mm. Use 75 mm ± 20 mm for proportioning purposes.

Step 6

Water Content. Figure 13-7 and Table 13-8A recommend that a 75-mm slump, air-entrained concrete made with 25-mm nominal maximum-size aggregate should have a water content of about 175 kg/m³. However, rounded gravel should reduce the water content of the table value by about 25 kg/m³. Therefore, the water content can be estimated to be:

$$175 \text{ kg/m}^3 - 25 \text{ kg/m}^3 = 150 \text{ kg/m}^3$$

In addition, the water reducer will reduce water demand by 10% resulting in an estimated water demand of:

$$150 \text{ kg} \times 10/100 = 15 \text{ kg}$$
$$150 \text{ kg} - 15 \text{ kg} = 135 \text{ kg}$$

Step 7

Cement Content. The cement content is based on the maximum water-cement ratio and the water content. Therefore:

$$135 \text{ kg} / 0.31 = 435 \text{ kg}$$

Step 8

Cement Type. Cement: Type IL, ASTM C595 specified. May consider additional SCMs such as fly ash or slag cement, if permitted. The relative density provided by the manufacturer is 3.00.

Step 9

Admixture Effects. For an 7% air content, the air-entraining admixture manufacturer recommends a dosage rate of 0.5 g per kg of cement. From this information, the amount of air-entraining admixture per cubic meter of concrete is:

$$0.5 \text{ g/kg} \times 435 \text{ kg} = 218 \text{ g or } 0.218 \text{ kg}$$

The water reducer dosage rate of 3 g per kg of cement results in:

$$3 \text{ g/kg} \times 435 \text{ kg} = 1305 \text{ g or } 1.305 \text{ kg}$$

Step 10

Fine-Aggregate Content. At this point, the amounts of all ingredients except the fine aggregate are known. In the absolute volume method, the volume of fine aggregate is determined by subtracting the absolute volumes of the known ingredients from 1 cubic meter. The absolute volume of the water, cement, and coarse aggregate is calculated by dividing the known mass of each ingredient by the product of their relative density (specific gravity) and the density of water. Volume computations are as follows:

MATERIAL	MASS, kg	DENSITY, kg/m³	VOLUME, m³
Cement	435	3000 = (3.0 × 1000)	0.145
Water	135	1000 = (1 × 1000)	0.135
Air	7.0%		0.070
Coarse Aggregate	1077	2680 = (2.68 × 1000)	0.402
Fine Aggregate		2640 = (2.64 × 1000)	
Total			1.000

The calculated absolute volume of fine aggregate is then:

$$1.0 \text{ m}^3 - (0.145 \text{ m}^3 + 0.135 \text{ m}^3 + 0.070 \text{ m}^3 + 0.400 \text{ m}^3) = 0.248 \text{ m}^3$$

The weight of dry fine aggregate is:

$$0.248 \text{ m}^3 \times 2640 \text{ kg/m}^3 = 655 \text{ kg}$$

The trial mixture then has the following proportions for one cubic meter of concrete:

MATERIAL	MASS, kg
Cement	435
Water	135
Coarse Aggregate	1072
Fine Aggregate	655
Total Yield	2297
Air-entraining admixture	0.218
Water reducer	1.305
Slump	75 mm (± 20 mm for Trial)

The liquid admixture volume is generally too insignificant to include in the water calculations. However, certain admixtures, such as shrinkage reducers, plasticizers, and corrosion inhibitors are exceptions: due to their relatively large dosage rates, their volumes should be included.

Step 11

Moisture Corrections. Tests indicate that for this example, coarse-aggregate moisture content is 2% and fine-aggregate moisture content is 6%.

With the aggregate moisture contents (*mc*) and absorptions (*abs*) indicated, the trial batch aggregate proportions become:

$$M_{batch} = M_{SSD} \times \frac{1 + mc}{1 + abs}$$

Coarse aggregate (*mc* = 2%, *abs* = 0.5%):

$$CA_{SSD} = 1072 \text{ kg} \times (1.02 / 1.005) = 1088 \text{ kg}$$

Fine aggregate (*mc* = 6%, *abs* = 0.7%):

$$FA_{SSD} = 655 \times (1.06/1.007) = 689 \text{ kg}$$

Note: (0.5% absorption ÷ 100) + 1 = 1.005;
(0.7% absorption ÷ 100) + 1 = 1.007

A correction also needs to be made to the mixing water for batching (batch water) to account for the water absorbed by aggregate that is less than SSD or water contributed by aggregate that is above SSD:

$$W_{corr} = M_{SSD} \times \frac{abs - mc}{1 + abs}$$

Coarse aggregate water correction:

$$CAW_{corr} = 1072 \text{ kg} \times (0.005\text{-}0.02) / 1.005 = 16 \text{ kg}$$

Fine aggregate water correction:

$$FAW_{corr} = 655 \text{ kg} \times (0.007\text{-}0.06) / 1.007 = 34 \text{ kg}$$

The correction for the batch water content then becomes:

$$W_{corr} = (1072 \text{ kg} - 1088 \text{ kg}) + (655 \text{ kg} - 689 \text{ kg}) =$$
$$-16 + -34 = -50 \text{ kg}$$

The estimated batch weights for one cubic meter of concrete are revised to include aggregate moisture as follows:

MATERIAL	MASS, kg	MOISTURE CORRECTIONS, kg	BATCH WEIGHTS, kg
Cement	435		435
Water	135	-50	85
Coarse Aggregate	1072	+16	1088
Fine Aggregate	655	+34	689
Total Yield	2297		2297
Air-entraining admixture	0.218		0.218
Water reducer	1.305		1.305

Step 12

Trial Mixtures. At this stage, the estimated batch weights should be checked by means of trial batches or by full-size field batches. Enough concrete must be mixed for appropriate air and slump tests and for casting the three cylinders required for 28-day compressive-strength tests, plus beams for flexural tests if necessary. For a laboratory trial batch it is convenient, in this case, to scale down the weights to produce 0.1 m³ of concrete as follows:

MATERIAL	MASS,* kg	CORRECTED TO 0.1 m³	TRIAL MIXTURE, kg
Cement	435	0.1	43.5
Water	85	0.1	8.5
Coarse Aggregate	1088	0.1	108.8
Fine Aggregate	689	0.1	68.9
Total Yield	2297		229.7
Air-entraining admixture	0.218		21.8 mL
Water reducer	1.305		130 mL

*Includes moisture corrections

The above concrete, when mixed, had a measured slump of 100 mm, an air content of 8%, and a density of 2274 kg/m³. During mixing, some of the premeasured water may remain unused or additional water may be added to approach the required slump. In this example, although 8.5 kg of water was calculated to be added, the trial batch used only 8.0 kg. The mixture excluding admixtures therefore becomes:

MATERIAL	TRIAL MIXTURE, kg
Cement	43.5
Water	8.0
Coarse Aggregate	108.8
Fine Aggregate	68.9
Total Yield	229.2

The yield of the trial batch is: 229.2 kg / 2274 kg/m³ = 0.10079 m³

The mixing water content is determined from the added water plus the free water on the aggregates, calculated as follows:

Water added	=	8.0 kg
Free water on coarse aggregate: 108.8 kg / 1.02 × 0.015%	=	1.61 kg
Free water on fine aggregate: 68.9kg / 1.06 × 0.053%	=	3.44 kg
		13.05 kg

Note: (2% *mc* − 0.5% *abs*) ÷ 100 = 0.015%
(6% *mc* − 0.7% *abs*) ÷ 100 = 0.053%

The mixing water required for a cubic meter of the same slump concrete as the trial batch is:

$$13.05 \text{ kg} /0.10079 \text{ m}^3 = 129 \text{ kg/m}^3$$

Batch Adjustments. The measured 100-mm slump of the trial batch is high (above initial 25 mm to 75 mm range, and ±20 mm tolerance), the yield was slightly high, and the 8.0% air content as measured in this example is also too high (more than 0.5% above 7.0%). Adjust the yield, reestimate the amount of air-entraining admixture required for 7% air content, and adjust the water to obtain a 75-mm slump. Increase the mixing water content by 3 kg/m³ for each 1% by which the air content is decreased from that of the trial batch and reduce the water content by 2 kg/m³ for each 10 mm reduction in slump. The adjusted mixture water for the reduced slump and air content is:

$$([3 \text{ kg/m}^3/\%] \times 1\%) - ([2 \text{ kg/m}^3/10 \text{ mm}] \times 25 \text{ mm}) + 129 \text{ kg} = 127 \text{ kg}$$

With less mixing water needed in the trial batch, less cement also is needed to maintain the desired water-cement ratio of 0.31. The new cement content is:

$$\frac{127 \text{ kg}}{0.31} = 410 \text{ kg}$$

The amount of coarse aggregate remains unchanged because workability is satisfactory. The adjusted batch weights based on the new cement and water contents are calculated after the following volume computations:

MATERIAL	MASS, kg/m³	DENSITY, kg/m³	VOLUME, m³
Cement	410	3000 = (3.0 × 1000)	0.137
Water	127	1000 = (1 × 1000)	0.127
Air			0.07
Coarse Aggregate	1072	2680 = (2.68 × 1000)	0.400
Fine Aggregate		2640 = (2.64 × 1000)	
Total			1.000

The calculated absolute volume of fine aggregate is then:

$$1.000 \text{ m}^3 - (0.137 \text{ m}^3 + 0.127 \text{ m}^3 + 0.070 \text{ m}^3 + 0.403 \text{ m}^3) = 0.266 \text{ m}^3$$

The weight of dry fine aggregate is:

$$0.266 \text{ m}^3 \times 2640 \text{ kg/m}^3 = 702 \text{ kg}$$

Coarse aggregate:

$$CA_{SSD} = (mc = 2\%, abs = 0.5\%) = 1072 \text{ kg} \times (1.02/1.005) = 1088 \text{ kg}$$

Fine aggregate:

$$FA_{SSD} = (mc = 6\%, abs = 0.7\%) = 702 \text{ kg} \times (1.06/1.007) = 739 \text{ kg}$$

Coarse aggregate water correction:

$$CAW_{corr} = 1072 \times (0.005 - 0.02)/1.005 = 16 \text{ kg}$$

Fine aggregate water correction:

$$FAW_{corr} = 702 \times (0.007 - 0.06)/1.007 = 37 \text{ kg}$$

Batch water correction:

$$W_{corr} = (1072 - 1088) + (702 - 739) = -16 + -37 = -53 \text{ kg}$$

The estimated batch weights for one cubic meter of concrete are revised to include aggregate moisture as follows:

MATERIAL	MASS, kg	MOISTURE CORRECTIONS, kg	BATCH WEIGHTS, kg
Cement	410		410
Water	127	-53	74
Coarse Aggregate	1072	+16	1088
Fine Aggregate	702	+37	739
Total	2311		2311
Air-entraining admixture	0.218		21.8 mL
Water reducer	1.305		130 mL

After checking these adjusted proportions in a trial batch, it was found that the concrete had the desired slump, air content, and yield. The 28-day test cylinders had an average compressive strength of 48 MPa, which exceeds the f'_{cr} of 43.5 MPa. Due to fluctuations in moisture content, absorption rates, and relative density (specific gravity) of the aggregate, the density determined by volume calculations may not always equal the density determined by ASTM C138 (AASHTO T 121). Occasionally, the proportion of fine-to-coarse aggregate is kept constant in adjusting the batch weights to maintain workability or other properties obtained in the first trial batch. After adjustments to the cementitious materials, water, and air content have been made, the volume remaining for aggregate is appropriately proportioned between the fine and coarse aggregates.

Additional trial concrete mixtures with water-cement ratios above and below 0.31 should also be tested to develop a strength to water-cement ratio relationship. From the data, a new more economical mixture with a compressive strength closer to f'_{cr} and a lower cement content can be proportioned and tested. The final mixture would probably look similar to the above mixture with a slump range of 25 mm to 75 mm and an air content of 5% to 8%. The amount of air-entraining admixture must be adjusted to field conditions to maintain the specified air content.

Example 2. Absolute Volume Method (Inch-Pound)

Conditions and Specifications. Concrete is required for a building foundation. A specified compressive strength, f'_c, of 3500 psi is required at 28 days using a Type I portland cement. The design calls for a minimum of 3 in. of concrete cover over the reinforcing steel. The minimum distance between reinforcing bars is 4 in. No statistical data on previous mixes are available. The materials available are as follows:

• Cement: Type I, ASTM C150, with a relative density of 3.15.

• Coarse aggregate: Well-graded ¾-in. nominal maximum-size gravel containing some crushed particles with an oven dry relative density (specific gravity) of 2.68, absorption of 0.5% (SSD), and oven dry rodded bulk density (unit weight) of 100 lb/ft³. The laboratory sample for trial batching has a moisture content of 2%.

• Fine aggregate: Natural sand with an oven dry relative density (specific gravity) of 2.64 and absorption of 0.7%. The laboratory sample moisture content is 6%. The fineness modulus is 2.80.

From this information, the task is to proportion a trial mixture that will meet the above conditions and specifications. This can be done by following the steps discussed earlier.

Step 1

Strength Requirements. Since no statistical data are available, and there are no durability exposure concerns from Table 13-1, then f'_{cr} from Table 13-3B is equal to f'_c + 1200. Therefore,

$$f'_{cr} = 3500 + 1200 = 4700 \text{ psi}$$

Step 2

Determining w/cm. Table 13-1 requires no maximum w/cm. The recommended water-cementing materials ratio for an f'_{cr} of 4700 psi is 0.51 interpolated from Figure 13-4 or Table 13-4B for non-air entrained concrete:

$$\text{w/cm} = \{(5000 - 4700)(0.57 - 0.48)/(5000 - 4000)\} + 0.48 = 0.51$$

Step 3

Coarse Aggregates. From the specified information, a ¾-in. nominal maximum-size aggregate is adequate as it is less than ¾ of the distance between reinforcing bars and between the reinforcement and forms (cover). The quantity of ¾-in. nominal maximum-size coarse aggregate can be estimated from Figure 13-5 or Table 13-5. The bulk volume of coarse aggregate

recommended when using sand with a fineness modulus of 2.80 is 0.62. Since it weighs 100 lb/ft³, the oven dry weight of coarse aggregate (CA_{OD}) for a cubic yard of concrete (27 ft³) is:

$$CA_{OD} = 100 \text{ lb/ft}^3 \times 27 \text{ ft}^3/\text{yd}^3 \times 0.62 = 1674 \text{ lb/yd}^3$$

and, the mass of coarse aggregate at SSD:

$$CA_{SSD} = CA_{OD} \times (1 + abs/100) =$$
$$CA_{SSD} = 1674 \text{ lb/yd}^3 \times (1 + [0.5/100]) = 1682 \text{ lb/yd}^3$$

Step 4

Air Content. No air is required for exposure condition, but could be considered to improve workability and reduce bleeding. For first trial batch, design the mixture for non-air entrained concrete, and estimate 2% from Table 13-7 for batch proportions.

Step 5

Workability. As no slump was specified, a slump of 5 in. is selected for proportioning purposes based on placement methods of chute or pump for a foundation wall.

Step 6

Water Content. Figure 13-7 and Table 13-8B recommend that a 5-in. slump, non-air-entrained concrete made with ¾-in. nominal maximum-size aggregate should have a water content of about 350 lb/yd³ (interpolated). However, gravel with some crushed particles should reduce the water content of the table value by about 35 lb/yd³. Therefore, the water content can be estimated to be about 350 lb - 35 lb, which is 315 lb.

Step 7

Cementitious Materials Content. The cement content is based on the maximum w/cm and the water content. Therefore, 315 lb of water divided by a water-cement ratio of 0.51 requires a cement content of 618 lb.

Step 8

Cementitious Material Types. No special requirements. Type I cement, ASTM C150 specified. Could consider additional SCMs such as fly ash or slag cement.

Step 9

Admixture Effects. None specified. Could consider addition of 2% to 4% air-entrainment or a water reducer for increased workability depending on trial batch results.

Step 10

Fine Aggregate. At this point, the amounts of all ingredients except the fine aggregate are known. In the absolute volume method, the volume of fine aggregate is determined by subtracting the absolute volumes of the known ingredients from 27 ft³ (1 yd³). The absolute volume of the water, cement, and coarse aggregate is calculated by dividing the known weight of each by the product of their relative density (specific gravity) and the density of water. Volume computations are as follows:

MATERIAL	MASS, lb	DENSITY, lb/yd³	VOLUME, yd³
Cement	618	5307 = (3.15 × 62.4 × 27)	0.116
Water	315	1685 = (1.00 × 62.4 × 27)	0.187
Air	2%		0.020
Coarse Aggregate	1682	4515 = (2.68 × 62.4 × 27)	0.373
Fine Aggregate		4448 = (2.64 × 62.4 × 27)	
Total			1.000

The calculated absolute volume of fine aggregate is:

$$1.000 \text{ yd}^3 - (0.116 \text{ yd}^3 + 0.187 \text{ yd}^3 + 0.020 \text{ yd}^3 + 0.373 \text{ yd}^3) = 0.304 \text{ yd}^3$$

The weight of dry fine aggregate is:

$$0.304 \text{ yd}^3 \times 4448 \text{ lb/yd}^3 = 1352 \text{ lb}$$

The mixture then has the following proportions before trial mixing for one cubic yard of concrete:

MATERIAL	MASS, lb
Cement	618
Water	315
Coarse Aggregate	1682
Fine Aggregate	1352
Total	3967
Slump	5 in (± 1 in. for Trial)
Air	2% entrapped air

Step 11

Moisture Corrections. Tests indicate that for this example, coarse-aggregate moisture content is 2% and fine-aggregate moisture content is 6%.

With the aggregate moisture contents (*mc*) and absorptions (*abs*) indicated, the trial batch aggregate proportions become:

$$M_{batch} = M_{SSD} \times \frac{1 + mc}{1 + abs}$$

Coarse aggregate:

$$CA_{SSD} = (mc = 2\%, abs = 0.5\%) = 1682 \text{ lb} \times (1.02/1.005) = 1707 \text{ lb}$$

Fine aggregate:

$$FA_{SSD} = (mc = 6\%, abs = 0.7\%) = 1352 \text{ lb} \times (1.06/1.007) = 1423 \text{ lb}$$

Note: (0.5% absorption ÷ 100) + 1 = 1.005;
(0.7% absorption ÷ 100) + 1 = 1.007

A correction also needs to be made to the batch water content to account for the water absorbed by aggregate that is less than SSD or water contributed by aggregate that is above SSD.

$$W_{corr} = M_{SSD} \times \frac{abs - mc}{1 + abs}$$

Coarse aggregate water correction:

$$CAW_{corr} = 1682 \text{ lb} \times (0.005 - 0.02)/1.005 = 25 \text{ lb}$$

Fine aggregate water correction:

$$FAW_{corr} = 1352 \text{ lb} \times (0.007 - 0.06)/1.007 = 71 \text{ lb}$$

The correction for the batch water content then becomes:

$$W_{corr} = (1682 \text{ lb} - 1707 \text{ lb}) + (1352 \text{ lb} - 1423 \text{ lb}) = -25 \text{ lb} + -71 \text{ lb} = -96 \text{ lb}$$

The estimated batch weights for one cubic yard of concrete are revised to include aggregate moisture as follows:

MATERIAL	MASS, lb	MOISTURE CORRECTIONS, kg	BATCH WEIGHTS, lb
Cement	618		618
Water	315	-96	219
Coarse Aggregate	1682	+25	1707
Fine Aggregate	1352	+71	1423
Total Yield	3967		3967

Step 12

Trial Mixtures. At this stage, the estimated batch weights should be checked by means of trial batches or by full-size field batches. Enough concrete must be mixed for appropriate air and slump tests and for casting the three cylinders required for compressive strength tests at 28 days. For a laboratory trial batch it is convenient, in this case, to scale down the weights to produce 2.0 ft^3 of concrete or 2/27 yd^3.

MATERIAL	MASS,* lb/yd^3	CORRECTED TO 2.0 ft^3	TRIAL MIXTURE, lb
Cement	618	2/27	45.78
Water	219	2/27	16.22
Coarse Aggregate	1707	2/27	126.44
Fine Aggregate	1423	2/27	105.41
Total Yield	3967		293.85

*Includes Moisture Corrections

In this example, the above concrete, when mixed, had a measured slump of 6 in., an air content of 2.5%, and a density (unit weight) of 141.49 lb/ft^3. During mixing, some of the premeasured water may remain unused or additional water may be added to approach the required slump. In this example, although 16.22 lb of water was calculated to be added, the trial batch used only 16.08 lb. The mixture therefore becomes:

MATERIAL	TRIAL MIXTURE, lb
Cement	45.78
Water	16.08
Coarse Aggregate	126.44
Fine Aggregate	105.41
Total Yield	293.71

The yield of the trial batch is:

$$293.71/141.49 = 2.076 \text{ ft}^3$$

The mixing water content is determined from the added water plus the free water on the aggregates, calculated as follows:

Water added	=	16.08 lb
Free water on coarse aggregate: 126.44 lb/1.02 × 0.015	=	1.86 lb
Free water on fine aggregate: 105.41 lb/1.06 × 0.053	=	5.27 lb
		23.21 lb

Note: $(2\% \, mc - 0.5\% \, abs) \div 100 = 0.015\%$
$(6\% \, mc - 0.7\% \, abs) \div 100 = 0.053\%$

The mixing water required for a cubic yard of the same slump concrete as the trial batch is:

$$23.21 \text{ lb} \times 27 \text{ ft}^3 /2.076 \text{ ft}^3 = 302 \text{ lb}$$

Batch Adjustments. The measured 6-in. slump of the trial batch is borderline unacceptable (1.0 in. above 5 in.), the yield was slightly high, and the 2.5% air content as measured in this example is also high (0.5% above 2%). Adjust the yield, and adjust the water to obtain a 5-in. slump. Increase the mixing water content by 2.5 lb for each 0.5% by which the air content is decreased from that of the trial batch and reduce the water content by 10 lb for each 1 in. reduction in slump. The adjusted mixture water for the reduced slump and air content is:

$$(2.5 \text{ lb/in.} \times 1 \text{ in.}) - (10 \text{ lb/in.} \times 1 \text{ in.}) + 302 \text{ lb} = 295 \text{ lb}$$

With less mixing water needed in the trial batch, less cement also is needed to maintain the desired water-cement ratio of 0.51. The new cement content is:

$$295 \text{ lb}/0.51 = 578 \text{ lb}$$

The amount of coarse aggregate remains unchanged because workability is satisfactory. The adjusted batch weights based on the new cement and water contents are calculated after the following volume computations:

MATERIAL	MASS, lb	DENSITY, lb/yd^3	VOLUME, yd^3
Cement	578	5307 = (3.15 × 62.4 × 27)	0.109
Water	295	1685 = (62.4 × 27)	0.175
Air	2%		0.020
Coarse Aggregate	1682	4515 = (2.68 × 62.4 × 27)	0.373
Fine Aggregate		4448 = (2.64 × 62.4 × 27)	
Total			1.000

The calculated absolute volume of fine aggregate is then:

$$1.000 - (0.109 + 0.175 + 0.020 + 0.373) = 0.323 \text{ yd}^3$$

The weight of dry fine aggregate is:

$$0.323 \times 4448 = 1436 \text{ lb}$$

Coarse aggregate:

$$CA_{SDD} = (mc = 2\%, abs = 0.5\%) = 1682 \text{ lb} \times (1.02/1.005) = 1707 \text{ lb}$$

Fine aggregate:

$$FA_{SSD} = (mc = 6\%, abs = 0.7\%) = 1436 \text{ lb} \times (1.06/1.007) = 1512 \text{ lb}$$

Coarse aggregate water correction:

$$CAW_{corr} = 1682 \text{ lb} \times (0.005-0.02)/1.005 = 25 \text{ lb}$$

Fine aggregate water correction:

$$FAW_{corr} = 1436 \text{ lb} \times (0.007-0.06)/1.007 = 76 \text{ lb}$$

Batch water correction:

$$W_{corr} = (1682 \text{ lb} - 1707 \text{ lb}) + (1436-1512) = -25 + -76 = -101 \text{ lb}$$

The estimated batch weights for one cubic yard of concrete are revised to include aggregate moisture as follows:

MATERIAL	MASS, lb	MOISTURE CORRECTIONS, kg	BATCH WEIGHTS, lb
Cement	578		578
Water	295	-101	194
Coarse Aggregate	1682	+25	1707
Fine Aggregate	1436	+76	1512
Total Yield	3991		3991

Upon completion of checking these adjusted proportions in a trial batch, it was found that the proportions were adequate for the desired slump, air content, and yield. The 28-day test cylinders had an average compressive strength of 4900 psi, which exceeds the f'_{cr} of 4700 psi.

Due to fluctuations in moisture content, absorption rates, and specific gravity of the aggregate, the density determined by volume calculations may not always equal the unit weight determined by ASTM C138 (AASHTO T 121). Occasionally, the proportion of fine to coarse aggregate is kept constant in adjusting the batch weights to maintain workability or other properties obtained in the first trial batch. After adjustments to the cement, water, and air content have been made, the volume remaining for aggregate is appropriately proportioned between the fine and coarse aggregates.

Additional trial concrete mixtures with water-cement ratios above and below 0.51 should also be tested to develop a strength curve. From the curve, a new more economical mixture with a compressive strength closer to f'_{cr}, can be proportioned and tested. The final mixture would probably look similar to the above mixture with a slump range of 4 in. to 6 in.

Water Reducers. Water reducers are used either as a plasticizer to increase workability without the addition of water, or to reduce the w/cm of a concrete mixture to improve permeability or other properties.

Using the final mixture developed in the last example, assume that the project engineer approves the use of a water reducer to increase the slump to 6 in. to improve workability for a difficult placement area.

Assuming that the water reducer has a manufacturer's recommended dosage rate of 4 oz per 100 lb of cement to increase slump 2 in., the admixture amount becomes:

$$578/100 \times 4 = 23.0 \text{ oz/yd}^3$$

If a water reducer was used to reduce the water-cement ratio, the water and sand content would also need adjustment.

Air Entrainment. Air entrainment can be used for freeze-thaw resistance and for other benefits such as improved workability.

Using the final mixture developed in the last example, assume that the project engineer approves the use of an air-entraining admixture to provide a total air content in the concrete mixture of 4%.

Assuming that the air entrainment has a manufacturer's recommended dosage rate of 1 oz per 100 lb of cement to increase slump 2 in., the admixture amount becomes:

$$578/100 \times 1 = 6.0 \text{ oz/yd}^3$$

If an air-entraining admixture was used to increase the volume of air in the concrete mixture, the sand content would also need adjustment.

SCMs. Supplementary cementing materials are sometimes added in addition to, or as a partial replacement for, cement to aid in workability and resistance to sulfate attack and alkali reactivity. If a pozzolan or slag were required for the above example mixture, it would have been entered in the first volume calculation used in determining fine aggregate content. For example:

Assume that 75 lb of fly ash with a relative density (specific gravity) of 2.5 were to be used in addition to the originally derived cement content. The ash volume would be:

$$75 \text{ lb} / (2.5 \times 62.4 \text{ lb/ft}^3 \times 27 \text{ ft}^3/\text{yd}^3) = 0.018 \text{ yd}^3$$

The water to cementing materials ratio would be w/cm:

$$\frac{w}{c+p} = \frac{295}{578 + 75} = 0.45 \text{ by weight}$$

The water to portland cement only ratio would still be w/c:

$$\frac{w}{c} = \frac{295}{578} = 0.51 \text{ by weight}$$

The fine aggregate volume would have to be reduced by 0.018 yd³ to allow for the volume of ash.

The pozzolan amount and volume computation could also have been derived in conjunction with the first cement content calculation using a water-to-cementing-materials ratio of 0.51 (or equivalent). For example, assume 15% of the cementitious material is specified to be a pozzolan and

$$w/cm \text{ or } w/(c + p) = 0.51$$

Then with:

$w = 295$ lb and $c + p = 578$ lb
$p = 578 \text{ lb} \times 15/100 = 87$ lb
and
$c = 578 \text{ lb} - 87 \text{ lb} = 491$ lb

Appropriate proportioning computations for these and other mix ingredients would follow.

Example 3. Absolute Volume Method Using Multiple Cementing Materials and Admixtures (Metric)

The following example illustrates how to develop a mixture using the absolute volume method when more than one cementing material and admixture are used.

Conditions and Specifications. Concrete with a structural design strength, f'_c, of 40 MPa is required for a bridge to be exposed to freezing and thawing, deicers, and very severe sulfate soils. A value not exceeding 1500 coulombs is required to minimize permeability to chlorides. Water reducers, air entrainers, and plasticizers are allowed. A shrinkage reducer is requested to keep shrinkage under 300 millionths. Some structural elements exceed a thickness of 1 m, requiring control of heat development. The concrete producer has a standard deviation of 2 MPa for similar mixtures to that required here. For difficult placement areas, a slump of 200 mm to 250 mm is required. The following materials are available:

- Cement: Type HS, silica fume modified portland cement, ASTM C1157. Relative density of 3.14. Silica fume content of 5%.

- Fly ash: Class F. Relative density of 2.60.

- Slag cement: Grade 120. Relative density of 2.90.

- Coarse aggregate: Well-graded 19-mm nominal maximum-size crushed rock with an oven dry relative density of 2.68, absorption of 0.5%, and oven dry density of 1600 kg/m³. The laboratory sample has a moisture content of 2.0%. This aggregate has a history of alkali-silica reactivity in the field.

- Fine aggregate: Natural sand with some crushed particles with an oven dry relative density of 2.64 and an absorption of 0.7%. The laboratory sample has a moisture content of 6%. The fineness modulus is 2.80.

- Air entrainer: Synthetic.

- Retarding water reducer: Type D. Dosage of 3 g per kg of cementing materials per manufacturer's recommendations.

- Plasticizer: Type 1. Dosage of 30 g/kg of cementing materials per manufacturer's recommendations.

- Shrinkage reducer: Dosage of 15 g/kg of cementing materials.

Step 1

Strength Requirements. The design strength of 40 MPa is greater than the 35 MPa required in Table 13-1 for the exposure condition, F3.

For a standard deviation (*S*) of 2.0 MPa, the f'_{cr} must be the greater of:

$$f'_{cr} = f'_c + 1.34S = 40 \text{ MPa} + 1.34 \, (2.0 \text{ MPa}) = 42.7 \text{ MPa}$$

or

$$f'_{cr} = 0.9 \, f'_c + 2.33S = 36 \text{ MPa} + 2.33(2.0 \text{ MPa}) = 40.7 \text{ MPa}$$

therefore $f'_{cr} = 42.7$ MPa.

Step 2

Determining w/cm. Past field records using these materials indicate that a w/cm of 0.35 is required to provide a strength of 42.7 MPa. For a deicer environment and to protect embedded steel from corrosion, Table 13-1 requires a maximum water-cementing materials ratio of 0.40 and a strength of at least 35 MPa. For a severe sulfate environment, Table 13-2 requires a maximum water-to-cementing-materials ratio of 0.40 and a strength of at least 35 MPa. Both the water-to-cementing-materials ratio requirements and strength requirements are met and exceeded using the above determined 0.35 water-cementing materials ratio and 40 MPa design strength.

Step 3

Coarse Aggregate. The quantity of 19-mm nominal maximum-size coarse aggregate can be estimated from Figure 13-5 or Table 13-5. The bulk volume of coarse aggregate recommended when using sand with a fineness modulus of 2.80 is 0.62. Since the coarse aggregate has a bulk density of 1600 kg/m^3, the oven dry mass of coarse aggregate for a cubic meter of concrete is:

$$CA_{OD}: 1600 \times 0.62 = 992 \text{ kg}$$

and, the mass of coarse aggregate at SSD:

$$CA_{SSD} = CA_{OD} \times (1 + abs/100) =$$
$$992 \text{ kg/m}^3 \times (1 + [0.5/100]) = 997 \text{ kg}$$

Step 4

Air Content. For a severe exposure, Figure 13-6 or Table 13-6 suggests a target air content of 6% for 19-mm aggregate. Therefore, design the mixture for 5% to 7% and use 7% for batch proportions. The trial batch air content must be within ±0.5 percentage points of the maximum allowable air content.

Step 5

Workability. Assume a slump of 50 mm without the plasticizer and a maximum of 200 mm to 250 mm after the plasticizer is added. Use 250 ± 20 mm for proportioning purposes.

Step 6

Water Content. Figure 13-7 and Table 13-A recommend that a 50-mm slump, air-entrained concrete with 19-mm aggregate should have a water content of about 168 kg/m^3.

Assume the retarding water reducer and plasticizer will jointly reduce water demand by 15% in this case, while achieving the 250-mm slump, this results in an estimated water demand of:

$$168 \text{ kg} \times 15/100 = 25 \text{ kg}$$
$$168 \text{ kg} - 25 \text{ lb} = 143 \text{ kg}$$

Step 7

Cementitious Materials Content. The amount of cementing materials is based on the maximum water-cementing materials ratio and water content. Therefore:

$$143 \text{ kg} / 0.35 = 409 \text{ kg}$$

Fly ash and slag cement will be used to help control alkali-silica reactivity and control temperature rise. Local use has shown that a fly ash dosage of 15% and a slag dosage of 30% by mass of cementing materials are adequate. Therefore, the suggested cementing materials for one cubic meter of concrete are as follows:

$$\text{Cement:} \quad 55\% \text{ of } 409 \text{ kg} = 225 \text{ kg}$$
$$\text{Fly ash:} \quad 15\% \text{ of } 409 \text{ kg} = 61 \text{ kg}$$
$$\text{Slag:} \quad 30\% \text{ of } 409 \text{ kg} = 123 \text{ kg}$$

These dosages meet the requirements of Table 13-10 (2.8% silica fume from the cement + 15% fly ash + 30% slag = 47.8% which is less than the 50% maximum allowed).

Step 8

Cementitious Material Types. Type HS, silica fume modified portland cement, ASTM C1157, Class F fly ash, ASTM C618 (AASHTO M 295), and Grade 120 slag cement, ASTM C989 (AASHTO M 302).

Step 9

Admixture Effects. For an 8% air content, the air-entraining admixture manufacturer recommends a dosage of 0.5 g per kg of cementing materials. The amount of air entrainer is:

$$0.5 \text{ g/kg} \times 409 \text{ kg} = 205 \text{ g} = 0.205 \text{ kg}$$

The retarding water reducer dosage rate is 3 g per kg of cementing materials. This results in:

$$3 \text{ g/kg} \times 409 \text{ kg} = 1227 \text{ g or } 1.227 \text{ kg}$$

The plasticizer dosage rate is 30 g per kg of cementing materials. This results in:

$$30 \text{ g/kg} \times 409 \text{ kg} = 12{,}270 \text{ g or } 12.270 \text{ kg}$$

The shrinkage reducer dosage rate is 15 g per kg of cementing materials. This results in:

$$15 \text{ g/kg} \times 409 \text{ kg} = 6135 \text{ g or } 6.135 \text{ kg}$$

Step 10

Fine Aggregate. At this point, the amounts of all ingredients except the fine aggregate are known. The volume of fine aggregate is determined by subtracting the absolute volumes of all known ingredients from 1 cubic meter. The absolute volumes of the ingredients are calculated by dividing the known mass of each by the product of their relative density and the density of water. Assume a relative density of 1.0 for the chemical admixtures. Assume a density of water of 997.75 kg/m3 as all materials in the laboratory are maintained at a room temperature of 22°C (Table 13-14). Volumetric computations are as follows:

MATERIAL	MASS, kg/m³	DENSITY, kg/m³	VOLUME, m³
Cement	225	3133 = (3.14 × 997.75)	0.072
Fly ash	61	2594 = (2.60 × 997.75)	0.024
Slag cement	123	2893 = (2.90 × 997.75)	0.043
Water	143	998 = (1.0 × 997.75)	0.143
Air	7%		0.070
Coarse Aggregate	997	2674 = (2.68 × 997.75)	0.371
Fine Aggregate		2634 = (2.64 × 997.75)	
Total			1.00

The calculated absolute volume of fine aggregate is:

$$1.000 - (0.072 \text{ m}^3 + 0.024 \text{ m}^3 + 0.043 \text{ m}^3 + 0.143 \text{ m}^3 + 0.07 \text{ m}^3 + 0.371 \text{ m}^3) = 0.277 \text{ m}^3$$

The weight of dry fine aggregate is:

$$0.277 \text{ m}^3 \times 2,634 \text{ kg/m}^3 = 730 \text{ kg}$$

The calculated absolute volume of fine aggregate is then:

$$1 - 0.731 = 0.269 \text{ m}^3$$

The mass of dry fine aggregate is:

$$0.269 \times 2.64 \times 997.75 = 709 \text{ kg}$$

The admixture volumes are:

MATERIAL	MASS PER CUBIC METER, kg/m³	DENSITY, kg/m³	VOLUME, m³
Air entrainer	0.205	(1.0 × 997.75)	0.0002
Water reducer	1.227	(1.0 × 997.75)	0.0012
Plasticizer	12.270	(1.0 × 997.75)	0.0123
Shrinkage reducer	6.135	(1.0 × 997.75)	0.0062

Total = 19.84 kg of admixture with a volume of 0.0199 m³
Consider the admixtures part of the mixing water, and
Batch water = mixing water minus admixtures =
143 − 19.84 = 123 kg

The mixture then has the following proportions before trial mixing for 1 cubic meter of concrete:

MATERIAL	MASS, kg
Cement	225
Fly ash	61
Slag cement	123
Water	123
Coarse Aggregate	997
Fine Aggregate	709
Air-entraining admixture	0.205
Water reducer	1.227
Plasticizer	12.270
Shrinkage reducer	6.135
Total Yield	2258
Slump	250 mm (± 20 mm for Trial)

Estimated concrete density using SSD aggregate (adding absorbed water) = 225 + 61 + 123 + 123 + 997 + 709 + 20 (admixtures) = 2258 kg/m³.

Step 11

Moisture Corrections. The dry batch weights of aggregates have to be increased to compensate for the moisture on and in the aggregates and the mixing water reduced accordingly. The coarse aggregate and fine aggregate have moisture contents of 2% and 6%, and absorptions of 0.5% and 0.7%, respectively. With the moisture contents (*mc*) and absorptions (*abs*) indicated, the trial batch aggregate proportions become:

$$M_{batch} = M_{SSD} \times \frac{1 + mc}{1 + abs}$$

Coarse aggregate:

$$CA_{SSD} = (mc = 2\%, abs = 0.5\%) = 997 \text{ kg} \times (1.02/1.005) = 1012 \text{ kg}$$

Fine aggregate:

$$FA_{SSD} = (mc = 6\%, abs = 0.7\%) = 709 \text{ kg} \times (1.06/1.007) = 746 \text{ kg}$$

Note: (0.5% absorption ÷ 100) + 1 = 1.005;
(0.7% absorption ÷ 100) + 1 = 1.007

A correction also needs to be made to the batch water content to account for the water absorbed by aggregate that is less than SSD or water contributed by aggregate that is above SSD.

$$W_{corr} = M_{SSD} \times \frac{abs - mc}{1 + abs}$$

Coarse aggregate correction:

$$CAW_{corr} = 997 \text{ kg} \times (0.005\text{-}0.02)/1.005 = 15 \text{ kg}$$

Fine aggregate correction:

$$FAW_{corr} = 709 \text{ kg} \times (0.007\text{-}0.06)/1.007 = 37 \text{ kg}$$

and

Water correction:

$$W_{corr} = (997 \text{ kg} - 1012 \text{ kg}) + (709 \text{ kg} - 746 \text{ kg}) = -15 \text{ kg} + -37 \text{ kg} = -52 \text{ kg}$$

The estimated batch weights for one cubic yard of concrete are revised to include aggregate moisture as follows:

MATERIAL	MASS, kg	MOISTURE CORRECTIONS, kg	BATCH WEIGHTS, kg
Cement	225		225
Fly ash	61		61
Slag cement	123		123
Water	123	-52	71
Coarse Aggregate	997	+15	1012
Fine Aggregate	709	+37	746
Air-entraining admixture	0.205		0.205
Water reducer	1.227		1.227
Plasticizer	12.270		12.270
Shrinkage reducer	6.135		6.135
Total Yield	2,258		2,258

Step 12

Trial Mixtures. The above mixture is tested in a 0.1 m³ batch in the laboratory (multiply above quantities by 0.1 to obtain batch quantities). The mixture had an air content of 7.8%, a slump of 240 mm, a density of 2257 kg/m³, a yield of 0.1 m³, and a compressive strength of 44 MPa. Rapid chloride testing resulted in a value of 990 coulombs (ASTM C1202, *Standard Test Method for Electrical Indication of Concrete's Ability to Resist Chloride Ion Penetration* [AASHTO T 277]). ASTM C1567, *Standard Test Method for Determining the Potential Alkali-Silica Reactivity of Combinations of Cementitious Materials and Aggregate (Accelerated Mortar-Bar Method)*, was used to evaluate the potential of the mix for alkali-silica reactivity, resulting in an acceptable expansion of 0.02%. Temperature rise was acceptable, and shrinkage was within specifications. The water-soluble chloride content was 0.06%, meeting the requirements of Table 13-13.

The following mix proportions meet all applicable requirements and are ready for submission to the project engineer for approval:

MATERIAL	MASS, kg/m³
Cement	225
Fly ash	61
Slag cement	123
Water	123 (143 kg including admixtures)
Coarse Aggregate (SSD)	997
Fine Aggregate (SSD)	709
Air-entraining admixture	0.205
Water reducer	1.227
Plasticizer	12.270
Shrinkage reducer	6.135
Total	2258
Slump	200 mm to 250 mm
Air Content	5% to 8%
w/cm	0.35

Example 4. Laboratory Trial Mixture Using the PCA Water-Cement Ratio Method (Metric)

With the following method, the designer develops the concrete proportions directly from laboratory trial batching rather than the absolute volume of the constituent ingredients. When the quality of the concrete mixture is specified by water-cementitious material ratio, the trial batch procedure consists essentially of combining a paste (water, cementing materials, and, generally, a chemical admixture) of the correct proportions with the necessary amounts of fine and coarse aggregates to produce the required slump and workability. Representative samples of the cementing materials, water, aggregates, and admixtures must be used.

Conditions and Specifications. Concrete is required for a plain concrete pavement. The pavement specified compressive strength, f'_c, is 35 MPa at 28 days. The standard deviation, S, of the concrete producer is 2.0 MPa. Type IP cement and 19-mm nominal maximum-size coarse aggregate is locally available. Proportion a concrete mixture for these conditions and check it by trial batch.

Durability Requirements. The pavement will be exposed to freezing and thawing, but not deicers and therefore should have a maximum water to cementitious material ratio of 0.45 (Table 13-1) and at least 335 kg of cement per cubic meter of concrete as required by the owner.

Strength Requirements. For a standard deviation of 2.0 MPa, the f'_{cr} (required compressive strength for proportioning from Equations 13-1 and 13-2) must be the larger of:

$$f'_{cr} = f'_c + 1.34S = 35 + 1.34(2.0) = 37.7 \text{ MPa}$$

or

$$f'_{cr} = f'_c + 2.33S - 3.45 = 35 + 2.33(2.0) - 3.45 = 36.2 \text{ MPa}$$

Therefore the required average compressive strength = 37.7 MPa.

Aggregate Size. The 19-mm maximum-size coarse aggregate and the fine aggregate are in saturated-surface dry condition for the trial mixtures.

Air Content. The target air content should be 6% (Table 13-6) and the range is set at 5% to 7%.

Slump. The specified target slump for this project is 40 (±20) mm.

Batch Quantities. For convenience, a batch containing 10 kg of cement is to be made. The quantity of mixing water required is 10 kg × 0.45 = 4.5 kg.

Representative samples of fine and coarse aggregates are measured in suitable containers. The values are entered as initial mass in Column 2 below.

All of the measured quantities of cement, water, and air-entraining admixture are used and added to the mixer. Fine and coarse aggregates, previously brought to a saturated, surface-dry condition, are added until a workable concrete mixture with a slump deemed adequate for placement is produced. The relative proportions of fine and coarse aggregate for workability can readily be judged by an experienced concrete technician or engineer.

Workability. Results of tests for slump, air content, density, and a description of the appearance and workability are noted below:

BATCH NO.	SLUMP, mm	AIR CONTENT, %	DENSITY, kg/m³	CEMENT CONTENT, kg/m³	FINE AGGREGATE, % of total aggregate	WORKABILITY
1	50	5.7	2341	346	28.6	Harsh
2	40	6.2	2332	337	33.3	Fair
3	45	7.5	2313	341	38.0	Good
4	36	6.8	2324	348	40.2	Good

*Water-cement ratio was 0.45.

The amounts of fine and coarse aggregates not used are recorded on the data sheet in Column 3, and mass of aggregates used (Column 2 minus Column 3) are noted in Column 4. If the slump when tested had been greater than that required, additional fine or coarse aggregates (or both) would have been added to reduce slump. Had the slump been less than required, water and cement in the appropriate ratio (0.45) would have been added to increase slump. It is important that any additional quantities be measured accurately and recorded.

Mixture Proportions. Mixture proportions for a cubic meter of concrete are calculated in Column 5 by using the batch yield (volume, V), mass (M), and density (unit weight, D). For example, the mass of cement in kilograms per cubic meter of concrete is determined by dividing one cubic meter by the volume of concrete in the batch and multiplying the result by the number of kilograms of cement in the batch.

$$V = M/D = 67.9 \text{ kg} / 2313 \text{ kg/m}^3 = 0.0294 \text{ m}^3$$
$$C = 1 \text{ m}^3 / V = 1 \text{ m}^3 / 0.0294 \text{ m}^3 = 34.06 \text{ batches}$$
$$\text{Cement} = 34.06 \text{ batches} \times 10 \text{ kg/batch} = 341 \text{ kg}$$

The percentage of fine aggregate by mass of total aggregate is also calculated. In this trial batch, the cement content was 341 kg/m³ and the fine aggregate made up 38% of the total aggregate by mass:

$$\%FA = FA / (FA + CA) \times 100 = 38\%$$

The air content and slump were acceptable. The 28-day strength was 39.1 MPa, greater than f'_{cr}. The mixture in Column 5, along with slump and air content limits of 40 mm ± 20 mm and 5% to 8%, respectively, is now ready for submission to the project engineer.

MATERIAL	INITIAL MASS, kg	FINAL MASS, kg	MASS USED (Col 2 - Col 3)	MASS PER m³, kg/m³ [No. of batches (C) × Col 4]
Cement	10	0	10.0	341
Water	4.5	0	4.5	153
Coarse Aggregate	37.6	17.3	20.3	691
Fine Aggregate	44.1	11.0	33.1	1128
Air-Entraining Admixture	10 ml			
Total Yield			67.9	2313

Example 5. Laboratory Trial Mixture Using the PCA Water-Cement Ratio Method (Inch-Pound)

With the following method, the mix designer develops the concrete proportions directly from a laboratory trial batch, rather than the absolute volume of the constituent ingredients.

Conditions and Specifications. Air-entrained concrete is required for a foundation wall that will be exposed to moderate sulfate soils. A compressive strength, f'_c, of 4000 psi at 28 days using Type II cement is specified. Minimum thickness of the wall is 10 in. and concrete cover over ½-in. diameter reinforcing bars is 3 in. The clear distance between reinforcing bars is 3 in. The w/cm versus compressive strength relationship based on field and previous laboratory data for the example ingredients is illustrated by Figure 13-10. Based on the test records of the materials to be used, the standard deviation is 300 psi. Proportion and evaluate by trial batch a mixture meeting the above conditions and specifications. Enter all data in the appropriate blanks on a trial-mixture data sheet.

Water-Cementitious Materials Ratio. For these exposure conditions, S1, Table 13-1 indicates that concrete with a maximum w/cm of 0.50 should be used and the minimum design strength should be 4000 psi.

The w/cm for strength is selected from a graph plotted to show the relationship between the w/cm and compressive strength for these specific concrete materials (Figure 13-10).

For a standard deviation of 300 psi, f'_{cr} must be the larger of:

$$f'_{cr} = f'_c + 1.34S = 4000 \text{ psi} + 1.34 (300 \text{ psi}) = 4402 \text{ psi}$$

or

$$f'_{cr} = f'_c + 02.33S - 500 \text{ psi} = 4000 + 2.33 (300 \text{ psi}) - 500 \text{ psi} = 4199 \text{ psi}.$$

Therefore, f'_{cr} = 4400 psi.

From Figure 13-10, the w/cm for air-entrained concrete is 0.55 for an f'_{cr} of 4400 psi. This is greater than the 0.50 permitted for the exposure conditions; therefore, the exposure requirements govern. A w/cm of 0.50 must be used, even though this may produce strengths higher than needed to satisfy structural requirements.

Aggregate Size. Assuming it is economically available, 1½-in. maximum-size aggregate is satisfactory; it is less than ⅕ the wall thickness and less than ¾ the clear distance between reinforcing bars and between reinforcing bars and the form. If this size were not available, the next smaller available size would be used. Aggregates are to be in a saturated surface-dry condition for these trial mixtures.

Air Content. Because of the exposure conditions and to improve workability, a moderate level of entrained air is needed. From Table 13-6, the target air content for concrete with 1½-in. aggregate in a moderate exposure is 4.5%. Therefore, proportion the mixture with an air content range of 4.5% ± 1% and aim for 5.5% ± 0.5% in the trial batch.

Slump. The recommended slump range for placing a reinforced concrete foundation wall is 1 in. to 3 in., assuming that the concrete will be consolidated by vibration. Batch for 3 in. ± 0.75 in.

Batch Quantities. For convenience, a batch containing 20 lb of cement is to be made. The quantity of mixing water is determined by the w/cm required: 20 lb × 0.50 = 10 lb. Representative samples of fine and coarse aggregates are weighed into suitable containers. The values are entered as initial weights in Column 2 of the trial-batch data sheet.

All of the measured quantities of cement, water, and air-entraining admixture are used and added to the mixer. Fine and coarse aggregates, previously brought to a saturated surface-dry condition, are added in proportions similar to those used in mixtures from which Figure 13-10 was developed. Mixing continues until a workable concrete with a 3-in. slump deemed adequate for placement is produced. The relative proportions of fine and coarse aggregate for workability can readily be judged by an experienced concrete technician or engineer.

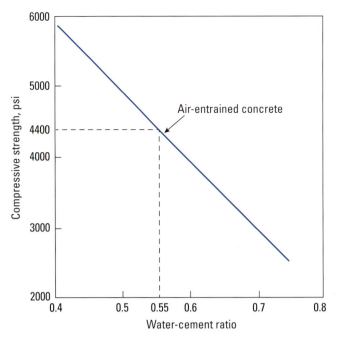

FIGURE 13-10. Relationship between strength and water to cement ratio based on field and laboratory data for specific concrete ingredients.

Workability. Results of tests for slump, air content, unit weight, and a description of the appearance and workability ("Good" for this example) are noted on the data sheet.

The amounts of fine and coarse aggregates not used are recorded on the data sheet in Column 3, and masses of aggregates used (Column 2 minus Column 3) are noted in Column 4. If the slump when tested had been greater than that required, additional fine or coarse aggregates (or both) would have been added to reduce slump. Had the slump been less than required, water and cement in the appropriate ratio (0.50) would have been added to increase slump. It is important that any additional quantities be measured accurately and recorded on the data sheet.

Mixture Proportions. Mixture proportions for a cubic yard of concrete are calculated in Column 5 by using the batch yield (volume) and density (unit weight). For example, the number of pounds of cement per cubic yard is determined by dividing 27 ft³ (1 yd³) by the volume of concrete in the batch and multiplying the result by the number of pounds of cement in the batch. The percentage of fine aggregate by weight of total aggregate is also calculated. In this trial batch, the cement content was 539 lb/yd³ and the fine aggregate made up 33.5% of the total aggregate by weight. The air content and slump were acceptable. The 28-day strength was 4950 psi (greater than f'_{cr}). The mixture in Column 5, along with slump and air content limits of 1 in. to 3 in. and 3.5% to 5.5%, is now ready for submission to the project engineer.

$$V = M/D = 144.3/144 = 1.0021 \text{ ft}^3$$
$$C = 1 \text{ yd}^3/V = 27/1.0021 = 26.943 \text{ batches}$$
$$\text{Cement} = 26.943 \text{ batches} \times 20 \text{ lb} = 539 \text{ lb}$$

Mixture Adjustments. To determine the most workable and economical proportions, additional trial batches could be made varying the percentage of fine aggregate. In each batch the water-cement ratio, aggregate gradation, air content, and slump should remain about the same. Results of four such trial batches are summarized below:

BATCH NO.	SLUMP, in.	AIR CONTENT, %	DENSITY, lb/ft³	CEMENT CONTENT, lb/yd³	FINE AGGREGATE, % of total aggregate	WORKABILITY
1	3	5.4	144	539	33.5	Good
2	2¾	4.9	144	555	27.4	Harsh
3	2½	5.1	144	539	35.5	Excellent
4	3	4.7	145	540	30.5	Excellent

*Water-cement ratio was 0.50.

MATERIAL	INITIAL MASS, lb	FINAL MASS, lb	MASS USED (Col 2 - Col 3)	MASS PER yd³ pcy [No. of batches (C) × Col 4]
Cement	20	0	20.0	539
Water	10	0	10.0	269
Coarse Aggregate	66.2	27.9	38.3	1032
Fine Aggregate	89.8	13.8	76.0	2048
Air-Entraining Admixture	0.3 oz			
Total Yield			144.3	3888

CONCRETE FOR SMALL JOBS

Although well-established ready mixed concrete mixtures are used for most construction, ready mix is not always practical for small jobs, especially those requiring one cubic meter (yard) or less. Small batches of concrete mixed at the site are required for such jobs.

If mixture proportions or mixture specifications are not available, Tables 13-15 and 13-16 can be used to select proportions for concrete for small jobs. Recommendations with respect to exposure conditions discussed earlier should be followed.

The proportions in Tables 13-15 and 13-16 are only a guide and may need adjustments to obtain a workable mixture with locally available aggregates (PCA 1988). Packaged, combined, dry concrete ingredients are also available (ASTM C387, *Standard Specification for Packaged, Dry, Combined Materials for Mortar and Concrete*).

TABLE 13-15A (SI). Proportions by Mass to Make One Tenth Cubic Meter of Concrete for Small Jobs

NOMINAL MAXIMUM SIZE COARSE AGGREGATE, mm	AIR-ENTRAINED CONCRETE				NON-AIR-ENTRAINED CONCRETE			
	CEMENT, kg	WET FINE AGGREGATE, kg	WET COARSE AGGREGATE, kg*	WATER, kg	CEMENT, kg	WET FINE AGGREGATE, kg	WET COURSE AGGREGATE, kg	WATER, kg
9.5	46	85	74	16	46	94	74	18
12.5	43	74	88	16	43	85	88	18
19.0	40	67	104	16	40	75	104	16
25.0	38	62	112	15	38	72	112	15
37.5	37	61	120	14	37	69	120	14

*If crushed stone is used, decrease coarse aggregate by 5 kg and increase fine aggregate by 5 kg.

TABLE 13-15B (Inch-Pound). Proportions by Mass to Make One Cubic Foot of Concrete for Small Jobs

NOMINAL MAXIMUM SIZE COARSE AGGREGATE, in.	AIR-ENTRAINED CONCRETE				NON-AIR-ENTRAINED CONCRETE			
	CEMENT, lb	WET FINE AGGREGATE, lb	WET COARSE AGGREGATE, lb*	WATER, lb	CEMENT, lb	WET FINE AGGREGATE, lb	WET COURSE AGGREGATE, lb	WATER, lb
3/8	29	53	46	10	29	59	46	11
1/2	27	46	55	10	27	53	55	11
3/4	25	42	65	10	25	47	65	10
1	24	39	70	9	24	45	70	10
1 1/2	23	38	75	9	23	43	75	9

*If crushed stone is used, decrease coarse aggregate by 3 lb and increase fine aggregate by 3 lb.

TABLE 13-16. Proportions by Bulk Volume* of Concrete for Small Jobs

NOMINAL MAXIMUM SIZE COARSE AGGREGATE, mm (in.)	AIR-ENTRAINED CONCRETE				NON-AIR-ENTRAINED CONCRETE			
	CEMENT	WET FINE AGGREGATE	WET COARSE AGGREGATE	WATER	CEMENT	WET FINE AGGREGATE	WET COURSE AGGREGATE	WATER
9.5 (3/8)	1	2 1/4	1 1/2	1/2	1	2 1/2	1 1/2	1/2
12.5 (1/2)	1	2 1/4	2	1/2	1	2 1/2	2	1/2
19.0 (3/4)	1	2 1/4	2 1/2	1/2	1	2 1/2	2 1/2	1/2
25.0 (1)	1	2 1/4	2 3/4	1/2	1	2 1/2	2 3/4	1/2
37.5 (1 1/2)	1	2 1/4	3	1/2	1	2 1/2	3	1/2

* The combined volume is approximately 2/3 of the sum of the original bulk volumes.

MIXTURE REVIEW

In practice, concrete mixture proportions will be governed by the limits of data available on the properties of materials, the degree of control exercised over the production of concrete at the plant, and the amount of supervision at the jobsite. It should not be expected that field results will be an exact duplicate of laboratory trial batches. An adjustment of the selected trial mixture is usually necessary on the job.

The mixture design and proportioning procedures presented here and summarized in Figure 13-11 are applicable to normal-weight concrete. For concrete requiring some special property, using special admixtures or materials – lightweight aggregates, for example – different proportioning principles may be involved.

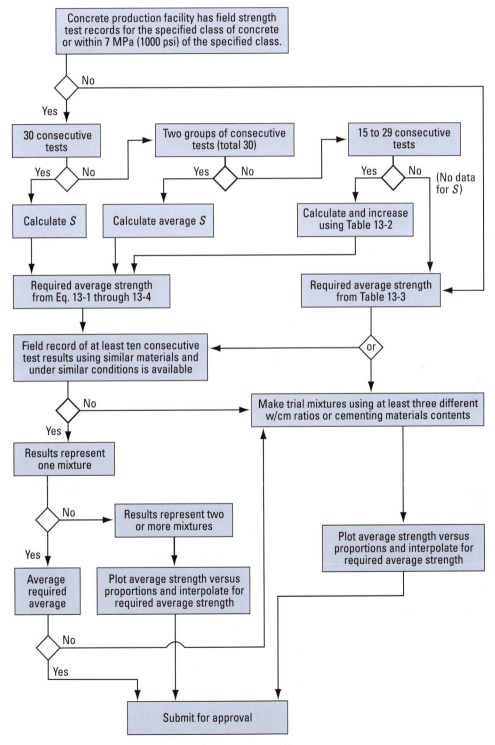

FIGURE 13-11. Flowchart for selection and documentation of concrete proportions.

REFERENCES

PCA's online catalog includes links to PDF versions of many of our research reports and other classic publications. Visit: cement.org/library/catalog.

Abrams, D.A., *Design of Concrete Mixtures, Lewis Institute, Structural Materials Research Laboratory*, LS001, Bulletin No. 1, Chicago, IL, 1918, 20 pages.

ACI Committee 201, *Guide to Durable Concrete*, ACI 201.2R-16, American Concrete Institute, Farmington Hills, MI, 2016, 86 pages.

ACI Committee 211, *Standard Practice for Selecting Proportions for Normal, Heavyweight and Mass Concrete*, ACI 211.1-91, American Concrete Institute, Farmington Hills, MI, 1991 (Reapproved 2009), 38 pages.

ACI Committee 211, *Guide for Selecting Proportions for High-Strength Concrete Using Portland Cement and Other Cementitious Material*, ACI 211.4R-08, American Concrete Institute, Farmington Hills, MI, 2008, 25 pages.

ACI Committee 211, *Guide for Submittal of Concrete Proportions*, ACI 211.5R-14, American Concrete Institute, Farmington Hills, MI, 2014, 14 pages.

ACI Committee 211, *Guide for Proportioning Concrete Mixtures with Ground Calcium Carbonate and Other Mineral Fillers*, ACI 211.7-20, American Concrete Institute, Farmington Hills, MI, 2020, 21 pages.

ACI Committee 213, *Guide for Structural Lightweight-Aggregate Con-crete*, ACI 213R-14, American Concrete Institute, Farmington Hills, MI, 2014, 53 pages.

ACI Committee 214, *Evaluation of Strength Test Results of Concrete*, ACI 214R-11, American Concrete Institute, Farmington Hills, MI, 2011 (Reapproved 2019), 16 pages.

ACI Committee 301, *Specification for Concrete Construction*, ACI 301-20, American Concrete Institute, Farmington Hills, MI, 2020, 73 pages.

ACI Committee 302, *Guide for Concrete Floor MI Slab Construction*, ACI 302.1R-15, American Concrete Institute, Farmington Hills, MI, 2015, 76 pages.

ACI Committee 318, *Building Code Requirements for Structural Concrete*, 318-19, American Concrete Institute, Farmington Hills, MI, 2019, 624 pages.

ACI Committee 329, *Report on Performance-Based Requirements for Concrete*, 329R-14, American Concrete Institute, Farmington Hills, MI, 2014, 46 pages.

Bentz, D., *Concrete Optimization Software Tool*, Computer software, National Institute of Standards and Technology (NIST), 2001.

Bureau of Reclamation, *Concrete Manual*, 8th ed., Denver, CO, revised 1981, 661 pages.

Caldarone, M.A.; Taylor, P.C.; Detwiler, R.J.; and Bhidé, S.B., *Guide Specification for High-Performance Concrete for Bridges*, EB233, Portland Cement Association, Skokie, IL, 2005, 64 pages.

Hover, K.C., "Graphical Approach to Mixture Proportioning by ACI 211.1-91," *Concrete International*, American Concrete Institute, Farmington Hills, MI, September 1995, pages 49 to 53.

Hover, K.C., "Concrete Design: Part 1, Finding Your Perfect Mix," *CE News*, September 1998, and "Concrete Design: Part 2, Proportioning Water, Cement, and Air," *CE News*, October 1998, and "Concrete Design: Part 3, Proportioning Aggregate to Finish the Process," *CE News*, November 1998.

Hover, K.C.; Bickley, J.; and Hooton, R. D., *Guide to Specifying Concrete Performance: Phase II Report of Preparation of a Performance-Based Specification for Cast-in-Place Concrete*, RMC Research and Education Foundation, Silver Spring, MD, March 2008, 53 pages.

Kosmatka, S.H., "Compressive versus Flexural Strength for Quality Control of Pavements," *Concrete Technology Today*, PL854, Portland Cement Association, Skokie, IL, 1985, pages 4 and 5.

NIST, *Internal Curing with Lightweight Aggregates*, Computer software, National Institute of Standards and Technology, 2010.

PCA, *Concrete for Small Jobs*, IS174, Portland Cement Association, Skokie, IL, 1988, 8 pages.

Shilstone, J.M., Sr., "Concrete Mixture Optimization," *Concrete International*, American Concrete Institute, Farmington Hills, MI, June 1990, pages 33 to 39.

Taylor, P.; Yurdakul, E.; Wang, X.; and Wang, X.; *Concrete Pavement Mixture Design and Analysis (MDA): An Innovative Approach to Proportioning Concrete Mixtures*, TPF-5(205), National Concrete Pavement Technology Center, Ames, IA, March 2015, 40 pages.

Taylor, P.; Van Dam, T.; Sutter, L.; and Fick, G., *Integrated Materials and Construction Practices for Concrete Pavement: A State-of-the-Practice Manual*, 2nd edition, TPF-5(286), Federal Highway Administration, Washington, DC, 2019, 338 pages.

Transtec Group, Inc., "COMPASS: Computer-Based Guidelines for Job-Specific Optimization of Paving Concrete," *Users Guide*, Austin, TX, 2010, 28 pages.

BATCHING, MIXING, TRANSPORTING, AND HANDLING CONCRETE

FIGURE 14-1. Concrete is most commonly batched and delivered in trucks as ready-mixed concrete (courtesy of NRMCA).

The production and delivery of concrete is achieved in different ways. Concrete is most commonly batched and delivered in trucks as ready-mixed concrete (Figure 14-1). ASTM C94, *Standard Specification for Ready-Mixed Concrete* (AASHTO M 157), establishes the standards for the manufacture and delivery of freshly mixed concrete and specifies many other materials and testing standards within ASTM.

ORDERING CONCRETE

The purchaser of ready-mixed concrete is typically the concrete contractor. In some cases, it may be the owner or specifier. When ready-mixed concrete is ordered, it is most important to communicate the requirements for fresh and hardened concrete to the concrete producer or manufacturer. Project specification requirements for concrete mixtures should be provided to the producer. This includes advanced communication regarding the day and time of placement, weather hold, quantity of concrete required, and the method and rate of placement.

ASTM C94 (AASHTO M 157) provides three options for ordering concrete with requirements for fresh concrete properties and compressive strength. If the concrete mixtures used on a project need special requirements for aesthetics, other hardened concrete properties, or durability, those need to be stated at the time concrete is ordered.

- *Option A* – performance-based order. The purchaser designates the compressive strength of the concrete, while the concrete producer selects the mixture proportions needed to obtain the required compressive strength, fresh concrete properties, or other performance parameters.

- *Option B* – prescription-based order. The purchaser states some of mixture proportions, such as cement, water, and admixture contents.

In this case, the purchaser controls the mixture proportions and therefore assumes responsibility for the performance of the mixture.

- *Option C* – combined option order. The purchaser designates the compressive strength and minimum cement content. The concrete producer selects the mixture proportions to comply with the requirements of the purchaser. The strength requirement governs: cement in excess of the minimum stated should be used if necessary to achieve the required strength.

For all ordering options, the purchaser should include requirements for slump, nominal maximum size of coarse aggregate, air content, and type of exposure anticipated. The concrete producer is typically not aware of where and how the concrete will be used without further information. There can be several types of concrete mixtures delivered to a single project. Mixtures are commonly assigned a designation that allows the purchaser to identify the type of mixture and location in the structure. This also establishes a link to the batch records for each load, which are retained at the plant.

The purchaser should also state delivery details based on the placement methods that will be used. Some placement methods require a continuous supply of concrete; alternatively, trucks should not be stacked up and waiting to discharge as this may impact the workability and quality, especially if excess water is added.

Ready-mixed concrete is sold by volume, in cubic meters or cubic yards. The basis for calculating the volume of concrete delivered in each load is described in ASTM C94 (AASHTO M 157). This is determined from the total weight of materials batched in that load and the measured density of freshly mixed concrete. Purchasers should estimate about 5% more concrete than the volume calculated from form dimensions to allow for testing and contingencies (waste and spillage, over-excavation, spreading forms, loss of entrained air, or settlement) during construction (NRMCA 2012b, 2016).

PRESCRIPTION VERSUS PERFORMANCE-BASED SPECIFICATIONS

Traditionally, the owner/agency develops the design requirements and establishes prescriptive provisions for the materials and proportions of concrete mixtures. These may include minimum cementitious materials content, and performance requirements for the concrete, such as the freeze-thaw resistance (air content) and strength. Contractors are then directed by the specifications to order the concrete mixture from the producer in accordance with the prescriptive provisions provided in the specifications. This procedure allows the owner to control the concrete mixture design and proportions. In turn, the risk associated with the performance of the mixture also resides with the owner. This method may lead to confusion over responsibility and authority if a problem with the concrete performance arises.

With the emergence of value engineering, design-build, performance specifications, and warranties, more owners are requiring the producer to develop the concrete mixture proportions based on performance criteria. This affords more flexibility and encourages innovation, but it also transfers more responsibility onto the concrete producer.

ACI 329R, *Report on Performance-Based Requirements for Concrete* (2014), outlines specific measures and considerations for specifying performance-based requirements for concrete construction. Prescriptive specifications will not guarantee how a concrete mixture performs. Significant reductions in project cost may be realized by allowing concrete suppliers to optimize mixtures for performance properties (Obla and Lobo 2006). See PCA EB221, *Specifiers Guide to Durable Concrete* (Detwiler and Taylor 2005), Caldarone and others (2005), Hover and others (2008), and NRMCA (2012a) for more information on specifying for concrete performance.

Sustainable construction attempts to control the environmental impact of construction products, including concrete mixtures. This may also result in prescriptive limits on concrete materials selection and proportions. As trends move from prescriptive specifications toward performance-based specifications and sustainable construction, concrete producers will become increasingly responsible for concrete mixture design and proportioning within the scope of their projects (see Chapter 13). Guides to understand and improve concrete specifications are given in NRMCA's *Guide to Improving Specifications for Ready Mixed Concrete* (2015), Daniel and Lobo (2014), Lamond and Pielert (2006), and CPMB 102, *Recommended Guide Specifications for Batching Equipment and Control Systems in Concrete Batch Plants* (2000).

BATCHING

Batching is the process of measuring quantities of concrete materials for mixtures by either mass or volume and introducing them into the mixer. To produce concrete of uniform quality, the ingredients must be measured accurately for each batch. Most ready-mixed concrete is produced by batching solid materials by mass and liquids by volume. This method of production is addressed in ASTM C94 (AASHTO M 157). ASTM C685, *Standard Specification for Concrete Made by Volumetric Batching and Continuous Mixing* (AASHTO M 241), covers volumetric batching and continuous mixing in equipment designed for volumetric production, commonly referred to as mobile mixers. Standards for batching equipment have been developed by the Concrete Plant Manufacturers Bureau (CPMB), Truck Mixer Manufacturers Bureau (TMMB), and Volumetric Mixer Manufacturers Bureau (VMMB) (CPMB 2018, TMMB 2005, and VMMB 2001).

TABLE 14-1. ASTM C94 Batching Tolerances

INGREDIENT	MEASURED QUANTITY	
Cementitious materials	±1% of required mass, and cumulative measured quantity of all cementitious materials shall also be within ±1% of the required cumulative mass at each intermediate weighing, if mass ≥ 30% of scale capacity	
	Up to +4%, and not less than required mass (-0%), if mass < 30% scale capacity*	
	INDIVIDUAL BATCHING	CUMULATIVE BATCHING
Aggregates	±2% of required mass[3] if mass ≥ 15% of scale capacity	±1% of required mass if mass ≥ 30% of scale capacity
	±0.3% of scale capacity if mass < 15% of scale capacity	±0.3% of scale capacity if mass < 30% of scale capacity
Batch Water	±1% of mixing water mass	
Admixtures	Greater of: ± 3% of required mass ± dosage for 50 kg (100 lb) of cement	

*Minimum batch size of 1 m³ (1 yd³)

ASTM C94 (AASHTO M 157) requires that materials should be measured within the percentages of accuracy shown in Table 14-1.

Scales and volumetric devices for measuring quantities of concrete ingredient materials should be verified for accuracy. The accuracy of scales and batching equipment should be checked periodically and adjusted when necessary (Figure 14-2). The typical frequency is at least once every 6 months. Scales are checked for accuracy by using a combination of certified field standard weights and product substitute loading. ASTM C94 (AASHTO M 157) requires that scales should be accurate to the greater of ±0.15% of the scale capacity or ±0.4% of the applied test load in all quarters of the scale capacity through its range of use. The minimum field standard weights required for this process is the greater of 500 kg (1000 pounds) or 10% of the scale capacity.

The Concrete Plant Manufacturers Bureau sets scale capacity limits in CPMB 100, *Concrete Plant Standards of the Concrete Plant Manufacturers Bureau* (2000).

Cementitious materials. Cement and supplementary cementitious materials are batched by mass and weighed in cumulative batchers or in separate batchers provided for each material. A vented dust seal is provided between the charging mechanism and the batcher. A vibrator may be fitted in ensure complete discharge. An overload removal port shall be provided. The sum of the capacities of a scale weighing cementitious materials shall not be less than 390 kg per cubic meter of rated capacity (660 lb per cubic yard).

Aggregates. Aggregates are batched by mass and weighed in a cumulative batcher or in separate batchers provided for each

material. There needs to be sufficient clearance above the batcher to permit removal of overload. The sum of the capacities of the scales weighing aggregates shall not be less than 1960 kg per cubic meter of rated capacity (3300 lb per cubic yard).

The standard size of batchers and minimum capacity requirements for cementitious materials and aggregates are shown in Table 14-2.

Water. Water is batched either by mass or volume. It is typically measured through water meters, in volumetric tanks, or in scales that measure the mass with discharging valves to prevent loss of water. Scales for water may be graduated either in kilograms (pounds) or liters (gallons), or both. Batchers shall have a volume capacity of not less than 200 L/m³ (40 gal/yd³) for a single batch. The sum of the capacities of a scale weighing water shall not be less than 190 kg per cubic meter or 190 L/m³ (320 lb per cubic yard or 40 gal/yd³) of rated capacity.

FIGURE 14-2. Batch control room in a typical ready-mixed concrete plant (courtesy of Ernst Concrete of Georgia).

Admixtures. Chemical admixtures are batched by weight or volume are typically charged into the mixture as aqueous solutions. Scales for admixtures may be graduated by weight or volume of the admixture required per 50 kg (100 lb) of cement. Admixtures that cannot be added in solution are commonly batched by mass or volume in bag quantities as directed by the manufacturer.

There are several types of admixture batching systems, including: a nutating disk, which is commonly used for liquid admixtures, one of several methods that can be used with manual, semiautomatic, and automatic systems, operated by remote control; manual dispensing systems, designed for low-volume concrete plants that depend on manual addition of the admixture into a calibration tube and discharged into the batch; and more sophisticated systems intended for automated high volume plants, which provide automatic fill and discharge.

Adding an admixture at the jobsite can also be accomplished with a tank and pressurized dispensing system. A calibrated holding tank and dispensing device should be part of the system so the plant operator can verify the proper amount of admixture is batched into the concrete mixer or truck-mounted tank.

TABLE 14-2. Standard Batch Sizes and Minimum Capacity Requirements for Cementitious Materials and Aggregates*

STANDARD METRIC			STANDARD INCH-POUND		
RATED CAPACITY OF PLANT, m³	MINIMUM VOLUME OF BATCHER, m³		**RATED CAPACITY OF PLANT, yd³**	MINIMUM VOLUME OF BATCHER, ft³	
	CEMENTITIOUS MATERIALS	**AGGREGATES**		**CEMENTITIOUS MATERIALS**	**AGGREGATES**
0.375	0.215	0.528	½	7.5	19
0.75	0.34	1.056	1	12	38
1	0.42	1.407	1½	16.5	57
1.5	0.59	2.11	2	21	76.0
2	0.76	2.81	3	30	114
3	1.09	4.22	4	39	152
3.5	1.26	4.93	4 ½	43.5	171
4	1.42	5.63	5	48	190
4.5	1.59	6.33	6	57	228
5	1.76	7.04	7	66	266
6	2.09	8.44	8	75	304
7	2.42	9.85	9	84	342
7.5	2.59	10.56	10	93	380
8	2.76	11.26	11	102	418
9	3.09	12.67	12	111	456
10	3.42	14.07	13	120	494
11	3.76	15.48	14	129	532
11.5	3.92	16.19	15	138	570
12	4.09	16.89	16	147	608

*Adapted from CPMB 100/100M-18

MIXING CONCRETE

All concrete should be mixed thoroughly until its ingredients are uniformly distributed. Mixers should not be loaded above their rated mixing capacities and should be operated at the mixing speed and for the period, either based on revolutions or time, recommended by the manufacturer. The rated mixing capacity of revolving drum truck mixers is limited to 63% of the gross internal volume of the barrel. For stationary plant mixers, the mixing capacity varies depending on the design. Increased output is typically obtained by double batching for a load, or using a larger mixer, or additional mixers, rather than by speeding up or overloading the equipment on hand. If the blades of a mixer become worn or coated with hardened concrete, mixing action will be less efficient. These conditions should be corrected by replacing the blades or removing the hardened concrete (CPMB100 2018).

If concrete has been adequately mixed, samples taken from different portions of a batch should have essentially the same strength, density, air content, slump, and coarse-aggregate content, with some allowance for testing variability. The process for evaluating mixers for their ability to achieve uniformly mixed concrete are addressed in the Annex of ASTM C94 (AASHTO M 157). Requirements for maximum allowable differences to evaluate mixing uniformity within a batch of freshly mixed concrete are given in Table 14-3.

Structural lightweight concrete can be mixed the same way as normal-density concrete when the aggregates have less than 10% total absorption by mass or when the absorption is less than 2% by mass during the first hour after immersion in water. For aggregates not meeting these limits, mixing procedures are described by Bohan and Ries (2008).

TABLE 14-3. Requirements for Uniformity of Concrete (ASTM C94 [AASHTO M 157])

TEST	REQUIREMENT*
Density calculated on an air-free basis	16 kg/m³ (1.0 lb/ft³)
Air content, volume of concrete	1.0%
Slump:	
If average slump is 100 mm (4 in.) or less	25 mm (1.0 in.)
If average slump is 100 to 150 mm (4 to 6 in.)	40 mm (1.5 in.)
Coarse aggregate content, portion by mass of each sample retained on 4.75 mm (No. 4) sieve	6.0%
Average compressive strength at 7 days for each sample,[†] based on average strength of all comparative test specimens	7.5%[‡]

* Expressed as maximum permissible difference in results of tests of samples taken from two locations in the concrete batch.

[†] Not less than 3 cylinders will be molded and tested from each of the samples.

[‡] Approval of the mixer shall be tentative, pending results of the 7-day compressive strength tests.

STATIONARY MIXERS

Concrete may be mixed in a stationary mixer (Figure 14-3). Stationary mixers are used at construction jobsites, manufacturing plants, and at ready-mixed concrete plants. Most ready-mixed plants outside the U.S. employ the use of stationary mixers at plants. They are available in sizes up to 9 m³ (12 yd³) and can be of the revolving drum tilting or nontilting type, reversing drum, or a horizontal shaft revolving blade or paddle type. Precast concrete plants typically use smaller capacity pan-type vertical shaft stationary mixers. All types may be equipped with loading skips and some are equipped with a swinging discharge chute. Stationary mixers should have timing devices that are set for a given mixing time and locked so that the batch cannot be discharged until the designated mixing time has elapsed.

Careful attention should be paid to the required mixing time. ASTM C94 (AASHTO M 157) states a default minimum mixing time of one minute plus 15 seconds for every cubic meter (yard), unless mixer performance tests demonstrate that shorter periods can achieve uniformly mixed concrete. The Concrete Plant Manufacturers Bureau qualifies certain mixers for a shorter mixing time if uniformity requirements are met with less time (CPMB 100 2018). ASTM C94 recognizes that longer mixing can result in loss of air and indicates that the mixing time should not be extended by more than 60 s of established mixing

FIGURE 14-3. Concrete can be mixed at the jobsite or plant in a stationary mixer (courtesy of S. Parkes, Votorantim Cimentos).

time for air entrained concrete. Short mixing times, less than what the mixer is designated for, can result in non-homogenous mixtures with improper consistency, poor distribution of air voids (resulting in poor freeze-thaw resistance), low strength gain, and early stiffening problems (see Chapter 9). The mixing period should be measured from the time all solid materials are in the mixer drum, provided all the batch water is added before one-fourth of the mixing time has elapsed (ASTM C94 [AASHTO M 157]).

The sequence that ingredients are introduced into the mixer has a large impact on concrete uniformity. Under usual mixing conditions, most of the batch water should be charged into the drum before the solid materials are added. Coarse aggregate should be charged initially to avoid head packs, or cement balls. The aggregate and cementitious materials are ribbon-fed into the mixer, with the cement ending before the aggregates. The last 10 to 25% of the batch water should be added after all other materials are in the mixer. When heated water is used in cold weather concreting (see Chapter 19), this order of charging may require some modification to prevent possible rapid stiffening when hot water comes into contact with cement. In this case, addition of the cementitious materials should be delayed until most of the aggregate and water have intermingled in the drum. Supplementary cementitious materials should be weighed and batched into the mixer with the cement.

If retarding or water-reducing admixtures are used, they should be added in the same sequence in the charging cycle each time the mixture is batched. If not, significant variations in the time of initial setting and percentage of entrained air may result between batches. Addition of the admixtures should be completed no later than one minute after addition of water to the cement has been completed or prior to the start of the last three-fourths of the mixing cycle, whichever occurs first. If two or more admixtures are used in the same batch of concrete, they

A - Cementitious material storage
B - Cement delivery
C - Water storage
D - Aggregate storage bins
E - Aggregate receiving hopper
F - Conveyor belt
G - Admixture storage
H - Weigh hopper
I - Central Mixer
J - Concrete loaded in ready-mix truck
K - Washout station
L - Settling pits
M - Reclaimed aggregates
N - Recycled water

FIGURE 14-4. Schematic of a ready-mix concrete plant.

should be added separately to avoid any interaction that might interfere with the efficiency of the admixtures and adversely affect the concrete properties. The sequence in which they are added to the mixture is also important depending on any adverse side effects they may exhibit (see Chapter 6).

READY-MIXED CONCRETE

Ready-mixed concrete is batched and mixed at a concrete plant and delivered to the project in a freshly mixed and unhardened state. Figure 14-4 illustrates a central mix ready-mix plant. Ready-mixed concrete can be manufactured by any of the following methods:

• *Central-mixed concrete* – mixed completely in a stationary mixer and is delivered either in a truck mixer (Figure 14-5), a truck agitator (Figure 14-6 left), or a nonagitating truck (Figure 14-6 right).

• *Shrink-mixed concrete* – mixed partially in a stationary mixer and mixing is completed in a truck mixer.

• *Truck-mixed concrete* – mixed completely in a truck mixer (Figure 14-7).

ASTM C94 (AASHTO M 157) notes that when a truck mixer is used for complete mixing, uniformly mixed concrete should be attained within 70 to 100 revolutions of the drum at mixing speed. Mixing speed is typically about 12 rpm to 18 rpm.

FIGURE 14-5. Central mixing in a stationary mixer of the tilting drum type with delivery by a truck mixer operating at agitating speed.

Excessive mixing along with the addition of water to maintain slump, can result in concrete strength loss, temperature rise, excessive loss of entrained air, and accelerated slump loss. The homogeneity of concrete is maintained after mixing and during delivery by turning the drum at agitating speed. Agitating speed is usually about 2 rpm to 6 rpm. Additional revolutions at mixing speed are often applied before concrete is discharged or if any adjustments are made to the mixture at the jobsite.

When truck mixers are used, ASTM C94 (AASHTO M 157) requires that the purchaser determine a time limit from the start of mixing to when the concrete discharge must be completed. If no time limit is stated by purchaser, the manufacturer will

FIGURE 14-6. (left) Truck agitators are also used with central-mix batch plants. Agitation mixing capabilities allow truck agitators to supply concrete to projects with slow rates of concrete placement and at distances greater than nonagitating trucks. (right) Nonagitating trucks are used with central-mix batch plants where short hauls and quick concrete discharge (by conveyor, in this example) allows the rapid placement of large volumes of concrete (courtesy of Gomaco).

establish a limit. The time limit to complete discharge will be stated on the delivery ticket. ASTM C94 also requires that the purchaser shall state any drum revolution limit as to when the concrete discharge must begin. If no drum revolution limit is stated by purchaser, the manufacturer shall determine and communicate the limit to the purchaser prior to delivery. The purchaser can waive the limits on time and revolutions if the concrete is of adequate consistency and air content.

Current admixture and concrete mixture technology can facilitate longer delivery durations. Mixers and agitators should always be operated within the limits for volume and speed of rotation designated by the equipment manufacturer. This information is noted on plates affixed to the equipment. Trucks may be equipped with advanced communication and measuring equipment that record concrete temperature, slump, water and/or admixture additions, age, and track location of the truck.

The National Ready Mixed Concrete Association (NRMCA) administers a certification program for concrete production

facilities that audits the management of materials, accuracy of measuring devices, measurement accuracy, automation capability, recording capability, and condition of stationary mixers and delivery vehicles. The inspection and certification ensures that the facility conforms to ASTM C94 (AASHTO M 157) and industry standards. Several state departments of transportation also perform inspections of concrete plants and truck mixers as part of the approval process to bid on state paving projects. Some invoke a requirement for NRMCA certification.

VOLUMETRIC (CONTINUOUS) MIXER CONCRETE

Volumetric mixers (Figure 14-8) batch concrete by volume, and continuously deliver a stream of dry ingredients, water, and admixtures into a mixing auger. The auger blends the materials into a fresh, homogenized mix as long as materials are being fed. The concrete is produced in accordance with ASTM C685 (AASHTO M 241). The concrete mixture is easily adjusted for different application needs and varying weather conditions.

These mixers have traditionally been used to produce smaller quantities of concrete or for remote locations (Figure 14-9). Recent use includes production of concretes that are avoided in truck mixers – such as rapid setting or latex modified concrete, as well as for use on projects where time requirements are a consideration. Applications include delivery into pumps, use on barges for underwater placement, mounting onto railroad cars for tunnel work, and for delivery into drum mixers at a small batch plant.

RETEMPERING CONCRETE

Specifications require that the slump or slump flow (for self-consolidating concrete) should be within a specific range when discharged. Tolerances for ready-mix concrete are stated in ASTM C94 (AASHTO M 157). Because of the various

FIGURE 14-7. Truck-mixed concrete is mixed completely in a truck mixer (courtesy of NRMCA).

FIGURE 14-8. Volumetric mixers measure materials by volume and continuously mix concrete as the dry ingredients, water, and admixtures are fed into a mixing auger at the rear of the vehicle.

FIGURE 14-9. Volumetric mixers are traditionally used for smaller quantities of concrete or in remote locations (courtesy of L. Szabo, Zimmerman Industries).

uncertainties associated with the production and delivery and the impact on slump, concrete producers may hold back some of the batch water for a concrete batch so that it can be used to retemper the load during transportation to the jobsite or upon arrival onsite to achieve target slump prior to discharge. ASTM C94 (AASHTO M 157) allows water or water-reducing admixture, or both, to be added to the concrete if the slump or slump flow is less than specified providing the following conditions are met:

- the maximum quantity of water and/or water-reducing admixture that can be added on site is determined by the manufacturer and does not exceed the allowable w/cm,

- adjusting the concrete mixture with water or water-reducing admixture is done prior to discharge of a significant portion of the load (a preliminary sample is permitted to measure slump or air content),

- the maximum allowable slump or slump flow is not exceeded,

- the maximum allowable mixing and agitating time (or drum revolutions) are not exceeded,

- concrete is remixed for a minimum of 30 revolutions at mixing speed to ensure the adjustment is properly distributed through the load, and

- the quantity of water or water-reducing admixture is recorded.

Water and water-reducing admixtures can be added to concrete by automated systems that monitor slump in the mixer and add water or water-reducing admixture as needed, or by the truck mixer operator at the jobsite. The automated system should be capable of making the additions to an accuracy of ±3% of the target quantity. For mixers that do not have an automated system, water is added from a tank on the truck mixer and the quantity is measured via a sight gage on the tank or a water meter. The recorded addition should be accurate to 1 L (1 gal). Indiscriminate addition of water in excess of that designed for the mixture to make concrete more fluid should not be allowed because this increases the w/cm and lowers the strength and quality of concrete.

ASTM C94 (AASHTO M 157) permits adjustment of the mixture with an air-entraining admixture if the air content at the jobsite is lower than the permitted range. Other adjustments may be the use of admixtures that modify setting characteristics of concrete. These adjustments should be made by the concrete producer or with their approval.

TRANSPORTING AND HANDLING CONCRETE

Good advanced planning can help select the appropriate method of transporting and handling of concrete. The objective is to safely install quality concrete in a productive manner. The greatest productivity will be achieved if the work is planned to optimize the personnel and equipment selected. Delays, segregation, early stiffening, and premature drying can all seriously affect the quality of the finished work and must be taken into consideration.

Planning and Scheduling. The objective in planning any project schedule is to produce the fastest work using a qualified labor force and the proper equipment for the project at hand. Schedules should be realistic, allowing adequate time for project managers, concrete suppliers, and workers to evaluate conditions that affect placements and to coordinate resources such as drivers, pumps, and traffic controls (see Chapter 15). Avoiding delivery delays, traffic, jobsite congestion, and managing travel time are important elements of delivery.

Early Stiffening and Drying Out. Concrete can be designed to meet unique and challenging requirements. One of the most critical requirements is for mixtures to maintain plasticity and adequate slump to allow for placement. Concrete begins to react as soon as the cementitious materials and water are mixed together. The degree of stiffening that occurs in the first 30 minutes is not usually a problem unless excessive slump loss occurs. Concrete that is properly proportioned and kept agitated can generally be placed and compacted successfully within the typical ASTM C94 (AASHTO M 157) point of delivery time limit of 1½ hours. In hot weather the time limits may be reduced

FIGURE 14-10. Ready-mixed concrete is often placed in its final location by direct chute discharge from a truck mixer.

FIGURE 14-11. In comparison to conventional rear-discharge trucks, front-discharge truck mixers provide the driver with more mobility and control for direct discharge into place.

or in cold temperatures the time limits may be extended (see Chapters 18 and 19).

When challenges occur and if the project requirements allow for it, admixture technology can considerably extend the delivery time prior to discharge. Proper planning should eliminate or minimize any variables that would allow the concrete to stiffen to the extent that consolidation is not achieved or finishing is impaired.

Segregation. Segregation is the tendency for coarse aggregate to separate from the mortar. Segregation of concrete can impair uniformity of concrete in the structure. The mixture should be proportioned to minimize segregation. The method and equipment used to transport and handle the concrete must not result in segregation of the concrete materials.

Mixtures with too much coarse aggregate and limited amount of paste are considered boney mixes. These types of mixtures can be difficult to pump and finish and they tend to segregate more readily than properly proportioned mixtures. On the other hand, mixtures with too much paste are "sticky" and may be difficult to finish.

Significant variation can occur in aggregate gradations. ASTM C33, *Standard Specification for Concrete Aggregates*, is a single standard for available materials in all regions. Gap-graded aggregate with fewer mid and smaller aggregates can lead boney mixes while too many fines can lead to sticky mixes. Too many fines in a mix can also lead to increased water demand and, in extreme cases, significantly lower concrete strengths. Various techniques can be used to analyze aggregate gradations and determine sources of problems (see Chapters 5 and 20).

Viscosity modifying admixtures may be used to reduce segregation of mixtures. These types of admixtures are used extensively with self-consolidating concrete mixtures (see Chapter 6).

FIGURE 14-12. Versatile power buggy can move all types of concrete over short distances.

FIGURE 14-13. Concrete is easily lifted to its final location by bucket and crane.

TABLE 14-4. Methods and Equipment for Transporting and Handling Concrete

EQUIPMENT	TYPE AND RANGE OF WORK FOR WHICH EQUIPMENT IS BEST SUITED	CONCRETE MIXTURE CONSIDERATIONS	ADVANTAGES	POINTS TO WATCH FOR
Belt conveyors—large truck mounted or smaller conveyors mounted to mixer trucks.	For conveying concrete horizontally or to a slightly higher or lower level. Usually positioned between main discharge point and secondary discharge point. Truck mounted conveyers are for high volume placements and mixer mounted conveyers are for lower volume placements.	Concrete mixture should promote cohesion of the concrete as it is conveyed on the belt. Effective for mix designs that cannot be pumped due to aggregate size larger than 37.5 mm (1½ in). Slump must be suitable for transport on the belt.	Belt conveyors have adjustable reach, traveling diverter, and variable speed both forward and reverse. Can place large volumes of concrete quickly when access is limited.	End-discharge arrangements needed to prevent segregation and leave no mortar on return belt. In adverse weather (hot, windy) long reaches of belt need cover.
Buckets	Used with cranes, cableways, and helicopters for construction of buildings and dams. Convey concrete directly from central discharge point to formwork or to secondary discharge point.	Slower setting mixtures may be desired. Bucket are available form ⅓ yd³ to 10 yd³, but due to crane lifting capacity buckets larger than 3 yd³ are not often used.	Enables full versatility of cranes, cableways, and helicopters to be exploited. Clean discharge. Wide range of capacities.	Select bucket capacity to conform to placement needs and crane capacity. Discharge should be controllable. Limited to one lift at a time. Careful scheduling between operations is needed.
Chutes on truck mixers	For conveying concrete to a lower level, usually below ground level, on all types of construction.	No special requirements.	Low cost and easy to maneuver. No power required; gravity does most of the work.	Slopes should range between 1 to 2 and 1 to 3 and chutes must be adequately supported in all positions. End-discharge arrangements (downpipe) needed to prevent segregation.
Dropchutes	Used for placing concrete in vertical forms of all kinds. Some chutes are one-piece tubes made of flexible rubberized canvas or plastic, others are assembled from articulated metal cylinders (elephant trunks).	No special requirements.	Dropchutes direct concrete into formwork and carry it to bottom of forms without segregation. Their use avoids spillage of grout and concrete on reinforcing steel and form sides, which may diminish formed surface finish aesthetics. They also will prevent segregation of coarse particles.	Dropchutes should have sufficiently large, splayed-top openings into which concrete can be discharged without spillage. The cross section of a dropchute should be chosen to permit inserting into the formwork without interfering with reinforcing steel.
Mobile batcher mixers	Used for intermittent production of concrete at jobsite, or where only small quantities are required.	Desirable to place fast setting (accelerated) concrete mixtures.	A combined materials transporter and mobile batching and mixing system for quick, precise proportioning of specified concrete. One-person operation.	Trouble-free operation requires good preventive maintenance program on equipment. Materials must be identical to those in original mix design.
Non-agitating trucks	Used to transport concrete on short hauls over smooth roadways.	Concrete slump should be limited to low slump or zero slump mixtures (paving).	Capital cost of nonagitating equipment is lower than that of truck agitators or mixers.	Possibility of segregation. Clearance is needed for high lift of truck body upon discharge.
Pneumatic guns (shotcrete)	Used where concrete is to be placed in difficult locations and where thin sections and large areas are needed.	Special mixtures designed to hang on sloped, vertical and overhead structures.	Ideal for placing concrete in freeform shapes, for repairing structures, for protective coatings, thin linings, and building walls with one-sided forms.	Quality of work depends on skill of those using equipment. Only experienced nozzlemen should be employed.

TABLE 14-4. Methods and Equipment for Transporting and Handling Concrete (continued)

EQUIPMENT	TYPE AND RANGE OF WORK FOR WHICH EQUIPMENT IS BEST SUITED	CONCRETE MIXTURE CONSIDERATIONS	ADVANTAGES	POINTS TO WATCH FOR
Pumps	Used to convey concrete directly from central discharge point at jobsite to formwork or to secondary discharge point.	Concrete mixture should have a pumpable consistency and without tendency to segregate. Pumpabilty is depended on concrete mixtures having enough paste and adequate slump. Typically, a mortar fraction of 0.52 – 0.54 and a slump of 125 mm (5 in.) at the point of placement. Aggregate size must be less than ⅓ the diameter of pumpline. For example, topsize aggregate must be less than 37.5 mm (1½ in.) for large pumps with 125 mm (5 in.) line.	Pipelines take up little space and can be readily extended and deliver concrete in continuous stream. Pumps can move concrete both vertically and horizontally. Truck-mounted boom pumps or trailer pumps can be mobilized based on project needs. Self-climbing or deck mounted placing booms provide means to distribute concrete on each floor of a high-rise structure.	Constant supply of freshly-mixed concrete is needed. Care must be taken in operating pipeline to ensure an even flow and to clean out at conclusion of each operation. Pumping vertically, around bends, and through flexible hose will considerably reduce the maximum pumping distance.
Screw spreaders	Used for spreading concrete over large flat areas, such as in pavements and bridge decks.	Typically used for low slump mixtures (paving).	With a screw spreader a batch of concrete discharged from a bucket or truck can be quickly spread over a wide area to a uniform depth. The spread concrete has good uniformity of compaction before vibration is used for final compaction.	Screw spreaders are normally used as part of paving train. They should be used for spreading before vibration is applied.
Tremies	For placing concrete in slurry applications, and under water.	Underwater concrete mixture should have a minimum cement content of 390 kg/m³ (658 lb/yd³), and slump of 150 mm to 230 mm (6 in. to 9 in.). Concrete must flow and consolidate without any vibration and not disperse in the water.	Can be used to funnel concrete through slurry or water.	Precautions are needed to ensure that the tremie discharge end is always buried in fresh concrete, so that the seal is preserved between water and concrete mass. Diameter should be 250 mm to 300 mm (10 in. to 12 in.) unless pressure is available.
Truck agitators	Used to transport concrete for all uses in pavements, structures, and buildings. Haul distances must allow discharge of concrete within 1½ hours, but limit may be waived under certain circumstances.	No special requirements.	Truck agitators usually operate from central mixing plants where quality concrete is produced under controlled conditions. Discharge from agitators is well controlled. There is uniformity and homogeneity of concrete on discharge.	Timing of deliveries should suit job organization. Concrete crew and equipment must be ready onsite to handle concrete.
Truck mixers	Used to transport concrete for use in most applications. Haul distances must allow discharge of concrete within 1½ hours, but limit may be waived under certain circumstances (ASTM C94 [AASHTO M 157]).	No special requirements.	No central mixing plant needed, only a batching plant, since concrete is completely mixed in truck mixer. Discharge is same as for truck agitator.	Timing of deliveries should suit job organization. Concrete crew and equipment must be ready onsite to handle concrete. Control of concrete quality is not as good as with central mixing.
Wheelbarrows and power buggies	For short flat hauls on all types of onsite concrete construction, especially where accessibility to work area is restricted.	No special requirements.	Very versatile and therefore ideal inside and on jobsites where placing conditions are constantly changing.	Slow and labor intensive.

FIGURE 14-14. (left) A truck-mounted pump and boom can conveniently move concrete vertically or horizontally to the desired location. (right) View of concrete discharging from flexible hose connected to rigid pipeline leading from the pump. Rigid pipe is used in pump booms and in pipelines to move concrete over relatively long distances. Up to 8 m (25 ft) of flexible hose may be attached to the end of a rigid line to increase placement mobility.

METHODS AND EQUIPMENT FOR TRANSPORTING AND HANDLING CONCRETE

The selection of methods and equipment used to convey concrete from the producer to the final placement impacts the final performance. Machines for transporting and handling concrete are vastly improved over previously available methods. The transporting and handling equipment should be chosen to minimize delays in concrete placement. Optimized results often include adjustment to mixtures based on the method of conveyance selected. Table 14-4 provides a list of conveyance methods commonly used. Regardless of the method of conveyance that is selected, equipment used must

be maintained in good operating condition and contingency planning should include means to recover from an equipment failure, such as available backup equipment.

There have been many advancements in the methods used to convey concrete over the last 80 years. Ready-mixed concrete trucks traditionally deliver concrete to the projects (Figures 14-10 and 14-11). The wheelbarrow and buggy, although still used on relatively smaller projects, evolved into the power buggy (Figure 14-12); the bucket hauled over a pulley wheel evolved into the bucket and crane (Figure 14-13); and high capacity concrete conveyors and boom pumps (Figure 14-14) are prevalent today. Direct placement from truck chutes is the preferred method as it is economical and effective. However, chutes are not always practical for larger placements or those that cannot be accessed by truck.

As concrete-framed buildings became taller, the need to hoist reinforcement and formwork as well as concrete to higher levels led to the development of the tower crane – a familiar sight on the building skyline today (Figures 14-13 and 14-15). The crane and bucket method of placing is much slower than pumping and is often used for smaller volume slab placements and columns. Utilizing construction methods as self-climbing formwork systems, reusable formwork liners, panelized horizontal formwork systems, prefabricated reinforcement cages, and state-of-the-art concrete pumping systems, allows for faster concrete placements. See Chapter 15 for more information on concrete placement.

FIGURE 14-15. The tower crane and bucket can easily handle concrete for tall-building construction (courtesy of Baker Concrete).

The first mechanical concrete pump was developed and used in the 1930s and the hydraulic pump was developed in the 1950s. Trailer or boom pumps are economical to use in placing both large and small quantities of concrete, depending on jobsite conditions. The advanced mobile pump with hydraulic placing boom (Figure 14-14) is probably the single most important innovation in concrete handling equipment. Modern booms range up to 60 m (200 ft) in length. On high rise buildings that exceed the range of boom pumps, risers are set up and independent placing booms that climb from floor to floor are utilized to place concrete. It is now possible to pump high-strength concrete in a single stage over 600 m (1968 ft), allowing for supertall applications. Pumping is economical to use in placing both large and small quantities of concrete, depending on jobsite conditions. For small to medium size projects, a combination of truck mixer and boom pump can be used to transport and place concrete.

For small projects, truck-mixer-mounted conveyor belts are also available (Figure 14-16) and for larger placements independent truck mounted conveyors can be used. The conveyor belt is an efficient, portable method of handling and transporting concrete. A dropchute at the end of the conveyor prevents concrete from segregating as it leaves the belt; a scraper prevents loss of mortar.

FIGURE 14-16. The conveyor belt is an efficient, portable method of transporting concrete.

FIGURE 14-17. A conveyor belt mounted on a truck mixer places concrete up to about 12 m (40 ft) without the need for additional transporting equipment.

FIGURE 14-18. The screw spreader quickly spreads concrete over a wide area to a uniform depth. Screw spreaders are used primarily in pavement construction.

Conveyor belts can be operated in series and on extendable booms of hydraulic cranes (Figure 14-17). The use of conveyors is preferred when concrete mixture proportions include large top size aggregates that cannot be pumped such as dams and mat foundations.

The screw spreader (Figure 14-18) has been very effective in placing and distributing concrete for pavements. Screw spreaders can place a uniform depth of concrete quickly and efficiently.

Shotcrete is concrete that is pneumatically projected onto a surface at high velocity (Figure 14-19). It may also be known as gunite or sprayed concrete. Shotcrete is used for both new construction and repair work. It is especially suited for curved or thin concrete structures and shallow repairs (see Chapter 15).

FIGURE 14-19. Shotcrete is pneumatically applied concrete (courtesy of American Shotcrete Association).

See ACI 304R, *Guide for Measuring, Mixing, Transporting, and Placing Concrete* (2000), and Panarese (1987) for extensive information on methods to transport concrete.

CHOOSING THE BEST METHOD OF CONCRETE PLACEMENT

When choosing the best method for concrete placement, initial considerations are the type of project, its physical size, the total amount of concrete to be placed, and the placement schedule. Further consideration will identify the amount of work that is below, at, or above ground level. This aids in selecting the concrete transporting and handling equipment necessary for placing concrete at the required levels for efficient placement without plugged lines or cold joints. Concrete supply must be continuous to achieve monolithic placements.

Concrete must be moved from the mixer to the point of placement as rapidly as possible without segregation or loss of ingredients. The transporting and handling equipment must have the capacity to move sufficient concrete so that cold joints are eliminated (see Chapter 16).

The specifications and performance of transporting and handling equipment are being continuously improved. The best results and lowest costs will generally be realized if the work is planned to get the most out of the equipment. Panarese (1987) is very helpful in deciding which method to use based on capacity and range information for various methods and equipment.

Work At or Below Ground Level

The largest concrete volume placements on a typical job are usually either below or at ground level and therefore can be placed by methods different than those employed on a superstructure. Concrete work below ground can vary enormously – from filling large-diameter bored piles or massive mat foundations to the intricate work involved in basement and subbasement walls. A crane can be used to handle formwork, reinforcing steel, and concrete. The concrete may be chuted directly from the truck mixer to the point needed. Chutes must not slope greater than 1:2 (vertical to horizontal) or less than 1:3 (vertical to horizonal). Long chutes, over 6 m (20 ft), or those not meeting slope standards, must discharge into a hopper before distribution to point of need. Belt conveyors are very useful for work near ground level. Since placing concrete below ground is frequently a matter of horizontal movement assisted by gravity, portable conveyors can be used for high output at relatively low cost.

Alternatively, a concrete pump can readily move the concrete to its final position. Pumps must be of adequate capacity and must be capable of moving concrete without segregation. The loss of slump caused by pressure forcing mix water into the aggregates as the concrete mixture travels from pump hopper to discharge

at the end of the pipeline must be minimal – not greater than 50 mm (2 in.). The air content generally should not be reduced by more than 2 percentage points during pumping. Air loss greater than this may be caused by a boom configuration that allows the concrete to free-fall. In view of this, specifications for both slump and air content may be required at the discharge end of the pump (point of placement). A tapered hose may be needed to keep from losing entrained air from the end of the pump (see Chapter 9). Pipelines must not be made of aluminum or aluminum alloys. These cause excessive entrainment of air; aluminum reacts with alkalies in the fresh concrete to form hydrogen gas. This can result in high air content and serious reduction in concrete strength. For more information on pumping concrete see the American Concrete Pumping Association (ACPA), concretepumpers.com.

Work Above Ground Level

Conveyor belt, crane and bucket, hoist, pump, or even a helicopter, can be used for lifting concrete to locations above ground level (Figure 14-20). The tower crane and pumping boom (Figure 14-21) are the right tools for tall buildings. The volume of concrete needed per floor as well as boom placement and length affect the use of a pump; large volumes minimize pipeline movement in relation to output.

FIGURE 14-20. For work aboveground or at inaccessible sites, a concrete bucket can be lifted by helicopter (courtesy of Paschal).

FIGURE 14-21. A pump boom mounted on a mast and located near the center of a structure can frequently reach all points of placement. It is especially applicable to tall buildings where tower cranes cannot be tied up with placing concrete. Concrete is supplied to the boom through a pipeline from a ground-level pump. Concrete can be pumped hundreds of meters (feet) vertically with these pumping methods.

RETURNED FRESH CONCRETE

Returned fresh is defined by ASTM C1798, *Specification for Returned Fresh Concrete for Use in a New Batch of Ready-Mixed Concrete* as fresh concrete not yet discharged from a ready-mixed concrete transportation unit when it is returned to the manufacturer. Returned fresh concrete in a quantity of less than 450 kg (1000 lb) or 0.2 m³ (0.25 yd³) is not subject to ASTM C1798.

The new batch of concrete containing returned fresh concrete should meet all requirements of a specification. If using ASTM C94 (AASHTO M 157), it is recommended the purchaser select Option A for concrete. In this option, the manufacturer has responsibility for making necessary adjustments to mixture proportions to ensure the new concrete batch containing returned fresh concrete will meet project requirements.

Any additional water that is added to the returned fresh concrete prior to batching of new concrete mixture shall be measured to an accuracy of ± 3 % of the amount added or 4 L (1.0 gal), whichever is greater.

Returned fresh concrete that has reached an age of 1½ hours after the original loading or first mixing of the returned fresh concrete, shall be treated with extended set control admixture.

Returned concrete having a temperature that exceeds 38°C (100°F) should not be reused in a new concrete mixture.

REFERENCES

PCA's online catalog includes links to PDF versions of many of our research reports and other classic publications. Visit: cement.org/library/catalog.

ACI Committee 304, *Guide for Measuring, Mixing, Transporting, and Placing Concrete*, ACI 304R-00 (Reapproved 2009), American Concrete Institute, Farmington Hills, MI, 2000, 48 pages.

ACI Committee 329, *Report on Performance-Based Requirements for Concrete*, ACI 329R-14, American Concrete Institute, Farmington Hills, MI, 2014, 52 pages.

Bohan, R.P.; and Ries, J., *Structural Lightweight Aggregate Concrete*, IS032, Portland Cement Association, Skokie, IL, 2008, 8 pages.

Caldarone, M.A.; Taylor, P.C.; Detwiler, R.J.; and Bhidé, S.B., *Guide Specification for High-Performance Concrete for Bridges*, EB233, Portland Cement Association, Skokie, IL, 2005, 64 pages.

CPMB, *Concrete Plant Standards of the Concrete Plant Manufacturers Bureau*, Concrete Plant Manufacturers Bureau, Publication No. 100-18, 16th Revision, Silver Spring, MD, 2018, 34 pages.

CPMB, *Recommended Guide Specifications for Batching Equipment and Control Systems in Concrete Batch Plants*, Concrete Plant Manufacturers Bureau, Publication No. 102-00, 2nd Revision, Silver Spring, MD, 2000, 16 pages.

Daniel, D.G.; and Lobo, C.L., *User's Guide to ASTM Specification C94 for Ready-Mixed Concrete*, MNL 49, Version 2, ASTM International, West Conshohocken, PA, 2014, 190 pages.

Detwiler, R.J.; and Taylor, P.C., *Specifier's Guide to Durable Concrete*, EB221, 2nd edition, Portland Cement Association, Skokie, IL, 2005, 72 pages.

Hover, K.C.; Bickley, J.; and Hooton, R.D., *Guide to Specifying Concrete Performance: Phase II Report of Preparation of a Performance-Based Specification For Cast-in-Place Concrete*, RMC Research and Education Foundation, Silver Spring, MD, March 2008, 53 pages.

Lamond, J.; and Pielert, J., *Significance of Tests and Properties of Concrete and Concrete-Making Materials*, STP169D, ASTM, West Conshohocken, PA, 2006, 645 pages.

NRMCA, *Guide Performance-Based Specification for Concrete Materials – Section 03300 for Cast-in-Place Concrete*, National Ready Mixed Concrete Association, Silver Spring, MD, 2012a, 27 pages.

NRMCA, "Concrete Yield," *Technology in Practice*, TIP 8, National Ready Mixed Concrete Association, Silver Spring, MD, 2012b.

NRMCA, *Guide to Improving Specifications for Ready Mixed Concrete*, Publication 2PE003, National Ready Mixed Concrete Association, Silver Spring, MD, 2015, 22 pages.

NRMCA, "Discrepancies in Yield," *Concrete in Practice*, CIP 8, National Ready Mixed Concrete Association, Silver Spring, MD, 2016.

Obla, K.; and Lobo, C., *Experimental Case Study Demonstrating Advantages of Performance Specifications*, Report to the RMC Research Foundation, Project 04-02, RMC Research Foundation, Silver Spring, MD, January 2006, 39 pages.

Panarese, W. C., *Transporting and Handling Concrete*, IS178, Portland Cement Association, Skokie, IL, 1987, 28 pages.

TMMB, *Truck Mixer, Agitator and Front Discharge Concrete Carrier Standards*, Truck Mixer Manufacturers Bureau, Publication No. 100-05, Silver Spring, MD, 2005, 21 pages.

VMMB, *Volumetric Mixer Standards*, Volume Mixer Manufacturers Bureau, Publication No. 100-01, Silver Spring, MD, 2001, 12 pages.

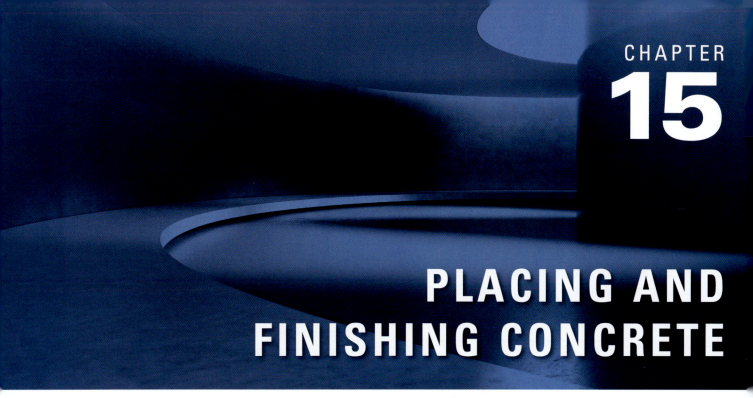

PLACING AND FINISHING CONCRETE

FIGURE 15-1. Placing and finishing concrete takes proper planning and communication to execute the project successfully (courtesy of Baker Concrete).

Concrete placements can range from a simple residential basement to a large-scale commercial project (Figure 15-1). No matter how basic or complex the project is, the quality of the completed project is dependent on appropriate design, quality of the materials used, and the knowledge and ability of the concrete contractor who places and finishes the concrete. There are many other considerations that play a role in the success of a concrete project, including safety, concrete mixture proportioning, joint performance, and acceptable tolerances. ACI 117, *Standard Specifications for Tolerances for Concrete Construction and Materials* (2010), gives tolerances for cast-in-place concrete.

PRECONSTRUCTION CONFERENCE AND CHECKLIST

Every construction project requires good planning, well-executed project documents, construction practices that follow accepted standards, and clear and consistent communication among team members. Many potential problems can be avoided well before the start of a concrete project by communicating responsibilities and fielding questions during a preconstruction conference. Preconstruction conferences should be held well in advance of the project start date to ensure there is sufficient time for all parties to fully understand their responsibilities. It is important that all members involved in the concrete placement are represented during the preconstruction conference (owner, architect, structural engineer, general contractor, concrete contractor, concrete supplier, testing agency, and inspectors).

The National Ready Mixed Concrete Association (NRMCA) and the American Society of Concrete Contractors (ASCC) provide a guide, *Checklist for the Concrete Pre-Construction Conference* (NRMCA/ASCC 2005), to assist decision makers and other participants with planning concrete projects. The checklist allocates responsibilities and establishes procedures related to concrete construction including: subgrade preparation, forming, concrete mixture proportioning, necessary equipment, ordering and scheduling materials and operations, placing, consolidating, finishing, jointing, curing and protection, testing and acceptance as well as safety and environmental issues.

As part of the prescription to performance initiative, NRMCA and ASCC have also published the *Checklist for Concrete Producer-Concrete Contractor Fresh Concrete Performance Expectations* (NRMCA/ASCC 2012). This checklist is intended to clearly define responsibilities of each party for performance of the fresh concrete and applies to properties of the fresh concrete that are typically outside of the domain of the contract documents. It is assumed that the mix design developed by the concrete producer will result in hardened concrete properties that meet the requirements in the contract documents (see Chapter 13).

There are many other preconstruction checklists available to assist in planning and execution of concrete placements, including: *Checklist for Ordering and Scheduling Ready Mixed Concrete* (NRMCA/ASCC 2007a), *Checklist for Pumping Ready-Mixed Concrete* (NRMCA/ASCC 2007b), *Concrete Inspector's Checklist – Site of Delivery* (FHWA 2017), and *Inspection-in-Depth: Major Structures – Concrete Plant and Placement* (FHWA 1999).

WORKING SAFELY WITH CONCRETE

Controlling Hazards

Construction equipment and tools, building materials, and the jobsite itself can all present constant hazards to busy construction personnel. It is important to note, that while personal protective equipment (PPE) is essential for each worker, it should never be considered the first or only line of defense. The hierarchy of controls (Figure 15-2) outlines the successive order in which various hazard control systems should be considered and implemented to best protect workers (NIOSH 2015).

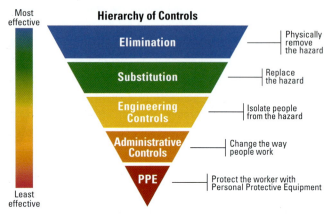

FIGURE 15-2. The Hierarchy of Controls helps determine how and in what order control systems can be implemented to decrease workers' exposure to hazards (adapted from NIOSH 2015).

Generally, methods at the top of the hierarchy are more effective and protective compared to those lower down the pyramid. In many cases, controls such as elimination or substitution are not feasible however, which should prompt exploration of engineering and/or administrative controls. Finally, although PPE is at the bottom of the hierarchy, it is absolutely essential and a worker's last line of defense. PPE is frequently used in conjunction with other control methods. Similarly, numerous control methods can be used for the same job, which increases the layers of safety.

Personal Protective Equipment

Hard hats are required on construction projects to protect from a variety of hazards – largely from projectile objects. Similarly, proper footwear is required and either composite or steel toe work boots can help protect from crippling crushing injuries. When choosing work boots, traction and ankle support should also be taken into consideration to help prevent slips, trips, and falls and ankle injuries.

Hearing protection is a key piece of PPE that must be taken seriously because once hearing is damaged, it is not repairable. There are many styles of hearing protection and it is imperative that workers choose the best style for them: one they will actually wear. In some circumstances where a noise level is either so great, or a worker is exposed for a long enough duration of time at certain decibels, then double hearing protection should be worn.

Some concrete jobs require working from heights. Falling from a height is one of the leading causes of fatalities in the construction industry and workers must exercise extreme caution and use appropriate equipment and PPE. This may include employing proper ladder safety and/or using fall protection including a harness and lanyard restraint.

Proper eye protection is essential when working with cement or concrete. Eyes are particularly vulnerable to blowing dust, splattering concrete, and other foreign objects. On some jobs it may be advisable to wear full-cover goggles or safety glasses with side shields. Actions that cause dust to become airborne should be avoided due to both physical safety and health concerns.

Local or general ventilation can help control exposures below applicable exposure limits. Respirators may be used in poorly ventilated areas, where exposure limits are exceeded, or when dust causes discomfort or irritation. Prior to use of a respirator, employees must be fit-tested (and re-tested annually) to ensure their respirator will be effective. In some cases, facial hair or other factors can cause the respirator to not seal properly to the face, which can allow contaminated air into the facepiece of a respirator. Respirable crystalline silica is a key driver of the use of respirators. OSHA updated its *Respirable Crystalline Silica Standard* (OHSA 2017), which decreased the permissible exposure limit (PEL) and updated health monitoring, sampling, and other requirements.

The Occupational Safety and Health Administration (OSHA) is the primary governing body that regulates safety and health at working places. The agency divides its regulations between general industry (29 CFR 1910) and construction (29 CFR 1926). More information on OSHA regulatory changes can be found at osha.gov.

Protect Your Back

All materials used to make concrete – portland cement, sand, coarse aggregate, and water – can be quite heavy, even in small quantities. When lifting heavy materials, the back should be straight, legs bent, and the weight between the legs as close to the body as possible. If a load requires two or more people, one person should not attempt moving it on their own. Mechanical equipment should be used to place concrete as close as possible to its final position. After the concrete is deposited in the desired area by chute, pump, or wheelbarrow, it should be pushed – not lifted – into final position with a shovel. A short-handled, square-end shovel is an effective tool for spreading concrete, but special concrete rakes or come-alongs can also be used. Excessive horizontal movement of the concrete should be avoided. Not only does it require extra effort, but it may also lead to segregation of the concrete ingredients.

Skin Safety

Fresh, moist concrete is alkaline in nature and can be extremely caustic. It is, therefore, important to avoid prolonged, direct contact between the skin and wet concrete or clothing saturated with wet concrete. Such contact can cause skin irritation and severe chemical burns (PCA 1998). Following contact, use fresh water to thoroughly wash the skin areas exposed to the fresh concrete.

A chemical burn from concrete occurs with very little warning because little or no heat is sensed by the skin. Seek **immediate** medical attention if irritation or inflammation begins on exposed skin areas or if you have persistent or severe discomfort.

If clothing areas become saturated from contact with fresh concrete, mortar, or grout, promptly rinse them with clean water to prevent continued contact with skin surfaces. Indirect contact through clothing can be as serious as direct contact. When working with fresh concrete, begin each workday by wearing clean clothing. To ensure all concrete has been removed, it is prudent to shower or bathe after the workday is complete.

Wear waterproof gloves, long-sleeve shirts, full length trousers, and waterproof leather or rubber boots when working with fresh concrete. Boots must be high enough to prevent fresh concrete from flowing into them. When finishing concrete, use waterproof pads between fresh concrete surfaces and knees. Wear gloves to protect hands. Also, wear proper eye protection when working with fresh concrete, mortar, or grout. If contact with the eyes is made, flush them immediately with fresh water and seek **prompt** medical attention.

When handling dry cement, grinding, or when working in an environment where concrete dust is present, wear proper eye protection and a National Institute for Occupational Safety and Health (NIOSH)-approved dust respirator. As with fresh concrete, if dry cement gets in your eyes, flush the eyes immediately and repeatedly with water and seek **prompt** medical attention.

PREPARATION BEFORE PLACING CONCRETE ON GROUND

Preparation prior to placing concrete for pavements or slabs on ground includes compacting, trimming, moistening the subgrade and subbase when necessary; setting formwork; and placing reinforcing steel and other embedded items securely in place.

Subgrade Preparation

Cracks, slab settlement, and structural failures can often be traced to an inadequately prepared and poorly compacted subgrade. The subgrade on which concrete is to be placed should be well drained, of uniform bearing capacity, proper moisture content, level or properly sloped (Figure 15-3), and free of sod, organic matter, and frozen soil.

FIGURE 15-3. A base course foundation for concrete pavement is shaped by an auto-trimmer to design grades, cross section, and alignment using automatic sensors that follow string lines.

The strength, or bearing capacity, of the subgrade should be adequate to support anticipated structural loads. The material should allow for proper drainage and should also be prepared to ensure the absence of hard or soft spots leading to inconsistent support conditions (see Tarr and Farny 2008). In general, undisturbed soil is superior to compacted material for supporting concrete.

The major causes of nonuniform support are:

- the presence of soft unstable saturated soils or hard rocky soils,
- backfilling without adequate compaction, and
- expansive soils. Uniform support cannot be achieved by merely dumping granular material on a soft area.

To prevent bridging and settlement cracking, soft or saturated soil areas and hard spots (rocks) should be excavated and replaced with soils similar to the surrounding subgrade or, when similar soil is not available, with granular material, such as crushed stone. All fill materials must be compacted to provide uniform support equal to the rest of the subgrade (Figure 15-4). Expansive, compressible, and potentially troublesome soils should be evaluated by a geotechnical engineer; a special foundation or slab design may be required. Proof rolling the subgrade using a fully-loaded dump truck or similar heavy equipment is commonly used to identify areas of unstable soils requiring additional attention (Figure 15-5).

FIGURE 15-4. (left) Adequate compaction of a base course foundation for concrete pavement can be achieved by using a vibratory roller. (right) Vibratory plate compactors are also used to prepare subgrades under slabs.

FIGURE 15-5. Proof roll of subgrade with full-loaded truck can help identify areas of unstable soils. The subgrade shown is not acceptable for placement.

The subgrade should be moistened with water in advance of placing concrete, but should not contain puddles or wet, soft, muddy spots during concrete placement (Figure 15-6). The modified proctor test, governed by ASTM D1557, *Standard Test Methods for Laboratory Compaction Characteristics of Soil Using Modified Effort*, and AASHTO T 180, *Standard Method of Test for Moisture–Density Relations of Soils Using a 4.54-kg (10-lb) Rammer and a 457-mm (18-in.) Drop*, should be performed to determine the optimum moisture content for a given soil type. For ideal compaction moisture content should be near the optimum percentage. Moistening the subgrade

FIGURE 15-6. Water trucks with spray-bars are used to moisten subgrades and base course layers to achieve adequate compaction and to reduce the amount of water drawn out of concrete as it's placed.

prevents the subgrade from drawing too much water from the concrete mixture and also increases the ambient air's relative humidity, which decreases the potential amount of evaporation from the concrete surface.

In cold weather, concrete must not be placed on frozen subgrade. Snow, ice, and other debris must be removed from within forms before concrete is placed. Care should be taken, in cold weather applications, to limit surface temperatures of supporting materials and the concrete to a temperature differential of less than 20°F (11°C) to avoid inconsistent setting, rapid moisture loss, and plastic shrinkage cracking (ACI 306R 2016 and Mustard and Ghosh 1979).

Where concrete is to be deposited on and bonded to rock or hardened concrete, all loose material must be removed. Where concrete is intended to act independently, granular fill, sand, sheet goods, or coating materials may be used as bond breakers. Cut faces defining the boundaries of concrete placements should be nearly vertical or horizontal rather than sloping.

Subbase. Concrete can be built without a subbase. However, a subbase is frequently placed on a subgrade to improve stability during construction, serve as a leveling course to correct minor surface irregularities, enhance uniformity of support, bring the site to the desired grade, and serve as a capillary break between the concrete and subgrade.

Where a subbase is used, the contractor should place and compact to near maximum density a 100 mm (4 in.) thick layer of granular material such as limestone trimmings (crusher fines), gravel, or crushed stone. If a thicker subbase is needed for achieving the desired grade, the material should be compacted in thin layers about 100 mm (4 in.) deep unless tests determine compaction of a thicker lift is possible (Figure 15-7). Subgrades and subbases can be compacted with small plate vibrators, vibratory rollers, or hand tampers. If a subbase is used, it must be well compacted.

FIGURE 15-7. Nuclear gauges containing radioactive sources used to measure soil density and moisture can determine if a subbase has been adequately compacted.

Moisture Control and Vapor Retarders

Many of the moisture problems associated with enclosed slabs on ground (floors) can be minimized or eliminated by:

• sloping the landscape away from buildings,

• using a 100 mm (4 in.) thick granular subbase to form a capillary break between the soil and the slab,

• providing drainage for the granular subbase to prevent water from collecting under the slab,

• installing foundation drain tile,

• installing a vapor retarder directly beneath a concrete slab, and

• use water reducer to lower water content of mixture.

A vapor retarder is not a vapor barrier. A vapor retarder slows the movement of water vapor by use of a 0.25 mm to 0.40 mm (10 mil to 15 mil) polyethylene film or polyolefin sheet that is overlapped approximately 150 mm (6 in.) at the edges and taped. A vapor retarder does not stop 100% of vapor migration; a vapor barrier stops nearly 100%. Vapor barriers are thick, rugged, multiple-ply-reinforced membranes that are sealed at

FIGURE 15-8. Vapor retarders and vapor barriers can be made from many types of materials, but all serve the same purpose: to reduce or virtually eliminate the passage of water vapor through concrete.

the edges. Vapor retarders are more commonly used. However, many of the following principles apply to vapor barriers as well. Newer polyolefin products are very low permeance and highly tear and puncture resistant (Figure 15-8).

A vapor retarder should be placed under all concrete floors on ground that are likely to receive an impermeable floor covering such as sheet vinyl tile. They should also be used for any purpose where the passage of water vapor through the floor might damage moisture-sensitive equipment or materials in contact with the floor (Kanare 2008).

Vapor retarders placed directly under concrete slabs may increase the time delay before final finishing due to longer bleeding times, particularly in cold weather. To minimize this effect, a minimum 75 mm (3 in.) thick layer of approved granular, self-draining compactible subbase material can be placed over the vapor barrier (or insulation if present) (ACI 302.2R 2006). However, slabs that will receive moisture-sensitive flooring should be placed directly on the vapor retarder (Kanare 2008). Sand over polyethylene sheeting is slippery, somewhat dangerous, and difficult to keep in place while concreting. This practice is not recommended. It often results in sand lenses protruding into the concrete. A 50-mm to 100-mm (2 in. to 4 in.)-thick subbase will alleviate this problem in areas where moisture vapor will not damage flooring or products stored on the slab. The subbase over a vapor retarder must be protected from rain, construction activities, or other external sources of moisture to prevent excessive vapor migration after the concrete slab is placed.

If concrete is placed directly on a vapor retarder, the water to cementitious materials ratio should be kept low (0.45 or less) because excess mixing water can only escape to the surface as bleed water. Because of a longer bleeding period, settlement cracking over reinforcement and shrinkage cracking is more likely. A flow chart to determine when and where a vapor retarder should be used is provided in Figure 15-9 (ACI 302.1R 2015).

Good quality, well-consolidated concrete at least 100 mm (4 in.) thick is practically impermeable to the passage of liquid water unless the water finds a path (crack) to travel through or is under considerable pressure. However, such concrete – even concrete several times as thick – is not impermeable to the passage of water vapor.

Water vapor that passes through a concrete slab transports soluble salts which can increase the surface pH and form a deposit (efflorescence) as the moisture evaporates. Floor coverings such as linoleum, vinyl tile, carpeting, wood, and synthetic surfacing effectively seal the moisture within the slab. Sufficiently high pH moisture condensate may eventually deteriorate the adhesives causing the floor covering to loosen, buckle, or blister.

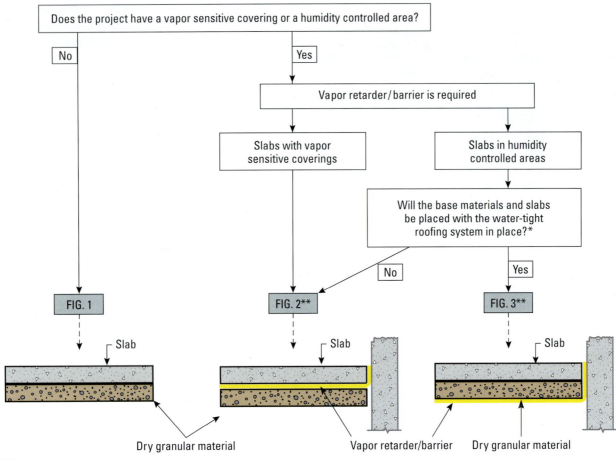

FIGURE 15-9. Flowchart for determination of location of vapor retarder in concrete slabs on ground construction (adapted from ACI 302.1R-15).

NOTES:
* If granular material is subject to future moisture infiltration, use Fig. 2. A reduced joint spacing, a low shrinkage mix design, or other measures to minimize slab curl will likely be required if Fig. 2 is used.
** At perimeter, vapor retarder/barrier should be turned up and sealed to adjacent concrete.

To prevent problems with floor covering materials caused by moisture within the concrete, the following steps should be taken:

- use a low water to cementitious materials ratio concrete,
- moist-cure the slab for 7 days,
- allow the slab a drying period of at least 2 months (Hedenblad 1997 and 1998), and
- test the slab moisture condition before installing the floor covering.

Flooring-material manufacturers publish the recommended test and specified moisture limits for installing their product. The flooring industry typically specifies testing the floor moisture condition by measuring the internal relative humidity (RH) within the slab depth (ASTM F2170, *Standard Test Method for Determining Relative Humidity in Concrete Floor Slabs Using in situ Probes*) versus capturing the moisture vapor emission rate (MVER) from a concrete slab (ASTM F1869, *Standard Test Method for Measuring Moisture Vapor Emission Rate of Concrete Subfloor Using Anhydrous Calcium Chloride*).

ASTM F2170 incorporates a relative humidity sensor sealed in a hole extending into the slab interior. Disposable RH probes are now available that combines the meter and the probe into a single unit (Figure 15-10). ASTM F1869 involves the use of a desiccant beneath a sealed plastic dome (Figure 15-11). Both of these tests require the building to be enclosed and maintained at operating ambient temperature and relative humidity. For more information and additional tests for water vapor transmission, see PCA EB075, *Concrete Floors on Ground* (Tarr and Farny 2008), PCA EB119, *Concrete Floors and Moisture* (Kanare 2008), ACI 302.2R (2006), and Chapter 20.

Insulation Under Concrete

Insulation is sometimes installed over the vapor retarder to assist in keeping the temperature of a concrete slab above the dew point; this helps prevent moisture in the air from condensing on the slab surface. Insulation below the concrete can provide energy savings for slabs containing heat piping or other heating provisions and is mandatory for cold storage, ice-skating rinks, and freezer floors. The insulation prevents

FIGURE 15-10. Single-unit disposable RH probe (courtesy of Wagner Electronics).

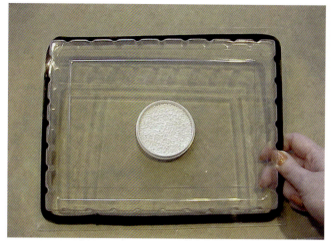

FIGURE 15-11. Moisture vapor emission rate is determined according to ASTM F1869 using commercially available calcium chloride kits that absorb moisture from a specific test area over a known length of time.

frost penetration into subbase and subgrade soils and frost heaving of these materials. For freezer floors and ice rinks, the insulation is located between the bottom of a concrete slab and a mudslab. Mudslabs are generally constructed by placing a lean concrete mixture. Heat return ducts are encased in the mudslab to prevent subgrade freezing and heaving that could occur in spite of the insulation below the floor. Mudslabs also provide a stable working surface.

In addition to functioning as a thermal barrier, under-slab insulation must provide support to the slab both during and after construction. Insulating properties are based on the material itself and the thickness used (Tarr and Farny 2008). Furthermore, insulation below floors should not be allowed to absorb water as the insulating quality is lost with increased moisture content.

Codes and specifications often require insulation at the perimeter of a slab. In cold climate regions, insulation below floors on ground is used at floor edges adjacent to exterior foundations or walls. The insulation is placed inward for a distance of about 600 mm to 900 mm (24 in. to 36 in.) and may also be used vertically along the foundation wall to aid in heat retention.

Formwork

Forms for paving and slabs on ground should be properly aligned, clean, tight, adequately braced, and constructed of materials that will impart the desired off-the-form finish to the hardened concrete. Form facing material for concrete can be constructed of lumber, metal, hardboard, or plastic. Edge forms used for flatwork and intermediate screeds should be set accurately and firmly to the specified elevation and contour for the finished surface. Slab edge forms are usually metal or wood braced firmly with wood or steel stakes to keep them in horizontal and vertical alignment. The forms should be straight and free from warping and have sufficient strength to resist concrete pressure without deforming. They should also be strong enough to support any mechanical placing and finishing

equipment used. Wood forms, unless oiled or otherwise treated with a form-release agent, should be moistened before placing concrete, otherwise they will absorb water from the concrete and may swell. Forms should be made for easy removal so as to minimize damage to the concrete. Avoid the use of large nails with wood forms, and minimize the quantity to facilitate removal and reduce damage to concrete. For architectural concrete, the form-release agent should be a nonstaining material. See Johnston (2014) and ACI 347R, *Guide to Formwork for Concrete* (2014) for more information on formwork.

Reinforcement and Embedments

Reinforcing steel should be clean of debris and free of excessive rust or mill scale when concrete is placed. Thin films of surface discoloration are common and not deemed significant. Reinforcing or embedded steel that is less than 2580 mm^2 (4 in.^{2}) in cross sectional area must be at least -12°C (10°F), and reinforcing or embedded steel that is greater than 2580 mm^2 (4 in.^{2}) in cross sectional area must be at least 0°C (32°F) (ACI 306R 2016 and ACI 301 2020). Mortar splattered on reinforcing bars from previous placements need not be removed from steel and other embedded items if the next lift is to be completed within a few hours. However, loose, dried mortar must be removed from items that will be encased by later lifts of concrete. See Chapter 8 for more details on reinforcement for concrete.

Embedments are items other than reinforcement that are located partially or completely within the concrete. This includes fully embedded circulation coils, such as coolant ducts and heating ducts; drains; and pipes that pass completely through the concrete. Embedments can be supported on chairs, slab bolsters, or small concrete bricks and blocks depending on the concrete dimensions. Supports should be strong enough and the spacing close enough to sustain imposed loads from the concrete placing crew without displacement of the embedded item.

FIGURE 15-12. Concrete should be placed as near as possible to its final position.

DEPOSITING CONCRETE

Continuous, uninterrupted placement of concrete is desirable. There are several ways to handle and deposit fresh concrete where it is needed, including directly from a truck chute, by buggy, by crane and bucket, by conveyor belt, or by pump (Figures 15-12 to 15-14). Concrete should be deposited continuously as near as possible to its final position without objectionable segregation. See Chapter 14 for more information on methods of transporting and handling concrete.

FIGURE 15-13. The swing arm on a conveyor belt allows fresh concrete to be placed evenly across a deck (courtesy of Baker Concrete).

In general, concrete should be placed in walls, thick slabs, or foundations in layers of uniform thickness and thoroughly consolidated before the next layer is placed. Lift heights are governed by formwork pressures and the economics of the form. Refer to ACI 347R-14 for formulas for calculating formwork pressures. The lift depth will depend on the width between forms, size of coarse aggregate used, the volume of steel reinforcement and consistency of the concrete mixture. Layers should be about 150 mm to 500 mm (6 in. to 20 in.) deep for reinforced members and 375 mm to 500 mm (15 in. to 20 in.) thick for mass work using large aggregates (more than 25 mm

FIGURE 15-14. Concrete deposited by boom pumps can aid in continuous rapid placements (courtesy of Baker Concrete).

[1 in.]) or stiff consistency concrete mixtures (slump ≤ 75 mm [3 in.]). Lift heights should be optimized to reduce segregation and cold joints. With light reinforcing and special considerations given to the mixture for resistance to segregation, lift heights may be extended up to 1.5 m (60 in.).

Flatwork

In slab construction, placing should start along the perimeter at one end of the slab with each batch discharged against previously placed concrete (Figure 15-15). The concrete should not be deposited in large piles and moved horizontally into final position. This practice results in segregation because mortar tends to flow ahead of the coarser material.

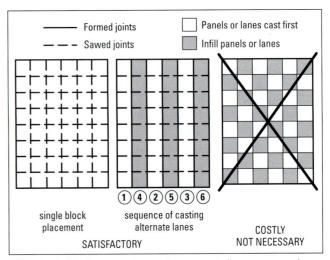

FIGURE 15-15. Sequence for casting concrete floors on ground.

Preparation Before Placing Concrete on Elevated and Vertical Formwork

Preparation prior to placing concrete for elevated slabs and vertical formwork includes preparing the supporting surface, erecting formwork and shoring, applying form release agent, and setting reinforcing steel and other embedded items securely in place. Once the supporting formwork is in place, many aspects of elevated and vertical placement will proceed in a manner consistent with placing concrete on ground (Figure 15-16).

FIGURE 15-16. Vertical formwork constructed for cast-in-place columns (courtesy of United Forming, Inc.).

Constructing Elevated and Vertical Formwork

Placing concrete for elevated concrete introduces multiple additional challenges beyond those encountered for placing concrete on ground. The bottom of the slab or other structural element (beam, drop panel) must now be appropriately supported, to ensure tolerances of ACI 117-10 are achieved while simultaneously providing a safe working surface for construction activities. Inadequate support of the formwork can affect the final tolerances. The grade which the formwork is installed over may be required to support loading in excess of that required for the final use of the structure. Similarly, for structures multiple levels in height, the loading applied for construction of any given level likely exceeds the capacity of the level below it, as such, loading applied to an existing elevated level must be shared across multiple levels, or transmitted directly to the ground below. Consult the contract documents for requirements related to formwork and shoring, the vertical elements used to support elevated formwork. Guidelines are presented in ACI 347R-14. For an in-depth discussion of a wide range of formwork topics as well as means to directly design formwork and shoring, see Johnston (2014). Sufficient consideration of both the vertical and lateral load paths are critical for any elevated formwork design. Consideration must also be given to the timing between concrete placements, ensuring sufficient strength of concrete prior to removal of concrete as well as consideration and coordination with the specifier to ensure the stiffness of the concrete is adequate for the demand on the structure at a given time.

Vertical Work

Walls less than 300 mm (12 in.) thick using appropriately sized aggregates and flowing concrete have been placed successfully with lifts of 900 mm to 1200 mm (36 in. to 48 in.), with 900 mm (36 in.) typically being the maximum lift height for cast-in place architectural concrete. The rate of placement should be rapid enough that previously placed concrete has not yet set when the next layer of concrete is placed upon it. Timely placement

and adequate consolidation will prevent flow lines, seams, and planes of weakness (cold joints) that result from placing freshly mixed concrete on top of concrete that is beyond its initial set (see Chapter 16). Typically, the formwork supplier will provide the allowable formwork pressure, and the contractor will then determine the rate of placement.

As with flatwork, to avoid segregation, concrete should not be moved horizontally over long distances as it is being placed in forms. In some work, such as placing concrete in sloping wingwalls or beneath window openings in walls, it may be necessary to move the concrete horizontally within the forms, but this should be kept to a minimum.

Where standing water is present, it should be pumped out prior to placement, otherwise concrete must be placed in a manner that displaces the water ahead of the concrete and does not allow the water to be mixed into the concrete. In all cases, water should be prevented from collecting in corners, at the ends, and along faces of forms. Care should be taken to avoid disturbing saturated subgrade soils so that they maintain sufficient bearing capacity to support structural loads.

Chutes on a ready mix truck provide an efficient means to deposit concrete near grade when truck access is available. Dropchutes are used to move concrete to lower elevations (such as in wall forms) without segregation and spattering of mortar on reinforcement and forms (Figure 15-17). Field studies indicate that free fall of concrete from heights of up to 46 m (150 ft) directly over reinforcing steel or at a high slump, does not result in segregation of the concrete ingredients nor reduce compressive strength so long as the material is confined (Suprenant 2001).

FIGURE 15-17. Dropchutes minimize the risk of segregation.

However, if a baffle is not used to control the flow of concrete onto sloped surfaces at the end of an inclined chute, segregation can occur.

Concrete is sometimes placed through openings (called placement windows) in the sides of tall, narrow forms. There is danger of segregation when a chute discharges directly through the opening without controlling concrete flow at the end of the chute. To decrease the tendency of segregation, a collecting hopper should be used outside the opening. This permits the concrete to flow more smoothly through the opening and within the reinforcement cage.

When concrete is placed in tall forms at a fairly rapid rate, some bleed water may collect on the top surface, especially with non-air-entrained concrete. Bleeding may be reduced by placing concrete more slowly and at a stiffer consistency.

To avoid cracks between structural elements in monolithic placement of deep beams, walls, or columns, concrete must be well consolidated and free of bleed water prior to continued placement of additional concrete on top of the first lift. The delay should be short enough to allow the next layer of concrete to join with the previous layer by vibration, thus preventing cold joints and honeycombing (see Chapter 16). Haunches and column capitals are considered part of the floor or roof slab and should be placed integrally at the same time.

The use of self-consolidating concrete (SCC) may allow for continuous placements if precautions are taken to account for form pressure increases due to the fluid behavior of SCC and for the heat of hydration in mass placements due to the high cementitious content of SCC (see section on **Special Placing Techniques**).

Depositing on Hardened Concrete

Recently placed hardened concrete requires little preparation prior to placing a lift of freshly mixed concrete on top of it. Care should be taken to remove the presence of any laitance (a thin layer of cement or concrete), dirt, or loose particles prior to next placement. Hardened concrete that has been in service for a period of time usually requires mechanical cleaning and roughening prior to placement of new concrete to produce a mechanical bond between the two layers.

Bonded Construction Joints in Structural Concrete. A bonded construction joint may be required between two structural concrete placements when bond is relied upon to provide adequate shear strength at the bond line between placements. Shear strength can also be achieved with reinforcing placed across the construction joint. ACI 318, *Building Code Requirements for Structural Concrete* (2019), includes provisions for determining when physical bond is required and requires the engineer to specify when bond is required. Bond may also be specified when a watertight joint is needed to minimize water infiltration.

The quality of a bonded joint therefore depends on the quality of the hardened concrete and on the preparation of its surface.

In columns and walls, the concrete near the top surface of a lift is often of inferior quality to the concrete beneath it. This may be due to poor consolidation or excessive laitance, bleeding, and segregation. Even in well-proportioned and carefully consolidated mixtures, some aggregate particle settlement and water gain (bleeding) at the top surface is unavoidable; this is particularly true with rapid placement rates and with taller structures.

Also, the encasing formwork prevents the escape of moisture from the fresh concrete. While formwork generally provides adequate curing as long as it remains in place, where there is no encasing formwork, the top surface may dry out too rapidly. Rapid drying may result in a weak porous layer unless protection and curing are provided.

Surface Preparation. When freshly mixed concrete is placed on recently hardened concrete, certain precautions must be taken to secure a well-bonded, watertight joint. The key to a successful bonded topping, as with a coating, is proper surface preparation.

In most cases it will be necessary to remove the entire surface down to sound concrete. Roughening and cleaning with lightweight chipping hammers, waterblasting, scarifiers, sandblasting (Figure 15-18), shotblasting, and hydrojetting are some satisfactory methods for exposing sound concrete.

The surface must be clean, sound, and rough with some coarse aggregate particles exposed. Any laitance, soft mortar, dirt, wood chips, form oil, or other foreign materials must be removed since they could interfere with proper bonding of the subsequent placement.

FIGURE 15-18. Sandblasting can clean any size or shape surface – horizontal, vertical, or overhead. Consult local environmental regulations regarding sandblasting.

The best surface cleaning method is vacuuming using a 51 mm (2 in.) line with a small brush attachment. The dust within the surface pores must be removed in order to achieve a well-bonded topping. Cleaning the prepared surface by brooming or blowing is unacceptable since the dust will ultimately settle back onto the surface. Sweeping compounds must also be avoided as they may leave an oily residue that can inhibit bond. Vacuum units should be located outdoors when possible. Even units with high filtering systems release dust back into the air when the units are emptied.

After thorough cleaning, the prepared surface should be moistened prior to receiving the concrete topping. Care must be taken to avoid contamination of the clean surface before a bonding grout and overlay concrete are placed.

Partially set or recently hardened concrete may only require stiff-wire brushing. In some types of construction such as dams, the surface of each concrete lift is cut with a high-velocity air-water jet to expose clean, sound concrete just before final set. This is usually done 4 hours to 12 hours after placing. The surface must then be protected and continuously cured until concreting is resumed for the next lift.

For two-course systems, the existing concrete must be scarified so that the coarse aggregate is exposed and the surface cleaned of dust and debris. A typical concrete surface profile (CSP) that might be required in this case would be a CSP 7 or above as described by the International Concrete Repair Institute's (ICRI) Surface Profile Plates (ICRI 1997). These plates, shown in Figure 15-19, range from a CSP 1 to a CSP 10 and rank the degree of surface roughness to be achieved.

The top surface of the base concrete can be roughened with a steel or stiff fiber broom just before it sets. The surface should be level, heavily scored, and free of laitance. It should be protected until it is thoroughly cleaned just before the grout coat and topping are placed. When placing a bonded topping on

concrete, the base should be cleaned of all curing compounds, laitance, dust, debris, grease, or other foreign substances using one of the following methods (consult the local jurisdiction concerning dust control):

- wet- or dry-grit sandblasting,
- high-pressure water blasting,
- mechanical removal by scabblers or scarifiers, and
- power brooming and vacuuming.

Hardened concrete may be left dry or moistened before new concrete is placed on it; however, the surface should be surface-saturated dry and free of standing water.

Bonding New to Previously Hardened Concrete. Care must be used when making horizontal construction joints in wall sections where freshly-mixed concrete is to be placed on hardened concrete. A good bond can be obtained by placing a rich concrete mixture (higher cement and sand content than normal) in the bottom 150 mm (6 in.) of the new lift and thoroughly vibrating the joint interface. Alternatively, a cement-sand bonding grout can be scrubbed into a clean surface immediately ahead of the concrete placement.

A topping concrete mixture can be bonded to the previously prepared base by one of the following procedures:

- **Portland cement-sand grouting.** A 1:1 cement-sand grout mixture having a water-cement ratio of not greater than 0.45, mixed to a creamlike consistency, is scrubbed into the prepared dry or damp (no free water) base slab surface. Bonding grout is placed just a short distance ahead of the overlay or top-course concrete (Figure 15-20). This method may also be applicable to horizontal joints in walls. The grout should not be allowed to dry out prior to the overlay placement; otherwise, the dry grout may act as a poor surface for bonding.

- **Latex.** A latex-bonding agent is added to the cement-sand grout and is spread in accordance with the latex manufacturer's direction.

FIGURE 15-19. Concrete Surface Profile (CSP) Plates developed by the International Concrete Repair Institute (ICRI 1997).

FIGURE 15-20. Application of a bonding grout just ahead of the overlay concrete. The grout must not dry out before the concrete is placed.

- **Epoxy.** An approved epoxy-bonding agent placed on the base concrete, prepared in accordance with the epoxy manufacturer's direction.

- **Placing topping directly onto a surface-saturated dry concrete substrate.** The surface of the base slab should have been prepared by one of the methods discussed previously. The bonding procedure should produce tensile bond strength with the base concrete in excess of 1.0 MPa (150 psi) or more for floors with heavy wheeled traffic.

The bonding method selected should also consider service conditions related to moisture transmission. Certain types of epoxy bonding agents are ideal for heavy duty industrial floor toppings but when use used in a parking garage structure, they may create a water vapor barrier that is detrimental. Guidance on placing concrete overlays on pavements is provided by Harrington (2014) and for concrete floors by Tarr and Farny (2008).

SPECIAL PLACING TECHNIQUES

Concrete may be placed by methods other than traditional cast-in-place. No matter what method is used, unless otherwise stated, the basics of mixing, placing, consolidating, finishing, and curing apply to most concrete applications.

Self-Consolidating Concrete

Self-consolidating concrete (SCC), also referred to as self-compacting concrete, flows and consolidates under its own weight. Self-consolidating concrete has unique properties of high workability without loss of stability (it is resistant to segregation), which has allowed for complex forms and rigorous construction schedules. SCC allows concrete to flow into densely reinforced areas, constricted spaces, or over long distances. The same characteristics have allowed contractors to improve placement techniques, and in some cases eliminate the use of cumbersome placing and consolidating equipment.

Figure 15-21 shows an example of mixture proportions used in self-consolidating concrete as compared to a conventional concrete mixture. Handling SCC requires tighter control because of SCC's inherently high fluidity and its dependence on maintaining homogeneous properties.

Transportation. Placing SCC in continuous pours is meant to reduce probabilities of surface blemishes or cold joints and may require that several truck loads be delivered at a rapid rate. This requires that coordination of scheduled deliveries take place before placement begins.

Transportation of SCC can be accomplished using any acceptable method for conventional concrete. SCC transportation concerns include leakage, sloshing, and timing. Given the high fluidity of SCC, leakage of paste or mortar can be a serious concern and can have a significant effect on the properties of the fresh mixture. Mixing equipment, including ready-mixed concrete trucks, typically will not leak, but concrete trucks may not be filled to capacity to prevent spillage, and onsite transportation containers, including buckets and formwork may require modifications to prevent leakage.

Sloshing of an SCC mixture can be significant because of SCC's high fluidity. Quick accelerations and decelerations of delivery vehicles should be avoided. Steep inclines in the transportation path needs to be considered and overfilling of the transportation equipment must be avoided (Bury 2002). While sloshing may not influence the fresh properties as much as leakage, the loss of concrete could be substantial. Typical load size may need to be reduced when transporting SCC in conventional concrete delivery vehicles.

Form Pressures. The fluidity of the SCC requires special considerations with regard to formwork pressure. The lateral pressure exerted upon the formwork increases as the mixture becomes more fluid. In addition, SCC is often placed in higher lifts than conventional concrete. SCC formwork pressures have been shown to reach full hydrostatic pressure (Alfes 2004, Assaad 2006, and Birch 2007). Accordingly, the formwork for SCC placements must be designed to withstand these pressures. Additionally, the formwork must be designed to prevent leakage of the highly fluid SCC mixture. Bottom-fed placements can produce significantly higher pressures than top-fed placements and may even exceed hydrostatic pressure (Figure 15-22) (Brameshuber 2003 and Leemann 2006).

Time is the single greatest factor influencing formwork pressure. The faster the SCC is discharged into the formwork, the more pressure it exerts (Brameshuber 2003, Tejeda-Dominguez and others 2005, and Assaad 2006). However, as soon as discharge is complete, formwork pressures quickly dissipate.

The mixture composition also influences formwork pressure. The use of supplementary cementitious materials has been shown to reduce formwork pressure due to a change in the thixotropic behavior caused by the alteration of the particle size distribution (Assaad 2003 and Gregori and others 2008).

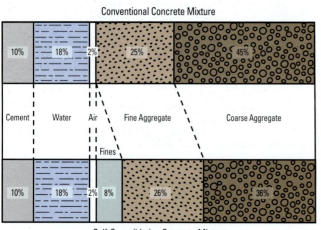

FIGURE 15-21. Examples of materials used in conventional concrete and self-consolidating concrete by absolute volume.

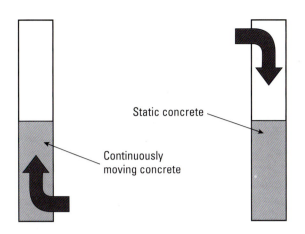

FIGURE 15-22. Bottom-fed placements increase the formwork pressure due to the fluidity of the moving concrete (Szecsy and Mohler 2009).

Static concrete

Continuously
moving concrete

FIGURE 15-23. A tremie pipe allows for placement of concrete under water.

Finishing and Curing SCC. ACI and PCI both recommend using normal finishing techniques for SCC (ACI 237R 2007 and PCI 2003). For flatwork applications, due to the significant reduction in bleeding, preventative measures should be taken to minimize evaporation from the surface before final finishing (Daczko 2006).

Placing Concrete Underwater

When concrete must be placed underwater, the work should be done under experienced supervision. The basic principles for normal concrete work apply to underwater concreting. Placing should be continuous with as little disturbance to the previously placed concrete as possible. The top surface should be kept as level as possible.

It is important that the concrete flow without segregation. Slump of concrete to be tremied should be specified at 150 mm to 230 mm (6 in. to 9 in.) and the mixture should have a maximum w/cm of 0.45. Generally, the concrete mixture will have a minimum cementitious materials content of 390 kg/m^3 (600 lb/yd^3). Antiwashout admixtures can be used to make concrete cohesive enough to be placed in limited depths of water, even without tremies. Using rounded aggregates, a higher percentage of fines, and entrained air may help to obtain the desired consistency.

Methods for placing concrete underwater include the following: tremie, pump, bottom-dump buckets, grouting preplaced aggregate, toggle bags, bagwork, and the diving bell. A tremie is a smooth, straight pipe long enough to reach the lowest point to be concreted from a working platform above the water (Figure 15-23). The diameter of the tremie pipe should be at least 8 times the diameter of the maximum size of aggregate. A hopper to receive the concrete is attached to the top of the pipe. The lower end of the tremie should be kept buried in the fresh concrete to maintain a seal below the rising top surface and to force the concrete to flow in beneath it under pressure.

Mobile concrete pumps with a variable radius boom make easy work of placing concrete underwater. Because the flexible hose on a concrete pump is similar to a tremie, the same placement techniques apply. Recommendations on the rate of concrete rise are generally in the range of 0.3 m/hr to 3 m/hr (1 ft/hr to 10 ft/hr).

See ACI 304R, *Guide for Measuring, Transporting, and Placing Concrete* (2000) and FHWA (1981 and 2010) for additional information on placing concrete underwater.

Preplaced Aggregate Concrete

When grouting preplaced aggregate, the forms are first filled with clean coarse aggregate, then grouted. Grouting preplaced aggregate is advantageous when placing concrete in flowing water. Concrete can be placed more quickly and economically than by conventional placement methods. However, the method is very specialized and must be performed by qualified experienced personnel.

Preplaced aggregate concrete is produced by first placing coarse aggregate in a form and later injecting a cement-sand grout, usually with admixtures, to fill the voids. Properties of the resulting concrete are similar to those of comparable concrete placed by conventional methods. However, considerably less thermal and drying shrinkage can be expected because of the point-to-point contact of aggregate particles.

Coarse aggregates should meet requirements of ASTM C33, *Standard Specification for Concrete Aggregates* (AASHTO M 80). In addition, most specifications limit both the maximum and minimum sizes; for example, 75 mm (3 in.) maximum and 12.5 mm (½ in.) minimum. Aggregates are generally graded to produce a void content of 35% to 40%. Fine aggregate used in the grout is generally graded to a fineness modulus of between 1.2 and 2.0, with nearly all of the material passing a 1.25 mm (No. 16) sieve.

The preplaced aggregate method has been used principally for restoration work and in the construction of reactor shields, bridge piers, and underwater structures. It has also been used in buildings to produce unusual architectural effects. Since the forms are completely filled with coarse aggregate prior to grouting, a dense and uniform exposed-aggregate facing is obtained when the surface is sandblasted, tooled, or retarded and wire brushed at an early age.

Tests for preplaced aggregate concrete are given in ASTM C937, *Standard Specification for Grout Fluidifier for Preplaced-Aggregate Concrete* and ASTM C943, *Standard Practice for Making Test Cylinders and Prisms for Determining Strength and Density of Preplaced-Aggregate Concrete in the Laboratory.* Preplaced aggregate concrete is discussed in more detail in ACI 304R-00, and ACI 304.1R, *Guide for the Use of Preplaced Aggregate Concrete for Structural and Mass Concrete Applications* (1992).

Shotcrete

Shotcrete is concrete that is pneumatically projected onto a surface at high velocity (Figure 15-24). Also known as gunite and sprayed concrete, shotcrete was patented in 1911.

FIGURE 15-24. Shotcrete is pneumatically projected onto concrete at high velocity.

Shotcrete is applied by a dry or wet process. In the dry process, a premixed blend of cement and damp aggregate is propelled by compressed air through a hose to a nozzle. Water is added to the cement and aggregate mixture at the nozzle and the intimately mixed ingredients are projected onto the surface. In the wet process, all the ingredients are premixed. The wet mixture is pumped through a hose to the nozzle, where compressed air is added to increase the velocity and propel the mixture onto the surface.

The concrete mixture is consolidated by the impact force and can be placed on vertical or horizontal surfaces without sagging. As the shotcrete mixture hits the surface, some coarser aggregates ricochet off the surface until sufficient paste builds up, providing a bed into which the aggregate can stick. To reduce overspray (mortar that attaches to nearby surfaces) and rebound (aggregate that ricochets off the receiving surface) the nozzle should be held at a 90° angle to the surface. The appropriate distance between nozzle and surface is usually between 0.6 m and 1.8 m (2 ft and 6 ft), depending on the material velocity and air volume.

Shotcrete is used for both new construction and repair work. It is especially suited for curved or thin concrete structures and shallow repairs, but can be used for thick members. The hardened properties of shotcrete are very operator dependent. Shotcrete has a density and compressive strength similar to normal- and high-strength concrete. Aggregate sizes up to 19 mm (¾ in.) can be used, however most mixtures contain aggregates only up to 9.5 mm (⅜ in.). Aggregate amounts of 25% to 30% pea gravel by volume are commonly used for wet shotcrete mixtures (Austin and Robbins 1995).

Supplementary cementitious materials can also be used in shotcrete. They improve workability, chemical resistance, and durability. The use of accelerating admixtures allows build-up of thicker layers of shotcrete in a single pass. They also reduce the time of initial set. However, using rapid-set accelerators often increases drying shrinkage and reduces later-age strength (Gebler and others 1992).

Steel fibers are used in shotcrete to improve flexural strength, ductility, and toughness. They can be used as a replacement for wire mesh reinforcement in applications like rock slope stabilization and tunnel linings (ACI 506.1R 2008). Steel fibers can be added up to 2% by volume of the total mix. Polypropylene fibers if used are normally added to shotcrete at a rate of 0.9 kg/m³ to 2.7 kg/m³ (1.5 lb/yd³ to 4.5 lb/yd³), but dosages up to 9 kg/m³ (15 lb/yd³) have also been used. See Chapter 7 for more information on fibers.

Trimming to specified contour lines or thickness and texturing finishes can be directly applied to shotcrete systems while the material is still fresh or after it has hardened. More information is available from the American Shotcrete Association, ACI 506R *Guide to Shotcrete* (2016), ASCE (1995), and Balck and others (2008).

CONSOLIDATION

Consolidation is the process of compacting fresh concrete; to mold it within the forms and around embedded items and reinforcement; and to eliminate stone pockets, honeycombs, and entrapped air (Figure 15-25). Consolidation should not remove significant amounts of intentionally entrained air.

FIGURE 15-25. Honeycombs and rock pockets are the results of inadequate consolidation.

Consolidation is accomplished by hand or by mechanical methods. The method chosen depends on the consistency of the mixture and the placing conditions, such as complexity of the formwork and amount and spacing of reinforcement. Generally, mechanical methods using either internal or external vibration are preferred (Whiting and others 1987).

Workable, flowing mixtures can be consolidated by hand rodding; inserting a tamping rod repeatedly into the concrete. The tamping rod should be long enough to reach the bottom of the form or lift and thin enough to easily pass between the reinforcing steel and the forms. Low-slump concrete can be transformed into flowing concrete through the use of plasticizers (water reducers) and without the addition of water to the concrete mixture (see Chapter 6).

Spading can be used to improve the appearance of formed surfaces. A flat, spadelike tool should be repeatedly inserted adjacent to the form. This forces the larger coarse aggregates away from form faces and assists entrapped air voids in their upward movement toward the top surface where they can escape. A mixture designed to be readily consolidated

by hand methods should not be mechanically consolidated; otherwise, the concrete is likely to segregate under intense mechanical action.

In highly reinforced elements, proper mechanical consolidation makes the placement of stiff mixtures with low water-cementitious materials ratios and high coarse-aggregate contents possible. Mechanical methods of consolidation include:

- centrifugation – used to consolidate moderate-to-high-slump concrete in making pipes, poles, and piles;
- shock or drop tables – used to compact very stiff low-slump concrete in the manufacture of architectural precast units; and
- vibration – internal and external, the most widely used method for consolidating concrete.

Vibration

When concrete is vibrated, the internal friction between the aggregate particles is temporarily disrupted and the concrete behaves like a liquid; it settles in the forms under the action of gravity and the large entrapped air voids rise more easily to the surface. Internal friction is reestablished in the mixture as soon as vibration stops.

Vibrators, whether internal or external, are usually characterized by their frequency of vibration, expressed as the number of vibrations per second (hertz), or vibrations per minute (vpm); they are also designated by the amplitude of vibration, which is the deviation in millimeters (inches) from the point of rest. The frequency of vibration can be measured using a vibrating reed tachometer.

When vibration is used to consolidate concrete, a backup should be on hand at all times in the event of a mechanical breakdown.

Internal Vibration. Internal or immersion-type vibrators, often called spud or poker vibrators (Figures 15-26 and 15-27), are commonly used to consolidate concrete in walls, columns,

FIGURE 15-26. Proper vibration makes possible the placement of stiff concrete mixtures, even in heavily-reinforced concrete members.

FIGURE 15-27. Internal vibrators are commonly used to consolidate concrete in walls, columns, beams, and slabs (courtesy of Baker Concrete).

TABLE 15-1. Range of Characteristics, Performance, and Applications of Internal* Vibrators

GROUP	DIAMETER OF HEAD, mm (in.)	RECOMMENDED FREQUENCY, VIBRATIONS PER MINUTE**	SUGGESTED VALUES OF			APPROXIMATE VALUES OF		APPLICATION
			ECCENTRIC MOMENT, mm • kg (in. • lb)	AVERAGE AMPLITUDE, mm (in.)	CENTRIFUGAL FORCE, kg (lb)	RADIUS OF ACTION, mm (in.)†	RATE OF CONCRETE PLACEMENT, m³/h (yd³/h)‡	
1	20-40 (0.75 -1.5)	9000-15,000	3.5-12 (0.03-0.10)	0.4-0.8 (0.015-0.03)	45-180 (100-400)	80-150 (3-6)	0.8-4 (1-5)	Plastic and flowing concrete in very thin members and confined places. May be used to supplement larger vibrators, especially in prestressed work where cables and ducts cause congestion in forms. Also used for fabricating test specimens.
2	30-60 (1.25 -2.5)	8500-12,500	9-29 (0.08-0.25)	0.5-1.0 (0.02-0.04)	140-400 (300-900)	130-250 (5-10)	2.3-8 (3-10)	Plastic concrete in thin walls, columns, beams, precast piles, thin slabs, and along construction joints. May be used to supplement larger vibrators in confined areas.
3	50-90 (2-3.5)	8000-12,000	23-81 (0.20-0.70)	0.6-1.3 (0.025-0.05)	320-900 (700-2000)	180-360 (7-14)	4.6-15 (6-20)	Stiff plastic concrete (less than 80 mm [3 in.] slump) in general construction such as walls, columns, beams, prestressed piles, and heavy slabs. Auxiliary vibration adjacent to forms of mass concrete and pavements. May be gang mounted to provide full-width internal vibration of pavement slabs.
4	80-150 (3-6)	7000-10,500	8-290 (0.70-2.5)	0.8-1.5 (0.03-0.06)	680-1800 (1500-4000)	300-510 (12-20)	11-31 (15-40)	Mass and structural concrete up to 50 mm (2 in.) slump deposited in quantities up to 3 m³ (4 yd³) in relatively open forms of heavy construction (powerhouses, heavy bridge piers, and foundations). Also used for auxiliary vibration in dam construction near forms and around embedded items and reinforcing steel.
5	130-150 (5-6)	5500-8500	260-400 (2.25-3.50)	1.0-2.0 (0.04-0.08)	1100-2700 (2500-6000)	400-610 (16-24)	19-38 (25-50)	Mass concrete in gravity dams, large piers, massive walls, etc. Two or more vibrators will be required to operate simultaneously to mix and consolidate quantities of concrete of 3 m³ (4 yd³) or more deposited at one time into the form.

  * Generally, extremely dry or very stiff concrete does not respond well to internal vibrators.

  ** While vibrator is operating in concrete.

  † Distance over which concrete is fully consolidated.

  ‡ Assumes the insertion spacing is 1½ times the radius of action, and that vibrator operates two-thirds of time concrete is being placed. These ranges reflect not only the capability of the vibrator but also differences in workability of the mixture, degree of deaeration desired, and other conditions experienced in construction. Adapted from ACI 309R-05.

beams, and slabs. A common type of internal vibrator is the flexible-shaft vibrator, which consists of a vibrating head connected to a driving motor by a flexible shaft. Inside the head, an eccentric weight connected to the shaft rotates at high speed, causing the head to revolve in a circular orbit. The motor can be powered by electricity, gasoline, or air. The vibrating head is usually cylindrical with a diameter ranging from 19 mm to 175 mm (¾ in. to 7 in.). Another common type is the motor-in-head vibrator, which have an electric motor built directly into the head generally at least 50 mm (2 in.) in diameter. The dimensions of the vibrator head as well as its frequency and amplitude in conjunction with the workability of the concrete mixture affect the performance of a vibrator.

Small-diameter vibrators have high frequencies ranging from 160 Hz to 250 Hz (10,000 vpm to 15,000 vpm) and low amplitudes ranging between 0.4 mm and 0.8 mm (0.016 in. and 0.03 in.). As

TABLE 15-2. Consistencies used in construction.

CONSISTENCY DESCRIPTION	SLUMP, mm (in.)	VEBE TIME, s	COMPACTING FACTOR AVERAGE	THAULOW DROP TABLE REVOLUTIONS
Extremely dry	–	32 to 18	–	112 to 56
Very stiff	–	18 to 10	0.7	56 to 28
Stiff	0 to 25 (0 to 1)	10 to 5	0.75	28 to 14
Stiff plastic	25 to 75 (1 to 3)	5 to 3	0.85	14 to 7
Plastic	75 to 125 (3 to 5)	3 to 0*	0.9	<7
Highly plastic	125 to 190 (5 to 7½)	–	–	–
Flowing	190+ (7½+)	–	0.95	–

*Test method is of limited value in this range.
Adapted from ACI 309R-05.

the diameter of the head increases, the frequency decreases and the amplitude increases. The effective radius of action of a vibrator increases with increasing diameter of the vibrator and slump of the concrete. For 100 mm (4 in.) slump concrete, vibrators with a diameter of 19 mm to 38 mm (¾ in. to 1½ in.) have a radius of action in freshly mixed concrete ranging between 75 mm and 150 mm (3 in. and 6 in.), whereas the radius of action for vibrators of 50 mm to 75 mm (2 in. to 3 in.) in diameter ranges between 175 mm and 350 mm (7 in. and 14 in.). Table 15-1 shows the range of characteristics and applications for internal vibrators for various applications. Table 15-2 describes characteristics of different concrete consistencies used in concrete construction, which can affect the choice of consolidation method.

Proper use of internal vibrators is critical for best results. Vibrators should not be used to move concrete horizontally, causing segregation. Whenever possible, the vibrator should be lowered vertically into the concrete at regularly spaced intervals and allowed to descend by gravity. It should penetrate to the bottom of the layer being placed and at least 150 mm (6 in.) into any previously placed layer. The vibrator is then withdrawn at roughly the same pace as the gravity insertion. The height of each layer will depend on the type of application and concrete properties (see **Depositing Concrete**).

In thin slabs, the vibrator should be inserted at an angle or horizontally in order to keep the vibrator head completely immersed. The vibrator should never be dragged around randomly in the slab. For slabs on ground, the vibrator should not make contact with the subgrade. The distance between insertions should be about 1½ times the radius of action so that the area visibly affected by the vibrator overlaps the adjacent previously vibrated area by a few centimeters (inches).

The vibrator should be held stationary until adequate consolidation is attained and then slowly withdrawn. An insertion time of 5 to 15 seconds will usually provide adequate consolidation. The concrete should move to fill the hole left by

the vibrator on withdrawal. If the hole does not refill, reinsertion of the vibrator at a nearby point should solve the problem.

Adequacy of internal vibration is judged by changes in the surface appearance of the concrete. Changes to watch for are the embedment of large aggregate particles, general batch leveling, the appearance of a thin film of mortar on the top surface, and the cessation of large bubbles of entrapped air escaping at the surface. Internal vibration may significantly affect the entrained-air-void system in concrete (Stark 1986 and Hover 2001).

Allowing a vibrator to remain immersed in concrete after paste accumulates over the head can result in nonuniformity. The length of time that a vibrator should remain in the concrete will depend on the workability of the concrete, the power of the vibrator, and the nature of the section being consolidated.

In heavily-reinforced sections where an internal vibrator cannot be inserted, it is sometimes helpful to vibrate the reinforcing bars by attaching a form vibrator to their exposed portions. This practice eliminates air and water trapped under the reinforcing bars and increases bond between the bars and surrounding concrete. This method should only be used if the concrete is still workable under the action of vibration. Internal vibrators should not be attached to reinforcing bars for this purpose because the vibrators may be damaged.

External Vibration. External vibrators can be form vibrators, vibrating tables, or surface vibrators such as vibratory screeds, plate vibrators, vibratory roller screeds, or vibratory hand floats or trowels. Form vibrators, designed to be securely attached to the outside of the forms, are especially useful for the following: consolidating concrete in members that are very thin or congested with reinforcement, stiff mixtures where internal vibrators cannot be used, and to supplement internal vibration.

Attaching a form vibrator directly to the concrete form is generally unsatisfactory. Rather, the vibrator should be attached to a steel plate that in turn is attached to steel

FIGURE 15-28. Vibratory screeds such as this truss-type unit reduce the work of strikeoff while consolidating the concrete.

I-beams or channels passing through the form stiffeners. Loose attachments can result in significant vibration energy losses and inadequate consolidation.

Form vibrators can be either electrically or pneumatically operated. They should be spaced to distribute the intensity of vibration uniformly over the form; optimum spacing is typically determined by trial and error. Sometimes it may be necessary to operate some of the form vibrators at a different frequency for better results; therefore, it is recommended that form vibrators be equipped to regulate their frequency and amplitude. The necessary duration of external vibration of the concrete is considerably longer than for internal vibration – generally between 1 minute and 2 minutes. A reed tachometer can not only determine frequency of vibration, but also give a rough estimate of amplitude of vibration by noting the oscillation of the reed at various points along the forms. This will assist in identifying dead spots or weak areas of vibration. A vibrograph

could be used if more reliable measurements of frequency and amplitude are needed.

Form vibrators should not be applied within the top meter (yard) of vertical forms. Vibration of the top of the form, particularly if the form is thin or inadequately stiffened, causes an in-and-out movement that can create a gap between the concrete and the form.

Vibrating tables are typically used in precasting plants. They should be equipped with controls so that the frequency and amplitude can be varied according to the size of the element to be cast and the consistency of the concrete. Stiffer mixtures generally require lower frequencies (below 6000 vpm) and higher amplitudes (over 0.13 mm [0.005 in.]) than more workable mixtures. Increasing the frequency and decreasing the amplitude as vibration progresses will improve consolidation.

Surface vibrators, such as vibratory screeds (Figures 15-28 to 15-30), are used to consolidate concrete in floors and other flatwork. Vibratory screeds give positive control of the strikeoff operation and save a great deal of labor. For slumps greater than 75 mm (3 in.), care should be taken because surface vibration of such concrete will result in an excessive accumulation of mortar and fine material on the surface; this may reduce wear resistance. For the same reason, surface vibrators should not be operated after the concrete has been adequately consolidated.

Because surface vibration of concrete slabs is least effective along the edges, an internal vibrator should be used along the edge forms immediately before the vibratory screed is applied.

Vibratory screeds are used for consolidating slabs up to 250 mm (10 in.) thick, provided such slabs are unreinforced or only lightly reinforced (welded-wire fabric). Internal vibration or a combination of internal and surface vibration is recommended for reinforced slabs. More detailed information regarding

FIGURE 15-29. Where floor tolerances are not critical, an experienced operator using a vibratory screed does not need screed rails supported by chairs to guide the screed. Instead, the operator visually matches elevations to forms or previous passes. This process is called wet screeding.

FIGURE 15-30. A laser level striking the sensors on this screed guides the operator as he strikes off the concrete. Screed rails and chairs are not needed and fewer workers are required to place concrete. Laser screeds interfaced with total station surveying equipment can also strike off sloped concrete surfaces.

internal and external vibration of concrete can be obtained from ACI 309R, *Guide for Consolidation of Concrete* (2005).

Consequences of Improper Vibration. Undervibration may cause honeycombing; excessive amount of entrapped air voids, often called bugholes; sand streaks; cold joints; placement lines; and subsidence cracking (see Chapter 16).

Defects from overvibration include:

• segregation as vibration and gravity causes heavier aggregates to settle while lighter aggregates rise;

• sand streaks;

• loss of entrained air in air-entrained concrete;

• excessive form deflections or form damage; and

• form failure caused by excessive pressure from vibrating the same location too long or placing concrete more quickly than the designed rate of placement.

Undervibration is much more prevalent than overvibration. For detailed guidance on proper vibration, see ACI 309R (2005).

FINISHING

Concrete can be finished in many ways, depending on the intended service use. Various colors and textures, such as exposed-aggregate or a pattern-stamped surface, may be specified. Some surfaces may require only strikeoff and screeding to proper contour and elevation, while other surfaces may specify a broomed, floated, or troweled finish. Details are given in ACI 302, *Guide for Concrete Slab and Floor Construction* (2015), PCA EB122, *Concrete Finisher's Guide* (Collins and others 2006), and Tarr and Farny (2008).

The mixing, transporting, and handling of concrete should be carefully coordinated with the finishing operations. Concrete should not be placed on the subgrade or into forms more rapidly than it can be spread, struck off, consolidated, and floated. Concrete placing and finishing crews should be staffed with an adequate number of proficient laborers and finishers

to correctly place, finish, and cure concrete slabs with due regard for the effects of concrete temperature and atmospheric conditions on the setting time of the concrete and the size of the placement to be completed.

> ACI 302.1R-15 defines the **Window of Finishability** as: *The time period available for finishing operations after the concrete has been placed, consolidated, and struck-off, and before final troweling.*

Finishing Flatwork

Tools for placing, consolidating, and finishing concrete flatwork are largely mechanized, though some amount of hand labor is generally necessary for striking off, compacting, and finishing. Large jobs require more equipment and small jobs may have restrictions based on access to the work area.

Screeding (Strikeoff). Screeding or strikeoff is the process of cutting off excess concrete to bring the top surface of a slab to proper grade. The manually used template is called a straightedge, although the lower edge may be straight or slightly curved, depending on the surface specified. It should be moved across the concrete with a sawing motion while advancing forward a short distance with each movement. There should be a surplus (surcharge) of concrete against the front face of the straightedge to fill in low areas as the straightedge passes over the slab. A 150 mm (6 in.) slab needs a surcharge of about 25 mm (1 in.). Straightedges are sometimes equipped with vibrators that consolidate the concrete and assist in reducing the strikeoff work. This combination straightedge and vibrator is called a vibratory screed. Vibratory screeds are discussed earlier in this chapter under **Consolidation**. Screeding, consolidating, and bullfloating must be completed before bleed water collects on the surface.

Bullfloating or Darbying. To eliminate high and low spots and to embed large aggregate particles, a bullfloat or darby (Figure 15-31 left) should be used immediately after strikeoff. Initial float

FIGURE 15-31. (left) Darbying brings the surface to the specified level and is done in tight places where a bullfloat cannot reach. (right) Bullfloating must be completed before any bleed water accumulates on the surface.

operations should level, shape, and smooth the surface and work up a slight amount of cement paste.

A long-handle bullfloat (Figure 15-31 right) is used on areas too large to reach with a darby. Highway straightedges are used to obtain very flat surfaces (Figure 15-32). For non-air-entrained concrete, these tools can be made of wood, aluminum, or magnesium alloy; for air-entrained concrete they should be of aluminum or magnesium alloy. It is important that the surface not be closed during this finishing phase as bleeding is typically still in progress and trapped bleed water or bleed air can lead to surface delaminations. Bullfloating or darbying must be completed before bleed water accumulates on the surface. Care must be taken not to overwork the concrete as this could result in a less durable surface. For more information on bleeding, see Chapter 9 and Kosmatka (2006).

FIGURE 15-32. Highway straightedges are used on highway pavement and floor construction where very flat surfaces are desired (courtesy of Baker Concrete).

Further finishing may not be required in all situations. However, on most slabs bullfloating or darbying is followed by edging, jointing, floating, troweling, and brooming. A slight hardening of the concrete is necessary before the start of any of these finishing operations. When the bleed water sheen has

evaporated, and the concrete will sustain foot pressure with only about 6 mm (¼ in.) indentation, the surface is ready for continued finishing operations.

> **Warning:** One of the principal causes of surface defects in concrete slabs is finishing while bleed water is present on the surface. If bleed water is worked into the surface, the water-cementitious materials ratio is significantly increased. This reduces strength, entrained air content, and watertightness of the concrete surface. Any finishing operation performed on the surface of a concrete slab while bleed water is present can cause crazing, dusting, or scaling (see Chapter 16).

Floating and troweling the concrete before the bleeding process is completed may also trap bleed water or air under the finished surface producing a weakened zone or void under the finished surface; this occasionally results in delaminations. For this reason, concrete with more than 3% entrained air should not be finished with a hard-trowel (ACI 302.1R 2015, ACI 301 2020, Tarr and Farny 2008, and PCA 2001).

Spreading dry cement on a wet surface to absorb excess water causes surface defects such as dusting, crazing, and mortar flaking. These wet spots should be avoided, if possible, by adjustments in aggregate gradation, mix proportions, and consistency. When wet spots do occur, finishing operations should be delayed until the water either evaporates or is removed with a rubber floor squeegee or by dragging a soft rubber garden hose to push the standing water off of the slab surface (ACI 302.1R 2015). If a squeegee or hose is used, extreme care must be taken so that excess cement paste is not removed with the water, and so that the surface is not damaged.

Edging and Hand Tooling Joints. Edging is required along all edge forms and isolation and construction joints in floors and outdoor slabs such as walks, drives, and patios. Edging densifies and compacts concrete next to the form where

FIGURE 15-33. Edging involves cutting concrete away from the forms and then running an edger to smooth and densify the corner.

FIGURE 15-34. Joints may be tooled in fresh concrete using a groover and a straight edge as a guide (left), or a bullfloat fitted with a jointing attachment for tooling contraction joints (right).

floating and troweling are less effective, making it more durable and less vulnerable to scaling, chipping, and popouts. Edging tools should be held flat across their width and the leading face of the edger slightly inclined to provide a flat edge which prevents the edger from catching on the concrete surface causing tearing and defects (Figure 15-33). First pass edging operations should be completed before the onset of bleeding; otherwise, the concrete should be cut away from the forms to a depth of 25 mm (1 in.) using a pointed mason trowel or a margin trowel to push coarse aggregate particles away from the form edge. Edging may be required after each subsequent finishing operation for exterior and interior slabs.

Proper jointing practices can minimize unsightly random cracks. Contraction joints, sometimes called control joints, can be formed with a hand groover or by inserting strips of plastic, wood, metal, or preformed joint material into the fresh concrete (Figure 15-34). When hand methods are used to form control joints in exterior concrete slabs, mark the forms to accurately locate the joints. Prior to bullfloating, the edge of a thin strip of wood or metal may be used to knock down the coarse aggregate where the joint will be hand tooled. The slab should then be jointed immediately after bullfloating or in conjunction with the edging operation. If insert strips are used, they should be kept vertical. Hand tooling and insert strips are not recommended for jointing industrial floors. For these slabs, contraction joints should be sawn into hardened concrete. Jointing is discussed further under **Jointing Concrete**.

Floating. After the concrete has been edged, joints may be hand-tooled, and the surface should be floated (Figure 15-35). Floating embeds aggregate particles just beneath the surface; removes slight imperfections, humps, and voids; compacts the mortar at the surface in preparation for additional finishing operations; and reestablishes the moisture content of the paste at the near surface where evaporation has its greatest impact.

FIGURE 15-35. Hand floating (left hand) the surface with a hand float held flat on the concrete surface and moved in a sweeping arc with a slight sawing motion. Troweling (right hand) with blade tilted is performed before moving the kneeboards (courtesy of Baker Concrete).

The concrete should not be overworked as this may bring an excess of water and fine material to the surface and result in subsequent surface defects.

Hand floats are made of fiberglass, magnesium, or wood. The metal float reduces the amount of work required because drag is reduced as the float slides more readily over the concrete surface. A magnesium float is essential for hand-floating air-entrained concrete because a wood float tends to stick to and tear the concrete surface. The light metal float also produces a smoother surface than the wood float. The hand float should be held flat on the concrete surface and moved with a slight sawing motion in a sweeping arc to fill in holes, cut off lumps, and smooth ridges.

Floating concrete is now done almost exclusively by machine. Typically, a gasoline-powered troweling machine equipped with float blade shoes is used (Figure 15-36). A power trowel can be

FIGURE 15-36. Power floating using walk-behind (left) and ride-on equipment (right). When the bleed water sheen has evaporated and the concrete will sustain foot pressure with only slight indentation, the surface is ready for floating and final finishing operations (courtesy of Baker Concrete).

either walk-behind or ride-on. Walk-behind machines usually have a single set of 3 or 4 blades and ride-on machines can have two or three sets of 4 or 5 blades. The diameter of power trowels ranges from 900 mm to 1200 mm (36 in. to 48 in.). Float shoes are relatively wide (typically 250 mm [10 in.]) and 350 mm to 450 mm (14 in. to 18 in.) long. They slip over the trowel blades on power equipment. Both the leading and trailing edges of float blades are turned up slightly. Another attachment available for use is the pan attachment, steel discs with the outer edge turned up. Pan attachments can distribute the load of the machine over a larger area (less contact pressure). Pans work the surface more, but they can significantly increase the flatness and allow the contractor to further delay initial finishing (Bimel 1998).

The floating produces a relatively even (but not smooth) texture that has good slip resistance and is often used as a final finish, especially for exterior slabs. Where a float finish is the desired final finish, it may be necessary to float the surface a second time after additional stiffening of the concrete surface. Marks left by hand edgers and groovers are normally removed during floating unless the marks are desired for decorative purposes; in such cases the edger and groover should be used again after final floating.

Troweling. Where a smooth, hard, dense surface is desired, floating should be followed by steel troweling. Troweling should only be done on a surface that has been previously floated.

Combination blades may be used between floating and final troweling. They are slightly smaller than float blades (typically 200 mm [8 in.] wide) and have only the leading edge turned up. The angle of upturn is important, however, and these blades must be used with care so as not to densify and seal the surface. When angled they act as trowels and can densify the surface. If the surface is densified prematurely, rising bleed water might be trapped during the floating period. In most cases, float blades are recommended to decrease this risk.

A power trowel is similar to a power float, except that the machine is fitted with smaller, individual steel trowel blades (typically 150 mm [6 in.] wide and 350 mm to 450 mm [14 in. to 18 in.] long). Trowel blades can be tilted to various degrees to exert higher pressure to the slab surface during final finishing (Figure 15-37). This increased pressure densifies and closes the slab surface. Generally, greater tilt will produce a smoother and denser surface. It is customary when hand-finishing large slabs to float and immediately trowel an area before moving the kneeboards. These operations should be delayed until the concrete has hardened sufficiently so that water and fine material are not brought to the surface. Premature floating and troweling can cause scaling, crazing, or dusting and produce a surface with reduced wear resistance. When power-troweling in a systematic pattern, it is important that concrete setting properties remain consistent. Otherwise, it is difficult to avoid areas that set slower or bleed longer to reach other areas that are ready for troweling.

Initial troweling may produce the desired surface free of defects. Two or more passes are frequently required to increase the compaction of fines at the surface and give greater resistance

FIGURE 15-37. Water spraying the surface should be avoided during finishing.

FIGURE 15-38. Adjustable blades on a power-trowel can be pitched to impart a greater polishing force to the concrete slab.

to wear. Surface smoothness, density, and wear resistance can all be improved by additional troweling passes. There should be a lapse of time between successive trowelings to permit the concrete to become harder. Too long a delay will result in a surface that is too hard to float and trowel. As the surface stiffens, each successive troweling should be made with smaller trowels, using progressively more tilt and pressure on the trowel blade (Figure 15-38). Each successive troweling should be made in a direction at right angles to the previous pass. The final pass should make a ringing sound as the trowel moves over the hardening surface. If necessary, tooled edges and joints should be rerun after troweling to maintain uniformity and true lines.

Exterior concrete should not be troweled for several reasons including; loss of entrained air caused by overworking the surface, and slip and fall accidents occurring due to smoother surfaces.

For safety reasons, floating and brooming are the preferred textures for safety for outdoor concrete. Traditionally, bullfloats have been used successfully without sealing the surface or removing substantial air from the surface. The risk of these occurrences increases with the use of power floating equipment.

Burnished concrete. Repeated hard steel troweling results in a somewhat polished (glossy) look, referred to as a burnished surface (Figure 15-39). This special finish provides added resistance to abrasion and wear. Achieving a burnished floor finish requires multiple passes with a trowel: at first, the trowel blades are flat to the surface, then tilt angle and rotation speed are increased on each of the following passes. When a floor has reached the burnished condition, the trowel blades make a sharp metallic ringing sound as they slide over the surface.

Plastic combination blades have been used at slower rotation speeds to create a mirror-like polished final finish after troweling.

Surface defects will be more noticeable on a burnished finish than on a surface that is not as smooth. A burnished finish will not be perfectly uniform and may exhibit dark colors, craze cracking, circular patterns from power troweling, and other surface blemishes. All of these conditions are cosmetic and do not affect floor performance. A specific color or consistency cannot be guaranteed, and some mottling should be expected. Because the final appearance is subjective, specifiers must clearly communicate their expectations. The contractor and concrete supplier can take precautions to avoid gross discoloration or mottling (see Chapter 16).

Material requirements for a burnished floor may include fine aggregate grading limits, minimum cement content, uniform w/cm, maximum slump, and avoiding calcium chloride admixtures.

FIGURE 15-39. A burnished finish is preferred for interior floors that do not require high slip-resistance.

Brooming. Brooming or tining should be performed before the concrete has thoroughly hardened, but the surface should be sufficiently hard to retain the scoring impression to produce a slip-resistant surface (Figures 15-40 and 15-41). Rough scoring

FIGURE 15-40. Brooming provides a slip-resistant surface mainly used on exterior concrete.

FIGURE 15-41. This machine is tining the surface of fresh concrete (left). Tining of pavements improves tire traction and reduces hydroplaning. Transverse tining maximizes traction; however, longitudinal tining is preferred to minimize tire noise (right).

or tining, can be achieved with a rake, a steel-wire broom, or a stiff, coarse, fiber broom; such coarse-textured brooming usually follows floating. If a finer texture is desired, the concrete should be floated to a smooth surface and then brushed with a soft-bristled broom.

Interior concrete could also be troweled before brooming. Pavements should be broomed transversely to the main direction of traffic. The magnitude of the texture is dependent on the degree of hard troweling – a light trowel will texture more easily than a denser finish. Best results are obtained with equipment that is specially made for texturing concrete.

Finishing Formed Surfaces

Many as-cast concrete surfaces require little or no additional treatment when they are carefully constructed with the proper forming materials. In accordance with ACI 301, *Specifications for Concrete Construction* (2020), these surfaces are divided into several classes; as-cast finishes, surface finish SF-1, surface finish SF-2, and surface finish SF-3. Table 15-3 provides more detail on each finish type.

Smooth-rubbed finish. A smooth, rubbed finish is produced on a newly hardened concrete surface no later than the day following form removal. The forms are removed and necessary patching completed as soon as possible. The surface is then wetted and rubbed with a carborundum brick or other abrasive until a satisfactory uniform color and texture are produced.

Grout-cleaned rubbed finish. A grout cleandown (sack-rubbed finish) can be used to impart a uniform color and appearance to a smooth surface. After defects have been repaired, the surface should be saturated thoroughly with water and kept wet at least one hour before finishing operations begin. Next, a grout should be prepared consisting of 1 part cement, 1½ to 2 parts of fine aggregate meeting the requirements of ASTM C144, *Standard Specification for Aggregate for Masonry Mortar* (AASHTO M 45) or ASTM C404, *Standard Specification for Aggregates for Masonry Grout* – typically passing a 600 μm (No. 30) sieve, and

TABLE 15-3. Specified Surface Finishes (ACI 301-20).

AS-CAST FINISHES
Produce as-cast formed finishes in accordance with Contract Documents.

SURFACE FINISH SF-1
No formwork facing material is specified.
Patch voids larger than 1½ in. wide or ½ in. deep.
Remove projections larger than 1 in.
Tie holes need not be patched.
Surface tolerance Class D as specified in ACI 117-10.

SURFACE FINISH SF-2
Patch voids larger than ¾ in. wide or ½ in. deep.
Remove projections larger than ¼ in.
Patch tie holes.
Surface tolerance Class B as specified in ACI 117-10.

SURFACE FINISH SF-3
Patch voids larger than ¾ in. wide or ½ in. deep.
Remove projections larger than ⅛ in.
Patch tie holes.
Surface tolerance Class A as specified in ACI 117-10.

sufficient water for a thick, creamy consistency. It should be preshrunk by mixing at least 15 minutes before use and then remixed without the addition of water and applied uniformly by brush, plasterer's trowel, or rubber float to completely fill all air bubbles and holes. Finally, the surface should be rubbed with clean, dry burlap to remove all excess grout. All air holes should remain filled, but no visible film of grout should remain after the rubbing. Any section cleaned with grout must be completed in one day as grout remaining overnight is difficult to remove.

Cork-floated finish. A cork-floated finished surface should be vigorously floated with a wood, rubber, or cork float immediately after applying the grout to fill any small air holes (bugholes) that

FIGURE 15-42. Patterned, textured, and colored concretes are very attractive.

are left; any remaining excess grout should be scraped off with a float. If the float pulls grout from the holes, a sawing motion of the tool should correct the difficulty. Any grout remaining on the surface should be allowed to stand undisturbed until it loses some of its plasticity but not its damp appearance. Final finish can be produced by a swirling motion of a cork float.

If possible, work should be done in the shade and preferably during cool, damp weather. During hot or dry weather, the concrete can be kept moist with a fine atomized fog spray.

The completed rubbed surface should be wet-cured by keeping the area damp for 36 hours following the clean down. When completely dry, the surface should have a uniform color and texture.

SPECIAL SURFACE FINISHES

A variety of patterns and textures, and colors can be used to produce decorative finishes. Patterns can be formed with divider strips or by scoring or stamping the surface just before the concrete hardens.

Textures can be produced with little effort and expense with floats, trowels, and brooms; more elaborate textures (Figure 15-42) can be achieved with special techniques (Kosmatka and Collins 2004).

Stamped Concrete

Pattern Stamping. Cobblestone, brick, and tile finishes in a variety of sizes and patterns can be impressed deeply into partially set concrete with special imprinting tools (Figure 15-43). The concrete also may be colored integrally, by the dry-

shake method, or both. The joints may be filled with plain or colored mortar to create any number of striking effects.

To receive a stamped pattern, concrete should contain small size coarse aggregate such as pea gravel, 9.5 mm ($^3/_8$ in.) top size. Finishing follows the normal procedures; however, the surface should not be troweled more than once. After the surface is troweled or floated to the desired texture, platform

FIGURE 15-43. Stamp molds can be used to provide pattern on concrete surfaces often with release powder to prevent the concrete from sticking to the stamp mold. Tampers are used to stamp the pattern into the surface.

stamping pads are used. One pad is placed next to the other so that the pattern is accurately aligned; at least two pads are required. The finisher simply steps from one pad to the next, stamping the design to depth of about 25 mm (1 in.).

A form release agent may be brushed on the pads to keep them free of mortar if the pad is in direct contact with the concrete. In addition to the weight of the worker, a hand-tamper is sometimes used to ensure adequate indentation of the stamping pad.

For many patterns, especially those with a smooth surface, a sheet of 0.025 mm to 0.05 mm (1 mil to 2 mil) polyethylene plastic is placed over the slab, then the stamping pads are placed on the plastic sheet. The plastic stretches during stamping, rounding the edges of the incised pattern. Also, the plastic minimizes the possibility of concrete sticking to the stamp and then tearing the surface when moved to its next location. Textured surfaces may require the use of powdered release agents for greater detail to the surface. Stamped impressions in the pattern (joints) are usually left open, although they can be grouted for contrast.

Pattern Rolling. Pattern rolling is similar to pattern stamping in appearance. Unlike traditional pattern stamping, which uses stationary imprinting tools, pattern rolling uses a hollow cylinder to impress a pattern on the surface. The cylinder, usually with a brick or cobblestone relief (Figure 15-44), is rolled across the surface to create the pattern. Water can be added inside the cylinder to adjust its weight to control depth of imprint or adjust to the stiffness of the concrete.

FIGURE 15-44. Pattern rolling is accomplished with a hollow cylinder that impresses a pattern on the concrete surface (courtesy of Marshall Concrete Products, Inc.).

Normal slab construction procedures are followed in preparing the subbase and slab. Integral color admixtures are added to the concrete or dry-shake coloring agents are applied to the surface to color the concrete. After the initial bleed water has evaporated and the surface has been floated, the slab is covered with a thin plastic sheet stapled to the wooden edge forms. If a rough texture is desired, especially at impressed

joints or edges, the cylinder should be applied directly to the concrete surface without the plastic sheet.

Before rolling, a chalk line or straightedge should accurately mark the rolling location on the plastic to keep the cylinder and pattern alignment. The size of the slab to be patterned should be a whole multiple of the cylinder width, usually 1 m (3 ft). Forms at locations where the last pass is less than a roller width should be set 6 mm (¼ in.) lower than the elevation of other forms. A 6 mm (¼ in.) wood strip is nailed to the top of this form to provide the proper slab elevation when screeding the slab. Prior to rolling, the wood strip is removed and the roller extends beyond the edge to provide a full, complete imprint. Rolling usually begins within ½ to 1 hour after floating.

Although only a limited number of patterns are available compared to traditional pattern stamping, a wide variety of colors and textures can be achieved. The primary advantage of pattern rolling is the speed at which a pattern can be imprinted.

Exposed-Aggregate Concrete

An exposed-aggregate finish provides a rugged, attractive surface in a wide range of textures and colors. Methods for obtaining an exposed aggregate surface include seeding, monolithic placements, and toppings.

Seeding. In the seeding technique, the aggregate is evenly distributed or seeded in one layer onto the concrete surface immediately after the slab has been bullfloated or darbied. Select aggregates are carefully chosen to avoid deleterious substances; they are usually of uniform size such as 9.5 mm to 12.5 mm (⅜ in. to ½ in.) or larger as shown in Table 15-4. They should be washed thoroughly before use to assure satisfactory bond. Flat or elongated aggregate particles should not be used since they are easily dislodged when the aggregate is exposed. Caution should be exercised when using crushed stone; it not only has a greater tendency to stack during the seeding operation (requiring more labor), but the sharp angular edges may be undesirable in some applications (pool decks, for example).

TABLE 15-4. Aggregate Sizes for Exposed-Aggregate Slabs

6.3 mm (¼ in.)	to	12.5 mm (½ in.)
9.5 mm (⅜ in.)	to	16 mm (⅝ in.)
12.5 mm (½ in.)	to	19 mm (¾ in.)
16 mm (⅝ in.)	to	22.4 mm (⅞ in.)
19 mm (¾ in.)	to	25 mm (1 in.)
25 mm (1 in.)	to	37.5 mm (1½ in.)
31.5 mm (1¼ in.)	to	50 mm (2 in.)

The particles must be completely embedded in the concrete. This can be done by lightly tapping with a hand float, a darby, or the broad side of a piece of lumber. Then, when the concrete can support a finisher on kneeboards, the surface should be hand-floated with a magnesium float or darby until the mortar completely surrounds and slightly covers all the aggregate particles.

Monolithic Placements and Toppings. In the monolithic technique, select aggregate, usually gap-graded, is mixed throughout the batch of concrete. Toppings are an alternative method in which the select exposed-aggregate is mixed into a topping that is placed over a base slab of conventional concrete. Exposing the aggregate is commonly performed by washing and brushing, using retarders, or scrubbing. When the concrete has hardened sufficiently, simultaneously brushing and flushing with water should expose the aggregate. In washing and brushing, the surface layer of mortar should be carefully washed away with a light spray of water and brushed until the desired exposure is achieved.

On large surface areas, a water-insoluble retarder can be sprayed or brushed on the surface immediately after floating (see Chapter 6). When the concrete becomes too hard to produce the required finish with normal washing and brushing, a dilute hydrochloric acid could be used. Surface preparation should be minimized and applicable local environmental laws should be followed.

To expose aggregate after the concrete has hardened; mechanical methods such as abrasive blasting, waterblasting, tooling or bushhammering, and grinding and polishing are effective once the concrete has a compressive strength of around 28 MPa (4000 psi). Abrasive blasting is best applied to a gap-graded aggregate concrete. The nozzle should be held perpendicular to the surface and the concrete removed to a maximum depth of about one-third the diameter of the coarse aggregate.

Waterblasting can also be used to texture the surface of hardened concrete, especially where local ordinances prohibit the use of sandblasting for environmental reasons. High-pressure water jets are used on surfaces with or without retarders (Figure 15-45).

In tooling or bushhammering, a layer of hardened concrete is removed and the aggregate is fractured at the surface. Resulting finishes can vary from a light scaling to a deep, bold texture obtained by jackhammering with a single-pointed chisel. Combs and multiple points can be used to produce finishes similar to those sometimes used on cut stone.

For more information on exposed aggregate finishes, see Kosmatka and Collins (2004), PCA (1972), and PCA (1995).

FIGURE 15-45. In exposing the aggregate, the paste may be removed by mechanical methods such as water blasting. High-pressure jets can be used on surfaces with or without retarding admixtures.

Polished Concrete

Polished concrete is a systematic mechanical abrasion of a concrete surface using progressively finer diamond polishing pads (Fields 2006). Polishing is an abrasive process. Grinding, honing, and polishing are the three levels of abrasion. In large facilities, heavy industrial floor polishers are used; smaller machines and hand-held devices are useful for small floors or tight locations. A dense, strong concrete provides the best results.

As with any abrasive process the material surface is ground, or in effect "sanded" starting with coarse grit diamond impregnated pads and progressively changing to finer grit pads until the final surface aesthetic is achieved. Polishing typically begins with planetary grinders holding 50- to 80-grit diamond pads to flatten the surface and remove high spots; it can also expose fines or expose aggregates depending on the grit. Each successive step approximately doubles the grit rating to achieve polished finishes as high as 3000 grit. Each pass increases smoothness and shine. A greater number of passes increases reflectivity. Honing, using a finer grit, removes scratches left by the preceding grind and creates a soft glow or stone-like appearance.

Polishing can be done either wet or dry, but improvements in dust control equipment and filters have increased the popularity of indoor dry grinding. One method of slab finishing that has gained popularity is diamond polishing (Nasvik 2007 and Thome 2004). Diamond polishing (or diamond grinding) involves the use of terrazzo-type surface grinders that impart a smooth, glossy finish that exposes the aggregate within the paste matrix. Usually, a thin layer, about 1 mm to 1.5 mm ($\frac{1}{32}$ in. to $\frac{1}{16}$ in.), of the surface is removed. The process involves successive passes using finer grinding pads. The pads are rated with grit values based on the relative surface smoothness they create in comparison to conventional grinding grits.

Burnishing and polishing achieve similar results, but burnishing is done as part of the finishing steps, and polishing is done after the concrete has hardened and gained sufficient strength to not be disturbed as it is rubbed with progressively finer pads. Polishing can allow for embedded materials to be exposed. It can be used on new or old concrete floors and can correct problem surfaces.

A polished surface finish is similar to terrazzo. Special aggregate may be used for aesthetic effects, although polishing may not expose aggregate to the same degree as terrazzo (Figure 15-46). Terrazzo, with its ground and polished surface, is also an exposed-aggregate concrete, which is primarily used indoors.

FIGURE 15-46. Polishing concrete.

Colored Finishes

Colored and white concrete finishes for decorative effects in both interior and exterior applications can be achieved by four different methods (Kosmatka and Collins 2004):

- one-course or integral method,
- two-course method,
- dry-shake method, and
- stains and paints.

Two-course method. In the two-course method, a base slab is placed and left with a rough texture to bond better to a colored topping layer. As soon as the base slab can support a finisher's weight, the topping course can be placed. If the base slab has hardened, prepare a bonding grout for the base slab prior to placing the topping mix. The topping mix is normally 13 mm (½ in.) to 25 mm (1 in.) thick, with a ratio of cement to sand of 1:3 or 1:4. The mix is floated and troweled in the prescribed manner. The two-course method is commonly used because it is more economical than the one-course method.

Dry-shake method. In the dry-shake method, a prepackaged dry-color material is cast onto the surface of a concrete slab or precast panel. After the slab has been bullfloated once, the dry coloring material should be broadcast evenly over the surface. The required amount of coloring material can usually be determined from previously cast sections. After the material has absorbed water from the fresh concrete, it should be floated into the surface. If recommended by the material manufacturer to apply in two coats, the rest of the material should be applied immediately, by casting onto the surface at right angles to the initial application and then floated to assure uniform color distribution. The slab should again be floated to work the remaining material into the surface. The surface can then be troweled at the same time as a typical slab. For exterior surfaces that will be exposed to freezing and thawing, little or no troweling followed by brooming with a soft bristle concrete broom is recommended. Curing should begin immediately after finishing; take precautions to prevent discoloring the surface. See Kosmatka and Collins (2004) and Harris (2004) for more information.

Stains, Paints, and Clear Coatings. Many types of stains, paints, and clear coatings can be applied to concrete surfaces. Among the principal paints used are portland cement base, latex-modified portland cement, and latex (acrylic and polyvinyl acetate) paints (PCA 1992). However, stains and paints are used only when it is necessary to color existing concrete (Figure 15-47). It is difficult to obtain a uniform color with dyes or stains; therefore, the manufacturer's directions should be closely followed. Stains and dyes are often used to even out colors on existing slabs or to create mottled or variegated effects.

FIGURE 15-47. Staining the surface of existing concrete can provide a pleasing aesthetic.

Portland cement-based paints can be used on either interior or exterior exposures. The surface of the concrete should be damp at the time of application and each coat should be dampened as soon as possible without disturbing the paint. Damp curing of conventional portland cement paint is essential.

On open-textured surfaces, such as concrete masonry, the paint should be applied with stiff-bristle brushes (scrub brushes). Paint should be worked well into the surface. For concrete with a smooth or sandy finish, whitewash or dutch-type calcimine brushes are best.

The latex materials used in latex-modified portland cement paints retard evaporation, thereby retaining the necessary water for hydration of the portland cement. When using latex-modified paints, moist curing is typically not required. Most latex paints are resistant to alkali and can be applied to new concrete after 10 days of good drying weather. The preferred method of application is by long-fiber, tapered nylon brushes 100 mm to 150 mm (4 in. to 6 in.) wide; however, roller or spray methods can also be used. The paints may be applied to damp, but not wet surfaces. If the surface is moderately porous, or if extremely dry conditions prevail, prewetting the surface is advisable.

Clear coatings are frequently used on concrete surfaces to:
• prevent soiling or discoloration of the concrete by air pollution,
• facilitate later cleaning of the surface,
• brighten the color of the aggregates, and
• render the surface water-repellent and thus prevent color change due to rain and water absorption.

Superior coatings often consist of methyl methacrylate forms of acrylic resin, as indicated by a laboratory evaluation of commercial clear coatings (Litvin 1968). The methyl methacrylate coatings should have a higher viscosity and solids content when used on smooth concrete, since the original appearance of smooth concrete is more difficult to maintain than the original appearance of exposed-aggregate concrete.

When a change of color would be objectionable, other materials, such as silane and siloxane penetrating sealers, are commonly used as water repellents for many exterior concrete applications.

Mock-Ups

A preconstruction mock-up (field sample) should be made for each finish to determine the timing and steps involved for special concrete placements and architectural concrete. Accepted field mockup will serve as the reference to which concrete will be compared for periodic and final acceptance. Field mockups should be constructed at an acceptable location on site. Field mockups should also provide a simulated repair area to demonstrate an acceptable repair procedure with an acceptable color and texture match.

Field mockups should be constructed using same placing and finishing procedures, equipment, materials, and curing methods that will be used for construction project. This should include variations in temperature during placement and any other considerations for repeating the same level of work as the approved sample. Mockups should be protected from physical damage and retained until final acceptance of structure.

JOINTING CONCRETE

The following three types of joints are common in concrete construction: isolation joints, contraction joints, and construction joints.

Isolation Joints

Isolation joints permit both horizontal and vertical differential movements at adjoining parts of a structure (Figure 15-48). They are used, for example, around the perimeter of a floor on ground, around columns; around machine foundations to separate the slab from the more rigid parts of the structure; and to isolate slabs-on ground from fixed foundation members.

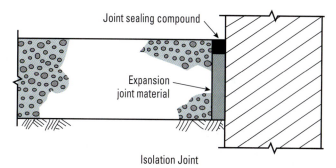

FIGURE 15-48. Isolation joints permit horizontal and vertical movements between abutting faces of a slab and fixed parts of a structure.

Isolation-joint material (often called expansion-joint material) can be as thin as 6 mm (¼ in.) or less, but 13 mm (½ in.) material is commonly used. Care should be taken to ensure that all the edges for the full depth of the slab are isolated from adjoining construction to minimize cracking stress potential.

Columns on separate footings are isolated from the floor slab either with a circular or square-shaped isolation joint. A square shape should be rotated to align its corners with control and construction joints.

Contraction Joints

Contraction joints provide for movement in the plane of a slab or wall. These joints induce controlled cracking caused by drying and thermal shrinkage at preselected locations (Figure 15-49). Contraction joints (also sometimes called control joints) should be constructed to permit transfer of loads perpendicular to the plane of a slab or wall. If no contraction joints are used, or if they are too widely spaced in slabs on ground or in lightly reinforced walls, random cracks may occur; cracks are most likely when drying and thermal shrinkage produce tensile stresses in excess of the concrete's tensile strength.

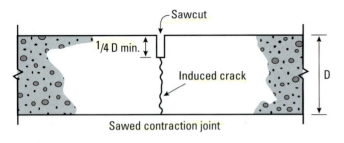

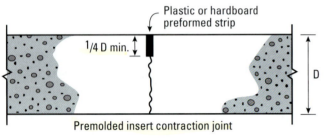

FIGURE 15-49. Contraction joints provide for horizontal movement in the plane of aslab or wall and induce controlled cracking caused by drying and thermal shrinkage.

Saw-Cut Joints. Contraction joints in slabs on ground can be created using several techniques. One of the most common methods is to saw a continuous straight slot in the top of the slab (Figure 15-50). When installed to a depth of ¼ the slab depth, these joints create planes of weakness at which cracks are expected to form. Vertical loads are transmitted across a contraction joint by aggregate interlock between the opposite faces of the crack provided the crack is not too wide and the spacing between joints is not too great.

Crack widths at saw-cut contraction joints that exceed 0.9 mm (0.035 in.) through the depth of the slab do not reliably transfer loads. However, the width of the crack at the surface cannot be used to evaluate potential load transfer capability as cracks are typically much wider at the drying surface than at the bottom. In addition, the effectiveness of load transfer by aggregate interlock depends on other variables than crack width. Other factors include: slab thickness, subgrade support, load magnitude, repetitions of load, and aggregate angularity.

FIGURE 15-50. Sawing a continuous cut in the top of a slab is one of the most economical methods of forming a contraction joint.

TABLE 15-5. Spacing of Contraction Joints in Meters (Feet)*

SLAB THICKNESS mm (in.)	MAXIMUM-SIZE AGGREGATE LESS THAN 19 mm (¾ in.)	MAXIMUM-SIZE AGGREGATE 19 mm (¾ in.) AND LARGER
125 (5)	3.0 (10)	3.75 (13)
150 (6)	3.75 (12)	4.5 (15)
175 (7)	4.25 (14)	5.25 (18)**
200 (8)	5.0 (16)**	6.0 (20)**
225 (9)	5.5 (18)**	6.75 (23)**
250 (10)	6.0 (20)**	7.5 (25)**

* If concrete cools at an early age, shorter spacings may be needed to control random cracking. A temperature difference of only 6°C (10°F) may be critical. For slump less than 100 mm (4 in.), joint spacing can be increased by 20%.

** When spacings exceed 4.5 m (15 ft), load transfer by aggregate interlock decreases markedly. If shrinkage is high or unknown, joints should not exceed 4.5 m (15 ft).

Smooth steel dowels may be used to increase load transfer at contraction joints when heavy wheel loads are anticipated. Dowels may be round, square, or plate shapes. Sizes and spacing of dowels, which are placed at the center of the slab depth, are shown in Tarr and Farny (2008). See ACI 302.1R (2015) and PCA (1982a) for further discussions on doweled joints.

Sawing must be coordinated with the setting time of the concrete. It should be started as soon as the concrete has hardened sufficiently to prevent aggregates from being dislodged by the saw (usually within 4 hours to 12 hours after the concrete hardens). Sawing should be completed before drying shrinkage stresses become large enough to produce cracking. The timing depends on factors such as mix proportions, ambient conditions, and type and hardness of aggregates. Dry-cut sawing techniques allow saw cutting to take place shortly after final finishing is completed. Generally, the slab should be cut before the concrete cools, after the concrete sets enough to prevent raveling or tearing while saw cutting, and before drying-shrinkage cracks start to develop.

Edging and Hand Tooling Joints. Contraction joints can also be formed in the fresh concrete with hand groovers (Figures 15-33 and 15-34) or by placing strips of wood, metal, plastic, or preformed joint material at the joint locations. The top of the strips should be flush with the concrete surface. Contraction joints, whether sawed, grooved, or preformed, should extend vertically into the slab to a depth of at least ¼ the slab thickness or a minimum of 25 mm (1 in.) deep. It is recommended that the joint depth not exceed ⅓ the slab thickness if load transfer from aggregate interlock is important. While industrial slabs subjected to hard-wheeled traffic should be sawcut, contraction joints may be formed using a combination of either grooving or inserts and sawing to achieve the final joint depth.

Other Considerations. Contraction joints in walls are also planes of weakness that permit differential movements in the plane of the wall. The thickness of the wall at a contraction joint should be reduced by a minimum of 25%, preferably 30%. Under the guidance of the design engineer, in lightly reinforced walls, half of the horizontal steel reinforcement should be cut at the joint. Care must be taken to cut alternate reinforcement bars precisely at the joint. A gap created by two cuts in the bar can lower the risk that the cut is not precisely located at the joint. At the corners of openings in walls where contraction joints are located, extra diagonal or vertical and horizontal reinforcement should be provided to control crack width. Contraction joints in walls should be spaced not more than about 6 meters (20 ft) apart. In addition, contraction joints should be placed where abrupt changes in wall thickness or height occur, and near corners – if possible, within 3 meters to 4 meters (10 ft to 15 ft). Depending on the structure, these joints may need to be caulked to prevent the passage of water through the wall. Instead of caulking, a waterstop can be used to prevent water from leaking through the crack that occurs in the joint.

The spacing of contraction joints in slabs on ground depends on slab thickness, shrinkage potential of the concrete, subgrade friction, service environment, and the absence or presence of steel reinforcement. Unless reliable data indicate that more widely spaced joints are feasible, the suggested intervals given in Table 15-5 should be used for well-proportioned concrete with aggregates having normal shrinkage characteristics. However, the risk of random cracking is reduced if the joint spacing is held to a maximum of 4.5 m (15 ft) regardless of the slab thickness. Joint spacing should certainly be decreased for concrete that may have high shrinkage characteristics. The panels created by contraction joints should be approximately square. Panels with excessive length-to-width ratio (more than 1½ to 1) are likely to crack near the center of the long dimension of the panel. In joint layout design it is also important to remember that contraction (control) joints should only terminate at a free edge or at an isolation joint. Contraction joints should never terminate at another contraction joint (T-intersection) as cracking will be induced from the end of the terminated joint into the adjacent panel. This is sometimes referred to as sympathetic cracking.

Construction Joints

Construction joints (Figure 15-51) are stopping points during the construction process. In structural building systems, a true construction joint should bond new concrete to existing concrete and prohibit movement. Deformed tiebars are often used in construction joints to restrict movement. For slabs on ground, construction joints are designed and built to function as contraction or isolation joints. For example, in a floor on ground, the construction joints align with columns and function as contraction joints, and are purposely made unbonded. The designer of suspended slabs should determine the location

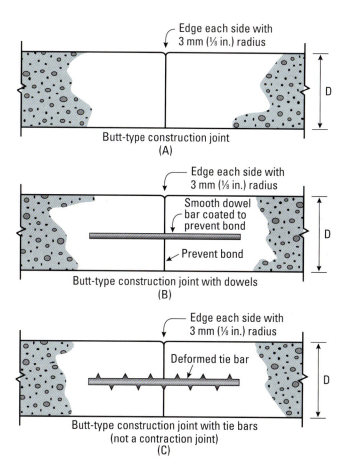

FIGURE 15-51. Construction joints are stopping points in the construction process. Construction-joint types (A) and (B) are also used as contraction joints.

of construction joints. Oils, form-release agents, and paints are used as debonding materials. In thick, heavily-loaded floors, unbonded doweled construction joints are commonly used. For thin slabs that are not loaded by heavy moving loads or subject to differential soil movement, a flat-faced butt-type joint will suffice.

On most structures it is desirable to have wall joints that will not detract from appearance. When properly made, wall joints can be inconspicuous or hidden by rustication strips. They can become an architectural as well as a functional feature of the structure. However, if rustication strips are used in structures exposed to deicing salts, such as bridge columns and abutments, care should be taken to ensure that the reinforcing steel has the required depth of concrete cover to protect from corrosion.

Horizontal Construction Joints

For making a horizontal construction joint in reinforced concrete wall construction, good results have been obtained by constructing the forms to the level of the joint, overfilling the forms a few centimeters (inches), and then removing the excess concrete just before hardening occurs; the top surface then can be manually roughened with stiff brushes. The procedure is illustrated in Figure 15-52.

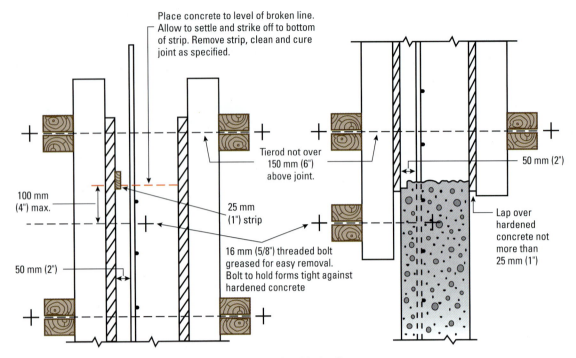

FIGURE 15-52. A straight, horizontal construction joint can be built using this detail.

In the case of vertical construction joints cast against a bulkhead, the concrete surface generally is too smooth to permit proper bonding. So, particular care should be given to removal of the smooth surface finish before re-erecting the forms for newer placements against the joint. Stiff-wire brushing may be sufficient if the concrete is less than three days old; otherwise, bushhammering or sandblasting may be needed. This should be followed by washing with clean water to remove all dust and loose particles.

Horizontal joints in walls should be made straight and horizontally placed at suitable locations. A straight horizontal construction joint can be made by nailing a 25 mm (1 in.) wide wood strip to the inside face of the form near the top (Figure 15-52). Concrete should then be placed to a level slightly above the bottom of the strip. After the concrete has settled and before it becomes too hard, any laitance that has formed on the top surface should be removed. The strip can then be removed and any irregularities in the joint leveled off. The forms are removed and then erected above the construction joint for the next lift of concrete. To prevent concrete leakage from staining the wall below, gaskets should be used where forms contact previously placed hardened concrete.

A variation of this procedure makes use of a rustication strip instead of the 25 mm (1 in.) wide wood strip to form a groove in the concrete for architectural effect (Figure 15-53). Rustication strips can be V-shaped, rectangular, or slightly beveled. If a V-shaped strip is used, the joint should be located at the point of the V. If rectangular or beveled, the joint should be made at the top edge of the inner face of the strip.

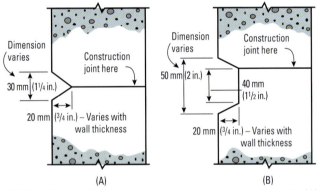

FIGURE 15-53. Horizontal construction joints in walls with (A) V-shaped and (B) beveled rustication strips.

Joint Layout for Floor Slabs

A typical joint layout for all three joint types – isolation, contraction, and construction – is illustrated in Figure 15-54. Isolation joints are provided around the perimeter of the floor where it abuts the walls and around all fixed elements that may restrain movement of the slab. This includes columns and machinery bases that penetrate the floor slab. These isolation joints will not be loaded by moving wheel loads. With the slab isolated from other building elements, the remaining task is to locate and correctly space contraction joints to eliminate random cracking. Construction joint locations are coordinated with the floor contractor to accommodate work schedules and crew size. Unbonded construction joints should coincide with the contraction joint pattern and act as contraction joints. Construction joints should be planned to provide long-strips for each placement rather than a checker-

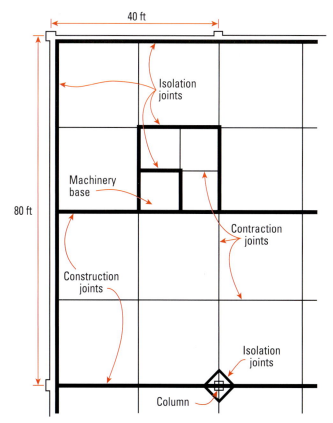

FIGURE 15-54. Typical joint layout for a 150 mm (6 in.) thick concrete floor on ground. The machinery base should be bordered with an isolation joint and load transfer devices (plates or dowels) should be provided, or the adjoining panel should be bordered with contraction joints. The interior contraction joints within the bay containing the machine base, can be eliminated if steel reinforcement is provided at the re-entrant corner to hold random cracks tightly closed.

board pattern. Contraction joints are then placed to divide the long-strips into relatively square panels, with panel length not exceeding 1.5 times the width. Contraction joints should stop at free edges or isolation joints. Avoid reentrant corners in joint layout whenever possible.

For more information on joints, see ACI 302.1R (2015), PCA (1982a), and Tarr and Farny (2008). For joints in walls, see PCA (1982a), PCA (1982b), PCA (1984), PCA (1984a), and PCA (1984b).

Treating Joints

There are three options for treating joints: they can be filled, sealed, or left open. The movement at contraction joints in a floor is generally very small. For light traffic floor slabs and some industrial and commercial uses, these joints can be left unfilled or unsealed. Where there are wet conditions or hygienic and dust-control requirements, joints should be sealed to prevent moisture penetration through joints or infiltration of dirt and debris. Joint sealants are flexible and can accommodate substantial joint movement. Exterior pavement joints are often sealed because they are only loaded by pneumatic tires. This type of loading imparts relatively low stresses that typically

do not cause spalling. However, when traffic by small, hard-wheeled vehicles will cross joints, such as for interior industrial slabs, filling is necessary. Joint fillers are semi-rigid so they can provide some compressive lateral support to the joint walls which protects the joints from spalling.

The difference between a filler and a sealant is the hardness of the material. Fillers are more rigid than sealants and provide support to joint edges. In many places where traffic loading is light, a resilient material such as a polyurethane elastomeric sealant is satisfactory. However, heavy-traffic areas require support for joint edges to prevent spalling at saw-cuts. In such cases, a good quality, semi-rigid epoxy or polyurea filler with a Shore Hardness of A-80 or D-50 (ASTM D2240, *Standard Test Method for Rubber Property – Durometer Hardness*) should be used. The material should be installed full depth in the saw cut, without a backer rod, and trimmed flush with the floor surface.

Isolation joints are intended to accommodate movement; thus a flexible, elastomeric sealant should be used to keep foreign materials out of the joint.

Unjointed Floors

An unjointed floor, or one with a limited number of joints, can be constructed when joints are unacceptable. Three unjointed floor methods are suggested:

Post-tensioning. A prestressed floor can be built using post-tensioning. With this method, steel strands in ducts are tensioned after the concrete hardens to produce compressive stress in the concrete. This compressive stress will counteract the development of tensile stresses in the concrete and provide a crack free floor. Large areas without intermediate joints, 1000 m² (10,000 ft²) and greater, can be constructed in this manner.

Continuous Reinforcement. Continuous reinforcement is often used in concrete pavements. At reinforcement percentages ranging from 0.5% to 0.7% of the slab cross-sectional area, the amount of steel is about 5 to 7 times greater than that used in conventionally reinforced slabs. Most floors that contain continuous reinforcement are designed with a minimum of 0.5% steel.

The high percentage of reinforcing steel does not prevent cracking. Instead, closely spaced, narrow cracks form. Crack spacings range from about 0.6 m to 1.2 m (2 ft to 4 ft). Because the cracks are narrow, load transfer occurs by aggregate interlock and the reinforcing steel which reduces the impact of curling and warping.

Special effort should be made to reduce subgrade friction in floors without contraction joints. Tarr and Farny (2008) discuss use of distributed steel in floors.

Shrinkage-Compensating Concrete (SCC). Concrete made with expansive cement and or shrinkage reducing admixtures can be used to offset the amount of drying shrinkage anticipated after curing. Contraction joints are not needed when construction joints are used at intervals of 10 m to 35 m (40 ft to 120 ft). Large areas without joints, up to 2000 m² (20,000 ft²), have been cast in this manner.

After concrete is placed, and initial set occurs, a bond develops between the concrete and the reinforcing steel and the shrinkage-compensating component in the concrete undergoes a slight expansion– this is a quasi-form of prestressing. When the concrete subsequently dries and shrinks, the tensile forces in the steel are diminished as the concrete remains in slight compression. When properly designed, SCC expands about 0.03 to 0.07%, which is comparable to conventional concrete, thus compensating for drying shrinkage.

The benefits of SCC are twofold:

- it eliminates cracking and the need for joints, and
- it eliminates warping potential thus improving long-term flatness.

See ACI 223, *Standard Practice for the Use of Shrinkage-Compensating Concrete* (2010), for more information on shrinkage compensating concretes.

CURING AND PROTECTION

All newly placed and finished concrete slabs should be cured and protected from drying, from extreme changes in temperature, and from damage by subsequent construction activities and traffic (see Chapter 17).

Initial curing should begin immediately after strike-off operations and continue until finishing operations are complete (Figure 15-55). Final curing is needed to ensure continued hydration of the cement, assure proper strength gain and durability of the concrete, and to minimize early drying shrinkage.

Special precautions are necessary when concrete work continues during periods of adverse weather. In cold weather, arrangements should be made in advance for heating, covering,

FIGURE 15-55. An excellent method of wet curing is to completely cover the surface with wet burlap and keep it continuously wet during the curing period.

insulating, or enclosing the concrete (see Chapter 19). Hot-weather work may require special precautions against rapid evaporation and drying and high temperatures (see Chapter 18).

Rain Protection

Prior to commencing placement of concrete, the owner and contractor should be aware of procedures to be followed in the event of rain during the placing operation. Protective coverings such as polyethylene sheets or tarpaulins should be available and onsite at all times.

When rain occurs, if practical, all batching and placing operations should stop and the fresh concrete should be covered to the extent that the rain does not indent the surface of the concrete or wash away the cement paste. When rain ceases, the covering should be removed and remedial measures taken such as surface retexturing or reworking in-place plastic concrete, before concrete placing resumes. For more information on protecting concrete surfaces from rain refer to Kozikowski and others (2019).

REMOVING FORMS

It is advantageous to leave forms in place as long as possible to continue the curing period. However, at times it may be necessary to remove forms as soon as possible. It is often necessary to remove forms quickly to permit their immediate reuse. Vertical forms are typically removed at the start of the next day provided that concrete has met minimum strength requirements. There are exceptions, such as curb or mechanical pad form faces where the forms are removed earlier so that formed surfaces can be finished while the concrete is still green.

Beam and floor slab forms and supports (shoring) should never be removed until the concrete is strong enough to satisfactorily carry the dead load of the structure and any imposed construction loads. The concrete should be strong enough so that the surfaces will not be damaged when reasonable care is used in removing forms. In general, for concrete temperatures above 10°C (50°F), the side forms of reasonably thick, supported sections can usually be removed 24 hours after placing. Shoring may be removed between 3 days and 21 days, depending on the size of the member and the strength gain of the concrete. For most conditions, it is better to rely on the strength of the concrete rather than arbitrarily selecting an age at which forms may be removed. Advice on reshoring is provided by ACI 347R (2014).

For form removal, the designer should specify the minimum strength requirements for various members. The age-strength relationship should be determined from representative samples of concrete used in the structure and field-cured samples under job conditions. However, it should be remembered that strengths are affected by the materials used, concrete temperature, curing temperature and other conditions. The time required for form removal, therefore, will vary from job to job.

Stripping should be started some distance away from, and then move towards, a projection. This relieves pressure against projecting corners and reduces the chance of edges breaking off. If it is necessary to wedge between the concrete and the form, only wooden wedges should be used. A pinch bar or other metal tool should not be placed against the concrete to pry away formwork.

Recessed forms require special attention. Wooden wedges should be gradually driven behind the form and the form should be tapped lightly to break it away from the concrete. Forms should not be pulled off rapidly after wedging has been started at one end; this is almost certain to break the edges of the concrete.

REFERENCES

PCA's online catalog includes links to PDF versions of many of our research reports and other classic publications.
Visit: cement.org/library/catalog.

ACI Committee 117, *Specifications for Tolerances for Concrete Construction and Materials*, ACI 117-10, American Concrete Institute, Farmington Hills, MI, 2010, reapproved 2015, 76 pages.

ACI Committee 223, *Standard Practice for the Use of Shrinkage-Compensating Concrete*, ACI 223-10, American Concrete Institute, Farmington Hills, MI, 2010.

ACI Committee 237, *Self-Consolidating Concrete*, ACI 237R-07, reapproved 2019, American Concrete Institute, Farmington Hills, MI, 2007, 30 pages.

ACI Committee 301, *Specification for Concrete Construction*, ACI 301-20, American Concrete Institute, Farmington Hills, MI, 2016.

ACI Committee 302, *Guide for Concrete Floor and Slab Construction*, ACI302.1R-15, American Concrete Institute, Farmington Hills, MI, 2015.

ACI Committee 302, *Guide for Concrete Slabs that Receive Moisture-Sensitive Flooring Materials*, ACI302.2R-06, American Concrete Institute, Farmington Hills, MI, 2006.

ACI Committee 303, *Guide to Cast-in-Place Architectural Concrete Practice*, ACI 303R-12, American Concrete Institute, Farmington Hills, MI, 2012.

ACI Committee 304, *Guide for Measuring, Mixing, Transporting, and Placing Concrete*, ACI 304R-00, American Concrete Institute, Farmington Hills, MI, 2000, 41 pages.

ACI Committee 306, *Cold-Weather Concreting*, ACI 306R-16, American Concrete Institute, Farmington Hills, MI, 2016.

ACI Committee 309, *Guide for Consolidation of Concrete*, ACI 309R-05, American Concrete Institute, Farmington Hills, MI, 2005.

ACI Committee 318, *Building Code Requirements for Structural Concrete and Commentary*, ACI 318-19, American Concrete Institute, Farmington Hills, MI, 2019.

ACI Committee 347, *Guide to Formwork for Concrete*, ACI 347R-14, American Concrete Institute, Farmington Hills, MI, 2014.

ACI Committee 506, *Guide to Fiber-Reinforced Shotcrete*, ACI 506.1R-08, American Concrete Institute, Farmington Hills, MI, 2008, 12 pages.

ACI Committee 506, *Guide to Shotcrete*, ACI 506R-16, American Concrete Institute, Farmington Hills, MI, 2016.

Alfes, C., "Fresh Concrete Pressure of Highly Flowable Concrete and Self-Compacting Concrete in Element Walls," *Betonwerk und Fertigteil-Technik/Concrete Plant and Precast Technology*, Vol. 70, No. 11, November 2004, pages 6 to 17.

ASCE, *Standard Practice for Shotcrete*, American Society of Civil Engineers, New York, NY, 1995, 65 pages.

Assaad, J.J.; and Khayat, K.H., "Effect of Casting Rate and Concrete Temperature on Formwork Pressure of Self-Consolidating Concrete," *Materials and Structures/Materiaux et Constructions*, Vol. 39, No. 3, April 2006, pages 333 to 341.

Assaad, J.J.; Khayat, K.H.; and Mesbah, H., "Variation of Formwork Pressure with Thixotropy of Self-Consolidating Concrete," *ACI Materials Journal*, Vol. 100, No. 1, January-February 2003, pages 29 to 37.

Austin, S.; and Robins, P., *Sprayed Concrete: Properties, Design and Application*, Whittles Publishing, U.K., 1995, 382 pages.

Balck, L.; Gebler, S.; Isaak, M.; and Seabrook, P., *Shotcrete for the Craftsman* (CCS-4), American Concrete Institute, Farmington Hills, MI, 2008, 85 pages.

Bimel, C., "Is Delamination Really a Mystery?," *Concrete International*, American Concrete Institute, Farmington Hills, MI, Vol. 20, No. 1, January 1998, page 29.

Birch, B.F.; Lange, D.; and Khayat, K.H., "Formwork Pressure," *Self-Consolidating Concrete: A White Paper by Researchers at The Center of Advanced Cement Based Materials*, Advanced Cement-Based Materials Center, Evanston, IL, 2007, pages 34 to 42.

Brameshuber, W.; and Uebachs, S., "Investigations on the Formwork Pressure Using Self-Compacting Concrete," *Self-Compacting Concrete, Proceedings of the 3rd International RILEM Symposium*, PRO33, RILEM Publications, Bagneux, France, 2003, pages 281 to 287.

Bureau of Reclamation, *Concrete Manual*, 8th ed., U.S. Bureau of Reclamation, Denver, CO, revised 1981, 661 pages.

Bury, M.A.; and Buehler, E., "Methods and Techniques for Placing Self-Consolidating Concrete—An Overview of Field Experiences in North American Applications," *Conference Proceedings: First North American Conference on the Design and Use of Self-Consolidating Concrete*, Advanced Cement-Based Materials Center, Evanston, IL, 2002, pages 281 to 286.

Collins, T.C.; Panarese, W.C.; and Bradley, B.J., *Concrete Finisher's Guide*, EB122, Portland Cement Association, 2006, 88 pages.

Daczko, J.A.; and Vachon, M., "Self-Consolidating Concrete (SCC)," *Significance of Tests and Properties of Concrete and Concrete-Making Materials*, STP169D, ASTM International, West Conshohocken, PA, 2006, pages 637 to 645.

FHWA, *Concrete Inspector's Checklist*, Federal Highway Administration, Washington, DC, 2017, 2 pages.

FHWA, *Inspection-in-Depth: Major Structures–Concrete Plant and Placement*, Federal Highway Administration, Washington, DC, 1999.

FHWA, *Tremie Concrete for Bridge Piers and Other Massive Underwater Placements*, Federal Highway Administration, Washington, DC, 1981.

Gebler, S.H.; Litvin, A.; McLean, W.J.; and Schutz, R., "Durability of Dry-Mix Shotcrete Containing Rapid-Set Accelerators," *ACI Materials Journal*, May-June 1992, pages 259 to 262.

Greening, N.R.; and Landgren, R., *Surface Discoloration of Concrete Flatwork*, RX023, Research Department Bulletin No. 203, Portland Cement Association, Skokie, IL, 1966, 24 pages.

Gregori, A.; Ferron, R. P.; Sun, Z.; and Shah, S. P., "Experimental Simulation of Self-Consolidating Concrete Formwork Pressure," *ACI Materials Journal*, Vol. 105, No. 1, January-February 2008, pages 97 to 104.

Harrington, D., *Guide to Concrete Overlays*, Third Edition, ACPA TB021.03P, National Concrete Pavement Technology Center, Ames, IA, 2014, 145 pages.

Harris, B., *Guide to Stamped Concrete*, Craftsman Book Company, Carlsbad, CA, November 2004, 141 pages.

Hedenblad, G., *Drying of Construction Water in Concrete-Drying Times and Moisture Measurement*, Swedish Council for Building Research, Stockholm, 1997, 54 pages.

Hedenblad, G., "Concrete Drying Time," *Concrete Technology Today*, PL982, Portland Cement Association, 1998, pages 4 and 5.

Hover, K.C., "Vibration Tune-up," *Concrete International*, American Concrete Institute, Farmington Hills, MI, September 2001, pages 31 to 35.

ICRI, *Selecting and Specifying Concrete Surface Preparation for Sealers, Coatings, and Polymer Overlays*, Guideline No. 03732, International Concrete Repair Institute, St. Paul, MN, 1997.

Johnston, D.W., *Formwork for Concrete*, ACI SP-4, 8th Edition, American Concrete Institute, Farmington Hills, MI, 2014.

Kanare, H.M., *Concrete Floors and Moisture*, EB119, Portland Cement Association, Skokie, IL, and National Ready Mixed Concrete Association, Silver Spring, MD, 2008, 176 pages.

Kosmatka, S.H., "Bleed Water," *Significance of Tests and Properties of Concrete and Concrete-Making Materials*, STP 169D, ASTM International, West Conshohocken, PA, 2006, pages 99 to 122.

Kosmatka, S.H.; and Collins, T.C., *Finishing Concrete with Color and Texture*, PA124, Portland Cement Association, Skokie, IL, 2004, 72 pages.

Kozikowski, R.L.; Tarr, S.M.; Rowswell, K.; and Suprenant, B.A., "Concrete Placements Exposed to Rain," *Concrete International*, Vol 41, No. 8, August 2019, pages 41 to 47.

Leemann, A.; Hoffmann, C.; and Winnefeld, F., "Pressure of Self-Consolidating Concrete on Formwork," *Concrete International*, Vol. 28, No. 2, February 2006, pages 27 to 31.

Litvin, A, *Clear Coatings for Exposed Architectural Concrete*, DX137, Portland Cement Association, Skokie, IL, 1968, 11 pages.

Mustard, J.N.; and Ghosh, R.S., "Minimum Protection and Thermal Stresses in Winter Concreting," *Concrete International*, V. 1, No. 1, January 1979, pages 96 to 101.

National Institute for Occupational Safety and Health, cdc.gov/niosh/topics/hierarchy, 2015.

NRMCA/ASCC, *Checklist for the Concrete Pre-Construction Conference*, National Ready-Mixed Concrete Association, Silver Spring, MD, and American Society of Concrete Contractors, St. Louis, MO, 2005, 18 pages.

NRMCA/ASCC, *Checklist for Concrete Producer-Concrete Contractor Fresh Concrete Performance Expectations*, National Ready-Mixed Concrete Association, Silver Spring, MD, and American Society of Concrete Contractors, St. Louis, MO, 2012, 4 pages.

NRMCA/ASCC, *Checklist for Ordering and Scheduling Ready Mixed Concrete*, National Ready-Mixed Concrete Association, Silver Spring, MD, and American Society of Concrete Contractors, St. Louis, MO, 2007a.

NRMCA/ASCC, *Checklist for Pumping Ready Mixed Concrete*, National Ready-Mixed Concrete Association, Silver Spring, MD, and American Society of Concrete Contractors, St. Louis, MO, 2007b.

OSHA, OSHA's *Respirable Crystalline Silica Standard for Construction*, OSHA Fact Sheet, DSG FS-3681, Occupational Safety and Health Administration, Washington, DC, December 2017, 2 pages.

PCA, *Building Movements and Joints*, EB086, Portland Cement Association, Skokie, IL, 1982a, 68 pages.

PCA, *Bushhammering of Concrete Surfaces*, IS051, Portland Cement Association, Skokie, IL, 1972.

PCA, *Color and Texture in Architectural Concrete*, SP021, Portland Cement Association, Skokie, IL, 1995, 36 pages.

PCA, *Concrete Slab Surface Defects: Causes, Prevention, Repair*, IS177, Portland Cement Association, Skokie, IL, 2001, 12 pages.

PCA, *Joints in Walls Below Ground*, CR059, Portland Cement Association, Skokie, IL, 1982b.

PCA, "Joints to Control Cracking in Walls," *Concrete Technology Today*, PL843, Portland Cement Association, Skokie, IL, September 1984, pages 4 and 5.

PCA, *Painting Concrete*, IS134, Portland Cement Association, Skokie, IL, 1992, 8 pages.

PCA, *Removing Stains and Cleaning Concrete Surfaces*, IS214, Portland Cement Association, Skokie, IL, 1988, 16 pages.

PCA, "Sealants for Joints in Walls," *Concrete Technology Today*, PL844, Portland Cement Association, Skokie, IL, December 1984a, pages 4 and 5.

PCA, "Why Concrete Walls Crack," *Concrete Technology Today*, PL842, Portland Cement Association, Skokie, IL, June 1984b, page 4.

PCA, *Working Safely with Concrete*, MS271, Portland Cement Association, Skokie, IL, 1998, 6 pages.

PCI, *Interim Guidelines for the Use of Self-Consolidating Concrete in Precast/Prestressed Concrete Institute Member Plants*, TR-6-03, Precast/Prestressed Concrete Institute, 1st Edition, Chicago, IL, 2003, 165 pages.

Stark, David C., *Effect of Vibration on the Air-Void System and Freeze-Thaw Durability of Concrete*, RD092, Portland Cement Association, Skokie, IL, 1986, 13 pages.

Suprenant, B.A., "Free Fall of Concrete," *Concrete International*, American Concrete Institute, Farmington Hills, MI, June 2001, pages 44 and 45.

Szecsy, R.; and Mohler, N., *Self-Consolidating Concrete*, IS546, Portland Cement Association, Skokie, IL, 2009, 24 pages.

Tarr, S. M.; and Farny, J. A., *Concrete Floors on Ground*, EB075, Portland Cement Association, Skokie, IL, 2008, 256 pages.

Tejeda-Dominguez, F.,D. A. Lange; and D'Ambrosia, M.D., "Formwork Pressure of Self-Consolidating Concrete in Tall Wall Field Applications," *Concrete Materials 2005*, Transportation Research Record No. 1914, Transportation Research Board, Washington, D.C., 2005, pages 1 to 7.

Whiting, D.; Seegebrecht, G.W.; and Tayabji, S. "Effect of Degree of Consolidation on Some Important Properties of Concrete," *American Concrete Institute*, Special Publication Volume 96, January 1, 1987, pages 125 to 160.

IMPERFECTIONS IN CONCRETE

FIGURE 16-1. Field mock-ups can serve as reference samples for concrete acceptance (courtesy of Baker Concrete).

It is to the credit of concrete that so few complaints are received on the vast amount of construction put in place. Great care is required throughout the entire construction process to ensure that the hardened concrete achieves its desired properties. Blemishes that appear on the surface of concrete may include: cracking, dusting, blisters, delaminations, crazing, popouts, scaling, spalling, bugholes, cold joints, honeycombing, discoloration, and/or efflorescence.

These issues, caused by specific factors that can be minimized or prevented by adhering to proper construction methods and use of clean, sound, durable materials. When troubleshooting concrete problems, it is important to relate the symptom to causes of distress and deterioration.

Differentiating whether a blemish is to be considered a defect in the concrete is essential to determining a remedy. Specification requirements, and consideration for future serviceability, durability, and aesthetics typically govern. Repairs of imperfections are not always required or necessary.

ACI 301, *Specifications for Concrete Construction* (2020), specifies surface requirements for cast-in-place concrete and finished slabs. Special finishes, including architectural concrete, may also be specified in accordance with ACI 301-20. If architectural concrete is specified, finish sample(s) exhibiting the desired surfaces, color, and texture and field mock-ups are required. An accepted field mockup will serve as the reference to which architectural concrete will be compared (Figure 16-1). Field mockups should also provide a simulated repair area to demonstrate an acceptable repair procedure with an acceptable color and texture.

CRACKING

Unexpected cracking of concrete is a frequent cause of complaints. Cracking can be the result of one or a combination of factors, such as drying shrinkage, thermal contraction, restraint (external or internal) from shortening, subgrade settlement, and applied loads. The following sections describe the most common types of non-structural cracks found in concrete structures.

Plastic Shrinkage Cracking

Plastic-shrinkage cracks are relatively short cracks that may occur before final finishing on days when wind, low humidity, and high temperatures occur (Figure 16-2). Plastic-shrinkage cracks have varying lengths, spaced from a few centimeters (inches) up to 3 m (10 ft) apart, and can penetrate to full depth of a slab.

FIGURE 16-2. Plastic shrinkage cracks are relatively short cracks that resemble tears on the surface of concrete.

When surface moisture evaporates faster than it can be replaced by rising bleed water, it causes the surface to shrink more than the interior concrete. As the interior concrete restrains shrinkage of the surface concrete, stresses develop that exceed the concrete's tensile strength, resulting in surface cracks.

Protective measures for preventing plastic shrinkage cracking includes initial curing procedures that prevent the surface from rapidly drying. Fogging the air above the concrete and erecting sun and windshades lessen the risk of plastic-shrinkage cracking (see Chapter 17).

Drying Shrinkage Cracking

Concrete contracts with moisture loss. This is known as drying shrinkage and may cause cracking (Figure 16-3). Typical concrete shrinkage has been measured at 400 to 800 millionths. A millionth is typical measure of shrinkage in concrete and is in terms of a millionth of a unit length per length of concrete. For example, shrinkage of 500 millionths in a concrete slab 3 meters (10 ft) long results in a change in length of 0.0015 m or 1.5 mm (0.005 ft or 0.06 in.). However, for some mixtures, shrinkage exceeding 1100 millionths has been documented (Tarr and Farny 2008). Cracking may occur as a result of normal

FIGURE 16-3. Typical drying shrinkage cracking.

concrete shrinkage combined with poor jointing practices. It may also be due to higher than normal shrinkage, lack of isolation joints, or improper subgrade support.

The major factor influencing the drying-shrinkage properties of concrete is the total water content of the concrete. As the water content increases, the amount of shrinkage increases proportionally. Large increases in sand content and significant reductions in size and amount of coarse aggregate increase shrinkage because total water is increased. Smaller size coarse aggregates provide less internal restraint to shrinkage. Use of aggregates exhibiting high-shrinkage (see Chapter 10) and some accelerating admixtures, including calcium chloride also increases total concrete shrinkage.

To accommodate this shrinkage and control the location of cracks, joints are placed at regular intervals. Experience has shown that contraction joints (induced cracks) should be spaced at about 24 to 30 times the thickness of the concrete. For example, a 200-mm (8-in) thick slab requires a spacing of between 4.8 m (16 ft) and 6.0 m (20 ft). To ensure activation of joints, joints should be cut to a minimum depth of 1/4 thickness of concrete. For more information on joints and volume change see Chapters 10 and 15.

Expecting a concrete structure to be completely free of random cracking is unreasonable; it is generally accepted that up to 3% of the panels formed by joints in unreinforced slabs will crack. Steel reinforcement can be installed at the 1/3 depth of the slab to hold random cracks that do occur tightly closed and to prevent further deterioration. Use of the appropriate amount of reinforcement (see Chapter 8) or synthetic macrofibers (see Chapter 7) can also help reduce and mitigate formation and propagation of drying shrinkage cracks. A typical amount of drying shrinkage might be about 200 to 300 millionths (0.02% to 0.03%) for concrete that contains an average amount of reinforcement. However, extending the reinforcing steel through sawcut control joints increases the risk of random cracking and may result in a far greater percentage of cracked panels without further design considerations (Figure 16-4).

FIGURE 16-4. Random cracking of a slab.

Thermal Cracking

Concrete expands when heated and contracts when cooled. Concrete has a coefficient of thermal expansion and contraction of about 10×10^{-6} per °C (5.5×10^{-6} per °F). Concrete placed during hot temperatures will contract as it cools during the night. A 22°C (40°F) drop in temperature between day and night – not uncommon in some areas – would cause about 0.7 mm (0.03 in.) of contraction in a 3-m (10-ft) length of concrete, sufficient to cause cracking if the concrete is restrained (Figure 16-5).

FIGURE 16-5. Cracking of a sidewalk due to thermal expansion of adjacent parking lot abutting the sidewalk without expansion joints.

Thermal expansion and contraction of concrete varies with factors such as aggregate type, cement content, water-cement ratio, temperature changes, concrete age, and relative humidity. Of these, aggregate type has the greatest influence.

Designers should give special consideration to structures in which some portions of the structure are exposed to temperature changes, while other portions are partially or completely protected. Allowing for movement by using properly designed expansion or isolation joints and correct detailing will help minimize the effects of temperature variations.

Settlement Cracking

Settlement cracks may develop in fresh concrete over embedded items, such as reinforcing steel, or adjacent to forms or hardened concrete as the concrete settles, or subsides (Figure 16-6). Settlement cracking results from insufficient consolidation (vibration), overly wet concrete (high water cementitious materials ratios), or a lack of adequate cover over embedded items. Concrete made with air entrainment, sufficient fine materials, and low water contents can help minimize subsidence cracking. Revibration while the concrete is still plastic may eliminate these cracks. Also, fibers have been reported to reduce subsidence cracking (Suprenant and Malisch 1999).

FIGURE 16-6. Subsidence cracking over reinforcement.

Other Types of Cracking

Insufficiently compacted subgrades and soils susceptible to frost heave or swelling can produce cracks in concrete. Overloading of concrete also results in flexural crack formation and possible failure.

Cracks can also be caused by freezing and thawing of saturated concrete, alkali-aggregate reactivity, sulfate attack, or corrosion of reinforcing steel. However, cracks from these sources may not appear for years.

Prevention of Cracking

Cracking in concrete can be reduced significantly or eliminated by observing the following practices:

- Use proper subgrade preparation, including uniform support and proper subbase material at adequate moisture content.

- Minimize the mixing water content by maximizing the size and amount of coarse aggregate and use a low-shrinkage, well-graded aggregate.

- Use the lowest amount of mixing water required for workability. Consider the use of water reducers if higher workability is needed.

- Avoid calcium chloride admixtures.

- Prevent rapid loss of surface moisture during initial finishing operations through use of fogging, spray-applied evaporation reducers, plastic sheets, and windbreaks to avoid plastic-shrinkage cracks.

- Provide contraction joints at recommended intervals, spaced at 24 to 30 times the slab thickness.

- Provide isolation joints to prevent restraint from adjoining elements of a structure.

- Protect concrete from extreme changes in temperature.

- If concrete is to be placed directly on a vapor barrier, use a mixture with a low overall water content.

- Properly place, consolidate, finish, and cure the concrete.

- Avoid using excessive amounts of cementitious materials.

- Consider using a shrinkage-reducing admixture to reduce drying shrinkage.

- Consider using fibers to help control plastic shrinkage cracks.

Refer to Chapter 10 for more recommendations on how to reduce shrinkage, control volume change, and related cracking in concrete. Proper mixture design and selection of suitable concrete materials can significantly reduce or eliminate the formation of cracks and deterioration related to freezing and thawing, alkali-aggregate reactivity, sulfate attack, and steel corrosion.

Repair of Cracking

The activity (dormant or moving, static or dynamic), moisture condition, size, and location of cracking will dictate whether repair is needed and what method to use. Tightly closed cracks and fine cracks subject only to light industrial traffic should be judiciously neglected, but kept under observation. They usually do not affect serviceability.

As concrete contracts (shrinks), cracks can open. If cracks open up or show signs of spalling, they should be repaired. Spalling occurs at widened (opened) cracks subjected to heavy loading or expansive forces. Unsupported edges at the crack cannot support loads, so they break off, or spall. If not repaired, the spall becomes wider and deeper. Thus, what is initially a relatively simple repair of routing and sealing can develop into a more complex and costly partial depth repair.

Inactive crack/joint repair. For crack repairs where there is no movement, the spalled concrete is removed by routing out the crack or joint with a saw and securing the two sections back together. Preparing concrete by routing includes removing all unsound concrete, dust, and debris. A structural epoxy can be used to bond the two crack faces or an epoxy concrete repair mortar can be installed in the routed-out space. If required by the epoxy mortar supplier, a primer is applied to the concrete repair surface prior to epoxy mortar placement.

Active crack/joint repair. If slab end movements are to be accommodated at the joint or crack, the repair is done with a semi-rigid epoxy. This will provide support to the vertical edges, but still allow expansion or contraction movements without damaging the repaired area. Routing in preparation to installation of the semi-rigid epoxy should be to a depth of approximately 25 to 40 mm (1 to 1½ in.) and width of 10 mm (⅜ in.).

Partial- and full-depth repairs. Deeper and wider spalls require deeper and wider repairs. In every case, the repair must extend to sound concrete. If depth to sound concrete is greater than one-half of the slab thickness, full-depth repairs must be made.

For more information on repairing cracks in concrete refer to ACI 224.1R, *Causes, Evaluation, and Repair of Cracks in Concrete Structures* (2007).

DUSTING

Dusting – the development of a fine, powdery material that easily rubs off the surface of hardened concrete – is the result of a thin, weak layer, called laitance, composed of cementitious materials and fine particles (Figure 16-7).

FIGURE 16-7. Dusting is the result of a thin, weak layer, called laitance on the surface of concrete.

Causes of Dusting

Fresh concrete is a cohesive mass, with the aggregates, cementitious materials and water uniformly distributed throughout. A certain amount of time must elapse before the cementing materials and water react sufficiently to develop hardened concrete. During this period, the cement and aggregate particles are partly suspended in the water. Because the cement, SCMs, and aggregates are more dense than water, they tend to sink. As they move downward, the displaced water moves upward and appears at the surface as bleed water.

The bleeding process on its own is not harmful to concrete, unless excessive. The surface bleed water should evaporate prior to proper completion of final finishing operations. When concrete mixtures have a higher water cement ratio or other mixture components that contribute to higher bleeding rates and capacities, the bleed water may not evaporate on its own and a layer of laitance may form. The laitance – the weakest, most permeable, and least wear-resistant concrete – is at the top surface, exactly where the strongest, most impermeable, and most wear-resistant concrete is needed.

Floating and troweling concrete with bleed water on it mixes the excess water back into the surface, elevating the surface w/cm, further weakening the concrete's strength and wear resistance and giving rise to dusting. Dusting may also be caused by:

- water applied during finishing,
- exposure to rainfall during finishing,
- spreading dry cement over the concrete surface,
- a low cementitious materials content,
- a high water content (high water-to-cementitious materials ratio),
- lack of proper curing (especially allowing rapid drying of the surface),
- carbonation during winter concreting (caused by unvented heaters),
- freezing of the surface, and
- dirty aggregate.

Prevention of Dusting

Take precautions to avoid the situations listed above. Protect the fresh surface of concrete from rain. Excessively high water contents in the concrete mixture should also be avoided. Heated enclosures are commonly used for protecting concrete when air temperatures are near or below freezing. The enclosures frequently are heated by oil- or propane-fired blowers or coke-burning salamanders (see Chapter 19). As these heaters produce a dry heat, care must be taken to prevent rapid drying

of the concrete surface, especially near the heater. The burning of fuel also produces carbon dioxide that will combine with the calcium hydroxide in fresh concrete to form a weak layer of calcium carbonate on the surface. When this occurs, the concrete surface will dust in response to abrasion. Carbon dioxide-producing heaters should not be used while placing and finishing concrete and during the first 24 to 36 hours of the curing period unless properly vented to outside the heated enclosure.

Remedy for Dusting

One way to correct a dusting surface is to grind off the thin layer of laitance to expose the solid concrete underneath. Another possible method is to apply a surface hardener. This treatment will not convert a poor concrete slab surface into a good one; it may improve wearability and reduce dusting of the surface.

The major ingredient in many floor-surface hardeners is sodium silicate (water glass) or a metallic silicofluoride (magnesium and zinc fluosilicates are also widely used). The application should follow manufacturer's instructions, and the treatment is usually applied in two or three coats, letting the surface dry between each application. More information on surface hardeners is available in PCA EB075, *Concrete Floors on Ground* (Tarr and Farny 2008).

BLISTERS

Blisters (Figure 16-8) of varying size, may appear on the surface of a concrete slab during finishing operations. These bumps appear when bubbles of entrapped air or water rising through the plastic concrete get trapped under an already sealed surface.

Causes of Blisters

Experienced concrete finishers attribute blistering to three principal causes:

- an excess amount of entrapped air held within the concrete,
- insufficient vibration during compaction, and
- finishing when the concrete is still spongy.

FIGURE 16-8. (left) Blisters are surface bumps that may range in size from 5 mm to 100 mm (¼ in. to 4 in.) in diameter with a depth of about 3 mm (⅛ in.) (courtesy of NRMCA). (right) A cross section of concrete showing a void trapped under a blister.

A high percentage of material passing the 600 µm, 300 µm, and 150 µm (No. 30, 50, and 100) sieves, may result in a sticky or tacky concrete that can become more easily sealed when floating or finishing at any early age. Sticky mixes have a tendency to crust under drying winds while the remainder of the concrete remains plastic and the entrapped air inside rises to the surface. Usually, all that is needed to relieve this condition is to reduce the amount of sand in the mix. A reduction of 60 kg/m³ to 120 kg/m³ of sand (100 lb/yd³ to 200 lb/yd³) may be enough. This is done by replacing the sand removed with a like amount of the smallest size coarse aggregate available. The slightly harsher mix should release most of the entrapped air with normal vibration. On days when surface crusting occurs, slightly different finishing techniques may be needed, such as the use of wood floats to keep the surface open and flat troweling to avoid enfolding air into the surface under the blade of a trowel.

Insufficient vibration during compaction may not adequately release entrapped air. Overuse of vibration methods can leave the surface with excessive fines, inviting crusting and early finishing.

Any tool used to compact or finish the surface will tend to force the entrapped air toward the surface. Blisters may not appear after the first finishing pass. However, as the work progresses (during the second or third pass), the front edge of the trowel blade is lifted to increase the surface density, and air under the surface skin is forced ahead of the blade until enough is concentrated (usually near a piece of large aggregate) to form blisters.

The use of SCMs or fibers may also contribute to blistering because they change the bleed rate and cohesive characteristics of the concrete mixture. The tendency may be to close (seal) the surface prior to completion of bleeding when using these materials. Blisters, which may be full of water and/or air, also can appear at any time and without apparent cause. Floating the concrete a second time helps to reduce blistering. Delayed troweling will depress the blisters even though it may not reestablish complete bond.

Prevention of Blisters

To avoid blisters, the following should be considered:

- Do not use concrete with a high water content, excessively high air content, or excess fines.
- Use cement contents in the range of 280 kg/m³ to 360 kg/m³ (470 lb/yd³ to 610 lb/yd³).
- Warm the subgrade before placing concrete on it during cold weather.
- Avoid overworking the concrete, especially with vibrating screeds, jitterbugs, or bullfloats. Overworking causes

aggregate to settle and bleed water and excess fines to rise. Properly vibrate to release entrapped air.

- Do not attempt to seal (finish) the surface too soon. Use a wood bullfloat on non-air-entrained concrete to avoid early sealing. Magnesium or aluminum tools should be used on air-entrained concrete.
- Use proper finishing techniques and proper timing during and between finishing operations. Flat floating and flat troweling are often recommended. Hand floating should be started when a person standing on a slab makes a 5-mm (¼-in.) imprint or about a 3-mm (⅛-in.) imprint for machine floating. If moisture is deficient, a magnesium float should be used. Proper lighting is also very important during finishing operations.
- Reduce evaporation over the slab by using a fog spray or slab cover.
- Avoid using air contents over 3% for hard-troweled slabs.
- Caution should be used placing a slab directly on polyethylene film or other vapor barriers.

Remedy for Blisters

A small patch can be made by drilling out the blistered surface and filling the void with a dry-pack mortar or other appropriate patch material. If the blistering is wide-spread, the recommended fix is grinding and overlaying with a new surface. Patching of blisters will bring more attention to their location.

DELAMINATION

Delaminations are a separation along a plane parallel to a concrete surface. The delaminated mortar thickness typically ranges from about 3 mm to 9 mm (⅛ in. to ⅜ in.) (Figure 16-9). The affected area can be anywhere from a few square centimeters (square inches) to a few square meters (square yards) or widespread damage.

Causes of Delaminations

Delaminations are similar to blisters in that delaminated areas of surface mortar result from bleed water and bleed air being trapped below the prematurely sealed (densified) mortar surface. The primary cause is finishing the surface before bleeding is complete. When concrete slabs are finished by hard troweling, a densified surface layer is created. A transition zone exists between the densified upper surface and the underlying body of the slab. Surface delaminations most often occur within this transition zone when the paste is weakened and fractures as the densified upper layer dries/shrinks relative to the underlying concrete. Weakening of the paste can be caused by trapping rising bleed water below the densified surface, which increases the water-cementitious ratio.

Another cause of delaminations is hard-troweling air-entrained

FIGURE 16-9. Surface delaminations are areas of loss of surface mortar ranging from about 3 mm to 9 mm (⅛ in. to ⅜ in.).

concrete (above 3%), which can result in elongated coalesced air voids in a lens within the transition zone.

Delaminations also may be the result of disruptive stresses from chloride-induced corrosion of steel reinforcement or of poorly bonded areas in two-course construction. The resulting delaminations are deeper than those caused by trapped air or bleed water.

Prevention of Delaminations

It is necessary to wait for a period of time after placing the concrete to allow air and water to escape from the concrete. The waiting period varies with the concrete mixture, mixing and placing procedures, and weather conditions. Delaminations are very difficult to detect during finishing and become apparent after the concrete surface has dried and the delaminated area is loaded. To avoid conditions that lead to delaminations, see the recommendations under the section on blisters.

Remedy for Delaminations

A delaminated area that has separated from the underlying concrete can leave a hole in the surface and resembles spalling. A delamination survey can be conducted by sounding – dragging a chain across the surface or tapping with a hammer and listening for hollow or drummy sounds. A hollow sound indicates delaminated areas, and a ringing sound indicates intact areas. This test is described in ASTM D4580, *Standard Practice for Measuring Delaminations in Concrete Bridge Decks by Sounding*. Nonstandard methods for detecting delaminated areas are acoustic impact, infrared thermography, and ground-penetrating radar.

As with blisters, delaminations can be repaired by patching or, if wide-spread, by grinding and overlaying with a new surface. Epoxy injection may also be beneficial in some applications.

CRAZING

Crazing, a network pattern of fine cracks that do not penetrate much below the surface (Figure 16-10). Crazing cracks are very fine and barely visible except when the concrete is drying after the surface has been wet. The cracks encompass small concrete areas less than 50 mm (2 in.) in dimension, forming a chicken-wire pattern. The term "map cracking" is often used to refer to cracks that are similar to crazing cracks, but map cracking is more visible and surrounds larger areas of concrete. Although crazing cracks may be unsightly and can collect

FIGURE 16-10. Crazing is a fine network of cracks at the surface of concrete. The image on the far right shows crazing due to alkali-aggregate reactivity in the fine aggregates (middle image courtesty of R.D. Hooton).

dirt, crazing does not compromise the structural integrity of the concrete or does not ordinarily indicate the start of future deterioration.

Causes of Crazing

Crazing is caused by minor surface shrinkage. When concrete is just beginning to gain strength, the climatic conditions, particularly the relative humidity during the drying period in a wetting and drying cycle, are an important cause of crazing. Low humidity, high air temperature, hot sun, or drying wind, either separately or in any combination, can cause rapid surface drying that encourages crazing. A surface into which dry cement has been cast will be more subject to crazing. The conditions that contribute to dusting, as described above, also will increase the tendency to craze. In rare cases, crazing can also be due to alkali-aggregate reactivity (AAR) in the coarse or fine aggregate (Figure 16-10 right).

Prevention of Crazing

To prevent crazing, curing procedures should begin early, within minutes after final finishing when weather conditions warrant. When the temperature is high and the sun is out, some method of curing with water should be used, since this will stop rapid drying and lower the surface temperature. The concrete should be protected against rapid changes in temperature and moisture wherever feasible.

Remedy for Crazing

Crazing cracks are a surface issue and, though unsightly, are unlikely to lead to structural or serviceability problems. There is no repair method, thus it is best to take precautions, as outlined above. If surface quality is of concern, a breathable sealer can be applied. Methyl methacrylate or ultra low viscosity epoxy sealers can be used to fill the narrow cracks where saturation and subsequent freezing is a concern. However, it should be noted that applying a sealer will highlight the crazing pattern by making the cracks appear darker.

FIGURE 16-11. Popouts are fractured aggregates that vary in size generally from 5 mm to 50 mm (¼ in. to 2 in.) but up to as much as 300 mm (1 ft) in diameter.

POPOUTS

A popout is a conical fragment that breaks out of the surface of the concrete leaving a hole that may vary in size generally from 5 mm to 50 mm (¼ in. to 2 in.) but up to as much as 300 mm (1 ft) in diameter (Figure 16-11). Usually a fractured aggregate particle will be found at the bottom of the hole, with part of the aggregate still adhering to the point of the popout cone. Popouts may also occur in fine aggregates (Figure 16-12).

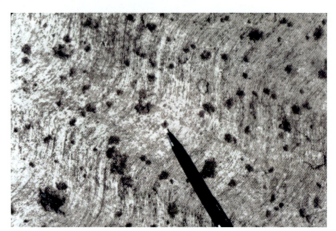

FIGURE 16-12. Popouts caused by ASR of sand-sized particles.

Causes of Popouts

The cause of a popout usually is a piece of porous rock having a high rate of absorption and relatively low specific gravity. As the porous aggregate absorbs moisture or freezing occurs under moist conditions, its swelling creates internal pressures sufficient to rupture the concrete surface. Oxidation of iron sulfate (pyrite) can also contribute to popouts and is usually accompanied by formation of rust. Pyrite, hard-burned dolomite, coal, shale, soft fine-grained limestone, or chert commonly cause popouts. Popouts may also occur to relieve pressure created by water uptake of expansive gel formed during the chemical reaction between the alkalies in the concrete and reactive siliceous aggregates, a phenomenon known as alkali-silica reactivity (ASR) (see Chapter 11 for more information on ASR). Popouts may also occur in the fine aggregate.

Most popouts appear within the first year after placement. Popouts caused by ASR may occur as early as a few hours to a few weeks, or even a year, after the concrete is placed (Landgren 2002). Popouts caused by moisture-induced swelling may occur shortly after placement due to the absorption of water from the plastic concrete, or they may not appear until after a season or year of high humidity or rainfall or after the concrete has been exposed to freezing temperatures. Popouts are considered a cosmetic detraction and generally do not affect the service life of the concrete.

Prevention of Popouts

The following steps can be taken to minimize or eliminate popouts:

- Use concrete with the lowest water content possible for the application.

- Use a durable crushed-stone or beneficiated-aggregate concrete.

- Use two-course construction with clean, sound rock in the topping, and the porous aggregates in the base slab, thus limiting the susceptible aggregate's exposure to excess moisture.

- Use a blended cement or a supplementary cementitious material such as fly ash (proven to control ASR) where popouts are caused by alkali-silica reactivity. Use of a low alkali loading in the concrete is also beneficial (see Chapter 11).

- Use air-entrained concrete.

- During hot, dry, and windy weather, cover the surface with plastic sheets after screeding and bullfloating to reduce evaporation before final finishing.

- Do not finish concrete with bleed water on the surface.

- Avoid hard-steel troweling where not needed, such as most exterior slabs.

- Use wet-curing methods such as continuous sprinkling with water, fogging, ponding, or covering with wet burlap soon after final finishing. Wet-cure for a minimum of 7 days, as wet cures can greatly reduce or eliminate popouts caused by ASR.

- If using aggregate with a susceptibility to ASR-induced popouts, use wet curing methods. Avoid plastic film, curing paper, and especially curing compounds as they don't dilute the alkalies at the surface. Flush curing water from the surface before final drying.

- Impervious floor coverings or membranes should be avoided as they can aggravate popout development. Avoid use of vapor barriers.

- Slope the slab surface to drain water properly.

Remedy for Popouts

Surfaces with popouts can be repaired. In general, repair of popouts is very labor intensive and therefore the blemishes are frequently tolerated. If repairs are deemed necessary, a small patch can be made by drilling out the spalled particle and filling the void with a dry-pack mortar or other appropriate repair material. As a general rule, mortar used for patching should be made from the same materials as the original concrete except that a proportion of off-white cement should be mixed with the original cement to lighten the color and thus better match the existing surface.

If the popouts in a surface are too numerous to patch individually, a thin-bonded concrete overlay may be used to restore serviceability.

SCALING AND MORTAR FLAKING

Scaling is local flaking or peeling away of the near-surface portion of hardened concrete or mortar. Light, medium, severe, and very severe scaling ranges from no exposure of coarse aggregate up to a loss of mortar and coarse aggregate particles to a depth of greater than 20 mm (0.8 in.). The aggregate is usually clearly exposed and often stands out from the concrete (Figure 16-13).

Mortar flaking over coarse aggregate particles, sometimes called popoffs, is another form of scaling that somewhat resembles a surface with popouts (Figure 16-14). However, mortar flaking usually does not result in freshly fractured aggregate particles and there are fewer, if any, conical voids such as those found in popouts. Aggregate particles with flat surfaces are more susceptible than round particles to this type of defect. Mortar flaking occasionally precedes more widespread surface scaling, but its presence does not necessarily lead to more extensive scaling.

Causes of Scaling and Mortar Flaking

Scaling is primarily a physical action caused by hydraulic pressure from water freezing within the concrete and not

FIGURE 16-13. Scaling is the general loss of surface mortar typically due to exposure to freeze-thaw and deicer salts.

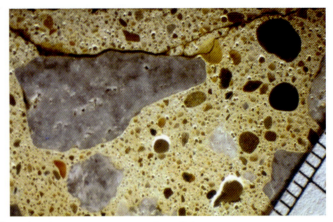

FIGURE 16-14. Mortar flaking is another form of scaling that consists of loss of paste over coarse aggregate particles.

usually caused by chemical action. When pressure due to ice formation exceeds the tensile strength of concrete, scaling can result if entrained-air voids are not present to act as internal pressure relief valves.

The presence of a deicer solution in water-soaked concrete during freezing causes an additional buildup of internal pressure (see Chapter 11).

Mortar flaking over coarse aggregate particles is caused essentially by the same actions that cause regular scaling. Excessive and early drying of the surface mortar alone can aggravate scaling.

However, the moisture loss is accentuated over aggregate particles near the surface. The lack of moisture necessary for cement hydration results in a mortar layer of lower strength and durability, higher shrinkage, and poorer bond with the aggregate.

Upon freezing in a saturated condition, this thin, weakened mortar layer breaks away from the aggregate. Poor finishing practices can also aggravate mortar flaking.

Prevention of Scaling and Mortar Flaking

Field service experience and extensive laboratory testing have shown that when properly spaced air voids are present, air-entrained concrete will have excellent resistance to surface scaling and mortar flaking due to freezing and thawing and the application of deicer chemicals, provided the concrete is properly proportioned, placed, finished, and cured.

Scaling resistance may decrease as the amount of certain Supplementary Cementitious Materials (SCMs) increase. ACI 318, *Building Code Requirements for Structural Concrete* (2019), limits the maximum dosage of fly ash, slag cement, and silica fume to 25%, 50%, and 10%, respectively, by mass of cementing materials, for deicer exposures. Total SCM content should not exceed 50% by mass of the cementitious materials. Certain transportation departments in northern states allow the use of up to 30% fly ash in pavements. However, concretes that are properly designed, placed, and cured, have demonstrated

good scaling resistance even when made with higher dosages of some of these materials. Formed and mechanically finished surfaces have shown less vulnerability to scaling than hand-finished sections (Thomas 2013). The selection of materials and dosages should be based on local experience and the durability should be demonstrated by field or laboratory performance.

When concrete in service will be exposed to cycles of freezing and thawing or deicing chemicals, consult local guidelines on allowable practices and use the following guidelines to ensure adequate concrete performance:

- A proper air content (a minimum of 5% to 8% for 19 mm [¾ in.] nominal size aggregate) with a satisfactory air void system (having a spacing factor $\leq$ 0.200 mm [0.008 in.] and a specific surface area of 24 mm²/mm³ [600 in.²/in.³] or greater).

- A low water-to-cementitious materials ratio ($\leq$0.45) for plain concrete (maximum of 0.40 for reinforced concrete) per ACI 318-19.

- A minimum compressive strength of 31 MPa (4500 psi) for plain concrete exposed to freezing and thawing cycles that will be in continuous contact with moisture and exposure to deicing chemicals (minimum of 35 MPa or 5000 psi for reinforced concrete per ACI 318-19).

- Fly ash, slag cement, and silica fume dosages not exceeding 25%, 50%, and 10%, respectively with combinations not exceeding 50%, by mass of cementing materials, for deicer exposures, unless otherwise demonstrated by local practice or testing.

- Proper finishing after bleed water has evaporated from the surface.

- A minimum of 7 days of moist curing at or above 10°C (50°F).

- A minimum 30-day drying period after moist curing prior to exposure to freeze-thaw cycles and deicers when saturated.

- Adequate drainage (1% minimum slope, 2% preferred).

- For additional protection, consider applying a breathable sealer after the initial drying period (see Chapter 17).

Remedy for Scaling and Mortar Flaking

If scaling or mortar flaking should develop, or if the concrete is suspected of having poor quality, a breathable surface treatment may be applied to help protect the concrete against further freeze-thaw damage. These sealer treatments are made with linseed oil, silane, siloxane, or other materials (see Chapter 17).

Impermeable materials, such as most epoxies, should not be used on slabs on ground or other concrete where moisture can collect and freeze under the coating. The freezing water can cause delaminations or blistering under the impermeable coating; therefore, a breathable surface treatment should be used.

Thin-bonded overlays or surface-grinding methods can usually remedy a scaled surface if sound, air-entrained concrete is present below the scaled surface.

SPALLING

A spall is a loss of section of concrete, detached from a larger mass. Spalling is a deeper surface defect than scaling, often appearing as circular or oval depressions on surfaces or as elongated cavities along joints (Figure 16-15). Spalls may be 25 mm (1 in.) or more in depth and 150 mm (6 in.) or more in diameter, although smaller spalls also occur.

Causes of Spalling

Spalls are caused by pressure or expansion within the concrete, bond failure in two-course construction, impact loads, fire, or weathering. Improperly constructed joints and corroded reinforcing steel are two common causes of spalls. If left unrepaired, spalls can accelerate concrete deterioration.

Prevention of Spalling

Spalls can be avoided by:

- Properly designing the concrete element – including joints – for the environment and anticipated service.

- Using proper concrete mixes and concreting practices.
- Taking special precautions where necessary.

The first line of defense against steel corrosion caused by chloride-ion ingress (for example, from deicers) should be the use of a low permeability concrete made with a water-cement ratio of 0.40 or less. If steel reinforcement is used, it should have adequate cover as outlined in ACI 318-19. For extreme conditions, the addition of SCMs including silica fume, or latex to a concrete mixture will dramatically lower its permeability.

Other examples of special precautions to reduce steel corrosion induced by chloride-ion ingress in extreme conditions are:

- use of epoxy-coated reinforcing steel (ASTM D3963, *Standard Specification for Fabrication and Jobsite Handling of Epoxy-Coated Steel Reinforcing Bars*),
- application of breathable surface sealers such as silane, siloxane, and methacrylate-based compounds,
- use of corrosion-inhibiting admixtures, and
- cathodic protection methods.

These methods may be combined for added protection.

Remedy for Spalling

Spalled areas of concrete can be repaired when no more than the top third of the concrete is damaged and the underlying concrete is sound. If more than the top third of the concrete is damaged, if steel reinforcement bars are uncovered, a full-depth repair may be required. The economics of partial-depth repair versus complete replacement should be considered, as it may be more cost effective (and more uniform-looking) to replace the entire area.

Concrete should be removed to a depth of at least 40 mm (1½ in.). The boundaries of the repair area should be determined by sounding the pavement for delaminated or unsound areas. Patch limits should be extended about 100 mm (4 in.) beyond the

FIGURE 16-15. Spalling is a loss of section of concrete typically caused by pressure of expansion within the concrete (left) or erosion from impact of slab edges and joints (right).

edges of the unsound areas. Spalled or delaminated concrete is removed by sawing and chipping or by milling. If jackhammers are used, they should sized to prevent concrete damage beyond the repair area. It is best to keep the area rectangular or square and provide vertical edges at the boundaries to contain the patch. The exposed concrete should be lightly sandblasted to clean and roughen the surface so that sufficient bond can be obtained with the repair material. Follow repair material manufacturer's recommendation for surface preparation.

Repair materials can be portland-cement-based, proprietary materials, or polymer concretes (epoxy, methyl methacrylate, and polyurethane). Bituminous materials have also been used, but are usually considered temporary. Repair materials should have a thermal expansion coefficient that is compatible with the underlying concrete. Some materials are formulated to work with a bonding agent; other materials require only a clean roughened surface for good bond. Refer to manufacturer's recommendation for substrate conditioning and surface preparation. The International Concrete Repair Institute (ICRI) has technical guidelines for surface preparation and repair of concrete (ICRI 2008). Patch materials are usually mixed on site in small quantities to ensure prompt placement (without segregation), consolidation, finishing, and curing. The air temperature at placement needs to be above 4°C (40°F) for all cement-based repair materials and many of the proprietary rapid-setting mixes. The polymer concretes can be applied at lower temperatures and to wet substrates, but perform better when placed under more favorable conditions. Wet curing methods (spraying, wet coverings) reduce shrinkage of repair materials more than sealed curing (sheet materials, curing compounds). When spall repair involves a joint, the joint must be restored to proper working condition to allow for thermal expansion of the slab. This is one of the most important steps to ensure that partial-depth repairs will function properly. Sealants placed in joints keep out non-compressible materials and allow free joint movement. Consult manufacturers for further recommendations on repair materials and placement methods.

BUGHOLES

Bugholes are surface voids that result from the migration of entrapped air (and to a lesser extent water) to the fresh concrete-form interface. These surface defects manifest themselves mostly in vertical surfaces (Figure 16-16).

During consolidation, the densification and subsequent shrinkage of the fresh concrete forces entrapped air voids and excess water out of the cementitious matrix. The water tends to migrate upward due to a density differential and become bleed water. The air bubbles, however, seek the nearest route to reach pressure equilibrium. When in a vertical form, the closest distance for the air bubbles' migration is often to the interior form surface. If these bubbles are not directed by sufficient vibration vertically to the free surface of the setting concrete, bugholes will be present after form removal.

Bugholes are found more frequently in the upper portion of the concrete structure or at angled form surfaces as a result of additive accumulation from escaping air voids along the height of the structure. These surface voids are primarily an aesthetic problem for exposed structural concrete. However, problems do arise if the concrete surface is to be painted or if the voids reach a larger diameter depending on minimum cover (typically greater than 25 mm [1 in.]).

Causes of Bugholes

Perhaps the most significant cause of bugholes is improper vibration. Consolidation, usually through vibration, sets the air bubbles and water into motion. Adequate vibration sends both entrapped air and excess water to the free surface of the concrete – either vertically winding through the matrix or laterally in a direct route to the form wall. Improper vibration will either insufficiently liberate the voids or over-consolidate the concrete resulting in segregation and bleeding. (See ACI 309, *Guide for Consolidation of Concrete* [2005], for a full description of consolidation using vibration.)

FIGURE 16-16. Bugholes are surface voids that result from the migration of entrapped air and water.

Another factor that promotes bughole formation is the form material itself. Nonpermeable forms (for example, steel or polymer impregnated wood) and the use of form-release agents can restrict the movement of the air voids between the concrete-form interface that is necessary for bughole reduction. It is imperative that when form-releasing agents are used, they are used according to the manufacturer's recommendations and used only with specified form material.

Mix design can also be considered a significant contributor to bughole formation. Mix designs vary widely in their use of aggregate type, size, and grading and their use of admixtures and air-entrainment. A sticky or stiff mixture that does not respond well to consolidation can be directly linked to increased surface void formation.

Prevention of Bugholes

Through careful selection of materials, quality workmanship, and dutiful supervision, surface voids can be minimized.

- Proper consolidation. Vibration should be completed with each lift of concrete placed. As successive lifts are placed, the vibrator should penetrate the previous lift, working the entrapped air towards the form and then vertically up the sides.

- Permeable Forms. When impermeable forms are used, more vibration is necessary to move the air voids to the free surface of the concrete. The use of permeable forms has been shown through research to reduce bugholes significantly by allowing escaping air to move through the form to the ambient air. Choosing the proper form-releasing agent in the proper amount can also improve the surface quality.

- Mix design. Workable, flowing mixtures are easier to place and consolidate and therefore reduce the risk of bughole formation. Concrete with an optimally graded aggregate properly proportioned cement content, and any admixture that provides increased flow, workability, or ease of consolidation contributes to bughole reduction. Self-consolidating concrete (SCC) is increasingly popular (in precast especially) to improve surface quality.

Remedy for Bugholes

Bugholes as surface defects are not detrimental to structural concrete from a durability standpoint. It is better to minimize the occurrence of bugholes than to try and repair the surface. Generally, unless the surface is going to be viewed from close range, the frequency and distribution of bugholes for a surface finish as specified in ACI 301-20, will be acceptable from an appearance point of view (see Chapter 15).

Any patching of bugholes is likely to emphasize them, as repair mortars will likely be a different color than the rest of the concrete surface. However, if the surface is to be painted, bugholes

FIGURE 16-17. Honeycombing in concrete.

may be able to be patched beforehand to provide a smoother appearance. See discussion under **Remedy for Honeycombing**.

HONEYCOMBING

Honeycombing occurs when mortar fails to fill the spaces between coarse aggregates (Figure 16-17). It usually becomes apparent when the formwork is stripped, revealing a rough and 'stony' concrete surface with air voids between the coarse aggregate. Sometimes, however, a surface skin of mortar masks the extent of the defect.

Honeycombing is always an aesthetic problem, and, depending on the depth and extent, may reduce both the durability performance and the structural strength of the member.

Causes of Honeycombing

Congested reinforcement, segregation, and insufficient fine aggregate contents can contribute to honeycombing. Holes and gaps in the formwork can allow some of the mortar to drain out of the concrete.

Prevention of Honeycombing

Higher concrete slumps and vibration may assist in preventing honeycombing by increasing the flowability of the concrete.

To minimize the incidence of honeycombed concrete:

- Ensure the mix has sufficient mortar to fill the voids between the coarse aggregate.

- Use a mix with appropriate workability (appropriate fines content and overall aggregate gradation) for the situation in which it is to be placed.

- Ensure the concrete is fully compacted and the placing methods minimize the risk of segregation.

- Ensure the reinforcement layout and the section shape will permit the concrete to flow around the reinforcement and completely fill the forms.

- Check that the formwork is rigid and well braced, the joints are watertight and any penetrations through the formwork, are properly sealed.

Remedy for Honeycombing

It is always better to avoid imperfections such as honeycombing in concrete rather than have to repair them. The extent and depth of the honeycombed area first needs to be defined. This can be done by chiseling out the affected area to expose sound concrete or by using non-destructive testing techniques.

If the honeycombed area is small in extent and depth, and does not significantly jeopardize the quality of the cover concrete protecting the reinforcement, then it can be repaired with mortar of a similar color and engineering properties of the base concrete. Any lightly attached stones should be removed before the mortar is worked into the spaces between the aggregate ensuring that it completely fills the honeycombed area. The area should be slightly over filled and screeded off to give a similar texture to the surrounding surface. The repair should then be cured.

Consideration needs to be given to the appearance of the repaired surface relative to adjacent untreated surfaces. As a general rule, mortar used for patching should be made from the same materials as the original concrete except that a proportion of white cement should be mixed with the cement to lighten the color and thus better match the existing surface.

If the honeycombing is extensive and penetrates down to the reinforcement or even deeper, then it is necessary to cut out the defective concrete and replace it with sound concrete. It is critical that the reinforcement be surrounded by sound concrete to protect it from corrosion.

The advice of a suitably qualified engineer should be obtained to check that the load-carrying capacity of the member, as repaired, will be satisfactory.

COLD JOINT

Cold joints are discontinuities in concrete members. Cold joints often result in "placement" lines visible on the surface, indicating the presence of a joint where one layer of concrete had hardened before subsequent concrete was placed (Figure 16-18). Aside from their appearance, cold joints can be a concern if they allow moisture penetration or if the loss of tensile strength of the concrete across the joint is deemed detrimental to the performance of the structure.

Causes of Cold Joints

Cold joints result primarily from a placement delay of sufficient duration to prevent a union of the material in two successive lifts. Placement lines can also result from improper consolidation between concrete lifts.

Prevention of Cold Joints

Cold joints may be reduced with the use of a set retarder which extends the period during which concrete remains plastic. This allows a large placement to be completed before setting occurs, which helps eliminates cold joints.

The transporting and handling equipment must have the capacity to move sufficient concrete so that cold joints are eliminated. For monolithic construction, the rate of placement should be rapid enough that previously placed concrete has not yet set and is still responsive to vibration. Lift heights should be optimized, layers should be sufficiently shallow to permit the two layers to be integrated by proper vibration to reduce segregation and cold joints. Timely placement and adequate consolidation will prevent flow lines, seams, and planes of weakness.

Remedy for Cold Joints

Coring over a cold joint and breaking the core sample in compression (direct shear test) is the best way to determine if there is a proper bond (Figure 16-19) (Volz and others 1997). If strengthening of a cold joint is required, the remedies discussed under **Cracking** can be considered. Cracks as narrow as 0.05 mm (0.002 in.) can be bonded by the injection of epoxy or cementitious grout.

FIGURE 16-18. Cold joint in concrete.

FIGURE 16-19. Coring through concrete to determine if visible line is a placement line or cold joint.

DISCOLORATION

Discoloration is a departure of color from that which is normal or desired. Surface discoloration of concrete flatwork can appear as gross color changes in large areas of concrete, spotted or mottled light or dark blotches on the surface (Figure 16-20), or early light patches of efflorescence. Although unsightly, discoloration typically does not indicate a concern with durability or strength.

FIGURE 16-20. Discoloration of concrete surface due to use of calcium chloride and plastic sheeting.

Causes of Discoloration

Laboratory studies to determine the effects of various concreting procedures and concrete materials show that no single factor is responsible for all discoloration. Factors found to influence discoloration are calcium chloride admixtures, alkalies in cementitious materials, hard-troweled surfaces, inadequate or inappropriate curing, a wet substrate, variation of the water-cement ratio at the surface, and changes in the concrete materials or proportions. Discoloration from these causes appears very soon after placing the concrete.

Discoloration at later ages may be the result of atmospheric or organic staining – simply stated, the concrete is dirty. This type of discoloration is usually removed by power washing with pressurized water and, possibly, chemical cleaners.

The use of calcium chloride in concrete may discolor the surface. Calcium chloride accelerates the hydration process but has a retarding effect on the hydration of the ferrite compound in portland cement. The ferrite phase normally becomes lighter with hydration; however, in the presence of calcium chloride the retarded, unhydrated ferrite phase remains dark.

Extreme discoloration can result from attempts to hard-trowel the surface after it has become too stiff to trowel properly. Vigorously troweling a surface to progressively compact it can reach the point where the water-cement ratio is drastically decreased in localized areas. This dense, low-water-cement-ratio concrete in the hard-troweled area is almost always darker than the adjacent concrete.

Waterproof paper and plastic sheets used to moist-cure concrete, especially those containing calcium chloride, have been known to give a mottled appearance to flat surfaces due to the difficulty in keeping the cover in complete contact with the surface over the entire area. The places that are in contact will tend to be lighter in color than those that are not.

Concrete materials and proportions affect concrete color. Individual cementitious materials may differ in color. Thus, substituting one cement or fly ash for another may change the color of concrete. Concretes containing significant amounts of supplementary cementitious materials – fly ash, silica fume, metakaolin, or slag, for example – may differ in color from those containing no mineral admixture. The color of the sand has an effect on the color of the concrete. High-strength concrete with a low water-cement ratio is darker in color than low-strength concrete with a high water-cement ratio.

A rare discoloration ranging in color from buff to red/orange has been reported in Wisconsin, Illinois, Louisiana, and a few other states. This type of discoloration is more likely to occur during periods of high relative humidity and high ambient temperature. This staining occurs more often with certain types and amounts of wet-curing. Fly ash aggravates the staining by intensifying the color.

Prevention of Discoloration

The discoloration of concrete can be avoided or minimized by:

• Avoiding the use of calcium chloride admixtures,

• Using consistent concrete ingredients, uniformly proportioned from batch to batch,

• Using proper and timely placing, finishing, and curing practices, and

• Being consistent in concreting practices.

Disruptions or changes in the concrete materials or proportions, formwork, finishing, or curing can result in significant and sometimes permanent discoloration. In particular, additional hydration of the ferrite compounds in cementitious materials leads to more reduced iron being available to oxidize and discolor the concrete.

Remedy for Discoloration

To eradicate discoloration, the first (and usually effective) remedy is an immediate, thorough flushing with water. Permit the slab to dry, then repeat the flushing and drying until the discoloration disappears. If possible, use hot water. Acid washing using concentrations of weaker acids such as 3% acetic acid (vinegar) or 3% phosphoric acid will lessen carbonation and mottling discoloration. Treating a dry slab with 10% solution of caustic soda (sodium hydroxide) gives some success in blending light spots into a darker background. Harsh acids should not be used, as they can harm the concrete surfaces and expose the aggregates.

One of the best methods to remove most discoloration, when other remedies have failed, is to treat a surface with a 20% to 30% water solution of diammonium citrate (Greening and Landgren 1966). Any chemical treatment should be approached with caution and tested in an inconspicuous area first to ensure it does not damage the concrete or create additional discoloration.

Staining discolored concrete with a chemical stain is another way to make color variations less noticeable. Usually, darker colors hide color variations more effectively. Chemical stains can be used on interior or exterior concrete.

The following chemicals are largely ineffective at removing buff to red/orange discoloration: hydrochloric acid (2%), hydrogen peroxide (3%), bleach, phosphoric acid (10%), diammonium citrate (0.2 M), and oxalic acid (3%) (Miller and others 1999, and Taylor and others 2000).

IRON (RUST) STAINING

Iron oxide staining leaves unsightly stains on the surface of concrete.

Causes of Iron Staining

Aggregates can occasionally contain particles of iron oxide or iron sulfides (for example, pyrite) that result in rust stains exposed concrete surfaces (Figure 16-21). Exposed reinforcement and tie wires can also leave rust stains on the surface of concrete. Shavings from iron tools used in placing or finishing concrete can also rust and cause staining.

FIGURE 16-21. Iron oxide stain caused by impurities in the coarse aggregate (courtesy of T. Rewerts).

Prevention of Iron Staining

Aggregates should meet the staining requirements of ASTM C330, *Standard Specification for Lightweight Aggregates for Structural Concrete* (AASHTO M 195), when tested according to ASTM C641, *Standard Test Method for Iron Staining Materials*

in Lightweight Concrete Aggregates; the quarry face and aggregate stockpiles should not show evidence of staining.

- Maintain cover for embedded steel to prevent corrosion of reinforcement. Also take time to assure tie wires are bulled away from form faces and slab surfaces.
- Take care to remove wires and patch tie holes to eliminate potential source of rusting.
- Care must be taken with tools. Do not use metal tools (including shovels) on concrete surfaces to avoid rust stains from iron shavings.

Remedy for Iron Staining

If staining is difficult to remove, commercial sodium bisulfate cleaners are somewhat successful in removing stains; however, the difference between cleaned and stained areas decreases over several weeks.

EFFLORESCENCE

Efflorescence can be considered a type of discoloration (Figure 16-22). It is a deposit, usually white in color, that occasionally develops on the surface of concrete, often just after a structure is completed. Although unattractive, efflorescence is usually harmless. In rare cases, excessive efflorescence deposits can occur within the surface pores of the material, causing expansion that may disrupt the surface.

FIGURE 16-22. Efflorescence is a deposit of soluble salts on the surface of concrete.

Causes of Efflorescence

Efflorescence is caused by a combination of circumstances: soluble salts in the material, available moisture to dissolve these salts, and evaporation or hydrostatic pressure that moves the solution toward the surface. Water in moist, hardened concrete dissolves soluble salts. This salt-water solution migrates to the surface by evaporation or hydraulic pressure where the water evaporates, leaving a salt deposit at the surface. Efflorescence is particularly affected by temperature, humidity, and wind. In the summer, even after long rainy periods, moisture evaporates so quickly that comparatively small amounts of salt are brought to the surface. Usually efflorescence is more common in the winter when a slower rate of evaporation allows migration of salts to the surface. If any of the conditions that cause efflorescence – water, evaporation, or salts – are not present, efflorescence will not occur.

All concrete materials are susceptible to efflorescence. Even small amounts of water-soluble salts (a few tenths of a percent) are sufficient to cause efflorescence when leached out and concentrated at the surface. In some cases, these salts come from beneath the surface, but chemicals in the concrete can react with chemicals in the atmosphere to form efflorescence. For example, hydration of cement produces soluble calcium hydroxide, which can migrate to the concrete surface with moisture flow, and combine with carbon dioxide in the air to form a white calcium carbonate deposit.

All concrete ingredients should be considered for soluble-salt content. Common efflorescence-producing salts are carbonates of calcium, potassium, and sodium; sulfates of sodium, potassium, magnesium, calcium, and iron; and bicarbonate of sodium or silicate of sodium.

Prevention of Efflorescence

To reduce or eliminate soluble salts:

- Never use unwashed sand. Use sand that meets the requirements of ASTM C33, *Standard Specification for Concrete Aggregates*.
- Use clean mixing water free from harmful amounts of acids, alkalies, organic material, minerals, and salts. Drinking water is usually acceptable. Do not use seawater (see Chapter 4).

Low absorption of moisture is the best assurance against efflorescence. Concrete will have maximum watertightness when made with properly graded aggregates, an adequate cement content, a low water-cement ratio, and thorough curing.

Remedy for Efflorescence

When there is efflorescence, the source of moisture should be determined and corrective measures taken to keep water out of the structure. Chloride salts are highly soluble in water, so the first rain often will wash them off the surface of concrete. With the passage of time, efflorescence becomes lighter and less extensive unless there is an external source of salt. Light-colored surfaces show the deposits much less than darker surfaces. Most efflorescence can be removed by dry brushing, water rinsing with brushing, light waterblasting or light sandblasting, followed by flushing with clean water. If this is not satisfactory, it may be necessary to wash the surface with a dilute solution of muriatic acid (1% to 10%). For integrally colored concrete, only a 1% to 2% solution should be used to prevent surface etching that may reveal the aggregate and change color and

texture. Always pretest the treatment on a small, inconspicuous area to be certain there is no adverse effect. Before applying an acid solution, always dampen concrete surfaces with clean water to prevent the acid from being absorbed deeply which may cause damage. The cleaning solution should be applied to no more than 0.4 m² (4 ft²) at one time to avoid surface damage. Wait about 5 minutes, then scour with a stiff bristle brush. Immediately rinse with clean water to remove all traces of acid. Neutralize the treated area with a solution of 0.5 kg (1 lb) of baking soda (sodium bicarbonate) dissolved in 20 L (5 gal) of water. The entire concrete element should be treated to avoid discoloration or mottled effects. Surfaces to be painted should be thoroughly rinsed with water and allowed to dry.

ANALYSIS OF IMPERFECTIONS IN CONCRETE

When troubleshooting concrete problems, it is important to relate the symptom to causes of future distress and deterioration. Deterioration in concrete is defined as a worsening of condition with time, which may result in a progressive reduction in the ability of the concrete to serve its intended function.

The cause of most concrete defects can be determined by experienced technicians, field inspectors, or engineers. There are many field and laboratory tests available to aid in determining mechanisms behind problematic symptoms. A petrographic (microscopical) analysis on samples of the concrete may assist with a field investigation. A petrographic analysis of concrete is performed in accordance with ASTM C856, *Standard Practice for Petrographic Examination of Hardened Concrete*. Samples for the analysis are usually, 100-mm (4-in.) diameter cores or saw-cut sections. Broken sections can be used, but cores or saw-cut sections are preferred because they are less apt to be disturbed. Samples should represent concrete from both problem and nonproblem areas. The petrographer should be provided with a description and photographs of the problem, plus information on mixture design, construction practices used, and environmental conditions.

The petrographic report often includes the probable cause of the problem, extent of distress, general quality of the concrete, and expected durability and performance of the concrete. Corrective action, if necessary, would be based to a great extent on the petrographic report.

For more information on properly identifying symptoms refer to PCA (2002), PCA (2007), and ACI 201.1R, *Guide for Conducting a Visual Inspection of Concrete in Service* (2008).

For more information on cleaning and repair of concrete surfaces, refer to PCA's *Effects of Substances on Concrete and Guide to Protective Treatments* (2007), *Removing Stains and Cleaning Concrete Surfaces* (1988), ACI 546.3R, *Guide to Materials Selection for Concrete Repair* (2014), ICRI (2018), and von Fay (2015).

REPAIRING AND CLEANING CONCRETE

As discussed earlier in the chapter, when appearance is important, specifications should clearly define vertical surface finish requirements and any special appearance or color uniformity requirements. Acceptable appearances should be judged at an agreed upon distance, generally at a distance of 6 m (20 ft) (ACI 303R 2012). Mock-ups may be required for acceptance. Mock-ups also provide a valuable measure of parameters in determining acceptable appearance including color, surface texture, as well as constructability.

After forms are removed, all undesired bulges, fins, and small projections can be removed by chipping or tooling, as specified. Undesired bolts, nails, ties, or other embedded metal can be removed or cut back to a depth of 13 mm (½ in.) from the concrete surface.

When required, the surface can be rubbed or ground to provide a uniform appearance (see Chapter 15). Any cavities such as tie rod holes should be filled unless they are intended for decorative purposes. Honeycombed areas should be repaired and stains removed in accordance with specified requirements. Patching can be minimized by exercising care in constructing the formwork and placing the concrete. The condition of form work impacts the consistency of the final surface. Form liners and use of form release agents should be monitored to yield the desired finish appearance. In general, repairs are easier and more successful if they are made as soon as practical, preferably as soon as the forms are removed. However, the procedures discussed below apply to both new and old hardened concrete.

Holes, Defects, and Overlays

Repaired areas usually appear darker than the surrounding concrete. White cement may be used in mortar or concrete for patching where appearance is important. Samples should be applied and cured in an inconspicuous location to determine the most suitable proportions of white and gray cements. Steel troweling should be avoided since this may darken the patch.

Bolt holes, tie rod holes, and other cavities that are small in area but relatively deep should be filled with a dry-pack mortar. The mortar should be mixed as stiff as is practical: use 1 part cement, 2½ parts sand passing a 1.25 mm (No. 16) sieve, and just enough water to form a ball when the mortar is squeezed gently in the hand. The cavity should be clean with no oil or loose material and kept damp with water for several hours. A neat-cement paste should be scrubbed onto the void surfaces, but not allowed to dry before the mortar is placed. The mortar should be tamped into place in layers about 13 mm (½ in.) thick. Vigorous tamping and adequate curing will ensure good bond and minimum shrinkage of the patch.

Concrete used to fill large areas and thin-bonded overlays should be as similar as possible to the concrete being repaired to ensure compatible thermal expansion and contraction.

Otherwise if the concrete is questionable in service, the repair material should have a low water-cement ratio, often with a cement content equal to or greater than the concrete to be repaired. Cement contents typically range from 360 kg/m³ to 500 kg/m³ (600 lb/yd³ to 850 lb/yd³) and the water-cementitious materials ratio is usually 0.45 or less. The aggregate size should be no more than one third the patch or overlay thickness. A 9.5-mm (⅜-in.) nominal maximum size coarse aggregate is commonly used. The fine aggregate proportion can be higher than usual, often equal to the amount of coarse aggregate, depending on the desired properties and application.

Before the repair material is applied, the surrounding concrete should be clean and sound (Figure 16-23). Abrasive methods of cleaning (sandblasting, hydrojetting, waterblasting, scarification, or shotblasting) are usually required.

FIGURE 16-23. Concrete prepared for patch installation.

For overlays, a cement-sand grout, a cement-sand-latex grout, or an epoxy bonding agent may be applied to the prepared surface with a brush or broom (see Chapter 15). Typical grout mix proportions are 1 part cement and 1 part fine sand and water or latex modified mixing water. The grout should be applied immediately before the new concrete is placed. The grout should not be allowed to dry before the freshly mixed concrete is placed; otherwise bond may be impaired. The existing concrete should be surface saturated dry when the grout is applied but not wet with free-standing water. There are many manufactured topping materials that are applied to feather down edge thickness, but concrete toppings are normally limited to a minimum of 20 mm (¾ in.). Some structures, like bridge decks, should have a minimum concrete topping thickness of 40 mm (1½ in.).

Honeycombed and other defective concrete should be cut out to expose sound material. If defective concrete is left adjacent to a patch, moisture may get into the voids; in time, weathering action will cause the patch to spall. The edges of the defective area should be cut or chipped straight and at right angles to the surface, or slightly undercut to provide a key at the edge of the patch. No feathered edges should be permitted (Figure 16-24). Based on the size of the patch, either a mortar or a concrete repair mixture should be used.

(A) Incorrectly installed patch repair. The feathered edges will break down under traffic or weather away.

(B) Correctly installed patch repair. The chipped area should be at least 20 mm (¾ in.) deep with the edges at right angles or undercut to the surface.

FIGURE 16-24. Concrete repair material installation.

Shallow repairs can be filled with a dry-pack mortar as described earlier. This should be placed in layers not more than 13 mm (½ in.) thick, with each layer given a scratch finish to improve bond with the subsequent layer. The final layer can be finished to match the surrounding concrete by floating, rubbing, or tooling. Formed surfaces are matched by pressing a section of form material against the repair patch while still plastic.

Deep repairs can be filled with concrete held in place by forms. Such patch repairs should be reinforced and doweled to the hardened concrete (Bureau of Reclamation 1981). Large, shallow vertical or overhead repairs may best be accomplished by shotcreting. Several proprietary low-shrinkage cementitious repair products are also available.

Curing Patch Repairs

Following repair, good curing is essential (Figure 16-25). Curing should be started immediately to avoid early drying. Wet burlap, plastic sheets, curing paper, tarpaulins, or a combination of these can be used. In locations where it is difficult to hold these materials in place, an application of two coats of membrane-curing compound is often the preferred method.

FIGURE 16-25. Good curing is essential to successful patch repairs. This patch is covered with polyethylene sheeting plus rigid insulation to retain moisture and heat for rapid hydration and strength gain.

Cleaning Concrete Surfaces

Where appearance is important, all surfaces should be cleaned after construction has progressed to the stage where further discoloration from subsequent construction activities is no longer a risk.

There are three techniques for cleaning concrete surfaces: water, chemical, and mechanical (abrasion). Water loosens dirt and rinses it from the surface. Chemical cleaners, usually mixed with water, react with dirt to separate it from the surface, and then the dirt and chemicals are rinsed off with clean water. Sandblasting is the most common mechanical method which removes dirt by abrasion.

Before selecting a cleaning method, it should be tested on an inconspicuous area to be certain it will be helpful and not harmful. If possible, identify the characteristics of the discoloration because some treatments are more effective than others in removing certain materials.

Water cleaning. Water cleaning methods include low-pressure washes, moderate-to-high-pressure waterblasting, and steam. Low-pressure washing is the simplest method, requiring only that water run gently down the concrete surface for a day or two. The softened dirt is then flushed using a slightly higher pressure rinse. Stubborn areas can be scrubbed with a nonmetallic-bristle brush and rinsed again. High-pressure waterblasting is used effectively by experienced operators. Steam cleaning must be performed by skilled operators using special equipment. Water methods are the least harmful to concrete, but they are not without risk. Serious damage may occur if the concrete surface is subjected to freezing temperatures while it is still wet; and water can bring efflorescence to the surface.

Chemical cleaning. Chemical cleaning is usually done with water-based mixtures formulated for specific concrete. An organic compound called a surfactant (surface-active agent), which acts as a detergent to wet the surface more readily, is included in most chemical cleaners. A small amount of acid or alkali is typically also included to separate the dirt from the surface.

Chemicals commonly used to clean concrete surfaces and remove discoloration include weak solutions (1% to 10% concentration) of hydrochloric, acetic, or phosphoric acid. Diammonium citrate (20% to 30% water solution) is especially useful in removing stains and efflorescence on formed and flatwork surfaces. Chemical cleaners should be used by skilled operators taking suitable safety precautions.

There can be problems related to the use of chemical cleaners. Their acid or alkaline properties can lead to unwarranted reactions between cleaner and concrete as well as mortar, painted surfaces, glass, metals, and other building materials. Since chemical cleaners are used in the form of water-diluted solutions, they too can liberate soluble salts from within the concrete to form efflorescence. Some chemicals can also expose the aggregate in concrete.

Mechanical cleaning. Mechanical cleaning includes sand-blasting, shotblasting, scarification, water blasting, power chipping, and grinding. These methods wear the dirt off the surface rather than separate it from the surface. They wear away both the dirt and some of the concrete surface; it is inevitable that there will be some loss of decorative detail, increased surface roughness, and rounding of sharp corners. Abrasive methods may also reveal defects (voids) hidden just beneath the formed surface. However, mechanical impact methods can bruise the concrete surface which compromises the quality of substrate and may reduce bond of repair materials.

Chemical and mechanical cleaning can each have an abrading effect on the concrete surface that may change the appearance of a surface when compared to that of an adjacent uncleaned surface.

REFERENCES

PCA's online catalog includes links to PDF versions of many of our research reports and other classic publications.
Visit: cement.org/library/catalog.

ACI Committee 201, *Guide for Conducting a Visual Inspection of Concrete in Service*, ACI 201.1R-08, American Concrete Institute, Farmington Hills, MI, 2008, 16 pages.

ACI Committee 224, *Causes, Evaluation, and Repair of Cracks in Concrete Structures*, ACI 224.1R-07, American Concrete Institute, Farmington Hills, MI, 2008, 22 pages.

ACI Committee 301, *Specification for Concrete Construction*, ACI 301-20, American Concrete Institute, Farmington Hills, MI, 2020, 73 pages.

ACI Committee 303, *Guide to Cast-in-Place Architectural Concrete Practice*, ACI 303R-12, American Concrete Institute, Farmington Hills, MI, 2012, 36 pages.

ACI Committee 318, *Building Code Requirements for Structural Concrete and Commentary*, ACI 318-19, American Concrete Institute, Farmington Hills, MI, 2019, 628 pages.

ACI Committee 546, *Guide to Materials Selection for Concrete Repair*, ACI 546.3R-14, American Concrete Institute, Farmington Hills, MI, 2014, 72 pages.

Greening, N.R.; and Landgren, R., *Surface Discoloration of Concrete Flatwork*, RX203, Portland Cement Association, Skokie, IL, 1966, 19 pages.

ICRI, *Guide for Selecting and Specifying Materials for Repair of Concrete Surfaces*, 320.2R-2018, International Concrete Repair Institute, Sterling, VA, 2018, 38 pages.

ICRI, *Guidelines for Surface Preparation for the Repair of Deteriorated Concrete Resulting from Reinforcing Steel Corrosion*, 310.1R-2008, International Concrete Repair Institute, Sterling, VA, 2008.

Kerkhoff, B., *Effects of Substances on Concrete and Guide to Protective Treatments*, IS001, Portland Cement Association, Skokie, IL, 2007, 36 pages.

Landgren, R.; and Hadley, D.W., *Surface Popouts Caused by Alkali-Aggregate Reaction*, RD121, Portland Cement Association, Skokie, IL, 2002, 18 pages.

Miller, F. M.; Powers, L.J.; and Taylor, P.C., *Investigation of Discoloration of Concrete Slabs*, SN2228, Portland Cement Association, Skokie, IL, 1999, 22 pages.

PCA, *Concrete Slab Surface Defects: Causes, Prevention, Repair*, IS177, Portland Cement Association, Skokie, IL, 2007, 16 pages.

PCA, *Removing Stains and Cleaning Concrete Surfaces*, IS214, Portland Cement Association, Skokie, IL, 1988, 16 pages.

PCA, *Types and Causes of Concrete Deterioration*, IS536, Portland Cement Association, Skokie, IL, 2002, 16 pages.

Tarr, S.M.; and Farny, J.A., *Concrete Floors on Ground*, EB075, 4th edition, Portland Cement Association, Skokie, IL, 2008, 254 pages.

Taylor, P.C.; Detwiler, R.J.; and Tang, F.J., *Investigation of Discoloration of Concrete Slabs* (Phase 2), SN2228b, Portland Cement Association, Skokie, IL, 2000, 22 pages.

Thomas, M., *Supplementary Cementing Materials in Concrete*, CRC Press, Boca Raton, FL, 2013, 210 pages.

von Fay, K.F., *Guide to Concrete Repair*, 2nd Edition, United States Department of the Interior, Bureau of Reclamation, Technical Service Center, Denver, CO, 2015, 390 pages.

Volz, J.S.; Olsen, C.A.; Osterle, R.G.; and Gebler, S.H., "Are They Pour Lines Or Cold Joints," *Concrete Construction*, Hanley-Wood, LLC, April 1997.

CURING CONCRETE

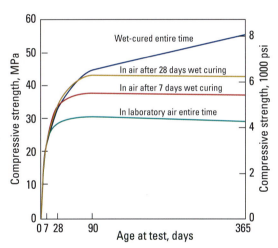

FIGURE 17-1. Effect of wet curing time on strength gain of concrete (Gonnerman and Shuman 1928).

Curing, by definition, is the maintenance of a satisfactory moisture content and temperature in concrete for a sufficient period of time during and immediately following placement so that the desired concrete properties may develop.

Proper curing promotes continued hydration of cementitious materials. The extent of hydration influences the strength and durability of concrete. As discussed in Chapter 9, when the relative humidity within the concrete drops to about 80% or the temperature of the concrete drops below -10°C (14°F), hydration and strength gain virtually stop (Powers 1948).

IMPORTANCE OF CURING

The need for adequate curing of concrete cannot be overemphasized. Curing has a strong influence on the final properties of hardened concrete. Curing improves strength, volume stability, permeability, and durability (including resistance to abrasion, freezing and thawing, and scaling). Cather (1992) defined the curing-affected zone as that portion of the concrete most influenced by curing measures. Exposed slab surfaces are especially sensitive to curing as strength development and durability of the top surface of a slab can be reduced significantly when curing is neglected (see Chapter 16). Loss of water will also cause the concrete to shrink, thus creating tensile stresses within the concrete. If these stresses develop before the concrete has attained adequate tensile strength, surface cracking can result (see Chapter 10).

With proper curing, concrete gains strength. The strength improvement is rapid at early ages but continues more slowly thereafter for an indefinite period provided there is a presence of water. Figure 17-1 shows the strength gain of concrete with age for different wet curing periods while Figure 17-2 shows the relative strength gain of concrete cured at different temperatures. Additional information on strength gain and hydration can be found in Chapter 9.

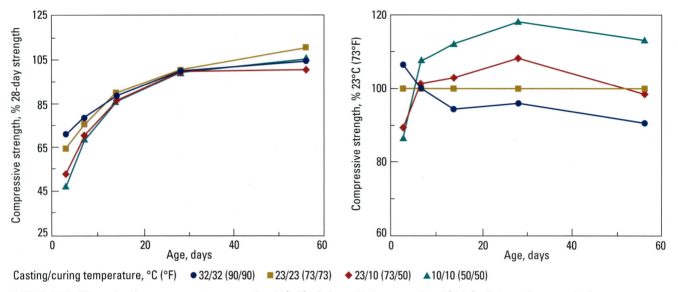

FIGURE 17-2. Effect of curing temperature on strength gain (left) relative to 28-day strength and (right) relative to the strength of concrete at 23°C (73°F) (Burg 1996). Note: Expanded y-axes.

TYPES OF CURING

The timing at which drying and the need for curing begins depends on the resulting rate of surface moisture evaporation. Surface moisture evaporation depends on the ambient environment (temperature, relative humidity, and windspeed) and the bleeding characteristics of the concrete. Curing often involves a series of procedures used at particular times as the concrete ages. These include initial curing, intermediate curing, external curing, internal curing, and accelerated curing.

Initial Curing

Initial curing includes procedures implemented anytime between placement and final finishing to reduce moisture loss from the concrete surface. Fogging is a method of initial curing and is applied to minimize plastic shrinkage cracking until finishing operations are complete.

Other methods to prevent the rapid loss of moisture from the concrete surface include evaporation retarders or possibly a temporary application of plastic sheeting to the surface. The plastic sheeting, when used, is typically installed immediately after strike-off and removed before finishing operations commence.

For some conditions including temperatures outside of desired curing ranges, it may also be necessary to modify the exposure to the environment by erecting windbreaks, sunshades, or enclosures (Figure 17-3).

Intermediate Curing

Along with initial curing measures, intermediate curing is sometimes necessary and refers to procedures implemented after finishing is completed but before the concrete surface has hardened sufficiently to prevent damage or marring due to the application of curing. Significant loss of moisture can occur when curing measures are delayed until final finishing is completed. The same techniques used for initial curing can be used for intermediate curing.

External Curing

External curing includes methods to maintain favorable moisture and temperature within concrete and at the concrete surface so the hydration of portland cement and reactions of supplementary cementitious materials can continue. External curing operations should begin during finishing operations as soon as the concrete stiffens enough to prevent marring or erosion of the surface.

If temperatures are favorable, hydration is relatively rapid during the first few days after concrete is placed. However, it is important for water to be retained within the concrete and at the concrete surface during this period and to prevent or substantially reduce evaporation. External curing methods may consist of supplying additional moisture to the concrete, retaining the mixing water, and/or modifying the concrete temperature.

FIGURE 17-3. Polyethylene plastic sheets admitting daylight are used to fully enclose a building frame. The temperature inside is maintained at 10°C (50°F) with ventilated space heaters.

Internal Curing

Internal curing refers to the process by which hydration continues because an internal water supply is available in addition to the mixing water. Saturated lightweight concrete aggregates and superabsorbent polymers (SAP) provide a source of moisture for internal curing. Lightweight aggregates used for internal curing purposes are specified by ASTM C1761, *Standard Specification for Lightweight Aggregate for Internal Curing of Concrete*. Internal curing must be accompanied by external curing methods to assure that exposed concrete surfaces are properly cured (ACI 308-213R 2013 and Lam 2005).

Internal curing allows the concrete to gain additional strength and also results in a reduction of permeability due to a significant extension in the curing time (Holm and Ries 2006). Internal curing can also help avoid early age cracking of concrete with high cementitious materials content typical of many high strength concretes. Likewise, early shrinkage of concrete caused by rapid drying due to self desiccation and autogenous shrinkage can also be diminished through internal curing (see Chapters 9 and 10).

For concretes with low water to cement ratios, 60 kg/m³ to 180 kg/m³ (100 lb/yd³ to 300 lb/yd³) of saturated lightweight fine aggregate can provide the additional moisture for internal curing. Super-absorbent polymer does not replace any aggregate, but instead is used as an admixture dosed by weight of cementitious materials. The superabsorbent polymers should be batched dry and blended with the sand, due to their tendency to clump when wet (Lam 2005). The quantified water which the SAP absorbs should be included as part of the mixing water.

In most cases, excess moisture will eventually diffuse out of the concrete. The time it takes for the concrete to dry must be taken into consideration if it is to be covered with a moisture-sensitive flooring system.

Accelerated Curing

Accelerated curing is any method by which the concrete is rapidly cured to achieve high-early strength. These techniques are especially useful in the precast concrete industry, where high-early strength enables the early removal of formwork, resulting in cost-saving benefits. This is usually accomplished with live steam, heating coils, or electrically heated forms or pads. A higher curing temperature provides earlier strength gain in concrete but it may decrease long-term strength gain. Regardless of the manner of accelerated curing, the maximum internal temperature of the concrete should not exceed 70°C (158°F) unless it is demonstrated to not cause damage by testing or historic field data (see PCI 1999 and Chapter 11).

CURING METHODS AND MATERIALS

Concrete can be kept wet either by supplying additional moisture or by retaining moisture. In some cases, curing also involves modifying the concrete temperature. The method or combination of methods chosen to cure concrete depends on factors such as availability of curing materials, size, shape, and intended use of concrete, project location or production facilities, aesthetic appearance, and economics.

Supplying Additional Moisture

Methods that provide additional sources of moisture replace water lost through evaporation or bound through chemical reaction and maintain hydration in the concrete during the important early hardening period. These include immersion, ponding, fogging, sprinkling, and saturated wet coverings. These methods provide some evaporative cooling, which is beneficial in hot weather concreting. The need for wet curing is greatest during placement and the first few hours after finishing. Wet coverings may create a mottled appearance. Where appearance of the concrete is important, the water used for curing must be free of substances that will stain or discolor the concrete.

Immersion. The most ideal method of water curing consists of total immersion of the finished concrete element. This method is commonly used in the laboratory for curing concrete test specimens.

Ponding. On horizontal surfaces, such as pavements and floors, concrete can be cured by ponding. Ponding is a particularly effective method for preventing loss of moisture from the concrete; it is also effective for maintaining a uniform temperature in the concrete. The temperature of curing water used shall be at least 10°C (50°F) and not be more than 20°C (35°F) colder than surface temperature of the concrete at the time the water and concrete come in contact to help prevent thermal stresses that result in cracking (ACI 301 2020).

Dikes can be built around the perimeter of the concrete surface to retain the ponded water. However, when earth or sand is used as the dike material it may also discolor the concrete surface. Since ponding requires considerable labor and supervision, the method is usually only used on smaller projects. Occasional checks to maintain continuous ponding are suggested especially in arid climates.

Fogging. Fogging the air above the concrete surface before and after final finishing is the most effective way to minimize evaporation and reduce plastic shrinkage cracking. A fine fog mist is applied through a system of nozzles or sprayers directed at a level approximately 1.5 m (5 ft) above the concrete surface to raise the relative humidity of the ambient air over flatwork (Figure 17-4). Fog nozzles atomize water using air pressure (Figure 17-5) to create a fog blanket, slowing evaporation from the concrete. They should not be confused with garden-hose nozzles, which leave liquid water on the slab. Fogging should be continued until a suitable curing material such as a curing compound, wet burlap, or curing paper can be applied.

FIGURE 17-4. Fogging cools the air and raises the relative humidity above flatwork to lessen rapid evaporation from the concrete surface.

FIGURE 17-5. Fog nozzle.

Sprinkling. Once the concrete has set sufficiently to prevent water erosion, ordinary lawn sprinklers are effective as a curing method if even coverage is provided and water runoff is controlled. Soaker hoses are useful on surfaces that are vertical or nearly so.

The cost of sprinkling may be a disadvantage. The method requires an ample water supply and careful supervision. If sprinkling is done at intervals, the concrete must be prevented from drying between applications of water by using burlap or similar materials (Figure 17-6); otherwise alternate cycles of wetting and drying can result in surface crazing or cracking.

FIGURE 17-6. Lawn sprinklers saturating burlap with water keep the concrete continuously wet. Intermittent sprinkling is acceptable if no drying of the concrete surface occurs.

Wet Coverings. Fabric coverings saturated with water, such as burlap, cotton mats, or other moisture-retaining fabrics, are commonly used for curing. These materials may be single use or reusable. Curing materials that are reusable must be properly maintained and stored so that they do not rot, tear, or crease, and all coverings must be free of any substance that is harmful to concrete or causes discoloration.

Burlaps that are fastened with polyethylene sheathing are a popular method of wet curing. Treated burlap that reflects light

and is resistant to rot and fire is also available. The requirements for burlap are described in AASHTO M 182, *Specification for Burlap Cloth Made from Jute or Kenaf and Cotton Mats,* and those for white burlap-polyethylene sheeting are described in ASTM C171, *Standard Specification for Sheet Materials for Curing Concrete* (AASHTO M 171). New burlap should be thoroughly rinsed in water to remove soluble substances that may stain the concrete and to make the burlap more absorbent prior to use on concrete.

For effective external curing, wet, moisture-retaining fabric coverings should be placed as soon as the concrete has hardened sufficiently to prevent surface damage. During the waiting period other intermediate curing methods may be used, such as fogging or the use of evaporation retarders. While applying the covering, care should be taken to cover the entire surface with wet fabric, especially at the edges of slabs where drying occurs on two or more adjacent surfaces. The coverings should be kept continuously wet so that a film of water remains on the concrete surface throughout the curing period. Use of polyethylene film over wet burlap is a good practice; it will eliminate the need for continuous watering of the covering. Periodically rewetting the fabric under the plastic with a soaker hose may be required so it does not dry out. Alternate cycles of wetting and drying during the early curing period may cause crazing of the surface.

Wet coverings of clean sand or sawdust may be useful on small projects. Wet hay or straw can be used to cure flat surfaces, although it's use is not preferred due to the many other viable options for curing. If used, it should be placed in a layer at least 150 mm (6 in.) thick and held down with wire screen, burlap, or tarpaulins to prevent the wind from blowing it away. A major disadvantage of wet sand, sawdust, hay, or straw coverings is the difficulty in removal and possibility of discoloring the concrete.

FIGURE 17-7. Impervious curing paper is an efficient means of curing horizontal surfaces.

FIGURE 17-8. Polyethylene film is an effective moisture barrier for curing concrete and easily applied to complex as well as simple shapes. To minimize discoloration, the film should be kept flat on the concrete surface.

Retaining Moisture

Covering the concrete with impervious paper or plastic sheeting or applying membrane-forming curing compounds reduces the loss of mixing water from the surface of the concrete.

Impervious Paper. Impervious paper for curing concrete consists of two sheets of kraft paper cemented together by a bituminous adhesive with fiber reinforcement. Such paper, conforming to ASTM C171 (AASHTO M 171), is an efficient means of curing horizontal surfaces and structural concrete of relatively simple shapes. An important advantage of this method is that periodic additions of water are not required. Curing with impervious paper enhances the hydration of cement by preventing loss of moisture from the concrete (Figure 17-7).

As soon as the concrete has hardened sufficiently to prevent surface damage, it should be thoroughly wetted and the widest paper available applied. Edges of adjacent sheets should be overlapped about 150 mm (6 in.) and tightly sealed with sand, wood planks, pressure-sensitive tape, mastic, or glue. The sheets must be weighed down to maintain close contact with the concrete surface during the entire curing period.

Impervious paper can be reused if it effectively retains moisture. Tears and holes should be repaired with curing-paper patches.

In addition to curing, impervious paper provides some protection to the concrete against damage from subsequent construction activity as well as protection from the direct sun. It should be light in color and non-staining to the concrete.

Plastic Sheets. Plastic sheet materials, such as polyethylene film, can be used to cure concrete (Figure 17-8). Polyethylene film is a lightweight, effective moisture retarder and is easily applied to complex as well as simple shapes. Its application is the same as described for impervious paper.

Curing with polyethylene film (or impervious paper) can cause patchy discoloration, especially if the concrete contains calcium chloride and has been finished by steel troweling.

This discoloration is more pronounced when the film becomes wrinkled. It can be difficult and time consuming on a large project to place sheet materials without wrinkles. Flooding the surface under the covering may prevent discoloration, but other means of curing should be used when uniform color is important.

Polyethylene film should conform to ASTM C171 (AASHTO M 171), which specifies a 0.10-mm (4-mil) thickness for curing concrete, and lists only clear and white opaque film. However, black film is available and is satisfactory under some conditions such as cold weather or for interior locations. White film should be used for curing exterior concrete during hot weather to reflect the sun's rays. Clear film has little effect on heat absorption.

ASTM C171 (AASHTO M 171) also includes a sheet material consisting of burlap impregnated on one side with white opaque polyethylene film. Combinations of polyethylene film bonded to an absorbent fabric such as burlap help retain moisture on the concrete surface.

Polyethylene film may also be placed over wet burlap or other wet covering materials to retain the water in the wet covering material. This procedure reduces the labor-intensive need to re-wet covering materials. There are also single-use plastic coverings available for use that help eliminate potential staining from re-use of coverings.

Liquid-Applied Evaporation Retarders. Temporary evaporation retarders (usually polymers) can be applied immediately after screeding to reduce water evaporation before final finishing operations and curing commence. These materials are sprayed onto the surface during finishing and should have no adverse effect on the concrete properties or inhibit the adhesion of membrane-curing compounds. Repeated applications of evaporation retarders followed by finishing are not recommended since these materials typically contain very high water contents (above 90%) which increase the surface w/cm.

Membrane-Forming Compounds. Liquid membrane-forming compounds consisting of waxes, resins, chlorinated rubber, and other materials can be used to retard or reduce evaporation of moisture from concrete. They are practical and the most widely used method for curing, not only for freshly placed concrete, but also for extending curing of concrete after removal of forms or after initial curing. Curing compounds should be able to maintain the internal relative humidity of the concrete surface above 80% for at least seven days to sustain the chemical reactions for continued hydration of the cementitious materials.

Membrane-forming curing compounds are of two general types: clear, or translucent; and white pigmented. Clear or translucent compounds may contain a fugitive dye that makes it easier to check visually for complete coverage of the concrete surface when the compound is applied. The dye fades away soon after application. Pigmented compounds that are reflective are recommended on sunny days as they control the concrete temperature by reducing solar-heat gain. Pigmented compounds should be agitated in the container prior to application to re-suspend pigment that may have settled out and provide uniform coverage.

Curing compounds should be applied immediately after final finishing of the concrete is completed (Figure 17-9). The concrete surface should be damp when the curing compound is applied. Application of a curing compound immediately after final finishing and before all free water on the surface has evaporated will help prevent the formation of crazing and plastic shrinkage cracks. Power-driven spray equipment is recommended for uniform application of curing compounds on large projects. Spray nozzles recommended by the product manufacturer or use of windbreaks should be arranged to prevent wind-blown loss of curing compound. Otherwise proper coverage application rates will not be achieved.

Curing compounds should be thoroughly mixed and uniformly applied and not prone to discoloration. They should not sag, run off high-spots, or collect in grooves. They should form a tough film to withstand early construction traffic without damage, with good moisture-retention properties.

Manufacturer's recommended application rates should be followed. An even single coat is applied at a typical rate of 3 m²/liter to 5 m²/liter (150 ft²/gallon to 200 ft²/gallon); but products may vary. If two coats are necessary for effective protection, the second coat should be applied perpendicular to the first. Complete coverage of the surface must be attained because even small pinholes in the membrane will result in loss of moisture from the concrete.

Note that bonding of subsequent materials might be inhibited by the presence of a curing compound even after the moisture retention characteristics of the compound have diminished. Most curing compounds are not compatible with adhesives used with floor covering materials. Consequently, they should either be tested for compatibility, or not used when bonding

FIGURE 17-9. Liquid membrane-forming curing compounds should be applied with uniform and adequate coverage over the entire surface and edges for effective, extended curing of concrete.

of overlying materials is necessary. For example, a curing compound should generally not be applied to the base of a two-course floor. Similarly, some curing compounds may affect the adhesion of paint to concrete floors or tilt-up wall panels. Curing compound and floor covering manufacturers should be consulted to determine if their products are suitable for the intended application (Kanare 2008).

Caution is necessary when using curing compounds containing solvents of high volatility in confined spaces or near sensitive occupied spaces such as hospitals because evaporating volatiles may cause respiratory problems. Applicable local environmental laws concerning volatile organic compound (VOC) emissions should be followed.

Curing compounds should conform to ASTM C309, *Standard Specification for Liquid-Membrane Forming Compounds for Curing Concrete* (AASHTO M 148). A method for determining the efficiency of curing compounds, waterproof paper, and plastic sheets is described in ASTM C156, *Standard Test Method for Water Loss [from a Mortar Specimen] Through Liquid Membrane-Forming Curing Compounds for Concrete* (AASHTO T 155). Curing compounds with sealing properties are specified under ASTM C1315, *Standard Specification for Liquid Membrane-Forming Compounds Having Special Properties for Curing and Sealing Concrete*. Reactive chemical surface treatments, generally forms of silicate compounds (such as sodium silicate), do not meet the requirements of ASTM C309 or ASTM C1315 (Tarr 2014 and ACI 302.1R 2015).

Forms Left in Place. Forms provide satisfactory protection against loss of moisture if the top exposed concrete surfaces are kept wet. A soaker hose is excellent for this application. The forms should be left on the concrete as long as practical.

Most form facing materials have a coating so as to avoid absorbing water from hardened concrete. Untreated, wood forms left in place have potential to absorb water as they dry out. They should be kept wet by sprinkling, especially during hot, dry weather. If this cannot be done, they should be removed as soon as practical and another curing method started without delay for the remaining duration of curing required. Color variations may occur from variations in timing of formwork removal and uneven curing of walls. When appearance dictates, as with decorative concrete, this should be qualified by an approved mock-up.

Modifying Concrete Temperature

Maintaining appropriate concrete temperatures is an important factor in the overall curing process to ensure the desired concrete properties may develop. Supplying heat to the concrete accelerates early strength gain. In hot and cold weather applications, concrete may need thermal curing measures (cooling or heating) for protection. Curing concrete in hot weather should follow the recommendations in Chapter 18 and ACI 305R, *Guide to Hot Weather Concreting* (2010). Recommendations for curing concrete in cold weather can be found in Chapter 19 and ACI 306R, *Guide to Cold Weather Concreting* (2016).

Steam Curing. Steam curing is advantageous when early strength gain in concrete is important or where additional heat is required to accomplish hydration, as in cold weather.

Two methods of steam curing are used: live steam at atmospheric pressure (for enclosed cast-in-place structures and large precast concrete units) and high-pressure steam in autoclaves (for small manufactured units). Only live steam at atmospheric pressure will be discussed.

A typical steam-curing cycle consists of: (1) an initial delay prior to steaming, (2) a period of increasing temperature, (3) a period for holding the maximum temperature constant, and (4) a period for decreasing the temperature. A typical atmospheric steam-curing cycle is shown in Figure 17-10.

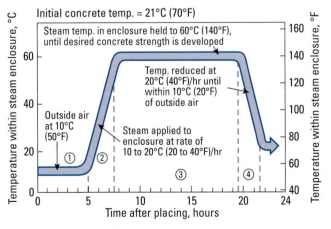

① Initial delay prior to steaming   3 to 5 hours
② Temperature increase period   2½ hours
③ Constant temperature period   6 to 12 hours*
④ Temperature decrease period   2 hours

    *Type III or high-early-strength cement,
    longer for other types

FIGURE 17-10. A typical atmospheric steam-curing cycle.

Steam curing at atmospheric pressure is generally done in an enclosure to minimize moisture and heat losses. Tarpaulins are frequently used to form the enclosure. Application of steam to the enclosure should be delayed to allow for some hardening of the concrete (typically at least 3 hours after placement of the concrete). However, a 3- to 5-hour delay period prior to steaming will achieve maximum early strength, as shown in Figure 17-11.

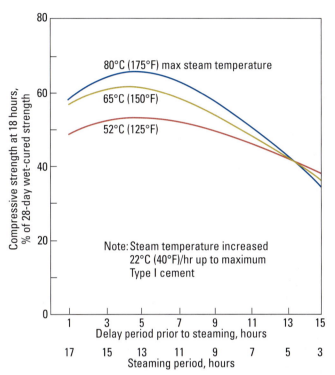

FIGURE 17-11. Relationship between strength at 18 hours and delay period prior to steaming. In each case, the delay period plus the steaming period totaled 18 hours (Hanson 1963).

Steam temperature in the enclosure should be kept at about 60°C (140°F) until the desired concrete strength has developed. Strength will not increase significantly if the maximum steam temperature is raised from 60°C to 70°C (140°F to 158°F). It is recommended that the internal temperature of concrete not exceed 70°C (158°F) to avoid delayed ettringite formation (DEF) and undue reduction in ultimate strength (see Chapter 11).

Monitoring air temperatures alone is not sufficient because the heat of hydration may cause the internal temperature of the concrete to exceed 70°C (158°F). Besides early strength gain, there are other advantages of curing concrete at temperatures of around 60°C (140°F); for example, there is reduced drying shrinkage and creep as compared to concrete cured at 23°C (73°F) for 28 days (PCI 2013, Klieger 1960, and Tepponen and Eriksson 1987).

To prevent damaging volume changes, temperatures in the enclosure surrounding the concrete should not be increased or decreased more than 20°C (36°F) per hour depending on the size and shape of the concrete element (PCI 1999).

The curing temperature in the enclosure should be maintained until the concrete has reached the desired strength. The time required will depend on the concrete mixture and steam temperature in the enclosure. See PCI (1999 and 2013) for more information.

Electrical, Oil, Microwave, and Infrared Curing. Electrical, hot oil, microwave, and infrared curing methods have been available for accelerated and normal curing of concrete for many years. Electrical curing methods include a variety of techniques: use of the concrete itself as the electrical conductor, use of reinforcing steel as the heating element, use of a special wire as the heating element, electric blankets, and the use of electrically heated steel forms (presently the most popular method). Electrical heating is especially useful in cold-weather concreting. Hot oil, hot water, or closed loop steam may be circulated through pipes surrounding the steel forms to heat the concrete. Infrared heating and microwave radiation have had limited use in accelerated curing of concrete. Concrete that is cured by infrared methods is usually under a covering or enclosed in steel forms. Electrical, oil, and infrared curing methods are used primarily in the precast concrete industry.

Insulating Blankets or Covers. Formwork can be economically insulated with commercial blanket (Figure 17-12) or batt insulation that has a tough moisture-proof covering. Suitable insulating blankets are manufactured of fiberglass, sponge rubber, cellulose fibers, mineral wool, vinyl foam, and open-cell polyurethane foam. Sprayed on insulation foam may be applied to steel or wood forms. When insulated formwork is used, care should be taken to ensure that concrete temperatures do not become excessive.

Layers of dry, porous material such as straw or hay can be used to provide insulation against freezing of concrete when temperatures fall below 0°C (32°F).

During cold weather, additional heat is often required to maintain favorable concrete curing temperatures of 10°C to 20°C (50°F to 70°F). Framed enclosures of canvas tarpaulins, reinforced polyethylene film, or other materials can be placed around the structure and heated by indirect fired or properly vented space heaters or steam. Portable hydronic heaters are used to thaw subgrades as well as heat concrete without the use of an enclosure. In all cases, care must be taken to avoid loss of moisture from the concrete. Use of supplemental heaters that are un-vented can lead to a build-up of carbon monoxide that can pose a dangerous health hazard to the construction personnel. Exposure of fresh concrete to heater or engine exhaust gases must be avoided as this can result in surface deterioration and dusting (see Chapter 19).

Self-Annealing Concrete. Self-annealing concrete refers to the process in which the mixing water and internal heat of hydration generated by concrete are retained within the formwork to

FIGURE 17-12. Concrete footing pedestal being covered with a tarpaulin to retain the heat of hydration.

accelerate the curing process. This accelerates early strength gain while maintaining moisture, then gradually allows concrete to cool to ambient temperature or to a point where the concrete strength can withstand thermal stresses prior to removing the formwork protection (Table 17-1).

Capturing and controlling the heat of hydration can be accomplished using either passive or active methods. Passive methods can utilize insulating materials placed within existing formwork in direct contact with concrete that stay in place after the conventional formwork is stripped until the desired properties have been achieved. Stay-in-place insulated concrete forms (ICFs) can also be used with similar results.

CURING DURATION AND TEMPERATURE

The period of time that concrete should be protected from freezing, abnormally high temperatures, and against loss of moisture depends upon a number of factors: the required strength and durability; type and quality of cementing materials used; mixture proportions; required strength, size and shape of the concrete member; ambient conditions; and future exposure conditions.

In accordance with ACI 308.1, *Specification for Curing Concrete* (2011), for concrete slabs on ground (floors, pavements, canal linings, parking lots, driveways, sidewalks) and for structural concrete (cast-in-place walls, columns, slabs, beams, small footings, piers, retaining walls, bridge decks), the length of the curing period for ambient temperatures above 10°C (50°F) should be a minimum of 7 days. The curing period may be shortened to 3 days for high-early-strength concretes, in accordance with ACI 301, *Specification for Concrete Construction* (2020). In cold weather concreting, additional time may be needed to attain 70% of the specified compressive or flexural strength. When the daily mean ambient temperature is 10°C (50°F) or lower, ACI 306R (2016) recommendations for curing and protection period should be followed to prevent damage by freezing (see Chapter 19).

TABLE 17-1. Compressive Strength Development of Different Curing Methods (Ciuperca 2013)

CONCRETE MIX DESIGN	FORM / CURING TYPE	COMPRESSIVE STRENGTH, MPa (psi)				
		7 DAYS	28 DAYS	58 DAYS	90 DAYS	14 MONTH
320 kg/m³ (540 lb/yd³) portland cement 71 kg/m³ (120 lb/yd³) fly ash (Class F)	Conventional form	22.3 (3240)	32.1 (4660)	38.9 (5640)	42.7 (6190)	46.9 (6810)
	Standard cured ASTM C39	21.8 (3170)	38.3 (5555)	41.0 (5960)	50.7 (7360)	N/A
	Self-annealed	42.6 (6180)	45.6 (6610)	47.3 (6860)	47.5 (6890)	55.0 (7980)

The curing period may be 3 weeks or longer for lean concrete mixtures used in massive structures such as dams; conversely, it may be only a few days for higher cement contents, especially if Type III or HE cement is used. Steam-curing periods are normally much shorter, ranging from a few hours to 3 days; but generally 24-hour cycles are used. Since all the desirable properties of concrete are improved by curing, the curing period should be as long as necessary and reasonable.

When wet curing is interrupted, the development of strength continues for a short period and then stops after the concrete's internal relative humidity drops below about 80%. However, if wet curing is resumed, strength development will be reactivated, but the original potential strength may no longer be achieved. Therefore, it is best to wet-cure the concrete continuously from the time it is placed and finished until it has gained sufficient strength, impermeability, and durability.

On hardened concrete and on flat concrete surfaces in particular, curing water should not be more than about 11°C (20°F) cooler than the concrete. This will minimize cracking caused by thermal stresses due to temperature differentials between the concrete and curing water.

The curing period should be prolonged for concretes made with cementing materials possessing slow-strength-gain characteristics, especially in cold weather. Since supplementary cementitious materials (SCMs) generally do not hydrate as rapidly as portland cements, concrete mixtures that include SCMs (the majority of modern concretes) may require longer curing to allow optimal properties to develop.

For mass concrete (large piers, locks, abutments, dams, heavy footings, and massive columns and transfer girders) in which no pozzolan is used as part of the cementitious material, curing of unreinforced sections should continue for at least 2 weeks. If the mass concrete contains an SCM, minimum curing time for unreinforced sections should be extended to at least 3 weeks. Heavily reinforced mass concrete sections should be cured for a minimum of 7 days.

Hydration proceeds at a much slower rate when the concrete temperature is low. Temperatures below 10°C (50°F) are unfavorable for the development of early strength; below 4°C (40°F) the development of early strength is greatly retarded; and at or below freezing temperatures, down to -10°C (14°F), little or no strength develops. A higher curing temperature provides earlier strength gain in concrete than a lower temperature but it may decrease 28-day strength as shown in Figure 17-13.

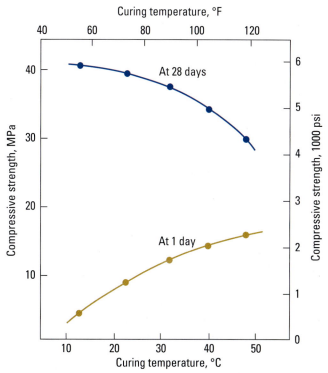

FIGURE 17-13. One-day strength increases with increasing curing temperature, but 28-day strength decreases with increasing curing temperature (Verbeck and Helmuth 1968).

Maturity

The concept of maturity can be used to estimate the development of strength when there is variation in the curing temperature of the concrete. The maturity method takes into account the influence of concrete temperature on hydration, using strength and temperature calibration data for mixtures over short time intervals at early ages. It follows that concrete should be protected so that its temperature remains favorable for hydration and moisture is not lost during the early hardening period. See Chapter 19 for more information on the Maturity Method.

Field Cured Cylinders

If strength tests are used to establish the time when curing can cease or forms can be removed, representative concrete test cylinders or beams may be fabricated in the field, located adjacent to the structure, and cured using the same methods. This practice is described by ASTM C31, *Standard Practice for Making and Curing Concrete Test Specimens in the Field* (AASHTO T 23). While the intent of field-cured cylinders is to replicate the curing conditions of the concrete in service, the difference in volume of the cylinders versus the structure dimensions will lead to differing results in temperature and moisture.

Match Curing

Equipment is available that can monitor internal concrete temperatures and match that temperature and relative humidity in a concrete curing box; this is the most accurate means of representing in-place concrete strengths.

Cores, cast-in-place removable cylinders, and nondestructive testing methods may also be used to determine the in-place strength of a concrete member. See Chapter 20 for more information.

SELECTING CURING METHOD BASED ON CONSTRUCTION APPLICATION

The most effective method for curing concrete depends on the ambient conditions during construction, materials used, method of construction, and the intended use of the hardened concrete. For special projects, final curing measures can involve many time-consuming options. Often, more than one type of curing method may be used. For example, fog spraying or plastic covered wet burlap can precede application of a curing compound. The effectiveness and appropriateness of a curing method should be evaluated for each concrete project.

Paving, Bridge Decks, and Floor Slabs

An exposed concrete surface is particularly susceptible to insufficient hydration because it dries first. All exposed surfaces, especially exposed edges and joints, must be protected against moisture evaporation. Initial curing and protective measures should be initiated immediately to reduce the chance for plastic shrinkage cracking and crazing.

Slabs should be cured by wet curing methods when feasible or by impervious paper or plastic sheets. Curing compounds should not be used with floor coverings unless approved by flooring manufacturer. Floors that will receive moisture sensitive floor coverings should not be wet cured if drying time is an issue with construction schedule.

Concrete mixtures with high cement contents and low water-cementing materials ratios (less than 0.40), such as that often used for high-performance bridge decks may require special curing needs. Membrane-forming curing compounds may not retain enough water in the concrete. Therefore, fogging and wet curing may become necessary to maximize hydration (Copeland and Bragg 1955).

For lower w/cm concrete mixtures, the permeability of the paste is normally so low that externally applied curing water will not penetrate far beyond the surface layer (ACI 308R 2016 and Meeks and Carino 1999). Therefore, bulk properties such as compressive strength can be considerably less sensitive to surface moisture conditions at lower w/cm; however other surface properties such as abrasion and scaling resistance can be markedly improved by wet-curing low w/cm concrete. Fogging during and after placing and finishing also helps minimize plastic cracking in concretes with very low water-cement ratios.

Cast-in-Place Concrete Structures

Formwork can be left in place as a method of curing cast-in-place concrete. Exposed surfaces need to be protected from excessive moisture loss. Care also must be taken to ensure concrete temperatures within the forms are within project specifications. This may require the use of soaker hoses, heating or cooling forms, or applying insulated coverings. Once forms are removed, additional curing methods, such as compounds and coverings, may be desired depending on curing duration.

Mass Concrete

Water curing is sometimes specified for mass concrete. Typically, this is inappropriate because it artificially cools the surface only, which increases the likelihood of thermal cracking. When water curing is specified for mass concrete, heated water must be used. The temperature of the curing water should be high enough that it does not cool the concrete. This can be dangerous and costly when the temperature difference limit of 20°C (35°F).

In most cases, water retaining curing methods are more appropriate for mass concrete. Such methods include, form curing, membrane curing, and the use of a curing compound. See Chapter 18; ACI 207.1R, *Guide to Mass Concrete* (2005); and Gajda (2007) for more information on thermal control and curing of mass concrete.

Precast Concrete

Precast concrete is typically prepared by accelerated curing to produce high early strengths. As with mass concrete, control of elevated-temperature curing process is required during the manufacture of precast concrete to prevent excessive internal temperatures, and minimize temperature gradients (see Chapter 11, ACI 308 2016, and PCI 2013 for more information). Other methods for curing precast concrete include conventional wet curing by methods that supply additional moisture and retain mixing water.

Decorative Concrete

For textured surfaces, care should be taken to allow the concrete to set sufficiently so that the texture is not marred during curing. The recommended curing method for new concrete placements to receive stains is unwrinkled, non-staining, high quality curing paper. Water curing may be appropriate for decorative concrete including pigments and stains, but there have been reports of efflorescence problems attributed to water transporting soluble salts to the slab surface. Curing compounds are incompatible with staining applications since these materials prevent stain penetration. For more information on curing decorative concrete see Kosmatka and Collins (2004), and ACI 310R, *Guide to Decorative Concrete* (2019).

OTHER CONSIDERATIONS

Sealing Compounds

Sealing compounds (sealers) are liquids applied to the surface of hardened concrete to reduce the penetration of liquids or gases such as water (Figure 17-14), deicing solutions, and aggressive chemicals to protect concrete from freeze-thaw damage, corrosion of reinforcing steel, alkali-silica reactivity, and acid attack. In addition, sealers used on interior floor slabs reduce dusting and the absorption of spills while making the concrete surface easier to clean.

Sealers are fundamentally different from curing compounds. The primary purpose of a curing compound is to reduce the loss of water from newly placed concrete and it is applied immediately after finishing. Sealers, on the other hand, retard the penetration of harmful substances into hardened concrete and are typically not applied until the concrete is of sufficient age and sufficiently dry to allow sealers to penetrate the surface of concrete. Surface sealers are generally classified as either film-forming or surface penetrating.

Film-Forming Sealers

Film-forming sealing compounds do not penetrate much beyond the surface of the concrete. The relatively large molecular structure of these compounds limits their ability to penetrate the surface. These materials can impede the penetration of water, and protect against mild chemicals. They may also prevent the absorption of grease and oil as well as reduce dusting under light traffic.

Film-forming surface sealers consist of acrylic resins, chlorinated rubber, urethanes, epoxies, and alpha methylstyrene. The effectiveness of film-forming sealers depends on the continuity of the layer formed. Film-forming sealers may alter the color and

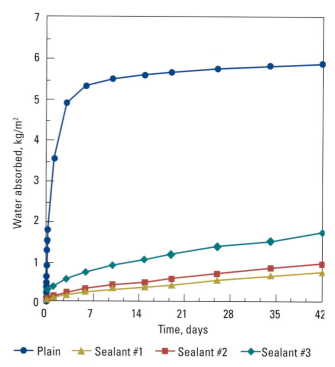

FIGURE 17-14. Sealants are shown to be effective at reducing the amount of water absorbed into concrete (Golias 2010).

sheen of the concrete surface. Since they are films, they are subject to wear by abrasion, grit, and traffic, which can cause damage requiring the reapplication of the material. Consult manufacturers' application recommendations because some of these materials are intended for interior use only and may yellow and deteriorate under exposure to ultraviolet light.

Surface Penetrating Sealers

Water-repellent penetrating sealers have a very small molecular size that allows penetration and saturation of the concrete as deep as 3 mm (⅛ in.). The two most common are silane and siloxane, compounds which are derived from the silicone family. These sealers allow the concrete to breathe, thus preventing a buildup of vapor pressure between the concrete and sealer that can occur with some film-forming materials. However, periodic reapplication is recommended. In northern states and coastal areas silanes and siloxanes are popular for protecting bridge decks and other exterior structures from corrosion of reinforcing steel caused by chloride infiltration from deicing chemicals or sea spray.

The penetrating sealer used most extensively historically has been a mixture of 50% boiled linseed oil and 50% mineral spirits (AASHTO M 233, *Standard Specification for Boiled Linseed Oil Mixture for Treatment of Portland Cement Concrete*). Due to volatile organic compound (VOC) limitations, and regulations, this type of sealer can no longer be used in some jurisdictions. Although this mixture is an effective sealer, it has two main disadvantages: it darkens the concrete, and periodic reapplication is necessary for long-term protection.

It is recommended that at least 28 days should be allowed to elapse before applying sealers to new concrete. Application of any sealer should only be done on concrete that is clean and allowed to dry for at least 24 hours at temperatures above 16°C (60°F). Penetrating sealers cannot fill surface voids if they are filled with water. Some surface preparation may be necessary if the concrete is dirty. Concrete placed in the late fall should not be sealed until spring because the sealer may cause the concrete to retain water that may exacerbate freeze-thaw damage.

The precautions outlined earlier regarding volatile solvents in curing compounds also apply to sealing compounds. The scale resistance provided by concrete sealers should be evaluated based on criteria established in ASTM C672, *Standard Test Method for Scaling Resistance of Concrete Surfaces Exposed to Deicing Chemicals*. For more information on surface sealing compounds, see AASHTO M 224, S*tandard Specification for Use of Protective Sealers for Portland Cement Concrete;* ACI 515.2R, *Guide to Selecting Protective Treatments for Concrete* (2013); and Kerkoff (2007).

Curing prior to applying sealers is an absolute must when using a sealer; curing is necessary to produce a lower permeability and higher durability of the concrete surface. Satisfactory freeze-thaw durability of exterior concrete still primarily depends on an adequate air-void system, sufficient strength, and the use of proper placing, finishing, and curing techniques (see Chapter 11).

REFERENCES

PCA's online catalog includes links to PDF versions of many research reports and other classic publications.
Visit: cement.org/library/catalog.

ACI Committee 207, *Guide to Mass Concret*e, ACI 207.1R-05, American Concrete Institute, Farmington Hills, MI, 2005, reapproved 2012, 30 pages.

ACI Committee 213, *Guide for Structural Lightweight Aggregate Concrete*, ACI 213R-14, American Concrete Institute, Farmington Hills, MI, 2014, 53 pages.

ACI Committee 301, *Specification for Concrete Construction*, ACI 301-20, American Concrete Institute, Farmington Hills, MI, 2020, 73 pages.

ACI Committee 305, *Guide Hot-Weather Concreting*, ACI 305R-10, American Concrete Institute, Farmington Hills, MI, 2010, 23 pages.

ACI Committee 306, *Guide to Cold Weather Concreting*, ACI 306R-16, American Concrete Institute, Farmington Hills, MI, 2016, 28 pages.

ACI Committee 308, *Specification for Curing Concrete*, ACI 308.1-11, American Concrete Institute, Farmington Hills, MI, 2011, 7 pages.

ACI Committee 308, *Guide to External Curing of Concrete*, ACI 308R-16, American Concrete Institute, Farmington Hills, MI, 2016, 36 pages.

ACI Committee 308 and Committee 213, *Report on Internally Cured Concrete Using Prewetted Absorptive Lightweight Aggregate*, ACI (308-213)R-13, American Concrete Institute, Farmington Hills, MI, 2013, 12 pages.

ACI Committee 310, *Guide to Decorative Concrete*, ACI 310R-19, American Concrete Institute, Farmington Hills, MI, 2019, 48 pages.

ACI Committee 515, *Guide to Selecting Protective Treatments for Concrete*, ACI 515.2R-13, American Concrete Institute, Farmington Hills, MI, 2013, 29 pages.

Bohan, R.P.; and Ries, J., *Structural Lightweight Aggregate Concrete*, IS032, Portland Cement Association, Skokie, IL, 2008, 8 pages.

Burg, R.G., *The Influence of Casting and Curing Temperature on the Properties of Fresh and Hardened Concrete*, Research and Development Bulletin RD113, Portland Cement Association, 1996, 20 pages.

Cather, R., "How to Get Better Curing," *Concrete, The Journal of the Concrete Society*, London, V. 26, No. 5, Sept.-Oct., 1992, pages 22 to 25.

Ciuperca, R., U.S. Patent No. 8,545,749, Oct 1, 2013, *Concrete Mix Composition, Mortar Mix Composition and Method of Making and Curing Concrete or Mortar and Concrete or Mortar Objects and Structures*.

Copeland, L.E.; and Bragg, R.H., *Self-Desiccation in Portland Cement Pastes*, Research Department Bulletin RX052, Portland Cement Association, 1955, 13 pages.

Gajda, J., *Mass Concrete for Buildings and Bridges*, EB547, Portland Cement Association, Skokie, IL, 2007, 44 pages.

Golias, M., *The Use of Soy Methyl Ester-Polystyrene Sealants and Internal Curing to Enhance Concrete Durability*, MS thesis, Purdue University, West Lafayette, IN, 2010, 137 pages.

Gonnerman, H.F.; and Shuman, E.C., "Flexure and Tension Tests of Plain Concrete," Major Series 171, 209, and 210, *Report of the Director of Research*, Portland Cement Association, November 1928, pages 149 and 163.

Hanson, J.A., *Optimum Steam Curing Procedure in Precasting Plants*, with discussion, Development Department Bulletins DX062 and DX062A, Portland Cement Association, 1963, 28 pages and 19 pages, respectively.

Holm, T.; and Ries, J., "Lightweight Concrete and Aggregates," Chapter 46, *Significance of Tests and Properties of Concrete and Concrete Making Materials*, STP169D, ASTM International, West Conshohocken, PA, 2006, pages 546 to 560.

Kanare, H.M., *Concrete Floors and Moisture*, EB119, Portland Cement Association, Skokie, Illinois, and National Ready Mixed Concrete Association, Silver Spring, MD, 2008, 176 pages.

Kerkhoff, B., *Effect of Substances on Concrete and Guide to Protective Treatments*, IS001, Portland Cement Association, 2007, 36 pages.

Klieger, P., *Some Aspects of Durability and Volume Change of Concrete for Prestressing*, Research Department Bulletin RX118, Portland Cement Association, 1960, 15 pages.

Kosmatka, S.H.; and Collins, T.C., *Finishing Concrete Slabs with Color and Texture*, PA124, Portland Cement Association, Skokie, IL, 2004, 72 pages.

Lam, H., *Effects of Internal Curing Methods on Restrained Shrinkage and Permeability*, SN2620, Portland Cement Association, Skokie, IL, 2005, 134 pages.

Meeks, K.W.; and Carino, N.J., *Curing of High-Performance Concrete: Report of the State-of-the-Art*, NISTR 6295, National Institute of Standards and Technology, Building and Fire Research Laboratory, Gaithersburg, MD, March 1999, 191 pages.

PCI, *Manual for Quality Control for Plants and Production of Structural Precast Concrete Products*, MNL 116-99. 4th Edition, Precast/Prestressed Concrete Institute, Chicago, IL, 1999, 328 pages.

PCI, *Manual for Quality Control for Plants and Production of Architectural Precast Concrete Products*, MNL 117-13. 4th Edition, Precast/Prestressed Concrete Institute, Chicago, IL, 2013, 334 pages.

Stanley, C., *How Hot is Your Concrete and Does it Really Matter? Heat Evolution and its Effects on Concrete Cast on Construction Sites in Asia*, 30th Conference on Our World in Concrete and Structures, Singapore, 23-24 August 2005, 11 pages.

Tarr, S., "A Condensed Look at Liquid Densifiers." *Concrete Construction*, January 2014, pages 41 to 16.

Tepponen, P.; and Eriksson, B., "Damages in Concrete Railway Sleepers in Finland," *Nordic Concrete Research*, No. 6, The Nordic Concrete Federation, Oslo, 1987, pages 199 to 209.

Verbeck, G.J.; and Helmuth, R.A., "Structures and Physical Properties of Cement Pastes," *Proceedings, Fifth International Symposium on the Chemistry of Cement*, Vol. III, The Cement Association of Japan, Tokyo, 1968, pages 9 to 41.

HOT WEATHER CONCRETING

FIGURE 18-1. Nighttime placement using liquid nitrogen to reduce concrete temperature.

Hot weather conditions can adversely influence concrete quality primarily by accelerating the rate of cement hydration and moisture evaporation. Projects placed in hot weather may require changes to the concrete mixture, modifying the concrete environment, or shifting the construction schedule for more favorable conditions to accommodate higher temperatures (Figure 18-1).

Detrimental hot weather conditions include:

- high ambient air temperature,
- high concrete temperature,
- low relative humidity,
- high wind speeds, and
- solar radiation.

Hot weather conditions can create problems in concrete, including lower later-age strength development (due to accelerated early-age strength development), and other issues such as:

- increased water demand,
- accelerated slump loss,
- difficulty controlling entrained air content,
- faster rate of setting,
- plastic shrinkage cracking and crazing,
- thermal cracking, and
- unplanned cold joints.

WHEN TO TAKE PRECAUTIONS

Ideally, concrete would be placed at temperatures of 10°C to 15°C (50°F to 60°F) to maximize beneficial concrete properties; however these temperatures are not always practical or possible. Concrete temperatures of 10°C to 27°C (50°F to 80°F) during placement are more reasonable to expect and are acceptable for the majority of concrete placements without detriment.

IMPACT OF HOT WEATHER ON CONCRETE PROPERTIES

Workability

Weather conditions at a jobsite rarely coincide with laboratory conditions in which concrete mixtures are designed and specimens are stored and tested in a controlled environment. As concrete temperature increases, slump loss is often compensated for by adding water to the concrete at the project site. At higher temperatures a greater amount of water is required to hold slump constant than is needed at lower temperatures. The addition of water results in a higher water-cementitious materials ratio, which lowers the strength and adversely affects other desirable properties of the hardened concrete. This is in addition to an adverse effect on strength at later ages due to the higher temperature, even without the addition of water. Retarding and water-reducing admixtures can be used to control slump loss.

As shown in Figure 18-2, if the temperature of freshly mixed concrete is increased from 10°C to 38°C (50°F to 100°F), about 20 kg/m³ (33 lb/yd³) of additional water is needed to maintain the same 75-mm (3-in.) slump. This additional water can reduce strength by 12% to 15% and produce results that may not comply with project specifications. Adjusting mixture proportions to meet the higher water demand while maintaining w/cm will improve the concrete strength, however concrete durability and resistance to cracking (due to volume change) will still be impacted by the higher water content. Also, the higher cementitious materials content necessary to maintain the w/cm with additional mixing water will further increase the concrete temperature and water demand.

Heat of Hydration

Heat generated during cement hydration raises the temperature of concrete in varying degrees depending on the size of the concrete placement, its surrounding environment, and the amount of cement in the concrete. As a general rule, a total temperature rise of about 2°C to 9°C per 45 kg (5°F to 16°F per 100 lb) of portland cement can be expected from heat of hydration depending on the thickness of the concrete. Hot-weather concrete work and mass concrete placements require preventative measures to address the generation of heat from cement hydration and related thermal volume changes (see **Thermal Control of Mass Concrete**).

Setting Time

High temperatures of freshly mixed concrete increase the rate of setting and shorten the length of time within which the concrete can be transported, placed, and finished. As a general rule of thumb, the setting time is reduced by about 33% for every 5°C (10°F) increase in the initial concrete temperature. For example, if the initial setting time of a particular concrete is 2 hours, and the fresh concrete temperature at initial set is 15°C (60°F), the same concrete mixture's initial setting time can be expected to be about 90 minutes when the temperature is 20°C (68°F). However, cementitious materials behave differently and do not always follow this generalization. Figure 18-3 shows that the setting time was reduced by 2 or more hours with a 10°C (18°F) increase in concrete temperature.

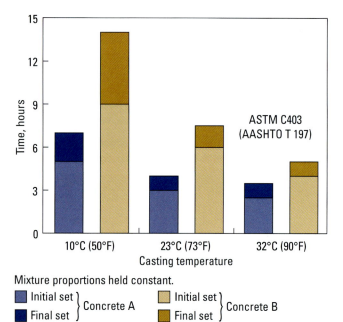

FIGURE 18-3. Mortar extracted from concrete. Effect of concrete temperature on setting time. "Concrete A" refers to mixtures of 356 kg/m³ (600 lb/yd³) of Type I cement at a water-cement ratio of 0.45, while "Concrete B" refers to mixtures of 335 kg/m³ (564 lb/yd³) of Type I/II cement with a water-cement ratio of 0.46 (Burg 1996).

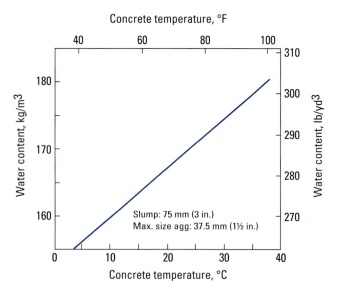

FIGURE 18-2. The water requirement of a concrete mixture increases with an increase in concrete temperature (Bureau of Reclamation 1981).

However, laboratory settings do not adequately demonstrate all of the variables involved in setting of concrete and timing of finishing operations in the field (Lee and Hover 2016). Concrete should remain plastic long enough so that it can be placed, consolidated, and finished to the desired surface quality without developing unwanted cold joints.

Strength Gain

Figure 18-4 and 18-5 show the effect of concrete temperatures on compressive strength for portland cement-only mixtures (no SCMs). The concrete temperatures at the time of mixing, casting, and curing were 23°C, 32°C, 41°C, and 49°C (73°F, 90°F, 105°F, and 120°F). After 28 days, the specimens were all wet-cured at 23°C (73°F) until the 90-day and one-year test ages. The tests, using identical concretes of the same water-cement ratio (0.45), show that while higher concrete temperatures produce higher early strengths than concrete at 23°C (73°F), at later ages concrete strengths are lower.

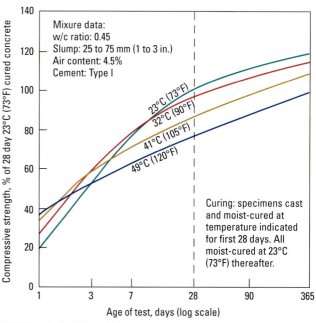

FIGURE 18-4. Effect of high concrete temperatures on compressive strength at various ages (Klieger 1958).

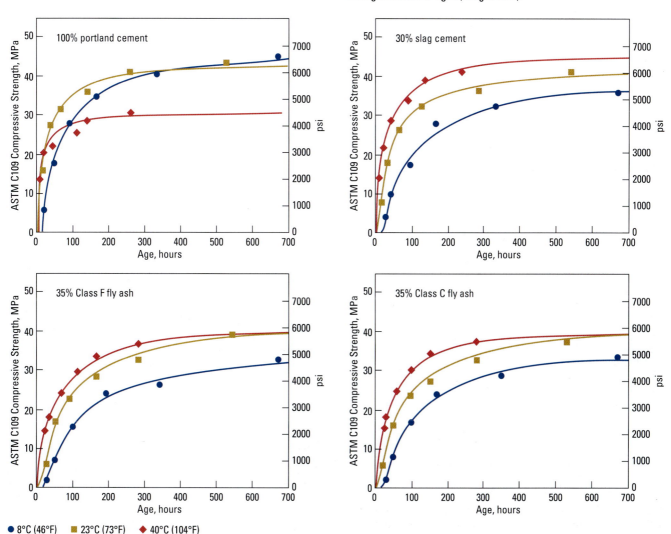

● 8°C (46°F) ■ 23°C (73°F) ◆ 40°C (104°F)

FIGURE 18-5. Compressive strength results for ASTM C109 mortars made with various cementitious materials and cured at different temperatures (Brooks and others 2007).

The proper fabrication, curing, and testing of strength test specimens during hot weather is critical. Steps should be taken to assure ASTM C31, *Standard Practice for Making and Curing Concrete Test Specimens in the Field* (AASHTO T 23) procedures are followed regarding initial curing of strength specimens for acceptance or quality control testing. Concrete mixtures with a specified strength less than 40 MPa (6000 psi) should have an initial curing temperature between 16°C and 27°C (60°F and 80°F). For concrete mixtures with a specified strength of 40 MPa (6000 psi) or greater, the initial curing temperature should be between 20°C and 26°C (68°F and 78°F). If the initial 24-hour curing is at 38°C (100°F), the 28-day compressive strength of the test specimens may be 10% to 15% lower than if cured at the required ASTM C31 (AASHTO T 23) curing temperatures (Gaynor and others 1985).

Specimens cured in the field in the same manner as the structure (match cured) more closely represent the actual strength of concrete in the structure at the time of testing. However, test specimens give little indication of whether a deficiency is due to the quality of the concrete as delivered or to the influence of handling and curing. On some projects, field-cured specimens are made in addition to standard cured specimens used for acceptance of concrete. Field cured cylinders do not always represent the concrete in service. Match cured cylinders are especially useful when the weather is unfavorable, to determine when forms can be removed, or when the structure can be put into service. For more information see Chapter 17.

Plastic Shrinkage Cracking and Crazing

In hot weather, there is an increased tendency for cracks to form in both the fresh and hardened concrete. Rapid evaporation of water from freshly placed concrete can cause plastic shrinkage cracks before the surface has hardened. Cracks may also develop in the hardened concrete because of increased drying shrinkage due to higher water contents or thermal volume changes as the concrete cools (see Chapter 10).

Concrete surfaces should dry out slowly after the curing period to reduce the possibility of surface crazing and cracking.

The following conditions, individually or collectively, increase the possibility of plastic shrinkage cracking and crazing:

- lower relative humidity,
- increased wind speed,
- increased concrete temperature,
- higher cementitious materials content,
- lower w/cm, and
- concrete mixtures that have low bleed rates (including silica fume mixtures).

Plastic shrinkage cracking and crazing (Figures 18-6 and 18-7) are often associated with hot weather concreting; however, they can occur any time ambient conditions produce rapid evaporation of moisture from the concrete surface (including cold weather concreting). Plastic shrinkage cracks and crazing occur when surface moisture evaporates faster than the rate at which bleed water reaches the surface (See Chapter 16 for more information on crazing due to plastic shrinkage and crazing).

The nomograph in Figure 18-8 is a graphical solution intended to estimate the rate of evaporation from a body of water that is often used to estimate rate of evaporation of bleed water from concrete (Kohler 1952, Menzel 1954, and NRMCA 1960).

Menzel (1954) adopted the Kohler (1952) equations, simply converting the units used to express vapor pressure and wind speed as shown in Equation 18-1. The values for the saturation vapor pressure of water, are themselves temperature dependent:

$$W = 0.315 \, (e_o - e_a)(0.253 + 0.060 V) \; (kPa)$$
$$W = 0.44 \, (e_o - e_a)(0.253 + 0.096 V) \; (psi)$$

EQUATION 18-1.

FIGURE 18-6. Crazing cracks are a network of fine cracks shown, compared to a drying shrinkage stress crack.

FIGURE 18-7. Typical plastic shrinkage cracks.

Uno (1998) used equations for saturation vapor pressure and combined them with the Kohler/ Menzel equation to produce Equation 18-2 that takes vapor pressure into account:

$$E = 5([T_c + 18]^{2.5} - r \cdot [T_a + 18]^{2.5})(V + 4) \times 10^{-6} \text{ (SI)}$$
$$E = (T_c^{2.5} - r \cdot T_a^{2.5})(1 + 0.4V) \times 10^{-6} \text{ (inch-pound)}$$

EQUATION 18-2.

Where (Equation 18-1 and 18-2):

W = mass of water evaporated in kg (lb) per m² (ft²) of water-covered surface per hour

e_o = saturation water vapor pressure in kPa (psi) in the air immediately over the concrete surface, at the concrete temperature. Obtain e_o from chemical handbooks (for example Weast [1987]), or Table 18-1.

e_a = water vapor pressure in kPa (psi) in the air surrounding the concrete obtained by multiplying the saturation vapor pressure at the temperature of the air surrounding the concrete by the relative humidity of the air. Air temperature and relative humidity are measured approximately 1.2 m to 1.8 m (4 ft to 6 ft) above the concrete surface on the windward side and shielded from direct sunlight.

V = average wind speed in km/h (mph), measured at 0.5 m (20 in.) above the concrete surface.

E = evaporation rate, kg/m²/h (lb/ft²/h)

T_c = concrete (water surface) temperature, °C (°F)

r = (relative humidity %)/100

T_a = air temperature, °C (°F)

When the rate of evaporation of bleed water exceeds 1 kg/m² (0.2 lb/ft²) per hour, precautionary measures such as windscreens or fogging are recommended. With some concrete mixtures, such as those containing certain supplementary cementing materials (SCMs), cracking is possible if the rate of evaporation exceeds 0.5 kg/m² (0.1 lb/ft²) per hour. Concrete containing silica fume is particularly prone to plastic shrinkage because bleeding rates are commonly as low as 0.25 kg/m² (0.05 lb/ft²) per hour, producing little to no surface bleed water. Therefore, for mixtures with low bleed rates protection from premature drying is essential even at lower evaporation rates.

One or more of the precautions listed below may minimize the occurrence of plastic shrinkage cracking. They should be considered during planning and execution:

• Keep the concrete temperature low by cooling aggregates and mixing water.

• Use (appropriate volume) of synthetic fibers in the concrete mixture.

• Prewet aggregates that are dry and absorptive prior to batching concrete.

• Dampen the subgrade and fog forms prior to placing concrete.

• Erect temporary windbreaks and sunshades to reduce wind velocity and solar radiation over the concrete surface.

• Fog the slab immediately after placing and before finishing, taking care to prevent the accumulation of water that may increase the w/cm at the surface. Finishing a wet surface will increase the w/cm and weaken the concrete surface.

• Protect the concrete with temporary coverings, such as polyethylene sheeting, during any extended delay between placing and finishing. See Chapter 17 for more information on initial and intermediate curing.

TABLE 18-1. Saturation water vapor pressure (e_o)*

SI UNITS		INCH-POUND UNITS	
CONCRETE TEMPERATURE, °C	SATURATION WATER VAPOR PRESSURE, kPa	CONCRETE TEMPERATURE, °F	SATURATION WATER VAPOR PRESSURE, psi
5	0.872	40	0.121
10	1.230	50	0.178
15	1.705	60	0.257
20	2.338	70	0.363
25	3.167	80	0.508
30	4.242	90	0.697
35	5.623	100	0.951
40	7.374	110	1.273
45	9.582	120	1.694

* Adapted from ACI 305R (2010).

To use these charts:

1. Enter with air temperature, move up to relative humidity.
2. Move right to concrete temperature.
3. Move down to wind velocity.
4. Move left: read approximate rate of evaporation.

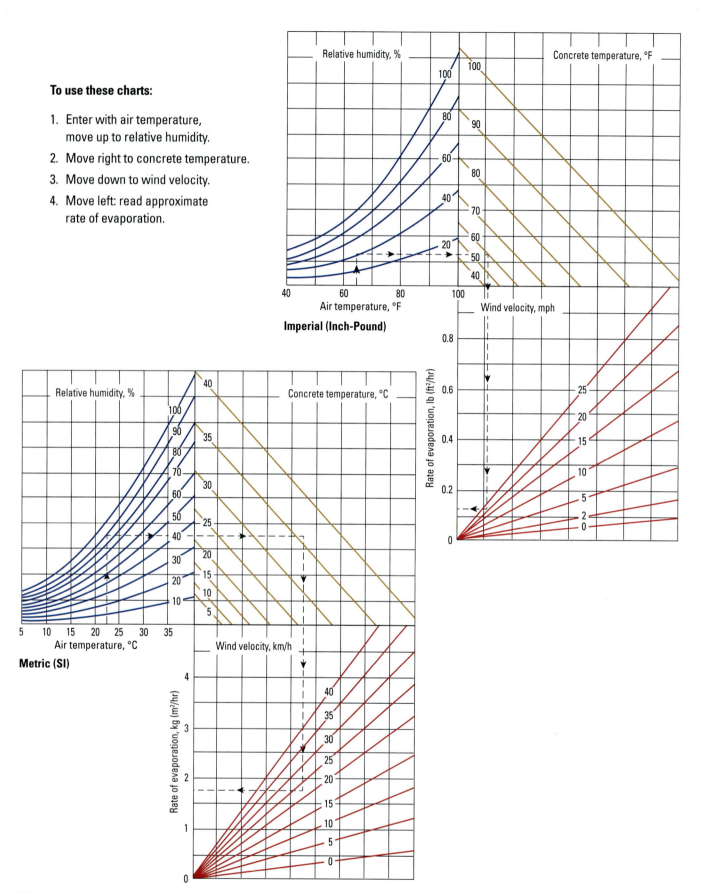

FIGURE 18-8. Effect of concrete and air temperatures, relative humidity, and wind velocity on rate of evaporation of surface moisture from concrete. Wind speed is the average horizontal air or wind speed in km/h (mph) measured at 500 mm (20 in.) above the evaporating surface. Air temperature and relative humidity should be measured at a level approximately 1.2 m to 1.8 m (4 ft to 6 ft) above the evaporating surface and on the windward side shielded from the sun's rays (Kohler 1952, Menzel 1954, and NRMCA 1960).

PREPARATION BEFORE CONCRETING DURING HOT WEATHER

In hot weather construction, it is impractical to recommend a single maximum ambient temperature. Atmospheric conditions, including air temperature, relative humidity and wind speed, in conjunction with site conditions influence the precautions needed. For example, flatwork done under a roof that blocks solar radiation with exterior walls in place that screen the wind could be successfully completed using a concrete with a high temperature. However, this same concrete could cause difficulty if placed outdoors because of direct exposure to sun and wind. Additionally, placement of concretes with a high surface-to-volume ratio, such as thin overlays with relatively high cementitious contents, require special care.

ACI 305R, *Guide to Hot Weather Concreting* (2010), recommends advanced planning for hot weather conditions for concrete placed in ambient conditions that fall between 24°C and 38°C (75°F and 100°F). Last-minute attempts to mitigate hot-weather concreting are rarely performed early enough to prevent damaging effects. If acceptable field data is not available, the maximum temperature limit established for conditions at the project site should be based on trial-batch tests at the anticipated placement temperature and for the typical concrete section thickness, rather than on ideal temperatures of 20°C to 30°C (68°F to 86°F) cited in ASTM C192, *Standard Practice for Making and Curing Concrete Test Specimens in the Laboratory* (AASHTO R 39). If possible, large trial batches should be made to establish the relationship for the properties of interest as a function of time at various concrete temperatures. This process will establish the maximum allowable time to discharge concrete after batching for various concrete temperatures.

ASTM C94, *Standard Specification for Ready-Mixed Concrete* (AASHTO M 157), states that the producer shall deliver the ready-mixed concrete during hot weather at concrete temperatures as low as practicable, subject to the approval of the purchaser. ASTM C94 (AASHTO M 157) also notes in some situations difficulty may be encountered when concrete temperatures approach 32°C (90°F). However, ASTM C94 (AASHTO M 157) does not mandate a maximum concrete temperature unless materials are heated. If heated water or aggregates are used, the maximum concrete temperature shall not exceed 32°C (90°F) during production or transportation.

Setting a maximum concrete temperature to achieve required strength and durability is specific to each concrete mixture and daily ambient conditions for a given project. For most applications it is too complex to simply limit the maximum temperature of concrete as placed; circumstances and concrete requirements vary too widely. For example, a temperature limit that would serve successfully at one project location (such as in a cooler climate) could be highly restrictive at another (such as in a warmer climate). The inverse could occur if concrete mixtures that were designed for hot-weather conditions were placed in cooler ambient conditions.

ACI 305.1, *Specification for Hot Weather Concreting* (2014); and ACI 301, *Standard Specification for Concrete Construction* (2020), require that the temperature of concrete as delivered shall not exceed 35°C (95°F), unless otherwise specified. For hot weather concrete specified by ACI 305.1, qualification of a concrete mixture with a maximum concrete temperature that exceeds 35°C (95°F) shall be supported by either past field experience or by preconstruction testing. As an optional requirement, the specified limits may be changed on the basis of evidence of satisfactory performance or specific project needs.

The following list of precautions will reduce the potential problems of hot-weather concreting:

- organize a preconstruction conference to discuss the precautions required for the project;
- use materials and mixture proportions that have proven field performance in hot-weather conditions;
- cool the concrete or one or more of its ingredients;
- use a concrete consistency (slump) that allows for rapid placement and consolidation;
- reduce the time of transport, placing, and finishing;
- schedule concrete placements to limit exposure to harsh atmospheric conditions;
- consider nighttime placements;
- consider protection methods to limit moisture loss during placing and finishing, such as sunshades and windscreens;
- apply initial curing procedures including fogging and/or liquid-applied evaporation reducers to control evaporation after strike-off and prior to finishing concrete.

Which precautions to use and when to use them will depend on many factors including: the type of member or construction application, characteristics of the materials being used, and the experience of the placing and finishing crew in dealing with the atmospheric conditions on the project site.

CONCRETE MIXTURES FOR HOT WEATHER

Concrete mixtures for hot weather conditions must be proportioned to produce a material that can be readily placed, economical, and have low heat development.

Cement

As discussed in Chapter 2, Types I and Type II portland cement are the most widely used cement in concrete placements. Where applicable, lower heat options are available. ASTM C150, *Standard Specification for Portland Cement* (AASHTO M 85), includes requirements for a low heat cement, Type IV cement, which is rarely produced in the U.S. However, if available, Type II(MH), cement with a moderate heat of hydration, could be used. This cement's heat of hydration is generally lower than that of either a Type I or a conventional Type II cement. ASTM C595, *Standard Specification for Blended Hydraulic Cements* (AASHTO M 240), includes provisions for moderate- and low-heat cements signified by (MH) and (LH) designations. Likewise, Type MH and Type LH cements specified under ASTM C1157, *Standard Performance Specification for Hydraulic Cement*, are moderate- and low-heat cement options. Before specifying these cement types, their local availability should be verified (see Chapter 2).

Supplementary Cementitious Materials

Supplementary cementitious materials (SCMs) are beneficial because they generally reduce the temperature rise of concrete. Some SCMs, such as metakaolin and silica fume, are often used to increase durability and strength; however, their use does not impact the final temperature of the concrete.

Chapter 3 discusses the influence of each SCM on water demand and heat of hydration. These materials generally slow both the rate of setting and the rate of slump loss. However, some caution regarding finishing is needed. The rate of bleeding can be slower than the rate of evaporation and plastic shrinkage cracking or crazing may result without taking proper measures when using SCMs. This is discussed in greater detail under **Plastic Shrinkage Cracking**.

Aggregates

Aggregates have a pronounced effect on the fresh concrete temperature because they represent 70% to 85% of the total volume of concrete. The density and specific heat of aggregates influences the overall temperature of the concrete. If different sources of aggregates are available, the use of aggregates with a lower coefficient of thermal expansion (CTE) can be beneficial. The use of an aggregate with a lower CTE reduces the volume change of concrete due to temperature changes, which also reduces stresses and the likelihood of thermal cracking (see Chapter 10).

Chemical Admixtures

Set-retarding admixtures and extended set-controlling admixtures, ASTM C494, *Standard Specification for Chemical Admixtures for Concrete* (AASHTO M 194), Types B and D, and Type S, can be beneficial in offsetting the accelerating effects of high temperature despite the potential for increased rate of slump loss resulting from their use.

Water-reducing admixtures, ASTM C494 (AASHTO M 194), Types A, D, F, and G, can also be used to potentially reduce the amount of cementitious material needed for the strength specified (by lowering the w/cm). This has the benefit of reducing the heat energy of the concrete without reducing the compressive strength.

Air entrainment is also affected in hot weather. At elevated temperatures, an increase in the dosage of air-entraining admixture is generally necessary to produce a given air content in concrete.

Admixtures should be tested under job conditions before construction begins; this will determine their compatibility with the other concrete ingredients. Trial mixtures that simulate the anticipated temperatures as well as haul and placement times while measuring slump loss over time have been helpful in verifying that certain dosages of admixtures will perform properly in the field.

COOLING CONCRETE MATERIALS

Because of the detrimental effects of high concrete temperatures, all operations in hot weather should be directed toward keeping the concrete temperature below the specified limits. During hot weather the most favorable concrete temperature for achieving high quality freshly mixed concrete is usually lower than can be obtained without artificial cooling.

A common method of cooling concrete is to lower the temperature of the concrete materials before mixing. One or more of the ingredients can be cooled. The contribution of each ingredient in a concrete mixture to the temperature of the freshly mixed concrete is related to the temperature, specific heat, and the quantity of each material.

The initial temperature of concrete can be estimated from the temperatures of its ingredients using Equation 18-3.

$$T = \frac{\sum C_j M_j T_j}{\sum C_j M_j}$$

EQUATION 18-3.

Where:

$\sum$ = indicates summation over concrete ingredients (*j*)

T = initial temperature of the freshly mixed concrete, °C (°F)

C_j = the specific heat of ingredient *j*, kJ/kg/°C (Btu/lb/°F)

M_j = mass of ingredient *j*, kg (lb), and

T_j = difference between temperature of ingredient *j* and the baseline temperature, °C (°F).

The temperature of the ingredients is relative to a baseline of 0°C (or 0°F), so the difference in temperature is the temperature of the ingredient. For a mixture with aggregates, cement and water, this becomes Equation 18-4.

$$T = \frac{C_a T_a M_a + C_c T_c M_c + C_w T_w M_w + C_w T_{wa} M_{wa}}{C_a M_a + C_c M_c + C_w M_w + C_w M_{wa}}$$

EQUATION 18-4.

Where the subscripts are: *a* for aggregates, *c* for cement, *w* for (batch) water, and *wa* is for water in the aggregates. Note that this equation applies when the aggregates are at the same temperature, and T_{wa} can generally be simplified to T_a. For a mixture with portland cement and limestone aggregates, using values in Table 18-2, Equation 18-4 can be simplified to Equation 18-5.

$$T = \frac{0.22\,T_a M_a + 0.18\,T_c M_c + T_w M_w + T_a M_{wa}}{0.22 M_a + 0.18 M_c + M_w + M_{wa}}$$

EQUATION 18-5.

TABLE 18-2. Specific heat of selected concrete ingredients

MATERIAL	SPECIFIC HEAT kJ/kg/°C (Btu/lb/°F)
Portland cement*	0.75 (0.18)
SCMs[†]	
fly ash	0.73 (0.18)
slag cement	0.63 (0.15)
silica fume	0.87 (0.21)
Aggregate[‡]	
Basalt	0.90 (0.22)
Granite	0.78 (0.19)
Limestone	0.91 (0.22)
Sandstone	0.77 (0.19)
Siliceous	0.77 (0.19)
Water[§]	
Liquid (23°C/73°F)	4.19 (1.00)
Ice (0°C/32°F)	2.11 (0.50)

* PCA 1937 and Bentz and others 1998. For blended cements, it is recommended to use a weighted average of the portland cement and SCM or limestone contents.
[†] Trinhztfy and others 1982.
[‡] Schindler 2002.
[§] Weast and others 1987. The heat of fusion of ice is 334 kJ/kg (144 Btu/lb).

Since the ratio of specific heats of aggregate and cement to water is constant whether Metric (SI) or Imperial (inch-pound) units are used, this equation is applicable for both units.

Example calculations for initial concrete temperature for a system with portland cement, aggregate, batch water, and water in the aggregates are shown in Table 18-3. This equation is also useful in estimating the impacts of changing the temperatures of various ingredients on the initial concrete temperatures as well.

Figure 18-9 can be developed from the equations above and graphically shows the effect of material temperature on

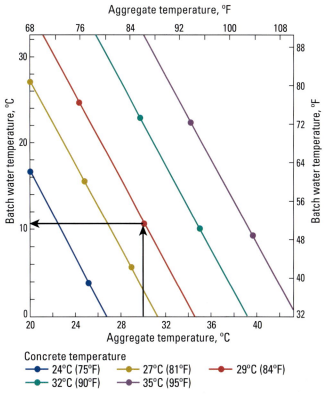

FIGURE 18-9. Temperature of freshly mixed concrete as affected by temperature of its ingredients. Aggregate (granite): 1360 kg (3000 lb); Moisture in aggregate: 27 kg (60 lb); Mixing water: 109 kg (240 lb); Portland cement at 66°C (150°F): 256 kg (564 lb). Black arrows illustrate the example in the text.

the temperature of fresh concrete. As an example of using Figure 18-9, to keep the initial concrete temperature for a mixture below 29°C (85°F), if the aggregate temperature is 30°C (86°F), then the water temperature should be below approximately 12°C (54°F) (see black arrows in the figure). It is evident that although concrete temperature is primarily dependent upon the aggregate temperature (due to quantity of material in the mixture), cooling the mixing water is another method to lower concrete temperature.

When more detailed information is available, additional terms can be added to the numerator and denominator of the temperature equations as shown in Equation 18-6, where subscripts are: *ca* for coarse aggregate, *fa* for fine aggregate, *wc* for water in coarse aggregate, and *wf* for water in fine aggregate. The fine and coarse aggregate may be potentially at different temperatures, be of different mineral types, and also have different moisture contents. This equation might also apply when only one aggregate is heated (see Chapter 19).

$$T = \frac{C_{ca} T_{ca} M_{ca} + C_{fa} T_{fa} M_{fa} + C_c T_c M_c + C_w T_w M_w + C_w T_{ca} M_{wc} + C_w T_{fa} M_{wf}}{C_{ca} M_{ca} + C_{fa} M_{fa} + C_c M_c + C_w M_w + C_w M_{wc} + C_w M_{wf}}$$

EQUATION 18-6.

TABLE 18-3A. Effect of Temperature of Materials on Initial Concrete Temperatures (SI).

MATERIAL	MASS, M, kg	SPECIFIC HEAT, C kJ/kg/°C	ENERGY TO VARY TEMPERATURE 1°C kJ	INITIAL TEMPERATURE T, °C	TOTAL ENERGY IN MATERIAL,* kJ
	(1)	(2)	(3) Col. 1 × Col. 2	(4)	(5) Col. 3 × Col. 4
Aggregate (total)	1839 (M_a)	0.78	1434	27 (T_a)	38,718
Portland cement	335 (M_c)	0.75	251	66 (T_c)	16,566
Batch water	123 (M_w)	4.19	515	27 (T_w)	13,905
Free water on aggregate	40 (M_{wa})	4.19	168	27 (T_a)	4,536
Concrete (sum)	2337		2368		73,725
Initial concrete temperature				**73,725 / 2368 = 31.1°C**	

*Total energy (kJ) is relative to a baseline of 0°C.

To achieve a 1 °C reduction in initial concrete temperature:
Cement temperature must be lowered by 2,368 / 251 = 9.4°C,
Or batch water temperature dropped 2,368 / 515 = 4.6°C,
Or aggregate temperature cooled 2,368 / 1434 = 1.7°C.

TABLE 18-3B. Effect of Temperature of Materials on Initial Concrete Temperatures (Inch-Pound).

MATERIAL	MASS, M, lb	SPECIFIC HEAT, C Btu/lb/°F	ENERGY TO VARY TEMPERATURE 1°F Btu	INITIAL TEMPERATURE T, °F	TOTAL ENERGY IN MATERIAL,* Btu
	(1)	(2)	(3) Col. 1 × Col. 2	(4)	(5) Col. 3 × Col. 4
Aggregate (total)	3100 (M_a)	0.19	589	80 (T_a)	47,120
Portland cement	564 (M_c)	0.18	102	150 (T_c)	15,300
Batch water	282 (M_w)	1.00	282	80 (T_w)	22,560
Free water on aggregate	88 (M_{wa})	1.00	88	80 (T_a)	7040
Concrete (sum)	4034		1061		92,020
Initial concrete temperature				**92,020 / 1061 = 86.7°F**	

*Total energy (Btu) is relative to a baseline of 0°F.

To achieve a 1 °F reduction in initial concrete temperature:
Cement temperature must be lowered by 1,061 / 102 = 10.4°F,
Or batch water temperature dropped 1,061 / 282 = 3.8°F,
Or aggregate temperature cooled 1,061 / 589 = 1.8°F.

Cementitious Materials

The temperature of hydraulic cements and other cementitious materials have only a minor effect on the temperature of the freshly mixed concrete because of their low specific heats and the relatively small amount of cementitious materials in a concrete mixture. For instance, a portland cement temperature change of 4°C (8°F) generally will change the concrete temperature by only 0.5°C (1°F).

Because hydraulic cements lose heat slowly during storage, they may still be hot when delivered (this heat is produced while grinding the cement clinker and other ingredients during manufacture). Cement temperatures in storage silos at ready-mix plants are often greater than 50°C (120°F) even in cold weather conditions.

Since the temperature of cementitious materials affects the temperature of the fresh concrete to some extent, some specifications place a limit on their temperature at the time of use. This limit varies from 66°C to 82°C (150°F to 180°F). However, it is more practical to specify a maximum temperature for freshly mixed concrete rather than place a temperature limit on individual ingredients (Lerch 1955).

Aggregate

There are several methods of keeping aggregates cool. Cooling effects are realized when stockpiles are shaded from the sun and kept moist by sprinkling. Since evaporation is a cooling process, sprinkling provides effective cooling, especially when the relative humidity is low. It is undesirable to cool aggregate stockpiles with seawater. Using seawater will contribute to acceleration of concrete setting time and corrosion of steel reinforcement.

Lowering the temperature of concrete 0.5°C (1°F) typically requires a 0.8°C to 1.1°C (1.5°F to 2°F) reduction in the temperature of the coarse aggregate.

Mixing Water

Water is the easiest ingredient in concrete to cool. Mixing water should be stored in insulated tanks or not exposed to the direct rays of the sun. Tanks and pipelines carrying mixing water should be buried, insulated, shaded, or painted to keep water as cool as practical.

Cooling the mix water temperature 2.0°C to 2.2°C (3.5°F to 4°F) will usually lower the concrete temperature about 0.5°C (1°F). Water can be cooled by refrigeration, liquid nitrogen, or ice. However, because mixing water is such a small percentage of the total mixture, it is difficult to lower concrete temperatures more than about 4°C (8°F) by cooling the batch water alone (Gajda and others 2005).

Ice Substitution

Substituting ice for batch water is more effective than using chilled batch water for cooling concrete. Ice both lowers the batch water temperature and lowers the mix temperature by extracting heat during the phase change that occurs as ice melts into liquid water (Gajda and others 2005).

When ice is used, the ice must be completely melted by the time mixing is completed and the concrete is discharged, otherwise the ice could create large voids in the concrete. The amount of water and ice must not exceed the total mixing-water requirements. Crushed or flaked ice is more effective than chilled water in reducing concrete temperature. When using crushed ice, care must be taken to store it at a temperature that will prevent the formation of lumps.

Figure 18-10 shows crushed ice being charged into a truck mixer prior to the addition of other materials. The volume of ice generally should not replace more than approximately 75% of the total batch water. The maximum temperature reduction from the use of ice is limited to about 11°C (20°F).

The approximate temperature of concrete when ice is used for a portion of the batch water can be calculated from the equations used previously with the inclusion of terms (M_i and H_i, respectively) to account for the mass of ice and its heat of fusion, as shown in Equation 18-7:

FIGURE 18-10. Substituting ice for part of the mixing water will substantially lower concrete temperature.

Where:

C_i = the specific heat of ice (kJ/kg/°C or Btu/lb/°F)

T_i = the difference between the initial temperature of the ice the reference temperature (°C or °F),

M_i = the mass in kg (lb) of ice,

T_{mi} = the difference between reference temperature and the melting point of ice (0°C or 32°F), and

H_i = the heat of fusion of ice in kJ/kg (Btu/lb).

All of the temperatures are referenced to a baseline of 0°C or 0°F. Since ice has a different specific heat than liquid water, two terms are used for ice. The term $C_i T_i M_i$ accounts for the energy in the ice as its temperature is raised to its melting point, after which the term for the specific heat of water applies for the (melted) ice ($C_w T_{mi} M_i$). In SI units, T_{mi} is zero, and that term drops out. By convention, since the ice melts completely, the term $C_i M_i$ in the denominator becomes $C_w M_i$. Inserting constants for the specific heat of portland cement (0.75 kJ/kg/°C), granitic aggregates (0.78 kJ/kg/°C), ice (2.11 kJ/kg/°C), and the heat of fusion of water (334 kJ/kg), simplifies to Equation 18-8A.

In inch-pound units, substituting values for C_w (1 Btu/lb/°F) and T_{mi} (32°F), the term for melted ice ($C_w T_{mi} M_i$) becomes 32 M_i. By including values for the specific heat of portland cement (0.18 Btu/lb/°F), and granitic aggregates (0.19 Btu/lb/°F), ice (0.5 Btu/lb/°F), the heat of fusion of water (144 Btu/lb), simplifies to Equation 18-8B.

Calculations in Tables 18-4A and 18-4B show the effect of 44 kg or 75 lb of ice in reducing the temperature of concrete. The impact of the reference temperature is shown in Table 18-4 as the energy contributed by the ice at 10°C

$$T = \frac{C_a T_a M_a + C_c T_c M_c + C_w T_w M_w + C_w T_a M_{wa} + C_i T_i M_i + C_w T_{mi} M_i - H_i M_i}{C_a M_a + C_c M_c + C_w M_w + C_w M_{wa} + C_i M_i}$$

EQUATION 18-7.

$$T = \frac{0.19T_aM_a + 0.18T_cM_c + T_wM_w + T_aM_{wa} + M_i(0.5T_i - 80)}{0.19M_a + 0.18M_c + M_w + M_{wa} + M_i} \quad \text{(SI)}$$

EQUATION 18-8A.

$$T = \frac{0.19T_aM_a + 0.18T_cM_c + T_wM_w + T_aM_{wa} + M_i(0.5T_i - 128)}{0.19M_a + 0.18M_c + M_w + M_{wa} + M_i} \quad \text{(inch-pound)}$$

EQUATION 18-8B.

(below the reference temperature), is -930 kJ in Table 18-4A, but 760 Btu in Table 18-4B, as the ice is at higher temperature (20°F) than the reference temperature of 0°F.

Liquid Nitrogen

Liquid nitrogen can be used to cool concrete, either by injecting it directly into a central mixer drum or the drum of a truck mixer, or by precooling materials, particularly aggregates, during batching. Liquid nitrogen can cool concrete to lower temperatures than ice can. Figure 18-11 shows liquid nitrogen added directly into a truck mixer near a ready-mix plant. This may also be added at the project site.

Liquid nitrogen precooling should be performed by trained professionals since liquid nitrogen is a cryogenic liquid. Liquid nitrogen vapors can displace oxygen from air and cause localized fog, which may reduce visibility. When dosing liquid nitrogen inside a mixer, care should be taken to prevent the liquid nitrogen from contacting the metal drum; the cryogenic temperature of the liquid nitrogen can crack the drum. The addition of liquid nitrogen does not in itself influence the amount of mixing water required except that lowering the concrete temperature can reduce water demand.

FIGURE 18-11. Liquid nitrogen is an effective method of reducing concrete temperature for mass concrete placements or during hot-weather concreting.

The initial concrete temperature, volume of concrete, and target final temperature will determine the necessary rate of injection of liquid nitrogen. In general, approximately 6 liters (1½ gallons) of liquid nitrogen are needed to cool 1 cubic meter (cubic yard) of concrete by 0.5°C (1°F). Liquid nitrogen has been successfully used to precool concrete to temperatures as low as 2°C (35°F) for specialized applications.

TABLE 18-4A. Effect of Ice on Initial Temperature of Concrete (SI)

MATERIAL	MASS, M kg	SPECIFIC HEAT, C kJ/kg/°C	ENERGY TO VARY TEMPERATURE 1°C kJ	INITIAL TEMPERATURE, T °C	TOTAL ENERGY IN MATERIAL* kJ
	(1)	(2)	(3) Col. 1 × Col. 2	(4)	(5) Col. 3 × Col. 4
Aggregate (total)	1839 (M_a)	0.78	1434	27 (T_a)	38,718
Portland cement	335 (M_c)	0.75	251	66 (T_c)	16,566
Batch water	79 (M_w)	4.19	331	27 (T_w)	8937
Free water on aggregate	40 (M_{wa})	4.19	167	27 (T_a)	4536
Ice	44 (M_i)	2.11	93	-10 (T_i)	-930
melted	44 (M_{mi})	4.19	184	0 (T_{mi})	0
Ice, heat of fusion	44 (M_i) × (-334 kJ/kg) =				-14,696
Concrete (sum)	2337		2368		53,131
Concrete temperature				**53,131 / 2368 = 22.4°C**	

*Total energy (kJ) is relative to a baseline of 0°C.

TABLE 18-4B. Effect of Ice on Initial Temperature of Concrete (Inch-Pound).

MATERIAL	MASS, M lb	SPECIFIC HEAT, C Btu/lb/°F	ENERGY TO VARY TEMPERATURE 1°F Btu	INITIAL TEMPERATURE, T °F	TOTAL ENERGY IN MATERIAL* Btu
	(1)	(2)	(3) Col. 1 × Col. 2	(4)	(5) Col. 3 × Col. 4
Aggregate (total)	3100 (M_a)	0.19	589	80 (T_a)	47,120
Portland cement	564 (M_c)	0.18	102	150 (T_c)	15,300
Batch water	207 (M_w)	1.00	207	80 (T_w)	16,560
Free water on aggregate	88 (M_{wa})	1.00	88	80 (T_a)	7040
Ice	75 (M_i)	0.50	38	20 (T_i)	760
melted	75 (M_{mi})	1.00	75	32 (T_{mi})	2400
Ice, heat of fusion	75(M_i) × (-144 Btu/lb)				-10,800
Concrete (sum)	4034		1061		78,380
Concrete temperature			78,380 / 1061 = 73.9°F		

*Total energy (Btu) is relative to a baseline of 0°F.

TRANSPORTING AND HANDLING CONCRETE IN HOT WEATHER

Before concrete is placed, precautions should be taken during hot weather to maintain or reduce concrete temperature. Mixers, chutes, conveyor belts, hoppers, pump lines, and other equipment for handling concrete should be shaded, painted, or covered with wet burlap to reduce the effect of solar heating.

Transporting and placing concrete should be completed quickly during hot weather. Delays contribute to slump loss and an increase in concrete temperatures. Sufficient labor and equipment must be available at the project site to handle and place concrete immediately upon delivery.

In hot weather it is important to discharge the concrete as quickly as possible. Many project specifications require that discharge of concrete be completed within 90 minutes of mixing. However, these restrictions may be extended under certain conditions (ACI 301 2020). If specific time limitations on the completion of discharge of the concrete are desired, they should be included in the project specifications. It is also reasonable to obtain test data from a trial batch simulating the batch to placement time, mixing conditions, and anticipated concrete temperatures to document, a reduction or extension of in the time limit, if necessary.

PLACING AND FINISHING IN HOT WEATHER

Forms, reinforcing steel, and subgrade should be wetted with cool water just before the concrete is placed (Figure 18-12). During placing and finishing operations, initial curing measures such as fogging can be directed above the concrete surface. This not only cools the concrete surfaces and surrounding air but also increases the relative humidity at, and adjacent to, the concrete member. The increase in relative humidity minimizes the rate of evaporation of water from the concrete after placement. For slabs on ground, it is a good practice to moisten subgrades that are dry before concreting. There should be no standing water or puddles in the forms or on the subgrade when concrete is placed.

FIGURE 18-12. Dampening the subgrade, yet keeping it free of standing water, will reduce drying of the concrete and reduce problems from hot weather conditions.

During extremely high temperatures, improvements may be obtained by restricting concrete placements to early morning, late evening, or nighttime hours , especially in arid climates. This practice has resulted in substantially less thermal shrinkage and cracking of thick slabs and pavements.

Since the setting of concrete is more rapid in hot weather, extra care must be taken with placement techniques to avoid cold joints. For placement of walls shorter lifts can be specified to assure enough time for consolidation with the previous lift. Temporary sunshades and windbreaks help to minimize cold joints in slabs. Finishing on dry and windy days requires additional precautions to minimize the potential for plastic shrinkage cracking and crazing.

Curing and Protection in Hot Weather

To prevent the drying of exposed concrete surfaces, wet curing should commence as soon as the surfaces are finished and continue for at least 72 hours. In hot weather, continuous wet curing for the entire curing period is preferred. Avoid cyclic wetting and drying of a surface during curing.

If wet curing cannot be continued beyond 24 hours, the concrete should be protected from drying with curing paper, heat-reflecting plastic sheets, or membrane-forming curing compounds. Pigmented curing compounds can be used on exposed concrete surfaces. Application of a curing compound during hot weather should be preceded by 24 hours of wet curing. If this is not practical, the curing compound should be applied immediately after final finishing (see Chapter 17).

Thermal Control of Mass Concrete

Mass concrete is not exclusively a hot weather issue. The thermal control of mass concrete placements require special considerations to reduce or control the heat of hydration and the resulting temperature rise to avoid damaging the concrete through excessive temperatures and temperature differences. Mass concrete includes not only large volumes of concrete used in dams and other massive structures, but also moderate-to high-cementitious content concrete in structural members of buildings, bridges, and other infrastructure. See ACI 207.1, *Guide to Mass Concrete* (2005) for more information on determining if the mixture and placement qualifies as mass concrete. These conditions can result in thermal cracking and later age strength reduction and damage of the concrete (Gajda 2007). Thermal shock cracking can also occur if appropriate temperatures are not achieved for a sufficient period of time.

In mass concrete, temperature rise (Figure 18-13) results from the heat of hydration of cementitious materials. As the interior concrete increases in temperature and thermally expands, the surface concrete may be cooling and thermally contracting. This causes tensile stresses that may result in thermal cracks at the surface if the temperature difference between the surface and center is too great (the resulting thermal stress exceed the developing tensile strength). The width and depth of cracks depends upon the temperature difference, physical properties of the concrete, and size, location, and amount of any reinforcing steel.

The temperature rise in a mass concrete placement is related to the initial concrete temperature (Figure 18-14), ambient temperature, size of the concrete element (its minimum dimension), and type and quantity of cementitious materials. Thinner concrete members with moderate amounts of cementitious materials dissipate heat more rapidly.

For resistance to delayed ettringite formation, the maximum internal temperature of concrete should be less than 70°C (158°F) (see Chapter 11).

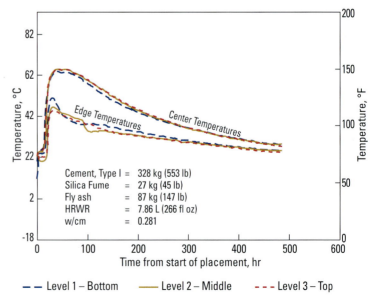

FIGURE 18-13. (left) A drilled pier (caisson), 3 m (10 ft) in diameter and 12.2 m (40 ft) in depth in which low-heat, high-strength concrete is placed and (right) temperatures of this concrete measured at the center and edge and at three different levels in the caisson (Burg and Fiorato 1999).

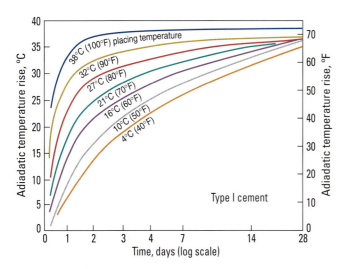

FIGURE 18-14. The effect of concrete-placing temperature on temperature rise in mass concrete with 223 kg/m³ (376 lb/yd³) of Type I portland cement. Higher placing temperatures accelerate temperature rise (ACI 207.2R 2007).

Internal concrete temperature gain can be controlled by a number of means for mass concrete structures, including some or all of the following considerations:

- using concrete with reduced cementitious contents,

- using concrete with increased SCM levels,

- precooling the concrete prior to placement,

- post cooling the placed concrete with embedded cooling pipes, and

- splitting the placement into multiple thinner placements with adequate time between placements to dissipate heat.

Massive structurally reinforced concrete members with high cement contents (300 kg/m³ to 600 kg/m³ or 500 lb/yd³ to 1000 lb/yd³) cannot use many of the placing techniques and controlling factors mentioned above to maintain the low temperatures needed to control thermal cracking. For these concretes (often used in bridges, foundations, and power plants), a good technique is to minimize external restraint from adjacent concrete elements, with the use of isolation joints. Where appropriate, place the concrete in multiple smaller sections with adequate time between placements to dissipate the temperature rise of the concrete, or control internal differential thermal strains between the surface and the center. The latter is done by properly designing the concrete mixture and either keeping the concrete surface warm through use of insulation, or reducing the internal concrete temperature by precooling the concrete or postcooling with internal cooling pipes. Regardless of season, insulation is necessary in mass concrete applications to limit the temperature difference between the center and surface of the element.

Studies and experience have shown that by limiting the maximum temperature difference between the interior and exterior surface of the concrete to less than about 20°C (35°F), surface cracking can be minimized or avoided (FitzGibbon 1977 and Fintel and Ghosh 1978). Some sources indicate that the maximum temperature differential (MTD) for concrete containing granite or limestone (aggregates with a low coefficient of thermal expansion) should be 25°C and 31°C (45°F and 56°F), respectively (Bamforth 1981). The actual MTD for a particular mass concrete placement with a particular concrete can be determined using equations in ACI 207.2R, *Report on Thermal and Volume Change Effects on Cracking of Mass Concrete* (2007), or by a stress/strain approach in PCA EB547, *Mass Concrete for Buildings and Bridges* (Gajda 2007).

In general, an MTD of 20°C (35°F) should be assumed unless a demonstration or calculations based on physical properties of the actual concrete mixture and the geometry of the concrete member demonstrate that higher MTD values are allowable.

Limiting the temperature difference from the time of placement through the time the concrete cools to within 20°C (35°F) of the average air temperature, allows the concrete to cool slowly to ambient temperature with little or no surface cracking. However, this is true only if the member is not restrained by continuous reinforcement crossing the interface of adjacent or opposite sections of hardened concrete. Restrained concrete may crack due to eventual thermal contraction after cooling.

Figure 18-15 illustrates the relationship between temperature rise, cooling, and temperature differences for a section of mass concrete. As can be observed, if the forms (which are providing insulation in this case) are removed too early, cracking will occur once the difference between interior and surface concrete temperatures exceeds the temperature difference of 20°C (35°F). If higher temperature differences are permissible, the forms can be removed sooner. For large concrete placements, surface insulation may be required for several weeks or longer.

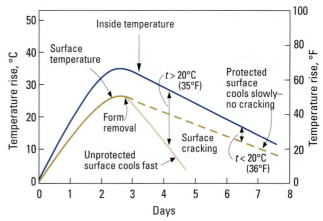

FIGURE 18-15. Potential for surface cracking after form removal, assuming a temperature difference limit, *t*, of 20°C (35°F) (Fintel and Ghosh 1978).

Temperatures and temperature differences in mass concrete placements can be more precisely predicted by various methods, including a method given by Gajda (2007), commercially-available finite element software, and proprietary mass concrete software. A thermal control plan is now commonly submitted by the contractor to the owner/engineer prior to a mass concrete placement. The contents of a thermal control plan are described in Gajda (2007) and ACI 301 (2020).

The maximum temperature rise can be approximated using the example shown in Equation 18-9. For concrete containing 300 kg/m³ to 600 kg/m³ (500 lb/yd³ to 1000 lb/yd³) of Type I or Type II portland cement and the least dimension of the member is no more than 1.8 m (6 ft), this approximation would be 15°C of temperature rise for every 100 kg/m³ (16°F for every 100 lb/yd³ of portland cement). For example, the maximum temperature, T_{max}, of such an element made with concrete having 400 kg/m³ (674 lb/yd³) of Type I cement and cast with an initial concrete temperature of 25°C (77°F) is shown in Equation 18-9.

$$T_{max} = 25°C + (15°C \times 400/100) \text{ or } 85°C \quad \text{(SI)}$$
$$T_{max} = (77°F + [16°F \times 674/100] \text{ or } 185°F) \quad \text{(inch-pound)}$$

EQUATION 18-9.

The slow rate of heat exchange between concrete and its surroundings is due to the heat capacity and thermal diffusivity of concrete. Heat escapes from uninsulated concrete at a rate that is inversely proportional to the square of its least dimension. A 150-mm (6-in.) thick wall cooling from both sides may take approximately 1½ hours to dissipate 95% of its developed heat. A 1.5-m (5-ft) thick wall would take an entire week to dissipate the same amount of heat under similar conditions (ACI 207.2R 2007). If internal cooling pipes were not used during the construction of the Hoover Dam, it would have taken more than 200 years to dissipate its internal heat.

Commercially available temperature sensors are commonly used to monitor temperature and temperature differences in mass concrete placements (Figure 18-16). Specific locations

FIGURE 18-16. Temperature sensors are used to monitor temperature and temperature differences in mass concrete placements (courtesy of J. Gajda).

for monitoring temperatures and temperature differences in typical mass concrete placements are described in Gajda (2007) and ACI 301 (2020).

REFERENCES

PCA's online catalog includes links to PDF versions of many of our research reports and other classic publications. Visit: cement.org/library/catalog.

ACI Committee 207, *Guide to Mass Concrete*, ACI 207.1-05, American Concrete Institute, Farmington Hills, MI, 2005, reapproved 2012, 31 pages.

ACI Committee 207, *Report on Thermal and Volume Change Effects on Cracking of Mass Concrete*, ACI 207.2R-07, American Concrete Institute, Farmington Hills, MI, 2007, 28 pages.

ACI Committee 301, *Specifications for Concrete Construction*, ACI 301-20, American Concrete Institute, Farmington Hills, MI, 2020, 73 pages.

ACI Committee 305, *Guide to Hot-Weather Concreting*, ACI 305R-10, American Concrete Institute, Farmington Hills, MI, 2010, 23 pages.

ACI Committee 305, *Specification for Hot-Weather Concreting*, ACI 305.1-14, American Concrete Institute, Farmington Hills, MI, 2014, 7 pages.

Bentz, D.P.; Waller, V.; and de Larrard, F., "Prediction of Adiabatic Temperature Rise in Conventional and High-Performance Concretes Using a 3-D Microstructural Model," *Cement and Concrete Research*, Vol. 28, No. 2, February 1998, pages 285 to 297.

Brooks, A.G.; Schindler, A.K.; and Barnes, R.W., "Maturity Method Evaluation for Various Cementitious Materials," *Journal of Materials in Civil Engineering*, Vol. 19, No. 12, December 2007, pages 1017 to 1025.

Burg, R.G., *The Influence of Casting and Curing Temperature on the Properties of Fresh and Hardened Concrete*, Research and Development Bulletin RD113, Portland Cement Association, 1996, 13 pages.

Burg, R.G.; and Fiorato, A.E., *High-Strength Concrete in Massive Foundation Elements*, Research and Development Bulletin RD117, Portland Cement Association, 1999, 22 pages.

Bureau of Reclamation, *Concrete Manual*, 8th ed., Denver, CO, revised 1981, 661 pages.

FitzGibbon, M.E., "Large Pours for Reinforced Concrete Structures," Current Practice Sheets No. 28, 35, and 36, *Concrete*, Cement and Concrete Association, Wexham Springs, Slough, England, March and December 1976 and February 1977.

Fintel, M.; and Ghosh, S.K., "Mass Reinforced Concrete Without Construction Joints," presented at the *Adrian Pauw Symposium on Designing for Creep and Shrinkage*, Fall Convention of the American Concrete Institute, Houston, TX, November 1978.

Gajda, J.; Kaufman, A.; and Sumodjo, F., "Precooling Mass Concrete", *Concrete Construction*, August 2005, pages 36 to 38.

Gajda, J., *Mass Concrete for Buildings and Bridges*, EB547, Portland Cement Association, Skokie, IL, 2007, 44 pages.

Gaynor, R.D.; Meininger, R.C.; and Khan, T.S., *Effect of Temperature and Delivery Time on Concrete Proportions*, NRMCA Publication No. 171, National Ready Mixed Concrete Association, Silver Spring, MD, June 1985, 20 pages.

Klieger, P., *Effect of Mixing and Curing Temperature on Concrete Strength*, Research Department Bulletin RX103, Portland Cement Association, 1958, 20 pages.

Kohler, M.A., "Lake and Pan Evaporation," *Water Loss Investigations: Lake Hefner Studies*, Geological Survey Circular 229, U.S. Government Printing Office, Washington, D.C., 1952, pages 127 to 158.

Krishnaiah, S.; and Singh, D.N., "Determination of Thermal Properties of Some Supplementary Cementing Materials Used in Cement and Concrete," *Construction and Building Materials*, Vol. 20, 2006, pages 193 to 198.

Lee, CH.; and Hover, K.C., "What Do We Mean by "Setting?" *Concrete International*, American Concrete Institute, Vol. 38, No. 8, August, 2016, pages 53 to 59.

Lerch, W., *Hot Cement and Hot Weather Concrete Tests*, IS015, Portland Cement Association, 1955, 11 pages.

Menzel, C.A., "Causes and Prevention of Crack Development in Plastic Concrete," *Proceedings of the Portland Cement Association*, 1954, pages 130 to 136.

NRMCA, "Plastic Cracking of Concrete," *Engineering Information*, National Ready Mixed Concrete Association, Silver Spring, Maryland, July 1960, 2 pages.

PCA, *Tables of Heat Content of Cement Clinker, Raw Materials and Combustion Gases*, Manufacturing Research Bureau, Portland Cement Association, Chicago, IL, 1937, 19 pages.

Trinhztfy, H.W.; Blaauwendraad, J.; and Jongendijk, J., "Temperature Development in Concrete Structures Taking Accounting of State Dependent Properties," *Proceedings*, RILEM International Conference on Concrete at Early Ages, Vol. 1, Paris, pages 211 to 218, 1982.

Uno, P.J., "Plastic Shrinkage Cracking and Evaporation Formulas," *ACI Materials Journal*, Vol. 95, No. 4, July-August 1998, pages 365 to 375.

Weast, R.C., ed., *CRC Handbook of Chemistry and Physics*, 68th edition, CRC Press, Boca Raton, FL, 1987.

COLD WEATHER CONCRETING

FIGURE 19-1. Closeup view of ice impressions in paste of frozen fresh concrete.

During cold weather, the concrete mixture and its temperature should be adapted to the construction procedures and ambient weather conditions and to the required properties of the concrete. Concrete gains strength at a slower rate when exposed to low temperatures, and is susceptible to freezing and surface quality issues.

Cold weather conditions can adversely affect concrete quality including:

- freezing of fresh paste,
- slower rate of setting,
- slower strength development (low early strength),
- dusting,
- plastic shrinkage cracking and crazing, and
- thermal cracking.

WHEN TO TAKE PRECAUTIONS

Cold weather is defined by ACI 306R, *Guide to Cold Weather Concreting* (2016), as existing when the air temperature has fallen to, or is expected to fall below 4°C (40°F). Given that ambient conditions can change quickly, and that time is required to install cold weather protection measures, the concrete producer and contractor must be adequately prepared to take action before conditions threaten to impair the quality of concrete. Freshly mixed concrete must be protected against the disruptive effects of freezing until the concrete attains a minimum compressive strength of 3.5 MPa (500 psi) (Powers 1962).

EFFECT OF FREEZING ON FRESH CONCRETE

Ice crystal formations occur as unhardened cement paste freezes. Ice crystal impressions (Figure 19-1) do not occur in hardened concrete. The disruption of a fresh paste matrix by freezing can cause reduced strength gain and increased porosity. Significant ultimate strength reductions, up to about 50%, can occur if concrete is frozen within a few hours after placement or before it attains a compressive strength of 3.5 MPa (500 psi) (McNeese 1952). Concrete exposed to deicers should attain a minimum compressive strength of 35 MPa (4500 psi) prior to repeated cycles of freezing and thawing (Klieger 1957 and Gebler and Klieger 1986).

The strength of concrete that has been frozen only once at an early age may be restored by providing favorable subsequent curing conditions. However, this concrete will likely be more permeable and less resistant to weathering. The critical period after which concrete is not seriously damaged by one or two freezing cycles is dependent upon the concrete mixture and conditions of placing, finishing, curing, and successive drying. See Chapter 11 for more information on freeze-thaw resistance.

IMPACT OF COLD WEATHER ON CONCRETE PROPERTIES

Workability

Workability of concrete is influenced by concrete materials, proportions, concrete temperature, and ambient air temperatures. As the temperature of concrete drops, the workability increases. Cold weather retards hydration which lowers water demand loss of concrete mixtures. Figure 19-2 illustrates the effects of casting temperature on slump.

Heat of Hydration

Concrete generates heat during transition from the plastic to the hardened state as a result of the chemical reactions. The heat generated is called heat of hydration. Dimensions of the concrete, placement ambient air temperature, initial concrete temperature, water-cement ratio, admixtures, and the type, composition, fineness, and amount of cementitious materials all affect heat generation and buildup (see Chapter 2).

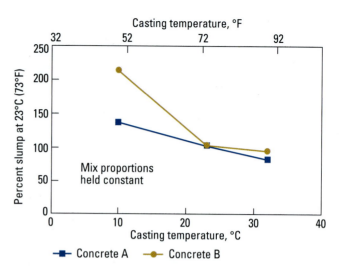

FIGURE 19-2. Slump characteristics as a function of casting temperature. Materials were stored at desired temperature for at least 24 hours prior to mixing. Laboratory was kept at casting temperature during mixing and while conducting fresh tests. "Concrete A" refers to mixtures of 356 kg/m³ (600 lb/yd³) of Type I cement at a water-cement ratio of 0.45, while "Concrete B" refers to mixtures of 335 kg/m³ (564 lb/yd³) of Type I/II cement with a water-cement ratio of 0.46 (Burg 1996).

Heat of hydration is useful in cold weather concreting as it contributes to the heat necessary to provide satisfactory curing temperatures, provided the heat is contained. This is particularly true in more massive elements. The heat liberated during hydration can offset some or all of the lost heat during placing, finishing, and early curing operations in thicker elements.

Setting Time

Temperature affects the rate at which hydration of cement occurs. Lower temperatures retard hydration and setting times of concrete. Figure 19-3 illustrates the effect of cooler temperatures on initial and final setting times of concrete

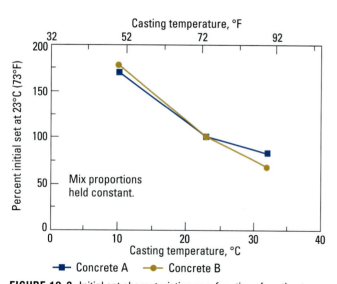

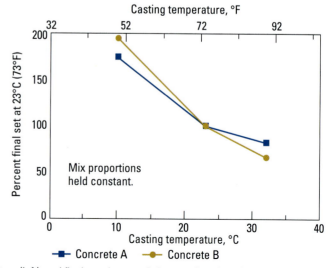

FIGURE 19-3. Initial set characteristics as a function of casting temperature (left), and final set characteristics as a function of casting temperature (right). "Concrete A" refers to mixtures of 356 kg/m³ (600 lb/yd³) of Type I cement at a water-cement ratio of 0.45, while "Concrete B" refers to mixtures of 335 kg/m³ (564 lb/yd³) of Type I/II cement with a water-cement ratio of 0.46 (Burg 1996).

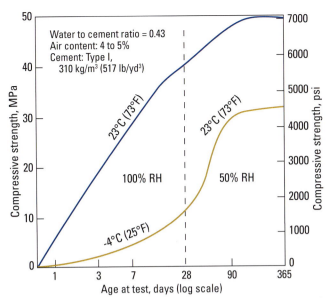

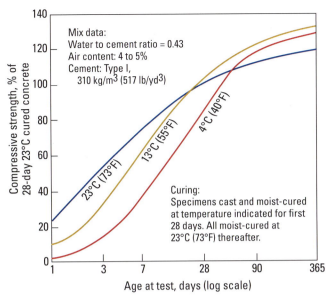

FIGURE 19-4. Effect of temperature conditions on the strength development of concrete. Concrete for the lower curve was cast at 4°C (40°F) and placed immediately in a curing room at -4°C (25°F). Both concretes received 100% relative humidity curing for first 28 days followed by 50% relative humidity curing (Klieger 1958).

FIGURE 19-5. Effect of low temperatures on concrete compressive strength at various ages. Note that for this particular mixture made with Type I cement, the best temperature for the first 28 days to develop the highest long-term strength (1 year) was 13°C (55°F) (Klieger 1958).

(tested in accordance with ASTM C403), *Standard Test Method for Time of Setting of Concrete Mixtures by Penetration Resistance*. As discussed in Chapter 18, laboratory setting times from tests on mortar may not be directly transferable to concrete setting times in the field.

Strength Gain

Low temperatures slow the rate of strength gain of concrete. If hardened concrete is frozen, but kept above about -10°C (14°F), it will still be able to gain strength slowly. However, below that temperature, cement hydration and concrete strength gain cease. Figures 19-4 and 19-5 show the age-compressive strength relationship for concrete that has been cast and cured at various temperatures. Note in Figure 19-5 that concrete cast and cured at 4°C (40°F) and 13°C (55°F) had relatively low strengths for the first week; but after 28 days – when all specimens were wet-cured at 23°C (73°F) – strengths for the 4°C (40°F) and 13°C (55°F) concretes grew faster than the 23°C (73°F) concrete and at one year they were slightly higher.

Concrete test cylinders must be maintained at a temperature between 16°C (60°F) and 27°C (80°F) at the jobsite for up to 48 hours until they are taken to a laboratory for curing (ASTM C31, *Standard Practice for Making and Curing Concrete Test Specimens in the Field* [AASHTO T 23]). For concrete mixtures with a specified strength of 40 MPa (6000 psi) or greater, the initial curing temperature shall be between 20°C and 26°C (68°F and 78°F). During this period, cylinders should be kept in a curing box and covered with a nonabsorptive cap or impervious plastic bag. The temperature in the curing box should be accurately controlled by a thermostat (Figure 19-6). When stored in an insulated curing box outdoors, the temperature of the cylinders

are more likely to be properly controlled. If kept in a trailer the heat may be inadvertently turned off at night or over a weekend or holiday in cold weather conditions. In that case, the cylinders would not be cured within the prescribed curing temperatures during this critical period.

Cylinders stripped from molds after the first 24 ± 8 hours must be wrapped tightly in plastic bags or laboratory curing started immediately.

In addition to laboratory-cured cylinders, field-cured cylinders may be specified in accordance with ASTM C31. As discussed in Chapter 17, field cured cylinders may not adequately replicate the curing conditions of concrete in service. It is sometimes difficult to find the right locations for field curing. Differences in the surface to volume ratios between cylinders and the structure, in conjunction with differences in mass, make correlations between field-cured cylinder strengths and in-place strengths difficult.

FIGURE 19-6. Insulated curing box with thermostat for curing test cylinders. Heat is supplied by electric rubber heating mats on the bottom. A wide variety of designs are available for curing boxes.

Cast-in-place cylinders (ASTM C873, *Standard Test Method for Compressive Strength of Concrete Cylinders Cast in Place in Cylindrical Molds*), nondestructive testing methods (see Chapter 20), as well as the maturity method (discussed later in this chapter) are helpful in monitoring in place concrete strength.

PREPARATION BEFORE CONCRETING DURING COLD WEATHER

Concrete can be placed safely without damage from freezing in cold climates if precautions are taken. Under these circumstances planning and preparation are required to protect concrete and support hydration. All materials and equipment needed for adequate protection and curing must be on hand and ready for use before concrete placement is started.

Preparations should be made to protect the concrete from low temperatures using enclosures, windbreaks, portable heaters, insulated forms, and blankets to maintain required concrete temperature (Figure 19-7). Concrete must be delivered at the required temperature and the temperature of forms, reinforcing steel, the ground, or other concrete on which the fresh concrete is cast must also be considered. Concrete should not be cast on frozen concrete or on frozen ground. Forms, reinforcing steel, and embedded fixtures must be free of snow and ice at the time concrete is placed. Thermometers and proper storage facilities for test cylinders should be available to monitor and verify that precautions are adequate.

CONCRETE MIXTURES FOR COLD WEATHER

High-Early Strength

Rapid early strength development is desirable in cold weather construction to reduce the length of time temporary protection is required. The additional cost of high-early-strength concrete is often offset by earlier reuse of forms and shores, savings in the shorter duration of temporary heating, earlier setting times that allow earlier finishing of flatwork, and earlier use of the structure. High-early-strength concrete can be obtained by using one or a combination of the following:

- Type III or HE high-early-strength cement,
- additional Type I/II portland cement (60 kg/m³ to 120 kg/m³ [100 lb/yd³ to 200 lb/yd³]),
- addition of silica fume or metakaolin, and
- set accelerating admixtures.

Principal advantages of using the options above occur during the first 7 days. At a 4°C (40°F) curing temperature, the advantages of Type III cement are more pronounced and persist longer than at the higher temperature (Figure 19-8).

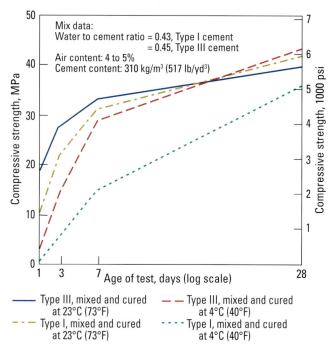

FIGURE 19-8. Early-age compressive-strength relationships for Type I and Type III portland cement concretes mixed and cured at 4°C (40°F) compared to 23°C (73°F) (Klieger 1958).

FIGURE 19-7. When suitable preparations to build enclosures and insulate equipment have been made, cold weather is no obstacle to concrete construction.

TABLE 19-1. Recommended Concrete Temperature for Cold-Weather Construction*

LINE	CONDITION		THICKNESS OF SECTIONS, mm (in.)			
			LESS THAN 300 (12)	300 TO 900 (12 TO 36)	900 TO 1800 (36 TO 72)	OVER 1800 (72)
1	Minimum temperature of fresh concrete as *mixed* for weather indicated.	Above -1°C (30°F)	16°C (60°F)	13°C (55°F)	10°C (50°F)	7°C (45°F)
2		-18°C to -1°C (0°F to 30°F)	18°C (65°F)	16°C (60°F)	13°C (55°F)	10°C (50°F)
3		Below -18°C (0°F)	21°C (70°F)	18°C (65°F)	16°C (60°F)	13°C (55°F)
4	Minimum temperature of fresh concrete *as placed and maintained.*[†]		13°C (55°F)	10°C (50°F)	7°C (45°F)	5°C (40°F)

* Adapted from Table 5.1 of ACI 306R-16.

[†] Placement temperatures listed are for normal-weight concrete in accordance with ASTM C94 (AASHTO M 157) and ACI 301-20. Lower temperatures can be used for lightweight concrete if justified by tests. For recommended duration of temperatures in Line 4, see Table 19-3.

The use of certain SCMs such as silica fume and metakaolin may improve early strength development. Other SCMs may lower early strength gain.

Accelerators

Set accelerating admixture can be used to decrease the setting time and may increase the early-age strength development of concrete in cold weather. Set accelerators containing chlorides should not be used where there is an in-service potential for corrosion, such as in concrete members containing steel reinforcement or where aluminum or galvanized inserts will be used. Chloride-based accelerators are not recommended for concretes exposed to soil or water containing sulfates or for concretes susceptible to alkali-aggregate reaction. Set accelerators must not be used as a substitute for proper curing and frost protection.

Air-Entrained Concrete

Entrained air is particularly desirable in any concrete that will be exposed to freezing weather while in service. Concrete that is not air entrained can suffer strength loss and internal as well as surface damage as a result of freezing and thawing (Figure 19-9). Air entrainment provides the capacity to relieve stresses due to ice formation within the concrete.

If concrete is non-air entrained, concrete should be protected during freezing months to ensure concrete work can be done under cover where there is no chance that rain, snow, or water from other sources can saturate the concrete and where there is no chance of freezing. Steel troweled floor finishes should not be specified when entrained air content above 3% is specified for slabs, because entrained air may promote blistering and delamination of the slab surface. See Chapters 6 and 9 for more information on air-entraining admixtures and their impact on concrete properties and Chapter 11 for information on freeze-thaw resistance of concrete.

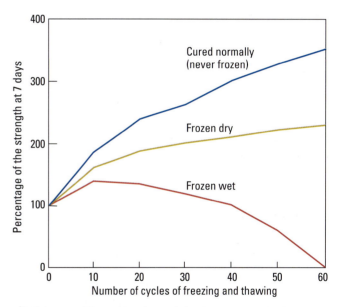

FIGURE 19-9. Effect of freezing and thawing on strength of concrete that does not contain entrained air (cured 7 days before first freeze) (Powers 1956).

HEATING CONCRETE MATERIALS

Temperature of Concrete as Mixed

For cold weather construction, it is recommended that the temperature of fresh concrete as mixed not be less than shown in Lines 1, 2, or 3 of Table 19-1 for the respective thickness of section. Note that lower concrete temperatures are recommended for more massive concrete sections because heat generated during hydration is dissipated less rapidly in thicker sections. Also note that at lower ambient air temperatures more heat is lost from concrete during transportation and placement. Therefore, the recommended concrete temperatures as mixed are higher for colder weather.

In cold weather conditions, there is little advantage in using fresh concrete at a temperature much above 21°C (70°F). Higher concrete temperatures do not afford proportionately longer protection from freezing because the rate of heat loss is greater. Also, high concrete temperatures are undesirable since they increase thermal shrinkage after hardening, require more mixing water for the same slump, and contribute to the possibility of plastic shrinkage cracking (caused by rapid moisture loss through evaporation). Therefore, the temperature of the concrete as mixed should not be more than 11°C (20°F) above the minimum recommended in Table 19-1. In accordance with ACI 306.1R-16 and ACI 301-20, normal concrete operations can be resumed if ambient temperatures return to above 10°C (50°F) during more than half of any 24-hour duration. The graph in Figure 19-10 is based on Equation 19-1 (see Chapter 18).

$$T = \frac{C_{ca}T_{ca}M_{ca} + C_{fa}T_{fa}M_{fa} + C_cT_cM_c + C_wT_wM_w + C_wT_{ca}M_{wc} + C_wT_{fa}M_{wf}}{C_{ca}M_{ca} + C_{fa}M_{fa} + C_cM_c + C_wM_w + C_wM_{wc} + C_wM_{wf}}$$

EQUATION 19-1.

Where:

T = temperature of the freshly mixed concrete, °C (°F);

C_{ca}, C_{fa}, C_c, C_w, C_{wc}, and C_{wf} = specific heats of the coarse aggregate, fine aggregate, cement, batch water, and water in the coarse aggregate and water in the fine aggregate, respectively, kJ/kg/°C (Btu/lb/°F) (See Table 18-2 for specific heats of common concrete ingredients);

T_{ca}, T_{fa}, T_c, T_w, T_{wc}, and T_{wf} = temperature of the coarse aggregate, fine aggregate, cement, batch water, and water in the coarse aggregate and water in the fine aggregate, respectively, °C (°F) (generally $T_{wf} = T_{fa}$ and $T_{wc} = T_{ca}$); and

M_{ca}, M_{fa}, M_c, M_w, M_{wc}, and M_{wf} = mass of the coarse aggregate, fine aggregate, cement, batch water, water in coarse aggregate, and water in the fine aggregate, respectively, kg (lb).

If the weighted average temperature of aggregates and cement is above 0°C (32°F), the proper mixing water temperature for the required concrete temperature can be selected from Figure 19-10. The range of concrete temperatures in the chart corresponds with the recommended values given in Lines 1, 2, and 3 of Table 19-1.

When the temperature of one or more of the aggregates is below 0°C (32°F) the free moisture on the aggregates will freeze, Equation 19-1 can be modified to account for the additional heat required to return the frozen free moisture to a thawed moist state.

SI — Substitute $M_{wf}(0.5T_{fa}-80)$ for $T_{wf}M_{wf}$
Substitute $M_{wc}(0.5T_{ca}-80)$ for $T_{wc}M_{wc}$
Inch-pound — Substitute $M_{wf}(0.5T_{fa}-128)$ for $T_{wf}M_{wf}$
Substitute $M_{wc}(0.5T_{ca}-128)$ for $T_{wc}M_{wc}$

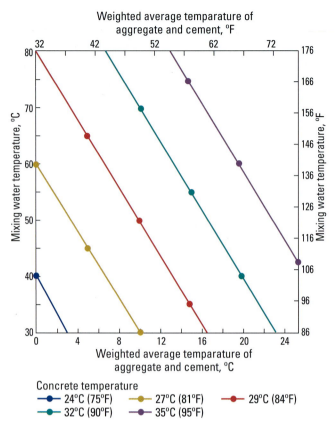

FIGURE 19-10. Temperature of mixing water needed to produce heated concrete of required temperature. Temperatures are based on the following mixture and are reasonably accurate for other typical mixtures: Aggregate (granite): 1360 kg (3000 lb); Moisture in aggregate: 27 kg (60 lb); Added mixing water: 109 kg (240 lb); Portland cement 256 kg (564 lb).

Aggregates

The temperature of aggregates varies with weather and type of storage. Aggregates usually contain frozen lumps and ice when the temperature is below freezing. Frozen aggregates must be thawed to avoid aggregate pockets in the concrete after batching, mixing, and placing. If thawing takes place in the mixer, the cooling effect caused by melting ice must be taken into consideration when estimating final concrete temperatures.

At air temperatures consistently above -4°C (25°F) it is seldom necessary to heat aggregates. The desired concrete temperature can usually be obtained by heating only the mixing water. At temperatures below freezing, in addition to heating the mixing water, often only the fine aggregate needs to be heated to produce concrete of the required temperature, provided the coarse aggregate is free of frozen lumps.

Three of the most common methods for heating aggregates are: storage in bins or weigh hoppers heated by steam coils or live steam; storage in silos heated by hot air or steam coils; and stockpiling over heated slabs, steam vents, or pipes. Although heating aggregates stored in bins or weigh hoppers is most commonly used, the volume of aggregate that can be heated is often limited and quickly consumed during production.

Circulating steam through pipes over which aggregates are stockpiled is a recommended method for heating aggregates. Stockpiles can be covered with tarpaulins to retain and distribute heat and to prevent formation of ice. Live steam, at pressures of 500 kPa to 900 kPa (75 psi to 125 psi), can be injected directly into the aggregate pile to heat it, but the resultant variable moisture content in aggregates might result in erratic mixing-water control and must be measured and accounted for in batching. To avoid hot spots in the aggregates, limit temperatures to below 100°C (212°F), or average temperatures to below 65°C (150°F).

On small jobs aggregates can be heated by stockpiling over controlled heating sources. Care should be taken to prevent scorching the aggregates.

Water

Of the ingredients used to make concrete, mixing water is the easiest and most practical to heat. The mass of aggregates and cement in concrete is much greater than the mass of water. However, water can store more than four times as much heat per unit weight than cement or the aggregate. For cement and aggregates, the specific heat (that is, energy required to raise the temperature 1°C/kg (1°F/lb) of material) varies with the lithology, but is generally about 0.77 kJ to 0.91 kJ (0.19 Btu to 0.23 Btu) for most aggregates (see Table 18-2), compared to 4.19 kJ/kg (1 Btu/lb) for water. Fluctuations in mixing-water temperature from batch to batch should be avoided. The temperature of the mixing water can be adjusted by blending hot and cold water.

To avoid the possibility of a rapid set of the concrete when either water or aggregates are heated to above 38°C (100°F), they should be combined in the mixer first, before the cement is added. If this mixer-loading sequence is followed, water temperatures up to the boiling point can be used, provided the aggregates are cold enough to reduce the final temperature of the aggregates and water mixture to appreciably less than 38°C (100°F).

DELIVERING CONCRETE IN COLD WEATHER

Temperature Loss During Delivery

Temperature loss during haul time may be an issue when delivery times approaching or greater than 1 hour are anticipated. The following equations may be used to estimate temperature loss. These equations include adjustments to the initial concrete temperature to assure minimum temperature requirements at delivery are met. Adjust the values proportionally for times less than or greater than 1 hour (ACI 306R 2016).

For revolving drum mixers	$T = 0.25\,(t_r - t_a)$
For covered-dump body	$T = 0.10\,(t_r - t_a)$
For open-dump	$T = 0.15\,(t_r - t_a)$

where:

t_r – Required delivery temperature (°C or °F)

t_a – Ambient air temperature (°C or °F)

PLACING AND FINISHING IN COLD WEATHER

Temperature of Concrete as Placed and Maintained

ACI 306.1, *Standard Specification for Cold Weather Concreting* (1990), and ACI 301, *Specifications for Concrete Construction* (2020), state if the average of highest and lowest ambient temperature from midnight to midnight is expected to be less than 4°C (40°F) for more than 3 successive days, then the concrete must be delivered to meet the minimum temperatures shown in Table 19-1, unless otherwise specified.

The concrete should be placed in the forms before its temperature drops below that given on Line 4 of Table 19-1. That temperature should be maintained for the duration of the protection period given in Chapter 17.

Concreting on Ground During Cold Weather

Concreting on ground during cold weather involves some extra preparation. Placing concrete on the ground involves different procedures than those used at an upper level. In particular, if frozen, the ground must be thawed before placing concrete. When the subgrade is frozen to a depth of approximately 80 mm (3 in.) or less, the surface region can be thawed by steaming; spreading a layer of hot sand, gravel, or other granular material where the grade elevations allow it; removing and replacing suitable compacted and with unfrozen fill; covering the subgrade with insulation for several days; or using hydronic heaters under insulated blankets. (See **Hydronic Systems** later in this chapter.) Placing concrete for floor slabs and exposed footings should be delayed until the ground is thawed and warmed sufficiently to ensure that it will not freeze again during the protection and curing period.

Construction of enclosures for slabs on ground is generally much simpler than elevated enclosures. Indirect vented, electric, or hydronic heaters are required if the area is enclosed, to protect personnel from CO_2 build up, and the concrete from carbonation. Cement hydration will furnish some of the curing heat and the use of insulating blankets may be sufficient for some slabs.

Once cast, footings should be backfilled as soon as concrete is strong enough to bear the soil pressure with unfrozen fill. Concrete should never be placed on a frozen subgrade or backfilled with frozen fill. Once these frozen materials thaw, uneven settlements may occur and cause cracking.

ACI 306R-16 requires that concrete not be placed on any surface that would lower the temperature of the concrete in place below the minimum values shown on Line 4 in Table 19-1. In addition, concrete placement temperatures should not be higher than these minimum values by more than 11°C (20°F) to reduce rapid moisture loss and the potential development of plastic shrinkage cracks.

ACI 301-20 requires contact surface temperatures be maintained above 32°F if placing concrete in contact with ground, subbase, or mud mat. ACI 301 also requires formwork contact surface temperature be maintained above 5°C (10°F) and insulated or heated as required to protect concrete from freezing.

Slabs can be cast on ground at ambient temperatures as low as 2°C (35°F) as long as the minimum concrete temperature as placed is not less than shown on Line 4 of Table 19-1. Although surface temperatures need not be higher than a few degrees above freezing, they also should preferably not be more than 5°C (10°F) higher than the minimum placement temperature. The duration of curing should not be less than that described in Chapter 16 for the appropriate exposure classification. Because of the risk of surface imperfections that might occur on exterior concrete placed in late fall and winter, many concrete contractors choose to delay concrete placement until spring. By waiting until spring, temperatures will be more favorable for cement hydration; this will help generate adequate strengths along with sufficient drying so the concrete can resist freeze-thaw damage.

Concreting Above Ground During Cold Weather

Working above ground in cold weather usually involves several different approaches in comparison to work at ground level:

- The concrete mixture may not need to be changed to generate more heat because portable heaters can be used to heat the undersides of floor and roof slabs. However, there are advantages to having a mixture that will produce a high strength at an early age; for example, artificial heat can be cut off sooner, and forms can be recycled faster.

- Enclosures must be constructed to retain the heat under floor and roof slabs.

- Portable heaters used to warm the underside of formed concrete can be direct fired heating units if the underside is exposed to fresh air and exhaust gases are vented to the outside. Position heaters to provide uniform temperature.

Position heaters to provide uniform temperature. Before placing concrete, the heaters under a formed deck should be turned on to preheat the forms and melt any snow or ice remaining on top. Care should be taken to monitor against overheating and potential for form fires. Temperature requirements for surfaces in contact with fresh concrete are the same as those outlined in **Concreting on Ground During Cold Weather**. Metallic embedments at temperatures below the freezing point may result in localized freezing that decreases the bond between concrete and steel reinforcement. ACI 306R-16 suggests that a reinforcing bar having a cross-sectional area of about 650 mm² (1 in.²) should have a temperature of at least -12°C (10°F) immediately before encasement in fresh concrete at a temperature of at least 13°C (55°F). ACI 301-20 requires that massive metallic embedded items in concrete, including piping and conduit shall be maintained above 5°C (10°F).

Windbreaks may also provide enclosures for above ground construction (Figure 19-11). When slab finishing is completed, insulating blankets or other insulation must be placed on top of the slab to ensure that proper curing temperatures are maintained. The insulation value (R) necessary to maintain the concrete surface temperature of walls and slabs above ground at 10°C (50°F) or above for 7 days may be estimated from Figure 19-12. To maintain a temperature for longer periods, more insulation is required. ACI 306-16 has additional graphs and tables for slabs placed on ground at a temperature of 2°C (35°F). Insulation can be selected based on R-values provided by insulation manufacturers or by using the information in Table 19-2. Insulation should be moisture proof or protected from moisture. Most commercial concrete insulating blankets consist of layers of cellular air cushioning or foam packeting sandwiched between plastic sheeting.

FIGURE 19-11. Finishing this concrete flatwork can proceed because a windbreak has been provided and there is adequate heat under the slab.

When concrete strength development is not determined, a conservative estimate can be made if adequate protection at the recommended temperature is provided for the duration of time found in Table 19-3. However, the actual amount of insulation and length of the protection period should be determined from the monitored in-place concrete temperature and the desired strength. A correlation between curing temperature, curing time, and compressive strength can be determined from laboratory testing of the particular concrete mixture used in the field (see **Maturity Concept**). Corners and edges are particularly vulnerable during cold weather and must be protected. As a result, the thickness of insulation for these areas, especially on columns, should be about three times the thickness that is required for walls or slabs. On the other hand, if the ambient temperature rises much above the temperature assumed in selecting insulation values, the temperature of the concrete may become excessive. To lower the probability of thermal shock and cracking when forms are removed. See **Terminating the Heating Period**. Temperature readings of insulated concrete

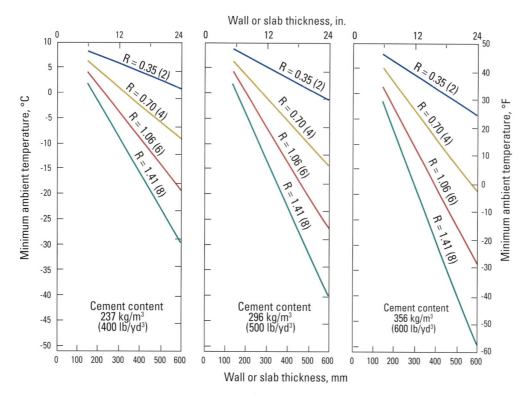

FIGURE 19-12. Thermal resistance (R) of insulation required to maintain the concrete surface temperature of walls and slabs above ground at 10°C (50°F) or above for 7 days. Concrete temperature as placed: 10°C (50°F). Maximum wind velocity: 24 km/h (15 mph). Note that in order to maintain a certain minimum temperature for a longer period of time, more insulation or a higher R-value is required (adapted from ACI 306R-16).

should therefore be taken at regular intervals and should not vary from ambient air temperatures by more than the values given in ACI 306R-16. The maximum temperature differential between the concrete interior and the concrete surface should be about 20°C (36°F) to minimize cracking. The weather forecast should be checked and appropriate action taken for expected temperature changes.

Columns and walls should not be cast on foundations at temperatures below 0°C (32°F) because chilling of concrete in the bottom of the column or wall will retard strength development. Concrete should not be placed on any surface that would lower the temperature of the as-placed concrete below the minimum values shown on Line 4 in Table 19-1.

MONITORING CONCRETE TEMPERATURE

Calibrated thermometers are needed to check the concrete temperatures as delivered, as placed, and as maintained. A Calibrated thermometer is shown in Figure 19-13. Other types of devices, such as infrared thermometers, can also be used. Commercial sensors (Figure 19-14) may be useful for monitoring in-place concrete temperature. (Note, infrared thermometers are not standardized for use in concrete.)

After the concrete has hardened, temperatures can be checked with an infrared thermometer or with an ordinary thermometer that is kept covered with insulating blankets.

FIGURE 19-13. A bimetallic pocket thermometer with a metal sensor suitable for checking fresh concrete temperatures.

FIGURE 19-14. Commercial sensors may be used for monitoring concrete temperatures (courtesy of Judlau – OHL Group and COMMAND Center).

TABLE 19-2. Insulation Values of Various Materials[†]

MATERIAL	DENSITY kg/m³ (lb/ft³)		THERMAL RESISTANCE, *R*, FOR 10-mm (1-in.) THICKNESS OF MATERIAL,* (m² • °C) / W ([°F • hr • ft²]/ Btu)	
Blanket and batt				
Glass-fiber batts	8	(0.47 to 0.51)	0.542	(3.08)
Glass-fiber batts	10 to 12	(0.61 to 0.75)	0.608	(3.45)
Glass-fiber batts	13 to 14	(0.79 to 0.85)	0.664	(3.77)
Glass-fiber batts	22	(1.40)	0.766	(4.35)
Rock and slag wool batts	32 to 37	(2.00 to 2.30)	0.690	(3.92)
Rock and slag wool batts	45	(2.80)	0.750	(4.26)
Mineral wool	16 to 48	(1.00 to 3.00)	0.629	(3.57)
Mineral wool	16 to 29	(1.00 to 1.80)	0.734	(4.17)
Board and slabs				
Cellular glass	120	(7.50)	0.608	(3.45)
Cement fiber slabs, portland cement binder	400 to 432	(25.0 to 27.0)	0.342	(1.94)
Cement fiber slabs, magnesia oxysulfide binder	352	(22.0)	0.308	(1.75)
Glass fiber board	24 to 96	(1.50 to 6.00)	0.750	(4.26)
Expanded rubber (rigid)	64	(4.00)	0.881	(5.00)
Extruded polystyrene, smooth skin, aged 180 days	22 to 58	(1.40 to 3.60)	0.881	(5.00)
Expanded polystyrene, molded beads	16 to 24	(1.00 to 1.50)	0.704	(4.00)
Expanded polystyrene, molded beads	29	(1.80)	0.766	(4.35)
Mineral fiberboard, wet felted	160	(10.0)	0.678	(3.85)
Rock wool board, floors and walls	64 to 128	(4.00 to 8.00)	0.734	(4.17)
Rock wool board, roofing	160 to 176	(10.0 to 11.0)	0.629	(3.57)
Acoustical tile	336 to 368	(21.0 to 23.0)	0.483	(2.74)
Perlite board	144	(9.00)	0.490	(2.78)
Polyisocyanurate, unfaced	26 to 37	(1.60 to 2.30)	1.067	(6.06)
Loose fill				
Cellulose fiber, densely packed	56	(3.50)	0.641	(3.64)
Perlite, expanded	32 to 64	(2.00 to 4.00)	0.608	(3.45)
Perlite, expanded	64 to 120	(4.00 to 7.50)	0.527	(2.99)
Perlite, expanded	120 to 176	(7.50 to 11.00)	0.451	(2.56)
Glass fiber	29 to 37	(1.80 to 2.30)	0.719	(4.08)
Rock and slag wool	64	(4.00)	0.629	(3.57)
Vermiculite, exfoliated	112 to 131	(7.00 to 8.20)	0.375	(2.13)
Vermiculite, exfoliated	64 to 96	(4.00 to 6.00)	0.400	(2.27)
Woods (12% moisture)				
Oak	657 to 753	(41.0 to 47.0)	0.148	(0.84)
Birch	689 to 721	(43.0 to 45.0)	0.148	(0.84)
Maple	641 to 705	(40.0 to 44.0)	0.155	(0.88)
Ash	609 to 673	(38.0 to 42.0)	0.160	(0.91)

R-values are the reciprocal of U values (conductivity).

* Values are from ASHRAE (2017).

[†] Values are for mean temperature of 75°F. Representative values for dry materials are intended as design (not specification) values for materials in normal use. Thermal values of insulating materials may differ from design values depending on in situ properties (e.g., density and moisture content, orientation, etc.) and manufacturing variability. For properties of specific product, use values supplied by manufacturer.

CURING AND PROTECTION IN COLD WEATHER

After concrete is in place, it should be protected and kept within the recommended temperatures listed on Line 4 of Table 19-1. These curing temperatures should be maintained until specified strength is gained to withstand exposure to low temperatures, anticipated environment, and construction and service loads. The length of protection required to accomplish this will depend on the cementitious materials types and amount, whether accelerating admixtures were used, and the loads that must be carried. Recommended minimum periods of protection are given in Tables 19-3A and 19-3B. The duration of heating structural

concrete requiring full service loading before forms and shores are removed should be based on the adequacy of in-place compressive strengths rather than an arbitrary time period.

Wet Curing

Strength gain stops when moisture required for hydration is no longer available. Concrete retained in forms or covered with insulation may retain enough moisture at 5°C to 13°C (40°F to 55°F) to maintain curing. However, a means of providing wet curing is needed to offset drying from low humidity in cold weather and from the dry air produced by heaters used in enclosures during cold weather.

TABLE 19-3A. Recommended Duration of Concrete Protection in Cold Weather*

SERVICE CATEGORY	CONVENTIONAL CONCRETE,† DAYS	HIGH-EARLY-STRENGTH CONCRETE,‡ DAYS
No load, not exposed‡ favorable moist-curing	2	1
No load, exposed, but later has favorable moist-curing	3	2
Partial load, exposed	6	4
Fully stressed, exposed	See Table 19-3B	

TABLE 19-3B. Recommended Duration of Protection for Fully Stressed Concrete Subjected to Freezing and Thawing

REQUIRED PERCENTAGE OF STANDARD-CURED 28-DAY STRENGTH	DAYS AT 10°C (50°F) TYPE OF HYDRAULIC CEMENT			DAYS AT 21°C (70°F) TYPE OF HYDRAULIC CEMENT		
	I OR GU	II OR MH	III OR HE	I OR GU	II OR MH	III OR HE
50	6	9	3	4	6	3
65	11	14	5	8	10	4
85	21	28	16	16	18	12
95	29	35	26	23	24	20

*Table 19-3A and Table 19-3B are adapted from ACI 306R-16. Cold weather is defined as when the temperature has, or is expected to, fall below 4°C (40°F). For recommended concrete temperatures, see Table 19-1. For concrete that is not air entrained, ACI 306R states that protection for durability should be at least twice the number of days listed in Table 19-3A.

Table 19-3B is adapted from ACI 306R-16. The values shown are approximations and will vary according to the thickness of concrete, mix proportions, and so on. They are intended to represent the ages at which supporting forms can be removed.

For recommended concrete temperatures, see Table 19-1.

† Made with ASTM C150 (AASHTO M 85) Type I, II, or C1157 GU, or MH hydraulic cement.

‡ Made with ASTM C150 (AASHTO M 85) Type III or C1157 HE hydraulic cement, an accelerator, or an extra 60 kg/m³ (100 lb/yd³) of cement.

Live steam exhausted into an enclosure around the concrete is an excellent method of curing because it provides both heat and moisture. Steam is especially useful in extremely cold weather because the moisture provided offsets the rapid drying that occurs when very cold air is heated.

Enclosures

Heated enclosures are effective for protecting concrete in cold weather. Enclosures can be constructed of wood, canvas tarpaulins, or polyethylene film (Figure 19-15). Prefabricated, rigid plastic enclosures are also available (Figure 19-16). Clear plastic enclosures that admit daylight are popular but temporary heat in these enclosures can prove expensive.

FIGURE 19-16. Even in the winter, an outdoor swimming pool can be constructed if a heated enclosure is used.

When enclosures are constructed below a deck, the framework can be extended above the deck to serve as a windbreak. Typically, a height of 2 m (6 ft) will protect concrete and construction personnel against biting winds that cause temperature drops and excessive evaporation. Wind breaks may be taller or shorter depending on anticipated wind velocities, ambient temperatures, relative humidity, and concrete placement temperatures.

Enclosures can be quickly transported using flying forms; more often, though, they must be removed so that the wind will not

FIGURE 19-15. Tarpaulin heated enclosure maintains an adequate temperature for proper curing and protection during prolonged winter weather.

interfere with maneuvering the forms into position. Similarly, enclosures can be built in large panels with the windbreak included; much like gang forms.

Insulating Materials

Tarpaulins and insulated blankets are often necessary to retain the heat of hydration more efficiently and keep the concrete warm. Thermometer readings of the concrete's temperature will indicate whether the covering is adequate. The heat generated during hydration will offset to a considerable degree the loss of heat during placing, finishing, and early curing operations.

Heat and moisture can be retained in the concrete by covering it with commercial insulating blankets (Figure 19-17). The effectiveness of insulation can be determined by placing a thermometer under it and in contact with the concrete. If the temperature falls below the minimum required on Line 4 in Table 19-1, additional insulating material, or material with a higher R-value, should be applied. Corners and edges of concrete are most vulnerable to freezing. In view of this, temperatures at these locations should be checked more often.

FIGURE 19-17. Stack of insulating blankets. These blankets trap heat and moisture in the concrete, providing beneficial curing.

The thermal resistance (R-values) for common insulating materials are given in Table 19-2. For maximum efficiency, insulating materials should be kept dry and in close contact with concrete or formwork.

Insulating blankets for construction are made of fiberglass, sponge rubber, open cell polyurethane foam, vinyl foam, mineral wool, or cellulose fibers. The outer covers are made of canvas, woven polyethylene, or other tough fabrics that will withstand rough handling. The R-value for a typical insulating blanket is about 1.2 m² • °C/W for 50 mm to 70 mm thickness, (7 [°F • hr • ft²]/Btu for 2 in. to 3 in.), but since R-values are not marked on the blankets, their effectiveness should be checked with a thermometer. If necessary, they can be used in multiple layers to attain the desired insulation.

Historically, concrete pavements have been protected from cold weather by spreading 300 mm (1 ft) or more of dry straw or hay

on the surface for insulation. If used, tarpaulins, polyethylene film, or waterproof paper should be used as a protective cover over the straw or hay to make the insulation more effective and prevent it from blowing away. The straw or hay should be kept dry or its insulation value will drop considerably.

Stay-in-place insulating concrete forms (ICF) became popular for cold-weather construction in the 1990s (Figure 19-18). Gajda (2002) showed that ICFs can be used to successfully place concrete in ambient temperatures as low as -29°C (-20°F). Forms built for repeated use often can be economically insulated with commercial blanket insulation. The insulation should have a tough moisture proof covering to withstand handling abuse and exposure to the weather. Rigid insulation can also be used (Figure 19-19).

FIGURE 19-18. Insulating concrete forms (ICF) permit concreting in cold weather.

FIGURE 19-19. With air temperatures down to -23°C (-10°F), concrete was cast in this insulated column form made of 19 mm (¾ in.) high-density plywood inside, 25 mm (1 in.) rigid polystyrene in the middle, and 13 mm (½ in.) rough plywood outside. R-value: 1.0 m² • °C/ W (5.6 [°F • hr • ft²] / Btu).

Heaters

Two types of heaters are used in cold weather concrete construction: direct fired and indirect fired (Figure 19-20). Indirect fired heaters are vented to remove the products of combustion. Where heat is supplied to the top surface of fresh concrete – for example, a floor slab – vented heaters are required.

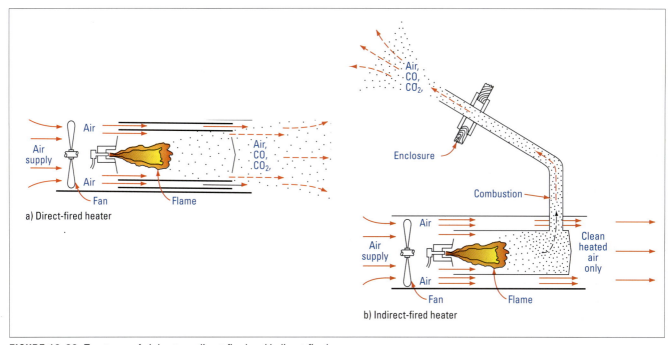

FIGURE 19-20. Two types of air heaters, direct-fired and indirect-fired.

Carbon dioxide (CO_2) in the exhaust must be vented to the outside and prevented from reacting with the fresh concrete (Figure 19-21). Direct fired units can be used to heat the enclosed space beneath concrete placed for a floor or a roof deck (Figure 19-22) only if adequate health and safety precautions are taken. Any heater burning a fossil fuel produces CO_2, this gas can combine with calcium hydroxide on the surface of fresh concrete to form a weak layer of calcium carbonate that interferes with cement hydration (Kauer and Freeman 1955). The result is a soft, chalky surface that will dust under traffic. Depth and degree of carbonation depend on concentration of CO_2, curing temperature, humidity, porosity of the concrete, length of exposure, and method of curing. Direct-fired heaters, therefore, should not be permitted to heat the air over concreting operations – until at least 24 hours have elapsed. In addition, the use of fossil fuel powered construction equipment should be restricted in enclosures during that time. If unvented heaters are used, immediate wet curing or the use of a curing compound will minimize the potential for carbonation.

A salamander is an inexpensive combustion heater without a fan that discharges its combustion products directly into the surrounding air; heating is accomplished by radiation from its metal casing. Salamanders are fueled by coke, oil, wood, or liquid propane. They are one form of a direct-fired heater. One disadvantage of salamanders is the safety risk associated with the high temperature of their metal casing. Salamanders should be placed so that they will not overheat formwork or enclosure materials. When placed on floor slabs, they should be elevated to avoid scorching the concrete.

> **Warning:** Carbon monoxide (CO), another product of combustion, can cause potentially serious health hazards: Four hours of exposure to 200 parts per million of CO will produce headaches and nausea. Three hours of exposure to 600 ppm can be fatal. The American National Standard A10.10, *Safety Requirements for Temporary and Portable Space Heating Devices and Equipment Used in the Construction Industry* (ANSI 2015), limits concentrations of CO to 50 ppm at worker breathing levels. The standard also establishes safety rules for ventilation and the stability, operation, fueling, and maintenance of heaters.

FIGURE 19-21. An indirect-fired heater. Notice vent pipe that carries combustion gases outside the enclosure.

FIGURE 19-22. A direct-fired heater installed through the enclosure.

Some heaters burn more than one type of fuel. The approximate heat values of fuels are as follows:

No. 1 fuel oil	37,700 kJ/L (135,000 Btu/gal)
Kerosene	37,400 kJ/L (134,000 Btu/gal)
Gasoline	35,725 kJ/L (128,000 Btu/gal)
Liquid-propane gas	25,500 kJ/L (91,500 Btu/gal)
Natural gas	37,200 kJ/m^3 (1,000 Btu/ft^3)

The output rating of a portable heater is usually the heat content of the fuel consumed per hour. A rule of thumb is that about 134,000 kJ are required for each 100 m^3 (36,000 Btu for 10,000 ft^3) of air to develop a 10°C (18°F) temperature rise.

Hydronic Systems

Hydronic systems transfer heat by circulating a glycol/water solution in a closed system of pipes or hoses (see Figure 19-23). These systems transfer heat more efficiently than forced air systems without the negative effects of exhaust gases and drying of the concrete from air movement. The specific heat of water/glycol solutions is more than six times greater than air. As a result, hydronic heaters can deliver large quantities of heat at low temperature differentials of 5°C (10°F) or less between the heat transfer hose and the concrete. Cracking and curling induced by temperature gradients within the concrete are nearly eliminated as is the danger of accidentally overheating the concrete and in potentially damaging long term strength gain.

Typical applications for hydronic systems include thawing and preheating subgrades. They are also used to cure elevated and on-grade slabs, walls, foundations, and columns. To heat a concrete element, hydronic heating hoses are usually laid on or hung adjacent to the structure and covered with insulated blankets and sometimes plastic sheets. Usually, construction of temporary enclosures is unnecessary. Hydronic systems can be used over areas much larger than would be practical to enclose. If a heated enclosure is necessary for other work, hydronic hoses can be sacrificed (left under a slab on grade) to make the slab a radiant heater for the structure built above (Grochoski 2000).

Electricity can also be used to cure concrete in winter. The use of large electric blankets equipped with thermostats is one method. The blankets can also be used to thaw subgrades or concrete foundations.

Electrical Resistance

Use of electrical resistance wires that are cast into the concrete is another method used during cold weather conditions. The power supplied is under 50 volts, and from 7.0 MJ to 23.5 MJ (1.5 kW/hr to 5 kW/hr) of electricity per cubic meter (cubic yard) of concrete is required, depending on the circumstances. Where electrical resistance wires are used, insulation should be included during the initial setting period. If insulation is removed before the recommended time, the concrete should be covered with an impervious sheet and the power supplied for the required time.

Radiant Heating and Steam

Steam is another source of heat for winter concreting. Live steam can be piped into an enclosure or supplied through radiant heating units. In choosing a heat source, it must be remembered

FIGURE 19-23. Hydronic system showing hoses (left) laying on soil to defrost subgrade and (right) warming the forms while fresh concrete is pumped in (courtesy of Groundheaters, Inc.).

that the concrete itself supplies heat through hydration of cement; this heat is often enough for curing needs provided the heat can be retained within the concrete using insulation.

TERMINATING THE HEATING PERIOD

To avoid cracking of the concrete due to sudden temperature change at the end of the curing period, ACI 306R-16 requires that the source of heat and cover protection be slowly removed. The maximum allowable temperature drop during the first 24 hours after the end of the protection is given in Table 19-4. The temperature drops apply to surface temperatures. Notice that the cooling rates for surfaces of mass concrete (thick sections) are lower than they are for thinner members.

TABLE 19-4. Maximum Allowable Temperature Drop During First 24 Hours After End of Protection Period*

SECTION SIZE, MINIMUM DIMENSIONS, mm (in.)			
LESS THAN 300 (12)	300 TO 900 (12 TO 36)	900 TO 1800 (36 TO 72)	OVER 1800 (72)
27°C (50°F)	22°C (40°F)	17°C (30°F)	11°C (20°F)

*Adapted from ACI 306R (2016) and ACI 301 (2020).

Gradual cooling can be accomplished by lowering the heat or by simply removing the heat source while leaving blankets or tarps and allowing the heat slowly to dissipate.

FORM REMOVAL AND RESHORING

Before shores and forms are removed, strength-critical structural concrete should be tested to determine if in-place strengths are adequate. If the curing temperatures listed on Line 4 of Table 19-1 are maintained, Table 19-3A can be used to determine the minimum time in days that vertical support for forms should remain in place.

In-place strengths can be monitored using one of the following: field-cured cylinders (ASTM C31 [AASHTO T 23]); probe penetration tests (ASTM C803, *Standard Test Method for Penetration Resistance of Hardened Concrete*); cast-in-place cylinders (ASTM C873, *Standard Test Method for Compressive Strength of Concrete Cylinders Cast in Place in Cylindrical Molds*); pullout testing (ASTM C900, *Standard Test Method for Pullout Strength of Hardened Concrete*); or maturity method (ASTM C1074, *Standard Practice for Estimating Concrete Strength by the Maturity Method* [AASHTO T 325]). Many of these tests are indirect methods of measuring compressive strength; they require correlation in advance with standard cylinders before estimates of in-place strengths can be made.

If in-place compressive strengths are not documented, Table 19-3B lists conservative time periods in days to achieve various percentages of the standard laboratory cured 28-day strength. The engineer issuing project drawings and specifications in cooperation with the formwork contractor must determine what percentage of the design strength is required (ACI 306R 2016). Vertical forms can be removed sooner than shoring and temporary falsework (ACI 347 2014).

Maturity Concept

The maturity concept is based on the principle that strength gain in concrete is a function of curing time and temperature. The maturity concept, as described in ASTM C1074 (AASHTO T 325) can be used to estimate strength development. Two maturity methods to estimate the in-place concrete strength are shown in Table 19-5. The first method is based on the Nurse-Saul function, also called Time-Temperature Factor method. A shortcoming of this method is that is fails to recognize the fact that maturity increases disproportionately at elevated temperatures depending on the type(s) of cementitious materials used and the water-to-cementitious materials ratio. The "Equivalent Age" maturity function is based on the Arrhenius equation; this function presents maturity in terms of equivalent age of curing at a specified temperature.

The Time-Temperature Factor method presents maturity in terms of °C•hr. Most maturity equipment uses a datum temperature of 0°C, which further simplifies the calculation. Given this simplification, maturity is typically calculated using metric units.

TABLE 19-5. Time-Temperature Factor and Equivalent Age Maturity Equations (ASTM C1074 [AASHTO T 325])

TIME-TEMPERATURE FACTOR		EQUIVALENT AGE	
$M = \sum_{0}^{t} (T - T_o) \Delta t$	M = maturity index, °C-hours T = average concrete temperature, °C, during the time interval Δt T_o = datum temperature (usually taken to be 0°C) t = elapsed time, hours Δt = time intervals, hours	$t_e = \sum e^{-\frac{E}{R}\left(\frac{1}{T} - \frac{1}{T_r}\right)} \Delta t$	t_e = equivalent age at the reference temperature E = apparent activation energy, J/mol (see ASTM C1074 for typical values) R = universal gas constant, 8.314 J/mol-K T = average concrete temperature, Kelvin, during the time interval Δt T_r = reference temperature, Kelvin Δt = time intervals, days or hours

FIGURE 19-24. Downloading data from a maturity sensor embedded in a pavement.

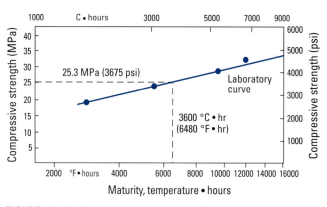

FIGURE 19-25. Maturity relation example (Nurse-Saul method).

To monitor the strength development of concrete in place using the maturity concept, the following information must be available:

- The strength-maturity relationship of the concrete used in the structure. The results of compressive strength tests at various ages on a series of cylinders made of a concrete mixture used in the structure; this must be done to develop a strength-maturity curve. These cylinders are cured in a laboratory at 23°C ± 2°C (73°F ± 3°F).

- A time-temperature record of the concrete in place. Temperature readings are obtained by placing expendable thermistors or thermocouples in the concrete (Figure 19-24).

See Figure 19-25 for an example using the maturity concept. Before construction begins a calibration curve is drawn by plotting the relationship between compressive strength and the maturity for a series of test cylinders (of the particular concrete mixture proportions) cured in a laboratory and tested for strength at successive ages.

The maturity concept is useful in checking the curing of concrete and in estimating strength in relation to time and temperature. It presumes that all other factors affecting concrete strength have been properly controlled. With these limitations in mind, the maturity method has gained greater acceptance as a surrogate for measuring the compressive strength of the concrete for removal of forms, or opening a pavement to traffic. It is no substitute for quality control and proper concreting practices (Gajda 2007, Kerkhoff and Gajda 2005, Malhotra 1974, and ACI 347 2014).

REFERENCES

PCA's online catalog includes links to PDF versions of many of our research reports and other classic publications. Visit: cement.org/library/catalog.

ACI Committee 301, *Specifications for Concrete Construction*, ACI 301-20, American Concrete Institute, Farmington Hills, MI, 2020, 74 pages.

ACI Committee 306, *Guide to Cold Weather Concreting*, ACI 306R-16, American Concrete Institute, Farmington Hills, MI, 2016, 29 pages.

ACI Committee 306, *Standard Specification for Cold Weather Concreting*, ACI 306.1-90, American Concrete Institute, Farmington Hills, MI, 1990, 6 pages.

ACI Committee 347, *Guide to Formwork for Concrete*, ACI 347-14, American Concrete Institute, Farmington Hills, MI, 2004, 32 pages.

ASHRAE Handbook of Fundamentals, American Society of Heating, Refrigerating, and Air-conditioning Engineers, Inc., New York, NY, 1977 and 1981.

ANSI, *Safety Requirements for Temporary and Portable Space Heating Devices Used in the Construction Industry*, A10.10, American National Standards Institute, New York, NY, 2015.

Burg, R.G., *The Influence of Casting and Curing Temperature on the Properties of Fresh and Hardened Concrete*, Research and Development Bulletin RD113, Portland Cement Association, Skokie, IL, 1996, 20 pages.

Gajda, J., *Cold-Weather Construction of ICF Walls*, SN2615, Portland Cement Association, Skokie, IL, 2002, 36 pages.

Gajda, J., *Mass Concrete for Buildings and Bridges*, EB547, Portland Cement Association, Skokie, IL, 2007, 44 pages.

Gebler, S.H.; and Klieger, P., *Effect of Fly Ash on Durability of Air-Entrained Concrete*, Research and Development Bulletin RD090, Portland Cement Association, Skokie, IL, 1986, 40 pages.

Grochoski, C., "Cold-Weather Concreting with Hydronic Heaters," *Concrete International*, American Concrete Institute, Farmington Hills, MI, April 2000, pages 51 to 55.

Kauer, J.A.; and Freeman, R.L., "Effect of Carbon Dioxide on Fresh Concrete," *Journal of the American Concrete Institute Proceedings*, Vol. 52, December 1955, pages 447 to 454. American Concrete Institute, Farmington Hills, MI, Discussion: December 1955, Part II, pages 1299 to 1304.

Kerkhoff, B.; and Gajda, J., "Maturity of Concrete: Concept, Tests, and Applications," CT053, *Concrete Technology Today*, Vol. 26, No. 3, Portland Cement Association, Skokie, IL, December, 2005, pages 4 and 5.

Klieger, P., *Curing Requirements for Scale Resistance of Concrete*, Research Department Bulletin RX082, Portland Cement Association, 1957, 17 pages.

Klieger, P., *Effect of Mixing and Curing Temperature on Concrete Strength*, Research Department Bulletin RX103, Portland Cement Association, 1958, 20 pages.

Malhotra, V.M., "Maturity Concept and the Estimation of Concrete Strength: A Review," Parts I and II, *Indian Concrete Journal*, Vol. 48, Associated Cement Companies, Ltd., Bombay, April and May 1974.

McNeese, D.C., "Early Freezing of Non-Air-Entrained Concrete," *Journal of the American Concrete Institute Proceedings*, Vol. 49, American Concrete Institute, Farmington Hills, MI, December 1952, pages 293 to 300.

Powers, T.C., *Resistance of Concrete to Frost at Early Ages*, Research Department Bulletin RX071, Portland Cement Association, Skokie, IL, 1956, 49 pages.

Powers, T.C., *Prevention of Frost Damage to Green Concrete*, Research Department Bulletin RX148, Portland Cement Association, Skokie, IL, 1962, 18 pages.

TEST METHODS

FIGURE 20-1. Quality control and testing concrete provides important feedback on compliance with project specifications.

Quality control and acceptance testing (Figure 20-1) are indispensable parts of the construction process. Test results provide important feedback on compliance with project specifications and also may be used to base decisions regarding any necessary adjustments to a concrete mixture. Past experience and sound judgment must be relied upon in evaluating test results.

Most specifications are a combination of prescriptive and performance requirements. However, specifiers are moving toward performance-based specifications that are concerned with the final performance of concrete rather than the process used to achieve the performance (Hover and others 2008, and ACI 329R 2014). Such specifications may not have acceptance limits for process control tests (such as slump or limits on the quantities of concrete ingredients) as with prescriptive specifications. Instead, physical tests are used to measure in-place concrete performance. These tests then become the basis for acceptance. Although process control tests may not be specified, a producer can use them to guide the project.

CLASSES OF TESTS

Project specifications may require a specific composition for the concrete mixture. These include maximum size of aggregate, aggregate proportions, and may also require minimum or maximum amount of cementitious materials. Properties of the freshly mixed and hardened concrete such as temperature, slump, density, air content, and compressive or flexural strength are also typically specified.

Cementitious materials are tested for their compliance with ASTM or AASHTO standard specifications to assure the potential for adequate strength development in concrete and avoid abnormal performance such as early stiffening, delayed setting, increased temperature or shrinkage, and low strength in concrete.

Aggregates are tested to determine their suitability for use in concrete by identifying harmful materials with chemical and petrographic examinations, and testing for abrasion resistance, cyclic freezing and thawing, and potential alkali-aggregate reactivity; and to assure uniformity with tests for moisture control, relative density or specific gravity, and gradation.

Fresh concrete is tested to evaluate the suitability of available materials, establish mixture proportions, and control concrete quality during construction. ASTM C94, *Standard Specification for Ready-Mixed Concrete* (AASHTO M 157), specifies that slump, air content, density, and temperature tests be performed when strength test specimens are made.

Hardened concrete is tested to prequalify a concrete mixture for certain properties including strength, modulus of elasticity, shrinkage, and durability; and may also be specified to be tested during construction to verify those properties in-place.

Following is a discussion of frequency of testing and descriptions of the common quality control tests to ensure uniformity of materials, desired properties of freshly mixed concrete, and required strength of hardened concrete. Special test methods are also described.

Lamond and Pielert (2006) and ASTM (2017 and 2019) provide extensive discussions of test methods for concrete and concrete ingredients.

FREQUENCY OF TESTING

The frequency of testing is a significant factor in the effectiveness of quality control of concrete. Specified testing frequencies are required also for acceptance of concrete or its components. Tests should be conducted at random locations within the quantity or time period specified by the testing standard or specification. Occasionally, specified testing frequencies may be insufficient to effectively control materials within specified limits during production. Therefore, process control tests (nonrandom tests) are often performed in addition to acceptance tests to document trends so that adjustments can be made to the concrete mixture before performing the required acceptance tests. The frequency of testing aggregates and concrete for typical batch-plant procedures depends largely upon the uniformity of materials, including the moisture content of aggregates, and the production process. Initially, it is advisable to make process control tests several times a day, but as work progresses and materials become more predictable, the testing frequency often can be reduced. ASTM C1451, *Standard Practice for Determining Variability of Concrete-Making Materials From a Single Source*, provides a standard practice for determining the uniformity of cementitious materials, aggregates, and chemical admixtures used in concrete.

Usually, aggregate moisture tests are conducted once or twice a day and monitored throughout the day with moisture meters. The first batch of fine aggregate in the morning may be overly wet because moisture can migrate overnight to the bottom of the storage bin. As fine aggregate is drawn from the bottom of the bin and additional aggregate is added, the moisture content should stabilize at a lower level and a moisture test can be conducted. It is important to obtain moisture samples representative of the aggregates being batched; a 1% change in moisture content of fine aggregate corresponds to approximately 8 kg/m³ (13 lb/yd³) of mixing water.

Slump, air content, density (unit weight), and temperature tests should be conducted for the first batch of concrete each day, whenever consistency of concrete appears to vary, and whenever strength-test specimens are made at the jobsite. Air-content tests should be made often enough at the point of delivery to ensure proper air content, particularly if temperature and aggregate grading change.

The number of strength tests will depend on the job specifications and the occurrence of variations in the concrete mixture. ACI 318, *Building Code Requirements for Structural Concrete* (2019) and ASTM C94 require that strength tests for each class of concrete placed each day be made at least once a day, at least once for each 115 m³ (150 yd³) of concrete, and if applicable, at least once for each 500 m² (5,000 ft²) of surface area for slabs or walls. In ACI 318-19, a strength test is defined as the average strength of two 150-mm × 300-mm (6-in. × 12-in.) or three 100-mm × 200-mm (4-in. × 8-in.) cylinders tested at 28 days or other age designated for f'_c. In addition, a 7-day test cylinder is often made and tested to provide an early indication of strength potential. As a rule of thumb, the 7-day strength is about 60% to 75% of the 28-day compressive strength, depending upon the type and amount of cementitious materials, water-cement ratio, initial curing temperature, and other variables. Additional specimens may be tested when high-strength concrete is involved or where structural requirements are critical. Specimens for strength tests are subjected to standard curing as defined in ASTM C31, *Standard Practice for Making and Curing Concrete Test Specimens in the Field* (AASHTO T 23). Strength of standard-cured specimens should not be used as an indication of the actual in-place concrete strengths.

In-place concrete strengths are typically estimated by testing specimens that have been field-cured in the same manner (as nearly as practicable) as concrete in the structure. ASTM C31 (AASHTO T 23) provides requirements for the handling and testing of field-cured specimens. Tests of field-cured specimens are commonly used to determine when forms and shores under a structural slab might be removed or when traffic will be allowed on new pavement.

PRECISION AND BIAS

A precision and bias statement for a test method should help the user in evaluating test results. The ASTM definition of precision is "the closeness of agreement between independent test results obtained under stipulated conditions." Precision depends on random errors and does not relate to the accepted reference value. Precision allows users to assess the general usefulness of a test result with respect to variability (ASTM E177).

A test determination is the value measured from a single test specimen. A test result can be a single determination or the average of two or more determinations, depending on how the test method is written or test results are used by other standards. For example, if three cylindrical specimens are made from the same sample of concrete for determining compressive strength, the strength measured on each cylinder is a test determination. The average of the three determinations is the test result that is used in concrete specifications to assess acceptability of the concrete.

Single-operator precision is defined as the closeness of agreement among test determinations obtained on identical test specimens by a single operator using the same apparatus in the same laboratory over a relatively short period of time.

Precision values are based on replicate testing of identical test specimens, which are defined as test specimens selected at random and made from a single quantity or batch of material that is as homogeneous as possible (ASTM C670). For materials like concrete, is it not possible to obtain test specimens that are exactly alike. For that reason, specimens are considered identical if they are made from a single quantity of material or batches of cementitious mixtures that are as homogenous as possible. In the previous example, the cylinders made from the same sample of concrete are considered identical specimens. If specimens are made at one facility from a single batch of concrete and distributed randomly to several laboratories, they are considered identical specimens for the purpose of determining precision indices from an interlaboratory study (ILS).

Figure 20-2 illustrates the concept of single-operator precision. It shows the values of test determinations obtained by operators in seven laboratories. The operator from each laboratory tested three identical specimens. Single-operator precision is the average scatter obtained by a single-operator when testing identical specimens using the same apparatus in the same laboratory. Single-operator precision is determined by calculating the pooled standard deviation from all the laboratories.

Multilaboratory precision is defined as the closeness of agreement among test results obtained with the same test method on identical test specimens in different laboratories with different operators using different equipment. Simply put, multilaboratory precision refers to the scatter in the average values obtained among the different laboratories testing identical specimens (Figure 20-3).

ASTM C802 provides the details for making the calculations for single operator precision and multilaboratory precision.

Two cases are commonly used in planning an ILS. In one case, test specimens will be made at one location and distributed to the participating laboratories. In the other case, homogenous batches of the same materials will be distributed to the laboratories with instructions on how to make the specimens. It can be expected that the latter case will result in greater multilaboratory variability because of the added source of variability among the laboratories associated with making the specimens.

Bias is "the difference between the expectation of the test results and an accepted reference value." Bias is the total systematic error as contrasted to random error and provides the relationship between typical test results under specific conditions and a related set of accepted reference values. A bias statement includes constant and variable components. Unknown and variable sources need to be identified. For the majority of ASTM test methods for cement and concrete, bias cannot be determined because there is no suitable reference material to serve as the true value, or the test method is the only

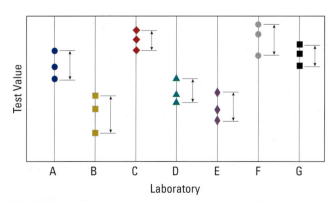

FIGURE 20-2. Single operator precision is determined by calculating the pooled standard deviation from all laboratories (courtesy of N. Carino).

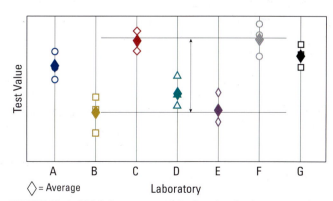

FIGURE 20-3. Multilaboratory precision is *related* to the scatter of the laboratory averages (courtesy of N. Carino).

way to measure the characteristic. ASTM C670 provides the methodology for determining the bias of a test method, and has recommended language for the bias statement for test methods for which bias cannot be determined.

There are two aspects of an operation of measurements, qualitative and quantitative (Shewhart and Deming 1939). This leads to two measurement concepts: a method with a written procedure, and a process that includes the method with a system of causes, repetition, and control. A measurement method cannot inherently be precise. This will vary in terms of individuals, equipment, and materials (Murphy 1961).

Repeatability is precision under conditions where independent test results are obtained with the same method on identical test items in the same laboratory by the same operator using the same equipment within short intervals of time. *Reproducibility* is the variability between single test results obtained under different measurement processes. This can include different operators within a single laboratory or among different laboratories.

THE CEMENT AND CONCRETE REFERENCE LABORATORY

The Cement and Concrete Reference Laboratory (CCRL) was established in 1929 (originally as the Cement Reference Laboratory, with concrete programs added in 1965) to aid in ensuring quality of construction materials. CCRL is a division of ASTM International and is an evaluation authority that provides laboratory inspections and proficiency sample services.

Laboratory inspections are a systemic review of laboratory practice, including procedures, practices, facilities and equipment to ensure that ASTM standards are met. Laboratory staff are observed conducting test methods and equipment dimensions are checked as well as operating characteristics to verify that they are within limits established in standards. A report is generated with any deficiencies noted, and laboratories take corrective actions. Although CCRL itself does not provide certification or accreditation services, other agencies often use CCRL's thorough inspection process as a basis for accreditation.

A review of a laboratory's quality system; including personnel qualifications, test records, equipment calibration and verification records, corrective actions taken, and quality system manual is another inspection service offered, using ASTM C1222, *Standard Practice for Evaluation of Laboratories Testing Hydraulic Cement*; C1077, *Standard Practice for Agencies Testing Concrete and Concrete Aggregates for Use in Construction and Criteria for Testing Agency Evaluation*; or C1093, *Standard Practice for Accreditation of Testing Agencies for Masonry*; as guidance. Typically, labs are inspected at intervals between 24- and 30-months.

The Proficiency Sample Program (PSP) is a separate service provided by CCRL. In this program, pairs of samples of materials are distributed, usually at intervals of 6 or 12 months, to laboratories

that then perform standard tests on the materials and report results back to CCRL. CCRL staff then compile and analyze the data to statistically identify testing bias or other issues (Youden 1959; Crandall and Blaine 1959). These programs help laboratories maintain a high degree of quality by identifying any issues with equipment or testing procedures between laboratory inspections. In addition, participation in an appropriate PSP is required by some quality standards (for example, ASTM C1222 for hydraulic cements, C1077 for concrete, and C1093 for masonry). Programs are available for portland cement, blended cement, masonry cement, pozzolans, concrete, masonry mortar, concrete masonry units, steel rebar, and potential ASR-reactivity of aggregates. Moore (2007) discusses utilizing PSP programs for cement laboratory quality control.

TESTING CEMENTITIOUS MATERIALS

Tests of the physical properties of cements should be used to evaluate the properties of the cement, rather than predicting the performance of concrete, as many other variables impact concrete performance. Cement specifications limit the properties with respect to the type of cement. Cement should be sampled in accordance with ASTM C183, *Standard Practice for Sampling and the Amount of Testing of Hydraulic Cement* (AASHTO T 127). During manufacture, portland and blended cements are continuously monitored for chemistry and the properties discussed below.

Compressive Strength

The strength development characteristics of hydraulic cements are determined by measuring the compressive strength of 50 mm (2-in.) mortar cubes tested in accordance with ASTM C109, *Standard Test Method for Compressive Strength of Hydraulic Cement Mortars (Using 2-in. or [50-mm] Cube Specimens)* (AASHTO T 106) (Figure 20-4). In summary, the method uses mortars with a ratio of 1 part cement to 2.75 parts graded standard sand (by mass). The water-cement ratio for mortars of portland cements and portland-limestone cements is 0.485 for non-air entrained cements, and 0.460 for air-entrained cements. The water content for other blended cements

FIGURE 20-4. 50-mm (2-in.) mortar cubes are cast (left) and crushed (right) to determine strength characteristics of cement.

and masonry cements is that which produces a flow of 110 ± 5 in 25 drops of the flow table. The mortars are compacted in 2 lifts into the cube molds. After curing 20 to 72 hours in the molds, the specimens are removed from the molds and stored in limewater until testing.

The minimum compressive strength requirements in U.S. hydraulic cement specifications are provided in Chapter 2. Cement strengths (based on mortar cube tests) cannot be used to predict concrete strengths with any degree of accuracy.

Procedures to evaluate strength activity indexes of supplementary cementitious materials (SCMs) are also based on ASTM C109 (AASHTO T 106). The strength activity of a slag cement is evaluated using C109 mortar cubes using a mixture of 50% (by mass) of a reference portland cement and 50% of the slag cement, while fly ash and natural pozzolans are evaluated using mixtures of 20% by mass of fly ash and 80% of a portland cement meeting ASTM C150. For silica fume, the mixture is 10% silica fume and 90% portland cement. The activity index of the SCM is a ratio of the compressive strength of the SCM mixtures divided by the compressive strength of the portland cement mortar, multiplied by 100. (Additional details are in ASTM C989, *Standard Specification for Slag Cement for Use in Concrete and Mortars* [AASHTO M 302]; ASTM C618, *Standard Specification for Coal Fly Ash and Raw or Calcined Natural Pozzolans for Use in Concrete* [AASHTO M 295]; and C1240, *Standard Specification for Silica Fume Used in Cementitious Mixtures* [AASHTO M 307].)

The strength uniformity of a cement from a single source may be determined by following the procedures outlined in ASTM C917, *Standard Test Method for Evaluation of Variability of Cement from a Single Source Based on Strength.* Uniformity of other concrete ingredients can be evaluated using procedures in ASTM C1451.

Setting Time

To determine if a cement sets according to the time limits specified in cement specifications, tests are performed (at laboratory temperatures) using the Vicat apparatus (ASTM C191, *Standard Test Methods for Time of Setting of Hydraulic Cement by Vicat Needle* [AASHTO T 131]) (Figure 20-5). In summary, a paste of normal consistency is prepared and a 1-mm diameter needle with a 300-g weight is periodically placed on the surface of the paste. When the needle penetrates 25 mm (may be interpolated between the times for two depths), the initial setting time (determined from the initial contact time of the cement and water) has been reached, and when the needle no longer leaves an impression on paste, final set has been reached.

As discussed in Chapter 2, setting times of concretes do not correlate directly with setting times of pastes.

Cements are tested for early stiffening using ASTM C451, *Standard Test Method for Early Stiffening of Hydraulic Cement (Paste Method)* (AASHTO T 186), and ASTM C359, *Standard Test Method for Early Stiffening of Hydraulic Cement (Mortar Method)* (AASHTO T 185), which use the penetration techniques of the Vicat apparatus. However, these tests do not address all the variables that can influence early stiffening as discussed in Chapter 2.

Consistency and Flow

Consistency refers to the relative mobility of a freshly mixed cement paste or mortar or to its ability to flow. During cement testing, some tests require pastes that are mixed to *normal consistency*, defined as the water content that permits a penetration of 10 mm ± 1 mm of the Vicat plunger (see ASTM C187, *Standard Test Method for Amount of Water Required for Normal Consistency of Hydraulic Cement Paste* (AASHTO T 129). For mortar mixed to a specified flow, the mortar is placed in the brass mold centered on the flow table. After the mold is removed and the table undergoes a succession of drops, the diameter of the pat is measured to determine consistency. A few trials may be required to obtain the appropriate water content (Figure 20-6).

FIGURE 20-6. Normal consistency determination for paste using the Vicat plunger (see Figure 20-5) and brass mold (inset).

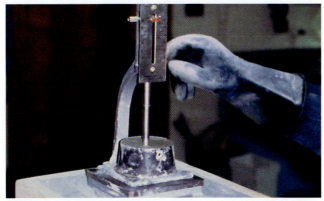

FIGURE 20-5. Setting of hydraulic cement is determined by the Vicat apparatus (ASTM C191 [AASHTO T 131]).

Other tests require mortars that are mixed to obtain either a fixed water-cement ratio or to yield a flow within a prescribed range. The flow is determined on a flow table following ASTM C1437, *Standard Test Method for Flow of Hydraulic Cement Mortar.* Both the normal consistency method and the flow test are used to regulate water contents of pastes and mortars, respectively, for use in subsequent tests; both allow comparing dissimilar ingredients with the same penetrability or flow.

Loss on Ignition and Insoluble Residue

The test for loss on ignition (LOI) of cement is performed in accordance with ASTM C114, *Standard Test Methods for Chemical Analysis of Hydraulic Cement* (AASHTO T 105)). Loss on ignition is determined by heating a cement sample of known weight up to a temperature between 900°C and 1000°C (about 1650°F and 1830°F) until a constant weight is obtained. The weight loss of the sample is then determined (Figure 20-7). A modified procedure known as split loss on ignition measures weight loss on a sample of cement sequentially heated to 550°C (1020°F) and then to about 950°C (1740°F). Typical ranges of LOI and variances between different cement types are discussed in Chapter 2.

FIGURE 20-7. Loss-on-ignition test of cement.

The insoluble residue is also determined using procedures in ASTM C114 (AASHTO T 105) and involves dissolving a cement sample in a strong acid, and then dissolving the remaining material in a strong base. This removes all soluble materials leaving an insoluble residue. ASTM C150 (AASHTO M 85) and ASTM C595 (AASHTO M 240) limits on insoluble residue are provided in Chapter 2.

Particle Size and Fineness

Fineness is usually measured by the Blaine air-permeability test, ASTM C204 (AASHTO T 153), *Standard Test Methods for*

FIGURE 20-8. Blaine test apparatus used for determining the fineness of cement.

Fineness of Hydraulic Cement by Air-Permeability Apparatus, that indirectly measures the surface area of the cement particles per unit mass (Figure 20-8). In summary, a powdered specimen is placed in a calibrated cell with a specific porosity and a fixed volume of air is passed through the sample. The time for the air to pass through the sample is recorded and the fineness is determined based on previous calibration with reference samples. ASTM C150 and AASHTO M 85 Blaine fineness limits are given in Chapter 2.

ASTM C430, *Standard Test Method for Fineness of Hydraulic Cement by the 45-μm (No. 325) Sieve* (AASHTO T 192) (Figure 20-9), and ASTM C1891, *Standard Test Method for Fineness of Hydraulic Cement by Air-Jet Sieving at 45-μm (No. 325),* provide fineness data in the form of the amount of cement passing a 45-μm (No. 325) sieve. X-ray or laser particle size analyzers (Figure 20-10) can also be used to determine values for fineness, although the results are not equivalent among different test methods.

FIGURE 20-9. Quick tests, such as washing cement over this 45-μm sieve, help monitor cement fineness during production. Shown is a view of the sieve holder with an inset top view of a cement sample on the sieve before washing with water.

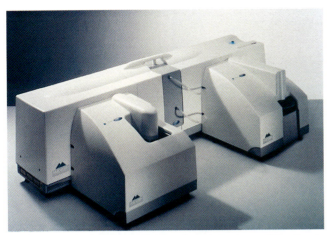

FIGURE 20-10. A laser particle analyzer uses laser diffraction to determine the particle size distribution of fine powders.

Density

The density of a cement, determined by ASTM C188 (AASHTO T 133), S*tandard Test Method for Density of Hydraulic Cement* (Figure 20-11) is defined as the mass of a unit volume of the solids or particles, excluding space between particles. It is reported as grams per cubic centimeter (g/cm³). It is not an indication of the cement's quality; rather, its principal use is in mixture proportioning calculations. More discussion on relative density and bulk densities of cement is found in Chapter 2.

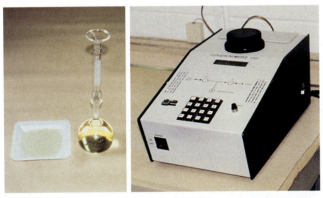

FIGURE 20-11. Density of cement can be determined by (left) using a Le Chatelier flask and kerosene or by (right) using a helium pycnometer.

Soundness

Soundness of some hydraulic cements is determined by the autoclave-expansion test, ASTM C151, *Standard Test Method for Autoclave Expansion of Hydraulic Cement* (AASHTO T 107). In summary, the method entails preparing pastes of normal consistency, molding as bars for 24 hours, then determining the length change after exposure to steam at 2 MPa (295 psi) for 3 hours. This method is no longer referenced in ASTM C150 (AASHTO M 85), ASTM C595 (AASHTO M 240), or ASTM C1157 due to a lack of correlation with field performance. More on soundness can be found in Chapter 2, Kabir and Hooton 2020, Gonnerman 1953, Helmuth 1998, Mehta 1978, and Klemm 2005.

Heat of Hydration

The heat of hydration is tested by isothermal conduction calorimetry according to ASTM C1702, *Standard Test Method for Measurement of Heat of Hydration of Hydraulic Cementitious Materials Using Isothermal Conduction Calorimetry*. This procedure measures the heat flow from a cementitious paste sample as it reacts with water in a sealed cell (compared to a reference cell) placed on a heat sink. The sample temperature is kept constant during the measurement.

ASTM C1679, *Standard Practice for Measuring Hydration Kinetics of Hydraulic Cementitious Mixtures Using Isothermal Calorimetry*, uses data on evolved heat over time. Indications of evolved heat over time provided by measurements of the changing temperatures of fresh cementitious mixtures can also be used in hydration evaluation, as described in ASTM C1753, *Standard Practice for Evaluating Early Hydration of Hydraulic Cementitious Mixtures Using Thermal Measurements*. Related thermal data compared from laboratory and field concrete mixtures can be used to estimate concrete strength development in the field by applying concrete maturity principles as described in ASTM C1074, *Standard Practice for Estimating Concrete Strength by the Maturity Method* (see Chapter 18).

Air Content

ASTM C185, *Standard Test Method for Air Content on Hydraulic Cement Mortar* (AASHTO T 137), determines the air content of a mortar in order to verify that the cement made with the mortar meets limits for air content in specifications. A maximum air content of 12% generally applies to cements for general concrete construction, and air-entraining cements are required to have air contents of between 16% and 22%. The method consists of preparing a mortar with fixed amounts of cement and a 20-30 standard sand, and an amount of water to provide a flow of 87½% +/- 7½% with 10 drops of a flow table. The mortar is weighed in a container of known volume and the air content determined from the densities of the ingredients and their proportions. The air content of a cement mortar cast in accordance with ASTM C185 does not correspond with air contents of concrete mixtures.

Air-Entraining Admixture Demand

ASTM C1827, *Standard Test Method for Determination of the Air-Entraining Admixture Demand of a Cementitious Mixture*, commonly known as the Foam Index Test, provides a rough indication of the required dosage of air-entraining admixture for various fly ashes relative to portland cement-only mixtures. This is known as the relative volume of AEA and although a good relative indicator, it cannot be used to estimate quantitatively an AEA dosage in a concrete mixture. Changes in the foam index value for a fly ash (or other SCM) can be used to anticipate the need to increase or decrease the dosage based on changes in materials characteristics.

TESTING AGGREGATES

Sampling Aggregates

Methods for obtaining representative samples of aggregates are given in ASTM D75, *Standard Practice for Sampling Aggregates* (AASHTO T 2). Accurate sampling is important. The location in the production process where samples will be obtained must be carefully planned. Sampling from a conveyor belt, stockpile, or aggregate bin may require special sampling equipment; caution must be exercised to obtain a sample free from segregation of individual particle sizes. The sample must be large enough to meet ASTM minimum sample size requirements. Samples obtained for moisture content testing should be placed in a sealed container or plastic bag as soon as possible to retain moisture until testing.

Reducing large field samples to small quantities for individual tests must be done in accordance with ASTM C702, *Standard Practice for Reducing Samples of Aggregate to Testing Size* (AASHTO R 76). For coarse aggregate, this is done by the quartering method or mechanical splitter. For the quartering method, the sample is thoroughly mixed and formed into a conical pile. The pile is flattened into a layer of uniform thickness and diameter (four to eight times the thickness). The flattened mass is divided into four equal parts, and two opposite quarters are discarded. This process is repeated until the desired size of sample remains to ensure the sample will be truly representative. A similar procedure is sometimes used for fine aggregate. Sample splitters are desirable for dry aggregate (Figure 20-12) but should not be used for fine aggregate samples with moisture contents that are above saturated surface dry. A sample splitter comprises chutes that empty into alternating directions so that one-half of the sample introduced into a hopper is diverted into one receptacle and the other half into another receptacle. The sample from one receptacle is reintroduced into the splitter as many times as necessary to obtain the required sample size.

FIGURE 20-12. Sample splitter commonly used to reduce coarse aggregate samples.

FIGURE 20-13. Performing a sieve analysis test of coarse aggregate in a laboratory.

Grading

Normal-weight aggregates should meet the requirements of ASTM C33, AASHTO M 6, or AASHTO M 80. The grading of an aggregate is determined by a sieve analysis in which the particles are divided into their various sizes as the sample passes through a stack of standard sieves arranged in order of progressively smaller openings (Figure 20-13).

The sieve analysis should be made in accordance with ASTM C136, *Standard Test Method for Sieve Analysis of Fine and Coarse Aggregates* (AASHTO T 27). Test results are presented typically in terms of percentage of material passing each sieve size. Size numbers (grading sizes) for coarse aggregates apply to the amounts of aggregate (by mass) in percentages that pass through an assortment of different size sieves. Specifications for the dimensions of openings in sieves and sieve frame sizes are provided in ASTM E11, *Standard Specification for Woven Wire Test Sieve Cloth and Test Sieves* (AASHTO M 92).

The grading requirements for concrete aggregate are shown in Chapter 5 and ASTM C33 (AASHTO M 6 and M 80). ASTM C33 sieves for fine aggregate have openings ranging from 75 μm (No. 200) to 9.5 mm (⅜ in.) (150 μm [No. 100] to 9.5 mm [⅜ in.] in AASHTO M 6). The standard sieves for coarse aggregate (ASTM C33/AASHTO M 80) have openings ranging from 300 μm (No. 50) to 100 mm (4 in.). For highway construction, ASTM D448, *Standard Classification for Sizes of Aggregate for Road and Bridge Construction* (AASHTO M 43) lists the same size numbers as noted in ASTM C33 (AASHTO M 6/M 80) plus additional coarse aggregate size numbers. Fine aggregate or sand has the same range of particle sizes for general construction and paving work.

FIGURE 20-14. Videograder for measuring size and shape of aggregate.

Particle Shape and Surface Texture

ASTM D4791, *Standard Test Method for Flat Particles, Elongated Particles, or Flat and Elongated Particles in Coarse Aggregate*, provides a method to determine flat and elongated particles. ASTM C1252 *Standard Test Methods for Uncompacted Void Content of Fine Aggregate (as Influenced by Particle Shape, Surface Texture, and Grading)*, provides an indirect method of comparing the impacts of shape and texture, as void contents increase with more angular, less rounded and rougher-surfaced particles (and vice-versa). ASTM C295, *Standard Guide for Petrographic Examination of Aggregates for Concrete*, provides procedures for the petrographic examination of aggregate which includes evaluation of particle shape and texture.

A number of automated test machines are available for rapid determination of the particle shape and size distribution of aggregate. These machines were designed to provide a faster alternative to the standard sieve analysis test. They can capture and analyze digital images of the aggregate particles to determine gradation. Figure 20-14 shows a videograder that measures size and shape of an aggregate using line-scan cameras wherein two-dimensional images are constructed from a series of line images. Other machines use matrix-scan cameras to capture two-dimensional snapshots of the falling aggregate. Maerz and Lusher (2001) developed a dynamic prototype imaging system that provides particle size and shape information by using a miniconveyor system to parade individual fragments past two orthogonally oriented, synchronized cameras. Measurement of aggregate shape properties using digital image analysis techniques has been standardized under AASHTO T 381, *Standard Method of Test for Determining Aggregate Shape Properties by Means of Digital Image Analysis* and AASHTO R 91, *Standard Practice for Determining Aggregate Source Shape Values from Digital Image Analysis Shape Properties*.

Densities

It is important to clearly communicate density values. The particle density is higher than the bulk density as the latter includes the volume of space between the particles. The particle density is used in mixture proportioning calculations, while the bulk density is used in characterizing aggregates.

Bulk Density (Unit Weight) and Void Content. Methods of determining the bulk density of aggregates and void content are given in ASTM C29, *Standard Test Method for Bulk Density ("Unit Weight") and Voids in Aggregate* (AASHTO T 19). In these standards, three methods are described for consolidating the aggregate in the container depending on the maximum size of the aggregate: rodding, jigging, and shoveling. The measurement of loose uncompacted void content of fine aggregate is described in ASTM C1252, *Standard Test Methods for Uncompacted Void Content of Fine Aggregate (as Influenced by Particle Shape, Surface Texture, and Grading)*.

Relative Density (Specific Gravity). Test methods for determining relative densities for coarse and fine aggregates are described in ASTM C127, *Standard Test Method for Relative Density (Specific Gravity), and Absorption of Coarse Aggregate* (AASHTO T 85) and ASTM C128, *Standard Test Method for Relative Density (Specific Gravity), and Absorption of Fine Aggregate* (AASHTO T 84).

(Particle) Density. The density of aggregate particles used in mixture proportioning computations (not including voids between particles) is determined by multiplying the relative density (specific gravity) of the aggregate times the density of water. An approximate value of 1000 kg/m³ (62.4 lb/ft³) is often used for the density of water.

Absorption and Surface Moisture

Several methods are used for determining the amount of moisture in aggregate samples. The total moisture content of fine or coarse aggregate can be measured in accordance with ASTM C566, *Standard Test Method for Total Evaporable Moisture Content of Aggregate by Drying* (AASHTO T 255). In this method a measured sample of the aggregate is dried either in a ventilated conventional oven, microwave oven, or over an electric or gas hotplate.

Using the mass measured before and after drying, the total moisture content can be calculated as shown in Equation 20-1:

$$P = 100(W - D)/D$$

EQUATION 20-1.

Where:

P = Total evaporable moisture content of sample, percent

W = Mass of original sample

D = Mass of dried sample.

The surface (free) moisture can be calculated if the aggregate absorption is known. Absorption refers to the increase in aggregate mass due to filling of permeable pores with water following a standard procedure. It is expressed as a percentage of the dry mass. The surface moisture content is equal to the total moisture content minus the absorption. Absorption can be determined using ASTM C127, *Standard Test Method for Relative Density (Specific Gravity) and Absorption of Coarse Aggregate* (AASHTO T 85), for coarse aggregate and ASTM C128, *Standard Test Method for Relative Density (Specific Gravity) and Absorption of Fine Aggregate* (AASHTO T 84), for fine aggregate.

If drying equipment is not available, a field or plant determination of surface (free) moisture in fine aggregate can be made in accordance with ASTM C70, *Standard Test Method for Surface Moisture in Fine Aggregate*. The same procedure can be used for coarse aggregate with appropriate changes in the size of sample and dimensions of the container. This test is based on displacement of water by a known mass of moist aggregate. The relative density (specific gravity) of the aggregate must be known accurately to determine the moisture content.

Electrical moisture meters are used in many concrete batching plants primarily to monitor the moisture content of fine aggregates, but some plants also use them to check coarse aggregates. They operate on the principle that the electrical resistance of damp aggregate decreases as moisture content increases, within the range of dampness normally encountered. The meters measure the electrical resistance of the aggregate between electrodes protruding into the batch hopper or bin. Moisture meters based on the microwave-absorption method are gaining popularity because they are more accurate than the electrical resistance meters. However, both methods measure moisture contents accurately and rapidly, but only at the location of the probes. These meters require frequent calibration and must be properly maintained. The variable nature of moisture contents in aggregates cause difficulty in obtaining representative samples for comparison with moisture meter readings. Several oven-dried moisture content tests should be performed to verify the calibration of these meters before trends in accuracy can be established.

Organic Impurities

Organic impurities in fine aggregate should be determined in accordance with ASTM C40, *Standard Test Method for Organic Impurities in Fine Aggregates for Concrete* (AASHTO T 21). A sample of fine aggregate is placed in a sodium hydroxide solution in a colorless glass bottle and shaken. The next day the color of the sodium hydroxide solution is compared with a glass color standard or standard color solution. If the color of the solution containing the sample is darker than the standard, the fine aggregate should not be used without further investigation.

Some fine aggregates contain small quantities of coal or lignite that give the solution a dark color. The quantity may be insufficient to reduce the strength of the concrete appreciably. If surface appearance of the concrete is not important, ASTM C33, *Standard Specification for Concrete Aggregates* and AASHTO M 6, *Standard Specification for Fine Aggregate for Hydraulic Cement Concrete*, state that fine aggregate is acceptable if the amount of coal and lignite does not exceed 1.0% of the total fine aggregate mass. A fine aggregate failing this limit may be used if, when tested in accordance with ASTM C87, *Standard Test Method for Effect of Organic Impurities in Fine Aggregate on Strength of Mortar* (AASHTO T 71), the 7-day strengths of mortar cubes made with the fine aggregate following ASTM C109 (AASHTO T 106), are at least 95% of the 7-day strengths of mortar made with the same fine aggregate, after it is washed in a 3% solution of sodium hydroxide and then thoroughly rinsed in water.

Staining Potential

When visible concrete staining is undesirable, such as in architectural applications, aggregates should meet the staining requirements of ASTM C330 (AASHTO M 195) when tested according to ASTM C641, *Standard Test Method for Iron Staining Materials in Lightweight Concrete Aggregates*; the quarry face and aggregate stockpiles should not show evidence of staining. As an additional aid in identifying staining particles, the aggregate can be immersed in a lime slurry (Midgley 1958; Mielenz 1964). If staining particles are present, a blue-green gelatinous precipitate will form within 5 to 10 minutes; this will rapidly change to a brown color on exposure to air and sunlight. The reaction should be complete within 30 minutes. If no brown gelatinous precipitate is formed when a suspect aggregate is placed in the lime slurry, there is little likelihood of any reaction taking place in concrete. These tests should be required when aggregates without a record of successful prior use are to be used in architectural concrete.

Objectionable Fine Material

Large amounts of clay and silt in aggregates can adversely affect durability, increase water requirements, and increase shrinkage. ASTM C33 (AASHTO M 6 and M 80) limits the amount of material passing the 75-μm (No. 200) sieve to 3% for concrete subjected to abrasion, or 5% in fine aggregate for other concrete, and to 1% or less in coarse aggregate. Testing for material finer than the 75-μm (No. 200) sieve should be done in accordance with ASTM C117, *Standard Test Method for Materials Finer than 75-μm (No. 200) Sieve in Mineral Aggregates by Washing* (AASHTO T 11).

Testing for clay lumps should be performed in accordance with ASTM C142, *Standard Test Method for Clay Lumps and Friable Particles in Aggregates* (AASHTO T 112).

Resistance to Freezing and Thawing

The performance of aggregates under exposure to freezing and thawing can be evaluated in two ways: past performance in the field, and laboratory freeze-thaw tests of concrete specimens, ASTM C1646, *Standard Practice for Making and Curing Test Specimens for Evaluating Resistance of Coarse Aggregate to Freezing and Thawing in Air-Entrained Concrete*. If similarly-sized aggregates from the same source have previously given satisfactory service when used in concrete, they may be considered suitable. Aggregates without a service record can be considered acceptable if they perform satisfactorily in air-entrained concretes subjected to freeze-thaw tests according to ASTM C1646. In these tests concrete specimens made with the aggregate in question are subjected to alternate cycles of freezing and thawing in water according to ASTM C666, *Standard Test Method for Resistance of Concrete to Rapid Freezing and Thawing* (AASHTO T 161). Deterioration is measured by the reduction in the dynamic modulus of elasticity, linear expansion, and weight loss of the specimens. A failure criterion of 0.035% expansion in 350 freeze-thaw cycles or less is used by a number of state highway departments to help indicate whether an aggregate is susceptible to D-cracking. Different aggregate types may influence the criteria levels and empirical correlations between laboratory freeze-thaw tests. Field service records should be kept to select the proper criterion (Grove and Vogler 1989).

Specifications may require that an aggregate's resistance to weathering be demonstrated by exposure to a sodium sulfate or magnesium sulfate solution ASTM C88, *Standard Test Method for Soundness of Aggregates by Use of Sodium Sulfate or Magnesium Sulfate* (AASHTO T 104). The test consists of a number of immersion cycles (wetting and drying) for a sample of the aggregate in a sulfate solution. This cycling creates a pressure through salt-crystal growth in the aggregate pores similar to that produced by freezing water. Upon completion of the cycling, the sample is then oven dried and the percentage of weight loss calculated. Unfortunately, this test is sometimes misleading. Aggregates behaving satisfactorily in the test might produce concrete with low freeze-thaw resistance; conversely, aggregates performing poorly might produce concrete with adequate resistance. This is attributed, at least in part, to the fact that the aggregates in the test are not confined by cement paste (as they would be in concrete) and the mechanisms of attack are not the same as in freezing and thawing. The test is most reliable for stratified rocks with porous layers or weak bedding planes.

An additional test that can be used to evaluate aggregates for potential D-cracking is the rapid pressure release method. An aggregate is placed in a pressurized chamber and the pressure is rapidly released causing the aggregate with a questionable pore system to fracture (Janssen and Snyder 1994). The amount of fracturing relates to the potential for D-cracking.

Abrasion and Skid Resistance

The most common test for abrasion resistance of coarse aggregate is the Los Angeles abrasion test (rattler method) performed in accordance with ASTM C131, *Standard Test Method for Resistance to Degradation of Small-Size Coarse Aggregate by Abrasion and Impact in the Los Angeles Machine* (AASHTO T 96) or ASTM C535, *Standard Test Method for Resistance to Degradation of Large-Size Coarse Aggregate by Abrasion and Impact in the Los Angeles Machine*. In this test a specified quantity of aggregate is placed in a steel drum containing steel balls, the drum is rotated, and the percentage of material worn away is measured. Specifications often set an upper limit on this mass loss. However, a comparison of the results of aggregate abrasion tests with the abrasion resistance of concrete made with the same aggregate do not generally show a clear correlation. Mass loss due to impact in the rattler is often as much as the mass loss caused by abrasion. The wear resistance of concrete is determined more accurately by abrasion tests of the concrete itself (see Chapter 11 and **Testing of Hardened Concrete**).

Alkali-Aggregate Reactivity

Certain constituents of some aggregates can react with alkalies in the pore solution of mortar or concrete leading to expansive reactions. These potential distress mechanisms are referred to as alkali-silica reaction (ASR) or alkali-carbonate reaction (ACR).

Test Methods for Identifying ASR-Susceptible Aggregates. The reactivity of an aggregate is classified according to Table 20-1. The most reliable test for aggregate reactivity is ASTM C1293, *Standard Test Method for Determination of Length Change of Concrete Due to Alkali-Silica Reaction (Concrete Prism Test)*,

TABLE 20-1. Classification of Aggregate Reactivity (ASTM C1778)

AGGREGATE-REACTIVITY CLASS	DESCRIPTION OF AGGREGATE REACTIVITY	ASTM C1293 1-YEAR EXPANSION %	ASTM C1260 14-DAY EXPANSION %
R0	Non-reactive	< 0.04	< 0.10
R1	Moderately reactive	≥ 0.04, < 0.12	≥ 0.10, < 0.30
R2	Highly reactive	≥ 0.12, < 0.24	≥ 0.30, < 0.45
R3	Very highly reactive	≥ 0.24	≥ 0.45

or CPT. Because this test takes one year to complete, a rapid assessment test, in the form of ASTM C1260 (AASHTO T 303), *Potential Alkali-Reactivity of Aggregates (Mortar-Bar Method)*, can be used to produce results in sixteen days. ASTM C1260 is less precise in determining aggregate reactivity, having been found to exhibit false positives and false negatives. These tests should not be used to disqualify potentially reactive aggregates, as reactive aggregates can be safely used with the careful selection of cementitious materials. See Chapter 11, ASTM C1778 (AASHTO R 80), and Farny and Kerkhoff (2007) for information on the mechanisms and control of ASR.

Test methods for identifying ACR aggregates. The three test methods commonly used to identify potentially alkali-carbonate reactive aggregate are:

• petrographic examination, ASTM C295, *Standard Guide for Petrographic Examination of Aggregates for Concrete*;

• rock cylinder method, ASTM C586, *Standard Test Method for Potential Alkali Reactivity of Carbonate Rocks as Concrete Aggregates (Rock-Cylinder Method)*; and

• concrete prism test, ASTM C1105, *Standard Test Method for Length Change of Concrete Due to Alkali-Carbonate Rock Reaction*.

Guidance on interpreting the results of these tests are provided in ASTM C1778 and AASHTO R 80).

Testing Aggregate Mineral Fillers

Aggregate mineral fillers (AMF) and ground calcium carbonate (GCC) are essentially a very fine aggregate used to broaden the size distribution of aggregate particles in concrete mixtures. Properties are specified under ASTM C1797, *Standard Specification for Ground Calcium Carbonate and Aggregate Mineral Fillers for use in Hydraulic Cement Concrete*. Particle size distribution measurements using either ASTM C136 for Types A and B, or ASTM C117 for Type C (see Chapter 5).

Other physical tests for AMF and GCC include a strength activity test, comparable to that for pozzolans in ASTM C311, *Standard Test Methods for Sampling and Testing Fly Ash or Natural Pozzolans* for use in Portland-Cement Concrete, (with 20% replacement of portland cement by filler) to evaluate whether the filler has any detrimental effect on concrete strength.

Several chemical tests for mineral fillers are provided in C1797: calcium carbonate ($CaCO_3$) content and magnesium carbonate ($MgCO_3$) contents for GCC, Types A and B, are determined by ASTM C25, *Standard Test Methods for Chemical Analysis of Limestone, Quicklime, and Hydrated Lime*. Testing for potentially harmful materials in AMF and GCC include clay content testing by AASHTO T 330, *Standard Method of Test for the Quantitative Detection of Harmful Clays of the Smectite Group in Aggregates Using Methylene Blue*, or ASTM C1777, *Standard Test Method*

for Rapid Determination of the Methylene Blue Value for Fine Aggregate or Mineral Filler Using a Colorimeter, and total organic carbon content by a method in an annex in the 2015 edition of ASTM C595 (subsequently removed from later versions).

Moisture content for GCC and AMF determined by ASTM C25 or ASTM C566, applies to pneumatically conveyed fillers. A maximum water requirement is determined using procedures in ASTM C311.

TESTING FRESHLY MIXED CONCRETE

Sampling Freshly Mixed Concrete

It is critical to obtain truly representative samples of freshly mixed concrete for control and acceptance tests. Unless the sample is representative, test results will be misleading. Samples should be obtained and handled in accordance with ASTM C172, *Standard Practice for Sampling Freshly Mixed Concrete* (AASHTO R 60). Except for routine slump and air-content tests performed for process control, ASTM C172 (AASHTO R 60) requires that sample size used for acceptance purposes be at least 28 L (1 ft³) and be obtained within 15 minutes between the first and final portions of the sample. The composite sample, made of two or more portions, should not be taken from the first or last portion of the batch discharge. The sample should be protected from sunlight, wind, contamination, and other sources of rapid evaporation during sampling and testing.

Consistency

The slump test described by ASTM C143, *Standard Test Method for Slump of Hydraulic-Cement Concrete* (AASHTO T 119), is the most generally accepted method used to measure the consistency of concrete. In this context, the term consistency refers to the relative fluidity of fresh concrete. The test equipment consists of a slump cone (a metal conical mold 300 mm [12 in.] high, with a 200 mm [8 in.] diameter base and 100 mm [4 in.] diameter top) and a steel rod 16 mm (⅝ in.) in diameter and not more than 600 mm (24 in.) long with hemispherically shaped tips. The dampened slump cone is placed upright on a flat, nonabsorbent rigid surface, and is filled in three layers of approximately equal volume. Therefore, the cone should be filled to a depth of about 70 mm (2½ in.) for the first layer, a depth of about 160 mm (6 in.) for the second layer, and overfilled for the third layer. Each layer is rodded 25 times. Following rodding, the last layer is struck off and the cone is slowly raised vertically 300 mm (12 in.) in 5 ± 2 seconds. The empty slump cone is inverted and placed next to the settled concrete. The tamping rod is placed on the inverted cone to provide a reference for the original height. The slump is the vertical distance the concrete settles, measured to the nearest 5 mm (¼ in.); a ruler is used to measure from the top of the mold to the displaced original center of the subsided concrete (see Figure 20-15).

FIGURE 20-15. Slump test for consistency of concrete.

A higher slump value is indicative of a more fluid concrete. The entire test through removal of the cone has to be completed within 2 ½ minutes from the start of filling the mold, as concrete will lose slump with time. If a portion of the concrete falls away or shears off while performing the slump test, another test is performed on a different portion of the sample. Shearing of the concrete mass may indicate that the mixture lacks cohesion.

Another test method for flow of fresh concrete involves the use of the K-Slump Tester (ASTM C1890, *Standard Test Method for K-slump of Freshly Mixed Concrete*). This is a probe-type instrument that is inserted into fresh concrete that has a depth of at least 175 mm (7 in.) of concrete and there is at least 60 mm (2⅜ in.) of concrete around the tester. The height of the mortar that has flowed through the openings into the tester provides a measure of fluidity; the higher the height, the more fluid the concrete. The K-slump tester is helpful in monitoring the loss in fluidity with time and for monitoring self-consolidating concrete.

Additional consistency tests include: the FHWA vibrating slope apparatus (Wong and others 2001 and Saucier 1966); British compacting factor test (BS 1881); Powers remolding test (Powers 1932); German flow table test (Mor and Ravina 1986); ASTM C1170, *Standard Test Method for Determining Consistency and Density of Roller-Compacted Concrete Using a Vibrating Table*; Kelly ball penetration test (ASTM C360, Withdrawn 1999) (Daniel 2006); Thaulow tester; Powers and Wiler plastometer (Powers and Wiler 1941); Tattersall (1971) workability device; BML viscometer (Wallevik 1996); BTRHEOM rheometer for fluid concrete (de Larrard and others 1993); ICAR rheometer (Koehler and Fowler 2004); free-orifice rheometer (Bartos 1978); delivery chute torque meter (US patent 4,332,158 [1982]); delivery-chute vane (US patent 4,578,989 [1986]); Angles flow box (Angles 1974); ring penetration test (Teranishs and others 1994); and the Wigmore consistometer (1948). The Vebe test and the Thaulow test are especially applicable to stiff and extremely dry mixes while the flow table is especially applicable to flowing concrete (Daniel 2006). The portable ICAR rheometer is suitable for field use and provides estimates of the fundamental rheological properties of fresh concrete (dynamic yield stress and plastic viscosity).

For self-consolidating concrete, ASTM C1611, *Standard Test Method for Slump Flow of Self-Consolidating Concrete*, can be used to evaluate consistency. After spreading has ceased, the average diameter of the concrete mass is measured and reported as the slump flow. After filling, the cone is lifted, and the diameter of the resulting pat is measured at two locations perpendicular to each other. The reported value being the average of the two measurements. There should be no segregation of water or aggregates (Figure 20-16).

FIGURE 20-16. ASTM C1611 slump flow test (courtesy of Master Builders Solutions).

Since SCC is characterized by special fresh concrete properties, many new tests have been developed to measure flowability, viscosity, blocking tendency, self-leveling, and stability of the mixture (Skarendahl and Peterson 1999 and Ludwig and others 2001). The J-Ring test (ASTM C1621, *Standard Test Method for Passing Ability of Self-Consolidating Concrete by J-Ring*) measures passing ability. The J-Ring consists of a ring of reinforcing bar such that it will fit around the base of a standard slump cone (Figure 20-17). The slump flow with and without the J-Ring is measured, and the difference calculated.

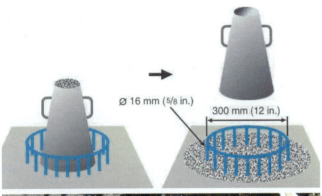

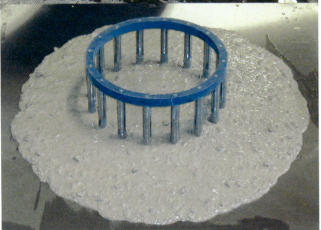

FIGURE 20-17. J-ring test (courtesy of Verein Deutscher Zementwerke).

Segregation Resistance

ASTM C1610, *Standard Test Method for Static Segregation of Self-Consolidating Concrete Using Column Technique*, evaluates static stability of a concrete mixture by quantifying aggregate segregation. A column is filled with concrete and allowed to sit after placement. The column is then separated into three pieces. The top and bottom sections (equal volumes) are carefully removed and then washed over a 4.75-mm (No. 4) sieve and the retained aggregate weighed. A non-segregating

mixture will have a consistent aggregate mass distribution in each section. A segregating mixture will have higher concentrations of aggregate in the lower sections.

An earlier assessment of the segregation resistance is ASTM C1712, *Standard Test Method for Rapid Assessment of Static Segregation Resistance of Self-Consolidating Concrete Using Penetration Test.* A 45 g hollow cylinder device is placed on top of an inverted slump cone mold containing SCC. The distance the weight sinks in 30 seconds correlates to the static segregation resistance of the mixture. If the penetration depth is less than 10 mm, the mixture is considered segregation resistant. A penetration value above 25 mm signals a mixture that is probably prone to segregation.

The Box Test was developed to evaluate low-slump concrete mixtures as used for slip-formed pavements (Cook and others 2014), and is a provisional standard, AASHTO TP 137, *Provisional Standard Test Method for Box Test in Slip Form Paving of Fresh Portland Cement Concrete*. The method investigates the concrete's response to internal vibration and the ability of the consolidated concrete to hold a vertical edge without formwork. Unconsolidated concrete is placed in a 28 L (1 ft³) collapsible wood form and vibrated by insertion and withdrawal (completed in 6 seconds) of a 25-mm (1-in.) square head vibrator. The wood forms are removed, and the sides of the concrete are inspected for excessive voids, with values of less than 10% to 30% or less indicating acceptable consolidation via the vibrator. If the sides have excessive voids, the mixture did not consolidate under the action of the vibrator and is not satisfactory. The edges of the concrete are inspected for edge slumping; a common target for slipform paving is an edge slump of 6 mm (0.25 in.) or less. Examples of satisfactory and unsatisfactory performance in the Box Test are shown in Figure 20-18. Based on comparing the workability of concrete mixtures made with different aggregate gradations, a new set of limits can be established for the Tarantula Curve discussed in Chapter 5.

FIGURE 20-18. (left) shows a mixture that showed good consolidation and no edge slumping. (center) shows a mixture with good consolidation and poor edge slumping. Vertical lines have been added to highlight the edge slumping. (right) shows a mixture with poor consolidation.

Temperature Measurement

Concrete temperature is measured in accordance with ASTM C1064, *Standard Test Method for Temperature of Freshly Mixed Hydraulic-Cement Concrete* (AASHTO T 309). Because of the important influence of the initial concrete temperature on the properties of freshly mixed and hardened concrete, many specifications place limits on the temperature of fresh concrete. Glass or armored thermometers are available (Figure 20-19). The thermometer should be accurate to ± 0.5°C (± 1°F) and should remain in a representative sample of concrete for a at least 2 minutes but not more than 5 minutes. At least 75 mm (3 in.) of concrete has to surround the sensing portion of the thermometer. Electronic temperature meters with digital displays are also available. Infrared thermometers are not approved for this standard as they only measure surficial temperatures. The temperature test is to be started within 5 minutes after obtaining the sample

FIGURE 20-19. A thermometer is used to take the temperature of fresh concrete.

Density and Yield

The density (unit weight) and yield of freshly mixed concrete (Figure 20-20) are determined in accordance with ASTM C138, *Standard Test Method for Density (Unit Weight), Yield, and Air Content (Gravimetric) of Concrete* (AASHTO T 121). The results may be used to determine the volume (yield) of concrete produced per batch (see Chapter 13). The test also can give indications of air content provided the relative densities of the ingredients are known. A balance or scale sensitive to 0.3% of the anticipated mass of the sample and container is required. For example, a 7-L (0.25-ft³) density container requires a scale sensitive to 50 g (0.1 lb). The size of the container used to determine density and yield varies with the size of aggregate: the 7-L (0.25-ft³) air meter container is commonly used with aggregates up to 25 mm (1 in.); a 14-L (0.5-ft³) container is used with aggregates up to 50 mm (2 in.). The volume of the container should be determined at least annually in accordance with ASTM C29, *Test Method for Bulk Density ("Unit Weight") and Voids in Aggregate*. Care is needed to consolidate the concrete adequately by either rodding or internal vibration. Strike off the top surface using a flat plate so that the container is filled to a flat smooth finish. The density is expressed in kilograms per cubic meter (pounds per cubic foot) and the yield in cubic meters (cubic feet). Yield is determined by dividing the total batch weight by the density.

FIGURE 20-20. Fresh concrete is weighed in a container of known volume to determine density (unit weight).

The density of unhardened as well as hardened concrete can also be determined by nuclear methods as described in ASTM C1040, *Standard Test Methods for In-Place Density of Unhardened and Hardened Concrete, Including Roller Compacted Concrete, By Nuclear Methods* (AASHTO T 271). The method requires establishing an empirical relationship between the device reading and the known densities of blocks of different materials.

Air Content

Several methods can be used to measure the air content of freshly mixed concrete. ASTM test methods include: ASTM C231, *Standard Test Method for Air Content of Freshly Mixed Concrete by the Pressure Method* (AASHTO T 152); ASTM C173, *Standard Test Method for Air Content of Freshly Mixed Concrete by the Volumetric Method* (AASHTO T 196); and ASTM C138 (AASHTO T 121). Although they measure only total air volume and not air-void characteristics, research has demonstrated that total air content is indicative of the adequacy of the air-void system. With any of the above methods, air-content tests should be started within 5 minutes after the final portion of the composite sample has been obtained.

The pressure method, ACTM C231 (AASHTO T 152), is based on Boyle's law, which relates pressure to volume of a gas. Many commercial air meters of this type are standardized to read air content directly when a predetermined pressure is applied to a fixed volume of concrete (Figure 20-21). The applied pressure compresses the air within the concrete sample, including the air in the pores of aggregates. For this reason, the pressure method is not suitable for determining the air content of concretes made with some lightweight aggregates or other very porous materials. Aggregate correction factors that compensate for air trapped in normal-weight aggregates are relatively constant and, though small, should be subtracted from the pressure meter gauge reading to obtain the correct air content. The instrument should be standardized for various elevations above sea level if it is to be used in localities having considerable differences in elevation. Some meters are based on the change in pressure of a known volume of air and are not affected by changes in elevation. Pressure meters are widely used because the mixture proportions and relative densities (specific gravities) of the concrete ingredients need not be known. Also, a test can be conducted in less time than is required for other methods.

FIGURE 20-22. Volumetric air meter.

FIGURE 20-21. Pressure meter (Type B) for determining air content.

The volumetric method (Figure 20-22) described in ASTM C173 (AASHTO T 196) is based on removing air from within a known volume of concrete by agitating the concrete in a fixed volume of a water-isopropyl alcohol mixture. This method can be used for concrete containing any type of aggregate, including lightweight or other porous aggregate. An aggregate correction factor is not necessary with this test. The volumetric test is not affected by atmospheric pressure, and the relative densities (specific gravities) of the concrete ingredients need not be known. Care must be taken to agitate the sample sufficiently to dislodge all air. The addition of 500 mL (1 pt) or more of alcohol accelerates the removal of air, thus shortening test times; it also dispels most of the foam and increases the accuracy of the measured air content, including tests performed on high-air-content or high-cement-content concretes.

The gravimetric method, ASTM C138 (AASHTO T 121), uses the same test equipment as used for determining the density (unit weight) of fresh concrete. The measured density of concrete is subtracted from the theoretical density as determined from the absolute volumes of the ingredients, assuming no air is

present. This difference, expressed as a percentage of the theoretical density, is the air content. Mixture proportions and relative densities (specific gravities) of the ingredients must be accurately known; otherwise results may be in error. Consequently, this method is suitable only if laboratory-type control is exercised, and it is not applicable for acceptance testing of air content.

Significant changes in density can, however, be a convenient way to detect variability in air content. AASHTO T 199, *Standard Method of Test for Air Content of Freshly Mixed Concrete by the Chace Indicator*, can be used as a quick check for the presence of low, medium, or high levels of air in concrete. It is not a substitute for the other more accurate methods. A representative sample of mortar from the concrete is placed in a cup and introduced into a graduated glass container (Figure 20-23). The container is filled with alcohol to the zero mark on the stem. A thumb is placed over the stem opening and the container is rotated repeatedly from vertical to horizontal end to dispel the air from the mortar. The drop in the alcohol level and the mortar content are used to estimate the air content of concrete.

FIGURE 20-23. Chace air indicator (AASHTO T 199).

Air-Void Analysis of Fresh Concrete

The conventional methods for analyzing air in fresh concrete, such as the pressure method noted above, only measure the total air content; consequently, they provide no information about the parameters that determine the quality of the air-void system. These parameters – the size and number of voids and spacing between them – can be measured on polished specimens of hardened concrete (ASTM C457, *Standard Test Method for Microscopical Determination of Parameters of the Air-Void System in Hardened Concrete*) (see **Testing Hardened Concrete, Air Content**); but the result of such analysis will only be available several days after the concrete already has set.

There are two test methods to determine the key air-void parameters in samples of fresh air-entrained concrete. Fresh concrete samples can be taken at the mix plant and on the jobsite. Testing concrete before and after placement into forms can verify how the applied methods of transporting, placing, and consolidation affect the air-void system. Because the samples are taken on fresh concrete, the air content and air-void system can be adjusted during production.

The air-void analyzer (AVA) test apparatus determines the volume and size distributions of entrained air bubbles (Figure 20-24). The measured data are used to estimate the spacing factor, specific surface, and total volume of entrained air. In this test method, air bubbles from a sample of fresh concrete rise through a viscous liquid, enter a column of water above it, then rise through the water and collect under a submerged pan that is attached to a sensitive balance (Figure 20-25). The viscous liquid retains the original bubble sizes. Large bubbles rise faster than small ones through the liquids. As air bubbles accumulate under the pan, the buoyancy of the pan increases. The balance measures this change in buoyancy, which is recorded as a function of time and can be related to the number of bubbles of different size.

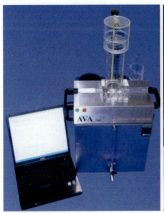

FIGURE 20-24. Equipment for the air-void analyzer.

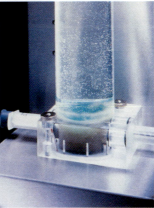

FIGURE 20-25. Air bubbles rising through liquids in column.

AASHTO adopted a provisional test method for the AVA (AASHTO T 348, *Standard for Air-Void Characteristics of Freshly Mixed Concrete by Buoyancy Change*). The AVA was not developed for measuring the total air-content of concrete, and because of the small sample size, may not give accurate results for this quantity. However, this method may be useful in assessing the quality of the air-void system; it gives good results in conjunction with traditional methods for measuring air content (Aarre 1998, Distlehorst and Kurgan 2007, and Petersen 2009).

The Super Air Meter (SAM) is a modified version of an ASTM C231 pressure meter (Figure 20-26). The meter uses two cycles of sequentially increasing pressures to determine a parameter termed the SAM number. The SAM number has been shown to correlate to the air-void spacing factor from hardened air-void analysis (ASTM C457) and resistance to rapid cycles of freezing and thawing (ASTM C666) (Welchel 2014 and Tanesi 2016). An AASHTO provisional test method, AASHTO TP 118, *Standard Method of Test for Characterization of the Air-Void System of Freshly Mixed Concrete by the Sequential Pressure Method,* has also been developed on this test method.

FIGURE 20-26. Super air meter.

Strength Specimens

Specimens molded in the field for strength tests are made and standard-cured in accordance with ASTM C31 (AASHTO T 23), and laboratory-molded specimens according to ASTM C192, *Standard Practice for Making and Curing Concrete Test Specimens in the Laboratory* (AASHTO R 39). Molding of strength specimens should be started within 15 minutes after the composite sample is obtained.

Traditionally, the standard test specimen for compressive strength of concrete with a nominal maximum aggregate size of 50 mm (2 in.) or smaller was a cylinder 150 mm (6 in.) in diameter by 300 mm (12 in.) high (Figure 20-27). In 2008, ACI 318 was revised to permit the use of cylinders 100 mm (4 in.) in diameter by 200 mm (8 in.) high. The smaller cylinders can only be used for concrete made with aggregate having a nominal maximum size of 25 mm (1 in.) or less. For larger aggregates, the diameter of the cylinder must be at least three times the nominal maximum size of aggregate and the height must be twice the diameter. Alternatively, it is permitted to wet-sieve fresh concrete with large aggregate using a 50-mm (2-in.) sieve in accordance with ASTM C172. While rigid metal molds are preferred, plastic, or other types of single-use molds conforming to ASTM C470, *Standard Specification for Molds for Forming Concrete Test Cylinders Vertically*, can be used. They should be placed on a flat, level, rigid surface and filled carefully to avoid distortion of their shape.

FIGURE 20-27. Preparing standard test specimens for compressive strength of concrete.

The smaller 100-mm (4-in.) diameter by 200-mm (8-in.) high cylinders have been commonly used with high-strength concrete containing up to 19 mm (¾ in.) maximum nominal size aggregate (Burg and Ost 1994, Forstie and Schnormeier 1981, and Date and Schnormeier 1984). The 100-mm × 200-mm (4-in. × 8-in.) cylinders are easier to cast, require less material, weigh considerably less than 150-mm × 300-mm (6-in. × 12-in.) cylinders and require less storage space for curing. In addition, the smaller cross-sectional area allows the use of smaller capacity testing machines to test high-strength concrete cylinders. The difference in indicated strength between the two cylinder sizes is insignificant as illustrated in Figure 20-28. The single-operator variability of 100 mm (4 in.) cylinders is reported to be slightly higher or similar to that for 150-mm (6-in.) cylinders (Detwiler and others 2001, Burg and others 1999, Pistilli and Willems 1993, and Carino and others 2004). The predominant size used in Canada is the 100-mm (4-in.) diameter cylinder (CSA A23.1). Consult job specifications for allowable cylinder sizes.

The smallest cross-sectional dimension of beams for flexural strength testing should be at least three times the nominal maximum size of aggregate. The length of beams should be at least three times the depth of the beam plus 50 mm (2 in.). For the same concrete, the flexural strength is affected by the cross-sectional dimensions of the beams. Therefore, the same beam size should be used for qualification and acceptance testing. Typically beams with cross sections of 150 mm × 150 mm (6 in. × 6 in.) have been used, but smaller 100 mm × 100 mm (4 in. × 4 in.) beams are acceptable provided the nominal

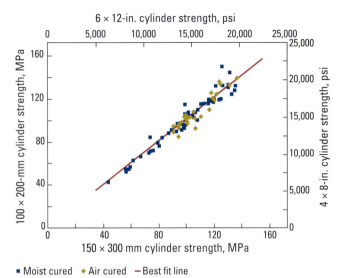

- ■ Moist cured ◆ Air cured — Best fit line

FIGURE 20-28. Comparison of 100 × 200-mm (4 × 8-in.) and 150 × 300-mm (6 × 12-in.) cylinder strengths (Burg and Ost 1994).

maximum aggregate is not greater than 35 mm (1⅜ in.). The specifier of tests should specify the beam size to be used.

ASTM C31 prescribes the method of consolidation and the number of layers to be used in making test specimens. For concrete with slump less than 25 mm (1 in.), consolidation is by vibration. For slump of at least 25 mm (1 in.), consolidation is by vibration or rodding. For cylinders to be consolidated by rodding, the molds are filled in two layers for 100-mm (4-in.) diameter cylinders and three layers for 150-mm (6-in.) diameter cylinders. Each layer is rodded 25 times and a smaller diameter rod is used for the smaller cylinders. The sides of the mold should be tapped lightly with the prescribed mallet or open hand. Cylinders to be vibrated should be filled in two layers with one insertion per layer for 100-mm (4-in.) diameter cylinders and two insertions per layer for 150 mm (6-in.) cylinders. Care should be used to avoid over-vibration. Usually vibration is sufficient when the top surface becomes relatively smooth and large air bubbles no longer break through the surface.

Beams up to 200 mm (8 in.) deep are molded using two layers if consolidated by rodding and using one layer if consolidated by vibration. Each layer is rodded once for each 1400 mm² (2 in.²) of top surface area. If vibration is used, the vibrator is inserted at intervals less than 150 mm (6 in.) along the center of the beam. For beams wider than 150 mm (6 in.), insertions of the vibrator is alternated along two lines. Internal vibrators should have a maximum width not more than ⅓ the width of beams or ¼ the diameter of cylinders. After casting, the tops of the specimens need to be protected against drying.

The strength of a test specimen can be greatly affected by jostling, changes in temperature, and exposure to drying, particularly within the first 24 hours after casting. Thus, test specimens should be cast in locations where subsequent movement is unnecessary and where protection is available.

FIGURE 20-29. Controlled moist curing in the laboratory for standard test specimens at a relative humidity of 95% to 100% and temperature of 23°C ± 2°C (73.5°F ± 3.5°F) (ASTM C511 or AASHTO M 201).

Specimens should be protected from rough handling at all ages. Identify specimens on the exterior of the mold to prevent confusion and errors in reporting.

Standard testing procedures require that specimens be cured under controlled conditions, either in the laboratory (Figure 20-29) or in the field. Standard curing gives an accurate indication of the quality of the concrete as delivered. After specimens are molded in the field, they are subjected to initial curing for up to 48 hours in accordance with ASTM C31. The temperature surrounding the specimens should be between 16°C and 27°C (60°F and 80°F) and moisture loss should be prevented. For concrete with a specified strength greater than 40 MPa (6,000 psi), the storage temperature should be between 20°C and 26°C (68°F and 78°F). Failure to comply with the initial curing requirements can result in low measured strength (Obla and others 2019). After initial curing and mold removal, specimens are subjected to final curing with free water maintained on their surfaces and at a temperature of 23.0°C ± 2.0°C (73.5°F ± 3.5°F). Specimens can be submerged in limewater or stored in a moist room. To prevent leaching of calcium hydroxide from concrete specimens, limewater must be saturated with hydrated lime, not agricultural lime (limestone), in accordance with ASTM C511, *Standard Specification for Mixing Rooms, Moist Cabinets, Moist Rooms, and Water Storage Tanks Used in the Testing of Hydraulic Cements and Concretes* (AASHTO M 201).

ASTM C31 also includes a field curing method in which specimens are cured in the field in the same manner as the structure. The strength of field-cured specimens is an indicator of the actual strength of concrete in the structure at the time of testing. However, they give little indication of whether a low strength test result is due to the quality of the concrete as delivered or to improper handling and curing of the specimens. On some projects, field-cured specimens are made in addition to those cast for standard curing; these are especially useful during cold weather, to determine when forms can be removed, or to determine when the structure can be put into use. Field-cured specimens may be required to check the adequacy of curing

and protection as discussed subsequently. The difference in volume of field cured cylinders versus the structure dimensions will lead to differing results in temperature and moisture. Cast-in-place (match cured) cylinders can provide a more accurate representation of the concrete in service. For more information see **Strength Tests of Hardened Concrete** and ASTM *Manual of Aggregate and Concrete Testing* (2020).

In-place concrete strength development can also be estimated by the maturity method (ACI 228.1R [2019] and ASTM C1074, *Standard Practice for Estimating Concrete Strength by the Maturity Method*), which is discussed further below and in Chapter 19. Other methods for estimating in-place strength are discussed subsequently under **Nondestructive Test Methods**.

Accelerated and Early-Age Compression Tests to Project Later Age Strength

ASTM C1768, *Standard Practice for Accelerated Curing of Concrete Cylinders*, can be used to speed up the rate of strength development so that a significant portion of the potential strength can be attained within a time period of 24 to 48 hours. Thus quality control tests can be completed within a few days of sampling and permit timely adjustments to the production process. Strength development of test specimens is accelerated using one of two curing procedures: submerged curing for 24 hours in warm water at 35°C ± 3°C (95°F ± 5°F), or autogenous curing for 48 hours in insulated containers. Later-age strengths can be estimated using previously established relationships between accelerated strength and standard 28-day compressive strength tests (Carino 2006), or they can be used along with ASTM C918 (AASHTO T 276).

ASTM C918, *Standard Test Method for Measuring Early-Age Compressive Strength and Projecting Later-Age Strength* (AASHTO T 276), uses the results of early-age strengths and the maturity method to estimate the later-age strength. This method requires monitoring the temperature of cylinders cured in accordance with ASTM C31 (AASHTO T 23) or ASTM C1768. Cylinders are tested at early ages beyond 24 hours, and the concrete temperature history is used to compute the maturity index at the time of test. A prediction equation relating strength to maturity index, is developed from laboratory or field data in accordance with ASTM C918 (AASHTO T 276). The prediction equation is used to project the strength at later ages based on the maturity index and measured strength of the specimens tested at early-age. See Carino (2006).

Chloride Content

The chloride content of fresh concrete should be checked to make sure it is below the specified limits, such as those given in ACI 318-19 for water-soluble chlorides, to avoid corrosion of reinforcing steel. An approximation of the chloride content of freshly mixed concrete can be made using a method developed by the National Ready Mixed Concrete Association

(NRMCA 1986). The chloride content of freshly mixed concrete may be estimated by summing up the total chloride contents of the individual constituents of the mixture. The NRMCA method provides a conservative approximation because it includes chlorides not soluble in water. If the calculated total chloride content is lower than the specified limit, no further action is needed. If the calculated total chloride content exceeds the limit, the water-soluble chloride content of the concrete can be determined in accordance with ASTM C1218. See also **Testing Hardened Concrete, Chloride Content**.

Portland Cement Content, Water Content, and Water-Cement Ratio

Test methods have been developed for estimating the portland cement and water content of freshly mixed concrete. Due to their complexity, however, they are rarely used for routine quality control. Nevertheless, results of these tests of fresh concrete can assist in determining the strength and durability potential of the concrete and can indicate if the required cement and water contents have been obtained. While not in current use, ASTM C1078, *Test Methods for Determining the Cement Content of Freshly Mixed Concrete* (withdrawn 1998), and ASTM C1079, *Test Methods for Determining the Water Content of Freshly Mixed Concrete* (withdrawn 1998), based on the Kelly-Vail method (Head and others 1983), can be used to determine cement content and water content. The disadvantage of these test methods is that they require sophisticated equipment and special operator skills, which may not be readily available.

Other tests for determining cement or water contents can be classified into four categories: chemical determination, separation by settling and decanting, nuclear methods, and electrical methods. The Rapid Analysis Machine (RAM) and nuclear cement gauge have been used to measure cement contents (Forester and others 1974 and PCA 1983). The microwave oven drying method (AASHTO T 318, *Standard Method of Test for Water Content of Freshly Mixed Concrete Using Microwave Oven*) and neutron-scattering methods have been used to estimate water contents. For an overview of these and other tests from all four categories, see Hime (2006). A combination of these tests can be run independently to determine either cement content or water content to calculate the water-cement ratio. None of these test methods, however, have the level of reliability required for use as acceptance tests.

Supplementary Cementitious Materials Content

Standard test methods are not available for determining the supplementary cementitious materials content of freshly mixed concrete. However, the presence of certain supplementary cementitious materials, such as fly ash, can be determined by washing a sample of the concrete's mortar over a 45-μm (No. 325) sieve and using a stereo microscope (150× to 250×) to view the residue retained (Figure 20-30). Fly ash particles appear as spheres of various colors. Sieving the mortar through a 150-μm

FIGURE 20-30. Fly ash particles retained on a 45-μm (No. 325) sieve after washing, as viewed through a microscope at 200×.

or 75-μm (No. 100 or 200) sieve before washing is helpful in removing sand grains.

Bleeding of Concrete

The bleeding tendency of fresh concrete can be determined by ASTM C232, *Standard Test Method for Bleeding of Concrete* (AASHTO T 158). A sample of fresh concrete is placed into a container with an inside diameter of about 255 mm (10 in.) and a height of about 280 mm (11 in.) and consolidated by rodding. The container is filled to a height of about 255 mm (10 in.). Bleed water is drawn off the concrete surface and recorded in regular intervals until cessation of bleeding. The bleeding tendency is expressed as the mass of bleed water expressed as a percentage of the mass of net mixing water in the test specimen. Typical values range from 0.1% to 2.5% of mixing water. The bleeding test is rarely used in the field, but it is useful for evaluating alternative mixtures in the laboratory (Figure 20-31).

FIGURE 20-31. ASTM C232 (AASHTO T 158) test for bleeding of concrete. Inset: The container needs to be covered during the test to prevent evaporation.

Time of Setting

ASTM C403, *Standard Test Method for Time of Setting of Concrete Mixtures by Penetration Resistance* (AASHTO T 197), is used to determine the time of setting of concrete by means of penetration resistance measurements made at regular time intervals on mortar sieved from the concrete mixture (Figure 20-32). The initial and final time of setting are determined as the times when the penetration resistance equals 3.4 MPa (500 psi) and 27.6 MPa (4,000 psi), respectively. Typically, initial setting occurs between 2 and 6 hours after batching and final setting occurs between 4 and 12 hours.

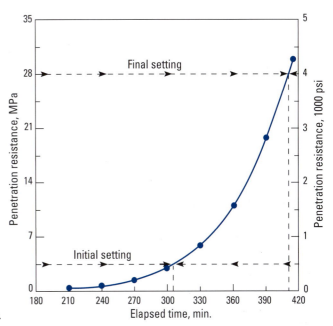

FIGURE 20-32. (left) Time of setting equipment. (right) Plot of test results.

Temperature, water-cementitious materials ratio, types of cementitious materials, and admixtures affect setting time.

TESTING HARDENED CONCRETE

Molded specimens prepared as described in the previous section **Strength Specimens** (ASTM C31 [AASHTO T 23], ASTM C192 [AASHTO R 39], or field specimens obtained in accordance with ASTM C873, *Standard Test Method for Compressive Strength of Concrete Cylinders Cast in Place in Cylindrical Molds*), or samples of hardened concrete obtained from construction (ASTM C42, *Standard Test Method for Obtaining and Testing Drilled Cores and Sawed Beams of Concrete* [AASHTO T 24]), can be used for tests of hardened concrete. Separate specimens should be obtained for each test performed because specimen preconditioning for certain tests can make the specimen unusable for other tests.

Strength Tests of Hardened Concrete

Strength tests of hardened concrete can be performed on the following: specimens molded from samples of freshly mixed concrete and then standard-cured in accordance with ASTM C31 or ASTM C192 (AASHTO T 23 or R 39) (Figure 20-33); in-situ specimens cored or sawed from hardened concrete in accordance with ASTM C42 (AASHTO T 24); or cast-in-place cylinders made using special molds and cured in the structure in accordance with ASTM C873 (Figure 20-34).

Cast-in-place cylinders can be used in concrete that is 125 mm to 300 mm (5 in. to 12 in.) in depth. The mold is filled in the normal course of concrete placement. The specimen is cured in place and in the same manner as the rest of the concrete section. The filled mold is removed from the structure and taken to a testing facility where the cylinder is prepared for compressive strength testing to determine the in-place concrete strength. If cast-in-place specimens have length-diameter ratios (L/D) less than 1.75, the strength correction factors given in ASTM C42 (AASHTO T 24) must be applied. This method is particularly applicable in cold-weather concreting, post-tensioning work, slabs, or any concrete work where a minimum in-place strength must be achieved before construction can continue to the next phase.

FIGURE 20-33. Compressive strength test cylinders and flexural strength specimens made in the field in accordance with ASTM C31 before being subjected to initial curing.

FIGURE 20-34. Concrete cylinders cast in place in cylindrical molds provide a means for determining the in-place compressive strength of concrete (ASTM C873).

For all methods, cylindrical specimens should have a diameter at least three times the nominal maximum size of coarse aggregate in the concrete and a length as close to twice the diameter as possible. Correction factors are available in ASTM C42 (AASHTO T 24) for specimens with length-to-diameter (L/D) ratios between 1.0 and 1.75. Refer to Ozyildirim and Carino (2006) for an explanation of why correction factors are needed for low L/D values. Cores and cylinders with a height of less than 95% of the diameter before capping should not be tested. A core diameter of at least 94 mm (3.70 in.) should be used if a L/D ratio greater than 1.0 is possible.

Cores should not be taken until the concrete can be drilled without disturbing the bond between the mortar and the coarse aggregate. For horizontal surfaces, cores should be taken vertically away from formed joints or edges. For vertical or sloped faces, cores should be taken perpendicular to the central portion of the concrete placement. Diamond-studded coring bits can cut through reinforcing steel. This should be avoided if possible when obtaining compression test specimens. The presence of bar reinforcement perpendicular to the core axis may reduce the measured compressive strength. There are, however, insufficient research data to develop appropriate correction factors to account for the presence of steel. If the specifier permits testing of cores with embedded reinforcement, engineering judgment is required to evaluate test results. A covermeter (electromagnetic device) or ground penetrating radar (GPR) can be used to locate reinforcing steel.

Length of a core drilled from a concrete structure should be determined in accordance with ASTM C1542, *Standard Test Method for Measuring Length of Concrete Cores*. For measuring member thickness, the use of ASTM C174, *Standard Test Method for Measuring Thickness of Concrete Elements Using Drilled Concrete Cores* (AASHTO T 148), may be stipulated.

Cores are obtained typically using water-cooled coring machines to prolong tool life and minimize damage. As a result, cores will absorb water leading to a moisture gradient from the exterior to the interior of the cores. It has been observed that the presence of a moisture gradient has a detrimental effect on the measured core strength. Figure 20-35 shows the effects of core conditioning on the strength of drilled cores. Forty-eight hour water immersion of the specimens prior to testing results in significantly lower test results than air-drying specimens for seven days before testing. The work by Fiorato and others (2000) showed that measured strengths varied by up to 25%, depending upon the time and type of conditioning prior to testing. Before testing, ASTM C42 requires that cores be kept in sealed plastic bags for at least 5 days after last being wetted. This is intended to provide reproducible moisture conditions and reduce the effects of moisture gradients introduced during specimen preparation.

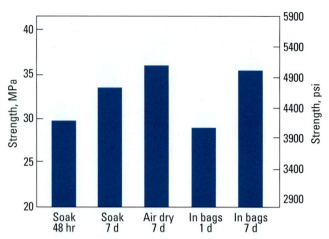

FIGURE 20-35. Effect of moisture conditioning on strength of drilled cores (Fiorato and others 2000).

Flexure test specimens that are saw-cut from in-place concrete are always immersed in lime-saturated water at 23.0°C ± 2.0°C (73.5°F ± 3.5°F) for at least 40 hours before testing.

Compressive strength test results are influenced by the condition of the ends of cylinders and cores. The ends of specimens should be ground or capped in accordance with the requirements of ASTM C617, *Standard Practice for Capping Cylindrical Concrete Specimens* (AASHTO T 231), or ASTM C1231, *Standard Practice for Use of Unbonded Caps in Determination of Compressive Strength of Hardened Cylindrical Concrete Specimens* (AASHTO T22). Various commercially available materials can be used to cap test specimens. ASTM C617 (AASHTO T 231) covers methods for making bonded caps of sulfur mortar, neat cement paste, or high-strength gypsum paste. Sulfur mortar caps must be allowed to harden at least 2 hours before the specimens are tested. For concrete strength of 35 MPa (5,000 psi) or greater, hardening time should be at least 16 hours unless data shows that shorter times are suitable. Sulfur mortar caps should be made as thin as is practicable for accurate test results (Lobo and others 1994). ASTM C617 requires that the average cap thickness not exceed 6 mm (¼ in.) for concrete compressive strengths less than 50 MPa (7000 psi) and not exceed 3 mm (⅛ in.) for higher concrete strength.

As an alternative to bonded caps, ASTM C1231 describes the use of neoprene caps that are not bonded to the ends of the specimen. This method of capping uses a disk-shaped 13 ± 2 mm (½ ± 1/16 in.) thick neoprene pad with a diameter slightly greater than the specimen diameter. The pad is placed in a cylindrical steel retainer with a cavity approximately 25 mm (1 in.) deep and with a diameter slightly larger than the diameter of the pad. This cap (retainer and pad) is placed on one or both ends of the cylinder, and the specimen is tested in accordance with ASTM C39, *Standard Test Method for Compressive Strength of Cylindrical Concrete Specimens* (AASHTO T 22). The test is stopped at 10% of the anticipated ultimate load to check that the axis of the cylinder is vertical within a tolerance of 0.5 degrees. The end of the specimen to receive an unbonded cap should not depart by more than 0.5 degrees from perpendicularity with the cylinder axis. If either the perpendicularity of the cylinder end, or the vertical alignment during loading are not met, the load applied to the cylinder may be concentrated on one side of the specimen. This can cause a short shear fracture in which the failure plane intersects the end of the cylinder. This type of fracture usually indicates the cylinder failed prematurely, at lower than the actual strength of the concrete. If end perpendicularity requirements are not met, the specimen can be saw-cut, ground, or capped with a sulfur mortar compound in accordance with ASTM C617 (AASHTO T 231). Premature shear fractures can also be reduced by: dusting the pad and end of cylinder with corn starch or talcum powder, preventing excess water from cylinders or burlap from draining into the retainer and below the pad, and checking bearing surfaces of retainers for planeness and indentations. The permitted hardness of the neoprene pads depends on the strength of the concrete being tested; the stronger the concrete, the harder the rubber that can be used. Unbonded neoprene caps are not permitted for acceptance testing of concrete with compressive strength above 80 MPa (12,000 psi). The neoprene pads are replaced after a specified number of tests, as stated in ASTM C1231.

Testing of specimens for strength (Figure 20-36) should be done in accordance with the following standard test methods:

- Compressive strength: ASTM C39 (AASHTO T 22),

- Flexural strength: ASTM C78, *Standard Test Method for Flexural Strength of Concrete (Using Simple Beam with Third-Point Loading)* (AASHTO T 97), or ASTM C293, *Standard Test Method for Flexural Strength of Concrete (Using Simple Beam With Center-Point Loading)* (AASHTO T 177), and

- Splitting tensile strength: ASTM C496, *Standard Test Method for Splitting Tensile Strength of Cylindrical Concrete Specimens* (AASHTO T 198).

For both concrete pavement thickness design and pavement concrete mixture proportioning, the modulus of rupture (flexural strength) should be determined by ASTM C78 (AASHTO T 97). However, ASTM C293 (AASHTO T 177) can be used for job control if empirical relationships to third-point loading (ASTM C78 [AASHTO T 97]) flexural strengths are determined before construction starts. For the same concrete, the measured flexural strength will, on average, be greater if ASTM C293 [AASHTO T 177] is used.

The moisture content of the specimen affects the resulting strength. Beams for flexural tests are especially vulnerable to moisture gradient effects and standard-cured beams need to be kept moist until the time of testing. Saturated specimens will show lower compressive strength and higher flexural strength compared with companion specimens tested dry. Cylinders used for acceptance testing for a specified strength must be cured in accordance with ASTM C31 (AASHTO T 23), to accurately represent the quality of the concrete made and cured under a standard condition. Cores are usually tested in the as-received condition, but rarely in a moist condition similar to standard-cured cylinders. In addition, cores are subjected to non-standard conditions of consolidation and curing. Therefore, cores and cylinders cannot be expected to have the same strengths at the same test age.

The variability in compressive-strength testing is less than for flexural-strength testing. To avoid the extreme care needed in flexural-strength acceptance testing, compressive-strength tests can be used to monitor concrete quality in projects designed on the basis of modulus of rupture. However, a laboratory-determined empirical relationship (Figure 20-37) must be developed between the compressive and flexural

FIGURE 20-36. Testing hardened concrete specimens: (left) cylinder, (right) beam.

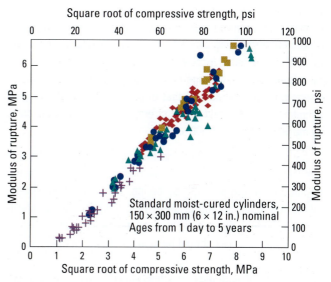

FIGURE 20-37. Long-term data show that flexural strength (measured by third-point loading) is proportional to the square root of compressive strength over a wide range of strength levels (Wood 1992).

strength of the concrete used (Kosmatka 1985a). Because of the robust empirical relationship between these two strengths and the economics of testing cylinders instead of beams, most state departments of transportation use compression tests of cylinders to monitor concrete quality for pavement and bridge projects.

Evaluation of Compression Test Results. ACI 318-19 states that the compressive strength of concrete can be considered satisfactory if the following conditions are met:

- Every average of any three consecutive strength tests equals or exceeds f'_c.

- No strength test (average of two or three cylinders) falls below f'_c by more than 3.5 MPa (500 psi) if f'_c is 34.5 MPa (5000 psi) or less; or by more than 0.10 f'_c if f'_c exceeds 34.5 MPa (5000 psi).

If the results of the cylinder tests do not meet these criteria, steps should be taken to increase the measured compressive strength. This would include review of the testing history to ensure that applicable standards were followed. In addition, if the second condition is not met, steps must be taken to ensure that the load capacity of the structure is not jeopardized. If the likelihood of low strength concrete is confirmed and the load carrying capacity is in question, the strength of the in-place concrete may be evaluated by drilled cores as discussed below.

In addition to the cylinders for acceptance testing, project specifications often require one or two 7-day cylinders and one or more "hold" cylinders. The 7-day cylinders monitor early strength gain to signal potential problems in meeting specified strength. Hold cylinders are commonly used to provide additional information in case the cylinders tested for acceptance are damaged or do not meet the required compressive strength. For low 28-day test results, the hold cylinders are typically tested at 56 days.

The licensed design professional or building official may require testing of field-cured cylinders to evaluate the adequacy of protection and curing procedures for the structure. According to ACI 318-19, such procedures are adequate if field-cured cylinders tested at the age designated for f'_c have an average strength of at least 85% of that of companion standard-cured cylinders, or if the field-cured strength at test age exceeds f'_c by more than 3.5 MPa (500 psi).

If low strength tests occur and the likelihood of low-strength concrete is confirmed, it may be necessary to evaluate the in-place concrete strength by coring. Three cores should be taken from each portion of the structure where the standard-cured cylinders did not meet acceptance criteria. Moisture conditioning of cores should be in accordance with ASTM C42 (AASHTO T 24). If the average strength of three cores is at least 85% of f'_c, and if no single core is less than 75% of f'_c, the concrete in the area represented by the cores is considered structurally adequate. If the results of properly conducted core tests are so low as to leave structural integrity in doubt, a strength evaluation of the structure in accordance with ACI 318-19 may be performed. Nondestructive or in-place test methods are not a substitute for core tests, but they can be used to confirm the likelihood of low strength concrete before drilling cores. Refer to NRMCA (1979), ACI 214R, *Guide to Evaluation of Strength Test Results of Concrete* (2011), ACI 228.1R, *Report on Methods for Estimating In-Place Concrete Strength* (2019), and ACI 318-19 for more information on evaluating concrete strength.

Air Content

The air content and air-void system parameters of hardened concrete can be determined using ASTM C457 to verify that the air-void system is adequate to resist damage in a freezing and thawing environment. The test is also used to determine the effects of different admixtures and methods of placement and consolidation on the air-void system. The test can be performed on molded specimens or samples removed from the structure. Using a polished section of a concrete sample, the air-void system is characterized with measurements using a microscope. The information obtained from this test method includes the combined volume of entrained and entrapped air, the specific surface (surface area of the air voids per unit volume of air), the spacing factor, and the number of voids per linear distance (Figure 20-38). See Chapter 11 for more information on resistance to freezing and thawing.

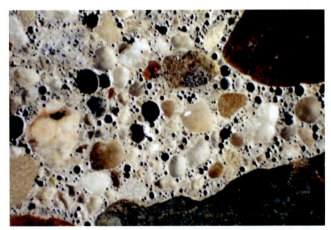

FIGURE 20-38. View of concrete air-void system under a microscope.

Density, Absorption, and Voids

The density, absorption, and voids content of hardened concrete can be determined in accordance with ASTM C642, *Standard Test Method for Density, Absorption, and Voids in Hardened Concrete*. Note that the boiling procedure in ASTM C642 can render the specimens useless for other tests, especially strength tests.

The saturated surface-dry (SSD) density can be determined by soaking the specimen in water for 48 hours and then determining its mass in air in the SSD condition and immersed in water. The SSD density is calculated as shown in Equation 20-2:

$$D_{SSD} = \frac{M_1 \rho}{M_1 - M_2}$$

EQUATION 20-2.

Where:

D_{SSD} is density in the SSD condition

M_1 is the SSD mass in air, kg (lb)

M_2 is the apparent mass immersed in water, kg (lb)

ρ is the density of water, approximately 1000 kg/m³ (62.4 lb/ft³) at 4°C (39.2°F)

The SSD density provides an approximation of the density of freshly mixed concrete. As was mentioned, the density of hardened concrete can also be determined using a nuclear gauge in accordance with ASTM C1040 (AASHTO T 271). This requires a pre-established relationship between the density of refence blocks and nuclear gauge measurements.

The rate of absorption (sorptivity) of water by hardened concrete can be determined using ASTM C1585, *Standard Test Method for Measurement of Rate of Absorption of Water by Hydraulic-Cement Concretes*. In this test, a concrete disk with a diameter of 100 mm (4 in.) and a height of 50 mm (2 in.) is preconditioned to control the internal relative humidity at the start of the test. The side of the disk and one end is sealed. The bare end is placed in water and the uptake of water is measured as a function of time. The water gain is plotted as a function of the square root of time, and the initial rate of water absorption is calculated through regression analysis. The rate of water absorption is affected by the degree of saturation and the structure of the pores in the paste, while the absorption determined by ASTM C642 is related to the volume of permeable pores.

Portland Cement Content

The portland cement content of hardened concrete can be determined by ASTM C1084, *Standard Test Method for Portland-Cement Content of Hardened Hydraulic-Cement Concrete* (AASHTO T 178). The test method includes two independent procedures: an oxide analysis procedure and a maleic acid extraction procedure. Each procedure requires substantial expertise in chemical testing. Cement content tests are valuable in determining the cause of lack of strength gain or poor durability of concrete. The user of these test methods should be aware that certain admixtures and aggregate types can interfere with test results. The presence of supplementary cementitious materials would also be reflected in the test results. However, they can be more accurately determined by complete chemical analysis using x-ray florescence spectrometry.

Supplementary Cementitious Material and Organic Admixture Content

The presence and amount of certain supplementary cementitious materials, such as fly ash and slag cement, can be determined by petrographic techniques (ASTM C856) as discussed below. A sample of the supplementary cementitious material used in the concrete is usually necessary as a reference to determine the type and amount of the supplementary cementitious material present in the concrete. The presence and possibly the amounts of organic admixtures (such as water reducers and air entraining agents) can be determined by infrared spectrophotometry (see Hime 2006).

Chloride Content

The water-soluble chloride-ion content of hardened concrete can be determined in accordance with ASTM C1218, *Standard Test Method for Water-Soluble Chloride in Mortar and Concrete*. In addition, ASTM C1152, *Standard Test Method for Acid-Soluble Chloride in Mortar and Concrete*, can be used to determine the acid-soluble chloride content of concrete which in most cases is equivalent to total chloride. ACI 318-19 places limits on the water-soluble chloride-ion content of concrete for different exposure categories and whether the concrete is prestressed. Tests of hardened concrete to demonstrate compliance with the limits are conducted at ages between 28 and 42 days.

The above tests for chloride-ion content use pulverized concrete samples. Thus, they also extract chloride ions from within fine and coarse aggregates particles that generally are not available and do not contribute to corrosion of reinforcing steel. ASTM C1524, *Standard Test Method for Water-Extractable Chloride in Aggregate (Soxhlet Method)*, can be used to investigate the amount of water-extractable chloride ions from aggregate particles. Because ASTM C1524 does not involve pulverizing the aggregate particles, it provides a more realistic measure of the chloride ions available for corrosion. Chlorides are extracted from a prepared aggregate sample of at least 30 g using a Soxhlet extractor, which subjects the sample to multiple extraction cycles every 20 ± 5 min for at least 70 cycles. The interpretation of information obtained from the Soxhlet procedure is a matter of debate.

Petrographic Analysis

Petrographic analysis uses microscopical techniques described in ASTM C856, *Standard Practice for Petrographic Examination of Hardened Concrete* (AASHTO T 299), to determine concrete constituents, characteristics, and distress mechanisms. The future performance of concrete elements can also be estimated. Some of the features that can be analyzed by a petrographic examination include paste, aggregate, fly ash, and air content; frost and sulfate attack; alkali-aggregate reactivity; degree of hydration; depth of carbonation; water-cement ratio; bleeding characteristics; fire damage; scaling; popouts;

effect of admixture; and other aspects. Almost any concrete distress or deterioration can be analyzed by petrography (St. John and others 1998). However, a standard petrographic analysis is sometimes accompanied by wet chemical analyses, infrared spectroscopy, x-ray diffractometry, x-ray florescence, scanning electron microscopy with attendant elemental analysis, differential thermal analysis, and other analytical tools. Some results from a petrographic examination depend on the subjective judgment and experience of the petrographer. ASTM C856 (AASHTO T 299) provides criteria for the qualifications of petrographers and technicians.

Several techniques are used by petrographers to estimate the *w/cm* (see Erlin 2006). Some of these are subjective and others require pre-establishing a correlation between *w/cm* and the characteristic used as an indicator of *w/cm*. However, ASTM C856 (2020) states: *"… there is no generally accepted standard procedure that employs microscopical methods for determining the w/c or w/cm of hardened concrete."*

Volume Changes – Length Change, Creep, and Elastic Modulus

Volume change limits are sometimes specified for certain concrete applications, such as industrial floor slabs, high-rise structures, pavements, and bridge decks. ASTM C157, *Standard Test Method for Length Change of Hardened Hydraulic-Cement Mortar and Concrete* (AASHTO T 160), determines length change in concrete due to drying shrinkage, chemical shrinkage, and causes other than externally applied forces and temperature changes. Early volume change of concrete before final setting can be determined using ASTM C827, *Standard Test Method for Change in Height at Early Ages of Cylindrical Specimens of Cementitious Mixtures.* ASTM C1698, *Standard Test Method for Autogenous Strain of Cement Paste and Mortar,* can be used to measure the autogenous strain of paste or mortar specimens from the time of final setting up to a specified age. Creep can be determined in accordance with ASTM C512, *Standard Test Method for Creep of Concrete in Compression.*

For some structures, it is critical to control deformations under service conditions. In such cases, a minimum value of modulus of elasticity of concrete may be specified. The static modulus of elasticity and Poisson's ratio of concrete can be determined using ASTM C469, *Standard Test Method for Static Modulus of Elasticity and Poisson's Ratio of Concrete in Compression.* Dynamic values of these parameters can be determined by using ASTM C215, *Standard Test Method for Fundamental Transverse, Longitudinal, and Torsional Frequencies of Concrete Specimens.* For the same concrete, the dynamic modulus of elasticity will be larger than the static value, but the difference decreases as the strength level of the concrete increases.

The cracking tendency of a mortar or concrete due to the combined effects of drying shrinkage, autogenous shrinkage,

heat of hydration, and stress relaxation under restrained conditions can be evaluated using ASTM C1581, *Standard Test Method for Determining Age at Cracking and Induced Tensile Stress Characteristics of Mortar and Concrete under Restrained Shrinkage.* The method was developed as a tool for screening alternative repair materials for concrete.

ASTM E1155, *Standard Test Method for Determining F_F Floor Flatness and F_L Floor Levelness Numbers,* can be used to evaluate the flatness and levelness of newly constructed floor slabs or the deformation of the slab surface due to curling and warping. Volume change is discussed further in Chapter 10.

Durability

In addition to tests for air content and chloride content previously described, the following tests are used to measure the durability performance of concrete (also see Chapter 11):

Resistance to Freezing-and-Thawing. The freeze-thaw resistance of concrete is usually determined in accordance with ASTM C666, *Standard Test Method for Resistance of Concrete to Rapid Freezing and Thawing* (AASHTO T 161). Prismatic specimens are monitored for changes in the dynamic modulus of elasticity, mass, and specimen length (optional) over a period of up to 300 cycles of freezing and thawing.

Since the development of internal microcracking is anticipated, freeze-thaw resistance in accordance with ASTM C666 is indirectly evaluated by changes in the relative dynamic modulus of elasticity (as measured by resonant frequency), indicating the degree of internal microcrack formation. A durability factor is then calculated according to Equation 20-3:

$$DF = \frac{P_n N}{M}$$

EQUATION 20-3.

where:

DF = durability factor of the specimen tested,

P_n = relative dynamic modulus at N cycles (%),

N = number of cycles at which the test specimen achieves the minimum specified value of Pc for discontinuing the test or the specified number of cycles of the test, whichever is less, and

M = specified number of cycles of the test (typically M = 300).

Concrete that will be exposed to deicers as well as freezing in a saturated condition can be tested for deicer-scaling resistance using ASTM C672-12, *Standard Test Method for Scaling Resistance of Concrete Surfaces Exposed to Deicing Chemicals* (withdrawn 2021). In ASTM C672, small slabs of concrete are used to evaluate scaling resistance using a visual damage rating. Although ASTM C672 requires that only surface scaling be monitored, many practitioners also measure mass loss, as is the current practice in Canada (Figure 20-39).

The Swedish Standard SS 13 72 44 (Marchand and others 1994, and Jacobsen and others 1996) correlates the scaling potential with the collected mass loss per unit area. Scaling of less than 1.0 kg/m^2 (0.2 lb/ft^2) after 50 freeze-thaw cycles in the presence of deicer salts indicates an acceptable scaling resistance (Gagné and Marchand 1993).

It is generally accepted that the ASTM C666 test is more severe than most natural exposures. Concrete mixtures that perform well in ASTM C666 (AASHTO T 161) do not always perform well in ASTM C672. ASTM C666 (AASHTO T 161) and ASTM C672 are often used to evaluate innovative designs of concrete mixtures, or new materials such as chemical admixtures, supplementary cementitious materials, and aggregates to determine their effect on resistance to freezing and thawing and deicers. Nmai (2006) presents a historical evolution of accelerated tests to assess frost resistance of concrete.

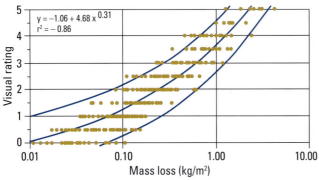

FIGURE 20-39. Correlation between mass loss per unit area and visual rating for each specimen tested according to ASTM C672 (Pinto and Hover 2001).

Sulfate Resistance. There is no standardized test for concrete sulfate resistance. ASTM C150 (AASHTO M 85) uses prescriptive limits on C$_3$A content to define sulfate resistant cement types (see Chapter 2) or an optional performance requirement for expansion, in accordance with ASTM C452, *Standard Test Method for Potential Expansion of Portland-Cement Mortars Exposed to Sulfate*. ASTM C452 fabricates mortars with sulfate added to bring the sulfate content up to 7% by mass, stores specimens in water for 14 days, and then measures the expansion. The optional limit in ASTM C150 (AASHTO M 85) is 0.040% expansion at 14 days.

ASTM C1012, *Standard Test Method for Length Change of Hydraulic-Cement Mortars Exposed to a Sulfate Solution*, is used in ASTM C595, *Standard Specification for Blended Hydraulic Cements* (AASHTO M 240), and ASTM C1157, *Standard Performance Specification for Hydraulic Cements*, to define cements as moderately- or highly-sulfate resistant. Mortar bar specimens are immersed in a 5% (33,800 ppm) sodium sulfate solution and length change is measured as a function of time. In this aggressive environment, expansions of less than 0.10% at 6 months indicate moderate sulfate resistance and

expansions less than 0.05% at 6 months (or less than 0.10% at 12 months), indicate high sulfate resistance. This test is valuable in assessing the sulfate resistance of concrete that will be continuously wet. It does not evaluate the more aggressive wetting and drying environment. The test can be modified to include wet-dry cycling. ASTM C1580, *Standard Test Method for Water-Soluble Sulfate in Soil*, and ASTM D516, *Standard Test Method for Sulfate Ion in Water* (AASHTO T 290), can be used to test soil and water for sulfate ion content to determine the severity of the sulfate exposure. Sulfate resisting cement do not guarantee sulfate resistant concrete, w/cm limits and adequate curing are also needed (see Chapter 11).

Alkali-Silica Reactivity. Evaluation of aggregates for potential alkali-silica reactivity (ASR), is covered under **Test Methods for Identifying ASR-Susceptible Aggregates**.

The efficacy of SCMs, such as fly ash and slag cement, to mitigate ASR should be evaluated by tests such as ASTM C1567, *Standard Test Method for Determining the Potential Alkali-Silica Reactivity of Combinations of Cementitious Materials and Aggregate (Accelerated Mortar-Bar Method)*, or ASTM C1293. Additionally, lithium admixtures to control ASR can be evaluated using a modified ASTM C1293 (Thomas and others 2008), CSA A23.2-28A, or CRD-C 662, *Determining the Potential Alkali-Silica Reactivity of Combinations of Cementitious Materials, Lithium Nitrate Admixture and Aggregate (Accelerated Mortar-Bar Method)*. ASTM C1567 follows a similar protocol as ASTM C1260, but with SCMs included in the cementitious materials. It is recommended that several levels of SCM be tested to determine the minimum content that will result in less than 0.10% expansion at 14 days. When ASTM C1293 is used to evaluate mitigation strategies, the test duration is extended to 2 years, and mitigation is demonstrated by the level of SCM(s) that result in less than 0.04% expansion.

A sequence of tests to evaluate aggregate reactivity, developed by Thomas and others (2008), has been adopted by ASTM C1778, *Standard Guide for Reducing the Risk of Deleterious Alali-Aggregate Reaction in Concrete* and AASHTO R 80, *Standard Practice for Determining the Reactivity of Concrete Aggregates and Selecting Appropriate Measures for Preventing Deleterious Expansion in New Concrete Construction*. The aggregates are first evaluated based on field history, then through petrographic examination (ASTM C295). Following the petrographic examination, the aggregates are tested according to ASTM C1260. If the expansion exceeds 0.10%, the aggregates are then tested according to ASTM C1293. Once the aggregates have been evaluated for reactivity, appropriate preventive measures can be prescribed (see Chapter 11). A drawback to this approach is that ASTM 1293 requires one year to complete the testing. An alternative to testing aggregate separately for potential reactivity is to test the concrete mixture using ASTM C1567 or ASTM C1293.

Samples from existing concrete structures can be evaluated for alkali-silica reaction using ASTM C856 (AASHTO T 299). There are also field methods for detecting the gel resulting from ASR. In one of these methods, a uranyl-acetate solution is applied to a broken or roughened concrete surface that has been dampened with distilled or deionized water (Natesaiyer and Hover 1988). After one minute, the solution is rinsed off and the treated surface is viewed under ultraviolet light. Areas of gel fluoresce bright yellow-green. Several materials not related to ASR in concrete can also fluoresce and interfere with an accurate indication of ASR gel. Materials that fluoresce like the ASR gel include: naturally fluorescent minerals, carbonated paste, opal, some other rock ingredients, and reaction products from fly ash, silica fume, and other pozzolans. Eye protection is required to prevent harm due to the ultraviolet light. In addition, the toxicity and radioactivity of uranyl acetate warrants special handling and disposal procedures regarding the solution and treated concrete. The other staining technique for detecting ASR gel uses benign solutions of sodium cobaltinitrite and rhodamine B to condition the specimen and produce a dark pink stain that corresponds to calcium-rich ASR gel (Guthrie and Carey 1997). Note that these rapid visual methods can identify evidence of ASR gel that was not the cause of damage to concrete. That is, ASR gel can be present when other mechanisms such as freezing and thawing, sulfate attack, and other deterioration mechanisms have caused the damage. These rapid methods for detecting the presence of ASR gel are useful but their limitations must be understood. Neither of the rapid procedures is a viable substitute for petrographic examination coupled with proper field inspection (Powers 1999).

Alkali-Carbonate Reactivity. Alkali-carbonate reactivity is relatively rare. Potential reactivity of dolomitic limestone aggregates can be evaluated by using ASTM C295, ASTM C586, *Standard Test Method for Potential Alkali Reactivity of Carbonate Rocks as Concrete Aggregates (Rock-Cylinder Method)*, and ASTM C1105, *Standard Test Method for Length Change of Concrete Due to Alkali-Carbonate Rock Reaction*. Existing concrete structures can be evaluated for alkali-carbonate reaction using ASTM C856 (AASHTO T 299).

Corrosion Activity. The corrosion activity of reinforcing steel in concrete exposed to chlorides or that has experienced excessive depth of carbonation can be evaluated using ASTM C876, *Standard Test Method for Corrosion Potentials of Uncoated Reinforcing Steel in Concrete*. The half-cell potential test provides an indication of the likelihood that steel corrosion is active, but it does not indicate the rate of corrosion. The linear polarization resistance (LPR) method, described in ACI 228.2R, *Report on Nondestructive Test Methods for Evaluation of Concrete in Structures* (2013), can be used to estimate corrosion rate. LPR provides an estimate of the corrosion rate at the time

of the test. This rate can change with changes in temperature, moisture content, and other factors. It is important to make LPR measurements several times a year to obtain a valid estimate of the mean corrosion rate.

Abrasion Resistance. Abrasion resistance of concrete is related strongly to the aggregate hardness (see discussion under **Testing Aggregates**). Over the years various test methods have been developed in an attempt to capture different abrasive actions. ASTM C779, *Standard Test Method for Abrasion Resistance of Horizontal Concrete Surfaces*, is the most commonly referenced concrete abrasion test method. ASTM C779 offers three loading regimes: revolving disks, dressing wheels, and ball bearings. There is little correlation between the different loading regimes in the test method, making it difficult to predict wear from one mechanism based on data from another test. The tests should be used for comparison purposes (employing the same test method) to select the best concrete mixture for abrasion resistance. Other available tests for abrasion and erosion include: ASTM C418, *Standard Test Method for Abrasion Resistance of Concrete by Sandblasting*, which uses the depth of wear under sandblasting; ASTM C944, *Standard Test Method for Abrasion Resistance of Concrete or Mortar Surfaces by the Rotating-Cutter Method*, which uses rotating cutters (more useful for smaller samples); ASTM C1138, *Standard Test Method for Abrasion Resistance of Concrete (Underwater Method)*, which simulates the effects of swirling water or cavitation; clamping concrete slabs inside a rotating drum filled with steel shot or aggregate; and application of rotating wire brushes.

Moisture Testing

Moisture related test methods fall into two general categories: qualitative or quantitative. Qualitative tests provide a gross indication of the presence or absence of moisture while quantitative tests measure the amount of moisture. Qualitative tests may give a strong indication that excessive moisture is present and the substrate is not ready for application of covering materials. Quantitative tests are performed to verify that a substrate is dry enough for these materials.

Qualitative moisture tests include: plastic sheet, mat bond, electrical moisture gauge tests. ASTM D4263, *Standard Test Method for Indicating Moisture in Concrete by the Plastic Sheet Method*, uses a 460-mm (18-in.) square sheet of clear plastic film that is taped to the slab surface and left for at least 16 hours. Water vapor from the slab will condense on the plastic film providing an indication of excess moisture in the concrete. The plastic sheet test has two limitations (ACI 302.2R-06): the duration of the test is too short to allow for potential moisture movement from the bottom of the slab; and the presence of moisture under the plastic sheet could be due to the slab being at the dew-point temperature rather than moisture migration. In the mat bond test, a 610-mm (2-ft) square sheet of floor covering is glued to the floor using the recommended adhesive and the edges are taped to the concrete for 72 hours. The effort needed to remove the flooring and the condition of the adhesive provide indications of the slab moisture condition. ASTM F2659, *Standard Guide for Preliminary Evaluation of Comparative Moisture Condition of Concrete, Gypsum Cement and Other Floor Slabs and Screeds Using a Non-Destructive Electronic Moisture Meter*, provides guidance on using electrical moisture meters. These are based on measuring electrical resistance or electrical impedance (resistance plus capacitance) and provide information on the moisture condition within approximately 25 mm (1 in.) of the surface. The electrical conductivity of concrete depends on moisture content as well as the pore structure of the hardened paste. Thus, while moisture meters display a moisture content, they should be used as indicators of relative moisture contents unless a concrete-specific correlation is developed. Sometime moisture meters are used in conjunction with the plastic sheet method to provide an objective indicator of the increase in surface moisture.

Quantitative test methods include: gravimetric moisture content, moisture vapor emission rate (MEVR), and relative humidity probe tests (Table 20-2). The most direct method for determining moisture content is to dry cut a specimen from the concrete element in question, place it in a moisture proof container, and transport it to a laboratory for testing. After obtaining the specimen's initial mass, dry the specimen in an oven at about 105°C (220°F) for 24 hours or until constant mass is achieved. The difference between the two masses divided by the dry mass, multiplied by 100, provides the moisture content in percent. ASTM F1869, *Standard Test Method for Measuring Moisture Vapor Emission Rate of Concrete Subfloor Using Anhydrous Calcium Chloride*, is a commonly used test for measuring the readiness of concrete for application of floor coverings. This involves placing a dish containing a fixed amount of calcium chloride under a flanged clear plastic cover that is sealed to the concrete surface. After 60 to 72 hours, the gain in mass of the calcium chloride is determined and the MVER is calculated from the mass gain, exposure area, and exposure time. The MVER is expressed in terms of pounds of moisture emitted from 1000 ft^2 in 24 hours. Converted to SI units, the rate is expressed as micrograms per square meter per second (µg/m^{2}s). See PCA EB119, *Concrete Floors and Moisture* (Kanare 2009), for more information.

ASTM F2420, *Standard Test Method for Determining Relative Humidity on the Surface of Concrete Floor Slabs Using Relative Humidity Probe Measurement and Insulated Hood*, uses a relative humidity probe under an insulated, impermeable box to trap moisture in a sealed air pocket above the floor

TABLE 20-2. Test Methods and Maximum Limits for Moisture Condition of Concrete

TEST METHOD	MAXIMUM LIMIT
ASTM F1869	3 lb/1000 ft^2 (170 µg/m^2) per 24 h
ASTM F2170	75%

Source: ASTM F710

FIGURE 20-40. Relative humidity hood test (ASTM F2420).

(Figure 20-40). The probe is allowed to equilibrate for at least 72 hours or until two consecutive readings at 24-hour intervals are within the accuracy of the instrument (typically ± 3% RH).

The maximum acceptable relative humidity limit for the installation of floor coverings ranges from a 60% to 90%. It can require several months of air-drying to achieve the desired relative humidity. A method for estimating drying time to reach a specified relative humidity based on water-cement ratio, thickness of structure, number of exposed sides, relative humidity, temperature, and curing conditions can be found in Hedenblad (1997), Hedenblad (1998), PCA EB075 (Tarr and Farny 2008), and Kanare (2009).

ASTM F2170, *Standard Test Method for Determining Relative Humidity in Concrete Floor Slabs Using In-situ Probes*, uses a relative humidity sensor installed in a hole drilled into the concrete. The hole is drilled to a prescribed depth below the surface that depends on whether the lab is drying from one surface or both surfaces. The hole is lined with a plastic sleeve, a humidity sensor is placed at the bottom of the hole, and the hole is capped. After 24 hours, the relative humidity at the bottom of the hole is measured. The liner ensures that the measured relative humidity is in equilibrium with the moisture content at the bottom of the hole. By making measurements with holes at different depths, the moisture profile in the slab can be established. See Kanare (2009) for more information.

Carbonation

The depth or degree of carbonation can be determined by petrographic techniques (ASTM C856) through the observation of calcium carbonate – the primary chemical product of carbonation. The depth of carbonation can also be determined by detecting where there is drastic change in the pH of the concrete. Uncarbonated concrete is highly alkaline with a pH above 12.5, while carbonated concrete has a pH of around 9 or less. The change in pH can be detected by spraying a newly fractured surface with a pH indicator solution, such as phenolphthalein. When a phenolphthalein solution is applied to a freshly fractured or freshly cut surface of concrete, areas with a pH between 8.3 and 10.0 turn red or purple while areas outside that range remain colorless (Figure 20-41). Other indicator solutions will display a range of colors depending on the pH value, and the depth where the color corresponds to a pH of 9 can be used as the approximate depth of carbonation. For more information, see **pH Testing Methods**, and Verbeck (1958), Steinour (1964), and Campbell and others (1991).

FIGURE 20-41. The depth of carbonation is determined by spraying phenolphthalein solution on a freshly broken concrete surface.

pH Testing Methods

The pH value of concrete is a measure of the acidic or basic nature of a surface. Adhesives used to bond flooring will not perform as required if they are applied to concrete surfaces that are highly basic. Thus, pH testing of concrete surfaces is required before applying flooring adhesives. Three methods have been used for measuring the surface pH of hardened concrete in the field. ASTM F710, *Standard Practice for Preparing Concrete Floors to Receive Resilient Flooring*, uses litmus paper designed for measuring the alkaline range of pH values. Place a few drops of distilled water on the concrete, wait 60 ± 5 seconds and immerse an indicator strip (litmus paper) in the water for 2 to 3 seconds. After removing the strip, compare it to the pH color scale supplied with the indicator strips. A second method uses a pH pencil. The pencil is used to make a 25 mm (1 in.) long mark on the surface after which 2 to 3 drops of distilled water are placed on the mark. After waiting 20 seconds, the color is compared to a color chart to judge the pH of the concrete.

The third method uses a wide-range liquid pH indicator applied to the surface of the concrete After several minutes, the resulting color is compared to a color chart to determine the pH of the concrete surface. See Kanare (2009) for more information.

These methods based on measuring the pH of drops of water placed on the concrete surface have been criticized as not providing a true measure of the pH of the surface concrete and that they are essentially measuring the pH of the water (Grubb and others 2007). It is argued placing a few drops of water on the surface is not sufficient to form a solution indicative of the pH of the concrete. It was suggested that a powder sample should be obtained from the concrete surface and placed in a small vial with water and measure the pH of that solution (Kakade 2014). ACI Committee 364 also recommends a procedure to determine the pH of a concrete surface (ACI 364.17T 2018).

Permeability and Diffusion

Both direct and indirect methods of measuring permeability are used. Historically, resistance to chloride-ion penetration was determined by ponding a chloride solution on a concrete surface and, at a later age, determining the chloride content of the concrete at particular depths using AASHTO T 259, *Standard Method of Test for Resistance of Concrete to Chloride Ion Penetration*. This test took 90 or more days to run. This lead to the development of ASTM C1202, *Standard Test Method for Electrical Indication of Concrete's Ability to Resist Chloride Ion Penetration* (AASHTO T 277), also called the rapid chloride permeability test (RCPT), which is often specified for testing concrete used in bridge decks. While ASTM C1202 is referred to as a "permeability" test, it is actually a test of electrical conductivity. The test measures the charge passed over a 6-hour period with 60 volts applied across the specimen. The test result is expressed in coulombs, and the higher the coulomb value the less resistance to chloride penetration. Conductivity is a good indicator of resistance to chloride pentation because both are affected the same factors, namely the structure of the pore system in the hardened paste. The RCPT takes 6 hours to complete after the specimen has been brought to a saturated condition. A more rapid test for electrical conductivity (or its inverse resistivity) than the RCPT was developed by the Florida Department of Transportation (FDOT 2004). This was subsequently adopted as AASHTO T 358, *Standard Method of Test for Surface Resistivity Indication of Concrete's Ability to Resist Chloride Ion Penetration*. This procedure uses the Wenner probe array method (Malhotra and Carino 2004) for measuring the resistivity along the surface of 100-mm × 200-mm (4-in. × 8-in.) cylinders. ACI 222R, *Guide to Protection of Metals in Concrete Against Corrosion* (2019), recommends using the Wenner probe method for assessing the resistivity of in-place concrete as it will affect corrosion rate. Because dry concrete has a high resistivity, test specimens need to be in a saturated condition to obtain a reliable indicator of the resistance to chloride ion penetration. The results of the electrical resistivity test have been correlated to the RCPT penetrability rating system (Smith 2006).

The surface resistivity method has some disadvantages related to probe spacing and specimen size dependency. ASTM C1876, *Test Method for Bulk Resistivity or Bulk Conductivity of Concrete*, uses electrodes on the ends of standard cylindrical specimens and overcomes the drawbacks of the surface method. The bulk resistivity, or inverse bulk conductivity, is a material property, whereas the charge passed in the ASTM C1202 test depends on the specimen size. There is, however, a simple theoretical relationship between electrical conductivity and charge passed (Weiss and others 2016).

The apparent chloride diffusion coefficient of hardened cementitious mixtures can be determined in accordance with ASTM C1556, *Standard Test Method for Determining the Apparent Chloride Diffusion Coefficient of Cementitious Mixtures by Bulk Diffusion*. Various absorption methods, including ASTM C642 and ASTM C1585, can be used as indicators of the relative resistance of concrete to the ingress of fluids.

Direct water permeability data can be obtained using CRD-C 163, *Test Method for Water Permeability of Concrete Using a Triaxial Cell*. A test method recommended by the American Petroleum Institute for determining the permeability of rock is also available. These methods, however, have limitations. Direct permeability testing using applied pressure is impractical for testing high quality concrete because of the time required to force measurable quantities of water through a specimen. For more information, see American Petroleum Institute (1956), Tyler and Erlin (1961), Whiting (1981), Pfeifer and Scali (1981), and Whiting (1988). For more information on permeability of concrete, see Chapter 9.

Thermal Properties

ASTM C177, *Test Method for Steady-State Heat Flux Measurements and Thermal Transmission Properties by Means of the Guarded-Hot-Plate Apparatus*, is used to determine values of thermal conductivity of concrete. This test method is the only appropriate method for determining the thermal conductivity of concrete from insulating to nomalweight concrete. Other methods, such as ASTM C518, *Standard Test Method for Steady-State Thermal Transmission Properties by Means of the Heat Flow Meter Apparatus*, do not allow adequate time for the concrete to come to thermal equilibrium which is needed to ensure thermal mass effects are not included with the thermal conductivity. Figure 20-42 shows an approximate relationship between thermal resistance and density for a particular lightweight insulating concrete mixture. The thermal conductivity of concrete increases with an increase in moisture content and density. See Brewer (1967) for additional density and conductivity relationships.

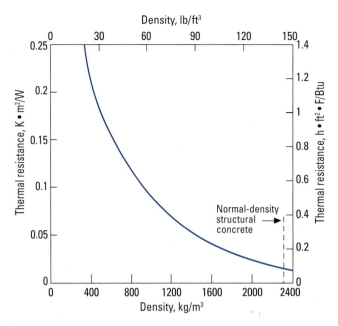

FIGURE 20-42. Thermal resistance of lightweight concrete versus density (PCA 1980).

Nondestructive Test Methods

Nondestructive tests (NDT) and other in-place tests can be used to evaluate the in-place concrete strength and locate internal defects in concrete members. The most widely used methods are the rebound hammer, probe penetration, pullout test, and several methods based on stress-wave propagation. Other techniques include: X-rays, gamma radiography, magnetic and eddy current cover meters, ground penetrating radar, and acoustic emissions. For more information on NDT and in-place methods used to evaluate quality of concrete, see ACI 228.1R (2019), ACI 228.2R (2013), Clifton (1985), Malhotra (1976 and 1984), and Malhotra and Carino (2004). Table 20-3 lists several NDT and in-place methods along with primary applications.

An NDT program may be undertaken for a variety of purposes regarding the strength or condition of hardened concrete, including:

- determination of in-place concrete strength,
- monitoring rate of concrete strength gain,
- location of defects, such as voids or honeycombing in concrete,
- determination of relative strength of comparable members,
- evaluation of concrete cracking and delaminations,
- evaluation of damage from mechanical or chemical actions,
- steel reinforcement location, size, and corrosion activity, and
- member dimensions.

Regardless of the type of NDT test used to estimate in-place strength, adequate and reliable correlation data with compressive strength data is necessary for a reliable estimate

TABLE 20-3. Nondestructive Test Methods for Concrete

PROPERTY OR CONDITION	PRIMARY NDT METHOD	SECONDARY NDT METHOD
In-place strength	Pullout test	Rebound hammer
	Pull-off test (for bond strength)	
	Probe penetration	
	Maturity method (new construction)	
General Quality and Uniformity	Visual inspection	Probe penetration
	Rebound hammer	
	Pulse velocity	
Thickness	Impact echo	Eddy-current thickness gage
	Ground penetrating radar	
	Ultrasonic echo	
Dynamic modulus of elasticity	Resonant frequency (on small specimens)	Pulse velocity
		Impact-echo
		Spectral analysis of surface waves
Density	Gamma radiometry (nuclear gauge)	
Reinforcement location	Covermeter	Radiography
	Ground penetrating radar	Ultrasonic echo
Reinforcement bar size	Covermeter	Radiography
Corrosion state of steel reinforcement	Half-cell potential	Radiography
	Polarization resistance	
Presence of near-surface defects	Sounding	Ground penetrating radar
	Infrared thermography	Radiography
Presence of internal defects	Impact echo, Ultrasonic echo, Impulse response	Pulse velocity
		Spectral analysis of surface waves, Radiography

of in-place strength. Correlation charts or graphs accompanying test instruments should not be relied upon unless they have been verified to provide accurate strength estimates for the concrete being evaluated. Verification of estimated in-place compressive strengths using drilled cores from one or two locations can provide guidance on the reliability of the NDT test results. The NDT method can then be used to survey larger portions of the structure. Care should be taken to consider the influence that other characteristic other than concrete strength can have on the NDT test results. Refer to ACI 228.1R-19 for guidance on developing correlations and how to use them for reliable estimates of in-place strength.

Rebound Hammer Tests. ASTM C805, *Standard Test Method for Rebound Number of Hardened Concrete*, is essentially a surface-hardness tester that provides a quick, simple means of checking concrete uniformity (Figure 20-43).

FIGURE 20-43. The rebound hammer gives an indication of the uniformity of concrete.

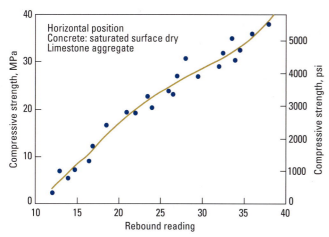

FIGURE 20-44. Example of a correlation curve for rebound number versus compressive strength.

It measures the rebound of a spring-loaded mass after it has struck a steel rod in contact with the concrete surface. The rebound number reading gives an indication of the relative compressive strength and elastic modulus of the concrete by use of a correlation curve (Figure 20-44). Two different concrete mixtures having the same strength, but different elastic moduli, will yield different readings. Carbonation of concrete will lead to higher rebound numbers, which may be mis-interpreted as concrete with higher quality than actual. An understanding of the factors influencing the rebound number is required.

Probe Penetration Tests. ASTM C803, *Standard Test Method for Penetration Resistance of Hardened Concrete*, is also called the Windsor probe test. The equipment consists of a powder-actuated gun that drives a hardened alloy probe into the concrete (Figure 20-45). The exposed length of the probe is measured and related to the compressive strength of the concrete by using a previously established correlation curve.

The results of the Windsor-probe test will be influenced greatly by the type of coarse aggregate used in the concrete. Therefore,

to improve accuracy of the estimated in place strength, a correlation curve for the particular concrete to be tested should be developed. This can be done using a cast slab for probe penetration tests and companion cores or cast cylinders for compressive strength (ACI 228.1R 2019).

When the probe is embedded into the concrete, a roughly conically-shaped region around the probe is subjected to extensive cracking and a small crater similar to a popout is created. Depending of the exposure of the test surface, the fractured concrete may have to be removed and the crater repaired.

Maturity Method. The principle of the maturity method is that strength gain of concrete is a function of time and temperature and samples of the same concrete will have equal strength if they have the same maturity index irrespective of their actual temperature histories. ASTM C1074, *Standard Practice for Estimating Concrete Strength by the Maturity Method*, provides procedures for measuring the maturity index of concrete based on the measured temperature history and using one of two maturity functions. ASTM C1074 also provides the procedure for developing the strength-maturity relationship for the specific concrete mixture to be used. Maturity meters include temperature probes embedded in the fresh concrete and control units that monitor and record the concrete temperature at regular intervals and calculate the maturity index. When a strength estimate is desired, the maturity index is read from the maturity meter and the strength is estimated from the strength-maturity relationship. ASTM C1074 also includes a procedure for establishing the most appropriate maturity function for the cementitious materials and admixtures in the concrete.

The maturity method assumes that the concrete in the structure has the same potential strength as the concrete used to develop the strength-maturity relationship in the laboratory. Thus, it is not able to account for errors in batching. For critical applications, such as formwork removal, the maturity method

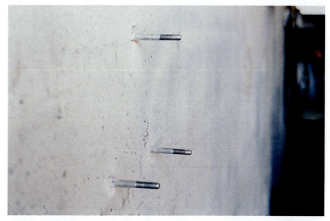

FIGURE 20-45. The Windsor-probe technique for determining the relative compressive strength of concrete. (left) Powder-actuated gun drives hardened alloy probe into concrete. (right) Exposed length of probe is measured and relative compressive strength of the concrete then determined from a correlation table.

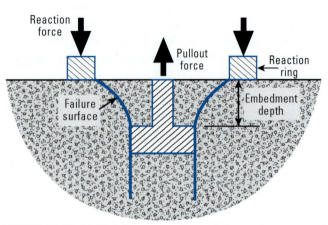

FIGURE 20-46. (right) Pullout test equipment being used to measure the in-place strength of concrete. (left) A schematic of the test configuration and the resulting failure surface (courtesy of N. Carino).

requires supplemental verification that the in-place concrete has the assumed strength potential. Methods for accomplishing this are provided in ASTM C1074.

Pullout Tests. ASTM C900, *Standard Test Method for Pullout Strength of Hardened Concrete*, involves casting the enlarged end of a steel insert in the concrete to be tested and then measuring the force required to pull it out (Figure 20-46). The applied tensile force reacts against a ring bearing on the surface of the concrete, which constrains the failure along a well-defined surface. The reaction ring affects the stresses in the concrete so that there is a band of high compression between the insert head and the bearing ring. As a result, the pullout strength has a strong correlation to compressive strength, and the relationship is little affected by factors such *w/cm* and type of materials used (Peterson 1997) (Figure 20-47). ASTM C900 also describes a procedure for performing pullout tests in existing concrete. In this case, special hardware is used to insert an expandable ring into an undercut created in the concrete by diamond studded router.

Pull-Off Tests. The pull-off test in accordance with ASTM C1583, *Standard Test Method for Tensile Strength of Concrete Surfaces and the Bond Strength or Tensile Strength of Concrete Repair and Overlay Materials by Direct Tension (Pull-off Method)*, can be used to evaluate the strength of a concrete surface before applying a repair material and for evaluating the bond strength of an overlay bonded to concrete. A metal disc is bonded to the surface using a fast-setting adhesive, a partial depth core is drilled to a specified depth, and the disc is pulled off by applying direct tensile force (Figure 20-48). Failure can occur in one of several potential failure planes: in the overlay, at the interface, or in the substrate. The failure load is divided by the cross-sectional area of the core and reported as the pull-off strength in MPa (psi).

Stress Wave and Vibration Tests. ASTM C597, *Standard Test Method for Pulse Velocity Through Concrete*, is based on the principle that the speed of sound (or stress waves) in a solid can be determined by measuring the travel time of short pulses of compressional stress waves through a test object. The pulse

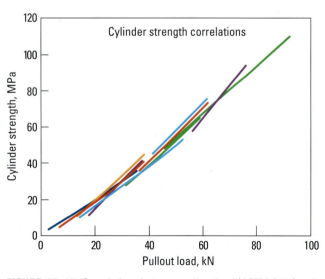
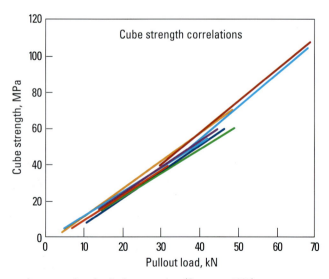

FIGURE 20-47. Correlations between pullout load (ASTM C900) and compressive strengths of cylinders or cubes (Petersen 1997).

FIGURE 20-48. Demonstration of pull-off test to measure the strength of the concrete substrate before application of an overlay or repair material (courtesy of N. Carino).

is introduced on one side of the test object by a transmitting transducer. The arrival of the pulse is measured by another transducer on the opposite side of the object. A control unit measures the time it takes for the pulse to travel between the two transducers. The distance between the transducers is measured and divided by the pulse travel time. The result is reported as the ultrasonic pulse velocity in m/s (ft/s), which in concrete typically ranges between 3500 m/s and 4500 m/s (11,500 ft/s to 15,000 ft/s). In general, high pulse velocities are indicative of sound concrete and low velocities are indicative of poor concrete. The pulse velocity, however, is affected greatly by the amount and type of aggregate, proximity of reinforcing steel, degree of saturation of the concrete, and to a minor degree by the w/cm. Therefore, care must be used when comparing the pulse velocities of different concretes. If internal defects (such as honeycombing voids or delaminations and cracking) are present in the test object, the actual pulse path length and pulse travel time will be increased. This results in a lower calculated pulse velocity. This test is useful for assessing the uniformity of the concrete in a structure.

ASTM C215, *Standard Test Method for Fundamental Transverse, Longitudinal, and Torsional Frequencies of Concrete Specimens*, covers the procedures to determine the fundamental resonant frequencies of prismatic and cylindrical test specimens. Resonant frequency is a function of the dynamic modulus of elasticity, Poisson's ratio, density, and geometry of the test specimen. Thus, if the geometry and mass of the specimen are known, the measured resonant frequencies can be used to calculate the dynamic modulus of elasticity and dynamic Poisson's ratio. In ASTM C215, two methods are presented to determine resonant frequency. In the *forced resonance method*, the specimen is forced to vibrate at continuously increasing frequencies until a maximum vibrational response is recorded.

In the *impact resonance method*, the specimen is struck with a small impactor and the resulting vibrational response is subjected to a frequency analysis. The fundamental transverse, longitudinal, and torsional frequencies of the test specimens are determined by changing the locations of the excitation and where the response is measured. Because the test method is nondestructive, it can be used to monitor changes in vibrational response as a function of time or as a function of the number of cycles of a disruptive action. Hence, resonant frequency testing is frequently used in laboratory durability tests such as freezing and thawing (ASTM C666 or AASHTO T 161).

Several other test methods based on stress waves (or sound) have been developed. ASTM D4580, *Standard Practice for Measuring Delaminations in Concrete Bridge Decks by Sounding*, employs simple hammer and chain drag soundings that are low-cost and accurate tests to identify delaminated areas of concrete. Hammer soundings can be used on either vertical or horizontal surfaces but are usually limited to small areas of delaminations. These areas are identified by striking the surface of the concrete with a hammer while listening for either a high-pitched ringing or low-pitched hollow sound. Another device uses bicycle sprocket wheels mounted on a handle to permit rapid sounding of vertical and overhead surfaces. Dragging either a single chain, in small areas, or for larger areas, a T-bar with four or more chains attached can be used to identify delaminated concrete. The sound emitted as the chains are dragged over the surface indicates whether the concrete is delaminated. Chain drag soundings are usually limited to horizontal surfaces that have a relatively rough texture. Smooth concrete may not bounce the chain links enough to generate adequate sound to detect delaminated areas. Note that corrosion of reinforcing bars in the area of delaminated concrete will probably extend beyond the boundary identified as delaminated.

A versatile technique based on stress-wave propagation is the impact-echo method. In this method, the surface of the test object is struck with a small impactor and a transducer located near the impact point monitors the arrival of the stress pulse after undergoing reflections from within the test object. The method is capable of detecting voids, cracks, and other defects within structural elements (Sansalone and Streett 1997). ASTM C1383, *Standard Test Method for Measuring the P-Wave Speed and the Thickness of Concrete Plates Using the Impact-Echo Method*, describes the use of the method to measure the thickness of plate-like concrete elements such as slabs, pavements, bridge decks, and walls. ASTM C1383 includes two procedures: Procedure A is used to measure the P-wave, and Procedure B uses that wave speed to measure the plate thickness. An advantage of the impact-echo method over the pulse velocity method is that it requires access to only one side of the structure and can determine the depth of internal defects.

Other stress-wave methods not covered in this review include: ultrasonic-echo, impulse response, and spectral analysis of surface waves (SASW). Information on these other methods may be found in Malhotra and Carino (2004) and ACI 228.2R-13.

Acoustic Emission Tests. Acoustic emission (AE) refers to the stress waves emitted from inside an object during a sudden release of stress. In concrete members, the most common source of acoustic emission is the formation of microcracks once the tensile capacity is reached. In acoustic emission monitoring, transducers are placed on the surface of the member, and these transducers record the arrival of stress waves caused by sudden release of stress. Compared with other stress-wave methods discussed previously, AE is a passive monitoring method as opposed to an active method for probing the interior conditions of a member. Common applications of AE include monitoring the response of a structure during load testing and detecting breaks in post-tensioning tendons. By using triangulation methods, it may be possible to estimate the location of the AE source within the member. Additional information on acoustic emission may be found in Malhotra and Carino (2004).

Nuclear (Radioactive) Tests. Nuclear (or radioactive) methods involve the use of high-energy electromagnetic radiation to gain information about the density or internal structure of the test object. These involve a source of penetrating electromagnetic radiation (x-rays or gamma rays) and a sensor to measure the intensity of the radiation after it has traveled through the object. If the sensor is in the form of photographic plate, the technique is called *radiography*. If the sensor is an electronic device that converts the incident radiation into electrical pulses, the technique is called *radiometry*. The use of x-rays for evaluating concrete members is limited due to the costly and potentially dangerous high-voltage equipment required as well as radiation hazards.

Gamma-radiography equipment can be used in the field to determine the location of reinforcement, honeycombing, and voids in structural concrete members (Mariscotti and others 2009). ASTM C1040 (AASHTO T 271) uses gamma radiometry to determine the density of fresh and hardened concrete in place. Users of nuclear-based methods require special training and certifications to ensure radiological safety.

Covermeters. Battery-operated devices, known as covermeters, are available to measure the depth of reinforcement in concrete and to estimate bar size. There are two types of covermeters: those based on magnetic reluctance and those based on eddy currents (ACI 228.2R 2013). Magnetic covermeters will only detect embedded magnetic materials, such as reinforcing steel. Eddy-current meters will detect any type of electrically conductive metal. Sophisticated instruments and software have been developed that permit reconstructing two-dimensional or three-dimensional images of the layout of embedded reinforcement. An eddy-current device has been developed to measure the thickness of concrete pavement slabs. This is done by placing a metal plate on the subgrade before placing the concrete. After the concrete has gained sufficient strength, a sensing head is passed over the plate and the measured response is converted to a slab thickness.

Ground-Penetrating Radar. Ground-penetrating (or short-pulse) radar (GPR) is analogous to sonar except that a pulse of electromagnetic energy is used instead of sound. An antenna placed on the surface sends out a short pulse of radiation in the microwave region of the electromagnetic spectrum. The pulse travels through the test object and a portion of the incident radiation is reflected when the pulse encounters any interfaces between materials of different electrical properties. At a concrete-metal interface there is 100% reflection of energy in the pulse. At a concrete-air interface only about 50% of the incident energy is reflected. GPR is effective in locating embedded metal and can probe deeper than electrical covermeters. It is also useful for evaluating grouting in reinforced concrete masonry construction. There are two test methods for evaluating bridge decks and pavements using GPR. ASTM D4748, *Standard Test Method for Determining the Thickness of Bound Pavement Layers Using Short-Pulse Radar*, can be used to measure the thickness of the upper layer of a multilayer pavement system. ASTM D6087, *Standard Test Method for Evaluating Asphalt-Covered Concrete Bridge Decks Using Ground Penetrating Radar*, can be used to evaluate the presence of delaminations due to corrosion of the top layer of reinforcement in concrete bridge decks.

Infrared Thermography. Infrared thermography (IRT) is a technique for locating near-surface defects by measuring surface temperature. It is based on two principles: 1) a surface emits infrared radiation with an intensity that depends on its temperature, and 2) the presence of a near-surface air gap will interfere with heat flow and alter the surface-temperature distribution. By measuring the surface temperature distribution using an infrared camera, the presence of a near surface defect can be inferred. This technique requires heat flow either into or out of the surface of the test object. This is accomplished by artificial heating with lamps, by solar radiation, or nighttime cooling. ASTM D4788, *Standard Test Method for Detecting Delaminations in Bridge Decks Using Infrared Thermography*, describes the application of IRT for performing a delamination survey of bridge decks. Another potential application is in the evaluation of structural repairs made using bonded fiber-reinforced polymer (FRP) sheets. It has been shown that active infrared thermography, in which heating lamps are used to create a transient heat flow condition, is effective in locating voids that may exist between FRP plies or at the FRP-concrete interface (Starnes and others 2003 and Levar and Hamilton 2003).

REFERENCES

PCA's online catalog includes links to PDF versions of many of our research reports and other classic publications.
Visit: cement.org/library/catalog.

Aarre, T., "Air-Void Analyzer," *Concrete Technology Today*, PL981, Portland Cement Association, Skokie, IL, April 1998, page 4.

ACI Committee 214, *Guide to Evaluation of Strength Test Results of Concrete*, ACI 214R-11, American Concrete Institute, Farmington Hills, MI, 2011, 16 pages.

ACI Committee 222, *Guide to Protection of Metals in Concrete Against Corrosion*, ACI 222R-19, American Concrete Institute, Farmington Hills, MI, 2019, 60 pages.

ACI Committee 228, *Report on Methods for Estimating In-Place Concrete Strength*, ACI 228.1R-19, American Concrete Institute, Farmington Hills, MI, 2019, 48 pages.

ACI Committee 228, *Report on Nondestructive Test Methods for Evaluation of Concrete in Structures*, ACI 228.2R-13, American Concrete Institute, Farmington Hills, MI, 2013, 82 pages.

ACI Committee 318, *Building Code Requirements for Structural Concrete and Commentary*, ACI 318-19, American Concrete Institute, Farmington Hills, MI, 2019, 629 pages.

ACI Committee 329, *Report on Performance-Based Requirements for Concrete*, ACI 329R-14, American Concrete Institute, Farmington Hills, MI, 2014, 46 pages.

ACI Committee 364, *TechNote: How to Measure pH of a Concrete Surface Prior to Installation of a Floor Covering*, ACI 364.17T-18, American Concrete Institute, Farmington Hills, MI, 2018, 5 pages.

American Petroleum Institute, *Recommended Practice for Determining Permeability of Porous Media*, API RP 27, American Petroleum Institute, Washington, D.C., 1956.

Angles, J., "Measuring Workability," *Concrete*, 1974, page 26.

ASTM, *Manual of Aggregate and Concrete Testing*, R0030, ASTM International, West Conshohocken, PA, 2019, 45 pages.

ASTM, *Manual of Cement Testing*, R0028, ASTM International, West Conshohocken, PA, 2017, 19 pages.

Bartos, P., "Workability of Flowing Concrete – Assessment by a Free Orifice Rheometer," *Concrete*, 1978, pages 28 to 30.

Bureau of Reclamation, *Concrete Manual*, 8th Edition, Denver, CO, 1981, page 11.

Burg, R.G.; Caldarone, M.A.; Detwiler, G.; Jansen, D.C.; and Willems, T.J., "Compression Testing of HSC: Latest Technology." *Concrete International*, American Concrete Institute, Farmington Hills, MI, August 1999, pages 67 to 76.

Burg, R.G.; and Ost, B.W., *Engineering Properties of Commercially Available High-Strength Concrete (Including Three-Year Data)*, RD104, Research and Development Bulletin, Portland Cement Association, Skokie, IL, 1994, 58 pages.

Campbell, D.H.; Sturm, R.D.; and Kosmatka, S.H., "Detecting Carbonation," *Concrete Technology Today*, PL911, Portland Cement Association, Skokie, IL, March 1991, pages 1 to 5.

Carino, N.J., "Prediction of Potential Concrete Strength at Later Ages," Chapter 14 in *Significance of Tests and Properties of Concrete and Concrete-Making Materials*, STP 169D, ASTM International, West Conshohocken, PA, 2006, pages 141 to 153.

Carino, N.J.; Guthrie, W.F.; and Lagergren, E.S., "Effects of Testing Variables on the Measured Compressive Strength of High-Strength (90 MPa) Concrete," ACI Special Publication SP-149, *American Concrete Institute*, Farmington Hills, MI, 2004, pages 589 to 632.

Clifton, J.R., "Nondestructive Evaluation in Rehabilitation and Preservation of Concrete and Masonry Materials," SP-85-2, *Rehabilitation, Renovation, and Preservation of Concrete and Masonry Structures*, ACI Special Publication SP-85, American Concrete Institute, Farmington Hills, MI, 1985, pages 19 to 29.

Cook, M.D.; Ghaeezadah; and Ley, M.T., "A Workability Test for Slip Formed Concrete Pavements," *Construction and Building Materials*, 68, 2014, pages 376 to 383.

Crandall, J.R.; and Blaine, R.L., "Statistical Evaluation of Interlaboratory Cement Tests," *Proceedings, American Society for Testing and Materials*, Vol. 59, 1959, pages 1129 to 1154.

CSA Standard A23.1-09/A23.2-09, *Concrete Materials and Methods of Concrete Construction/ Test Methods and Standard Practices for Concrete*, Canadian Standards Association, Toronto, Canada, 2009.

Daniel, D.G., "Factors Influencing Concrete Workability," Chapter 8 in *Significance of Tests and Properties of Concrete and Concrete-Making Materials*, STP 169D, ASTM International, West Conshohocken, PA, 2006, pages 59 to 72.

Date, C.G.; and Schnormeier, R.H., "Day-to-Day Comparison of 4 and 6 Inch Diameter Concrete Cylinder Strengths," *Concrete International*, American Concrete Institute, Farmington Hills, MI, August 1984, pages 24 to 26.

de Larrard, F.; Szitkar, J.; Hu, C.; and Joly, M., "Design and Rheometer for Fluid Concretes," *Proceedings, International RILEM Workshop*, Paisley, Scotland, March 2-3, 1993, pages 201 to 208.

Detwiler, R.J.; Thomas, W.; Stangebye, T.; and Urahn, M., "Variability of 4×8 in. Cylinder Tests," *Concrete International*, American Concrete Institute, Farmington Hills, MI, May 2001.

Distlehorst, J.A.; and Kurgan, G. J., *Development of Precision Statement for Determining Air Void Characteristics of Fresh Concrete with Use of Air Void Analyzer*, Transportation Research Record Number 2020, Transportation Research Board, Washington, D.C., 2007.

Eisenhart, C., "Realistic Evaluation of the Precision and Accuracy of Instrument Calibration Systems," *Journal of Research of the National Bureau of Standards-C.*, Vol. 67C, No. 2, 1963. (Reprinted in Ku and others, 1969.)

Erlin, B., "Petrographic Examination," Chapter 20 in *Significance of Tests and Properties of Concrete and Concrete-Making Materials*, STP 169D, ASTM International, West Conshohocken, PA, 2006, pages 207 to 244.

FDOT, *Florida Method of Test for Concrete Resistivity as an Electrical Indicator of its Permeability*, FM 5-578, Florida Department of Transportation, January 27, 2004.

Fiorato, A.E.; Burg, R.G.; and Gaynor, R.D., "Effects of Conditioning on Measured Compressive Strength of Concrete Cores," *Concrete Technology Today*, CT003, Portland Cement Association, Skokie, IL, 2000, pages 1 to 3.

Forester, J.A.; Black, B.F.; and Lees, T.P., *An Apparatus for Rapid Analysis Machine, Technical Report*, Cement and Concrete Association, Wexham Springs, Slough, England, April 1974.

Forstie, D.A.; and Schnormeier, R., "Development and Use of 4 by 8 Inch Concrete Cylinders in Arizona," *Concrete International*, American Concrete Institute, Farmington Hills, MI, July 1981, pages 42 to 45.

Gagné, R.; and Marchand, J., "La résistance à l'écaillage des bétons à haute performance : état de la question," *Atelier international sur la résistance des bétons aux cycles de gel-dégel en présence de sels fondants, CRIB/RILEM*, 30-31 août, Sainte-Foy, 1993, pages 21 to 48.

Gebler, S.H.; and Klieger, P., *Effects of Fly Ash on the Air-Void Stability of Concrete*, RD085, Research and Development Bulletin, Portland Cement Association, Skokie, IL, 1983.

Graves, R.E., "Grading, Shape, and Surface Texture," *Significance of Tests and Properties of Concrete and Concrete-Making Materials*, STP 169D, ASTM International, West Conshohocken, PA, 2006, pages 337 to 345.

Grubb, J.A.; Limaye, H.S.; and Kakade, A.M., "Testing pH of Concrete," *Concrete International*, April 2007, pages 78 to 83.

Guthrie, G.D.; and Carey, J.W., "A Simple Environmentally Friendly, and Chemically Specific Method for the Identification and Evaluation of the Alkali-Silica Reaction," *Cement and Concrete Research*, Vol. 27, No. 9, 1997, pages 1407 to 1417.

Hedenblad, G., *Drying of Construction Water in Concrete – Drying Times and Moisture Measurement*, LT229, Swedish Council for Building Research, Stockholm, 1997, 54 pages.

Hedenblad, G., "Concrete Drying Time," *Concrete Technology Today*, PL982, Portland Cement Association, 1998, pages 4 and 5.

Hime, W.G., "Analyses for Cement and Other Materials in Hardened Concrete," Chapter 27 in *Significance of Tests and Properties of Concrete and Concrete-Making Materials*, STP 169D, ASTM International, West Conshohocken, PA, 2006, pages 309 to 313.

Hime, W.G.; Mivelaz, W.F.; and Connolly, J.D., *Use of Infrared Spectrophotometry for the Detection and Identification of Organic Additions in Cement and Admixtures in Hardened Concrete*, RX194, Research Department Bulletin, Portland Cement Association, 1966, 22 pages.

Hover, K.C.; Bickley, J.; and Hooton, R. D., *Guide to Specifying Concrete Performance: Phase II Report of Preparation of a Performance-Based Specification For Cast-in-Place Concrete*, RMC Research and Education Foundation, Silver Spring, MD, USA, March 2008, 53 pages.

Janssen, D.J.; and Snyder, M.B., *Resistance of Concrete to Freezing and Thawing*, SHRP-C-391, Strategic Highway Research Program, Washington, D.C., 1994, 201 pages.

Kakade, A.M., "Measuring Concrete Surface pH-A Proposed Test Method," *Concrete Repair Bulletin*, March/April 2014, pages 16 to 20.

Kanare, H.M., *Concrete Floors and Moisture*, 2nd edition, EB119, Portland Cement Association, Skokie, IL, and National Ready Mixed Concrete Association, Silver Spring, MD, 2009, 176 pages.

Koehler, E.P.; and Fowler, D.W., "Development of a Portable Rheometer for Fresh Portland Cement Concrete," *International Center for Aggregate Research*, Report, ICAR 105-3F, Austin, TX, 2004, 321 pages.

Kosmatka, S.H., "Compressive versus Flexural Strength for Quality Control of Pavements," *Concrete Technology Today*, PL854, Portland Cement Association, 1985a, pages 4 and 5.

Ku, H.H., editor. *Precision Measurement and Calibration: Selected NBS Papers on Statistical Concepts and Procedures*, NBS Special Publication 300, vol. 1, 1969.

Lamond, J.F.; and Pielert, J.H., *Significance of Tests and Properties of Concrete and Concrete-Making Materials*, STP 169D, ASTM International, West Conshohocken, PA, 2006, 664 pages.

Levar, J.M.; and Hamilton, H.R., "Nondestructive Evaluation of Carbon Fiber-Reinforced Polymer-Concrete Bond Using Infrared Thermography," *ACI Materials Journal*, Vol. 100, No. 1, January-February 2003, pages 63 to 72.

Lobo, C.L.; Mullings, G.M.; and Gaynor, R.D., "Effects of Capping Materials and Procedures on the Compressive Strength of 100×200 mm (4×8 in.) High Strength Concrete Cylinders," *Cement, Concrete, and Aggregates*, Vol. 16, No. 2, December 1994, pages 173 to 180.

Malhotra, V.M., *Testing Hardened Concrete*, Nondestructive Methods, ACI Monograph No. 9, American Concrete Institute-Iowa State University Press, Farmington Hills, MI, 1976.

Malhotra, V.M., *In Situ/Nondestructive Testing of Concrete*, ACI Special Publication SP-82, American Concrete Institute, Farmington Hills, MI, 1984.

Malhotra, V.M.; and Carino, N.J., editors. *Handbook on Nondestructive Testing of Concrete*, 2nd Edition, CRC Press, Boca Raton, Florida, and ASTM International, West Conshohocken, PA, 2004.

Marchand, J.; Sellevold, E.J.; and Pigeon M., "The Deicer Salt Scaling Deterioration of Concrete – An Overview," *Proceedings of the Third CANMET/ACI International Conference on Durability of Concrete*, Nice, France, 1994, pages 1 to 46.

Mariscotti, M.A.J.; Jalinoos, F.; Frigerio, T.; Ruffolo, M.; and Thieberger, P., "Gamma-Ray Imaging for Void and Corrosion Assessment," *Concrete International*, Vol. 31, No. 11, November 2009, pages 48 to 53.

Midgley, H.G., "The Staining of Concrete by Pyrite," *Magazine of Concrete Research*, Vol. 10, No. 29, August 1958, pages 75 to 78.

Mielenz, R.C., *Reactions of Aggregates Involving Solubility, Oxidation, Sulfates or Sulfides*, Research Report No. 43, Highway Research Board, Washington, D.C., 1964, pages 8 to 18.

Moore, C.W., *Control of Portland Cement Quality*, EB121, Portland Cement Association, Skokie, Illinois, 2007, 172 pages.

Mor, A.; and Ravina, D., "The DIN Flow Table," *Concrete International*, American Concrete Institute, Farmington Hills, MI, December 1986.

Murphy, R.B., "On the Meaning of Precision and Accuracy," *Materials Research and Standards*, American Society for Testing and Materials, April, 1961. (Reprinted in Ku and others, 1969).

Natesaiyer, K.; and Hover, K.C., "Insitu Identification of ASR Products in Concrete," *Cement and Concrete Research*, Vol. 18, No. 3, 1988, pages 455 to 463.

Nmai, C.K, "Freezing and Thawing," *Significance of Tests and Properties of Concrete and Concrete-Making Materials*, ASTM STP 169D, edited by Lamond, J.F. and Pielert, J.H., American Society for Testing and Materials, PA, 2006, pages 154 to 163.

NRMCA, *In-Place Concrete Strength Evaluation – A Recommended Practice*, NRMCA Publication 133, revised 1979, National Ready Mixed Concrete Association, Silver Spring, MD, 1979.

NRMCA, "Standard Practice for Rapid Determination of Water Soluble Chloride in Freshly Mixed Concrete, Aggregate and Liquid Admixtures," *NRMCA Technical Information Letter No. 437*, National Ready Mixed Concrete Association, March 1986.

Ozyildirim, C.; and Carino, N.J., "Concrete Strength Testing," Chapter 13 in *Significance of Tests and Properties of Concrete and Concrete Making Materials*, STP 169D, ASTM International, West Conshohocken, PA, 2006, pages 125 to 140.

Obla, K.H.; Werner, O.R.; Hausfeld, J.L.; MacDonald, K.A.; Moody, G.D.; and Carino, N.J., "Who is Watching Out for the Cylinders?" *Concrete International*, Vol. 40, No. 8, August 2018.

Parry, J.M., *Wisconsin Department of Transportation QC/QA Concept*, Portland Cement Concrete Technician I/IA Course Manual, Wisconsin Highway Technician Certification Program, Platteville, Wisconsin, 2000, pages B-2 to B-4.

PCA, "Rapid Analysis of Fresh Concrete," *Concrete Technology Today*, PL832, Portland Cement Association, Skokie, IL, June 1983, pages 3 and 4.

Petersen, C.G., "Air Void Analyser (AVA) for Fresh Concrete, Latest Advances," *Ninth ACI International Conference on Superplasticizers and Other Chemical Admixtures in Concrete*, Seville, Spain, October 2009.

Petersen, C.G., "LOK-Test and CAPO-Test Pullout Testing: Twenty Years Experience," *Conference on Non-Destructive Testing in Civil Engineering*, British Institute of Non-Destructive Testing, Liverpool, UK, April 1997.

Pfeifer, D.W.; and Scali, M.J., *Concrete Sealers for Protection of Bridge Structures*, NCHRP Report 244, Transportation Research Board, National Research Council, 1981.

Pinto, R.C.A.; and Hover, K. C., *Frost and Scaling Resistance of High-Strength Concrete*, RD122, Research and Development Bulletin, Portland Cement Association, Skokie, IL, 2001, 70 pages.

Pistilli, M.F.; and Willems, T., "Evaluation of Cylinder Size and Capping Method in Compression Testing of Concrete," *Cement, Concrete and Aggregates*, ASTM International, West Conshohocken, PA, Summer 1993.

Powers, L.J., "Developments in Alkali-Silica Gel Detection," *Concrete Technology Today*, PL991, Portland Cement Association, Skokie, IL, April 1999, pages 5 to 7.

Powers, T.C., "Studies of Workability of Concrete," *Journal of the American Concrete Institute*, Volume 28, American Concrete Institute, Farmington Hills, MI, 1932, page 419.

Powers, T.C.; and Wiler, E.M., "A Device for Studying the Workability of Concrete," *Proceedings of ASTM*, Vol. 41, American Society for Testing and Materials, Philadelphia, PA, 1941.

Sansalone, M.; and Streett, W.B., *Impact-Echo: Nondestructive Evaluation of Concrete and Masonry*, Bullbrier Press, Jersey Shore, PA, 1997.

Saucier, K.L., *Investigation of a Vibrating Slope Method for Measuring Concrete Workability*, Miscellaneous Paper 6-849, U.S. Army Engineer Waterways Experiment Station, Vicksburg, Mississippi, 1966.

Shewhart, W.A.; and Deming, W.E., *Statistical Method from the Viewpoint of Quality Control*, Graduate School of the Dept. of Agriculture, Washington, D.C., 1939, 155 pages.

Smith, D., *The Development of a Rapid Test for Determining the Transport Properties of Concrete*, SN2821, Portland Cement Association, Skokie, IL, 2006, 125 pages.

St. John, D.A.; Poole, A.W.; and Sims, I., *Concrete Petrography – A Handbook of Investigative Techniques*, Arnold, London, 485 pages.

Starnes, M.A.; Carino, N.J.; and Kausel, E.A., "Preliminary Thermography Studies for Quality Control of Concrete Structures Strengthened with FRP Composites," *ASCE Journal of Materials in Civil Engineering*, Vol. 15, No. 3, May/June 2003, pages 266 to 273.

Steinour, H.H., *Influence of the Cement on Corrosion Behavior of Steel in Concrete*, RX168, Research Department Bulletin, Portland Cement Association, Chicago, IL, 1964, 22 pages.

Tanesi, J.; Kim, H.; Beyene, M.; and Ardani, A., "Super Air Meter for Assessing Air-Void System of Fresh Concrete," *Advances in Civil Engineering Materials*, Vol. 5, No. 2, 2016, pages 22 to 37.

Tarr, S.M.; and Farny, J.A., *Concrete Floors on Ground*, EB075, Portland Cement Association, Skokie, IL, 2008, 256 pages.

Tattersall, G.H., *Measurement of Workability of Concrete*, East Midlands Region of the Concrete Society of Nottingham, England, 1971.

Teranishs, K.; Watanabe, K.; Kurodawa, Y.; Mori, H.; and Tanigawa, Y., "Evaluation of Possibility of High-Fluidity Concrete," *Transactions of the Japan Concrete Institute (16)*, Japan Concrete Institute, Tokyo, 1994, pages 17 to 24.

Thomas, M.; Fournier, B.; Folliard, K.; Ideker, J.; and Shehata, M., *Test Methods for Evaluating Preventive Measures for Controlling Expansion Due to Alkali-Silica Reaction in Concrete*, ICAR 302-1, International Center for Aggregates Research, 2006, 62 pages.

Thomas, M.D.A.; Fournier, B.; and Folliard, K.J., *Report on Determining the Reactivity of Concrete Aggregates and Selecting Appropriate Measures for Preventing Deleterious Expansion in New Concrete Construction*, FHWA-HIF-09-001, Federal Highway Administration, Washington, D.C., April 2008, 28 pages.

Tyler, I.L.; and Erlin, B., *A Proposed Simple Test Method for Determining the Permeability of Concrete*, RX133, Research Department Bulletin, Portland Cement Association, Chicago, IL, 1961.

Verbeck, G.J., *Carbonation of Hydrated Portland Cement*, RX087, Research Department Bulletin, Portland Cement Association, Chicago, IL, 1958.

Wallevik, O., "The Use of BML Viscometer for Quality Control of Concrete," *Concrete Research*, Nordic Concrete Research, Espoo, Finland, 1996, pages 235 to 236.

Weiss, W.J.; Barrett, T.J.; Qiao, C; and Todak, H, "Toward a Specification for Transport Properties of Concrete Based on the Formation Factor of a Sealed Specimen," *Advances in Civil Engineering Materials*, Vol. 5, Issue 1, July 2016.

Weiss, W.J.; Yang, W.; and Shah, S.P., "Influence of Specimen Size and Geometry on Shrinkage Cracking," *Journal of Engineering Mechanics Division*, American Society of Civil Engineering, Vol. 126, No. 1, pages 93 to 101.

Welchel, D., *Determining the Air Void Distribution of Fresh Concrete with the Sequential Pressure Method*, MS Thesis, Oklahoma State University, Stillwater, OK, December 2014.

Whiting, D., *Rapid Determination of the Chloride Permeability of Concrete*, FHWA-RD-81-119, Federal Highway Administration, Washington, D.C., 1981.

Whiting, D., "Permeability of Selected Concretes," *Permeability of Concrete*, SP-108, American Concrete Institute, Farmington Hills, MI, 1988.

Wigmore, V.S., "Consistometer," *Civil Engineering* (London), Vol. 43, No. 510, December 1948, pages 628 to 629.

Wong, G. S.; Alexander, A. M.; Haskins, R.; Poole, T.S.; Malone, P.G.; and Wakeley, L., *Portland-Cement Concrete Rheology and Workability: Final Report*, FHWA – RD-00-025, Federal Highway Administration, Washington, D.C., 2001, 117 pages.

Wood, S.L., *Evaluation of the Long-Term Properties of Concrete*, RD102, Research and Development Bulletin, Portland Cement Association, Skokie, IL, 1992, 99 pages.

Youden, W.J., "Statistical Aspects of the Cement Testing Program", *Proceedings, American Society for Testing and Materials*, Vol. 59, 1959, pages 1120 to 1128.

PAVING

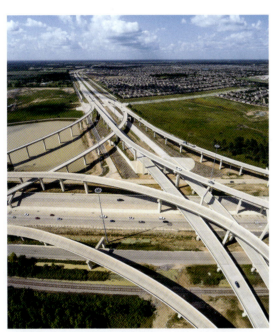

FIGURE 21-1. Concrete pavements are long life, low maintenance, and low life-cycle cost providing a safe, durable, and more fuel-efficient transportation network.

Concrete pavements have been a mainstay of infrastructure since the 1920s. In the United States, one of the first sections of concrete pavement was placed in Bellefontaine, Ohio in 1891. This road is still in use today with access to light traffic.

It is estimated that there are approximately 690,000 lane kilometers (430,000 lane miles) of concrete pavements in the U.S. Concrete pavements generally provide the longest life, least maintenance, and lowest life-cycle cost of all alternatives.

PAVEMENT TYPES

Overview

Concrete can be used for new pavements (Figure 21-1), reconstruction, resurfacing, restoration, or rehabilitation. Fundamentally, a concrete pavement needs to be designed to withstand the fatigue caused by traffic loads and the environment in which it is constructed. Pavement design includes the selection of slab thickness, joint dimensions, reinforcement and load-transfer requirements, and other features including durability requirements, constructability issues, and target service life. Both traditional and innovative approaches can be used and include the following pavement types:

- Jointed plain concrete pavement
- Continuously reinforced concrete pavement
- Concrete overlays
- Pervious concrete pavement
- Precast concrete pavement
- Roller-compacted concrete pavement

Jointed Plain Concrete Pavement

Jointed plain concrete pavements (JPCP) have regularly spaced transverse sawn contraction joints that allow the concrete to move longitudinally with changes in temperature and moisture as shown in Figure 21-2 (IMCP 2019).

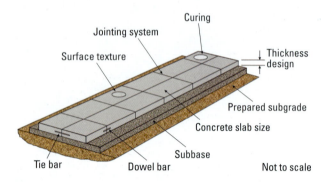

FIGURE 21-2. Typical JPCP design details (courtesy of ACPA).

Vertical movement between the transverse joints (joint faulting) is typically controlled by smooth dowels, which assist with transferring load between slabs, reducing the required slab thickness. The dowels are purposely not bonded to the concrete, so they allow horizontal movements. Longitudinal joints are most often held together with deformed tie bars that prevent the slabs from moving apart in the transverse direction. Joint spacing is based on slab thickness and the potential for curling (due to differential temperatures) and warping (due to differential moisture). The fundamental aim is to force cracks to form in controlled locations (contraction joints) rather than in random locations. JPCP transverse contraction joints are typically spaced at 4.6 m to 6.1 m (15 ft to 20 ft) in order to control cracking. Longitudinal construction joints are placed between lanes and shoulders.

Jointed plain concrete pavement construction is the most common type of concrete pavement, useful in most applications. Concrete pavement thickness is not significantly affected by the stiffness of the foundation system. The critical characteristics of a good base and subgrade are uniform and stable support with adequate drainage.

Critical design inputs for calculating thickness include the estimated traffic loading during the design life, failure criteria, concrete strength, load transfer between panels, and drainage characterization of supporting layers. For design of JPCP, many agencies use the American Association of State Highway and Transportation Officials (AASHTO) *Design Guide* (1993), although adoption of a *Mechanistic-Empirical Pavement Design Guide* (AASHTO 2020b) is increasing as states develop local calibrations. For more information on JPCP, see IMCP (2019), PavementDesigner.org (2018), Garber and others (2011), and AASHTO (1993).

Continuously Reinforced Concrete Pavement

Continuously reinforced concrete pavements (CRCP) contain continuous longitudinal and transverse reinforcement through the entire pavement. The reinforcement is designed to control transverse crack widths and hold cracks tightly together and is not designed to help carry traffic loads. As a result, CRCP systems do not include transverse contraction joints, but do include transverse and longitudinal construction joints. Longitudinal reinforcement is typically 0.70% to 0.75% of the cross-sectional area (lower in milder climates, higher in harsher climates). Transverse cracks should form at a spacing of about 0.6 m to 1.8 m (2 ft to 6 ft) as shown in Figure 21-3, and are generally narrow (about 0.5 mm [0.02 in.]), reducing the risk of contaminants and aggressive solutions from penetrating the system (IMCP 2019).

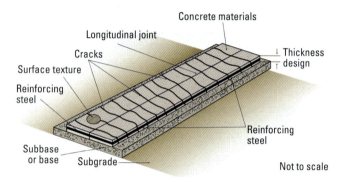

FIGURE 21-3. Typical CRCP details (courtesy of ACPA).

In addition to slab thickness and width, CRCP design involves selecting longitudinal and transverse reinforcement in order to provide desired performance for the design life. Factors affecting steel details include: crack spacing, crack width, steel stress, and bond development length. The most common type of reinforcement is deformed steel bars with a yield strength of about 420 MPa (60,000 psi). Epoxy or other protective coatings for reinforcement may be specified if the environment will promote corrosion. Coated reinforcement will reduce bond with the paste, potentially affecting the amount of steel required.

The free end (terminal) of a CRCP slab may move up to 50 mm (2 in.) due to environmental variation, depending on the nature of the underlying layer. Terminal joints are therefore placed where a CRCP butts against a structure or another pavement that will allow free horizontal movement of the slab while maintaining vertical alignment.

Detailed information on CRCP pavements is given in PavementDesigner.org (2018), Rasmussen (2011), and Roesler and others (2016).

BONDED OVERLAY OPTION _(Preventive Maintenance/Minor Rehabilitation)_	**UNBONDED OVERLAY OPTION** _(Minor/Major Rehabilitation)_
In general, bonded resurfacing is used to eliminate surface distress when the existing pavement is in good structural condition.	In general, unbonded resurfacing is highly reliable, with longer life than rehabilitation with asphalt.
Bonding is essential, so thorough surface preparation is necessary before resurfacing.	Minimal presurfacing repairs are necessary for unbonded resurfacing.

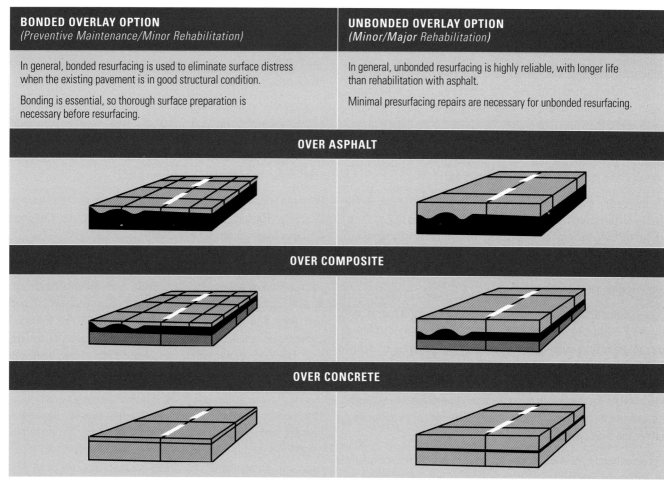

OVER ASPHALT

OVER COMPOSITE

OVER CONCRETE

FIGURE 21-4. Bonded and unbonded overlay options (adapted from Harrington 2014).

Concrete Overlays

Concrete overlays are a means of extending or improving the life of a pavement, while making the most use of the system already in place. Concrete overlays are able to restore the structural capacity and functional characteristics of a pavement and can be designed to deliver a range of service life, depending on the needs. The construction of a new surface results in improved surface characteristics including rideability, noise levels, and friction.

Concrete overlays may be classified as bonded and unbonded (Figure 21-4). Both overlay types can be constructed over concrete, flexible, and composite pavements. The choice between using a bonded or unbonded overlay is strongly dependent on the condition of the existing pavement. The pavement must be carefully reviewed to assess the amount of damage present, whether it can be readily repaired before overlaying, and whether the underlying cause of failure (such as alkali-silica reaction) will continue. These factors must be balanced with a decision on the desired service life of the overlay.

Bonded Overlays. Bonded overlays are relatively thin and constructed directly on top of existing pavements. The composite action of the new and old pavements means that less

thickness is required, but the condition of the existing paving system must be moderate or better. The existing surface should be roughened using shotblasting or milling. A bonding grout or epoxy is not required. Following surface preparation, the surface should be thoroughly cleaned and dried.

The design of a bonded concrete overlay is based on the assumption that the overlay and existing pavement will be one monolithic structure. Typical thicknesses range from 50 mm to 125 mm (2 in. to 5 in.). For high-traffic roads, a 150 mm (6 in.) thick, or greater, bonded overlay can be constructed.

Joints in the existing concrete pavement must be matched in the overlay. Transverse joints in bonded overlays of concrete pavements should be full depth plus 12.5 mm (0.5 in.). Similarly, longitudinal joints are cut either full depth or no less than half of the pavement thickness. Transverse and longitudinal joints in bonded overlays of hot-mix asphalt (HMA) and composite pavements should be cut to a depth of not less than one-third of the pavement thickness (T/3).

Reinforcement such as tie bars, dowel bars, and other embedded steel products are typically not used for overlays less than 150 mm (6 in.) thick. Continuously reinforced bonded overlays have been constructed, and in these cases steel is placed at sufficient depth to provide a minimum of 75 mm (3 in.) of concrete cover.

Unbonded Overlays. Unbonded concrete overlays are typically thicker than bonded overlays and often require a separation layer. Unbonded overlays can be placed over existing pavements that are moderately to significantly deteriorated. The separation layer minimizes reflection of existing pavement damage to the overlay, but also means that the new slab has to be thick enough to carry the loads on its own.

An unbonded overlay design assumes there is little to no bond between the bottom surface of the overlay and the top surface of the existing pavement. The design thickness of an unbonded overlay typically ranges from 150 mm to 275 mm (6 in. to 11 in.), but can be as thin as 100 mm (4 in.) for lower-volume roads or when loading is light.

If a full-depth concrete patch has been constructed in an existing asphalt or composite pavement to be overlaid, it must be isolated. One technique for isolating the patch is to apply a debonding agent or material (for example, asphalt emulsion coating) to the surface of the patch before the construction of the overlay. If a geotextile is used as the separation layer, it must be connected to a drainage system to prevent water collecting under the new concrete.

For an unbonded overlay of concrete, designers may intentionally place transverse joints in the overlay at an offset from those in the existing pavement. Joint spacings are designed based on the thickness of the overlay. Unbonded overlays of concrete, HMA, or composite pavements should include transverse and longitudinal joints that, when using conventional saws, should be cut to a depth of at least one-third of the pavement thickness (T/3).

Dowels are used in joints when the overlay's design thickness is greater than 200 mm (8 in.) and expected to carry heavy truck traffic.

There are many tools available for designing concrete overlays including: PavementDesigner.org (2018); *Bonded Concrete Overlay on Asphalt (BCOA) Thickness Designer* (ACPA 2012) and *BCOA-ME Design Guide* (Vandenbossche 2013); and *Mechanistic-Empirical Pavement Design Guide* (AASHTO 2020).

In general, a well-designed and well-constructed bonded concrete overlay requires relatively little maintenance over the pavement's life cycle, other than joint sealing maintenance, where applicable. Most common failures are due to insufficient bond in thin bonded overlays on concrete.

For more information on the design of overlays see the National Concrete Paving Technology Center's *Guide to Concrete Overlays* (Harrington and Fick 2014).

Pervious Concrete Pavement

Pervious concrete is used to provide a permeable pavement system that allows rainwater to pass through the surface and percolate into the underlying layer (Figure 21-5). Pervious concrete pavements reduce stormwater runoff, flash flooding, and standing water, and can reduce or even eliminate the need for on-site holding ponds or buried stormwater retention structures.

The mixture comprises specially graded coarse aggregate, cement, water, and little, if any, fine aggregate. The materials used for pervious concrete mixtures are the same as those used for conventional concrete, except that far less fine aggregate is used in the mixture. The gradation of the coarse aggregate is kept to a narrow band such as ASTM C33 No. 67, No. 8, or No. 89. Pervious concrete mixes contain minimal amounts of water, with typical water-to-cementitious materials (w/cm) ratios around 0.30 (see ACI 522.1 2020).

The result is a material with interconnected voids of between 15% to 25% by volume, leading to flow rates around 0.34 cm/s (480 in./hr., which is 200 L/m^2/min or 5 gal/ft^2/min). The exposed aggregate surface of pervious concrete can provide

FIGURE 21-5. Pervious concrete's key characteristic is the open-pore structure that allows high rates of water transmission.

FIGURE 21-6. Pervious concrete pavement in the rain.

enhanced traction for both pedestrians and vehicles while the potential for vehicle hydroplaning during wet weather conditions is also reduced (Figure 21-6). The low mortar content and high porosity tends to reduce strength, but sufficient strength is normally achieved for most low volume traffic paving and parking applications (see Table 21-1).

TABLE 21-1. Typical Values for Material Properties of Pervious Concrete (Garber and others 2011)

PROPERTY	TYPICAL VALUES
Unit weight	70% - 80% of conventional concrete mixtures
Density	1600 kg/m³ - 2000 kg/m³ (100 lb/ft³ - 125 lb/ft³)
Percent voids	15% - 25%
Permeability	2.5 m/hr - 50 m/hr (100 in./hr - over 2000 in./hr)
Compressive strength	3.5 MPa - 28 MPa (500 psi - 4000 psi) averaging 17 MPa (2500 psi)

Pervious concrete pavements systems can be designed using a standard pavement procedure. Typical designs include 125 mm to 150 mm (5 in. to 6 in.) of pervious concrete with 150 mm to 300 mm (6 in. to 12 in.) of a drainable aggregate base. The base layer should allow a percolation rate of 12.5 mm/hr (0.5 in./hr) if no overflow piping is installed. In freeze-thaw environments, a minimum of 300 mm (12 in.) of a drainable aggregate base, such as 25 mm (1 in.) of crushed stone, is typically constructed. A thicker pavement system may be required if heavier loads and higher traffic are anticipated, or if the percolation rate of the base layer is inadequate. If pervious concrete is placed over impermeable subgrades, such as clayey soils, then provisions need to be made for adequate storage and retention time to reduce peak flow and percolation into the subgrade.

Hydraulic Design. Hydraulic designs for pervious concrete pavement surfaces must consider: the total amount and peak rate of rainfall, nature of the pavement, and drainage onto and away from the pavement. The flow rate through the subbase and subgrade may be the parameter that controls the amount of water leaving the system. The total storage capacity of the pervious concrete system must therefore include the capacity of the pervious concrete pavement, the capacity of any subbase used, and the amount of water that leaves the system by infiltration into the underlying soil. In cold zones pervious concrete systems should not be designed to store water in the concrete itself because of water expansion upon freezing. Detailed guidance is available in perviouspavement.org/design/hydrological.html.

Placement techniques for pervious pavements are similar to hand placed conventional pavements using conventional formwork. Placement of pervious concrete should be continuous and rapid using vibrating and manual screeds to

avoid premature drying from the large surface area (Figure 21-7). Strike off should be 13 mm to 19 mm (½ in. to ¾ in.) above the forms to allow for compaction. Compaction and finishing is generally accomplished by a weighted roller-screed that spins as it is pulled across the fresh concrete (Figure 21-8). Pervious pavements are generally not finished beyond the initial effort required for compaction in order to prevent closing the surface.

FIGURE 21-7. Pervious concrete is usually placed and then struck off with a vibratory screed.

FIGURE 21-8. Compaction of pervious concrete with a steel roller.

Joints should be made soon after consolidation using a rolling joint tool (Figure 21-9). Recommended joint spacing of approximately 6 m (20 ft) is suggested. Joints may not be needed if random cracking is visually acceptable.

FIGURE 21-9. Joint roller, commonly referred to as a "pizza cutter."

Pervious concrete is cured immediately after finishing with moisture-retaining covers such as plastic sheeting or impervious paper (see Chapter 17). The primary maintenance requirement is the removal of soil and other debris materials that may clog the system. Periodic vacuuming, sweeping, or power spraying may be required to restore permeability. Pervious pavements have been shown to perform well in freeze-thaw and sulfate environments when designed and constructed properly. The ability for pervious concrete pavements to quickly drain water minimizes the saturation within the voids that could otherwise freeze and cause damage. However, air-entrained concrete should be used for pervious concrete pavements in freeze-thaw exposures.

American Concrete Institute (ACI) Committee 522 has developed recommendations (522R 2010) and specifications (522.1 2020) for pervious concrete pavement. Further information on pervious concrete is available from perviouspavement.org.

Precast Concrete Pavement

Precast concrete pavements are constructed using prefabricated concrete slabs installed over a prepared subbase or existing pavement. Precast panels are fabricated and cured off site at precast concrete facilities prior to construction (Figure 21-10), then transported and installed on site and then opened to traffic. A primary benefit of prefabrication is the high degree of quality control along with the controlled environment under which the panels can be produced in precast concrete facilities.

FIGURE 21-10. Precast panel fabrication.

Panels are installed normally by crane (Figure 21-11). Levelling is achieved by placing directly on the base, or supporting the slab on embedded threaded levelers, bolts, or a strongback beam until a grout or foam is pumped underneath for bearing.

Precast panels can be used for isolated full-depth repairs, such as joint replacements, for single or multiple consecutive slab replacements, or for total reconstruction of an entire section (Tayabji 2013). Repair or reconstruction of distressed pavements can be accomplished during overnight or weekend closures, thus reducing user delay and associated costs. This is especially appealing on heavily trafficked roadways.

FIGURE 21-11. Placing a precast panel.

Strength requirements are typically governed by the need to recycle forms in a short period, rather than structural design factors. Typical precast pavement mixtures are designed for average 28-day compressive strengths ranging from 28 MPa to 41 MPa (4000 psi to 6000 psi) with higher strengths usually achieved. For prestressed precast panels, compressive strengths of 21 MPa to 28 MPa (3000 psi to 4000 psi) are typically required for release of pre-tensioning.

Depending on the time between casting and placement, a major proportion of shrinkage may be completed before the slab is put into place. Prestressed systems should also be less prone to effects of warping.

Precast sections may be jointed or prestressed. Load transfer between panels is provided by dowels in jointed systems and by post-tensioning in prestressed systems. While the ride quality of the surface as installed may be acceptable for opening to traffic, if needed, diamond grinding can be used to ensure the final riding surface meets the requirements for high-speed roadways. Several proprietary systems are available that include details on dowel slots and undersealing approaches.

Jointed precast pavement systems are typically designed to replicate conventional JPCP in thickness. Precast panels are typically a minimum of 200 mm (8 in.) thick, but they can be adjusted as necessary to match the thickness and cross-section of the existing pavement. For repair or reconstruction projects, the precast panels are designed to match the thickness of the slab being removed minus 6 mm to 13 mm (¼ in. to ½ in.) to accommodate irregularities in the base.

Prestressed concrete (see Chapter 8) panels are typically designed with minimal non-prestressed reinforcement. This is because prestressing helps to minimize or eliminate cracking. Jointed precast panels are typically heavily reinforced with two mats of mild steel reinforcement to prevent any cracks that may form during handling or from widening over the life of the pavement. For jointed systems, doweled joints are used similar to conventional JPCP. However, it should be noted that there is no contribution to load transfer from aggregate interlock. Grout or mortar is typically used to ensure full bedding of the dowel bars in the precast panels after installation and to fill the joint between the panels.

For prestressed systems, there are two types of joints: the intermediate joints between the individual panels, and the expansion joints at the ends of each post-tensioned section of precast panels. The intermediate joints are non-functional joints that are post-tensioned together and sealed with epoxy applied to the abutting faces of the precast panels during construction. The expansion joints are designed to accommodate or absorb movement due to expansion and contraction of the post-tensioned slabs.

Detailed information on precast pavements is available in *Precast Concrete Pavement Technology* (Tayabji and Bush 2013 and Roesler and others 2016).

Roller-Compacted Concrete Pavement

Roller-compacted concrete (RCC) is a low workability mixture that is placed without internal vibration but is compacted using heavy vibratory steel drum and rubber-tired rollers (Harrington 2010).

RCC has similar strength properties and consists of the same basic ingredients as conventional concrete but has different mixture proportions. The largest difference between RCC mixtures and conventional concrete mixtures is that RCC is a dry mixture with a higher percentage of fine aggregates, which allows for tight packing and consolidation when compacted by vibratory rollers.

The 28-day unconfined compressive strength of RCC is comparable to that of conventional concrete, typically ranging from 28 MPa to 41 MPa (4000 psi to 6000 psi), with flexural strength ranging in values from 3.4 MPa to 6.9 MPa (500 psi to 1000 psi).

The goal in RCC mixture proportioning is to provide enough paste to cover the aggregates and fill the voids. RCC pavements generally are not air entrained, have a lower water and paste content, and require a larger fine aggregate content in order to produce a combined aggregate that is well graded and stable under the action of a vibratory roller (Figure 21-12). Aggregates constitute up to 85% of the volume of RCC and play an influential role in achieving the required workability, specified density, and strength. Roller-compacted concrete usually has a nominal maximum size of aggregate not greater than 19 mm (¾ in.) in order to minimize segregation and produce a relatively smooth surface texture. Since the paste content in RCC is lower than that of conventional concrete, there is less concrete shrinkage and associated cracking.

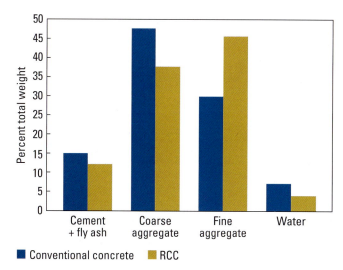

FIGURE 21-12. Typical mixture proportions (Harrington 2010).

Fresh RCC is stiffer than typical zero-slump conventional concrete. Its consistency is stiff enough to remain stable under vibratory rollers, yet wet enough to permit adequate mixing and distribution of paste without segregation.

The minimum thickness of an RCC pavement is typically 100 mm (4 in.), with a single-lift maximum thickness of 250 mm (10 in.). Design methodologies are available from PavementDesigner.org.

Unlike conventional concrete pavements, RCC pavements are constructed without forms, dowels, or reinforcing steel. Joint sawing is not required, but when sawing is specified, transverse joints are spaced farther apart than with conventional concrete pavements. When RCC is allowed to crack naturally, aggregate interlock usually provides adequate load transfer across the cracks. Cracks typically occur at 6 m to 18 m (20 ft to 60 ft) intervals, depending on the RCC's properties and pavement thickness. The primary reason joints are used in RCC pavements is to initiate crack locations or to improve aesthetics in parking lots, access roads, or areas of channelized traffic at speeds greater than 48 kph (30 mph). Dowels or tie bars are not used in RCC pavements.

RCC is typically placed with a high-density asphalt-type paver equipped with a high-density screed, and final compaction is achieved by a vibrating steel drum roller. When using a high-density screed, rolling compaction is reduced, thus minimizing variations in surface tolerance and improving surface smoothness. During placement, it is important that the subbase and/or subgrade be uniformly moist and firm. For pavement thicknesses greater than 250 mm (10 in.), two equal lifts are used. The operations from mixing through finished compaction should be completed within one hour (Figure 21-13).

FIGURE 21-13. Roller-compacted concrete (RCC) is typically placed with asphalt pavers and consolidated with steel wheel rollers.

RCC pavements are able to carry high, concentrated wheel loads making them suitable for heavy industrial, military, and mining applications. Roller-compacted concrete can also be used as a pavement surface layer for low-speed roads and industrial facilities where surface smoothness and appearance are not a major concern. Without diamond grinding, RCC's profile and smoothness may not be desirable for pavements carrying high-speed traffic. However, the pavement can support light vehicle traffic shortly after placement, making it suitable for applications that require immediate use of the pavement (Figure 21-14).

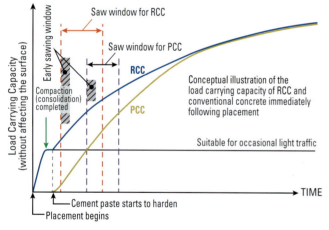

FIGURE 21-14. Strength development of RCC (Harrington 2010).

Detailed information on RCC pavements is available in the *Guide for Roller-Compacted Concrete Pavements* (Harrington and others 2010), ACI 325.10R, *Report on Roller-Compacted Concrete Pavements* (Reapproved 2001), and PCA EB215 *Roller-Compacted Concrete Quality Control Manual* (Arnold and Zamensky 2000).

PAVING MATERIALS AND MIXTURES

The large quantities of material needed for any given project mean that the cost of a paving mixture can impact its acceptability. As such, conventional paving mixtures tend to be relatively simple, despite having to meet a wide range of challenging demands for workability and durability.

Recommended w/cm ratios are in the range 0.38 to 0.42 for most concrete pavements, although higher values are not uncommon in areas that do not need freeze-thaw resistance and other durability resistance. These values are sought to provide reduced permeability and sufficient durability rather than strength. This is especially critical in cold regions where deicing salts are applied to the surface.

Portland and blended cements may be used in concrete paving applications. Supplementary cementing materials (SCMs) including slag cement, fly ash, and other pozzolans, either in binary or ternary combinations, are commonly used in paving mixtures. Dosage is based on balancing availability, cost, and required durability parameters.

Aggregates should be well graded, durable, and not prone to D-cracking nor alkali-carbonate reaction (ACR) (see Chapter 5). If aggregates are susceptible to alkali-silica reaction (ASR), they may be used with appropriate mitigation measures (see Chapter 11). It is becoming more common to recycle crushed concrete as a part of the aggregate system (Snyder and others 1994, and Reza and Wilde 2017), as base course material, or both.

Chemical admixtures, if used, generally consist of low- or mid-range water reducing admixtures and air-entraining admixtures. When needed for freeze-thaw durability, an adequate air void system is required. Typical specifications, for a 25 mm (1 in.) aggregate in severe environments, require at least 5% air content after the paver.

The key characteristic of concrete mixtures used for slipform concrete pavements is that they are stiff when fresh, and responsive to vibration (Figure 21-15). This behavior, along with minimizing segregation, is strongly affected by the gradation of the aggregate system, particularly the percentage of aggregate retained on the 2.36-mm and 1.18-mm (No. 8 and No. 16) sieves. It is important for the contractor to select an aggregate combination that provides a desirable overall gradation. Work by Cook and others (2013) has suggested combinations that fall within the Tarantula Curve increase the probability of meeting this need (see Chapter 5).

FIGURE 21-15. Vibrators on a typical slipform paving machine.

TABLE 21-2. Performance Engineered Mixture Concrete Properties and Current Test Methods*

PROPERTY	PURPOSE	TEST METHOD
Concrete Strength	Flexural Strength	ASTM C78 (AASHTO T 97)
	Compressive Strength	ASTM C39 (AASHTO T 22)
Shrinkage	Standard shrinkage	ASTM C157 (AASHTO T 160)
	Restrained shrinkage (cracking age)	AASHTO T 334
	Restrained shrinkage (stress/strength) dual ring	AASHTO T 336
Freeze-Thaw Durability	Air Content- Pressure Method	ASTM C231 (AASHTO T 152)
	Air Content- Volumetric Method	ASTM C173 (AASHTO T 196)
	Air Void System- Sequential Pressure Method	AASHTO TP 118
	Use of Protective Sealers	AASHTO M 224
	Quantifying Calcium Oxychloride Formation Potential	AASHTO T 365
	Rate of Absorption of Water	ASTM C1585
Transport Properties	Bulk Resistivity	AASHTO TP 119
	Surface Resistivity	AASHTO T 358
Aggregate Stability	Freeze-Thaw Resistance of Coarse Aggregate	ASTM C1646
	Rapid Freeze-Thaw Test for Concrete	ASTM C666 (AASHTO T 161)
	Determining Reactivity of Concrete Aggregates and Mitigating AAR	ASTM C1778 (AASHTO R 80)
Workability	Kelly Ball	AASHTO TP 129
	Box Test	AASHTO TP 137

*Adapted from AASHTO PP 84 (2020).

When performance-based measures are selected, the importance of quality control increases; compliance with the acceptance criteria is predicated on a well-designed and executed QC program (see Chapter 20) that includes processes, production, and construction control (AASHTO 2020a).

Desired pavement properties and suggested tests for performance engineered mixtures (PEMs) are shown in Table 21-2.

FIGURE 21-16. Central batch plant showing stationary mixer loading into dump trucks.

PAVEMENT CONSTRUCTION

For large paving projects, a dedicated batch plant is normally erected on site and concrete is mixed in a stationary mixer (Figure 21-16). The concrete is typically transported to the point of placement in dump trucks, which means that adjustments cannot be made to the batch after mixing. For smaller projects a ready-mixed concrete plant may be used to batch concrete, and then deliver it in truck mixers. In this case more attention is needed to ensure that the concrete delivered is uniform from batch to batch and that allowance is made for haul times (see Chapter 14).

Concrete can be placed using slipform or fixed-form paving methods, depending upon the nature of the project. Fixed-form paving is generally used in small or irregular sections where slipform paving is not practical (Figure 21-17). Fixed-form paving operations rely on a higher slump mixture that will flow easily to fill the forms. Vibration is is often provided using roller screeds while finishing and texturing is by hand.

Slipform paving is generally for placements that require high production rates, such as mainline paving (Figure 21-18). Slipform paving operations require a low-slump mixture that will not slough or lose its shape after it is vibrated and extruded from the paving machine. A limited amount of hand finishing may also be performed.

FIGURE 21-17. Fixed-form paving operation (courtesy of M. Ayers).

FIGURE 21-18. Slipform paving operation.

Dowels can be placed in prefabricated dowel baskets and secured in place prior to paving, or they can be inserted into the fresh concrete during placement using a dowel bar inserter (Figure 21-19).

Tie bars are commonly pre-placed prior to concrete placement, or a tie bar inserter may be used to place tie bars between lanes when two lanes are constructed at the same time. For lane additions bent tie bars can also be inserted during paving

and pulled straight after the pavement has hardened, or single-piece tie bars can be drilled and epoxied into the hardened concrete, and new concrete placed around it.

The construction process for CRCP is again similar to that for JPCP, except that a haul or access road to the side is required, along with a means of moving the concrete onto the alignment, such as a belt placer to allow room for the reinforcing steel (Figure 21-20). Attention needs to be paid to placing the steel in the correct final location. Currently, reinforcing steel is placed manually, either on chairs or on a transverse bar assembly. It is not recommended to use tube feeding of reinforcing steel as it has been found that steel location is much too variable.

FIGURE 21-20. Construction of CRCP.

Texture is applied to the surface of the concrete after placement and before curing. The purpose of texturing is to increase friction in wet weather, and to reduce noise. Two commonly used wet texture techniques include tining and drag. Tined surfaces are applied using a steel rake longitudinally (to reduce noise) (Figure 21-21). Drag textures are applied by dragging a piece of artificial turf or heavy burlap on the surface. Additional textures include diamond grinding and grooving. If used, grinding and grooving are done after curing and once the pavement is able to withstand the weight of the machine.

FIGURE 21-19. Dowel bar inserter.

FIGURE 21-21. Longitudinal tining the surface of concrete pavement has been shown to reduce noise.

Following finishing operations, curing compounds are sprayed on the surface to reduce water loss from the mixture and promote hydration of cementitious materials. Curing compounds should be applied to the surface immediately after finishing operations.

Contraction joints should be cut before random cracking occurs. Conventional sawing should be cut to a depth of one-third of the pavement thickness (T/3), while early entry saws that have a shoe to hold the concrete in place can be run somewhat earlier.

MAINTENANCE OF PAVEMENTS

In general, a well designed and constructed concrete pavement will require relatively little attention for up to 40 years. Typically, agencies require that concrete pavements remain crack free and smooth over the entire design life. Distress is often first observed at joints or cracks due to: warping and curling, loss of load transfer, loss of support, freeze-thaw related issues, aggregate expansion, or chemical attack. As a result, careful attention has to be paid to maintaining joint sealants if they are used, because if water does penetrate past the sealant, distress will be accelerated.

There are a variety of approaches to repairing such problems including: slab stabilization, slab jacking, partial-depth repair, full-depth repair, retrofitted edge drains, dowel bar retrofit, cross stitching, diamond grinding, diamond grooving, joint resealing, crack sealing, and thin concrete overlays. Selection of which approach to use is dependent on the cause and extent of distress, and is discussed in detail in the National Concrete Pavement Technology Center's *Preservation Guide* (Smith and others 2014).

SUBGRADES AND BASES/SUBBASES

Concrete support systems may typically be considered to have two layers: the *subgrade*, which is the natural soil layer on which a pavement is constructed; and the *base/subbase*, which is the layer upon which the pavement rests (Figure 21-22). It is not always necessary that concrete pavements have both a base and subbase layer. Those handling heavy loads (such as aircraft pavements) often have both (Kohn and others 2003). The primary objective of these layers is to provide uniform and stable support that will remain in place throughout a pavement's service life.

Due to its rigid nature, a concrete pavement distributes the pressure from applied loads over a larger area of the supporting material. As a result, deflections are small and pressures on the subgrade are low (ACPA 2007). Concrete pavements do not typically require especially strong foundation support, and the thickness of concrete required is not strongly affected by the stiffness of the support. More important than strength, however, is uniformity. Non-uniform support increases localized deflections and causes stress concentrations in the pavement that can lead to premature failures, including fatigue cracking, faulting, and pumping.

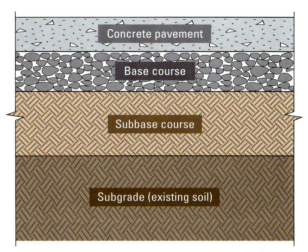

FIGURE 21-22. Typical pavement cross-section showing location of subgrade, subbase, base, and concrete pavement (courtesy of FHWA).

Requirements for subgrade preparation may vary considerably, depending on soil type, environmental conditions, drainage, and amount of heavy traffic. Subgrades may be modified by mixing in a small amount of cement, cement-kiln-dust, lime, lime-kiln-dust, fly ash, or some combination of these. Pavement subgrades may need improvement for a number of reasons: to increase strength and stiffness of poor quality soil, reduce swelling potential, and improve construction conditions. Different approaches in design and construction are available to ensure that a stable, stiff support system can be provided based on the existing materials available on site (Garber 2011 and ACPA 2007).

Base and subbases are typically made up of aggregates and may be untreated, using the materials as received, or stabilized by the addition of cementitious materials. A wide variety of materials and gradings have been used successfully for untreated bases or subbases. These materials include crushed stone, crushed concrete, bank-run sand-gravels, sands, soil-stabilized gravels, local materials such as sand-shell mixtures, and slag.

A principal criterion for base and subbase materials is to limit the amount of fines passing a 75-µm (No. 200) sieve because increasing fines reduces permeability. Soft aggregates should be avoided because fines may be created due to the abrasion or crushing action of compaction equipment and construction traffic. Generally, aggregates having less than 50% loss in the Los Angeles abrasion test (ASTM C131/AASHTO T 96) are satisfactory.

Subbase materials should prevent pumping of subgrade soils. This is achieved by using a minimum depth of 100 mm (4 in.) and specifying a minimum density for untreated bases and subbase soils (ASTM D698/AASHTO T 99) for heavily traveled projects.

Failing granular-base pavements, with or without their old bituminous mats, can be salvaged, strengthened, and reclaimed as soil-cement pavements. This is an efficient, economical way of rebuilding pavements. Since approximately 90% of the material used is already in place, handling, and hauling costs are cut to a minimum.

A cement-treated or lean concrete base provides a strong and uniform support for the pavement and joints, offers an all-weather working platform, and contributes to smoother pavements by giving firm support to the forms or paver during construction.

Soil-cement is a highly compacted mixture of soil (aggregate), cement, and water. It is widely used as a low-cost pavement base/subbase for roads, residential streets, parking areas, airports, shoulders, and materials-handling and storage areas. For more information see ACI 230.1R, *Report on Soil Cement* (2009); PCA EB003, *Soil-Cement Construction Handbook* (PCA 1995); PCA EB052, *Soil-Cement Laboratory Handbook* (PCA 1992); and PCA SR1007P, *Guide to Cement-Stabilized Subgrade Soils* (PCA 2020).

Soil-cement pavement products can be divided up into three main categories – each with their own unique contribution to a pavement structure. They include cement-modified soil (CMS) and cement-stabilized subgrade (CSS) soil, cement-treated base (CTB), and full-depth reclamation (FDR).

Cement-Modified Soil and Cement-Stabilized Subgrade

Cement-modified soil (CMS) describes a soil material that has been treated with a relatively small proportion of portland cement to provide a stable working platform. The improvements offered by the treatment include reducing the plasticity and shrink/swell potential of unstable, highly plastic, wet, or expansive soils, and increasing the bearing capacity. Cement-stabilized subgrade (CSS) not only provides all the benefits of CMS but also substantially increases soil stiffness and strength to the point where the treatment provides structural benefits to pavements (Figure 21-23).

FIGURE 21-23. Cement-modified soil and cement-stabilized subgrade still function essentially as soil with improved engineering properties.

Cement-modified soils and cement-stabilized subgrades are principally used to improve fine-grained, high plasticity soils such as silts and clays, and granular soils having a high plasticity and/or fines content. Typically, a cement content of 2% to 4% for CMS, and 3% to 6% for CSS will reduce the Plasticity Index (PI) (ASTM D4318, *Standards Test Methods for Liquid Limit, Plastic Limit, and Plasticity Index* [AASHTO T 90]) sufficiently, increase the California Bearing Ratio (CBR) (ASTM D1883, *Standard Test Method for California Bearing Ratio (CBR) of Laboratory-Compacted Soils* [AASHTO T 193]), and soil resistance values. The design requirements will vary, depending on existing soil properties, however a thickness of 150 mm (6 in.) has generally been shown to provide an adequate working platform.

Cement-modified soils and cement-stabilized subgrade construction consists of prewetting dry soils, if necessary; then distributing portland cement by mechanical spreader or in slurry form; mixing the cement into the soil using mixed-in-place methods; and then compacting with a tamping (sheepsfoot) roller (Figure 21-24). Surface compaction is completed with a steel drum, pneumatic tire, or other appropriate type of roller. The area is then shaped with a motor grader to final crown and grade, and the surface is sealed with a pneumatic-tire roller. All operations, including compaction, should be completed in the same day. CMS/CSS is not often cured; however, curing with a light water spray or bituminous coating will provide the maximum benefit from the cement.

FIGURE 21-24. Sheepsfoot roller is used to compact CMS/CSS.

Cement-Treated Base

A cement-treated base (CTB) consists of native soils, gravels, or manufactured aggregates blended with measured amounts of portland cement and water that hardens after curing. Cement-treated base is often used in highways, commercial, and heavy industrial applications. Cement-treated base is also common in airfield applications, particularly large airport runway applications. The Federal Aviation Administration requires a stabilized base, such as CTB, for runways that experience high volumes of heavy aircraft traffic.

Thicknesses for CTB are less than those required for unstabilized granular bases carrying the same traffic. The design of a concrete surface layer must take into account increased friction values between the layers. In some instances, an interlayer such as a geotextile or a layer of asphalt is used to reduce restraint stresses and the potential for cracking. Approaches for determining CTB thickness include the AASHTO *Design Guide* (AASHTO 1993) or the AASHTO *Mechanistic-Empirical Pavement Design Guide* (AASHTO 2020b). Some critical design inputs for calculating CTB thickness include the estimated traffic loading during the design life, subgrade strength, and CTB strength.

The mixture should be designed based on strength and resistance to freeze-thaw and wet-dry environments. The objective is to obtain maximum compaction by mixing an aggregate/granular material with the correct quantity of cement and water.

The cement content for CTB depends on the type of aggregate material used; however, it usually ranges from 3% to 6% for most applications. In general, a cement content that will provide a seven-day unconfined compressive strength of between 2.1 MPa and 2.8 MPa (300 psi and 400 psi) is satisfactory for most CTB applications.

The aggregate material used in a CTB mixture can be a variety of materials including existing or borrowed stone, gravel, sand, silt, and caliche. Recycled concrete aggregates (RCA) and recycled asphalt pavement (RAP) can also be incorporated. A well-graded sandy and gravelly aggregate material blend typically requires the least amount of portland cement for adequate hardening, while an aggregate material that is classified as either a poorly graded sandy material deficient in fines or a silty and/or clayey material requires more cement for hardening.

General ranges for engineering properties of CTB are listed in Table 21-3.

TABLE 21-3. Typical Engineering Properties of Cement-Treated Base

PROPERTY	7-DAY VALUE	
Compressive strength	2.1 MPa - 4.1 MPa	(300 psi - 600 psi)
Modulus of rupture (flexural strength)	0.7 MPa - 1.4 MPa	(100 psi - 200 psi)
Modulus of Elasticity	4,100 MPa - 6,900 MPa	(600,000 psi - 1,000,000 psi)
Poisson's ratio	0.15	

Construction of CTB includes initial preparation, processing, compaction, finishing, and curing. Initial preparation includes

shaping area to crown and grade and correcting unstable subgrade areas. There are two methods for processing CTB: mixed-in-place or central-plant-mixed. For CTB mixed-in-place, cement is placed dry onto the surface of the in-place aggregate using mechanical spreader equipment (Figure 21-25). The cement may also be placed on the surface in slurry form. If necessary, water is applied on the surface or directly into the mixing chamber. The single-shaft mixer then mixes the cement, water, and aggregate until a uniform material is achieved.

FIGURE 21-25. Placing dry cement for construction of cement-treated base.

The central-plant-mixed method requires mixing cement, aggregate material, and water in a stationary plant using pugmills or rotary-drum mixers. After mixing at the plant, the CTB material is deposited into trucks, hauled to the site, and spread evenly over the area. An aggregate spreader is commonly used to place the CTB mixture over the subgrade.

The CTB mixture is then compacted by using vibratory-steel rollers, tamping rollers, or pneumatic-tire rollers depending on the type of aggregate material used. Finishing operations include a combination of fine grading shaping, lightly applying water, and rolling with a pneumatic steel drum roller.

After finishing, the CTB must be adequately cured, allowing cement to hydrate and the cement-aggregate mixture to harden. The newly constructed base should be kept continuously moist (by lightly watering or misting) for a 3- to 7-day period. Alternately, a moisture-retaining cover or curing compound can be placed over the CTB soon after completion.

Full-Depth Reclamation

Full-depth reclamation (FDR) with cement is a technique in which an existing hot mix asphalt (HMA) pavement and granular base material is reclaimed (pulverized as necessary), combined with cementitious materials, and then recompacted to create a new base (Figure 21-26).

FIGURE 21-26. Full-depth reclamation of hot mix asphalt pavement.

Full-depth reclamation may be considered when the existing pavement or base and/or subgrade is damaged and inadequate for the current or future traffic. If rehabilitation of the pavement would otherwise require full-depth patching over more than 15% to 20% of the surface area, or the pavement cannot be rehabilitated with simple resurfacing methods, FDR is a viable option.

The amount of water and cement required in the mix will depend upon the project-specified strength and gradation of the final blend obtained from pulverizing the HMA during construction and mixing it with the base material. Typical specifications for pulverizing call for a minimum of 100% passing the 75-mm (3-in.) sieve, 95% passing the 50-mm (2-in.) sieve, and 55% passing the 4.75-mm (No. 4) sieve. In general, a cement content that provides a seven-day unconfined compressive strength between 2.1 MPa to 2.8 MPa (300 psi to 400 psi) is satisfactory for most FDR applications. Laboratory testing and trial mixes are required to select the best proportions for the job.

Thickness design for FDR is similar to a CTB and is calculated based on strength of the material, strength and stiffness characterizations of additional layers, anticipated loads, and performance requirements (for example, service life, serviceability, and reliability) (AASHTO 1993). Typical thickness values for FDR range from 150 mm to 300 mm (6 in. to 12 in.).

Full-depth reclamation requires a reclaimer mixer, grader, cement spreader, water truck, and roller. A reclaimer machine typically makes an initial pass over the existing flexible pavement, pulverizing the HMA surface and blending it with the base and/or subgrade material. A schematic of the mixing process is shown in Figure 21-27. Water may be added during this mixing stage to bring the material up to optimum moisture content. The material is then graded accordingly. Next, cement is spread either dry or in slurry form in a controlled manner onto the surface.

The reclaimer then mixes the cement into the pulverized material until the material is thoroughly mixed. This is followed by compaction, final grading, curing, and surfacing. Smooth-wheeled rollers are used for compaction. The FDR should be cured for from three to seven days.

For more information on FDR, see Reeder and others (2019).

SUSTAINABILITY OF CONCRETE PAVEMENTS

According to the FHWA, a sustainable pavement is one that achieves its specific engineering goals, while, it also meets basic human needs, uses resources effectively, and preserves and/or restores its surrounding ecosystems (Muench and Van Dam 2014). In addition, concrete pavement longevity, use of recycled materials, increased surface albedo, and lower fuel consumption all contribute to the sustainability of concrete pavements. Concrete pavements that have been well designed

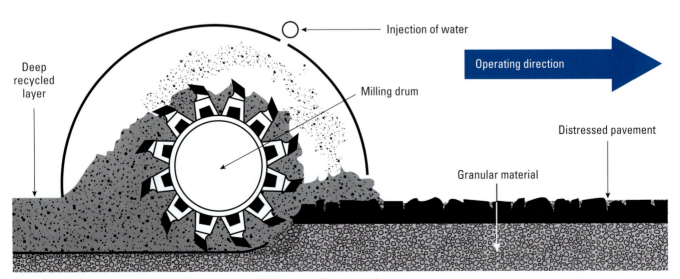

FIGURE 21-27. Schematic of full-depth reclamation (FDR) process (Garber 2011).

and constructed using quality materials have the service potential to last 40 years or more. This longevity can significantly reduce environmental and societal impacts due to rehabilitation activities over the pavement life cycle.

Stronger subgrade and base designs including cement-modified soils, cement-stabilized subgrades, cement-treated bases, and full-depth reclamation can provide a stronger and more stable support system relative to unbonded and untreated layers. These layer types can also be used to optimize pavement designs and reduce layer thicknesses.

Pervious concrete permits land owners to increase the effective use of land while still providing sufficient on-site water storage without the need for retention ponds or similar structures. In addition, the water that passes through the pavement serves to replenish groundwater levels.

The use of supplementary cementitious materials (SCMs) reduces the volume of materials in landfills while at the same time increasing potential durability of concrete pavements. Use of recycled aggregates and recycled concrete lowers the environmental impacts related to the processing, purchasing, and transportation of virgin aggregates. Existing materials are reused, reducing the exploitation of virgin material. Truck traffic is reduced, resulting in fuel savings and lower emissions.

Light colored concrete surfaces can increase albedo and reduce the amount of artificial lighting required in a given setting and reduce heat island effects. Vehicle fuel consumption can be reduced on rigid concrete surfaces, particularly for heavy vehicles. Additional benefits include reduced noise from tire-pavement interaction.

Concrete overlays make use of the existing pavement, eliminating the need for removal and disposal, as well as using the existing pavement as part of the new structure. Concrete overlays can be constructed and opened to traffic within a day, reducing user costs and accessibility issues.

For more information on sustainability of concrete paving, see Snyder and others (2016) and Chapter 12.

REFERENCES

PCA's online catalog includes links to PDF versions of many of our research reports and other classic publications. Visit: cement.org/library/catalog.

AASHTO, *Standard Practice for Developing Performance Engineered Concrete Pavement Mixtures*, PP 84-20, American Association of State Highway and Transportation Officials, Washington, DC, April 2020, 15 pages.

AASHTO, *Guide for Design of Pavement Structures, 4th Edition*, American Association of State Highway and Transportation Officials, Washington, DC, 1993, Compatible Software at acpa.org/WinPAS.

AASHTO, *Mechanistic-Empirical Pavement Design Guide – A Manual of Practice*, American Association of State Highway and Transportation Officials, 3rd edition, Washington, DC, 2020b, 264 pages.

ACI Committee 230, *Report on Soil Cement*, ACI 230.1R-09, American Concrete Institute, Farmington Hills, MI, 2009, 28 pages.

ACI Committee 325, *Report on Roller-Compacted Concrete Pavements*, ACI 325.10R-95, reapproved 2001, American Concrete Institute, Farmington Hills, MI, 1995, 32 pages.

ACI Committee 522, *Report on Pervious Concrete*, ACI 522R-10, American Concrete Institute, Farmington Hills, MI, 2010, 38 pages.

ACI Committee 522, *Specification for Construction of Pervious Concrete Pavement*, ACI 522.1-20, American Concrete Institute, Farmington Hills, MI, 2020, 13 pages.

ACPA, *Bonded Concrete Overlay on Asphalt (BCOA) Thickness Designer*, American Concrete Paving Association, Rosemont, IL, 2012.

ACPA, *StreetPave*, American Concrete Paving Association, Rosemont, IL, 2012.

ACPA, *Subgrades and Subbases for Concrete Pavements*, EB204P, American Concrete Pavement Association, Rosemont, IL, 2007.

Arnold, T.; and Zamensky, G., *Roller-Compacted Concrete: Quality Control Manual*, EB215, Portland Cement Association, 2000, 58 pages.

Cook, D.; Ghaeezah, A.; Ley, T.; and Russell, B., *Investigation of Optimized Graded Concrete for Oklahoma–Phase 1.* Oklahoma Department of Transportation, 2013.

Garber, S.; Rasmussen, R.O.; and Harrington, D., *Guide to Cement-Based Integrated Pavement Solutions*, SR035, National Concrete Pavement Technology Center, Ames, IA, August 2011, 82 pages.

Halsted, G.E.; Luhr, D.R.; and Adaska, W.S., *Guide to Cement-Treated Base (CTB)*, EB236, Portland Cement Association, Skokie, IL 2007, 20 pages.

Harrington, D.; Abdo, F.; Adaska, W.; and Hazaree, C., *Guide for Roller-Compacted Concrete Pavements*, National Concrete Pavement Technology Center, Ames, IA, August, 2010, 114 pages.

Harrington, D.; and Fick, G., *Guide to Concrete Overlays: Sustainable Solutions for Resurfacing and Rehabilitating Existing Pavements* (3rd edition), National Concrete Pavement Technology Center, Ames, IA, May 2014, 163 pages.

IMCP, *Integrated Materials and Construction Practices for Concrete Pavement: A State-of-the-Practice Manual*, 2nd edition, National Concrete Pavement Technology Center, Iowa State University, Ames, IA, Eds: Peter Taylor, Tom Van Dam, Larry Sutter, and Gary Fick, May 2019, 338 pages.

Luhr, D.R.; Adaska, W.S.; Halsted, G.E., *Guide to Full-Depth Reclamation (FDR) with Cement*, EB234, Portland Cement Association, Skokie, IL, 2008, 15 pages.

Muench, S.; and Van Dam, T., *Pavement Sustainability*, Tech Brief, FHWA-HIF-14-012, Federal Highway Administration, Washington, DC., 2014, 12 pages.

PavementDesigner.org, pavementdesigner.org, 2018.

PCA, *Cement-Stabilized Subgrade Soils*, SR1007P, Portland Cement Association, 2020, 59 pages.

PCA, *Soil-Cement Construction Handbook*, EB003, Portland Cement Association, Skokie, IL, 1995, 40 pages.

PCA, *Soil-Cement Laboratory Handbook*, EB052, Portland Cement Association, Skokie, IL, 1992, 60 pages.

Rasmussen, R.O.; Rogers, R.; and Ferragut, T.R., *Continually Reinforced Concrete Pavement*, Concrete Reinforcing Steel Institute, Chicago, IL, August 2011, 173 pages.

Reeder, G.; Harrington, D; Ayers, M; and Adaska, W., *Guide to Full-Depth Reclamation (FDR) with Cement*, SR1006P, National Concrete Pavement Tech Center, Ames, IA, 2017 (reprinted 2019), 104 pages.

Roesler, J.R.; Hiller, J.E.; and Brand, A.S., *Continuously Reinforced Concrete Pavement Manual: Guidelines for Design Construction, Maintenance, and Rehabilitation*, FHWA-HIF-16-026, Federal Highway Administration, August 2016, 129 pages.

Smith, K.; Harrington, D.; Pierce, L.; Ram, P.; and Smith, K., *Concrete Pavement Preservation Guide, Second Edition*, National Concrete Pavement Technology Center, Ames, IA, September 2014, 307 pages.

Snyder, M.B.; Van Dam, T.; Roesler, J.; and Harvey, J., *Strategies for Improving the Sustainability of Concrete Pavements*, FHWA-HIF-16-013, Federal Highway Administration, April 2016, 28 pages.

Tayabji, S.; and Buch, N., *Precast Concrete Pavement Technology*, Strategic Highway Research Program, Washington D.C., SHRP 2 Report S2-R05-RR-1, February 2013, 179 pages.

Vandenbossche, "BCOA-ME Design Guide," engineering.pitt.edu/Vandenbossche/BCOA-ME, 2013.

FIGURE 22-1. Veteran's Memorial, Columbus, Ohio. Concrete structures provide beautiful, durable and resilient building solutions (courtesy of Infinite Impact and Baker Concrete).

Concrete structures are the backbone of modern society (Figure 22-1). Worldwide, 70% of the population lives in concrete structures (Hanley Wood 2013). Concrete structures provide strong and solid construction with systems that can be easy to design and build and are versatile to suit any architectural style.

Concrete is used in residential, commercial, public, and industrial applications including:

• Buildings
• Bridges
• Dams and hydraulic structures
• Concrete pipe

One-, two-, and three-story stores, restaurants, schools, hospitals, commercial warehouses, terminals, and other industrial buildings are built with concrete because of its durability and ease of construction. Concrete dams, roads, bridges, and pipe are also part of contemporary infrastructure, providing durable and resilient structures that support modern society and the economy.

PERFORMANCE ASPECTS OF CONCRETE STRUCTURES

Architectural Concrete

Concrete has the inherent ability to be molded into virtually any shape and surface texture. When given an aesthetic finish, concrete features are referred to as decorative, or architectural concrete. Achieving a decorative appearance often requires that the surface involve special finishing techniques like exposing aggregates or using special ingredients like integral coloring pigments (see Chapter 15). However, many times the final appearance is purposely left as originally cast to showcase the natural beauty of concrete architecture (Figure 22-2). For more information on decorative concrete, see Chapter 17 and ACI 310 (2019).

FIGURE 22-2. Concrete has the ability to be molded into virtually any shape as shown in features of the Florida Polytechnic University Innovation, Science and Technology Building, constructed in 2014 (courtesy of Baker Concrete).

Durability

The longevity of concrete structures is readily apparent. Ancient structures demonstrate that concrete has an extremely long-life expectancy, and more contemporary buildings exhibit performance that exceeds current design life criteria. For example, the Hoover Dam was completed in 1936 (Figure 22-3), the Glenfinnan Viaduct in Scotland was completed in 1901, and the Pantheon in Rome was completed around 125 AD. All are still in use today.

FIGURE 22-3. The Hoover Dam, completed in 1935, is a testament to the durability of concrete structures.

Depending on the application, the design service life of building interiors is often 30 years. However, the concrete portion of structures often lasts 100 years and longer, more than triple the service life. When properly designed and maintained, concrete structures can be reused or repurposed several times. Reusing concrete buildings conserves future materials and resources and reduces the construction time that new structures may require (see Chapter 12). Concrete requires different degrees of durability depending on the exposure environment and the properties desired. Ingredient properties and compatibilities, proportions, placing and curing practices, and the service environment determine the durability and service life of the concrete. For detailed information on durability and deterioration mechanisms and related recommendations for proportioning concrete mixtures, see Chapter 11.

Resilience

Resilience is defined as the ability to prepare and plan for, absorb, recover from, and more successfully adapt to adverse events (National Research Council 2013). Resilience is growing in importance as tornadoes, hurricanes, floods, and wildfires have destroyed vast areas of development. Large portions of the U.S. are affected as the social and economic impact of damage from natural disasters continues to rise. The Federal Emergency Management Agency (FEMA) provides technical guidance to encourage the construction of shelters and other safe spaces. Concrete structures can provide occupants safety from all kinds of weather resistance to natural disasters, such as tornadoes, hurricanes, and storm surges (Figure 22-4).

FIGURE 22-4. Aerial photo of Pass Christian, Mississippi, shows the only surviving home in a neighborhood that was destroyed by an 8.5 m (28 ft) tall storm surge during Hurricane Katrina in 2005. Foundations are all that remain of most homes. The surviving concrete home was built using FEMA standards (courtesy of John Fleck).

Flood Resistance. Concrete is capable of resisting storm surge and is not damaged by water. Concrete submerged in water absorbs very small amounts of water over long periods of time, and the concrete is not damaged. In flood-damaged areas, concrete buildings are often salvageable. Concrete dams and levees are used for long-lasting flood control.

Concrete will only contribute to moisture problems in buildings if it is enclosed in a system that traps moisture between the concrete and other building materials. For instance, a vinyl wall covering in hot and humid climates will act as a vapor retarder and moisture can get trapped between the concrete and the wall covering. For this reason, impermeable wall coverings (such as vinyl wallpaper) should not be used on concrete walls without a pathway for vapor drying.

Safe Rooms. Concrete safe rooms and community shelters provide protection from high wind events when properly designed, detailed, and built. The 2021 *International Building Code* (IBC) requires storm shelters in emergency operation centers, call stations, fire, rescue, ambulance and police stations, and all K-12 schools located where the design wind speed for tornadoes is 112 m/s (250 mph). Storm shelters must comply with ICC/NSSA 500-2020 ICC/NSSA, *Standard on the Design and Construction of Storm Shelters*. FEMA P-361, *Safe Rooms for Tornadoes and Hurricanes: Guidance for Community and Residential Safe Rooms* (2015), describes the criteria for any safe room, private or public. Safe rooms are hardened structures specifically designed to meet FEMA criteria that can provide a very high probability of avoiding injury or death in extreme weather events (Figure 22-5). Concrete buildings are often used to provide this type of protection.

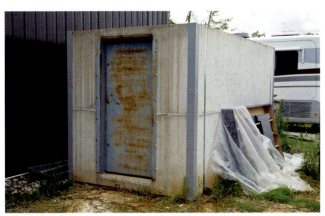

FIGURE 22-5. In June 2014, forty-two people rode out a tornado in this safe room in Saltillo, Mississippi (courtesy of FEMA/ Yuisa Rios).

Blast Resistance. Disaster impact conditions are not limited to mother nature. With proper reinforcement, concrete can also be made blast resistant (Smith and others 2009). Explosive blast loads can be resisted by properly designed reinforced concrete structures. This includes detailing reinforcement and splices for columns, beams, slabs, walls, and joints. Consideration should also be given to protecting structures against disproportionate collapse.

High performance concrete can be designed to have improved blast resistant properties. These concretes often have a compressive strength exceeding 100 MPa (14,500 psi) and contain steel fibers (see Chapters 7 and 23). These blast-resistant structures are often used in bank vaults and military applications. For more information, see ASCE 59, *Blast Protection of Buildings* (2011).

Non-Combustible Construction

Concrete has inherent fire resistance. Building codes (ICC and NFPA) establish fire ratings for different structural members and different building occupancies. Extensive research and testing over many years has established performance based on aggregate type and member thickness so that accurate estimates of concrete fire resistance can be determined. PCI 124, *Specification for Fire Resistance of Precast/Prestressed Concrete* (2018) contains procedures for designing fire resistance of precast/prestressed concrete members. ACI/TMS 216.1, *Code Requirements for Determining Fire Resistance of Concrete and Masonry Construction Assemblies* (2019), contains design and analytical procedures for determining the fire resistance of concrete and masonry members and building assemblies.

Concrete provides fire protection to the reinforcing steel and fire separation between building areas. At a minimum, fire ratings are intended to allow the structure to remain standing long enough to allow for safe egress of building occupants during a fire. For mid- and high-rise buildings, this is an important consideration. For some applications, like hospitals and nursing homes, non-combustible construction is important because occupants often have limited mobility. Because it is non-combustible, concrete often performs well enough to protect property as well as save lives, meaning it may be possible to repair the building following a fire rather than demolish the building. For a given thickness of concrete, replacing a portion of the normal weight aggregate with lightweight aggregate increases its fire resistance.

Typical building codes permit fire-resistant rated non-combustible construction of any height or area. Deviating from the specified fire resistance ratings reduces the permissible heights and floor areas. Buildings with non-combustible exterior loadbearing elements, and any type of interior construction permissible by the code, tend to be limited to six stories in height.

Seismic Resistance

Reinforced concrete systems provide resistance to help resist seismic demands. Appropriate design of reinforcement and other detailing requirements are intended to ensure that a structure can adequately carry seismic loads. For buildings in regions of low seismic risk, seismic provisions of ACI 318, *Building Code Requirements for Structural Concrete* (2019) are not required. For buildings in regions with moderate seismic risk, only the beams, columns, and slabs that contribute to the lateral force resisting system need to be specially designed. For buildings in regions with high seismic risk, all building components that are part of the lateral force resisting system, including foundations, should be designed to resist seismic loads.

A summary of the seismic detailing requirements contained in ACI 318 is provided by Fanella (2007). Reinforcement details are provided for beams, columns, two-way slabs, walls and foundations, flexural members of special moment frames, special moment frame members subjected to bending and axial load, joints of special moment frames, special reinforced concrete structural walls and coupling beams, and structural diaphragms and trusses.

The National Earthquake Hazards Reduction Program (NEHRP) publishes numerous guides and tools for earthquake resistant design and construction for both new and existing buildings. These materials find their way into building codes and practice. NEHRP resources are located online at nehrp.gov.

Sustainability

Energy Performance. A major appeal of insulated concrete walls is reduced energy to heat and cool the building. Although concrete is a poor thermal insulator, buildings made with it can be energy efficient. Insulation, thermal mass, and low air infiltration contribute to the energy saving. Typical R-value for EPS and XPS foams are, respectively, 4 and 5 per 25 mm (1 in.) of thickness. Thermal mass acts like a storage battery to store heat or cold, moderating temperature swings via damping and lag effects (see Chapter 12). In addition to saving energy and money associated with heating and cooling, concrete walls also provide more consistent interior temperatures for occupants, increasing their comfort.

Concrete is resistant to air passage. Cast-in-place systems have solid wall surfaces and panelized systems have few joints. Both situations limit the movement of air in or out of the structure. This air tightness prevents loss of conditioned air to the outside and infiltration of unconditioned air to the inside. Systems that incorporate continuous layers of insulation further enhance this effect.

The combination of thermal mass and reduced air infiltration results in tight building envelopes with steady temperatures. Insulation further improves this. It is common to downsize heating, ventilation, and air conditioning (HVAC) equipment and increase mechanical ventilation in concrete buildings. Also, air exchangers may need to be installed to keep adequate fresh air supplied to the interior. Heat or energy recovery exchangers can condition incoming air supplies to reduce energy losses.

Indoor Air Quality. Concrete contains low to negligible levels of volatile organic compounds (VOCs) (Budac 1998). Concrete encapsulates its ingredients and binds them into an inert matrix and does not off-gas. Decorative concrete flooring, wall, and ceiling assemblies can replace carpeting, wood flooring, and painted finishes which can be a source of VOCs and other air emissions. Exterior decorative finishes eliminate the need for siding materials while also providing a hard, durable exterior surface. Interior decorative finishes eliminate the need for gypsum wallboard and also provide a durable inside surface. These finishes minimize materials for construction and reduce construction waste (VanGeem 2008).

Acoustic Performance. Because of their inherent mass and stiffness concrete systems effectively block unwanted noise. Concrete minimizes sound transmission through a wall or floor, which is a desirable characteristic for most types of buildings.

Concrete has also been found to readily satisfy vibration acceptance criteria for human comfort under typical service conditions and occupancies. The acceptance criteria for walking structure borne sound excitations are easily satisfied for all types of reinforced concrete floor systems by providing member sizes based on only the appropriate minimum thickness requirements in ACI 318 (CRSI 2016).

Sound transmission class, STC, is a single number rating (in decibels, dB) to provide an estimate of performance of a wall in certain common sound insulation applications. Similarly, outdoor-indoor transmission class, OITC, is a single number rating to provide an outdoor-indoor transmission class.

Using field or laboratory testing or a calculation, walls are rated for STC between interior spaces or for OITC. To provide privacy between neighbors, building codes generally require an STC rating of 50 for interior-interior spaces. The *International Residential Code* (IRC) contains similar requirements, but with a minimum STC rating of 45 dB when tested in accordance with ASTM E90 for walls and floor/ceiling assemblies separating dwelling units.

Typical STC ratings for various concrete wall and roof/ floor assemblies range from 50 dB to 55 dB (CRSI 2016). An increase of 10 in the STC translates roughly to a decrease in the perceived noise by one half. Along with the use of concrete, using normal density concrete masonry units will generally result in STCs from the low 40s to the low 60s even before wall finishes are applied (TMS 302, *Standard Method for Determining Sound Transmission Ratings for Masonry Assemblies* [2018]).

STRUCTURAL COMPONENTS

Reinforced concrete construction for structures provides inherent stiffness, mass, and ductility. The design of reinforced concrete structures takes into consideration axial forces, bending moment, shear, torsion, and combinations of these (Wang and others 2007 and Kamara and Novak 2011).

Foundations

Concrete foundations range from very simple slab-on-grade to structural footings, to more complex deep foundation systems.

Shallow foundation. A majority of buildings are constructed with concrete footings, which distribute loads to the underlying soil. They are commonly strip footings or rectangular pads, but may be a variety of other shapes. They can also be cast as part of the concrete slab. Footings are constructed either against removable forms or sometimes directly against excavated earth (Figure 22-6).

FIGURE 22-6. Concrete footings constructed against removable forms (courtesy of M. Sheehan).

With a few exceptions, building codes require footings to be placed at a minimum depth of 300 mm (12 in.) below undisturbed ground or below the local frost level. This requirement prevents potentially damaging frost heave.

Structural loads are calculated and footings are sized to carry the load based on soil conditions at the site. Most soils have a bearing capacity of 7000 kg/m² to 20,000 kg/m² (1500 lb/ft² to 4000 lb/ft²). Footings may contain longitudinal and transverse reinforcement or may be unreinforced (plain). Their thickness depends on the magnitude of the dead load (from the structure), the size of the system (wall) they support, the presence or lack of reinforcement, and the bearing capacity of soil. Common sizes range from a minimum thickness of 150 mm (6 in.) for residential structures to well over 900 mm (36 in.) for larger structures. Most codes require minimum widths of approximately 300 mm (12 in.) although some residential codes may permit narrower widths.

Piers and Piles. Piers and piles are placed into the ground as part of a structural foundation and behave similar to columns. They may be referred to as caissons, foundation piers, bored piles, drilled shafts, sub-piers, and drilled piers (ACI 336.3R 2014). These members rely on skin friction along their sides, on end bearing, or a combination of both, to transfer the building's loads to the soil. The IBC distinguishes piers as relatively short in comparison to their width and piles as relatively slender, with length-to-horizontal dimension ≤ 12 for piers and ≥ 12 for piles (IBC 2021). Though the terms may be used somewhat interchangeably, concrete piers are typically cast-in-place in drilled holes and concrete piles may be cast-in-place in drilled holes or may be precast members driven into the ground. Concrete piers may be constructed with belled ends to connect multiple piers or piles and increase their end bearing capacity. Pier caps and pile caps generally use conventional compressive strength concretes.

The size of concrete drilled piers is often determined by the subsurface conditions and not by the stresses in the concrete. Consequently, concrete stresses are low and conventional concrete strengths may be used.

Caisson piles are cast-in-place in holes that have been excavated into the soil and extend to bedrock. They are used for very large or very tall buildings to transmit heavy loads to the ground. A rebar cage is placed into the hole before fresh concrete is placed (Figure 22-7). Concrete for piles and piers can either be freely dropped down the shaft or placed by means of a trunk. Studies show that the free-fall method does not reduce concrete quality (PCA 1995, Turner 1970, STS Associates 1994), with depths up to 45.7 m (150 ft) resulting in no appreciable segregation or loss of concrete strength.

Piers and piles, as well as their caps, can have a large diameter, so mixture proportions and temperature control procedures used for mass concrete may need to be considered (see Chapter 18). In addition, caps, piles and piers may contain congested reinforcement; consequently, the concrete may have to be flowable. The concrete should have a slump of a least 150 mm (6 in.) after 4 hours (ACI 336 2014).

FIGURE 22-7. Construction of mass concrete piles for the west approach to the San Francisco–Oakland Bay Bridge.

Precast concrete piles come in different sizes and shapes ranging from 254-mm (10-in.) square piles to 1.68-m (66-in.) diameter cylindrical piles (Dick 2005). Minimum concrete strengths should be 24 MPa (3500 psi) at time of stress transfer for pretensioned piles and 28 MPa (4000 psi) at time of post-tensioning piles unless higher strengths are required by the structural design (ACI 543 2012, and PCI 2019). Design compressive strengths are usually at least 34 MPa (5000 psi) and can be as high as 59 MPa (8500 psi).

For more information about the planning, design and construction of deep foundations and excavations, see the Deep Foundations Institute (dfi.org/default.asp).

Beams and Columns

Cast-in-place concrete may be used in non-prestressed concrete beams with short span lengths or in post-tensioned concrete beams for longer span lengths and unusually high loads. Strengths specified for short span beams are in the range of 24 MPa to 34 MPa (3500 psi to 5000 psi).

The design of most precast, prestressed concrete members is generally based on a concrete compressive strength at 28 days of at least 34 MPa to 41 MPa (5000 psi to 6000 psi). However, the concrete mix proportions are generally dictated by concrete strengths at time of transfer of the prestressing force. As a result, concrete strengths at 28 days are frequently in excess of the 28-day specified value.

Strengths for longer span post-tensioned beams for application such as bridges may be as high as 55 MPa (8000 psi) at 28 days. In addition, a minimum concrete strength at the time of post-tensioning is also specified and may control the concrete mixture proportions. In recent years, concrete with specified strengths up to 69 MPa (10,000 psi) has been specified by some specifications to achieve longer span lengths, wider beam spacing, or the use of shallower sections. In such cases, the concrete strength may be specified at 56 days to take advantage of the strength gain between 28 days and 56 days. The higher strengths are generally achieved through the use of higher cementitious materials content, lower water-cementitious materials ratio, and supplementary cementitious materials (see Chapter 23).

Columns act as vertical supports to beams and slabs, and to transmit the loads to the foundations (Figure 22-8 and Figure 22-9). Columns are primarily compression members, although they may also have to resist bending moment transmitted by beams. Columns may be classified as short or slender, braced or unbraced, depending on various dimensional and structural factors.

FIGURE 22-8. Concrete placement of a column (courtesy of Baker Concrete).

FIGURE 22-9. Concrete columns provide vertical supports to beams and slabs and transmit loads to the foundations (courtesy of United Forming, Inc.).

Typically, shorter structures tend to have lighter loads, so their design may not require interior columns. Instead, floors can be used to span between walls and transmit loads to them. Conversely, steel or concrete columns may be used in interior locations with exterior walls supporting loads and providing shear resistance.

Floors

Floors on ground. A concrete floor on ground is a common element of concrete construction. Floors on ground can be unreinforced (plain) or reinforced with mild steel or post-tensioning. The inclusion of reinforcement improves the flexural load-carrying ability of the floor and helps to hold cracks tight. When proper joint spacing is maintained, floors can be constructed without reinforcement.

Many floors on ground are slabs of uniform thickness cast between footings (Figure 22-10). Some floors are cast integrally with the footings, tapering from a thinner floor section to a thickened edge beneath the walls. Still other floors can be designed like a floating raft (reinforced) to bridge over poor soil conditions or to prevent damage from expansive soils. In all cases, the thickness of a concrete floor is designed for the magnitude of loads it will carry and the number of repetitions of the loads. The minimum practical floor thickness is 100 mm (4 in.). Industrial floors might be as thick as 350 mm (14 in.). Detailed information on the design and construction of concrete floors on ground is available in various references (Tarr and Farny 2008, ACI 360R, *Guide to Design of Slabs-on-Ground* (2010), and ACI 302.1R, *Guide to Concrete Floor and Slab Construction* (2015).

Elevated concrete floor systems, either cast-in-place (CIP) or precast, must be reinforced with metal decking, reinforcement or post-tensioned tendons in order to span between supports and transmit loads to the foundation. For CIP, the floors are

FIGURE 22-10. Finishing a concrete floor on ground with a laser screed.

either cast against removable formwork or corrugated metal decks. These surfaces act as a form to hold the fresh concrete but stay in place after the floor has hardened (Figure 22-11). Precast floors are typically hollow core panels (Figure 22-12), with the cores saving material, reducing dead load/weight, or double tees for longer spans.

An elevated floor must be designed for structural considerations and also provide fire resistance. By using structural lightweight concrete, either normal strength or high strength, dead weight loads for floors can be reduced. This is significant in high-rise construction as it lowers foundation costs for the building. Concrete floors on metal decks are used with steel beams/frames. Sections are manufactured in various widths and lengths from steel having a thickness ranging from 0.38 mm to 3.42 mm (0.0149 in. to 0.1345 in., or 28 to 10 Gage) (SDI 2014). The metal sheets are laid across beams as stay-in-place formwork for the concrete slab and can be designed to function as an integrated structural element with the slab. When the metal section serves as both reinforcement and formwork, they are composite steel floor deck-slabs, and when welded wire fabric (WWF) or rebar are needed to help carry loads, they

are non-composite steel floor decks. Concrete placed on cold-formed steel deck should have a minimum strength of 21 MPa (3000 psi) and not contain chloride admixtures or other corrosive materials. The slab depth for these systems ranges from 100 mm to 300 mm (4 in. to 12 in.) and is based on the load, any required fire resistance rating, or durability. Design manuals are available from the Steel Deck Institute (SDI 2010, SDI 2012) or sdi.org.

For concrete frames, or where longer spans or heavier loads are considered, a cast-in-place floor system should be used. Various concrete floor systems (Fanella 2000), most built with steel reinforcement, can provide economical column-free space without concerns for vibration and with low floor-to-floor heights. These systems allow for easier renovations to interior spaces should that become desirable in the future.

Many factors have contributed to the evolution of economical long-span concrete floor systems, including improved structural design methods, increases in the strength of both concrete and reinforcing steel, the development of flying forms, advanced production methods for precast members, the use of post-tensioning for CIP systems, and the implementation of more efficient construction techniques (Fanella and Munshi 2000).

The concrete for a cast-in-place floor is often assumed to have a density of 2400 kg/m^3 (150 lb/ft^3). In general, a strength of 28 MPa (4000 psi) results in the least expensive system. Formwork economy is important because it is one of the costliest aspects of concrete construction. Using readily available standard form sizes, repeating those sizes and shapes of members when possible (bay-bay and floor-floor), and using simple formwork keeps forming costs to a minimum (Fanella 2000).

Except for one-way and two-way joist systems, the minimum slab thickness necessary to satisfy structural requirements will normally provide a floor system that has at least a 2-hour fire resistance rating. It is important to note that the ACI 318-19 minimum cover requirements (see Chapter 11) for

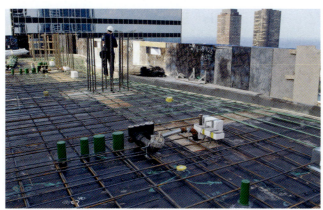

FIGURE 22-11. Sheets of coated plywood are used to support and form the underside of a concrete balcony during construction of the Aqua Tower, Chicago.

FIGURE 22-12. Hollow core panel placement (courtesy of PCI).

main reinforcement also generally result in a fire resistance rating greater than or equal to 2 hours. If the thickness necessary to satisfy fire resistance requirements exceeds that required for structural purposes, consideration should be given to using a different type of aggregate that provides higher fire resistance for the same thickness.

BUILDINGS

Modern buildings are divided into three categories. Low-rise is generally 1 to 3 stories; mid-rise 4 to 7; and high-rise is typically structures beyond 7 stories. The IBC defines high-rise construction as buildings with occupied floors more than 23 m (75 ft) above the lowest level of fire department vehicle access. While tall buildings and other prominent structures garner a lot of attention in the U.S. (see Chapter 23), more than 90% of buildings constructed are low-rise and small-area buildings, in the 1- to 3-story height range (Kamara and Novak 2011) (Figure 22-13).

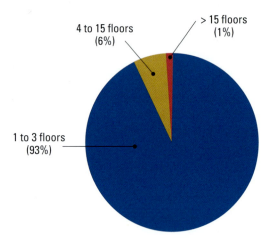

FIGURE 22-13. Building area by height in the U.S. (Kamara and Novak 2011).

ACI 318-19 applies to all structures ranging from single-story buildings to high-rises; buildings having complex shapes and more simple geometrics. It also applies to buildings requiring structurally intricate or innovative framing systems in contrast to those of more conventional or traditional systems of construction. For guidance, PCA EB204, *Simplified Design of Reinforced Concrete Buildings* (Kamara and Novak 2011) offers an alternative design process for primary framing members of a reinforced concrete building. When using the simplified approach, a designer will need to verify that all of the requirements of ACI 318-19 are met.

High-Rise Buildings

The first reinforced concrete high-rise in the U.S. was the 16-story Ingalls Building, completed in Cincinnati in 1903. Greater building height became possible as concrete strength increased. In the 1950s, 34 MPa (5000 psi) was considered high strength; by 1990, two high-rise buildings were constructed

in Seattle using concrete with strengths of up to 131 MPa (19,000 psi). Ultra-high-strength concrete is now manufactured with strengths in excess of 150 MPa (21,750 psi) (see Chapter 23).

High-rise buildings are commonly constructed with concrete frames – a system of columns and beams – with reinforced concrete floors to span between them. Many high-rise buildings specify modulus, E, rather than compressive strength, f_c'. Occupants of concrete towers are less likely to perceive building motions than occupants of comparable tall buildings with non-concrete structural systems.

A major economic consideration in high-rise construction is reducing the floor-to-floor height. Using a reinforced concrete flat plate system, the floor-to-floor height can be minimized while still providing high floor to ceiling heights. As a result, concrete has become the material of choice for many tall, slender towers.

Placement of concrete to the floor under construction may be accomplished by pump, bucket and crane, or buckets via elevator. Tower cranes are quite popular for high-rise construction. They are usually erected in the center core of the building, where they have full 360-degree access to deliver materials to all areas of the site (Figure 22-14).

FIGURE 22-14. Photo of tower crane in a concrete core (courtesy YoChicago.com).

Mid- and Low-Rise Buildings

In the U.S., residential concrete building systems were introduced in the early 1900s when Thomas Edison began experimenting with cast-in-place concrete. In parts of the U.S., concrete and masonry construction is the material of choice for single-family homes. Designers select concrete for one-, two-, and three-story stores, restaurants, schools, hospitals, commercial warehouses, terminals, and industrial buildings because of its durability, excellent acoustic properties, inherent fire resistance, and ease of construction.

Concrete for low- and mid-rise buildings is an economical choice because the load-bearing concrete exterior walls serve not only as a weather-tight enclosure, but may also carry roof and wind loads, eliminating the need for separate cladding and structural systems. Using concrete for frames, walls, and floors, designers and owners can achieve non-combustible performance, thermal mass, and sound attenuation in one system.

A concrete or concrete-and-steel composite frame may be used for mid- and low-rise buildings, or the structure may rely on wall panels to carry vertical loads. Cranes are typically used for mid- and low-rise buildings to place precast or tilt-up wall panels, columns, beams, and floor and roof panels (Figure 22-15).

FIGURE 22-15. A crane at ground level is used to place precast concrete panels for a mid-rise structure (courtesy of NCC/ Dukane Precast).

Building Systems

For mid- and low-rise buildings, six categories of concrete building systems include the majority of types in current use:

- removable forms (also known as cast-in-place),
- insulating concrete forms (ICFs),
- concrete masonry, panel systems like precast and tilt-up,
- autoclaved aerated concrete (AAC), and
- shotcrete.

Some differences between the various concrete systems are shown in Table 22-1 and are discussed in the following sections.

Removable Forms/Cast-in-Place. Removable forms or cast-in-place (CIP) concrete walls are constructed with ready-mix concrete placed into temporary forms erected on site. This has been one of the most common methods of concrete wall construction, from below grade to the top of a building. The techniques are similar at all levels, though different equipment may be needed.

High-rise tunnel forms, were introduced in the 1960s. It relies on reusable factory-made steel formwork to cast walls and slabs in one operation. It is highly systematic, and due to the repetitive nature, is well suited to projects with uniform room sizes, such as hotels, student housing, apartments, and commercial developments.

Newer forming techniques and modular systems make erecting and reusing forms much easier than ever, speeding cast-in-place construction. Flying forms are used on high-rise buildings to repeat floor forming, which is not only faster, but generally safer, too. The formwork system and protective scaffolding are moved in one continuous operation between floors. Concrete also saves time in the construction schedule by eliminating the added step of applying fire proofing necessary for wood or steel systems.

For one- and two-family dwellings and for homes up to two stories, the IBC includes prescriptive design and construction criteria for cast-in-place concrete walls for foundation, below grade walls, and above grade walls.

Stay-in-Place Form Systems. For one- and two-family dwellings, the *International Residential Code* (IRC) considers ICFs and other stay-in-place systems similar to cast-in-place for foundations, below grade walls, and above grade walls for homes up to two stories. For larger buildings like multi-family and commercial structures, a licensed engineer is typically required for structural design and an Evaluation Service Report documenting approval of the ICF for the type of construction mandated for the project will often be needed to finalize approval.

The two faces of forming material can be connected by ties of plastic, metal, or insulation. Some systems have hinged ties that allow pre-assembled forms to fold flat for easy, less costly shipping. Some systems use a ladder-type tie to connect narrow, tall vertical panels of material for field assembly.

Insulating concrete panels (ICPs) are another variation on forming concrete with insulation material. These systems consist of larger foam panels that form a post and beam concrete wall configuration, with wider spacing of the concrete elements than traditional ICFs, so they have more insulation and less concrete.

Both the thickness and density of the insulation can be varied with ICPs. Rigid EPS (expanded polystyrene) foam can be manufactured in different foam densities, such as 16 kg/m^3, 20 kg/m^3, and 24 kg/m^3 (1 lb/ft^3, 1¼ lb/ft^3, and 1½ lb/ft^3). Lighter densities would typically be used for residential walls up to 3.7 m (12 ft) tall, with heavier density foam used for taller walls and commercial applications.

Precast and Concrete Systems. There are two main types of conventionally formed panelized systems used for concrete walls: precast concrete and tilt-up concrete. By designing buildings with one predominant panel size and configuration, panel manufacturing becomes a repetitive process and gains efficiency, saving time and money. Precast panels can be sandwich or hollow core construction. Sandwich panels are popular because they provide two durable faces encasing insulation (Figure 22-16). The two whythes may act as composite or non-composite. A half-sandwich panels contain just one concrete face and therefore require additional insulation and finishing after erection.

FIGURE 22-16. An insulated precast sandwich wall with insulation (courtesy of Dukane Precast).

Hollow core panels can be used for walls, but are especially well suited to floors. Precast concrete panels are manufactured in factory settings. Not only is this conducive to improved quality control, it's also possible to manufacture year round and in any type of weather.

Precast panels must be shipped to the project, so their weight and size are limited by shipping and erection limitations. A sandwich panel that is 3 m × 6 m × 200 mm thick (10 ft × 20 ft × 8 in. thick) with 75-mm (3-in.) voids for insulation weighs more than 5400 kg (6 tons). Cranes and other lifting equipment should be appropriately sized to handle the panels.

Precast panels are typically cast horizontally. Horizontal casting is more convenient as it is faster to lay out the panels, openings, and reinforcement. Following casting, panels are placed into curing chambers to develop strength and durability. With horizontal casting, one face is cast separate from the other face.

For single-family homes, panel manufacturers usually provide pre-engineered plans for their panels. On larger buildings, it's common to custom engineer the building. The panel producer provides engineering data and specifications for their panels, which designers use in their analysis. Precasters are a resource for design detailing and to answer questions from building departments.

More information on precast design and construction is available from the Precast/Prestressed Concrete Institute at pci.org.

Tilt-up concrete. Tilt-up concrete panels are cast horizontally, typically on site, and then after sufficient strength has developed, lifted or tilted with a crane to form the walls of buildings (Figure 22-17). Panels can be produced in many shapes and sizes, including flat and curved sections. Concrete placement is fast and easy because it is done on the ground at the project location. Floor surfaces serve as casting beds, although sometimes separate casting beds are built. Panels can be stacked on top of each other if space is limited. Because the panels are cast on site, their size is not restricted to transportation considerations. Panels are typically large resulting in relatively few joints in buildings.

Tilt-up construction was traditionally most economical for larger panel sizes and buildings with larger footprints. Advancements and planning have made it more cost competitive even for buildings as small as 460 m² (5000 ft²). Tilt-up offers speed of construction with minimal capital investment.

Lift inserts, weld plates, and other hardware are all termed "embeds" and are placed in advance of concrete. Advancements in lifting inserts means that there are fewer limitations on panel size: 15.2 m (50 ft) tall panels are common, although the floor footprint may be one constraint on size. Cranes should have a safety margin of three to one for lifting. Bond breakers are used on the casting slab to facilitate lifting panels. Casting bed thickness must be adequate to support the crane during erection of panels. Operations are scheduled for lifting the panels in quick succession. Walls must be braced until panels are tied together, usually by welding. Vertical joints are then filled with sealants.

Tilt-up construction is covered by the 2020 IBC in the reinforced concrete section. Tilt-up is not addressed in the *International Residential Code* (IRC). ACI publishes ACI 551.2R, *Design Guide for Tilt-Up Concrete Panels* (2010), which presents information that expands on the provisions of ACI 318 to make it more specific to tilt-up construction.

FIGURE 22-17. Tilt-up wall panel is attached to the concrete slab and braced into position (courtesy of Tilt-up Concrete Association).

TABLE 22-1. Six Common Concrete Building Systems for Low-Rise Construction

APPLICATION	REMOVABLE FORM/ CAST-IN-PLACE	INSULATING CONCRETE FORM (ICF)	PRECAST	TILT-UP	MASONRY	AUTOCLAVED AERATED CONCRETE (AAC)
For single-family residential, most common:	Footings and foundation walls	Foundation walls, above grade exterior walls, floor decks	Walls, above grade exterior walls, floor decks	Above grade exterior walls	Foundation walls bearing and non-bearing interior and exterior above grade walls	Above grade exterior walls, floor decks
For multifamily residential, most common:	Footings, foundations, above grade structural elements, floor and roof decks	Foundations, above grade exterior walls and structural elements, non-load bearing above grade exterior walls, floor and roof decks	Foundations, above grade exterior walls and structural elements, non-load bearing above grade architectural panels, floor and roof decks	Above grade exterior walls and structural elements	Foundation walls, bearing and nonbearing interior and exterior above grade walls	Above grade exterior walls and structural elements, floor and roof decks
For commercial and Industrial, most common:	Footings, foundations, above grade structural elements, floor and roof decks	Foundations, above grade exterior wall and structural elements, non-load bearing above grade exterior walls, floor and roof decks	Foundations above grade bearing and nonbearing interior and exterior walls and structural elements, non-load bearing above grade architectural panels, floor and roof decks	Above grade exterior walls and structural elements	Foundations, bearing and nonbearing above grade interior and exterior wall and structural elements	Above grade exterior walls and structural elements, floor and roof decks
Codes and reference standards	ACI 332, IBC, IRC, PCA 100, ACI 318	ACI 332, IBC, IRC, PCA 100, ACI 318	IBC, IRC, ACI 318	ACI 318, IBC	TMS 402/602, IBC, IRC	Manufacturer assistance
In-situ forming systems	Removable forms (wood, reusable metal forms)	Stay-in-place forms (units or panels, flat, waffle-grid, and screen- grid systems)		Wood forms on a casting bed	None. Units set in mortar and cores grouted as required by design	None. AAC units set in thin-bed mortar
Insulation methods and energy considerations	Internal or external sheets / boards, internal furring plus insulation walls	Units/panels have R = 4 or 5 per 25 mm (in.) of thickness Typical R-value of walls about 20	Sandwich panel construction or applied sheet insulation R depends on void space and insulation properties	Primarily sandwich panel construction Typically: Uninsulated, R-2 Insulated, R-32	Integral insulation (in cells), internal, integral (between wythes) or external sheet/board sheet insulation, or lightweight units or internal furring and insulation	R = 1 per 25 mm (in.) of thickness
Typical sizes	Height: 2.4-2.7 m (8-9 ft) tall walls	Stackable units 406 mm (16 in.) tall	Height: Single-story: 2.4-3 m (8-10 ft), Two-story: 4.6-6 m (15-20 ft), Length: 6-9 m (20-30 ft)	Up to 15.2 m (50 ft)	200 × 200 × 400 mm (8 × 8 × 16 in.) nominal unit sizes (H × W × L)	Block height 200 mm (8 in.) Panels up to 610 mm (24 in.) tall
Typical wall thickness	100-610 mm (4-24 in.). Uninsulated walls are most commonly 150-200 mm (6-8 in.)	Block up to 1220 mm (48 in.) long. Wall cavity thickness up to 300 mm (12 in.), but typically 150 or 200 mm (6 or 8 in.). Overall wall thickness 340 mm (13¼ in.)	200 mm (8 in.) with a 75 mm (3 in.) void for insulation	Uninsulated 180-300 mm (7-12 in.). Insulated (sandwich) outer face 50-75 mm (2-3 in.). Inner face 50-150 mm (2-6 in.). Inner face 180-300 mm (7-12 in.)	200 mm (8 in.) nominal	100-300 mm (4-12 in.)
Common floor and ceiling assemblies*	Formed with wall forms or plywood	Proprietary ICF floor forms or combined with cast-in-place or precast floor systems	Precast panels for floors are similar to wall panels.		Proprietary steel joist systems can be used to build floor and roof decks using grouted reinforced CMU or combined with cast-in- place or precast floor systems	Proprietary steel joist systems with AAC floor units
Finishes	Off the form, form liners, plaster, or mechanically attached finish materials	Mechanically attached finish materials	Off the form, form liners, plaster, or mechanically attached finish materials	Off the form, form liners, plaster, or mechanically attached finish materials	Architectural masonry units, plaster, or adhered or mechanically attached finish materials, paints, or stains	Proprietary finish coats, or adhered or mechanically attached finish materials

* Wood or light gage steel joists can be used with any concrete wall system. Precast concrete floor or roof members for low-rise systems are often hollow-core panels. The cores can be used for utility chases or HVAC ducts.

More information on tilt-up design and construction is available from the Tilt-Up Concrete Association at tilt-up.org.

Concrete Masonry. The tough exterior of exposed units provides a durable finish in demanding environments, like schools and gymnasiums. Plain units can be sealed or painted and decorative masonry units can serve as a final finish, including providing extra durability in moist exposures, such as locker rooms, pools, and laundry facilities.

There are several ways to insulate masonry systems, such as using units with integral insulation, placing lightweight fill in the cells of units, or adding sheet insulation between wythes in multi-wythe systems (Figure 22-18) or on interior or exterior wall surfaces. Concrete masonry units for both load bearing and non-load bearing applications can also be formulated with lightweight mixtures to improve the thermal resistance of masonry walls. Recent changes to concrete masonry unit specifications allow for more innovative web configurations, including smaller web areas. These changes reduce thermal bridges across the wall section.

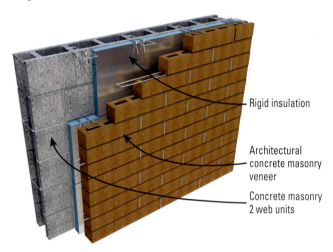

Rigid insulation

Architectural concrete masonry veneer

Concrete masonry 2 web units

FIGURE 22-18. A two-wythe masonry wall with rigid insulation in the cavity is very energy efficient.

More information on masonry design and construction is available in Farny and others (2008) and from the National Concrete Masonry Association at ncma.org.

Autoclaved Aerated Concrete. Autoclaved aerated concrete (also called autoclaved cellular concrete, or AAC) is a special type of lightweight building material. It is typically manufactured from a mortar consisting of pulverized siliceous material (sand, slag, or fly ash), cement or lime, and water. A gas forming admixture, for example aluminum powder, is added to the mixture. The chemical reaction of aluminum with the alkaline concrete mixing water forms hydrogen bubbles. The hydrogen expands in the mortar to form macropores with a diameter of 0.5 mm to 1.5 mm (0.02 in. to 0.06 in.). The material is then pressure steam cured (autoclaved) over a period of 6 to 12 hours

using a temperature of 190°C (374°F) and a pressure of 1.2 MPa (174 psi). This forms a hardened mortar matrix, which essentially consists of calcium silicate hydrates. This type of concrete contains up to 80% air. Because it has lower strength than most concrete products or systems, in load-bearing applications, it must typically be reinforced.

AAC is a porous mineral building material has densities between 300 kg/m³ and 1,000 kg/m³ (19 lb/ft³ and 63 lb/ft³) and compressive strengths between 2.5 MPa and 10 MPa (300 psi and 1500 psi). Due to the high macropore content – up to 80% by volume – autoclaved cellular concrete has a thermal conductivity of only 0.15 W/(m • K) to 0.20 W/(m • K) (1 Btu • in./[h • ft² • °F] to 1.4 Btu • in./[h • ft² • °F]). Autoclaved cellular concrete is produced in block or panel form for construction of residential or commercial buildings (Figure 22-19).

FIGURE 22-19. (top) Residential building constructed with autoclaved cellular concrete blocks. (bottom) Autoclaved cellular concrete block floating in water.

Cured blocks or panels of autoclaved aerated concrete are joined with thin bed mortar. Components can be used for walls, floors, and roofs. The lightweight material offers excellent sound and thermal insulation, and like all cement-based materials, is fire resistant (Table 22-2). A 203-mm (8-in.) thick AAC wall has a 4-hour fire rating, although actual performance will likely meet test requirements for up to 8 hours. Because it is noncombustible, it will not burn or give off toxic fumes. AAC walls require some type of applied finish, such as a polymer-modified stucco, natural or manufactured stone, or siding. Additional information can be found in ACI 523.2R, *Guide for Precast Cellular Concrete Floor, Roof, and Wall Units* (1996).

TABLE 22-2. Properties of Autoclaved Aerated Concrete (AAC)

PROPERTY	VALUE IN SI UNITS	VALUE IN INCH-POUND UNITS
Density (light enough to float in water)	320 to 800 kg/m³	20 to 50 lb/ft³
Compressive strength	2.1 to 6.2 MPa	300 to 900 psi
Allowable shear stress	0.06 to 0.20 MPa	8 to 22 psi
Thermal resistance (R-value)	0.03 to 0.05 per mm of thickness	0.8 to 1.25 per in. of thickness
Sound transmission class (STC)	40 for 102 mm thickness 45 for 203 mm thickness	40 for 4 in. thickness 45 for 8 in. thickness

BRIDGES

More than 70% of the bridges in the U.S. are constructed of concrete. These bridges perform year-round in a wide variety of climates and geographic locations. A popular method to accelerate bridge construction is to use prefabricated systems and elements. These are fabricated off-site or adjacent to the actual bridge site ahead of time, and then moved into place as needed, resulting in a shorter duration for construction. These systems are constructed with concrete – reinforced, pretensioned, or post-tensioned (or a combination thereof). Engineered to meet specific needs, high-performance concrete (HPC) is often used for bridge applications, including high-durability mixtures, high-strength mixtures, self-consolidating concrete, and ultra-high performance concrete (see Chapter 23).

Cast-In-Place Concrete Bridge Components

The majority of concrete bridges being built today consist of cast-in-place concrete decks supported on precast, prestressed concrete beams, which in turn are supported on pier caps and abutments. The pier caps are supported by concrete columns, which are supported either by directly piles or through an intermediate pile cap on top of the piles (Figure 22-20). Drilled piers may also be used instead of piles. In addition, there are trends towards the use of partial- or full-depth precast, prestressed concrete deck panels as a means of accelerating bridge construction.

In general, concretes for bridge decks have specified compressive strengths in the conventional range of 28 MPa to 35 MPa (4000 psi to 5000 psi). A more important consideration for bridge decks is to have a concrete with low permeability and low cracking potential. Some states specify a maximum value for permeability in terms of charge passed per ASTM C1202, *Standard Test Method for Electrical Indication of Concrete's Ability to Resist Chloride Ion Penetration* (AASHTO T 277). Most specified values are in the range of 1000 coulombs to

FIGURE 22-20. Cast-in-place concrete decks supported on precast, prestressed concrete beams, which in turn are supported on pier caps and abutments.

2000 coulombs at 28 days (Russell 2013). Although the goal is to have a low shrinkage concrete in bridge decks to reduce cracking, performance requirements for shrinkage are rarely specified. The use of higher strength concretes in bridge decks results in an increase in the amount of cracking. Consequently, concrete compressive strengths above 34 MPa (5000 psi) are not necessary unless required for structural purposes. Bridge deck concrete is generally pumped onto the bridge deck.

One important aspect of concrete bridge deck construction is the curing of the deck surface. Most states now require wet curing to begin within a certain distance or for a short time after the final finishing and to continue for a minimum of 7 days under a waterproof barrier (Russell 2013). Some states also require that a curing compound be applied at the end of the wet curing period. This allows the concrete to dry out more slowly and leads to slower development of tensile stresses from drying shrinkage and less cracking.

Precast, Prestressed Concrete Bridge Components

Precast bridges include post-tensioned, spliced precast, pretensioned girders, precast segmental box girders, cable-stayed structures, and arches. The Marc Basnight Bridge (Figures 22-21 and 22-22), in Dane County, North Carolina, was designed for a 100-year service life and constructed in 2019. The 2.8 mile-long bridge features the third-longest continuous segmental box girder superstructure in North America with 11 spans of lengths up to 350 feet. The remaining spans consist of precast, prestressed girders.

FIGURE 22-21. Concrete bridges provide beauty and durability to the transportation infrastructure. Shown is the Marc Basnight Bridge, Dane County, North Carolina, designed for a 100-year service life.

The requirements for some of these bridges and their components may be the same, however, because bridges are exposed to the outdoor environment, and the use of deicing chemicals in many regions of the U.S., durability of the concrete and the reinforcing system are major considerations in selecting concrete mixture proportions.

For precast, prestressed concrete components, the engineer generally specifies minimum strengths at time of transfer of the prestressing force and at 28 days; although other ages may be used. The *AASHTO LRFD Bridge Design Specifications* (AASHTO 2020) requires that the specified compressive strength for prestressed concrete shall not be less than 28 MPa (4000 psi). The engineer may also specify a minimum compressive strength at time of beam erection or a minimum compressive strength at time of post tensioning if a combination of pretensioning and post tensioning is utilized. In addition, the precast concrete producer prefers to achieve the transfer strength at an early age so that the components can be produced with a 24-hour production cycle. For this reason, Type III cement and heat curing of the components are frequently used to accelerate the strength gain. Some producers have adopted the use of self-consolidating concrete to improve the

surface appearance of the concrete, increase plant safety, and eliminate noise from vibrators.

Partial-depth precast concrete deck panels are used to span between adjacent beams as stay-in-place formwork. The stay-in-place panels become composite with the cast-in-place concrete deck to support additional dead loads and live loads. The panels are generally pretensioned in the direction of their span. Specified concrete compressive strengths are generally in the range of 34 MPa to 55 MPa (5000 psi to 8000 psi).

Full-depth precast concrete panels span over several beams and may span the complete width of the bridge. Each panel is typically pretensioned in the transverse direction of the bridge and multiple panels are post-tensioned together longitudinally before the panels are made composite with the supporting beams.

Concrete Specifications for Bridges

Specifications for concrete to be used in bridge structures in the U.S. have traditionally been prescriptive in nature. More recently, some states have begun to adopt performance specifications, which designate the required end results rather than the means to achieve them (Russell 2013). State specifications for bridges generally include standard specifications, supplemental specifications, and special provisions. The standard specifications document is used for all projects. It may be updated periodically through the use of supplemental specifications. Special provisions are those issued to define the requirements for specific projects. For larger projects, the special provisions are used to specify hardened concrete properties other than compressive strength. These have included specified values for modulus of elasticity, creep, and shrinkage.

The state standard specifications are generally based on the *AASHTO LRFD Bridge Design Specifications* (AASHTO 2020) and the *AASHTO LRFD Bridge Construction Specifications*

FIGURE 22-22. Construction of the Basnight Bridge, Dane County, North Carolina. The bridge featured the third-longest continuous segmental box girder superstructure in North America.

(AASHTO 2017 and 2019), which are published by the American Association of State Highway and Transportation Officials (AASHTO). These two documents are adopted by the state highway authorities with fine tuning for local conditions. Table 22-3 shows the different classes of normalweight concrete listed in the *AASHTO LRFD Bridge Construction Specifications* (2019). The class of concrete to be used in each part of the structure is then specified in the contract documents:

- Class A concrete is generally used for all elements of structures, except when another class is more appropriate, and specifically for concrete exposed to salt water.
- Class B concrete is used in footings, pedestals, massive pier shafts, and gravity walls.
- Class C concrete is used in thin sections, such as reinforced railings less than 4 in. thick, for filler in steel grate floors, etc.
- Class P concrete is used when strengths in excess of 28 MPa (4000 psi) are required. For prestressed concrete, consideration

should be given to limiting the nominal aggregate size to 19 mm (¾ in.).

- Class S concrete is used for concrete deposited underwater in cofferdams to seal out water.
- Class P(HPC) and Class A(HPC) are intended to be used for high-performance concretes as described in Chapter 23. For concrete Classes A, A(AE), and P used in or over salt water, the water/cementitious materials ratio shall be specified not to exceed 0.45.

DAMS AND HYDRAULIC STRUCTURES

Dams have played a dominant role in the advancement in the quality of life in the United States in providing water for irrigation, drinking water, and hydropower. The National Inventory of Dams includes over 80,000 dams in the inventory. Every state, except Alabama, has a Dam Safety Program that regulates the construction of new dams and the rehabilitation of unsafe existing dams. The Corps of Engineers, the Bureau of

TABLE 22-3. Classification of Normalweight Concrete for Bridge Construction

CLASS OF CONCRETE	MINIMUM CEMENT CONTENT, kg/m³ (lb/yd³)	MAXIMUM W/CM	AIR CONTENT RANGE, %	SIZE OF COARSE AGGREGATE PER AASHTO M 43 (ASTM D448), NOMINAL SIZE	SIZE NUMBER *	SPECIFIED COMPRESSIVE STRENGTH, MPa (psi) at days
A	363 (611)	0.49	—	25.0 mm to 4.75 mm (1.0 in. to No. 4)	57	28 (4000) at 28
A(AE)	363 (611)	0.45	6 ± 1.5	25.0 mm to 4.75 mm (1.0 in. to No. 4)	57	28 (4000) at 28
B	307 (517)	0.58	—	50 mm to 25.0 mm (2.0 in. to 1.0 in.) and 25.0 mm to 4.75 mm (1.0 in. to No. 4)	3 57	17 (2400) at 28
B(AE)	307 (517)	0.55	5 ± 1.5	50 mm to 25.0 mm (2.0 in. to 1.0 in.) and 25.0 mm to 4.75 mm (1.0 in. to No. 4)	3 57	17 (2400) at 28
C	390 (658)	0.49	–	12.5 mm to 4.75 mm (½ in. to No. 4)	7	28 (4000) at 28
C(AE)	390 (658)	0.45	7 ± 1.5	12.5 mm to 4.75 mm (½ in. to No. 4)	7	28 (4000) at 28
P	335 (564)	0.49	–[†]	25.0 mm to 4.75 mm (1.0 in. to No. 4) or 19.0 mm to 4.75 mm (¾ in. to No. 4)	7 67	≤ 41 (≤ 6000) at [†]
S	390 (658)	0.58	–	25.0 mm to 4.75 mm (1.0 in. to No. 4)	7	–
P(HPC)	–[‡]	0.40	–[†]	≤ 19.0 mm (≤ ¾ in.)	67	> 41 (> 6000) at [†]
A(HPC)	–[‡]	0.45	–[†]	–[‡]	–[‡]	≤ 41 (≤ 6000) at [†]

Notes:
*As noted in AASHTO M 43 (ASTM D448).
[†]As specified in the contract documents.
[‡]Minimum cementitious materials content and coarse aggregate size to be selected to meet other performance criteria specified in the contract.
(Adapted from AASHTO 2017 and 2019.)

Reclamation (USBR), and the Natural Resources Conservation Service have active dam safety programs that regulate over 13,000 dams. With this highly regulated market working to reduce the instances of unsafe dams and the green benefits of an expanding hydropower market, the use of multiple types of cementitious products in dam and hydraulic structures construction is relied upon to provide strength and durability. While the large scale concrete dam building era in the United States has slowed and water conservation has reduced the per capita consumption, the need for raw water storage for drinking and irrigation is still in demand. The climate change predictions suggest that longer drought periods and larger flood events will be more frequent which will result in the need for new dams for water storage and larger spillways to accommodate extreme flood events.

The most well known dam in the United States is the Hoover Dam (Figure 22-23). Concrete placement was started in 1933. The final block of concrete was placed and topped off at 221 m (726 ft) above the canyon floor in 1935. See usbr.gov/lc/hooverdam.

The world's largest hydroelectric dam is China's Three Gorges Dam. It is 2.3 km (1.4 mi) wide and 185 m (607 ft) in height. Construction began in 1994 and the dam started generating electricity in 2003. The dam provides hydroelectric power generation with 26 turbines producing 18,200 MW annually, equivalent to about 20% of China's power consumption. See ctgpc.com.

New Dam Construction and Modification to Existing Dams

New large concrete dam projects are rare in the United States. However, the San Vicente Dam in Southern California was raised by 36 m (117 ft) using 459,000 m³ (600,000 yd³) of roller-compacted concrete (RCC) in 2014. This is the tallest increase of any existing dam in the United States. The Army Corps of Engineers and Bureau of Reclamation dedicated a new auxiliary spillway for Folsom Dam in 2017 (Figure 22-24). It includes a 1100 ft-long approach channel beginning in Folsom Reservoir, a concrete control structure with six bulkhead and six radial gates, a 3100 ft-long auxiliary spillway chute, and a stilling basin that totaled to 230,000 m³ (300,000 yd³) of mass concrete. Outside of the U.S. where large dam building is common, the Longtan Dam in China used 7.5 million m³ (9.8 million yd³) of RCC for the 216 m (710 ft) high gravity dam.

Various types of cementitious products are used extensively in both new dam construction and modifications to existing dams and spillways. Considerations for mass concrete and thermal control of high volume placements and high cementitious contents should be given (see Chapter 18). Dams built in certain geologic areas where rock, such as limestone, exist have experienced significant seepage issues. Robust foundation grouting programs have been used utilizing cementitious grout as well as concrete diaphragm walls to address the seepage issues. Inadequate spillway capacity is commonplace with many older dams. Dams classified as high hazard structures typically require that the probable maximum precipitation (PMP) rainfall event be used as the inflow design storm. Dams that have reservoirs that are surrounded by development and cannot be raised to increase flood storage are often modified to add auxiliary spillways.

Overtopping Protection. The science of designing for overtopping protection of earthen dams started in 1980 when RCC was used at Ocoee Dam No. 2, located in Ocoee, Tennessee. The dam was a 9.1 m (30 ft) high, 137 m (450 ft) long rockfilled timber crib structure. Four years later, the design was implemented at Brownwood Country Club dam in Texas. Since then, over 135 dams in the United States have used

FIGURE 22-23. Hoover Dam, constructed from 1933 to 1935.

FIGURE 22-24. Folsum Dam (left, built in 1948) and Auxiliary Spillway (right, completed in 2017) (courtesy of USBR).

RCC for overflow spillways over all or a concrete or portion of earthen dams. A recent development has been attempting to air entrain RCC for facing systems. Historically, it has been found that if RCC is well compacted, there is moderate resistance to freeze thaw damage. Research has found that high paste RCC mixes can be air entrained and provide significant freeze thaw durability (Wu and others 2020).

Typical design criteria for overtopping protection systems set the minimum storm event for initiating flow over the spillway at above the 50-year storm and often time the 100-year event. This criterion often requires that the primary spillway be enlarged to accommodate these more common flood events. In addition, many older conventional concrete principal spillways have been damaged due to freeze thaw or alkali aggregate reactivity and need to be replaced. With improvements in conventional concrete mixes, new reinforced concrete spillways such as those with labyrinth cycles at the control section are becoming more common. Because many spillways have unique loading conditions caused by ice and because gates experience more severe exposure conditions, designers are using ACI 350-06 as the design code. The Army Corps of Engineers EM 1110-2-2104, *Strength Design for Reinforced Concrete Hydraulic Structures* (2016), USBR - *Reclamation Design Standards*, and the Natural Resources Conservation Service NEH Part 628 are two other design codes often used for hydraulic structures.

Gravity Dams. New gravity dams and the buttressing of existing gravity dams are often constructed using RCC due to its speed of construction. Since the construction of the first RCC gravity dam in 1980 (Willow Creek Dam in Oregon) many different mix proportions have been used along with facing systems and construction means and methods. Early, or first generation, RCC dams concentrated on using lean mixes while today medium to high paste mixes are common with Vebe times between 15 and 25 seconds, as determined per ASTM C1170, *Standard Test Method for Determining Consistency and Density of Roller-Compacted Concrete Using a Vibrating Table.* Fly ash contents today range between 30% and 60% of the total cementitious material content. Upstream and downstream facing systems have used precast panels, conventional concrete, flexible membranes, and more recently grout enriched RCC. ACI 207.5R, *Report on Roller Compacted Mass Concrete* (2011), and the Army Corps of Engineers' Engineering Manual EM 1110-2-2006, *Roller Compacted Concrete* (2000), are very useful as design aids and construction quality control requirements.

Slope Protection. Soil-cement has also found a niche as a viable option for slope protection on embankment dams, reservoir liners and upstream slope erosion protection, and for flood control structure such as open channels and stream bank protection. Soil-cement is a material produced by blending in-situ or borrowed soil and/or aggregates with predetermined proportions of portland cement, possibly supplementary cementitious materials, and water. A wide variety of soils can be used to make durable soil-cement for water resources applications. Typical mixtures are designed in the laboratory to produce a compressive strength of approximately 5 MPa (700 psi) at 7 days. The USBR practice is to perform compressive strength, freeze-thaw, and wet-dry tests on one set of cylinders prepared at the estimated cement content and two other sets prepared at ± 2% the estimated cement content (USBR 2013).

Depending on the application, the soil-cement is placed and compacted according to either the plating or the stair-step methods of construction. The former is suitable for placing and compacting soil-cement on flat and up to 3 to 1 (horizontal to vertical) slope. For steeper slopes, the stair-step method is used (Figure 22-25). For reservoir upstream slope protection, it is common to use the plating method for areas below the normal pool level and the stair-step method at higher elevation. Stair steps at the higher elevations offer added protection against abrasion and erosion from wave action and optimize the design by reducing the freeboard height required to contain wave run-up during storm events. For more detailed information on the use of soil-cement for water resources applications, see PCA EB203, *Soil-Cement Guide for Water Resources Applications* (Richards and Hadley 2006), and ACI 230, *Report on Soil Cement.* Another concrete product used in certain applications for slope protection is the articulating concrete block system. Several block sizes are available and are typically cabled together for ease of installation.

FIGURE 22-25. Soil cement stair-step slope protection (courtesy of R. Bass).

Concrete Pipe

From the early Roman aqueducts and sewer systems which still have some portions in service today, to the modern day networks of underground sewers crisscrossing our cities and neighborhoods, concrete pipe has a proven history of meeting the need to provide a healthy environment, transport goods, and improve agricultural development around the world (Figure 22-26). Reinforced concrete pipe (RCP) provides modern sewer and drainage systems including sanitary sewers, storm sewers, and roadway culverts.

FIGURE 22-26. Concrete pipe is a durable and essential part of the drainage system.

FIGURE 22-27. Early 20th century demonstration of the strength of concrete pipe.

History of Concrete Pipe

Concrete pipe used in North America for agricultural drain tile and engineered sewer systems can be traced back to the early 19th century. The oldest recorded concrete pipe sanitary sewer installation was in 1842 in Mohawk, New York. There are also several installations of concrete pipelines that were installed in the late 1800s that are still in service today.

In 1904, ASTM Committee C-4 on *Clay and Cement-Concrete Sewer Pipe* was established (Figure 22-27). ASTM Committee C-4 was made up of manufacturers and users of both clay and concrete sewer pipe and was the forerunner of the current ASTM Committee C13 on Concrete Pipe. ASTM C13's scope includes RCP specifications, testing procedures, and definitions.

Concrete Pipe Design and Fabrication

Reinforced concrete pipe is designed to resist the compressive stress and the reinforcement is designed to resist the tensile stress. Cage machines wrap circumferential wire in a helix around longitudinal wires while welding the intersecting wires automatically. Wire rollers utilize welded wire reinforcement which is rolled, cut, and welded or clipped to form the cage. A number of different reinforcing cage configurations are utilized in RCP production depending on the strength requirements of the application.

Concrete used in the production of RCP can be dry cast, wet cast, or self-consolidating concrete (SCC). Dry cast concrete is most commonly used. Dry cast concrete has a low water-to-cement ratio and a zero slump. The inherent strength in the compacted concrete allows for immediate removal of the forms, enabling a single form set to be used multiple times throughout a production day. The most common fabrication methods used in the production of RCP are the dry cast method and the packerhead method. The dry cast method utilizes a low-frequency, high-amplitude vibration system to consolidate the concrete inside the forms. The packerhead method consolidates the concrete using the packing pressure generated by a rotating rollerhead. The formwork used for dry cast production is typically more rigid due to the heavy vibration and/or packing pressures needed to consolidate the concrete.

The freshly made RCP is cured in a kiln or designated production area until the concrete reaches the desired strength for handling and yarding operations. There are many kiln systems and curing methods utilized by the RCP industry; many RCP manufacturers use a low-pressure steam system that allows the curing process to be accelerated. Regardless of the system or method utilized, maintaining proper temperature and moisture conditions are keys to successful curing of the RCP. When the RCP has attained the desired strength, it is tipped out and moved to the storage area in the yard. A number of quality control checks are performed throughout the production process and the finished product is tested for conformance to the applicable specifications, for example, ASTM C76, *Standard Specification for Reinforced Concrete Culvert, Storm Drain and Sewer Pipe* (AASHTO M 170).

Reinforced concrete pipe is available in a number of different shapes (Figure 22-28): circular, horizontal elliptical, vertical elliptical, and arch. Reinforced concrete pipe has many end uses, including sanitary sewers, storm drains, culverts, irrigation distribution systems, low-pressure sewer force mains, low-pressure water supply systems, treatment plant piping, outfalls, utility tunnels, groundwater recharge systems, jacked or tunneled installations, cattle pass tunneling, and trenchless installations; the applications are practically endless. See concretepipe.org for more information.

FIGURE 22-28. Concrete pipe in a storage yard.

REFERENCES

PCA's online catalog includes links to PDF versions of many of our research reports and other classic publications.
Visit: cement.org/library/catalog.

AASHTO, *AASHTO LRFD Bridge Construction Specifications*, 4th Edition, American Association of State Highway and Transportation Officials, Washington, DC, 2017 and 2020 Interim Revisions, 2019, 40 pages.

AASHTO, *AASHTO LRFD Bridge Design Specifications*, 9th Edition, American Association of State Highway and Transportation Officials, Washington, DC, 2020.

ACI Committee 207, *Report on Roller Compacted Mass Concrete*, ACI 207.5R, American Concrete Institute, Farmington Hills, MI, 2011, 75 pages.

ACI Committee 216, *Code Requirements for Determining Fire Resistance of Concrete and Masonry Construction Assemblies*, ACI/TMS 216.1-14 (19), American Concrete Institute joint with The Masonry Society, Farmington Hills, MI, 2019, 28 pages.

ACI Committee 230, *Report on Soil Cement*, ACI 230.1, American Concrete Institute, Farmington Hills, MI, 2009, 28 pages.

ACI Committee 302, *Guide for Concrete Floor and Slab Construction*, ACI 302.1R-15, American Concrete Institute, Farmington Hills, MI, 2015, 80 pages.

ACI Committee 310, *Guide to Decorative Concrete*, ACI 310R-19, American Concrete Institute, Farmington Hills, MI, 2019, 52 pages.

ACI Committee 318, *Building Code Requirements for Structural Concrete and Commentary*, ACI 318-19, American Concrete Institute, Farmington Hills, MI, 2019, 628 pages.

ACI Committee 332, *Residential Code Requirements for Structural Concrete and Commentary*, ACI 332-20, American Concrete Institute, Farmington Hills, MI, 2020, 80 pages.

ACI Committee 336, *Report on Design and Construction of Drilled Piers*, ACI 336.3R-14, American Concrete Institute, Farmington Hills, MI, 2014, 30 pages.

ACI Committee 350, *Code Requirements of Environmental Engineering Concrete Structures*, ACI 350-06, American Concrete Institute, Farmington Hills, MI, 2006, 492 pages.

ACI Committee 360, *Guide to Design of Slabs-on-Ground*, ACI 360R-10, American Concrete Institute, Farmington Hills, MI, 2010, 76 pages.

ACI Committee 523, *Guide for Precast Cellular Concrete Floor, Roof, and Wall Units*, 523.2R-96, American Concrete Institute, Farmington Hills, MI, 1996.

ACI Committee 543, *Guide to Design, Manufacture, and Installation of Concrete Piles*, ACI 543R-12, American Concrete Institute, Farmington Hills, MI, 2012, 64 pages.

ACI Committee 551.2R, *Design Guide for Tilt-Up Concrete Panels*, ACI 551.2R-15, American Concrete Institute, Farmington Hills, MI, 2015, 76 pages.

ASCE Committee 59, *Blast Protection of Buildings*, ASCE 59-11, American Society of Civil Engineers, Reston, VA, 2011, 128 pages.

CRSI, *Vibration and Sound Control in Reinforced Concrete Buildings*, CRSI Technical Note ETN-B-3-16, Schaumburg, IL, 2016, 8 pages.

Dick, J.S., "Precast Concrete Piles Provide Durability, Versatility," *PileDriver*, Pile Driving Contractors Association, Orange Park, FL, Quarter 3 2005, pages 28 to 33.

Fanella, D.A., *Concrete Floor Systems: Guide to Estimating and Economizing*, Second Edition, SP041.02, Portland Cement Association, Skokie, IL, 2000, 52 pages.

Fanella, D.A., *Seismic Detailing of Concrete Buildings*, Second Edition, SP382, Portland Cement Association, Skokie, IL, 2007, 82 pages.

Fanella, D.A.; and Munshi, J.A., *Long-Span Concrete Floor Systems*, SP339, Portland Cement Association, Skokie, IL, 1990, 108 pages.

Farny, J.A.; Melander, J.M.; and Panarese, W.C., *Concrete Masonry Handbook for Architects, Engineers, Builders*, Sixth Edition, EB008, Portland Cement Association, Skokie, IL, 2008, 308 pages.

FEMA 2015, *Safe Rooms for Tornadoes and Hurricanes: Guidance for Community and Residential Safe Rooms*, Third Edition, FEMA P-361, Federal Emergency Management Agency, Washington, D.C., March 2015, 187 pages.

Hanley-Wood, "Your Business," *Concrete Construction*, Hanley-Wood, Media Inc., Chicago, IL, June 2013.

ICC, *2021 International Building Code*, International Code Council, Country Club Hills, IL, 2021, 736 pages.

ICC, *2021 International Residential Code for One- and Two-Family Dwellings*, International Code Council, Country Club Hills, IL, ICC, 2021, 932 pages.

ICC/NSSA 500, *Standard for the Design and Construction of Storm Shelters*, International Code Council, Inc., Country Club Hills, IL, 2020.

Kamara, M.E.; and Novak, L.C., *Simplified Design of Reinforced Concrete Buildings*, Fourth Edition, EB204, Portland Cement Association, Skokie, IL, 2011, 336 pages.

National Research Council, "Disaster Resilience A National Imperative Committee on Increasing National Resilience to Hazards and Disasters", *The National Academies, Policy and Global Affairs, Committee on Science, Engineering, and Public Policy*, The National Academies Press, Washington D.C., 2013.

PCA 1995, "Troubleshooting Concrete: Effect of Free Fall on Concrete," *Concrete Technology Today*, PL951, Skokie, IL, March 1995, pages 7 to 8.

PCI 124, *Specification for Fire Resistance of Precast/Prestressed Concrete*, Precast/Prestressed Concrete Institute, 2018.

PCI, "Recommended Practice for Design, Manufacture, and Installation of Prestressed Concrete Piling", *PCI Journal*, July - August 2019.

Richards, D.L.; and Hadley, H.R., *Soil-Cement Guide for Water Resources Applications*, EB203, Portland Cement Association, Skokie, IL, 2006, 84 pages.

Russell, H.G., *NCHRP Synthesis of Highway Practice 441: High Performance Concrete Specifications and Practices for Bridges*, National Research Council, Transportation Research Board, Washington, DC, 2013, page 73.

SDI 2010, *American National Standards Institute/Steel Deck Institute NC-2010, Standard for Non-Composite Steel Floor Deck*, Steel Deck Institute, Fox River Grove, IL, sdi.org, 2010.

SDI 2012, *American National Standards Institute/Steel Deck Institute C-2011, Standard for Composite Steel Floor Deck-Slabs*, Steel Deck Institute, Fox River Grove, IL, sdi.org, 2012.

SDI 2014, *Steel Deck Institute, SDI Code of Standard Practice 2014 - No. COSP14*, Steel Deck Institute, Fox River Grove, IL, sdi.org, July 2014.

Smith, S.; McCann, D.; and Kamara, M., *Blast Resistant Design Guide for Reinforced Concrete Structures*, EB090, Portland Cement Association, Skokie, IL, 2009, 152 pages.

STS Consultants Ltd., "The Effect of Free Fall Concrete in Drilled Shafts," *A Report to the Federal Highway Administration*, 1994. Available from the International Association of Foundation Drilling, P.O. Box 280379, Dallas, TX 75228.

Tarr, S.M.; and Farny, J.A., *Concrete Floors on Ground*, Fourth Edition, EB075, Portland Cement Association, Skokie, IL, 2008, 252 pages.

TMS, S*tandard Method for Determining the Sound Transmission Class Rating for Masonry Walls*, TMS 302-18, The Masonry Society, Fort Collins, CO, 2018.

TMS, *Building Code Requirements and Specification for Masonry Structures*, TMS 402/602 2013/2016, The Masonry Society, Fort Collins, CO, 2013/2016.

Turner, C.D., "Unconfined Free Fall of Concrete," *Journal of the American Concrete Institute*, American Concrete Institute, Detroit, MI, December, 1970, pages 975-976.

USACE, *Roller Compacted Concrete*, EM 1110-2-2006, US Army Corps of Engineers, Washington, DC, 2000, 77 pages.

USACE, *Strength Design for Reinforced Concrete Hydraulic Structures*, EM 1110-2-2104, US Army Corps of Engineers, Washington, DC, 2016, 138 pages.

USBR, *Design Standards No. 13 – Embankment Dams*, Chapter 17, Soil-Cement Slope Protection, U.S. Bureau of Reclamation, Denver, August. 2013.

USBR, *Reclamation Design Standards*, usbr.gov/tsc/techreferences/designstandards.html, March 2021.

Wang, C.K.; Salmon, C.G.; and Pincheira, J.A., *Reinforced Concrete Design*, 7th ed., John Wiley & Sons, Hoboken, NJ, 2007, 960 pages.

Wu, Z.; Libre, N.A.; and Khayat, K.H., "Factors Affecting Air-Entrainment and Performance of Roller Compacted Concrete," *Construction and Building Materials*, Volume 259, October 30, 2020.

HIGH-PERFORMANCE CONCRETE

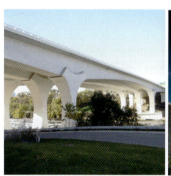

FIGURE 23-1. High-performance concrete is often used in bridges and tall buildings. (left) I-35W St. Anthony Falls bridge in Minneapolis, Minnesota. (right) 311 W. Wacker Drive, Chicago, Illinois.

High-performance concrete (HPC) exceeds the properties and constructability of normal concrete. Normal and special materials are used to make these specially designed concretes that must meet a combination of performance requirements. Special mixing, placing, and curing practices may be needed to produce and handle high-performance concrete. Performance tests are usually required to demonstrate compliance with specific project needs (ASCE 1993, Russell 1999, and Bickley and Mitchell 2001). High-performance concrete has been primarily used in bridges and tall buildings for its durability, strength, and high modulus of elasticity (Figure 23-1). It has also been used in shotcrete repair, poles, tunnels, parking garages, and agricultural applications.

High-performance concrete characteristics are defined, categorized, or developed for particular applications and environments (Goodspeed and others 1996, and Russell and Ozyildirim 2006). Some of the characteristics that may be required include enhanced engineering properties:

• High strength

• High early strength

• High modulus of elasticity

• Volume stability

• High durability

High-performance concretes are made with carefully selected high-quality ingredients and optimized mixture designs. These are batched, mixed, placed, compacted and cured to the highest industry standards. Typically, they will have low water-cementitious materials ratios of 0.20 to 0.45. High-range water reducers (or superplasticizers) are usually used to make these concretes fluid and more workable at lower w/cm.

High-performance concrete almost always has greater durability than normal concrete. This greater durability may be accompanied by normal

TABLE 23-1. Selected Properties of High-Performance Concrete

PROPERTY	TEST METHOD	CRITERIA THAT MAY BE SPECIFIED
High compressive strength	ASTM C39 (AASHTO T 22)	55 MPa to 140 MPa (8000 psi to 20,000 psi) at 28 days to 91 days
High-early compressive strength	ASTM C39 (AASHTO T 22)	20 MPa to 41 MPa (3000 psi to 6000 psi) at 3 hours to 18 hours, or 1 day to 3 days
High-early tensile strength	ASTM C78 (AASHTO T 97)	2 MPa to 4 MPa (300 psi to 600 psi) at 3 hours to 12 hours, or 1 day to 3 days
High modulus of elasticity	ASTM C469	34 GPa to more than 48 GPa (5 million psi to more than 7 million psi)
Low shrinkage	ASTM C157	Less than 800 millionths (microstrain) to less than 400 millionths (microstrain)
Low creep	ASTM C512	70 microstrain/MPa to less than 30 microstrain/MPa (0.52 microstrain/psi to less than 0.21 microstrain/psi)

strength or it may be partnered with high strength. Note that strength is not always the primary required property. For bridge decks, a normal strength concrete with very high durability and very low permeability is considered high performance concrete (Lane 2010). Table 23-1 lists properties that can be selected for high-performance concrete. Typical mix designs and member properties for many bridges can be found in the FHWA report compiling results for HPC bridges (Russell and others 2006). Not all properties can be achieved concurrently.

High-performance concrete specifications should be performance oriented. However, many specifications are a hybrid of performance requirements (such as permeability or strength limits) and prescriptive requirements (such as air content limits or dosage of SCMs) (Ferraris and Lobo 1998, and Caldarone and others 2005). Table 23-2 provides examples of high-performance concrete mixtures used in a variety of structures.

HIGH-STRENGTH CONCRETE

The definition of high strength concrete (HSC) changed historically as concrete strength used in the field increased. This publication considers high-strength concrete as a strength significantly beyond that used in normal practice. About 90% of ready mixed concrete has a 28-day specified compressive strength ranging from 20 MPa (3000 psi) to 40 MPa (6000 psi), with most of it between 20 MPa (3000 psi) and 35 MPa (5000 psi).

Most high-strength concrete applications are designed for compressive strengths of 70 MPa (10,000 psi) or greater as shown in Table 23-2. For bridges, the AASHTO *LRFD Specifications* (2020) state that the minimum allowable compressive strength for bridge decks and prestressed concrete members is 28 MPa (4000 psi). High strength concrete in bridge design applications has a minimum design strength of at least 55 MPa (8000 psi). For high-strength concrete, stringent

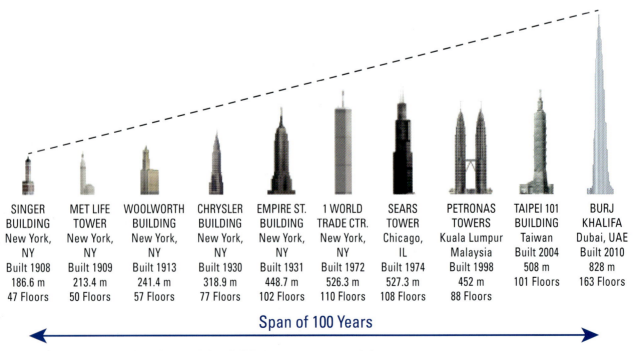

SINGER BUILDING	MET LIFE TOWER	WOOLWORTH BUILDING	CHRYSLER BUILDING	EMPIRE ST. BUILDING	1 WORLD TRADE CTR.	SEARS TOWER	PETRONAS TOWERS	TAIPEI 101 BUILDING	BURJ KHALIFA
New York, NY	New York, NY	New York, NY	New York, NY	New York, NY	New York, NY	Chicago, IL	Kuala Lumpur Malaysia	Taiwan	Dubai, UAE
Built 1908	Built 1909	Built 1913	Built 1930	Built 1931	Built 1972	Built 1974	Built 1998	Built 2004	Built 2010
186.6 m	213.4 m	241.4 m	318.9 m	448.7 m	526.3 m	527.3 m	452 m	508 m	828 m
47 Floors	50 Floors	57 Floors	77 Floors	102 Floors	110 Floors	108 Floors	88 Floors	101 Floors	163 Floors

Span of 100 Years

FIGURE 23-2. Holders of the title of "World's Tallest Building" over a 100-year period.

TABLE 23-2. High-Performance Concrete Mixtures Used in Various Structures

MIXTURE COMPONENTS AND PROPERTIES	MIXTURE NUMBER				
	1	2	3	4	5
Water, kg/m³ (lb/yd³)	130 (219)	145 (244)	151 (254)	160 (270)	108 (183)
Cement, kg/m³ (lb/yd³)	513 (865)	398 (671)*	311 (525)	400 (675)	356 (600)
Fly ash, kg/m³ (lb/yd³)	–	45 (76)	31 (53)	100 (168)	125 (210)
Slag cement, kg/m³ (lb/yd³)	–	–	47 (79)	–	56 (95)
Silica fume, kg/m³ (lb/yd³)	43 (72)	32 (54)*	16 (27)	50 (84)	24 (40)
Coarse aggregate, kg/m³ (lb/yd³)	1080 (1820)	1030 (1736)	1068 (1800)	847 (1428)	1068 (1800)
Fine aggregate, kg/m³ (lb/yd³)	685 (1155)	705 (1188)	676 (1140)	830 (1399)	775 (1306)
Water reducer, L/m³ (oz/yd³)	–	1.7 (47)	1.6 (41)	–	–
Retarder, L/m³ (oz/yd³)	–	–	–	–	–
Shrinkage-Reducing Admixture, L/m³ (oz/yd³)	–	–	–	–	–
Hydration Stabilizer, L/m³ (oz/yd³)	–	–	–	–	1.8 (48)
Air, %	–	5 - 8	7 ± 1.5	–	0 - 2.0
HRWR or plasticizer, L/m³ (oz/yd³)	16.2 (420)	3.2 (83)	2.1 (55)	3.0-3.5 (77-90)	6.6 (170)
Water to cementitious materials ratio	0.25	0.30	0.37	0.32	0.20
Slump or Slump Flow, mm (in.)	–	–	–	650 ± 75 (25 ± 3)	660-740 (26-29)
Comp. strength at 28 days, MPa (psi)	119 (17,250)	–	59 (8590)	60 (8700)	–
Comp. strength at 56 days MPa (psi)	–	–	–	80 (12,000)	80 (12,000)
Comp. strength at 91 days MPa (psi)	145 (21,000)	60 (8700)	–	–	–
Required E modulus at 28 days, GPa (million psi)	–	–	–	37.6 (5.45)	–
Required E modulus at 56 days, GPa (million psi)	41 (6)	–	–	41 (6)	–
Required E modulus at 91 days, GPa (million psi)	–	–	–	–	48 (7)

1. Two Union Square, Seattle, 1988.
2. Confederation Bridge, Northumberland Strait, Prince Edward Island/New Brunswick, 1997.
3. Wacker Drive Bi-Level Roadway, Chicago, 2001.
4. Burj Khalifa Tower, Dubai, United Arab Emirates, 2010.
5. The St. Regis, Chicago, 2020.
* Originally used a blended cement containing silica fume. Portland cement and silica fume quantities have been separated for comparison purposes.

application of the best practices is required. Compliance with the guidelines and recommendations for preconstruction laboratory and field-testing procedures described in ACI 363.2R, *Guide to Quality Control and Testing of High-Strength Concrete* (2011), are essential.

Concrete has become the material of choice for many tall towers. Buildings have progressed from the tall to the super-tall realm (Figure 23-2). The world's tallest building, the Burj Khalifa in Dubai, U.A.E., built in 2010, stands at 828 m (2,717 ft)

tall (Figure 23-3). The tower is primarily a concrete structure, with concrete construction utilized for the first 155 stories, above which exists a structural steel spire. High performance concrete is utilized for the tower (Table 23-2, Mixture No. 4), with wall and column concrete strengths ranging from 60 MPa to 80 MPa cube strength (7000 psi to 9300 psi cylinder strength). Additionally, the 80 MPa wall and column concrete was specified as a high-modulus concrete, in order to provide increased stiffness to the system (Baker and Pawlikowki 2012).

FIGURE 23-3. The Burj Khalifa in Dubai, U.A.E. is currently the world's tallest building. The tower is primarily a concrete structure, with concrete construction utilized for the first 155 stories, above which exists a structural steel spire.

As shown in Table 23-2, Mixture No. 1, the Two Union Square building in Seattle used concrete with a designed compressive strength of 131 MPa (19,000 psi) in its steel tube and concrete composite columns (Figure 23-4). High-strength concrete was used to meet a design criterion of 41 GPa (6 million psi) modulus of elasticity.

For tall buildings, carrying heavy loads is only one benefit of high compressive strengths; smaller member sizes decrease the building's dead load and the material has a higher modulus of elasticity, which increases the building's stiffness and its resistance to wind loads.

High-strength concrete is specified where reduced weight is important or where architectural considerations require smaller load-carrying elements. In high-rise buildings, HSC helps to achieve more efficient floor plans through smaller vertical members. HSC has also often proven to be the most economical alternative by reducing both the total volume of concrete and the amount of steel required for a load-bearing member. Since formwork typically accounts for 50%-60% of the cost of concrete construction, smaller column sizes that reduce the amount of formwork result in additional cost savings.

Traditionally, the specified strength of concrete has been based on 28-day test results. However, in high-rise concrete structures, the process of construction is such that the structural elements in lower floors are not fully loaded for periods of a year or more. For this reason, compressive strengths based on 56- or 91-day test results are commonly specified in order to achieve significant economy in material costs.

For bridges, the specified strength of concrete has also been based on 28-day test results. However, because of the use of fly ash and slag cement, which hydrate slower than the portland cement, 56-day strengths have also been specified on bridge projects. When later ages are specified, supplementary cementitious materials are usually incorporated into the concrete mixture and often this produces additional benefits because of the reduced peak heat generation during hydration.

FIGURE 23-4. The Two Union Square building in Seattle used concrete with a designed compressive strength of 131 MPa (19,000 psi) in its steel tube and concrete composite columns. High-strength concrete was used to meet a design criterion of 41 GPa (6 million psi) modulus of elasticity.

With use of low-slump or no-slump mixtures, high compressive-strength concrete is produced routinely under careful control in precast and prestressed concrete plants. These stiff mixtures are placed in ruggedly-built forms and consolidated by prolonged vibration or shock methods. However, cast-in-place concrete uses more fragile forms that do not permit the same compaction procedures. Hence, more workable concretes are necessary to achieve the required compaction and to avoid segregation and honeycombing. Superplasticizing admixtures are typically added to HPC mixtures to produce workable and often flowable mixtures.

Production of high-strength concrete may or may not require the purchase of special materials. The producer must know the factors affecting compressive strength and know how to vary those factors for best results. Each variable should be analyzed separately in developing a mix design. When an optimum or near optimum is established for each variable, it should be fixed as the remaining variables are studied. An optimized mixture design is then developed keeping in mind the economic advantages of using locally available materials. Many of the materials considerations discussed below also apply to most other high-performance concretes.

Hydraulic Cement

Selection of hydraulic cements for high-strength concrete should not be based only on mortar-cube tests but should also include tests of comparative strengths of concrete at 28, 56, and 91 days. Cement that yields the highest concrete compressive strength at extended ages (91 days) is preferable. For high-strength concrete, the cement should produce a minimum 7-day mortar-cube strength of approximately 30 MPa (4350 psi). Mortar cubes are tested according to ASTM C109, *Standard Test Method for Compressive Strength of Hydraulic Cement Mortars (Using 2-in. or [50 mm] Cube Specimens)*.

Trial mixtures with cement contents between 400 kg/m³ and 550 kg/m³ (between 675 lb/yd³ and 930 lb/yd³) should be made for each cement being considered for the project. Amounts will vary depending on target strengths. Other than decreases in sand content, with increasing cement content, the trial mixtures should be as nearly identical as possible.

Supplementary Cementitious Materials

Fly ash, silica fume, or slag cement are frequently used and are sometimes mandatory in the production of high-performance concrete. Other supplementary cementing materials (SCMs) such as metakaolin and other natural pozzolans may also be specified. The strength gain obtained with these SCMs cannot be attained by using additional cement alone, and SCMs greatly reduce permeability and improve durability. Supplementary cementitious materials are usually added at dosage rates of 5% to 25% or higher by mass of cementitious material. Some specifications only permit use of up to 10% silica fume, unless evidence is available indicating that concrete produced with a larger dosage rate will have satisfactory strength, durability, and volume stability. The water-to-cementitious materials ratio should be adjusted so that equal workability becomes the basis of comparison between trial mixtures. For each set of materials, there will be an optimum cement-plus-supplementary cementitious materials content at which strength does not continue to increase with greater amounts and the mixture becomes too sticky to handle. Blended cements can be used to make high-strength concrete with or without the addition of supplementary cementitious materials.

Aggregates

In high-strength concrete, careful attention must be given to aggregate size, grading, shape, surface texture, lithology, and cleanness. For each source of aggregate and concrete strength required there is an optimum-size aggregate that will yield the most compressive strength per weight of cementitious materials. To find the optimum size, trial batches should be made with 19 mm (¾ in.) and smaller coarse aggregates and varying cement contents. Many studies have found that 9.5 mm to 12.5 mm (⅜ in. to ½ in.) nominal maximum-size aggregates give optimum strength. Combining single sizes of aggregate to produce the required grading is recommended for close control and reduced variability in the concrete.

In high-strength concretes, the strength of the aggregate itself and the bond or adhesion between the paste and aggregate become important factors. Tests have shown that crushed-stone aggregates produce higher compressive strength in concrete than gravel aggregate using the same size aggregate and the same cementitious materials content. This is likely caused by superior aggregate-to-paste bond when using rough, angular, crushed material. For specified concrete strengths of 70 MPa (10,000 psi) or higher, the potential of the aggregates to meet design requirements must be established prior to use.

Coarse aggregates used in high-strength concrete should be free from detrimental coatings of dust and clay. Removing dust is important since it may affect the quantity of fines and consequently the water demand of a concrete mixture. Clay may affect the aggregate-paste bond. Washing of coarse aggregates may be necessary.

The quantity of coarse aggregate in high-strength concrete should be the maximum consistent with required workability. Because of the high percentage of cementitious material in high-strength concrete, an increase in coarse aggregate content beyond values recommended in standards for normal-strength mixtures is often necessary and allowable.

Due to the high amount of cementitious material in high-strength concrete, the role of the fine aggregate (sand) in providing workability and good finishing characteristics is not as critical as in conventional strength mixtures. Sand with a fineness modulus (FM) of about 3.0 – considered a coarse sand – has been found to be satisfactory for producing good workability and high compressive strength. For specified strengths of 70 MPa (10,000 psi) or greater, FM should be between 2.8 and 3.2 and not vary by more than 0.10 from the FM selected for the duration of the project. Finer sand, typically with a FM of between 2.5 and 2.7, may produce lower-strength, sticky mixtures.

High performance lightweight concrete has been used for bridges and other structures such as drilling platforms. This concrete typically uses normal-weight sand and lightweight coarse aggregate. Its lower mass also makes it an attractive option in seismic regions (Murugesh 2008 and Gilley 2008).

Admixtures

The use of chemical admixtures such as water reducers, retarders, high-range water reducers, or superplasticizers is typically necessary for HSC. They make more efficient use of the large amount of cementitious material in high-strength concrete and help to obtain the lowest practical water-to-cementing materials ratio. Chemical admixture efficiency must be evaluated by comparing strengths of trial batches. Also, compatibility between cement and SCM, as well as water-reducing and other admixtures, must be investigated by trial batches. From these trial batches, it will be possible to determine the workability, setting time, and amount of water reduction for given admixture dosage rates and times of addition.

The use of air-entraining admixtures where durability in a freeze-thaw environment is required is mandatory. However, air is not necessary nor desirable in high-strength concrete protected from the weather, such as interior columns and shear walls of high-rise buildings. Because air entrainment decreases concrete strength, testing to establish optimum air contents and spacing factors may be required. Certain high-strength concretes may not need as much air as normal-strength concrete for equivalent frost resistance. Pinto and Hover (2001) found that non-air-entrained, high-strength concretes had good frost and deicer-scaling resistance at a water-to-portland-cement ratio of 0.25. Burg and Ost (1994) found good frost resistance with non-air-entrained concrete containing silica fume at a water-to-cementitious materials ratio of 0.22.

HIGH MODULUS OF ELASTICITY CONCRETE

In high-rise buildings and in bridges, the stiffness of the structure is an important structural concern. On certain projects a minimum static modulus of elasticity has been specified as a means of increasing the stiffness of a structure. This represents the amount of force per unit area (stress) vs. the ratio of length change to original length (strain). It identifies how much a material can deform before failure. The stress-strain relationship and modulus of elasticity of concrete are important design parameters, especially for tall structures (see Chapter 9).

The modulus of elasticity is not necessarily proportional to the compressive strength of a concrete. There are code formulas for normal-strength concrete and suggested formulas for high-strength concrete. The modulus achievable is affected significantly by the properties of the aggregate, and also by the mixture proportions (Baalbaki and others 1991). If an aggregate has the ability to produce a high modulus, then the optimum modulus in concrete can be obtained by using as much of this aggregate as practical, while still meeting workability and cohesiveness requirements. If the coarse aggregate used is a crushed rock, and manufactured fine aggregate of good quality is available from the same source, then a combination of the two can be used to obtain the highest possible modulus.

Although several equations are available for estimating the modulus of elasticity based on compressive strength, many factors affect this property and designers should verify the value using trial field batching or documented performance (ACI 363.2R-2011). The St. Regis Chicago, spanning 363 m (1191 ft) and completed in 2020 (Figure 23-5), was designed for a static modulus target of 48 GPa (7 million psi) at 91 days for the bottom third of the structure (Table 23-2, Mixture No. 5). The required modulus value was the average required, with individual results allowed at 10% below specified. Ultimately, three different modulus levels were used for the structural system, decreasing with elevation. The average modulus in-place was 50.3 GPa (7.30 million psi) through level 30, 45.0 GPa (6.52 million psi) from level 30 to 80, and 43.7 GPa (6.34 million psi) from level 80 to the roof (level 95), well in excess of the minimum design values.

FIGURE 23-5. The St. Regis Chicago was designed for a high modulus of elasticity (courtesy of McHugh Construction Co.).

HIGH-EARLY-STRENGTH CONCRETE

High-early-strength concrete, also called fast-track concrete, achieves its specified strength at an earlier age than normal concrete. The time period in which a specified strength should be achieved may range from a few hours (or even minutes) to several days. High early strength can be attained using traditional concrete ingredients and concreting practices, although sometimes special materials or techniques are needed.

High early strength can be obtained using one or a combination of the following, depending on the age at which the specified strength must be achieved and on job conditions:

- Type III, Type HE, or other high-early-strength cements;
- high cementitious materials content (400 kg/m³ to 600 kg/m³ [675 lb/yd³ to 1000 lb/yd³]);
- low water-to-cementitious materials ratio (0.20 to 0.45 by mass);
- higher freshly mixed concrete temperature;
- higher curing temperature (Note: Keep internal member temperature under 70°C [158°F] to help prevent delayed ettringite formation);
- chemical admixtures (primarily accelerators and high-range water reducers);
- silica fume (or certain other supplementary cementitious materials);

TABLE 23-3. Strength Data for a Fast-Track Bonded Overlay

AGE	COMPRESSIVE STRENGTH MPa (psi)	FLEXURAL STRENGTH MPa (psi)	BOND STRENGTH MPa (psi)
4 hours	1.7 (252)	0.9 (126)	0.9 (120)
6 hours	7.0 (1020)	2.0 (287)	1.1 (160)
8 hours	13.0 (1883)	2.7 (393)	1.4 (200)
12 hours	17.6 (2546)	3.4 (494)	1.6 (225)
18 hours	20.1 (2920)	4.0 (574)	1.7 (250)
24 hours	23.9 (3467)	4.2 (604)	2.1 (302)
7 days	34.2 (4960)	5.0 (722)	2.1 (309)
14 days	36.5 (5295)	5.7 (825)	2.3 (328)
28 days	40.7 (5900)	5.7 (830)	2.5 (359)

Adapted from Knutson and Riley 1987.

- steam or autoclave curing;
- insulation to retain heat of hydration; and
- special rapid hardening cements.

High-early-strength concrete is used for prestressed concrete to allow for early stressing, precast concrete for rapid production of elements, high-speed cast-in-place construction, rapid form reuse, cold-weather construction, rapid repair of pavements (to reduce traffic downtime), fast-track paving, and several other uses.

In fast-track paving, use of high-early-strength mixtures allows traffic to open just hours after concrete is placed. An example of a fast-track concrete mixture used for a bonded concrete highway overlay consisted of 380 kg/m³ (640 lb/yd³) of Type III cement, 42 kg/m³ (70 lb/yd³) of Class C fly ash, 6.5% air,

a water reducer, and a water-to-cementitious materials ratio of 0.40. Strength data for this 40-mm (1½-in.) slump concrete are given in Table 23-3.

Figures 23-6 and 23-7 illustrate early strength development of concretes designed to open to traffic within 4 hours after placement. Figure 23-8 illustrates the benefits of blanket curing to develop early strength for patching or fast-track applications.

When designing early-strength mixtures, strength development is not the only criteria that should be evaluated; durability, early stiffening, camber of pretensioned members, autogenous shrinkage, drying shrinkage, temperature rise, and other properties also should be evaluated for compatibility with the project. Special curing procedures, such as fogging, may be needed to control plastic shrinkage cracking.

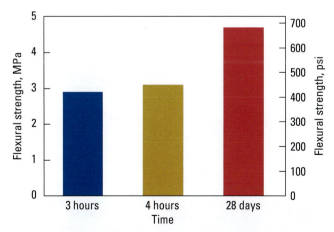

FIGURE 23-6. Strength development of a high-early strength concrete mixture using 390 kg/m³ (657 lb/yd³) of rapid hardening cement, 676 kg/m³ (1140 lb/yd³) of sand, 1115 kg/m³ (1879 lb/yd³) of 25 mm (1 in.) nominal max. size coarse aggregate, a water to cement ratio of 0.46, a slump of 100 mm to 200 mm (4 in. to 8 in.), and a plasticizer and retarder. Initial set was at one hour (Pyle 2001).

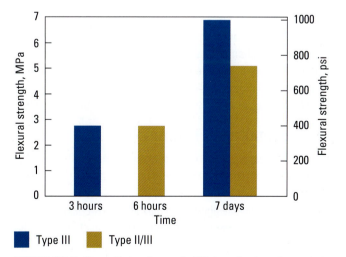

FIGURE 23-7. Strength development of high-early strength concrete mixtures made with 504 kg/m³ to 528 kg/m³ (850 lb/yd³ to 890 lb/yd³) of Type III or Type II/III cement, a nominal maximum size coarse aggregate of 25 mm (1 in.), a water-to-portland cement ratio of 0.30, a plasticizer, a hydration control admixture, and an accelerator. Initial set was at one hour (Pyle 2001).

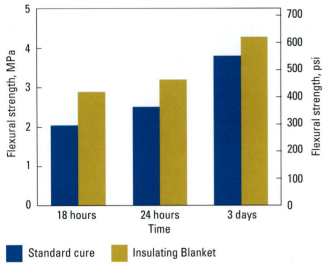

FIGURE 23-8. Effect of blanket insulation on fast-track concrete. The concrete had a Type I cement content of 421 kg/m³ (710 lb/yd³) and a water to cement ratio of 0.30 (Grove 1989).

VOLUME STABLE CONCRETE

As a larger number of buildings continue breaking height records, the importance of time-dependent shortening of concrete columns and shear walls becomes more critical (Fintel and Khan 1969). Differential shortening can affect the serviceability of high-rise buildings, leading to such undesirable conditions as sloping floors; cracking of structural members and interior partitions; buckled elevator guide rails, misaligned elevator stops relative to floors, and damage to façade elements and plumbing risers (Carreira and Poulos 2007, Baker and others 2007). In buildings more than 30 stories tall (or about 120 m or 400 ft) and for shorter buildings of hybrid or mixed construction, structural engineers typically design for differential shortening to minimize these effects.

Shortening occurs as a result of both drying shrinkage and creep. Drying shrinkage of concrete is due mainly to the evaporation of chemically uncombined water. Creep of concrete is the dimensional change or increase in strain with time due to a sustained stress. For more information on designing high-performance concrete for volume stability refer to Chapter 10.

HIGH-DURABILITY CONCRETE

HPC is often used in severe environments where a high degree of durability is needed and extended service life is specified. The Wacker Drive Bi-Level project in Chicago (see Table 23-2, Mixture No. 3) was designed as a biaxially post-tensioned, high-performance concrete slab structure. The project included the development and testing of high-performance concrete for both the precast and cast-in-place superstructures. Prototype segments were constructed and tested to validate the design of

FIGURE 23-9. The Wacker Drive Bilevel reconstruction specified a HPC concrete slab structure with a service life in excess of 75 years (courtesy of WJE).

the structure and evaluate maintenance methods to provide a service life in excess of 75 years (Figure 23-9).

The Confederation Bridge across the Northumberland Strait between Prince Edward Island and New Brunswick (Figure 23-10), Canada, has a 100-year design life (see Table 23-2, Mixture No. 2). This bridge contains HPC designed to efficiently protect the embedded reinforcement (Dunaszegi 1999 and Phipps 2008). The concrete had a diffusion coefficient of 4.8×10^{-13} m²/s at six months (a value 10 to 30 times lower than that of conventional concrete). The electrical resistivity was measured at 470 Ω-m to 530 Ω-m, compared to 50 Ω-m for conventional concrete. The design required that the concrete permeability, as determined using ASTM C1202, *Standard Test Method for Electrical Indication of Concrete's Ability to Resist Chloride Ion Penetration*, be rated at less than 1000 coulombs. The high concrete resistivity in itself will result in a rate of corrosion that is potentially less than 10% of the corrosion rate for conventional concrete. Refer to Chapter 11 for an in-depth review of durability issues that high-performance concrete can address.

FIGURE 23-10. The Confederation Bridge (shown here under construction) is a concrete bridge in a severe environment with a 100 year design life.

HIGH-PERFORMANCE CONCRETE CONSTRUCTION

Proportioning

The trial mixture approach is best for selecting proportions for high-performance concrete. To obtain high performance, it is necessary to use a low water to cementitious materials ratio and, often, a high portland cement content. The unit strength obtained for each unit of cement used in a cubic meter (yard) of concrete can be plotted as strength efficiency to evaluate mixture designs.

The water requirement of concrete increases as the fine aggregate content is increased for any given size of coarse aggregate. Because of the high cementitious materials content of these concretes, the fine aggregate content can be kept low. However, even with well-graded aggregates, a low water-cementitious materials ratio may result in concrete that is not sufficiently workable for the job. If a superplasticizer is not used, the design should be revised. A slump of around 200 mm (8 in.) will provide adequate workability for most applications. See Chapter 13, ACI 211.4, *Guide for Selecting Proportions for High-Strength Concrete Using Portland Cement and Other Cementitious Materials* (2008), PCA EB114, *High-Strength Concrete* (Farny and Panarese 1994), Nawy (2001), and Caldarone (2009) for additional guidance on proportioning concrete mixtures.

Trial Mixtures

High-performance concrete has been successfully mixed in transit mixers and central mixers. However, many of these concretes that have higher cementitious materials contents tend to be sticky and may cause build-up in these mixers, especially if silica fume is used. Where dry, uncompacted silica fume has been batched into a mixture, "balling" of the mixture has occurred and mixing has been incomplete. In these instances it has been necessary to experiment with the charging sequence, and the percentage of each material added at each step in the batching procedure. Batching and mixing sequences should be optimized during the trial mix phase. When truck mixing is unavoidable, the best practice is to reduce loads to 90% of the rated capacity of the trucks.

When there is no recent history of HPC mixtures that meet specified requirements, it is essential to first make laboratory trial mixtures to establish optimum proportions. At this stage, the properties of the mixture, such as workability, air content, density, strength, and modulus of elasticity can be determined. It is also important to determine how admixtures interact and their effects on concrete properties. Once laboratory mixture proportions have been determined, field trials using full loads of concrete are essential. They should be delivered to the site or to a mock-up to establish and confirm the suitability of the batching, mixing, transporting, and placing systems to be used.

For large projects or a mass concrete structure, a trial member may be required. One or more loads of the proposed mixture is cast into a trial member or mock-up. The fresh concrete is tested for slump, air content, temperature, and density. Casting the trial member or mock-up provides an opportunity to assess the suitability of the mixture for placing, compaction, and temperature gain. The trial member or mock-up can be instrumented to record temperatures and temperature gradients. It can also be cored and tested to provide correlation with standard cylinder test results. The cores can be tested to provide the designer with in-place strength and modulus values for reference during construction. The heat characteristics of the mixture can also be determined using embedded sensors (Figure 23-11) and computer programs (Gajda 2007), and the data used to determine how thermal control should be applied to the project.

FIGURE 23-11. Installation of embedded sensors for monitoring heat characteristics of the concrete mixture (courtesy of J. Gajda).

Placing, Consolidation, Finishing, and Curing

Close liaison between the contractor and the concrete producer allows concrete to be discharged rapidly after arrival at the jobsite. Final adjustment of the concrete should be supervised by the concrete producer's technicians at the site, by a concrete laboratory, or by a consultant familiar with the performance and use of high performance concrete.

Delays in delivery and placing must be eliminated. Sometimes it may be necessary to reduce batch sizes if placing procedures are slower than anticipated. Rigid surveillance must be exercised at the jobsite to prevent any addition of water exceeding the specified w/cm. Increases in workability should only be achieved by the addition of a superplasticizer.

Consolidation is very important in achieving the potential of high-performance concrete. Concrete must be vibrated as quickly as possible after placement in the forms. High-frequency vibrators should be small enough to allow sufficient clearance between the vibrating head and reinforcing steel. Over-vibration of workable normal-strength concrete often results in segregation, loss of entrained air, or both. On the other hand, high-performance concrete without a superplasticizer will be relatively stiff and contain little air. Consequently, inspectors should be more concerned with under-vibration rather than over-vibration. Most high-performance concrete, particularly very high-strength-concrete, is placed at slumps of 180 mm to 220 mm (7 in. to 9 in.). Even at these slumps, some vibration is required to ensure compaction. The amount of compaction should be determined by onsite trials.

High-performance concrete can be difficult to finish. High cementitious materials contents, large dosages of admixtures, low water contents, and air entrainment all contribute to concrete sticking to the trowels and other finishing equipment. When this occurs, finishing activities should be minimized. The finishing sequence should be modified to include the use of a fresno trowel in place of a bullfloat.

Curing of high-performance concrete is even more important than curing normal-performance concrete. Providing adequate moisture and favorable temperature conditions are recommended for a prolonged period, particularly when 56- or 91-day concrete strengths are specified.

Additional curing considerations apply with HPC. Where very low water-cementitious materials ratios are used in flatwork (slabs and overlays), and particularly where silica fume is used in the mixture, there will be little if any bleeding before or after finishing. In these situations it is imperative that fog curing or evaporation retarders be applied to the concrete immediately after the surface has been struck off. This is necessary to avoid plastic shrinkage cracking of horizontal surfaces and to minimize crusting. Fog curing, followed by 7 days of wet curing, has proven to be very effective. Internal curing can be very beneficial to HPC.

It is inevitable that some vertical surfaces, such as columns, may be difficult to cure effectively. Where projects are fast-tracked, columns are often stripped at an early age to allow raising of self-climbing form systems. Concrete is thus exposed to early drying, sometimes within eleven hours after casting. Because of limited access, additional curing is difficult and impractical.

In one project, tests were conducted on column concrete to determine if such early exposure and lack of curing have any harmful effects. The tests showed that for a portland cement-slag cement-silica fume mixture with a specified strength of 70 MPa (10,000 psi), the matrix was sound and a very high degree of impermeability to water and chloride ions had been achieved (Bickley and others 1994). Nevertheless, the best curing possible is recommended for all HPC.

The temperature history of HPC is an integral part of its curing process. Advantage should also be taken of recent developments in curing technology. Temperature increases and gradients that will occur in a concrete placement can be predicted by procedures that provide data for this purpose. With these techniques, measures to heat, cool, or insulate a concrete placement can be determined and then applied to significantly reduce both micro- and macro-cracking of the structure and assure durability. The increasing use of these techniques will be required in most structures using HPC to assure that the cover concrete provides long term protection to the steel, to meet the intended service life of the structure.

Thermal Control

The quality, strength, and durability of HPC are highly dependent on its temperature history from the time of delivery to the completion of curing. In principle, favorable construction and placing methods will enable: a low temperature at the time of delivery; the smallest possible maximum temperature after placing; minimum temperature gradients after placing; and a gradual reduction to ambient temperature after maximum temperature is reached. Excessively high temperatures and gradients can cause excessively fast hydration and micro- and macro-cracking of the concrete. Keeping member temperature under 70°C (158°F) internal curing temperature helps prevent delayed ettringite formation (DEF) (see Chapter 11 and Gajda 2007).

It has been a practice on major high-rise structures incorporating concrete with specified strengths of 70 MPa to 85 MPa (10,000 psi to 12,000 psi) to specify a maximum delivery temperature of 18°C (64°F) (Ryell and Bickley 1987). In hot weather conditions it is possible that this limit could only be met using liquid nitrogen to cool the concrete (see Chapter 19). Experience with very-high-strength concrete suggests that a delivery temperature of no more than 25°C (77°F), preferably 20°C (68°F), should be allowed. The specifier should state the required delivery temperature.

In HPC applications such as high-rise buildings, column sizes are large enough to be classified as mass concrete. Normally, excessive heat generation in mass concrete is controlled by using a low cementitious materials content. When high cementitious materials content HPC mixtures are used under these conditions, other methods of controlling maximum concrete temperature must be employed. Burg and Ost (1994) recorded temperature rise for 1220-mm (4-ft) concrete cubes. A maximum temperature rise of 9.4°C to 11.7°C for every 100 kg of cement per cubic meter of concrete (10°F to 12.5°F for every 100 lb of cement per cubic yard of concrete) was measured. Burg and Fiorato (1999) monitored temperature rise in high-performance concrete caissons; they determined that in-place strength was not affected by temperature rise due to heat of hydration.

Quality Control

A comprehensive quality-control program is required at both a concrete plant and onsite to guarantee consistent production and placement of high-performance concrete. Inspection of concreting operations, from stockpiling of aggregates through completion of curing, is important. Closer production control than is normally obtained on most projects is necessary. Also, routine sampling and testing of all materials is particularly necessary to control uniformity of the concrete.

While tests on concrete should always be made in strict accordance with standard procedures, some additional requirements are recommended, especially where specified strengths are 70 MPa (10,000 psi) or higher. In testing high-strength concrete, some changes and more attention to detail are required. For example, cardboard cylinder molds, which can cause lower strength-test results, should be replaced with plastic or reusable steel molds. Capping of cylinders must be done with great care using appropriate capping compounds. Lapping (grinding) the cylinder ends is a preferred alternative to capping. For specified strengths of 70 MPa (10,000 psi) or greater, end grinding to a flatness tolerance of 0.04 mm (0.0016 in.) is recommended (Caldarone and Burg 2009).

The physical characteristics of a testing machine can have a major impact on the result of a compression test. It is recommended that testing machines be extremely stiff, both longitudinally and laterally.

The quality control necessary for the production of high compressive strength concrete will, in most cases, lead to low variance in test results. Strict vigilance in all aspects of quality control on the part of the producer and quality testing on the part of the laboratory are necessary on high-strength concrete projects. For concretes with specified strengths of 70 MPa (10,000 psi) or greater, the coefficient of variation is the preferred measure of quality control.

ULTRA-HIGH PERFORMANCE CONCRETE

Ultra-high performance concrete (UHPC) is characterized by high strength and very low permeability, obtained by optimized particle packing and by a very low water cement ratio. UHPC stems from work done in the early 1970s to study high strength cement paste relying on water-cementitious materials ratios in the range of 0.20 to 0.30 (Yudenfreund and others 1972, Odler and others 1972, and Brunauer and others 1973). These mixtures achieved compressive strengths of 200 MPa (29,000 psi) under standard curing conditions.

Today's UHPC mixtures incorporate high doses of fiber reinforcement, typically in the range of 1.5% to 3% by volume.

FIGURE 23-12. Freshly-mixed ultra-high performance concrete.

The key role fibers play in UHPC the material may also be referred to as Ultra-High Performance Fiber-Reinforced Concrete or UHPFRC.

These mixtures generally rely on high cementitious binder contents often in the range of 900 kg/m³ to 1150 kg/m³ (2000 lb/yd³ to 2550 lb/yd³) and have an upper size limit of aggregate of 8 mm (⁵⁄₁₆-in) or less. Many UHPC mixtures are 100% powder. The critical factor in these mixtures is particle packing. Because UHPC mixtures rely on a high percentage of reactive materials the lower limit of particle size will be in the sub-micron range. Particle packing combined with HRWR chemical admixtures for w/cm ratios ranging between 0.15 to 0.25 yield highly flowable, exceedingly dense concrete mixtures (Figure 23-12).

Due to their complexity UHPC mixtures require precise control of component materials. Research has shown that with careful oversight, optimized non-proprietary mixtures can be successfully made by skilled concrete producers (Batista and others 2019). In addition, there are several proprietary pre-blended mixtures commercially available that, when combined with fibers and mixed according to manufacturer's recommendations, will yield UHPC mixtures with predictable performance characteristics.

Extended mixing times and relatively short placement windows are often associated with UHPC mixtures. While some UHPC mixtures can be produced in traditional drum mixers as with transit mixers or central mixers, the majority of UHPC is produced using high-shear planetary-type mortar or pan-type concrete mixers.

The compressive strength of UHPC is typically around 200 MPa (29,000 psi), but can be produced with compressive strengths up to 810 MPa (118,000 psi) (Semioli 2001). Depending on the mixture it is not uncommon for these materials to reach 36,000 psi (250 MPa). In their 1995 work, Richard and Cheyrezy show that specimens heat cured under pressure are capable of reaching strengths of roughly 94,000 psi (650 MPa), and when combined with steel aggregate approximately 116,000 psi (800 MPa).

Unlike traditional concrete that will resist a strain up to the point of cracking then rapidly lose strain resistance, UHPC will continue to resist a given strain responding with a series of fine cracks and delayed strain-softening prior to yielding. Essentially this provides ductility and enables UHPC concrete elements to "bend" which in turn can lead to smaller cross sections, reduced dead loads, and greater spans. However, the low comparative tensile strength may require prestressing reinforcement in severe structural service. Table 23-4 compares hardened concrete properties of UHPC with those of conventional concrete.

Architectural applications of UHPC tends to leverage its reliance on fiber reinforcement to yield very elaborate geometrical shapes and very thin sections (Figure 23-13).

The Federal Highway Administration (FHWA) studied multiple properties of UHPC, namely compressive and tensile strengths, creep and shrinkage, chloride ion penetration, and freeze-thaw durability. The 28-day compressive strengths ranged from 126 MPa to 193 MPa (18,000 psi to 28,000 psi), depending upon whether or not a secondary heat treatment was used to further develop compressive strength. The tensile strength was approximately 6.2 MPa (900 psi) without secondary heat treatment and 9.0 MPa (1300 psi) after secondary heat treatment. The UHPC showed excellent resistance to chloride ion penetration, exhibited good long-term creep and shrinkage behavior, and held up well in freeze-thaw testing (Graybeal 2006).

UHPC has been used in the beams and decks of U.S. bridges. The first bridge in the U.S. to use UHPC was the Mars Hill Bridge in Wapello County, Iowa. This 33.5-m (110-ft.) bridge used three 107-cm (42-in.) modified Iowa bulb-tee girders, and opened to traffic in 2006. The second bridge was the Cat Point Creek Bridge in Richmond County, Virginia. This ten-span bridge

opened to traffic in 2008 and contained one span with five UHPC bulb-tee girders. The third bridge was the Jakway Park Bridge in Buchanan County, Iowa. This 15.7-m (51.5-ft.) long bridge used UHPC pi-girders (Graybeal 2009 and Bierwagen 2009).

Other uses for UHPC in bridges include waffle deck panels in Iowa and cast-in-place UHPC connections between full-depth deck panels in New York. UHPC has also been used in pedestrian bridges (Bickley and Mitchell 2001 and Semioli 2001).

The US Army Corps of Engineers (USACE) is undergoing research on the use of UHPC panels for the lock wall repairs (USACE 2020). See Li and Li (2010) for other repair applications and Li (2010) for infrastructure applications.

FIGURE 23-13. Architectural UHPC staircase (photo by Thomas Mølvig, courtesy of Hi-CON).

TABLE 23-4. Comparison of Conventional Concrete and UHPC

MATERIAL CHARACTERISTIC	CONVENTIONAL CONCRETE MPa (psi)	UHPC MPa (psi)
Compressive strength	20 to 40 (3000 to 6000)	150 to 250 (22,000 to 36,000)
Direct tensile strength	1 to 3 (150 to 440)	6 to 12 (900 to 1700)
Elastic modulus (ASTM C469)	25,000 to 30,000 (3,600,000 to 4,400,000)	40,000 to 50,000 (6,000,000 to 7,200,000)

Adapted from ACI 239R (2018).

The low porosity of UHPC also gives excellent durability and transport properties, which makes it a suitable material for the storage of nuclear waste (Matte and Moranville 1999). A low-heat type of UHPC has been developed to meet needs for mass concrete pours for nuclear reactor foundation mats and underground containment of nuclear wastes (Gray and Shelton 1998).

REFERENCES

PCA's online catalog includes links to PDF versions of many of our research reports and other classic publications.
Visit: cement.org/library/catalog.

AASHTO, *LRFD Bridge Design Specifications*, 9th Ed., American Association of State Highway and Transportation Officials, Washington, D.C., 2020, 1912 pages.

ACI Committee 211, *Guide for Selecting Proportions for High-Strength Concrete Using Portland Cement and Other Cementitious Materials*, ACI 211.4-08, American Concrete Institute, Farmington Hills, MI, 2008, 25 pages.

ACI Committee 239, *Ultra-High-Performance Concrete: An Emerging Technology Report*, ACI 239R-18, American Concrete Institute, Farmington Hills, MI, 2018, 21 pages.

ACI Committee 363, *Guide to Quality Control and Testing of High-Strength Concrete*, 363.2R-11, American Concrete Institute, Farmington Hills, MI, 2011.

ASCE, *High-Performance Construction Materials and Systems*, Technical Report 93-5011, American Society of Civil Engineers, New York, NY, April 1993.

Baalbaki, W.; Benmokrance, B.; Chaallal, O.; and Aïtcin, P.C., "Influence of Coarse Aggregates on Elastic Properties of High Performance Concrete," *ACI Materials Journal*, Vol. 88, No. 5, 1991, pages 449 to 503.

Baker, W.F.; and Pawlikowski, J.J., "Higher and Higher: The Evolution of the Buttressed Core", *Civil Engineering*, American Society of Civil Engineers, October 2012, 8 pages.

Batista, C.; Rowland, L.; and Nielsen E., "The Next Generation of Ultra-High-Performance Concrete," *2nd International Interactive Symposium on Ultra-High Performance Concrete*, Albany, NY, 2019.

Bickley, J.A.; and Mitchell, D., *A State-of-the-Art Review of High Performance Concrete Structures Built in Canada: 1990 - 2000*, Cement Association of Canada, Ottawa, Ontario, May 2001, 122 pages.

Bickley, J.A.; Ryell, J.; Rogers, C.; and Hooton, R.D., "Some Characteristics of High-Strength Structural Concrete: Part 2," *Canadian Journal of Civil Engineering*, National Research Council of Canada, December 1994.

Bickley, J.A.; Sarkar, S.; and Langlois, M., "The CN Tower," *Concrete International*, American Concrete Institute, Farmington Hills, MI, August 1992, pages 51 to 55.

Brunauer, S.; Yudenfreund, M.; Odler, I.; and Skalny, J., "Hardened Portland Cement Pastes of Low Porosity, VI. Mechanism of the Hydration Process," *Cement and Concrete Research*, Vol. 3, No. 2, 1973, pages 129 to 147.

Burg, R.G.; and Ost, B.W., *Engineering Properties of Commercially Available High-Strength Concretes (Including Three-Year Data)*, Research and Development Bulletin RD104, Portland Cement Association, Skokie, IL, 1994, 62 pages.

Caldarone, M.A., *High-Strength Concrete: A Practical Guide*, Taylor and Francis, London, 2009, 272 pages.

Caldarone, M.A.; and Burg, R.G., "Importance of End Surface Preparation when Testing High Strength Concrete Cylinders," *HPC Bridge Views*, Issue 57, September/ October 2009.

Caldarone, M.A.; Taylor, P.C.; Detwiler, R.J.; and Bhide, S.B., *Guide Specification for High Performance Concrete for Bridges*, EB233, 1st edition, Portland Cement Association, Skokie, IL, 2005, 64 pages.

Dunaszegi, L., "HPC for Durability of the Confederation Bridge," *HPC Bridge Views*, Federal Highway Administration and National Concrete Bridge Council, Portland Cement Association, Skokie, IL, September/ October 1999, page 2.

Farny, J.A.; and Panarese, W.C., *High-Strength Concrete*, EB114, Portland Cement Association, Skokie, IL, 1994, 60 pages.

Goodspeed, C.H.; Vanikar, S.; and Cook, R.A., "High-Performance Concrete Defined for Highway Structures," *Concrete International*, American Concrete Institute, Farmington Hills, MI, February 1996, pages 62 to 67.

Gray, M.N.; and Shelton, B.S., "Design and Development of Low-Heat, High-Performance Reactive Powder Concrete," *Proceedings, International Symposium on High-Performance and Reactive Powder Concretes*, Sherbrooke, Quebec, August 1998, pages 203 to 230.

Graybeal, B., *Development of a Field-Cast Ultra-High Performance Concrete Composite Connection Detail for Precast Concrete Bridge Decks*, Report No. FHWA-HRT-12-042, Federal Highway Administration, McLean, VA, 2012.

Graybeal, B., "UHPC Making Strides," *Public Roads*, Vol. 72, Number 4, January/February 2009.

Grove, J.D., "Blanket Curing to Promote Early Strength Concrete," *Transportation Research Record*, Paper No. 880193, IDOT Research Project MLR-87-7, Transportation Research Board, Washington, D.C., 1989.

Knutson, M.; and Riley, R., "Fast-Track Concrete Paving Opens Door to Industry Future," *Concrete Construction*, Addison, IL, January 1987, pages 4 to 13.

Lane, S.N., "HPC Lessons Learned and Future Directions," *IABMAS 2010: The Fifth International Conference on Bridge Maintenance, Safety and Management*, International Association on Bridge Maintenance and Safety, Philadelphia, PA, July 11 to 15, 2010.

Li, M.; and Li, V.C., "High-Early-Strength ECC for Rapid Durable Repair - Material Properties," *ACI Materials Journal*, American Concrete Institute, Farmington Hills, MI, June 2010.

Li, V.C., "High-Ductility Concrete for Resilient Infrastructure," *Advanced Materials Journal*, July 2010.

Matte, V.; and Moranville, M., "Durability of Reactive Powder Composites: Influence of Silica Fume on Leaching Properties of Very Low Water/Binder Pastes," *Cement and Concrete Composites*, Vol. 21, 1999, pages 1 to 9.

Nawy, E.G., *Fundamentals of High-Performance Concrete*, John Wiley and Sons, New York, 2001.

Odler, I.; Yudenfreund, M.; Skalny, J.; and Brunauer, S., "Hardened Portland Cement Pastes of Low Porosity, III. Degree of Hydration.

Expansion of Paste, Total Porosity," *Cement and Concrete Research*, Vol. 2, No. 4, 1972, pages 463 to 480.

Phipps, A.R., "HPC for 100-Year Life Span," *HPC Bridge Views*, Federal Highway Administration and National Concrete Bridge Council, Issue 52, November/December 2008.

Pinto, R.C.A.; and Hover, K.C., *Frost and Scaling Resistance of High-Strength Concrete*, Research and Development Bulletin RD122, Portland Cement Association, 2001, 75 pages.

Richard, P.; and Cheyrezy, M., "Composition of Reactive Powder Concretes," *Cement and Concrete Research*, Vol. 25, No. 7, Oct. 1995, pages 1501 to 1511.

Russell, Henry G., "ACI Defines High-Performance Concrete," Concrete International, American Concrete Institute, Farmington Hills, Michigan, February 1999, pages 56 to 57.

Ryell, J.; and Bickley, J.A., "Scotia Plaza: High Strength Concrete for Tall Buildings," *Proceedings International Conference on Utilization of High Strength Concrete*, Stavanger, Norway, June 1987, pages 641 to 654.

Semioli, W.J., "The New Concrete Technology," *Concrete International*, American Concrete Institute, Farmington Hills, MI, November 2001, pages 75 to 79.

USACE, *Ultra-High Performance Concrete (UHPC) Panels for the Repair of Navigation Structures Subject to Impact and Abrasion, Building Strong Fact Sheet*, US Army Corps of Engineers, Engineering Research and Development Center, 18-19 August 2010.

Yudenfreund, M.; Odler, I.; and Brunauer, S., "Hardened Portland Cement Pastes of Low Porosity, I. Materials and Experimental Methods," *Cement and Concrete Research*, Vol. 2, No. 3, May 1972, pages 313 to 330.

INNOVATIONS IN CONCRETE TECHNOLOGY

FIGURE 24-1. The Pilazzo Italia Expo Milano 2015, was constructed with more than 700 self-consolidated, nano-engineered concrete panels, made from photocatalytic cement (courtesy of Italcementi).

Concrete technology is ever-changing. Due to its inherent flexibility and beauty, concrete stirs the imagination more than any other building material. While concrete embodies the fundamentals of strength, stability and safety, it also inspires the drive to innovate (Figure 24-1).

Concrete's advancement is dependent on any new technology's usefulness and adaptability by end users. Barriers to implementing new technologies include initial cost and lack of industry standards. Also, the goals of the developers and researchers may not match those of end users.

Concrete can be transformed into objects that defy most conceptions of the material that is taken for granted in everyday construction. Innovations in concrete technology include the following recent advancements:

- Translucent concrete
- Photocatalytic concrete
- Engineered cementitious composites
- Self-healing concrete
- Additive manufacturing
- Geosynthetic cementitious composite mats
- Robotics
- Drones
- Artificial intelligence
- Virtual cement hydration
- Nanotechnology
- Lunar concrete

TRANSLUCENT CONCRETE

Transparent concrete was first mentioned in a 1935 Canadian patent by Saint-Gobain, a glass manufacturer (Long 1935). However, similar technology was not explored until much later in the 1990s and early 2000s by Price (Hart 2005) and Losonczi (Graydon 2004). Further advances on light-transmitting concrete technology evolved into manufacturing translucent blocks, sheets, and panels.

Light transmitting or translucent concrete is now increasingly used in fine architecture and cladding for interiors. Translucent concrete is based on the concept of nano-optics, where embedded optical elements act as conduits to transmit light from one side of the surface to another. The optical materials allow light, shadows, and even colors to project through the concrete (Figure 24-2).

Materials

Translucent concrete is made by combining cementitious ingredients and fine aggregates and optical fibers, rods, or fabric. Cementitious ingredients often include conventional portland cement and other supplementary cementing materials. Some mixtures are also based on a polymeric matrix including epoxy-based mortars. Translucent concrete does not typically contain coarse aggregates as they may damage the fiber strands and stop light from passing through the concrete. Accelerators and rapid setting cements are often used in the mixture. Because the fibers are small in diameter and the proportion of fibers to concrete is relatively low, translucent concrete delivers the comparable strength and durability as conventional concrete.

Translucent Technology

A translucent material is made up of components with different indices of refraction, and optical fibers conduct light from artificial and natural sources.

The transparency of the concrete is characterized by the following factors:

- Transmittance – the ratio of the light energy falling on a body to that transmitted through it.
- Haze – an atmospheric phenomenon in which particulates obscure the clarity.
- Birefringence – maximum difference between refractive indices, dependent on the polarization and direction of light.
- Refractive index – a dimensionless number that describes how fast light travels (and how much light is refracted) through a material.
- Dispersion – the phenomenon of light scattering dependent on its frequency and length.

A general translucent concrete can be produced by adding 4% to 5% by volume of preferentially oriented optical fibers to the concrete mixture. The optical fibers used in the formulation of translucent concrete are fine glass or plastic threads (ranging from 0.25 mm to 2 mm) that guide the light directionally. The thinner and smaller the layer of fibers is, the more light it allows to pass through.

Due to bends in the fibers and roughness on the cut surfaces of the fibers, light transmission is generally a bit less than half the incident light on the fibers, so given 5% fibers, about 2% of the incident light is transmitted. The human eye's response to light is non-linear, so this can still provide a significant impact. In theory, the fibers could carry light around corners and over a distance of tens of meters, with the rate of loss increasing with length depending on the type of fiber and how it is bent.

Manufacture

There are three different layers in the optical fibers – the buffer coating, cladding, and the core. The buffer coating is an outer

FIGURE 24-2. Translucent concrete allows light, shadows, and colors to project through concrete through the use of optical fibers (right photo courtesy of Lucem).

layer that protects the cladding and core and the cladding provides a reflective surface to minimize light loss as it is transmitted through the core. Using acrylic rods, optical fibers, or translucent fabric, the materials are cast layer by layer into fine-grained concrete. After setting, the concrete is polished, and then cut to plates or stones with standard machinery.

To ensure that enough light is available, walls and other mounting systems can be equipped with some form of lighting, designed to achieve uniform illumination on the full plate surface. Usually mounting systems similar to those used for precast and natural stone panels are used.

Since optical fibers are an expensive material, the production of translucent concrete is expensive compared to traditional concrete. Several companies produce translucent concrete with very different production systems including translucent blocks (Litracon 2002), sheets that include light-conducting fabric (Luccon 2007), handcrafted and individually laid fibers for punctual transmission (Lucem 2007), and insulated panels (Zospeum 2017).

Applications

Translucent concrete is used in fine architecture as a façade material and for cladding of interior walls. In addition to providing light in otherwise dark places or windowless areas like subways and basements, it is used to construct sidewalks and speed bumps that illuminate with artificial light at night and provide increased safety for pedestrians and roadside traffic.

Translucent concrete has also been applied to various design products. It was used in the public square in Stockholm, Sweden. By day, the square's sidewalk looks as though it were made of ordinary concrete, but the translucent surface lights up at night when the colored lights beneath the surface illuminate. The Stuttgart City Library in Germany has a cube-shape translucent roof that allows natural light to illuminate the area. Other well-known projects include a large scale application of translucent concrete in the renovated headquarters of the Bank of Georgia which hosts almost 300 m^2 of light transmitting concrete.

PHOTOCATALYTIC (SELF-CLEANING)

Concrete has the ability to neutralize pollution and self-clean by incorporating a process known as photocatalysis. As sunlight hits a concrete surface that contains a photocatalyst, most organic and some inorganic pollutants are neutralized and can be washed away. The advantage of using solar light and rainwater as driving force has opened a new approach for environmentally-friendly building materials. Photocatalytic concrete was researched in the mid 1990s, and the first patent was applied for in Italy in 1998 (Cassar and Pepe 1998). The technology made its debut in the U.S. soon after. The application

of self-cleaning concrete has great potential in the field of degradation of pollutants, deodorization, sterilization, and energy conservation.

Materials

Self-cleaning concrete includes titanium dioxide (TiO_2) additions that can be applied to white or grey portland cement. Other products spray water-based titanium films (99% water, 1% TiO_2) onto the surface of concrete. Titanium dioxide is a naturally occurring compound that is used in many household products. It can absorb UV light without being consumed. When incorporating TiO_2 into concrete, it performs like any other portland cement and can be used in all varieties of concrete, including plaster.

Photocatalytic Technology

In the 1960s, Fujishima and Honda (1972) carried out the first research yielding the potential for practical applications of photcatalysis using TiO_2. When activated by the ultraviolet radiation in sunlight or artificial sources, TiO_2 oxidizes and reacts with external substances, decomposing organic compounds (Figure 24-3). The hydroxyl radicals are strong oxidizers that break down organic materials including dirt (soot, grime, oil, and particulate matter) and biological organisms (bacteria, viruses, algae, mold, and fungi), as well as odor-producing chemicals and then produce benign molecules of carbon dioxide, water, or other harmless substances (Chusid 2005).

The titanium-based catalyst is not consumed as it breaks down pollution. Because rain continuously washes away any pollution from the concrete surface, structures do not collect dirt and do not require chemical applications that are potentially harmful to the environment.

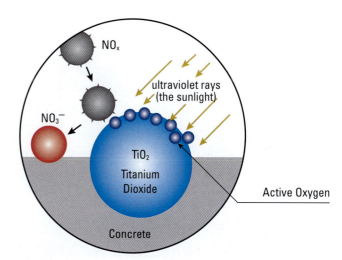

FIGURE 24-3. In the presence of light and air, titanium dioxide (TiO_2) breaks down harmful elements into relatively benign molecules by oxidation.

Photocatalytic concrete is light in color and reflects much of the sun's energy, which through albedo effects, reduces the heat gain of the concrete surfaces, reduces the air temperature in urban environments, and as a result, reduces the amount of overall smog. Both laboratory and field experiments reveal that adding TiO_2 to the surface of pavement and building materials can significantly reduce air pollution, by up to 60% in some applications (Giussani 2006).

Applications

These products provide value through unique architectural and environmental performance capabilities. The self-cleaning property of TiO_2, combined with its photocatalytic pollution breakdown abilities, and the beauty of white cement, make it an ideal additive to building surfaces in urban environments. Maintenance costs are significantly reduced.

Concrete products that are continually exposed to sunlight, like precast building panels, cast-in-place walls, pavers, and roof tiles, are especially well-suited for photocatalytic concrete.

Pavements made with photocatalytic concrete or pavers, are capable of reducing the pollution at a primary source – automotive tailpipes. A study conducted in the Netherlands used photocatalytic concrete pavers on a section of a busy roadway and monitored the air quality 0.5 m to 1.5 m (19.5 in. to 58.5 in.) above the pavement at both a control area with normal pavers and in a test section. It was observed that the NO_X levels were reduced by 25% to 45% (Ballari and others 2010).

Buildings in highly polluted locations have been constructed with photocatalytic concrete – noted applications include the Air France headquarters at Roissy-Charles de Gaulle International Airport near Paris, the Jubilee Church in Rome (Figure 24-4).

FIGURE 24-4. The Jubilee Church in Rome, constructed in 2000, is made from precast concrete coated with a layer of photocatalytic cement (courtesy of G. Basilico, Italcementi).

Three dimensional photocatalytic tiles with high surface to volume ratios have been constructed into modular forms for adhering to buildings or as standalone sculptures that can absorb air pollutions in urban environments (Manaugh 2006) such as with the Milan Expo in 2015.

ENGINEERED CEMENTITIOUS COMPOSITES (BENDABLE CONCRETE)

Engineered cementitious composites (ECC), also known as bendable concrete, were first researched in the early 1990s at the University of Michigan (Li 2019). More than twenty years later, use of ECC has applications in Japan, China, Singapore, Germany, South Korea, Denmark, and the United States. New developments are finding bendable concrete to be a promising technology that can be sourced with locally available materials (Zeiba 2019).

Materials

ECC belongs to the broad class of fiber-reinforced concrete (FRC), as it contains fiber in a cementitious matrix. The design which gives bendable concrete, or engineered cementitious composite, its impressive ductility is based off nacre, the substance that coats the inside of abalone shells. Nacre is composed of small aragonite platelets that are held together by natural polymers, allowing it to be both hard and flexible as platelets are able to slide side to side under stress. This effect is mimicked in bendable concrete by dispersing tiny fibers throughout a binder.

Quartz sand, air pores, water, and cementitious materials make up the ECC binder. Larger river sand with sizes up to 2.36 mm (No. 8) has been successfully used in ECC. Because of the tendency of matrix toughness to increase with sand particle size, the presence of larger particles requires appropriate adjustments of the binder. Viscosity modifying admixtures are commonly used to aid in dispersion of fibers. Other chemical admixtures are incorporated, as needed, to achieve desired concrete properties (Chapter 8).

An optimal fiber content high enough to achieve composite strain-hardening while still meeting workability and cost constraints, and a theoretical model to conduct composite optimization is needed when designing ECC mixtures.

ECC Technology

ECC represents a family of materials with the common feature of being ductile, with tensile strain capacity typically beyond 2%. Bendable concrete can deform up to 3% to 5% in tension before it fails (Figure 24-5), which gives it 300 to 500 times more tensile strain capacity than normal concrete (Li 2019). It is the ability to tolerate tensile strain that makes bendable concrete unique and highly resistant to cracking.

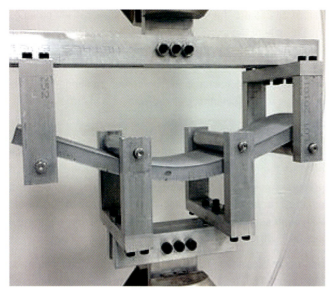

FIGURE 24-5. Engineered cement composite, also known as bendable concrete provides 300 to 500 times more tensile strain capacity than normal concrete (courtesy of V. Li).

Fiber properties (aspect ratio, mechanical properties), matrix properties (mechanical, initial flaw size distributions), and fiber/matrix interfacial properties (chemical bond, frictional bond, and other fiber/matrix interaction properties) are deliberately engineered to interact with one another in a certain prescribed manner, when the composite is loaded.

The design basis of ECC is also significantly different from those of high strength concrete (HSC) or UHPC (see Chapter 23). In general, UHPC has tensile strain capacity of 0.2% or less. HSC and UHPC are designed based on dense particle packing. Instead, the material microstructure of ECC is adjusted for synergistic interactions between the microstructural components, based on a body of knowledge known as ECC micromechanics (Li 2019).

Bendable concrete also has self-healing capabilities. Because bendable concrete keeps cracks relatively small, natural reactions within the hardened concrete generate healing products through mineralization (carbonation reactions) and continuous hydration which repairs the cracks and restores the durability of the concrete.

ECC can also be made photocatalytic to neutralize pollutants by incorporating embedded nano-titanium particles, thereby helping to maintain clean air in urban environments.

Manufacture

One of the essential challenges of processing ECC material is the uniform dispersion of fibers throughout the matrix. A variety of processing methods are available for ECC, the most common is casting into forms with mixtures designed as self-consolidating concrete (SCC). These can be cast-in-place or precast self-consolidating concrete (SCC) mixtures. The choice of processing method, including SCC casting, or other options such as spraying (shotcreting) or extrusion, depends on the application. In each case, the fresh properties must be appropriately adjusted to suit the application method.

Applications

In most cases, the adoption of ECC over normal concrete is driven by the significant improvement in structural resilience and durability. Bendable concrete presents an efficient alternative primarily in the construction and maintenance of infrastructure, where concrete is subject to harsh weather conditions and extreme loading. The much higher ductility suggests various potential applications.

Ideal for structures in high seismic regions, engineered cement composites have been applied in several bridge projects, as well as high-rise structures. Applications in service include the dampers on the Seisho Bypass Viaduct in Japan, which is roughly 28 km (17.4 mi) long. ECC was also used for the 27 story Glorio Tower in Central Tokyo, the 41 story Nabule Yokohama Tower in Yokohama, and the 60 story Kitahama Tower in Osaka, Japan. These structures incorporated ECC coupling beams in their building cores (Figure 24-6) (Li 2019).

FIGURE 24-6. ECC coupling beams in precast plants showing protruding reinforcing steel for connection to core wall (courtesy of V. Li).

In roads as well as other paved surfaces which must bear repeated loading of heavy vehicles, bendable concrete would crack less often, minimizing damage primarily from road salts which can corrode steel reinforcement. The first ECC link-slab project was completed in the summer of 2005, involving a link-slab 225 mm (8.9 in.) thick, 5.5 m (18 ft) long in the traffic direction, and 20.25 m (66.5 ft) across, on a bridge-deck in southern Michigan (Figure 24-7).

FIGURE 24-7. Construction of link-slab with self consolidating ECC mixture (courtesy of V. Li).

SELF-HEALING CONCRETE

Concrete has the inherent ability to self-heal microcracks directly related to the amount of unhydrated cement available and in contact with a source of water. However, this process is variable and unpredictable. To increase the crack-healing potential of concrete, specific healing agents can be incorporated in the concrete matrix. Self-healing concrete was developed in the Netherlands in 2006 by Henk Jonkers, a microbiologist and professor from Delft University (Jonkers 2007). Full scale research expanded shortly thereafter (Jonkers 2011).

Materials

The materials added to the concrete to enhance self-healing characteristics consist of two-component bio-chemical self-healing agents embedded in porous expanded clay particles with separate nitrogen, phosphorous, and nutrients (Figure 24-8). The expanded clay particles act as reservoirs and replace part of regular concrete aggregates. Protection of the bacterial spores by immobilization inside porous expanded clay particles

before addition to the concrete mixture substantially prolongs the life-time of the spores (Jonkers 2011). Replacing too high a fraction of sand and gravel for expanded clay has consequences for the concrete's compressive strength. The presence of the matrix-embedded bacteria and precursor compounds should not negatively affect other desired concrete characteristics.

Technology and Manufacture

The principle mechanism of bacterial crack healing is that the bacteria themselves act largely as a catalyst, and transform a precursor compound to a suitable filler material. The bacteria generate crystals that enclose their cells, and when mixed with other secretions (like proteins and sugar), they generate a glue-like substance.

Upon crack formation the two-component bio-chemical agent consisting of bacterial spores and calcium lactate are triggered and released from the clay particle by water entering the cracks. Subsequent bacterially mediated calcium carbonate formation results in physical closure of micro cracks. If the concrete cracks, the bacteria seal gaps by forming calcite.

Healing of up to 0.46 mm (0.02 in.) wide cracks in bacterial concrete was noted after 100 days submersion in water in a study conducted by Jonkers (2011). In this study, the crack-healing potential due to metabolic activity of bacteria was supported by oxygen profile measurements which revealed oxygen consumption by bacteria-based specimens, and not the control specimens.

Another possible mechanism involves a technique based on the external application of mineral-producing bacteria. In these studies, efficient sealing of surface cracks by mineral precipitation was observed when bacteria-based mixtures were sprayed or applied onto damaged surfaces or manually inserted into cracks (Van Tittelboom 2010).

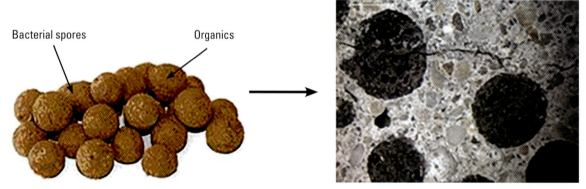

FIGURE 24-8. Self healing admixture composed of expanded clay particles (left) loaded with a two-component healing agent of 5% alkali-resistant bacterial spores and 95% organic bio-mineral precursor compound (calcium lactate). When embedded in the concrete matrix (right) the loaded expanded clay particles represent internal reservoirs containing the healing agent consisting bacterial spores and a suitable bio-mineral precursor (Jonkers 2011).

Applications

The bio-chemical self-healing agent shows potential for particularly increasing durability aspects of concrete construction in wet environments. The bacteria to be used as self-healing agent in concrete should be able to perform long-term effective crack sealing, preferably over the entire service life of the concrete.

ADDITIVE MANUFACTURING (3D PRINTING)

Additive manufacturing (AM) also known as 3D printing, uses computer-aided design to build objects layer by layer. Additive manufacturing grew from developments in both topography and photosculpture in the late 19th century. These parallel developments continued separately until they intersected with a patent (Swainson 1977) which described the method, medium, and apparatus for producing a three-dimensional product. A number of other patents quickly followed suit, and 3D printing transitioned from a research area into a new manufacturing technology.

Cementitious 3D construction printing, or 3D printing for concrete, has been explored since the 1990s (Khoshnevis 1998). Additive manufacturing for concrete is used to fabricate buildings or construction components in completely new shapes not previously possible with traditional concrete formwork.

Materials

Concrete materials used for 3D printing must include a rheology that permits the concrete flow sufficiently for the manufacturing system to place it, and then for the materials cease flowing almost immediately upon deposition, with enough strength to have the material hold its shape until setting and strength development can take place. Potential approaches to cementitious ingredients might include higher alite contents, lower sulfate addition levels, and incorporation of flocculating agents such as clay to provide dimensional stability and rigidity while the concrete sets. Other approaches might include heating or cooling the mixture during the deposition process to accelerate or retard the setting (Bohan 2016). The second requirement for 3D printed concretes is that they must be non-thixotropic materials, so they don't flow excessively after placement over time, or with vibration, before the concrete has set. In essence, the concrete must be flowable immediately prior to exiting the deposition device, yet nearly solid almost immediately after deposition. Shearing of the plastic concrete material must be eliminated.

Technology and Manufacture

Additive manufacturing typically uses computer-aided designs which are then translated into a numerical control programming language. That language translates the design into a series of commands executed by the printer. Those commands include three-dimensional toolpaths that direct the deposition nozzle to the correct spatial location along with commands that determine the amount of material deposited and the rate of deposition.

First and foremost, a concrete 3D printer must be quite substantial in size. This is a major challenge that has somewhat hindered the mainstream adoption of concrete 3D printing. Deposition of 3D printed concrete necessitates the use of mobile devices to achieve the size and scale required for commercial building construction when printing on site. Other systems preprint components and assemble them on site.

Current 3D printing technologies using concrete as a material include extrusion technology and shotcrete technology. Additional requirements for 3D printed concrete will depend on the nozzle diameter, printing speed, and time required for the concrete to support its own weight and that of any successive layers (Figure 24-9).

FIGURE 24-9. 3D concrete printer in operation. No-slump concrete leaves the nozzle as a relatively stiff continuous filament (courtesy of Skidmore, Owings, and Merrill).

An alternative to working with single, large robots was introduced by the Institute of Advanced Architecture of Catalonia in 2014 (IAAC 2014). While applying a similar extrusion technique, the deposition instrument is not a single large robot operating in a predefined space, but rather a group of small robots working together, with using sensors to track their relative positions.

Advantages of 3D printing technology include the reduction of raw material and replication rates and tolerances. Traditional processes for concrete placement require the use of molds for casting, which 3D printing eliminates. Final products are potentially identical to an original prototype, with strict tolerances. Because the 3D printing process builds successive layers without traditional machining, the only limitations upon the surface tolerances are those inherent to the scanned prototype or computerized image used and the method of deposition; there is virtually no distortion of the material.

Researchers have looked at methods of using reinforcing bars, fiber reinforcement, as well as several other alternative reinforcing methods (Asprone 2018). Many of the alternative reinforcing methods may not be practical for the early adoption of the technology.

The design of structures using 3D printing for concrete will require changes in design methodology. The traditional beam, column, slab, and frame elements with their corresponding shear, bending moment, and torsion capacity may no longer provide the optimal structure. Instead, continuously placed additive manufactured concrete structures may require design methodologies more similar to shells where shear stresses govern.

Applications

There are several advantages of 3D printing for concrete. Recent developments in the field of additive manufacturing have demonstrated that there is an untapped potential in quickly building affordable homes and entire communities. At the present time, the technology would be most useful for low-income housing, and emergency reconstruction by disaster relief agencies.

Research studies exploring the feasibility of the use of reinforced 3D printed concrete technology for government housing were conducted on the grounds of US Army Engineer Research and Development Center-Construction Engineering Research Laboratory (ERDC-CERL) in Champaign, Illinois. The building in Figure 24-10 used conventional cast-in-place concrete foundations, conventional roof structures, with 3D printed walls. These structures were modelled on commonly used military housing (Kreiger and others 2019). The studies showed the feasibility of both 3D printed concrete formwork and reinforced 3D printed concrete walls for different geometries.

FIGURE 24-10. Reinforced walls were constructed with 3D printed concrete at the US Army Engineer Research and Development Center-Construction Engineering Research Laboratory (ERDC-CERL).

Computer-aided design exemplifies the versatility and significant aesthetic potential that 3D concrete printing has for use in large scale structures. This is an area of active interest for many organizations, including the Center for Rapid

Automated Fabrication Technologies (CRAFT) at the University of Southern California, the University of Loughborough, Swiss Federal Institute of Technology (ETH Zurich), Shanghai-based contractor Winsun, Total Kustom in Minnesota, and the company ICON out of Austin, Texas, (Figure 24-11).

FIGURE 24-11. Noteworthy examples of projects include a 3D printed home located in Austin, Texas at Community First! Village using ICON's Vulcan construction system (courtesy of Philip Cheung).

ASTM Committee F42 on Additive Manufacturing Technologies was formed in 2009 and includes subcommittees on ceramics and construction. ASTM International also launched its Additive Manufacturing Center of Excellence (amcoe.org) with Auburn University, NASA, manufacturing technology organization EWI, and the UK-based Manufacturing Technology Centre (MTC) in July 2020. The American Concrete Institute formed a new committee in 2017 to address 3D printing – ACI Committee 564, *3-D Printing with Cementitious Materials*.

GEOSYNTHETIC CEMENTITIOUS COMPOSITE MATS

Geosynthetic Cementitious Composite Mats (GCCM) are an engineered concrete fabric that consists of a three-dimensional fiber matrix containing a dry high early-strength concrete mixture that hardens with the addition of moisture. Enhancing geosynthetic fabric with cement paste can provide both strength and durability along with puncture and abrasion resistance to be used for slope stabilization and other applications (Jongvivatsakul and others 2018). No special tools, mixing equipment, or machinery is required for installation. GCCM was patented by based on research at Imperial College and the Royal College of Art in London in 2005 (Crawford 2019).

Materials and Technology

The material construction for GCCM typically consists of three layers (Figure 24-12):

• A hydrophilic fibrous top layer. This layer consists of an elastomeric (typically polyester) yarn that aids in hydration by wicking water into the underlying fiber matrix.

• A three-dimensional fiber reinforced matrix. This layer is filled with a specially formulated dry, high-early strength

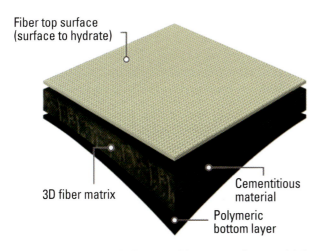

FIGURE 24-12. Goesythetic cementitious composite mats fabric consist of a three-dimensional fiber matrix containing a high early-strength concrete mixture (courtesy of Concrete Canvas).

cementitious blend. Fibers may consist of polypropylene, coated glass fibers, polyethylene, and PVC fibers. A mixture of these fibers can be used.

• A polymeric membrane. The polymeric bottom layer contains the dry cementitious mixture during transportation provides waterproofing upon installation.

A damp-proof layer may be used for applications requiring moisture proofing such as barriers or tunneling. The final fabric thickness will vary depending on selection of products and application and typically ranges from 2 mm to 70 mm (0.08 in. to 2¾ in.).

The flexural strength of GCCMs provides the best overall indication of material performance and ASTM D8058, *Standard Test Method for Determining the Flexural Strength of a Geosynthetic Cementitious Composite Mat (GCCM)*, was recently developed to test these materials in 2017. The final product is a 35 MPa to 50 MPa (5000 psi to 7250 psi) durable concrete shell with long service life depending on the final application and condition of exposure.

Manufacture

The GCCM fabric typically comes in rolls that can be quickly cut and secured in place. The rolls can also be manufactured as man-portable batched rolls for applications on sites with limited access. The fabric can be hung vertically, laid in trenches, or cut and formed into various shapes. The material is cut to size, secured using pegs and screws, as needed, and then hydrated thoroughly. The material can by hydrated by the addition of water or be placed in a location where it will come into contact with water.

The cloth remains flexible for up to two hours and is easily manipulated to fit the contours of irregular installations. The rate at which the fabric can be laid is around 200 m²/hour (240 yd²/hour).

Applications

GCCM is ideal for reinforcing or rehabilitating a culvert or drainage ditch, abutment, berm or other structure. GCCM has also been used a material for architectural sculptures including concrete furniture.

GCCMs are most commonly used as erosion control linings for water channels, either for drainage or irrigation and as facings for slopes as a replacement for non-structural concrete (Figure 24-13). The high early-strength concrete mix also enables GCCMs to have excellent abrasion resistance and chemical performance with a typical design life of well over 50 years.

FIGURE 24-13. GCCM is ideal or reinforcing or rehabilitating a culvert or drainage ditch, abutment, berm or other structure (courtesy Concrete Canvas).

Thin-walled concrete structures can be formed rapidly using GCCM. With this technology, GCCM has been used for rapid construction of secure shelters (Figure 24-14). These structures are constructed in four stages: delivery, inflation, hydration, and setting. The GCCM surface is bonded to the outer surface of a plastic inner that can bear the weight of the fabric.

FIGURE 24-14. GCCM has been used for rapid construction of secure shelters (courtesy of Concrete Canvas).

Similar technology to GCCM includes fabric bags of cementitious ingredients stacked for soil stabilization.

ROBOTICS

While automated construction tests began early in the industrial age, robotics significantly improved during mid-20th century, including the development of extrusion machines in the 1950s, and pumped concrete in the 1960s. Advances in 3D printing methods have claimed recent success in concrete placements (see **Additive Manufacturing**). Other breakthroughs in construction planning and design include Building Information Modeling (BIM) combined with automated construction processes. Part of the solution to delivering these projects is through leveraging robotics and automation to multiply construction productivity and improve worker safety.

Formwork Modeling and Automated Systems

The selection of formwork layout can be developed incorporating the actual site characteristics and imparting any changes during construction, by utilizing a 3D structural model and BIM (Kannan and Santhi 2013). Designing formwork with BIM and 3D modelling provides an automated sequential progress of the formwork construction from the start to finish. It can also be used to consider the repetition characteristics of reusing formwork systems in the model during simulation.

Automated climbing and gliding formwork systems are available for vertical high-rise construction that at the push of a button, raise all working platforms, and the concrete placing boom, along with interior and exterior formwork for an entire floor by hydraulic leverage. These can be an effective solution for repetitive buildings (such as towers) that increases jobsite production and reduces labor costs for buildings with more than 20 floor levels. A similar technology known as jump forms is used for construction on buildings of five stories or more.

Self-climbing form systems with working platforms allowed crews to cycle the vertical and horizontal formwork at the same speed in order to meet the tight schedule for the construction of the Regalia in South Florida (Figure 24-15) (Wright 2013).

Automated Reinforcement Fabrication

The placement of reinforcement including tying rebar and construction of cages can result in roughly 20% of the total cost of a project and is labor intensive and time consuming (Lab 2007).

Mechanizing reinforcement fabrication allows full-scale reinforced concrete projects typically placed conventionally cost advantages with elimination of labor and material. Automated rebar-tying was presented in the early 1990s (Altobelli 1991). An initial device enabled a worker to tie rebar from a standing position, thereby eliminating the grueling aspect of the task, bending over for long periods of time.

Advancements in technology now has robots handling the placement of rebar for construction (Toggle.is). The technology

FIGURE 24-15. Self-climbing form systems with working platform allowed crews to cycle the vertical and horizontal formwork for the construction of the Regalia, South Florida (courtesy of Doka USA).

has a robot lifting, and manipulating the bars, while a human may be part of the final wire tying. Current technology also provides automated pre-tying of cage assemblies on site. Additive manufacturing could also allow the continuous placement of reinforcement in a manner similar to the unspooling of consumable welding wire.

Automated Finishing

Automated concrete finishing operations were developed in the late 1980s in Japan (Altobelli 1991). A screeding robot was developed that propelled itself on wheels by riding on the reinforcement or form edges (Nomura and others, 1988). Remote controlled power trowels were implemented near the same time by Takenaka Corporation (Kikuchi and others 1988) and Shimizu Corporation (Kajioka and Fujimori 1990) completing over two million square feet of floor area in 1987. Radio controlled concrete finishers available now are lighter than traditional walk-behind trowels and can be operated from up to 0.8 km (½ mile) away.

Profile-3D-Printing is an additive and subtractive manufacturing process that combines deposition of concrete for rough layup with precision tooling for surface finishing of architectural building components commonly found in the architectural precast industry (Bard and others 2018).

DRONES

Remote piloted aircrafts, also called unmanned aerial vehicles (UAV) technology (commonly known as drones), are used to assist in construction operations. Flying over a construction site, drones can provide a unique perspective (Figure 24-16). The technology is used to map terrain, measure stockpiles, document project progress, and inspect infrastructure, among other tasks. Drones can also seek out and prevent safety hazards in a much quicker and cost-efficient manner than traditional methods.

A close fly-by of a structure can provide valuable visual data when equipped with high-definition photo and video equipment. The use of drones can assist in ascertaining structural integrity and project performance. Drones can help with the communication on sites and drones can fly in and out of places that are inaccessible to project construction units. Automatic crack identification for large-scale infrastructure has been proposed by adopting the UAV with image processing (Hallermann and Morgenthal 2014, and Kim and others 2015). Drones have also been employed with the use of laser scans to automatically measure concrete floor waviness and flatness (Turkan 2021).

In more mature markets, remote piloted aircraft have been deployed as delivery services. While it is difficult to gauge to what extent drones could help the supply chain, the idea is that they could help deliver supplies to active projects throughout a region.

Note, in the U.S., a drone can only be legally operated with an exemption from the Federal Aviation Administration (faa.gov/uas).

ARTIFICIAL INTELLIGENCE (AI)

Artificial intelligence (AI) uses forms of machine learning for computer systems to be able to perform tasks that normally require human intelligence, such as visual perception, language processing, and decision-making. Machine learning provides AI systems with the ability to gain insights into this data, and automatically learn and improve from experiences without being overtly programmed. This is done by analyzing algorithms and through pattern recognition. The primary aim is to allow the computers to learn automatically without human intervention or assistance and adjust actions accordingly (Russell and Norvig 2021).

In the construction industry, artificial intelligence is largely being used to help make document management and project management systems more accessible. For example, to easily scan a PDF document and locate information quickly.

FIGURE 24-16. Drones are used to map terrain, measure stockpiles, document project progress and safety, and inspect infrastructure (right photo courtesy of Ernst Concrete of Georgia).

Traditionally, the collection of data has been a manual process. AI technology is also being used to gather data collected by drones, recording devices, and sensors. The data sourced can be stored in a cloud and viewed via virtual reality (VR) platforms.

AI's main benefits include modeling and pattern detection, prediction, and optimization. AI may be used to predict possible issues like structural risk, safety hazards, and other conflicts (Li and others 2015, Yeum and Dyke 2015, and Fletcher and others 2017). Construction companies are now working to combine drone images and image classification algorithm to improve the accuracy, safety, and cost effectiveness of infrastructure repairs.

The advent of 3D printing combined with AI can provide the ability to print objects with the ability to change properties with time according to the environment (4D printing). Automation combined with BIM can facilitate asset management over the lifecycle of a structure, improving efficiencies, and minimizing contractual problems using decentralized blockchains (Pan and Zhang 2020).

VIRTUAL CEMENT HYDRATION

Computer technology now allows simulation of cement hydration (Figure 24-17), microstructure development, and physical properties. Combinations of materials, cement phases, or particle size distributions can be modeled to predict cement performance. Just some of the properties that can be simulated and predicted include: heat of hydration, adiabatic heat signature, compressive strength, setting time, rheology (yield stress and viscosity), percolation, porosity, diffusivity, thermal conductivity, electrical conductivity, carbonation, elastic properties, drying profiles, susceptibility to degradation mechanisms, autogenous shrinkage, and volumes of hydration reactants and products as a function of time. The effects of varying sulfate and alkali contents can also be examined, along with interaction with supplementary cementitious materials and chemical admixtures.

Computer modeling predicts performance without the expense and time requirements of physical testing. Garboczi and others (2019) provide a web-based computer model for hydration and microstructure development.

NANOTECHNOLOGY

While the meaning of nanotechnology varies, nanotechnology is commonly defined as the manipulation of matter with at least one dimension sized form 1 nm to 100 nm (NNI 2016). This definition reflects the importance of quantum mechanical effects.

Nanotechnology is already being used in concrete materials with a wide range of applications. The two main areas of nanotechnology in concrete research are nanoscience and nano-engineering. Nanoscience deals with the measurement and characterization of the nano and microscale structure of cement-based materials (Figure 24-18). Nano-engineering encompasses the techniques of manipulation of the structure at the nanometer scale to develop new cementitious composites with superior mechanical performance and durability.

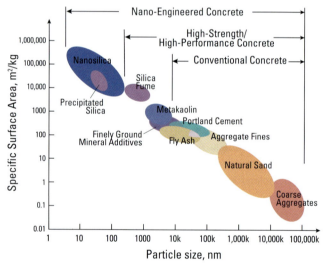

FIGURE 24-18. Particle size and specific surface area related to concrete materials (Sobolev and Ferrada-Gutiérrez 2005).

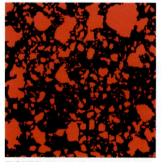 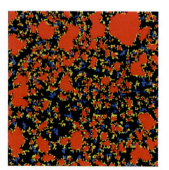 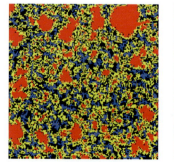 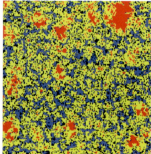

FIGURE 24-17. Example two-dimensional digital microstructures showing four stages of hydration in a microstructural model of C_3S hydration. The degrees of hydration are: A – 0%, B – 20%, C – 50%, D – 87%. In these images, colors are assigned as follows: Red = unreacted cement, blue = CH, yellow = C-S-H, and black = porosity (Garboczi and others 2019).

Advances in nanoscience has brought forth instrumentation including atomic force microscopy and molecular level modeling to better understand how the microstructure of C–S–H affects macroscale properties and performance of concrete (Dejong and Ulm 2007).

Concrete can be nano-engineered by the incorporation of nanosized objects (nanoparticles and nanotubes) to control material behavior and add novel properties; or by the grafting of molecules onto cement particles, cement phases, aggregates, and additives (including nanosized additives) to promote specific interfacial interactions (Sanchez and Sobolev 2010).

Examples of technologies in concrete include a range of novel properties such as those discussed previously: self-cleaning (photocatalytic), high ductility (engineered cementitious composites), self-control of cracks (self-healing), and self-sensing capabilities (sensors).

The current challenge facing the wider use of nanotechnology in construction is the expense and understanding of these innovations. As nanotech develops further and shows its full potential in improving concrete durability and impacting concrete sustainability, these technologies are likely to be more readily adopted.

LUNAR CONCRETE

NASA has challenged the concrete industry to aid in development of feasible options for constructing a permanent moon base. The U.S. space program established the goal and began research in the early 1980s. The aim of the study is to develop a lunar concrete that can be made entirely of materials existing on the Moon without incurring the enormous expense and logistical obstacles of shipping construction materials used on Earth to the Moon.

Materials

The Moon is a rich source of minerals from which metal, ceramic, cement, and oxygen can be obtained. Conceptually, the basic ingredients for lunar concrete would be the same as those for terrestrial concrete: aggregate, water, and cement. In the case of lunar concrete, the aggregate would be lunar regolith. The cement would be manufactured by beneficiating lunar rock that had a high calcium content. Water could potentially be supplied from off the Moon, or by combining oxygen with hydrogen produced from lunar soil (NASA 1988), or possibly from a water source present on the Moon.

The chemistry of lunar regolith is different from terrestrial soil (Figure 24-19). The Moon is very dry and is exposed to the bombardment of micrometeorites and solar wind irradiation.

Minerals with water as part of their structure such as clay, mica, and amphiboles are absent from the Moon. The lunar crust is chemically reduced, rather than being significantly oxidized like the Earth's crust and the bombardment of micrometeorites breaks up soil particles and melts portions forming irregular clusters of agglutinates. Solar wind includes large quantities of hydrogen and helium and other elements (NASA 2003).

FIGURE 24-19. Lunar regolith has notable differences to soil from terrestrial materials.

NASA has confirmed the presence of water on the moon's sunlit surface, a breakthrough that suggests water could be distributed across more parts of the lunar surface than the ice that has previously been found in dark and cold areas (NASA 2020).

Advancements in Lunar Concrete Technology

In 1986 the National Aeronautics and Space Administration (NASA) awarded the Construction Technology Laboratories (now CTLGroup) 40 g (1.4 oz) of lunar soil collected during the Apollo 16 mission for the determination of physical properties of concrete made with the lunar sample through a series of destructive and non-destructive tests (Figure 24-20) (Lin and others 1986). Portions of the lunar sample were examined by optical and electron microscopy to determine characteristics. Lunar soil was mixed with high alumina cement (which closely approximates a cement that could be manufactured by lunar soil) and distilled water. The lunar concrete was cured by using steam on a dry aggregate/cement mixture. Lin proposed that the water for such steam could be produced by mixing hydrogen with lunar ilmenite at 800°C (1470°F), to produce titanium oxide, iron, and water. It was capable of withstanding compressive pressures of 75 MPa (10,000 psi) and lost only 20% of that strength after repeated exposure to a vacuum chamber.

FIGURE 24-20. Sample of lunar concrete made from lunar soil collected during the Apollo 16 mission (courtesy of CTLGroup).

In an effort to promote lunar concrete technology, ACI formed Committee 125, *Lunar Concrete*, in 1988. The objective of the committee was to develop new technologies and correlate new and existing knowledge and to formulate recommendations for construction on the Moon. In 1991, ACI published SP-125: *Lunar Concrete Exploring the Production and Use of Concrete on the Moon* (ACI SP-125 1991).

A waterless lunar concrete, using moon dust as the aggregate, and a sulfur binder purified from lunar soil has been researched by University of Alabama in Huntsville and NASA (Toutanji and Grugel 2008). Only comparatively small amounts of lunar rock have been transported to Earth. To test the properties of lunar concrete, Toutanji and Grugel used a simulated lunar soil. They added 35 g of purified sulfur to every 100 g of dust. To melt the sulfur, it was heated to between 130°C and 140°C (266°F and 284°F) (Barras 2008). Once cooled, the mixture quickly hardened, forming a simulated lunar concrete. The samples were cast into small cubes and then exposed to 50 cycles of severe temperature changes from -27°C (-16.6°F) and room temperature. The simulated lunar concrete could withstand compressive pressures of 17 MPa (2500 psi). When the material was reinforced with silica, the strength was increased to around 20 MPa (2900 psi).

Artemis and Future Lunar Concrete

NASA is preparing to return to the moon before the end of the next decade with the intent of building outposts and paving the way for a permanent base (nasa.gov/specials/artemis).

A recent investigation on the International Space Station examined cement solidification in microgravity to help answer questions on how changes in gravity may affect the chemistry of concrete (NASA 2019 and Neves and others 2019). For the project, researchers mixed 12 pastes of cementitious powders and additives which were delivered to the space station via a capsule. Each sample included a bag with a burst pouch full of water. Astronauts ruptured the water pouch then massaged the liquid into the cement powder sample and mixed it with a spatula for 90 seconds. Then, for some samples, isopropyl alcohol was added to arrest the hydration of the cement. The samples were allowed to cure for various times, then sent back to Earth for analysis aboard a later mission. The study explored whether solidifying cement in microgravity would result in unique microstructures and provided a first comparison of cement samples processed on the ground and in microgravity.

On first evaluation, the samples processed on the space station show measurable changes in the cement microstructure compared to those processed on Earth. A primary difference was increased porosity, or the presence of more open spaces (Neves and others 2019). The next steps are to find binders that are specific for space and variable levels of gravity and consider the impacts of mixing in open space environment. There is still a lot to understand before astronauts start placing concrete in large scale for permanent structures in space.

REFERENCES

PCA's online catalog includes links to PDF versions of many of our research reports and other classic publications.
Visit: cement.org/library/catalog.

ACI Committee 125, *Lunar Concrete*, ACI SP-125, American Concrete Institute, Farmington Hills, MI, 293 pages.

Altobelli, F.R., *An Innovative Technology in Concrete Construction: Semi-Automated Rebar Tying*, M.S. Thesis, Massachusetts Institute of Technology, 1991, 75 pages.

Asprone, D.; Menna, C.; Bos, F.P.; Salet, T.A.M.; Mata-Falcón, J.; and Kaufmann, W., "Rethinking Reinforcement for Digital Fabrication with Concrete," *Cement and Concrete Research*, Vol. 112, 2018, pages 111 to 121.

Ballari, M.; Yu, Qingliang; and Brouwers, H.J.H., "Experimental Study of the NO and NO₂ Degradation by Photocatalytically Active Concrete," *Catalysis Today*, available online, 2010.

Barras, C., "Astronauts Could Mix DIY Concrete for Cheap Moon Base," *New Scientist*, October 17, 2008.

Bard, J.; Cupkova, D.; Washburn, N.; and Zeglin, G., "Robotic Concrete Surface Finishing: A Moldless Approach to Creating Thermally Tuned Surface Geometry for Architectural Building Components Using Profile-3D-Printing," *Construction Robotics*, December 14, 2018, pages 53 to 65.

Bohan, R.P., "Additive Manufacturing using Concrete: A Bridge Too Far?," *Proceedings of the 2016 IEEE-IAS/PCA Cement Industry Technical Conference*, Dallas, TX, 2016, pages 1 to 10.

Bos, F.; Wolfs, R.; Ahmed, Z.; and Salet, T., "Additive Manufacturing of Concrete in Construction: Potentials and Challenges of 3D Concrete Printing," *Virtual and Physical Prototyping*, Volume 11, Issue 3, August 2016.

Cassar, L.; and Pepe, C., *Hydraulic Binder and Cement Compositions Containing Photocatalyst Particles*, European Patent No. WO 9805601, February 12, 1998.

Chen, J.; and Poon, C., "Photocatalytic Activity of Titanium Dioxide Modified Concrete Materials – Influence of Utilizing Recycled Glass Cullets as Aggregates," *Journal of Environmental Management*, Vol. 90, No. 11, 2009, pages 3436 to 3442.

Chusid, M., "Next Step: Self-Cleaning Concrete! Photocatalysts Can Keep Concrete Clean and Reduce Air Pollution," *New Technology*, August/September 2005.

Crawford, W.; and Kujawsk, M., "Geosynthetic Cementitious Composite Mats – Essential Characteristics and Properties," *Proceedings of the XVII European Conference on Soil Mechanics and Geotechnical Engineering*, Reykjavík, Iceland, September 1 to 6, 2019, 8 pages.

DeJong M.J.; and Ulm, F.J., "The Nanogranular Behavior of C–S–H at Elevated Temperatures (up to 700°C)," *Cement and Concrete Research*, Vol. 37, 2007, pages 1 to 12.

"Drone/AI Collaboration Aims to Improve Finding Bridge-Surface Defects," ForConstructionPros.com, June 12, 2019.

Fletcher, E.; Abouelleil, A; and Rasheed, H., *FE-ANN Based Modeling of 3D Simple Reinforced Concrete Girders for Objective Structural Health Evaluation*, Iowa State University, Midwest Transportation Center, June 2017.

Fujishima, A.; and Honda, K., "Electrochemical Photolysis of Water at a Semiconductor Electrode," *Nature*, Vol. 238, 1972, pages 37 to 38.

Gaget, L., "Concrete 3D printer: The New Challenge of Construction Business," *Sculpteo Blog*, January 3, 2018.

Garboczi, E.J.; Bentz, D.P.; Snyder, K.A.; Martys, N.S.; Stutzman, P.E.; Ferraris, C.F.; and Bullard, J.W., *An Electronic Monograph: Modeling and Measuring the Structure And Properties of Cement-Based Materials*, National Institute of Standards and Technology, Gaithersburg, MD, ciks.cbt.nist.gov/monograph, 2019.

Graydon, O., *Concrete Casts New Light in Dull Rooms*, optics.org, March 11, 2004.

Hallermann, N.; and Morgenthal, G. "Unmanned Aerial Vehicles (UAV) for the Assessment of Existing Structures," *IABSE Symposium: Long Span Bridges and Roofs – Development, Design and Implementation*, Kolkata, India, 24-27 September 2013.

Hassan, M.M.; Dylla, H.; Mohammad, L.N.; and Rupnow, T., "Evaluation of the Durability of Titanium Dioxide Photocatalyst Coating for Concrete Pavement," *Construction and Building Materials*, Vol. 24, Iss. 8, 2010, pages 1456 to 1461.

Hart, S., "Concrete Gets Glamorous in the 21st Century," *Architectural Record*, January 2005, pages 175 to 182.

IAAC, *Minibuilders*. Available from: iaac.net, June 2014.

Jongvivatsakul, P.; Ramdit, T.; Phong Ngo, T.; and Likitlersuang, S. "Experimental Investigation on Mechanical Properties of Geosynthetic Cementitious Composite Mat (GCCM)," *Construction and Building Materials*, Vol. 166, 2018, pages 956 to 965.

Jonkers, H.M. "Self-Healing Concrete: A Biological Approach," In: S. Van der Zwaag, Ed., *Self-Healing Materials: An Alternative Approach to 20 Centuries of Material Science*, Springer, Inc., The Netherlands, 2007, pages 195-204.

Jonkers, H.M., "Bacteria-Based Self-Healing Concrete," *HeronJounal*, Vol. 56, No.1/2, 2011, 12 pages.

Kannan, M.; and Santhi, M. "Automated Construction Layout and Simulation of Concrete Formwork Systems using Building Information Modeling," *The 4th International Conference of European Asian Civil Engineering Forum*, National University of Singapore, June, 2013.

Katzman, L., "Building Toward a Cleaner Environment: A New Role for an Existing Product, TiO₂, " with support from a Sasaki Green RED grant for the Port of Los Angeles, Wilmington Waterfront Development Program- Buffer Design, December 2006.

Khoshnevis, B., "Innovative Rapid Prototyping Process Making Large Sized, Smooth Surface Complex Shapes an a Wide Variety of Materials," *Materials Technology*, 13, 1998, pages 52 to 63.

Kikuchi, K., Furuta, S., and Imai, T., "Development and the Result of Practical Works of Concrete Floor Finishing Robot," *The 5th International Symposium on Robotics in Construction*, Tokyo, Japan, June, 1988.

Kim, H.; Sim, S.H.; and Cho, S., "Unmanned Aerial Vehicle (UAV)-Powered Concrete Crack Detection based on Digital Image Processing," *6th International Conference on Advances in Experimental Structural Engineering, 11th International Workshop on Advanced Smart Materials and Smart Structures Technology*, University of Illinois, Urbana-Champaign, IL, August 1-2, 2015.

Kreiger, E.L.; Kreiger, M.A.; and Case, M.P., "Development of the Construction Processes for Reinforced Additively Constructed Concrete," *Science Direct*, February 25, 2019, 11 pages.

Lab, R.H., "Think Formwork – Reduce Costs," *Structure Magazine*, 2007, pages 14 to 16.

Long, B., *Light Transmitting Concrete Structure*, Canadian Patent CA353849, October, 29, 1935.

Li, F.; Maghareh, A.; and Dyke, S.J., "Experimental Implementation of Predictive Indicators for Configuring a Real-time Hybrid Simulation," *Engineering Structures*, 2015.

Li, V.C., *Engineered Cementitious Composites (ECC), Bendable Concrete for Sustainable and Resilient Infrastructure*, Springer-Verlag GmbH Germany, part of Springer Nature, 2019, 428 pages.

Li, V.C., "On Engineered Cementitious Composites (ECC). A Review of the Material and its Applications," *Journal of Advanced Concrete Technology*, Vol. 1, No. 3, 2003, pages 215 to 230.

Lin, T.D.; Love, H.; and Stark, D. "Physical Properties of Concrete Made with Apollo 16 Lunar Soil Sample," B. Faughnan and G. Maryniak, editors, *Space Manufacturing 6: Proceedings of the Eighth Princeton/AIAA/SSI Conference*, American Institute of Aeronautics and Astronautics, October 1987, pages 361 to 366.

Litracon, litracon.hu, March 2021.

Luccon, luccon.de, March 2021.

Lucem, lucem.com, March 2021.

NASA, "A Concrete Advantage for Space Explorers," nasa.gov, September 3, 2019.

NASA, Artemis. nasa.gov/specials/artemis

NASA, *The Second Conference on Lunar Bases and Space Activities of the 21st Century*, NASA Conference Publication 3166, Vol. 2, Houston, TX, April 5-7, 1988.

Neves, J.M.; Collins, P.J.; Wilkerson, R.P.; Gugel, R.N.; and Radlinska, A., "Microgravity Effect on Microstructural Development of Tri-calcium Silicate (C3S) Paste," *Frontiers in Materials*, April 24, 2019.

NNI, *The National Nanotechnology Initiative – Strategic Plan*, Executive Office of the President of the United States; December 2016.

Pan, Y.; and Zhang, L., "Roles of Artificial Intelligence in Construction Engineering and Management: A Critical Review and Future Trends," *Automation in Construction*, Volume 122, February 2021.

Paul, S.; and Dutta, A., "Translucent Concrete," *International Journal of Scientific and Research Publications*, ijsrp.org, Volume 3, Issue 10, October 2013.

Russell, S.J.; and Norvig, P., *Artificial Intelligence: A Modern Approach* (4th ed.), Pearson Education, Inc. Hoboken, NJ, 2021.

Sanchez, F.; and Sobolev, K., "Nanotechnology in Concrete – A Review," *Construction and Building Materials*, Vol. 24, May 2010.

Sharma, S.; and Reddy, O.P., "Transparent Concrete," *International Journal of Engineering Sciences & Research Technology*, ijesrt.com, March 2017.

Sikkema, J.K., *Photocatalytic Degradation of NO_x by Concrete Pavement Containing TiO_2*, Iowa State University (2003).

Sobolev K; and Ferrada-Gutiérrez, M., "How nanotechnology can change the concrete world: Part 1," *American Ceramic Society Bulletin*, 2005.

Swainson, W.K., *Method, medium and apparatus for producing three-dimensional figure product*, US4041476A, 1971.

Swint, D.O., Col., United States Air Force. Discussions. Spring, 1989.

Topličić-Ćurčić, G.; Jevtic, D.; Grdic, D.; Ristic, N.; Grdic, Z., *Photocatalytic Concrete – Environment Friendly Material*, 5th International Conference on Contemporary Achievements in Civil Engineering, Subotica, Serbia, April 2017.

Toutanji, H.A.; and Grugel, R.N., "Unconventional Approach," *Civil Engineering Magazine*, Vol 78, Issue 10, American Society of Civil Engineers, October 2008.

Turkan, Y., "Use of Laser Scans and Drones to Measure Concrete Floor Waviness/Flatness Automatically," *Strategic Development Council Technology Forum 49*, February 9, 2021.

Van Tittelboom, K.; De Belie, N.; De Muynck, W.; and Verstraete, W., "Use of Bacteria to Repair Cracks in Concrete," *Cement Concrete Research*, Vol. 40, 2010, pages 157 to 166.

Wiktor, V.; and Jonkers, H.M., "Quantification of Crack-Healing in Novel Bacteria-Based Self-Healing Concrete," *Cement and Concrete Composites*, Vol. 33, Iss. 7, August 2011, pages 763 to 770.

Wright, J., "Climb as You Build Formwork," *Concrete Construction Magazine*, March 14, 2013.

Yeum, C.M.; and Dyke, S.J., "Vision-Based Automated Crack Detection for Bridge Inspection," *Computer Aided Civil and Infrastructure Engineering*, 2015.

Zospeum, zospeum.com, March 2021.

APPENDIX

GLOSSARY

The intent of this Glossary is to clarify terminology used in concrete construction, with special emphasis on those terms used in *Design and Control of Concrete Mixtures*. Additional terminology that may not be in this book is included in the Glossary for the convenience of our readers. Other sources for terms include ACI CT-20: *Concrete Terminology* and ASTM terminology standards, including ASTM C125, *Standard Terminology Relating to Concrete and Concrete Aggregates*, and ASTM C219, *Standard Terminology Related to Hydraulic and Other Inorganic Cements*.

A

Absorption – see *Water absorption*.

Abrasion damage – wearing away of a surface by rubbing and friction.

Accelerating admixture – admixture that speeds the rate of hydration of hydraulic cement, shortens the normal time of setting, or increases the rate of hardening, of strength development, or both.

Addition – substance that is interground or blended in limited amounts into a hydraulic cement during cement manufacturing either as a "processing addition" to aid in manufacture and handling of the cement or as a "functional addition" to modify the properties of the cement.

Additive manufacturing – process of joining materials to make parts from three-dimensional model data, usually layering, as opposed to subtractive manufacturing and formative manufacturing methodologies. (Also known as 3D printing.)

Admixture – material, other than water, aggregate, hydraulic cement, and fibers used as an ingredient and added to the batch immediately before or during mixing.

Aggregate – granular mineral material such as natural sand, manufactured sand, gravel, crushed stone, and air-cooled blast-furnace slag. (See also *Lightweight aggregate* and *Heavyweight aggregate*.)

Air content – total volume of air voids, both entrained and entrapped, in cement paste, mortar, or concrete.

Air-entraining admixture – admixture for paste, mortar, grout, or concrete, that will cause air in the form of minute bubbles, to be incorporated during mixing, usually to increase a material's freeze-thaw resistance.

Air-entrained cement – cement containing an air-entraining addition added during its manufacture.

Air void – entrapped air pocket or an entrained air bubble in concrete, mortar or grout, or paste. Entrapped air voids usually are larger than 1 mm in diameter; entrained air voids are smaller.

Alite – a form of tricalcium silicate that is the principal phase in portland-cement clinker. (See also *Belite*.)

Alkali-aggregate reactivity – a reaction between aggregates containing certain forms of silica or carbonates and alkali hydroxides in concrete that produces expansive gel.

Alkali loading – a characteristic of concrete determined by multiplying the equivalent alkali content of a portland cement by the portland cement content of the concrete. When a blended cement is used, the alkali loading of the concrete is determined by multiplying the portland cement portion of the blended cement by the alkali content of the portland cement portion of the blended cement.

Architectural concrete – concrete that will be permanently exposed to view and which requires special care in selection of concrete ingredients, forming, placing, consolidating, and finishing to obtain the desired appearance.

Artificial intelligence – complex software functions or algorithms that allow for computers to learn and solve difficult tasks.

Autoclaved cellular concrete – concrete cured at high temperature and pressure in an autoclave resulting in high air content and low density.

Autogenous healing – a natural process of filling and sealing dormant cracks in concrete, or in mortar, when material is kept damp.

Autogenous shrinkage – the reduction in bulk volume that occurs during hydration due to chemical shrinkage within a sealed cementitious mixture that is not subjected to external forces and under constant temperature.

B

Batch water – see *Mixing water*.

Batching – process of weighing or volumetrically measuring and introducing into the mixer the ingredients for a batch of concrete, mortar, grout, or plaster.

Belite – a form of dicalcium silicate that occurs as a phase in portland-cement clinker. (See also *Alite*.)

Blast-furnace slag – nonmetallic byproduct of iron manufacturing, consisting essentially of silicates and aluminum silicates of calcium that are developed in a molten condition simultaneously with iron in a blast furnace. (See also *Slag cement*.)

Bleeding – flow of mixing water, or its emergence from, a newly placed concrete mixture caused by the settlement of the solid materials within the mixture.

Blended cement – a hydraulic cement containing combinations of portland cement, limestone, pozzolans, and/or slag cement.

Blisters – the irregular raising of a thin layer of mortar at the surface of concrete occurring during or soon after completion of finishing operations.

Bugholes – surface voids resulting from the migration of entrapped air or water to the interface between fresh concrete and formwork, typically occurring in vertical surfaces.

Bulking – increase in volume of a quantity of sand when in a moist condition compared to its volume when in a dry state.

Bulk density – the mass of a material per unit volume including voids between particles.

C

Calcination – the heating of solids to a controlled high temperature for the purpose of removing volatile substances, or oxidizing a portion of mass, rendering them friable. The root of the word calcination refers to its most prominent use, which is to remove carbon dioxide from limestone through combustion to yield calcium oxide (quicklime). Calcium oxide is a crucial ingredient in portland cement clinker manufacture.

Calcined clay – clay heated to high temperature to alter its physical properties for use as a pozzolan or supplementary cementing material in concrete.

Calcined shale – shale heated to high temperature to alter its physical properties for use as a pozzolan or supplementary cementing material in concrete.

Calcium aluminate cement – High aluminate content cement used in special applications for early strength gain, resistance to high temperatures, and resistance to sulfates, weak acids, and seawater.

Carbonation – reaction between carbon dioxide and a hydroxide or oxide to form a carbonate.

Calorimeter – an instrument for measuring heat exchange during a chemical reaction.

Cast-in-place concrete – concrete that is deposited and hardens in its final place within the completed structure.

Cellular concrete – high air content or high void ratio concrete resulting in low density.

Cement – see *Portland cement, Blended cement,* and *Hydraulic cement*.

Cement kiln dust – a fine grained, solid, highly alkaline particulate material chiefly composed of oxidized, anhydrous, micron-sized particles collected from electrostatic precipitators during the production of cement clinker.

Cement paste – constituent of concrete, mortar, grout, and plaster consisting of cementitious materials and water.

Cementitious material (cementing material) – any material having cementing properties (hydraulic or pozzolanic) that contribute to the formation of hydrated calcium silicate compounds.

Central-mixed concrete – concrete mixed in a stationary concrete mixer from which the freshly mixed concrete is then transported to the point of delivery.

Chemical admixture – see *Admixture*.

Chemical bond – bond between materials resulting from cohesion and adhesion developed by chemical reaction.

Chloride – chemical compounds containing chloride ions, which promote the corrosion of steel reinforcement. Chloride deicing chemicals are primary sources.

Clinker – a partially fused product of a cement kiln, which is ground with other ingredients to make hydraulic cement.

Coarse aggregate – natural gravel, crushed stone, or iron blast-furnace slag, usually larger than 5 mm (0.2 in.) and commonly ranging in size between 9.5 mm and 37.5 mm (⅜ in. to 1½ in.).

Coefficient of thermal expansion – change in linear dimension per unit length per degree of temperature change.

Cohesion – mutual attraction by which elements of a substance are held together.

Cold joint – a joint or discontinuity resulting from a delay in placement of sufficient duration to preclude bonding of the material.

Colored concrete – concrete containing either white cement and/or mineral oxide pigments to produce colors other than the normal gray hue of traditional gray cement concrete.

Compaction – process of inducing a closer arrangement of the solid particles in freshly mixed and placed concrete, mortar, or grout by reduction of voids, usually by vibration, tamping, rodding, puddling, or a combination of these techniques. Also called consolidation.

Compressive strength – maximum resistance that a concrete, mortar, or grout specimen will sustain when loaded axially in compression in a testing machine at a specified rate; usually expressed as force per unit of cross-sectional area, such as megapascals (MPa) or pounds per square inch (psi).

Consistency – relative mobility or ability of freshly mixed concrete, mortar, or grout to flow. (See also *Slump* and *Workability*.)

Construction joint – interface between concrete placements intentionally created to facilitate the construction process.

Continuous mixer – a mixer into which the ingredients of the mixture are fed without stopping, and from which the mixed product is discharged in a continuous stream. (See also *Volumetric mixing*.)

Contraction joint – weakened plane made by grooving, sawing, or forming a joint in the concrete to control cracking due to volume change in the structure. Also known as a "Control joint."

Corrosion – deterioration of material, typically metal, by chemical, electrochemical, or electrolytic reaction.

Crazing – network pattern of fine cracks that form on the near surface of concrete.

Creep – time-dependent deformation of concrete due to a sustained load.

Curing – process of maintaining moisture and a favorable temperature for a suitable period of time so that the desired properties of the material can develop.

Curling – out-of-plane deformation of the corners, edges, and surface of a pavement, slab, or wall panel from its original shape predominantly caused by temperature differences. (See also *Warping*.)

D

Dampproofing – treatment of concrete, mortar, grout, or plaster to retard the passage or absorption of water or water vapor.

D-cracking – a series of cracks in concrete near and roughly parallel to joints and edges resulting from use of coarse aggregate that is susceptible to damage during cycles of freezing and thawing.

Delamination – separation along a plane parallel to the surface of a material.

Delayed ettringite formation – a form of sulfate attack by which mature hardened concrete is damaged by internal expansion caused by the late formation of ettringite, not likely to occur unless the concrete has been exposed to temperatures during curing of 70°C (158°F) or greater. (See also *Ettringite*.)

Deleterious – causing, or with the potential to cause, harm or damage.

Diffusion – movement of species (ions, gas, or vapor) from an area of higher concentration to an area of lower concentration independent of the bulk motion of a fluid.

Discoloration – departure of color from that which is normal or desired.

Density – mass per unit volume; the weight per unit volume in air, expressed, for example, in kg/m³ (lb/ft³).

Dry process – in the manufacture of cement, the process in which the raw materials are ground, conveyed, blended, and stored in a dry condition. (See also *Wet process.*)

Dry-rodded density – mass per unit volume of dry aggregate, which includes the volume of the particles and the voids between particles, compacted by rodding under standardized conditions.

Drying shrinkage – shrinkage resulting from loss of moisture.

Durability – ability to resist weathering action and other conditions of service, such as chemical attack, freezing and thawing, and abrasion.

Durability factor – a measure of the change in a material property over a period of time as a response to exposure to a treatment that can cause deterioration, usually expressed as a percentage of the value of the property before exposure.

Dusting – the development of a powdered substance, or laitance, at the surface of hardened concrete.

E

Early stiffening – rapid development of rigidity in freshly mixed hydraulic cement paste, mortar, grout, plaster, or concrete.

Edging – the operation of tooling the edges of a fresh concrete slab to provide a rounded corner.

Efflorescence – a crystalline or powdery deposit the surface of a porous material (concrete, masonry, or plaster substrates) formed by migration of salts that crystallize into solid deposits through evaporation.

Engineered Cement Composite – a ductile, bendable concrete, with tensile strain capacity typically beyond 2%. ECC belongs to the broad class of fiber reinforced concrete (FRC) and can deform up to 3% to 5% in tension before it fails, which gives it 300 to 500 times more tensile strain capacity than normal concrete.

Entrained air – spherical microscopic air bubbles – usually 10 μm to 1000 μm in diameter – intentionally incorporated into concrete to provide freezing and thawing resistance and/or improve workability.

Entrapped air – irregularly shaped, unintentional air voids in fresh or hardened concrete 1 mm or larger in size.

Epoxy resin – class of organic chemical bonding systems used in the preparation of special coatings or adhesives for concrete or masonry or as binders in epoxy-resin mortars and concretes.

Equilibrium density – the density reached by structural lightweight concrete after exposure to relative humidity of 50% ± 5% and a temperature of 73.5°F ± 3.5°F (23°C ± 2°C) for a period of time sufficient to reach a density that changes less than 0.5% in a period of 28 days.

Equivalent alkalies – a chemical characteristic of cement determined using the formula % Na_2Oeq = % Na_2O + 0.658 % K_2O.

Ettringite – needle-like crystalline hydrate produced by the reaction of C_3A, gypsum, and water within a portland cement concrete.

Exothermic reaction – a chemical reaction that occurs with the evolution of heat.

Expansion joint – a separation provided between adjacent parts of a structure to allow movement.

Expansive cement – a cement that when mixed with water produces a paste that after setting increases in volume to a greater degree than does portland-cement paste.

Exposure class – designation used to describe environmental conditions to which concrete will be exposed.

Extended set-control admixture – an admixture that can predictably reduce the hydration rate of cement for applications requiring extended time of setting followed by normal strength development. (Also known as hydration control admixture.)

F

False set – the rapid development of rigidity in a freshly mixed portland cement paste, mortar, or concrete, in which rigidity can be dispelled and plasticity regained by further mixing without addition of water. (See also *Flash set.*)

Faulting – differential displacement of a slab or wall along a joint or crack.

Ferrocement – one or more layers of steel or wire reinforcement encased in portland cement mortar creating a thin-section composite material.

Fibers – thread or thread-like material ranging from 0.05 mm to 4 mm (0.002 in. to 0.16 in.) in diameter and from 10 mm to 150 mm (0.5 in. to 6 in.) in length and made of steel, glass, synthetic (plastic), carbon, or natural materials.

Fiber-reinforced concrete – concrete containing randomly oriented fibers in two or three dimensions throughout the concrete matrix.

Final set – the degree of stiffening of a cementitious mixture greater than initial set, generally stated as an empirical value indicating the time required for the cementitious mixture to stiffen sufficiently to resist, to an established degree, the penetration of a weighted test device. (See also *Initial set.*)

Fine aggregate – aggregate that passes the 9.5-mm (⅜-in.) sieve, almost entirely passes the 4.75-mm (No. 4) sieve, and is predominantly retained on the 75-μm (No. 200) sieve.

Fineness modulus (FM) – factor obtained by adding the cumulative percentages of material in a sample of aggregate retained on each of a specified series of sieves and dividing the sum by 100.

Finishability – ease of performing finishing operations to achieve specified surface characteristics.

Finishing – mechanical operations that establish the final appearance of any concrete surface that include screeding, consolidating, floating, troweling, or texturing.

Fire resistance – that property of a building material, element, or assembly to withstand fire or give protection from fire; it is characterized by the ability to confine a fire or to continue to perform a given structural function during a fire, or both.

Flash set – the rapid development of rigidity in a freshly mixed portland cement paste, mortar, or concrete, characteristically with the evolution of considerable heat, which rigidity cannot be dispelled nor can the plasticity be regained by further mixing without addition of water (also called quick setting). (See also *False set.*)

Flexural strength – The resistance of a specimen to bending and reported as modulus of rupture. (See also *Modulus of rupture.*)

Floating – smoothing and subsequent compaction and consolidation of an unformed fresh concrete surface using a tool made usually of wood, aluminum, or magnesium to impart a relatively even but still open texture.

Fly ash – residue from coal combustion, which is carried in flue gases, and is used as a pozzolan or cementing material in concrete.

Forms – temporary supports for keeping fresh concrete in place until it has hardened to such a degree as to be self supporting (when the structure is able to support its dead load).

Freeze-thaw resistance – ability of concrete to withstand cycles of freezing and thawing. (See also *Entrained air* and *Air-entraining admixture.*)

Fresh concrete – concrete that has been recently mixed and is still workable and plastic.

G

Gap-graded aggregate – aggregate intentionally graded so that portions of aggregate retained on specified intermediate sieves are substantially absent.

Geopolymer – an alternative cementitious material produced from alumino-silicate precursors that form nonhydrated reaction products by alkali activation.

Grading – size distribution of aggregate particles, determined by separation with standard screen sieves.

Grout – mixture of cementitious material with or without aggregate or admixtures to which sufficient water is added to produce a pouring or pumping consistency without segregation of the constituent materials.

Gypsum – common name for the mineral calcium sulfate dihydrate ($CaSO_4 \cdot 2H_2O$).

H

Hardened concrete – concrete that is in a solid state and has developed a certain strength.

Hardener – a chemical (including certain fluosilicates or sodium silicate) applied to concrete floors to reduce wear and dusting; or, in a two-component adhesive or coating, the chemical component that causes the resin component to cure.

Harsh mixture – a concrete mixture that lacks desired workability due to a deficiency of mortar or aggregate fines.

Heat of hydration – heat evolved during the chemical reaction of a hydraulic cement with water.

High-density concrete (heavyweight concrete) – concrete of very high density; normally designed by the use of heavyweight aggregates.

High-performance concrete – concrete meeting special combinations of performance and uniformity requirements that cannot always be achieved routinely using conventional constituents and normal mixing, placing, and curing practices.

High-strength concrete – concrete with a design strength of at least 70 MPa (10,000 psi).

Honeycomb – term that describes the failure of mortar to completely surround coarse aggregates in concrete, leaving empty spaces (voids) between them.

Hot cement – newly manufactured cement that has not had an opportunity to cool after grinding of the clinker.

Hydrated lime – dry powder obtained by treating quicklime with sufficient water to satisfy its chemical affinity for water; consists essentially of calcium hydroxide or a mixture of calcium hydroxide and magnesium oxide or magnesium hydroxide, or both.

Hydration – in concrete, mortar, grout, and plaster, the chemical reaction between hydraulic cement and water in which new compounds with strength-producing properties are formed.

Hydration control admixture – See *Extended set-control admixture*.

Hydraulic cement – cement that sets and hardens by chemical reaction with water, and is capable of doing so under water. (See also *Portland cement* and *Blended cement*.)

I

Initial curing – deliberate action taken between placement and final finishing of concrete to reduce the loss of water from the surface of the concrete.

Initial set – the degree of stiffening of a cementitious mixture less than final set, generally stated as an empirical value indicating the time required for the cementitious mixture to stiffen sufficiently to resist, to an established degree, the penetration of a weighted test device. (See also *Final set*.)

Internal curing – hydration of cement from the availability of internal water that is not part of the mixing water.

Isolation joint – separation that allows adjoining parts of a structure to move freely to one another, both horizontally and vertically.

J

Joint – see *Construction joint, Contraction joint, Isolation joint,* and *Expansion joint*.

Joint filler – compressible material used to fill a joint to prevent the infiltration of debris and provide support for sealants applied to the joint.

Joint sealant – compressible material used to exclude water and solid foreign materials from joints.

Joint spall – a spall adjacent to a joint.

K

Keyway – a recess or groove in one lift or placement of concrete that is filled with concrete of the next lift, giving shear strength to the joint.

Kiln – rotary furnace used in cement manufacture to heat and chemically combine raw inorganic materials, such as limestone, sand, and clay, into portland cement clinker.

L

Laitance – a layer of weak material derived from cementitious material and aggregate fines either carried by bleeding to the surface or to internal cavities of freshly placed mixture, or separated from the mixture and deposited on the surface or internal cavities during placement of the mixture.

Lean concrete – concrete of low cementitious material content.

Licensed design professional – an engineer or architect who is licensed to practice structural design as defined by the statutory requirements of the professional licensing laws of a state or jurisdiction, who is responsible for the structural design of a particular project, historically known as the *Engineer of record*.

Lightweight aggregate – low-density aggregate used to produce lightweight (low-density) concrete including expanded or sintered clay, slate, shale, perlite, vermiculite, or slag; natural pumice, scoria, volcanic cinders, tuff, or diatomite; sintered fly ash or industrial cinders.

Lightweight concrete – low-density concrete compared to normal-density concrete typically accomplished using lightweight aggregates.

Lime – general term that includes the various chemical and physical forms of quicklime, hydrated lime, and hydraulic lime. It may be high-calcium, magnesian, or dolomitic.

Loss on ignition – the percentage loss in mass of a sample ignited to constant mass at a specified temperature, usually 900°C to 1000°C (1650°F to 1830°F).

M

Map cracking – intersecting cracks that extend below the surface of hardened concrete that vary in width from fine and barely visible to open and well-defined (also called pattern cracking).

Masonry – concrete masonry units, clay brick, structural clay tile, stone, terra cotta, and the like, or combinations thereof, bonded with mortar, dry-stacked, or anchored with metal connectors to form walls, building elements, pavements, and other structures.

Masonry cement – hydraulic cement, primarily used in masonry and plastering construction, consisting of a mixture of portland or blended hydraulic cement and plasticizing materials (such as limestone, hydrated, or hydraulic lime) together with other materials introduced to enhance one or more properties such as setting time, workability, water retention, and durability.

Mass concrete – any volume of structural concrete in which a combination of dimensions of the member being cast, the boundary conditions, the characteristics of the concrete mixture, and the ambient conditions can lead to undesirable thermal stresses, cracking, deleterious chemical reactions, or reduction in the long-term strength as a result of elevated concrete temperature due to heat of hydration.

Maturity concept – the cumulative product of the age of the concrete and its average curing temperature above a certain base temperature. Used to perform an estimation of concrete strength based on time of curing and variations in curing temperature. (See also *Temperature-time factor.*)

Maximum size of aggregate – the smallest sieve opening through which the entire amount of aggregate is required to pass. (See also *Nominal maximum size of aggregate.*)

Metakaolin – highly reactive pozzolan made from heat-treated kaolin clays.

Mineral filler – a finely divided mineral product at least 65% of which passes the 75-μm (No. 200) sieve.

Mixing Water – the water in freshly-mixed cementitious mixtures, exclusive of any previously absorbed by the aggregate. (Also called *Batch water.*)

Modulus of elasticity – ratio of normal stress to corresponding strain for tensile or compressive stress below the proportional limit of the material; also referred to as elastic modulus, Young's modulus, and Young's modulus of elasticity; denoted by the symbol *E*.

Mohs scale – relative scale of the hardness of minerals ranging from 1 through 10, with a soft material like talc having a Mohs hardness of 1 and material like diamond having a Mohs hardness of 10.

Moist curing – curing with moist air (no less than 95% relative humidity) at atmospheric pressure and a temperature of about 23°C (73°F).

Mortar – mixture of cementitious materials, fine aggregate, and water, which may contain admixtures, and is usually used to bond masonry units.

Mortar cement – hydraulic cement, primarily used in masonry construction, consisting of a mixture of portland or blended hydraulic cement and plasticizing materials (such as limestone, hydrated, or hydraulic lime) together with other materials introduced to enhance one or more properties such as setting time, workability, water retention, and durability. Specifications for mortar cement typically require lower air contents and a flexural bond strength. (See also *Masonry cement.*)

Mixture design – the process by which the concrete performance characteristics are defined by reviewing particular conditions of use and establishing specific mixture characteristics.

Mixture proportioning – selecting proportions of available materials to produce concrete of required properties, with the greatest economy.

N

Natural cement – a hydraulic cement produced by calcining an argillaceous limestone at a temperature below the sintering point and then grinding to a fine powder.

Natural pozzolan – a raw or calcined natural material that has pozzolanic properties.

Neat cement paste – a mixture of hydraulic cement and water.

Nominal maximum size of aggregate – in specifications for and in descriptions of aggregate, the smallest sieve opening through which the majority of the aggregate (typically 85% to 95%) is required to pass. (See also *Maximum size of aggregate.*)

Normal weight concrete – class of concrete made with normal density aggregates, usually crushed stone or gravel, having a density of approximately 2400 kg/m³ (150 lb/ft³). (See also *Lightweight concrete* and *High-density concrete.*)

No-slump concrete – concrete having a slump of less than 6 mm (¼ in.).

O

Oil-well cement – see *Well cement.*

Oven-dry – the condition resulting from having been dried to essentially constant mass in an oven at a fixed temperature.

Overlay – layer of concrete or mortar placed on or bonded to the surface of an existing pavement or slab.

P

Parge – to coat with plaster, particularly foundation walls and rough masonry.

Particle-size distribution – the range of particle sizes characterizing a powder, such as hydraulic cement, or a granular material. For aggregate particle size distribution, see *Grading.*

Pavement (concrete) – highway, road, street, path, or parking lot surfaced with concrete.

Pea gravel – screened gravel, most of the particles of which pass a 9.5-mm (⅜-in.) sieve and are retained on a 4.75 mm (No. 4) sieve.

Permeability – property of allowing passage of liquids or gases.

Pervious concrete (no-fines or porous concrete) – concrete containing insufficient fines or no fines to fill the voids between aggregate particles in a concrete mixture. The coarse aggregate particles are coated with a cement and water paste to bond the particles at their contact points. The resulting concrete contains an interconnected pore system allowing storm water to drain through the concrete to the subbase below.

Petrographic analysis – laboratory study of concrete and mortar samples to determine various characteristics including, but not limited to, w/cm, paste-aggregate bond, and air content.

pH – chemical symbol for the logarithm of the reciprocal of hydrogen ion concentration, used to express the acidity or alkalinity (base) of a solution on a scale of 0 to 14, where less than 7 represents acidity, and more than 7 alkalinity.

Photocatalytic concrete – concrete that has the ability to neutralize pollution and self-clean by incorporating a process known as photocatalysis. (also known as Self-cleaning concrete.)

Pitting – development of relatively small cavities in a surface; in concrete, localized disintegration, such as a popout; in steel, localized corrosion evident as minute cavities on the surface.

Plain concrete – structural concrete with no reinforcement or with less reinforcement than the minimum amount specified for reinforced concrete in the applicable building code.

Plastic cement – special hydraulic cement product manufactured for plaster and stucco application. One or more inorganic plasticizing agents are interground or blended with the cement to increase the workability and molding characteristics of the resultant mortar, plaster, or stucco.

Plasticity – that property of freshly mixed cement paste, concrete, mortar, grout, or plaster that determines its workability, resistance to deformation, or ease of molding.

Plasticizer – admixture that increases the plasticity of portland cement concrete, mortar, grout, or plaster.

Plastic shrinkage cracking – crack that occurs in concrete before initial set.

Polymer-portland cement concrete – fresh portland cement concrete to which a polymer is added for improved durability and adhesion characteristics, often used in overlays for bridge decks; also referred to as polymer-modified concrete and latex-modified concrete.

Popout – shallow depression in a concrete surface resulting from the breaking away of pieces of concrete aggregates due to internal pressure.

Portland blast-furnace slag cement – hydraulic cement consisting of an intimate and uniform mixture of portland cement and granulated blast-furnace slag or slag cement produced by intergrinding, blending, or a combination of intergrinding and blending portland cement clinker or portland cement and slag cement or granulated blast furnace slag, in which the amount of the granulated blast-furnace slag or slag cement constituent is within specified limits.

Portland cement – Calcium silicate-based hydraulic cement produced by pulverizing portland cement clinker, and usually containing calcium sulfate and other ingredients. (See also *Hydraulic cement.*)

Portland cement plaster – a combination of portland cement-based cementitious material(s) and aggregate mixed with a suitable amount of water to form a plastic mass that will adhere to a surface and harden, preserving any form and texture imposed on it while plastic. (See also *Stucco.*)

Portland-limestone cement – hydraulic cement consisting of an intimate and uniform mixture of portland cement and limestone produced by intergrinding, blending or a combination of intergrinding and blending portland cement clinker or portland cement and limestone, in which the amount of the limestone constituent is within specified limits.

Portland-pozzolan cement – hydraulic cement consisting of an intimate and uniform mixture of portland cement and pozzolan produced by intergrinding, blending, or a combination of intergrinding and blending portland cement clinker or portland cement and pozzolan, in which the amount of the pozzolan constituent is within specified limits.

Portlandite – a crystalline calcium hydroxide ($Ca(OH)_2$).

Post-tensioning – method of prestressing in which prestressing steel is tensioned after concrete has hardened. (See *Prestressed concrete.*)

Pozzolan – siliceous or siliceous and aluminous materials, like fly ash or silica fume, which in itself possess little or no cementitious value but which will, in finely divided form and in the presence of moisture, chemically react with calcium hydroxide at ordinary temperatures to form compounds possessing cementitious properties.

Precast concrete – concrete cast in forms in a controlled environment and allowed to achieve a specified strength prior to placement on location.

Prestressed concrete – concrete in which compressive stresses are induced by high-strength steel tendons or bars in a concrete element before loads are applied to the element which will balance the tensile stresses imposed in the element during service. (See *Post-tensioning* and *Pre-tensioning*.)

Pre-tensioning – a method of prestressing in which the tendons are tensioned before the concrete is placed.

Q

Quality assurance – actions taken by an organization to provide and document that what is being done and what is being provided are in accordance with the contract documents and standards of good practice for the work.

Quality control – actions taken by a producer or contractor to provide control over what is being done and what is being provided so that applicable standards of good practice for the work are followed.

Quicklime – calcium oxide (CaO).

R

Reactive-powder concrete – high-strength, low-water content, and low-porosity concrete with high silica content and aggregate particle sizes of less than 0.3 mm (0.01 in.).

Ready-mixed concrete – concrete manufactured for delivery to a location in a fresh state.

Recycled concrete – hardened concrete that has been processed for reuse, usually as an aggregate.

Reinforced concrete – concrete to which tensile bearing materials such as steel rods or metal wires are added for tensile strength.

Reinforcement – bars, wires, strands, fibers, or other slender elements that are embedded in a matrix such that they act together to resist forces.

Relative density – a ratio relating the mass of a volume of material to that of water; also called specific gravity.

Relative humidity – the ratio of the quantity of water vapor actually present in the atmosphere to the amount of water vapor present in a saturated atmosphere at a given temperature, expressed as a percentage.

Resilience – longevity for continued use and adaptability to future use. Concepts of functional resilience include robustness, durability, enhanced disaster resistance, and longevity of which some components have been referred to as passive survivability.

Retarder – an admixture that delays the setting and hardening of concrete.

Retemper – to add water and remix a cementitious mixture to restore workability to a condition in which the mixture is placeable or usable.

Rodding – consolidation of concrete by means of a tamping rod.

Roller-compacted concrete (RCC) – a zero slump mixture of aggregates, cementitious materials and water that is consolidated by rolling with vibratory compactors; typically used in the construction of dams, industrial pavements, storage and composting areas, and as a component of composite pavements for highways and streets.

Rubbed finish – a finish obtained by using an abrasive to remove surface irregularities from concrete.

S

Sand streak – a streak of exposed fine aggregate in the surface of formed concrete caused by bleeding.

Saturated surface-dry – moisture condition of an aggregate particle or other porous solid when the permeable pores are filled with water and no water is on the exposed surfaces.

Scaling – disintegration and flaking of a hardened concrete surface, frequently due to repeated freeze-thaw cycles and application of deicing chemicals.

Screed – to strike off a cementitious mixture lying beyond the desired plane or shape; or the tool for striking off the cementitious mixture surface, sometimes referred to as a strikeoff.

Segregation – separation of the components (aggregates and mortar) of fresh concrete, resulting in a nonuniform mixture.

Self-consolidating concrete – fresh concrete that can flow around reinforcement and consolidate within formwork under its own weight without vibration.

Self-desiccation – the reduction in the internal relative humidity of a sealed cementitious mixture, due to chemical reactions between cementitious materials and water, that may reduce the rate of hydration or stop hydration.

Self-healing concrete – concrete with increased crack-healing potential due to alkali-resistant bacterial spores in the concrete matrix.

Set – a chemical process that results in a gradual development of rigidity of a cementitious mixture, adhesive, or resin.

Shear – an internal force tangential to the plane on which it acts.

Shoring – props or posts of timber or other material in compression used for the temporary support of excavations, formwork, or unsafe structures.

Shotcrete – mortar or small-aggregate concrete that is conveyed by compressed air through a hose and applied at high velocity to a surface. Also known as gunite and sprayed concrete.

Shrinkage – decrease in either length or volume of a material resulting from changes in moisture content, temperature, or chemical changes.

Shrinkage-compensating concrete – concrete containing expansive cement, or an admixture, which produces expansion during hardening and thereby offsets the contraction occurring later during drying (drying shrinkage).

Shrink-mixed concrete – ready-mixed concrete mixed partially in a stationary mixer and then mixed in a truck mixer.

Sieve – a metallic plate or sheet, woven-wire cloth, or other similar device with regularly spaced apertures of uniform size mounted in a suitable frame or holder for use in separating granular material according to size.

Sieve size – nominal size of openings between cross wires of a testing sieve.

Silica fume – very fine noncrystalline silica which is a byproduct from the production of silicon and ferrosilicon alloys in an electric arc furnace; used as a pozzolan in concrete.

Slag cement – hydraulic cement produced by finely grinding granulated blast-furnace slag.

Slump – measure of the consistency of freshly mixed concrete, equal to the immediate subsidence of a specimen molded with a standard slump cone.

Slump loss – the amount by which the slump of freshly mixed concrete changes (stiffens) during a period of time after an initial slump test was made on a sample or samples thereof.

Slurry – thin mixture of an insoluble substance, such as portland cement, slag cement, or clay, with a liquid, such as water.

Soil cement – mixture of soil and measured amounts of portland cement and water compacted to a high density; primarily used as a base material under pavements; also called cement-stabilized soil.

Soundness – the freedom of a solid from cracks, flaws, fissures, or variations from an accepted standard; in the case of a cement, freedom from excessive volume change after setting; in the case of aggregate, the ability to withstand the aggressive action to which concrete containing it might be exposed, particularly that due to weather.

Spacing factor – an index related to the maximum distance of any point in a cement paste or in the cement paste fraction of mortar or concrete from the periphery of an air void (also called Powers' spacing factor).

Spalling – loss of section of concrete, typically caused by pressure of expansion or erosion from impacts on edges and joints.

Specific gravity – see *Relative density*.

Stress – force per unit area.

Stability – ability of concrete to resist separation (segregation).

Stucco – portland cement plaster and stucco are the same material. The term "stucco" is widely used to describe the cement plaster used for coating exterior surfaces of buildings. However, in some geographical areas, "stucco" refers only to the factory-prepared finish coat mixtures. (See also *Portland cement plaster*.)

Subbase – the layer in a pavement system between the subgrade and the base course, or between the subgrade and the pavement.

Subgrade – the soil prepared and compacted to support a structure or a pavement system.

Sulfate attack – chemical attack on concrete caused by sulfates in the groundwater or soil manifested by slow expansion and disintegration of the concrete.

Superplasticizer (plasticizer) – admixture that increases the flowability of a fresh concrete mixture.

Supplementary cementitious (cementing) materials – cementitious material other than portland cement or blended cement. (See also *Cementitious material*.)

Surface moisture – free water retained on surfaces of aggregate particles and considered to be part of the mixing water in concrete, as distinguished from absorbed moisture.

Sustainable development – development that meets the needs of the present without compromising the ability of future generations to meet their own needs.

T

Temperature reinforcement – reinforcement designed to carry stresses resulting from temperature changes; also the minimum reinforcement for areas of members that are not subjected to primary stresses or necessarily to temperature stresses.

Temperature-time factor – product of temperature multiplied by time for a specific interval. (See also *Maturity concept*.)

Tensile strength – stress up to which concrete is able to resist cracking under axial tensile loading.

Ternary mixture – concrete containing three different types of cementitious materials.

Thermal conductivity – the ability of a homogeneous material to conduct heat, measured as the steady state heat flow per unit area through a body of unit thickness with one degree temperature difference between the surfaces.

Thermal contraction – contraction caused by decrease in temperature.

Thermal expansion – expansion caused by increase in temperature.

Thermal shock – the subjection of newly hardened concrete to a rapid change in temperature that may cause surface cracking.

Translucent concrete – a combination of glass and concrete used together in precast and prestressed panels.

Tremie – a pipe or tube through which concrete is deposited under water, having at its upper end a hopper for filling and a bail for moving the assemblage.

Trial batch – a batch of concrete prepared to establish acceptable proportions of the constituents.

Troweling – smoothing and compacting the unformed surface of fresh concrete by strokes of a trowel.

U

Ultra-high performance concrete – concrete that has a minimum specified compressive strength of 150 MPa (22,000 psi) with specified durability, tensile ductility, and toughness requirements; fibers are generally included to achieve specified requirements.

Uniformity – the homogeneity of a mixture, as characterized by the dispersion of its constituents.

Unit weight – density of fresh concrete or aggregate, normally determined by weighing a known volume of concrete or aggregate (bulk density of aggregates includes voids between particles).

V

Vapor barrier – membrane located under a concrete floor slab that is placed on the ground to retard transmission of water vapor.

Vapor retarder – a membrane that impedes the transmission of gas molecules.

Vibration – high-frequency agitation of freshly mixed concrete through mechanical devices, for the purpose of consolidation.

Viscosity – a measure of the resistance of a fluid to deform under a shear stress.

Volumetric mixing – measurements based on the volumes of the ingredients fed into a container that continually agitates and combines those ingredients for the production of concrete (also called *Volumetric measuring* and *Continuous-mixing*)

Volume change – either an increase or a decrease in volume due to any cause, such as moisture changes, temperature changes, or chemical changes. (See also *Creep* and *Shrinkage*.)

W

Warping – out-of-plane deformation of the corners, edges, and surface of a pavement, slab, or wall panel from its original shape predominantly caused by moisture differences. (See also *Curling*.)

Water absorption – the process by which a liquid (water) is drawn into and tends to fill permeable pores in a porous solid.

Water demand – water content required in a concrete or mortar mixture to achieve a desired consistency or workability.

Water-cement ratio (water-to-cement ratio and w/c) – ratio of mass of water to mass of cement in concrete.

Water cementitious materials ratio (water-to-cementing materials ratio and w/cm) – ratio of mass of water to mass of cementing materials in concrete, including portland cement plus supplementary cementitious materials.

Water reducer – admixture whose properties permit a reduction of water required to produce a concrete mix of a certain slump, reduce water-cement ratio, reduce cement content, or increase slump.

Weigh batching – measuring the constituent materials for mortar or concrete by mass.

Welded-wire reinforcement – a series of longitudinal and transverse wires arranged approximately at right angles to each other and welded together at all points of intersection.

Well cement – hydraulic cement suitable for use under high pressure and temperature in sealing water and gas pockets and setting casings during the drilling and repair of wells, and often contains retarders to meet the requirements of use.

Wet process – in the manufacture of cement, the process in which the raw materials are ground, blended, mixed, and pumped while mixed with water. (See also *Dry process*.)

White portland cement – cement manufactured from raw materials of low iron content.

Workability – that property of freshly mixed concrete, mortar, grout, or plaster that determines its working characteristics, that is, the ease with which it can be mixed, placed, molded, and finished. (See also *Slump* and *Consistency*.)

X

X-ray diffraction – the diffraction of X-rays by substances having a regular arrangement of atoms. Often used in laboratories to characterize materials at the microscopic level.

X-ray fluorescence – characteristic secondary radiation emitted by an element as a result of excitation by X-rays, used to yield chemical analysis of a sample.

Y

Yield – volume per batch of concrete expressed in cubic meters (cubic feet).

Yield strength – the stress at which a material exhibits a specific limiting deviation from the proportionality of stress to strain.

Z

Zero-slump concrete – concrete without measurable slump. (See also *No-slump concrete*.)

ACRONYMS

AAC	Autoclaved Aerated Concrete
AAR	Alkali-Aggregate Reactivity
ACR	Alkali-Carbonate Reaction
AM	Additive Manufacturing
ASR	Alkali-Silica Reaction
BCOA	Bonded Concrete Overlays on Asphalt
CIP	Cast-In-Place (Concrete)
CMS	Cement Modified Soil Subgrades
CRCP	Continuously Reinforced Concrete Pavement
CSP	Concrete Surface Profile
CTB	Cement Treated Bases
DEF	Delayed Ettringite Formation
DSC	Differential Scanning Calorimetry
ECC	Engineered Cementitious Composite
ECR	Epoxy Coated Rebar
EPD	Environmental Product Declaration
FDR	Full Depth Reclamation
FM	Fineness Modulus
FRC	Fiber Reinforced Concrete
FRP	Fiber Reinforced Polymers
GFRC	Glass Fiber Reinforced Concrete
GRP	Ground Penetrating Radar
HPC	High Performance Concrete
HRWR	High Range Water Reducer
ICF	Insulated Concrete Forms
ICP	Insulated Concrete Panels
JPCP	Jointed Plain Concrete Pavement
LCA	Life-Cycle Assessment
LCCA	Life-Cycle Cost Analysis
LCI	Life-Cycle Inventory
LEED	Leadership In Energy And Environmental Design

LID	Low Impact Development
LOI	Loss on Ignition
MTD	Maximum Temperature Differential
MVER	Moisture Vapor Emission Rate
NDT	Nondestructive Testing
OD	Oven-Dry
OPC	Ordinary Portland Cement
PCR	Product Category Rules
PRAHs	Permeability Reducing Admixtures Hydrostatic
PRAs	Permeability Reducing Admixtures
PVA	Polyvinyl Alcohol (fibers)
RCC	Roller Compacted Concrete
RCP	Reinforced Concrete Pipe
RH	Relative Humidity
RHA	Rice Husk Ash
SCC	Self Consolidating Concrete
	Shrinkage Compensating Concrete
	Self-Compacting Concrete
SCM	Supplementary Cementitious Material
SRAs	Shrinkage Reducing Admixtures
SSD	Saturated Surface Dry
TGA	Thermogravimetric Analysis
UHPC	Ultra High Performance Concrete
VMA	Viscosity Modifying Admixture
VOC	Volatile Organic Compounds
w/c	Water to Cement Ratio
w/cm	Water to Cementing Materials Ratio
WRA	Water Reducing Admixture
WWR	Welded Wire Reinforcement
XRD	X-Ray Diffraction

ASTM STANDARDS

ASTM International (ASTM) documents related to aggregates, cement, and concrete that are relevant to or referred to in the text are listed as follows and can be obtained at astm.org.

IEEE/ASTM SI 10	American National Standard for Metric Practice
A184/A184M	Standard Specification for Welded Deformed Steel Bar Mats for Concrete Reinforcement
A416/A416M	Standard Specification for Low-Relaxation, Seven-Wire Steel Strand for Prestressed Concrete
A421/A421M	Standard Specification for Stress-Relieved Steel Wire for Prestressed Concrete
A615/A615M	Standard Specification for Deformed and Plain Carbon-Steel Bars for Concrete Reinforcement
A648	Standard Specification for Steel Wire, Hard-Drawn for Prestressed Concrete Pipe
A704/A704M	Standard Specification for Welded Steel Plain Bar or Rod Mats for Concrete Reinforcement
A706/A706M	Standard Specification for Deformed and Plain Low-Alloy Steel Bars for Concrete Reinforcement
A722/A722M	Standard Specification for High-Strength Steel Bars for Prestressed Concrete
A767/A767M	Standard Specification for Zinc-Coated (Galvanized) Steel Bars for Concrete Reinforcement
A775/A775M	Standard Specification for Epoxy-Coated Steel Reinforcing Bars
A779/A779M	Standard Specification for Steel Strand, Seven-Wire, Uncoated, Compacted for Prestressed Concrete
A820/A820M	Standard Specification for Steel Fibers for Fiber-Reinforced Concrete
A821/A821M	Standard Specification for Steel Wire, Hard-Drawn for Prestressed Concrete Tanks
A881/A881M	Standard Specification for Steel Wire, Indented, Low-Relaxation for Prestressed Concrete
A882/A882M	Standard Specification for Filled Epoxy-Coated Seven-Wire Steel Prestressing Strand
A884/A884M	Standard Specification for Epoxy-Coated Steel Wire and Welded Wire Reinforcement
A886/A886M	Standard Specification for Steel Strand, Indented, Seven-Wire Stress-Relieved for Prestressed Concrete
A910/A910M	Standard Specification for Uncoated, Weldless, 2-Wire and 3-Wire Steel Strand for Prestressed Concrete
A911/A911M	Standard Specification for Low-Relaxation Steel Bars for Prestressed Concrete Railroad Ties
A934/A934M	Standard Specification for Epoxy-Coated Prefabricated Steel Reinforcing Bars
A944	Standard Test Method for Comparing Bond Strength of Steel Reinforcing Bars to Concrete Using Beam End Specimens
A951/A951M	Standard Specification for Steel Wire for Masonry Joint Reinforcement
A955/A955M	Standard Specification for Deformed and Plain Stainless-Steel Bars for Concrete Reinforcement
A970/A970M	Standard Specification for Headed Steel Bars for Concrete Reinforcement
A981/A981M	Standard Test Method for Evaluating Bond Strength for 0.600-in. [15.24-mm] Diameter Steel Prestressing Strand, Grade 270 [1860], Uncoated, Used in Prestressed Ground Anchors
A996/A996M	Standard Specification for Rail-Steel and Axle-Steel Deformed Bars for Concrete Reinforcement
A1022/A1022M	Standard Specification for Deformed and Plain Stainless Steel Wire and Welded Wire for Concrete Reinforcement
A1032	Standard Test Method for Hydrogen Embrittlement Resistance for Steel Wire Hard Drawn Used for Prestressed Concrete Pipe
A1034/A1034M	Standard Test Methods for Testing Mechanical Splices for Steel Reinforcing Bars
A1035/A1035M	Standard Specification for Deformed and Plain, Low-Carbon, Chromium, Steel Bars for Concrete Reinforcement
A1044/A1044M	Standard Specification for Steel Stud Assemblies for Shear Reinforcement of Concrete

A1055/A1055M	Standard Specification for Zinc and Epoxy Dual-Coated Steel Reinforcing Bars
A1060/A1060M	Standard Specification for Zinc-Coated (Galvanized) Steel Welded Wire Reinforcement, Plain and Deformed, for Concrete
A1061/A1061M	Standard Test Methods for Testing Multi-Wire Steel Strand
A1064/A1064M	Standard Specification for Carbon-Steel Wire and Welded Wire Reinforcement, Plain and Deformed, for Concrete
A1078/A1078M	Standard Specification for Epoxy-Coated Steel Dowels for Concrete Pavement
A1081/A1081M	Standard Test Method for Evaluating Bond of Seven-Wire Steel Prestressing Strand
A1094/A1094M	Standard Specification for Continuous Hot-Dip Galvanized Steel Bars for Concrete Reinforcement
C10/C10M	Standard Specification for Natural Cement
C29/C29M	Standard Test Method for Bulk Density ("Unit Weight") and Voids in Aggregate
C31/C31M	Standard Practice for Making and Curing Concrete Test Specimens in the Field
C33/C33M	Standard Specification for Concrete Aggregates
C39/C39M	Standard Test Method for Compressive Strength of Cylindrical Concrete Specimens
C40/C40M	Standard Test Method for Organic Impurities in Fine Aggregates for Concrete
C42/C42M	Standard Test Method for Obtaining and Testing Drilled Cores and Sawed Beams of Concrete
C70	Standard Test Method for Surface Moisture in Fine Aggregate
C78/C78M	Standard Test Method for Flexural Strength of Concrete (Using Simple Beam with Third-Point Loading)
C87/C87M	Standard Test Method for Effect of Organic Impurities in Fine Aggregate on Strength of Mortar
C88	Standard Test Method for Soundness of Aggregates by Use of Sodium Sulfate or Magnesium Sulfate
C91/C91M	Standard Specification for Masonry Cement
C94/C94M	Standard Specification for Ready-Mixed Concrete
C109/C109M	Standard Test Method for Compressive Strength of Hydraulic Cement Mortars (Using 2-in. or [50-mm] Cube Specimens)
C110	Standard Test Methods for Physical Testing of Quicklime, Hydrated Lime, and Limestone
C114	Standard Test Methods for Chemical Analysis of Hydraulic Cement
C117	Standard Test Method for Materials Finer than 75-µm (No. 200) Sieve in Mineral Aggregates by Washing
C123/C123M	Standard Test Method for Lightweight Particles in Aggregate
C125	Standard Terminology Relating to Concrete and Concrete Aggregates
C127	Standard Test Method for Relative Density (Specific Gravity) and Absorption of Coarse Aggregate
C128	Standard Test Method for Relative Density (Specific Gravity) and Absorption of Fine Aggregate
C131/C131M	Standard Test Method for Resistance to Degradation of Small-Size Coarse Aggregate by Abrasion and Impact in the Los Angeles Machine
C136/C136M	Standard Test Method for Sieve Analysis of Fine and Coarse Aggregates
C138/C138M	Standard Test Method for Density (Unit Weight), Yield, and Air Content (Gravimetric) of Concrete
C142/C142M	Standard Test Method for Clay Lumps and Friable Particles in Aggregates
C143/C143M	Standard Test Method for Slump of Hydraulic-Cement Concrete
C144	Standard Specification for Aggregate for Masonry Mortar
C150/C150M	Standard Specification for Portland Cement
C151/C151M	Standard Test Method for Autoclave Expansion of Hydraulic Cement

C156	Standard Test Method for Water Loss [from a Mortar Specimen] Through Liquid Membrane-Forming Curing Compounds for Concrete
C157/C157M	Standard Test Method for Length Change of Hardened Hydraulic-Cement Mortar and Concrete
C170/C170M	Standard Test Method for Compressive Strength of Dimension Stone
C171	Standard Specification for Sheet Materials for Curing Concrete
C172/C172M	Standard Practice for Sampling Freshly Mixed Concrete
C173/C173M	Standard Test Method for Air Content of Freshly Mixed Concrete by the Volumetric Method
C174/C174M	Standard Test Method for Measuring Thickness of Concrete Elements Using Drilled Concrete Cores
C177	Standard Test Method for Steady-State Heat Flux Measurements and Thermal Transmission Properties by Means of the Guarded-Hot-Plate Apparatus
C183/C183M	Standard Practice for Sampling and the Amount of Testing of Hydraulic Cement
C184	Standard Test Method for Fineness of Hydraulic Cement by the 150-μm (No. 100) and 75-μm (No.200) Sieves (Withdrawn 2002)
C185	Standard Test Method for Air Content of Hydraulic Cement Mortar
C186	Standard Test Method for Heat of Hydration of Hydraulic Cement (Withdrawn 2019)
C187	Standard Test Method for Amount of Water Required for Normal Consistency of Hydraulic Cement Paste
C188	Standard Test Method for Density of Hydraulic Cement
C191	Standard Test Methods for Time of Setting of Hydraulic Cement by Vicat Needle
C192/C192M	Standard Practice for Making and Curing Concrete Test Specimens in the Laboratory
C204	Standard Test Methods for Fineness of Hydraulic Cement by Air-Permeability Apparatus
C215	Standard Test Method for Fundamental Transverse, Longitudinal, and Torsional Resonant Frequencies of Concrete Specimens
C219	Standard Terminology Relating to Hydraulic and Other Inorganic Cements
C226	Standard Specification for Air-Entraining Additions for Use in the Manufacture of Air-Entraining Hydraulic Cement
C230/C230M	Standard Specification for Flow Table for Use in Tests of Hydraulic Cement
C231/C231M	Standard Test Method for Air Content of Freshly Mixed Concrete by the Pressure Method
C232/C232M	Standard Test Method for Bleeding of Concrete
C233/C233M	Standard Test Method for Air-Entraining Admixtures for Concrete
C235-68	Method of Test for Scratch Hardness of Coarse Aggregate Particles (Withdrawn 1976)
C243	Standard Test Method for Bleeding of Cement Pastes and Mortars (Withdrawn 2001)
C260/C260M	Standard Specification for Air-Entraining Admixtures for Concrete
C266	Standard Test Method for Time of Setting of Hydraulic-Cement Paste by Gillmore Needles
C270	Standard Specification for Mortar for Unit Masonry
C293/C293M	Standard Test Method for Flexural Strength of Concrete (Using Simple Beam With Center-Point Loading)
C294	Standard Descriptive Nomenclature for Constituents of Concrete Aggregates
C295/C295M	Standard Guide for Petrographic Examination of Aggregates for Concrete
C305	Standard Practice for Mechanical Mixing of Hydraulic Cement Pastes and Mortars of Plastic Consistency
C309	Standard Specification for Liquid Membrane-Forming Compounds for Curing Concrete

C311/C311M	Standard Test Methods for Sampling and Testing Fly Ash or Natural Pozzolans for Use in Portland Cement Concrete
C330/C330M	Standard Specification for Lightweight Aggregates for Structural Concrete
C331/C331M	Standard Specification for Lightweight Aggregates for Concrete Masonry Units
C332	Standard Specification for Lightweight Aggregates for Insulating Concrete
C341/C341M	Standard Practice for Preparation and Conditioning of Cast, Drilled, or Sawed Specimens of Hydraulic- Cement Mortar and Concrete Used for Length Change Measurements
C342-97	Standard Test Method for Potential Volume Change of Cement-Aggregate Combinations (Withdrawn 2001)
C348	Standard Test Method for Flexural Strength of Hydraulic-Cement Mortars
C349	Standard Test Method for Compressive Strength of Hydraulic-Cement Mortars (Using Portions of Prisms Broken in Flexure)
C360-92	Test Method for Ball Penetration in Freshly Mixed Hydraulic Cement Concrete (Withdrawn 1999)
C359	Standard Test Method for Early Stiffening of Hydraulic Cement (Mortar Method)
C387/C387M	Standard Specification for Packaged, Dry, Combined Materials for Concrete and High Strength Mortar
C403/C403M	Standard Test Method for Time of Setting of Concrete Mixtures by Penetration Resistance
C404	Standard Specification for Aggregates for Masonry Grout
C418	Standard Test Method for Abrasion Resistance of Concrete by Sandblasting
C430	Standard Test Method for Fineness of Hydraulic Cement by the 45-µm (No. 325) Sieve
C441/C441M	Standard Test Method for Effectiveness of Pozzolans or Ground Blast-Furnace Slag in Preventing Excessive Expansion of Concrete Due to the Alkali-Silica Reaction
C451	Standard Test Method for Early Stiffening of Hydraulic Cement (Paste Method)
C452/C452M	Standard Test Method for Potential Expansion of Portland-Cement Mortars Exposed to Sulfate
C457/C457M	Standard Test Method for Microscopical Determination of Parameters of the Air-Void System in Hardened Concrete
C465	Standard Specification for Processing Additions for Use in the Manufacture of Hydraulic Cements
C469/C469M	Standard Test Method for Static Modulus of Elasticity and Poisson's Ratio of Concrete in Compression
C470/C470M	Standard Specification for Molds for Forming Concrete Test Cylinders Vertically
C490/C490M	Standard Practice for Use of Apparatus for the Determination of Length Change of Hardened Cement Paste, Mortar, and Concrete
C494/C494M	Standard Specification for Chemical Admixtures for Concrete
C495/C495M	Standard Test Method for Compressive Strength of Lightweight Insulating Concrete
C496/C496M	Standard Test Method for Splitting Tensile Strength of Cylindrical Concrete Specimens
C511	Standard Specification for Mixing Rooms, Moist Cabinets, Moist Rooms, and Water Storage Tanks Used in the Testing of Hydraulic Cements and Concretes
C512/C512M	Standard Test Method for Creep of Concrete in Compression
C513/C513M	Standard Test Method for Obtaining and Testing Specimens of Hardened Lightweight Insulating Concrete for Compressive Strength
C518	Standard Test Method for Steady-State Thermal Transmission Properties by Means of the Heat Flow Meter Apparatus
C535	Standard Test Method for Resistance to Degradation of Large-Size Coarse Aggregate by Abrasion and Impact in the Los Angeles Machine
C563	Standard Test Method for Approximation of Optimum SO3 in Hydraulic Cement

C566	Standard Test Method for Total Evaporable Moisture Content of Aggregate by Drying
C567/C567M	Standard Test Method for Determining Density of Structural Lightweight Concrete
C586	Standard Test Method for Potential Alkali Reactivity of Carbonate Rocks as Concrete Aggregates (Rock- Cylinder Method)
C595/C595M	Standard Specification for Blended Hydraulic Cements
C596	Standard Test Method for Drying Shrinkage of Mortar Containing Hydraulic Cement
C597	Standard Test Method for Pulse Ultrasonic Velocity Through Concrete
C617/C617M	Standard Practice for Capping Cylindrical Concrete Specimens
C618	Standard Specification for Coal Ash and Raw or Calcined Natural Pozzolan for Use in Concrete
C637	Standard Specification for Aggregates for Radiation-Shielding Concrete
C638	Standard Descriptive Nomenclature of Constituents of Aggregates for Radiation-Shielding Concrete
C641	Standard Test Method for Iron Staining Materials in Lightweight Concrete Aggregates
C642	Standard Test Method for Density, Absorption, and Voids in Hardened Concrete
C666/C666M	Standard Test Method for Resistance of Concrete to Rapid Freezing and Thawing
C670	Standard Practice for Preparing Precision and Bias Statements for Test Methods for Construction Materials
C671-94	Standard Test Method for Critical Dilation of Concrete Specimens Subjected to Freezing (Withdrawn 2003)
C672/C672M	Standard Test Method for Scaling Resistance of Concrete Surfaces Exposed to Deicing Chemicals (Withdrawn 2021)
C682-94	Standard Practice for Evaluation of Frost Resistance of Coarse Aggregates in Air-Entrained Concrete by Critical Dilation Procedures (Withdrawn 2003)
C685/C685M	Standard Specification for Concrete Made by Volumetric Batching and Continuous Mixing
C688	Standard Specification for Functional Additions for Use in Hydraulic Cements
C702/C702M	Standard Practice for Reducing Samples of Aggregate to Testing Size
C778	Standard Specification for Standard Sand
C779/C779M	Standard Test Method for Abrasion Resistance of Horizontal Concrete Surfaces
C786/C786M	Standard Test Method for Fineness of Hydraulic Cement and Raw Materials by the 300-μm (No. 50), 150-μm (No. 100), and 75-μm (No. 200) Sieves by Wet Methods
C796/C796M	Standard Test Method for Foaming Agents for Use in Producing Cellular Concrete Using Preformed Foam
C802	Standard Practice for Conducting an Interlaboratory Test Program to Determine the Precision of Test Methods for Construction Materials
C803/C803M	Standard Test Method for Penetration Resistance of Hardened Concrete
C805/C805M	Standard Test Method for Rebound Number of Hardened Concrete
C806	Standard Test Method for Restrained Expansion of Expansive Cement Mortar
C807	Standard Test Method for Time of Setting of Hydraulic Cement Mortar by Modified Vicat Needle
C823/C823M	Standard Practice for Examination and Sampling of Hardened Concrete in Constructions
C827/C827M	Standard Test Method for Change in Height at Early Ages of Cylindrical Specimens of Cementitious Mixtures
C845/C845M	Standard Specification for Expansive Hydraulic Cement
C856	Standard Practice for Petrographic Examination of Hardened Concrete
C869/C869M	Standard Specification for Foaming Agents Used in Making Preformed Foam for Cellular Concrete
C873/C873M	Standard Test Method for Compressive Strength of Concrete Cylinders Cast in Place in Cylindrical Molds

C876	Standard Test Method for Corrosion Potentials of Uncoated Reinforcing Steel in Concrete
C878/C878M	Standard Test Method for Restrained Expansion of Shrinkage-Compensating Concrete
C881/C881M	Standard Specification for Epoxy-Resin-Base Bonding Systems for Concrete
C882/C882M	Standard Test Method for Bond Strength of Epoxy-Resin Systems Used With Concrete By Slant Shear
C884/C884M	Standard Test Method for Thermal Compatibility Between Concrete and an Epoxy-Resin Overlay
C900	Standard Test Method for Pullout Strength of Hardened Concrete
C917	Standard Test Method for Evaluation of Variability of Cement from a Single Source Based on Strength
C918/C918M	Standard Test Method for Measuring Early-Age Compressive Strength and Projecting Later-Age Strength
C928/C928M	Standard Specification for Packaged, Dry, Rapid-Hardening Cementitious Materials for Concrete Repairs
C937	Standard Specification for Grout Fluidifier for Preplaced-Aggregate Concrete
C938	Standard Practice for Proportioning Grout Mixtures for Preplaced-Aggregate Concrete
C939	Standard Test Method for Flow of Grout for Preplaced-Aggregate Concrete (Flow Cone Method)
C940	Standard Test Method for Expansion and Bleeding of Freshly Mixed Grouts for Preplaced-Aggregate Concrete in the Laboratory
C941	Standard Test Method for Water Retentivity of Grout Mixtures for Preplaced-Aggregate Concrete in the Laboratory
C942	Standard Test Method for Compressive Strength of Grouts for Preplaced-Aggregate Concrete in the Laboratory
C943	Standard Practice for Making Test Cylinders and Prisms for Determining Strength and Density of Preplaced-Aggregate Concrete in the Laboratory
C944/C944M	Standard Test Method for Abrasion Resistance of Concrete or Mortar Surfaces by the Rotating-Cutter Method
C953	Standard Test Method for Time of Setting of Grouts for Preplaced-Aggregate Concrete in the Laboratory
C995-01	Standard Test Method for Time of Flow of Fiber-Reinforced Concrete Through Inverted Slump Cone (Withdrawn 2008)
C979/C979M	Standard Specification for Pigments for Integrally Colored Concrete
C989/C989M	Standard Specification for Slag Cement for Use in Concrete and Mortars
C1005	Standard Specification for Reference Masses and Devices for Determining Mass and Volume for Use in the Physical Testing of Hydraulic Cements
C1012/C1012M	Standard Test Method for Length Change of Hydraulic-Cement Mortars Exposed to a Sulfate Solution
C1017/C1017M	Standard Specification for Chemical Admixtures for Use in Producing Flowing Concrete (Withdrawn 2022)
C1018-97	Standard Test Method for Flexural Toughness and First-Crack Strength of Fiber-Reinforced Concrete (Using Beam With Third-Point Loading) (Withdrawn 2006)
C1038/C1038M	Standard Test Method for Expansion of Hydraulic Cement Mortar Bars Stored in Water
C1040/C1040M	Standard Test Methods for In-Place Density of Unhardened and Hardened Concrete, Including Roller Compacted Concrete, By Nuclear Methods
C1059/C1059M	Standard Specification for Latex Agents for Bonding Fresh To Hardened Concrete
C1064/C1064M	Standard Test Method for Temperature of Freshly Mixed Hydraulic-Cement Concrete
C1067	Standard Practice for Conducting a Ruggedness Evaluation or Screening Program for Test Methods for Construction Materials
C1073	Standard Test Method for Hydraulic Activity of Slag Cement by Reaction with Alkali
C1074	Standard Practice for Estimating Concrete Strength by the Maturity Method

C1077	Standard Practice for Agencies Testing Concrete and Concrete Aggregates for Use in Construction and Criteria for Testing Agency Evaluation
C1078-87	Test Methods for Determining Cement Content of Freshly Mixed Concrete (Withdrawn 1998)
C1079-87	Test Methods for Determining the Water Content of Freshly Mixed Concrete (Withdrawn 1998)
C1084	Standard Test Method for Portland-Cement Content of Hardened Hydraulic-Cement Concrete
C1090	Standard Test Method for Measuring Changes in Height of Cylindrical Specimens of Hydraulic-Cement Grout
C1105	Standard Test Method for Length Change of Concrete Due to Alkali-Carbonate Rock Reaction
C1107/C1107M	Standard Specification for Packaged Dry, Hydraulic-Cement Grout (Nonshrink)
C1116/C1116M	Standard Specification for Fiber-Reinforced Concrete
C1138M	Standard Test Method for Abrasion Resistance of Concrete (Underwater Method)
C1140/C1140M	Standard Practice for Preparing and Testing Specimens from Shotcrete Test Panels
C1141/C1141M	Standard Specification for Admixtures for Shotcrete
C1150-96	Standard Test Method for The Break-Off Number of Concrete (Withdrawn 2002)
C1152/C1152M	Standard Test Method for Acid-Soluble Chloride in Mortar and Concrete
C1157/C1157M	Standard Performance Specification for Hydraulic Cement
C1170/C1170M	Standard Test Method for Determining Consistency and Density of Roller-Compacted Concrete Using a Vibrating Table
C1176/C1176M	Standard Practice for Making Roller-Compacted Concrete in Cylinder Molds Using a Vibrating Table
C1181	Standard Test Methods for Compressive Creep of Chemical-Resistant Polymer Machinery Grouts
C1202	Standard Test Method for Electrical Indication of Concrete's Ability to Resist Chloride Ion Penetration
C1218/C1218M	Standard Test Method for Water-Soluble Chloride in Mortar and Concrete
C1222	Standard Practice for Evaluation of Laboratories Testing Hydraulic Cement
C1231/C1231M	Standard Practice for Use of Unbonded Caps in Determination of Compressive Strength of Hardened Cylindrical Concrete Specimens
C1240	Standard Specification for Silica Fume Used in Cementitious Mixtures
C1252	Standard Test Methods for Uncompacted Void Content of Fine Aggregate (as Influenced by Particle Shape, Surface Texture, and Grading)
C1260	Standard Test Method for Potential Alkali Reactivity of Aggregates (Mortar-Bar Method)
C1293	Standard Test Method for Determination of Length Change of Concrete Due to Alkali-Silica Reaction
C1315	Standard Specification for Liquid Membrane-Forming Compounds Having Special Properties for Curing and Sealing Concrete
C1328/C1328M	Standard Specification for Plastic (Stucco) Cement
C1329/C1329M	Standard Specification for Mortar Cement
C1339	Standard Test Method for Flowability and Bearing Area of Chemical-Resistant Polymer Machinery Grouts
C1356	Standard Test Method for Quantitative Determination of Phases in Portland Cement Clinker by Microscopical Point-Count Procedure
C1365	Standard Test Method for Determination of the Proportion of Phases in Portland Cement and Portland- Cement Clinker Using X-Ray Powder Diffraction Analysis
C1383	Standard Test Method for Measuring the P-Wave Speed and the Thickness of Concrete Plates Using the Impact-Echo Method
C1385/C1385M	Standard Practice for Sampling Materials for Shotcrete

C1399/C1399M	Standard Test Method for Obtaining Average Residual-Strength of Fiber-Reinforced Concrete
C1435/C1435M	Standard Practice for Molding Roller-Compacted Concrete in Cylinder Molds Using a Vibrating Hammer
C1436	Standard Specification for Materials for Shotcrete (Withdrawn 2022)
C1437	Standard Test Method for Flow of Hydraulic Cement Mortar
C1438	Standard Specification for Latex and Powder Polymer Modifiers for use in Hydraulic Cement Concrete and Mortar
C1439	Standard Test Methods for Evaluating Latex and Powder Polymer Modifiers for use in Hydraulic Cement Concrete and Mortar
C1451	Standard Practice for Determining Variability of Concrete-Making Materials From a Single Source
C1480/C1480M	Standard Specification for Packaged, Pre-Blended, Dry, Combined Materials for Use in Wet or Dry Shotcrete Application (Withdrawn 2021)
C1506	Standard Test Method for Water Retention of Hydraulic Cement-Based Mortars and Plasters
C1524	Standard Test Method for Water-Extractable Chloride in Aggregate (Soxhlet Method)
C1542/C1542M	Standard Test Method for Measuring Length of Concrete Cores
C1550	Standard Test Method for Flexural Toughness of Fiber Reinforced Concrete (Using Centrally Loaded Round Panel)
C1556	Standard Test Method for Determining the Apparent Chloride Diffusion Coefficient of Cementitious Mixtures by Bulk Diffusion
C1565	Standard Test Method for Determination of Pack-Set Index of Portland and Blended Hydraulic Cement
C1567	Standard Test Method for Determining the Potential Alkali-Silica Reactivity of Combinations of Cementitious Materials and Aggregate (Accelerated Mortar-Bar Method)
C1579	Standard Test Method for Evaluating Plastic Shrinkage Cracking of Restrained Fiber Reinforced Concrete (Using a Steel Form Insert)
C1580	Standard Test Method for Water-Soluble Sulfate in Soil
C1581/C1581M	Standard Test Method for Determining Age at Cracking and Induced Tensile Stress Characteristics of Mortar and Concrete under Restrained Shrinkage
C1582/C1582M	Standard Specification for Admixtures to Inhibit Chloride-Induced Corrosion of Reinforcing Steel in Concrete
C1583/C1583M	Standard Test Method for Tensile Strength of Concrete Surfaces and the Bond Strength or Tensile Strength of Concrete Repair and Overlay Materials by Direct Tension (Pull-off Method)
C1585	Standard Test Method for Measurement of Rate of Absorption of Water by Hydraulic-Cement Concretes
C1600/C1600M	Standard Specification for Rapid Hardening Hydraulic Cement
C1602/C1602M	Standard Specification for Mixing Water Used in the Production of Hydraulic Cement Concrete
C1603	Standard Test Method for Measurement of Solids in Water
C1604/C1604M	Standard Test Method for Obtaining and Testing Drilled Cores of Shotcrete
C1608	Standard Test Method for Chemical Shrinkage of Hydraulic Cement Paste
C1609/C1609M	Standard Test Method for Flexural Performance of Fiber-Reinforced Concrete (Using Beam With Third-Point Loading)
C1610/C1610M	Standard Test Method for Static Segregation of Self-Consolidating Concrete Using Column Technique
C1611/C1611M	Standard Test Method for Slump Flow of Self-Consolidating Concrete
C1621/C1621M	Standard Test Method for Passing Ability of Self-Consolidating Concrete by J-Ring
C1622/C1622M	Standard Specification for Cold-Weather Admixture Systems

C1646/C1646M	Standard Practice for Making and Curing Test Specimens for Evaluating Resistance of Coarse Aggregate to Freezing and Thawing in Air-Entrained Concrete
C1679	Standard Practice for Measuring Hydration Kinetics of Hydraulic Cementitious Mixtures Using Isothermal Calorimetry
C1688/C1688M	Standard Test Method for Density and Void Content of Freshly Mixed Pervious Concrete (Withdrawn 2023)
C1697	Standard Specification for Blended Supplementary Cementitious Materials
C1698	Standard Test Method for Autogenous Strain of Cement Paste and Mortar
C1701/C1701M	Standard Test Method for Infiltration Rate of In Place Pervious Concrete
C1702	Standard Test Method for Measurement of Heat of Hydration of Hydraulic Cementitious Materials Using Isothermal Conduction Calorimetry
C1708/C1708M	Standard Test Methods for Self-leveling Mortars Containing Hydraulic Cements
C1709	Standard Guide for Evaluation of Alternative Supplementary Cementitious Materials (ASCM) for Use in Concrete
C1712	Standard Test Method for Rapid Assessment of Static Segregation Resistance of Self-Consolidating Concrete Using Penetration Test
C1723	Standard Guide for Examination of Hardened Concrete Using Scanning Electron Microscopy
C1738/C1738M	Standard Practice for High-Shear Mixing of Hydraulic Cement Pastes
C1740	Standard Practice for Evaluating the Condition of Concrete Plates Using the Impulse-Response Method
C1741	Standard Test Method for Bleed Stability of Cementitious Post-Tensioning Tendon Grout
C1747/C1747M	Standard Test Method for Determining Potential Resistance to Degradation of Pervious Concrete by Impact and Abrasion (Withdrawn 2022)
C1749	Standard Guide for Measurement of the Rheological Properties of Hydraulic Cementious Paste Using a Rotational Rheometer
C1753	Standard Practice for Evaluating Early Hydration of Hydraulic Cementitious Mixtures Using Thermal Measurements
C1754/C1754M	Standard Test Method for Density and Void Content of Hardened Pervious Concrete (Withdrawn 2021)
C1757	Standard Test Method for Determination of One-Point, Bulk Water Sorption of Dried Concrete (Withdrawn 2022)
C1758/C1758M	Standard Practice for Fabricating Test Specimens with Self-Consolidating Concrete
C1760-12	Standard Test Method for Bulk Electrical Conductivity of Hardened Concrete (Withdrawn 2021)
C1761/C1761M	Standard Specification for Lightweight Aggregate for Internal Curing of Concrete
C1768/C1768M	Standard Practice for Accelerated Curing of Concrete Cylinders
C1777	Standard Test Method for Rapid Determination of the Methylene Blue Value for Fine Aggregate or Mineral Filler Using a Colorimeter
C1778	Standard Guide for Reducing the Risk of Deleterious Alkali-Aggregate Reaction in Concrete
C1792	Standard Test Method for Measurement of Mass Loss versus Time for One-Dimensional Drying of Saturated Concretes (Withdrawn 2023)
C1797	Standard Specification for Ground Calcium Carbonate and Aggregate Mineral Fillers for use in Hydraulic Cement Concrete
C1810/C1810M	Gide for Comparing Performance of Concrete-Making Materials using Mortar Mixtures
C1827	Standard Test Method for Determination of the Air-Entraining Admixture Demand of a Cementitious Mixture
C1866	Standard Specification for Ground-Glass Pozzolan for Use in Concrete
C1872	Standard Test Method for Thermogravimetric Analysis of Hydraulic Cement

C1874	Standard Test Method for Measuring Rheological Properties of Cementitious Materials Using Coaxial Rotational Rheometer
C1875	Standard Practice for Determination of Major and Minor Elements in Aqueous Pore Solutions of Cementitious Pastes by Inductively Coupled Plasma Optical Emission Spectroscopy (ICP-OES)
C1876	Test Method for Bulk Electrical Reistivity or Bulk Conductivity of Concrete
C1890	Standard Test Method for K-slump of Freshly Mixed Concrete
C1891	Standard Test Method for Fineness of Hydraulic Cement by Air Jet Sieving at 45-μm (No. 325)
C1897	Standard Test Methods for Measuring the Reactivity of Supplementary Cementitious Materials by Isothermal Calorimetry and Bound Water Measurements
C1904	Standard Test Methods for Determination of the Effects of Biogenic Acidification on Concrete Antimicrobial Additives and/ or Concrete Products
D75/D75M	Standard Practice for Sampling Aggregates
D98	Standard Specification for Calcium Chloride
D140/D140M	Standard Practice for Sampling Asphalt Materials
D395	Standard Test Methods for Rubber Property – Compression Set
D412	Standard Test Methods for Vulcanized Rubber and Thermoplastic Elastomers – Tension
D448	Standard Classification for Sizes of Aggregate for Road and Bridge Construction
D511	Standard Test Methods for Calcium and Magnesium in Water
D512	Standard Test Methods for Chloride Ion in Water
D513	Standard Test Methods for Total and Dissolved Carbon Dioxide in Water
D516	Standard Test Method for Sulfate Ion in Water
D531	Standard Test Method for Rubber Property – Pusey and Jones Indentation
D546	Standard Test Method for Sieve Analysis of Mineral Filler for Asphalt Paving Mixtures
D575	Standard Test Methods for Rubber Properties in Compression
D558	Standard Test Methods for Moisture-Density (Unit Weight) Relations of Soil-Cement Mixtures
D624	Standard Test Method for Tear Strength of Conventional Vulcanized Rubber and Thermoplastic Elastomers
D632	Standard Specification for Sodium Chloride
D692/D692M	Standard Specification for Coarse Aggregate for Bituminous Paving Mixtures
D814	Standard Test Method for Rubber Property – Vapor Transmission of Volatile Liquids
D817	Standard Test Methods of Testing Cellulose Acetate Propionate and Cellulose Acetate Butyrate
D832	Standard Practice for Rubber Conditioning for Low Temperature Testing
D857	Standard Test Method for Aluminum in Water
D858	Standard Test Methods for Manganese in Water
D859	Standard Test Method for Silica in Water
D871	Standard Test Methods of Testing Cellulose Acetate
D888	Standard Test Methods for Dissolved Oxygen in Water
D914	Standard Test Methods for Ethylcellulose
D945	Standard Test Methods for Rubber Properties in Compression or Shear (Mechanical Oscillograph)
D979/D979M	Standard Practice for Sampling Bituminous Paving Mixtures
D991	Standard Test Method for Rubber Property – Volume Resistivity Of Electrically Conductive and Antistatic Products

D1053	Standard Test Methods for Rubber Property – Stiffening at Low Temperatures: Flexible Polymers and Coated Fabrics
D1067	Standard Test Methods for Acidity or Alkalinity of Water
D1068	Standard Test Methods for Iron in Water
D1073	Standard Specification for Fine Aggregate for Asphalt Paving Mixtures
D1126	Standard Test Method for Hardness in Water
D1139/D1139M	Standard Specification for Aggregate for Single or Multiple Bituminous Surface Treatments
D1179	Standard Test Methods for Fluoride Ion in Water
D1229	Standard Test Method for Rubber Property – Compression Set at Low Temperatures
D1246	Standard Test Method for Bromide Ion in Water
D1253	Standard Test Method for Residual Chlorine in Water
D1292	Standard Test Method for Odor in Water
D1329	Standard Test Method for Evaluating Rubber Property – Retraction at Lower Temperatures (TR Test)
D1343	Standard Test Method for Viscosity of Cellulose Derivatives by Ball-Drop Method
D1349	Standard Practice for Rubber – Standard Conditions for Testing
D1415	Standard Test Method for Rubber Property – International Hardness
D1426	Standard Test Methods for Ammonia Nitrogen In Water
D1429	Standard Test Methods for Specific Gravity of Water and Brine
D1439	Standard Test Methods for Sodium Carboxymethylcellulose
D1456	Standard Test Method for Rubber Property – Elongation at Specific Stress
D1460	Standard Test Method for Rubber Property – Change in Length During Liquid Immersion
D1557	Standard Test Methods for Laboratory Compaction Characteristics of Soil Using Modified Effort (56,000 ft-lbf/ft^3 [2,700 kN-m/m^3])
D1687	Standard Test Methods for Chromium in Water
D1688	Standard Test Methods for Copper in Water
D1691	Standard Test Methods for Zinc in Water
D1695	Standard Terminology of Cellulose and Cellulose Derivatives
D1696	Standard Test Method for Solubility of Cellulose in Sodium Hydroxide
D1795	Standard Test Method for Intrinsic Viscosity of Cellulose
D1883	Standard Test Method for California Bearing Ratio (CBR) of Laboratory-Compacted Soils
D1886	Standard Test Methods for Nickel in Water
D1926	Standard Test Methods for Carboxyl Content of Cellulose
D1971	Standard Practices for Digestion of Water Samples for Determination of Metals by Flame Atomic Absorption, Graphite Furnace Atomic Absorption, Plasma Emission Spectroscopy, or Plasma Mass Spectrometry
D1976	Standard Test Method for Elements in Water by Inductively-Coupled Plasma Atomic Emission Spectroscopy
D2136	Standard Test Method for Coated Fabrics – Low-Temperature Bend Test
D2137	Standard Test Methods for Rubber Property – Brittleness Point of Flexible Polymers and Coated Fabrics
D2240	Standard Test Method for Rubber Property – Durometer Hardness
D2363	Standard Test Methods for Hydroxypropyl Methylcellulose

D2364	Standard Test Methods for Hydroxyethylcellulose
D2419	Standard Test Method for Sand Equivalent Value of Soils and Fine Aggregate
D2632	Standard Test Method for Rubber Property – Resilience by Vertical Rebound
D2929	Standard Test Method for Sulfur Content of Cellulosic Materials by X-Ray Fluorescence
D2940/D2940M	Standard Specification for Graded Aggregate Material For Bases or Subbases for Highways or Airports
D2972	Standard Test Methods for Arsenic in Water D3082-09 Standard Test Method for Boron in Water
D3042	Standard Test Method for Insoluble Residue in Carbonate Aggregates
D3183	Standard Practice for Rubber – Preparation of Pieces for Test Purposes from Products
D3223	Standard Test Method for Total Mercury in Water
D3319	Standard Practice for Accelerated Polishing of Aggregates Using the British Wheel
D3352	Standard Test Method for Strontium Ion in Brackish Water, Seawater, and Brines
D3372	Standard Test Method for Molybdenum in Water
D3373	Standard Test Method for Vanadium in Water
D3385	Standard Test Method for Infiltration Rate of Soils in Field Using Double-Ring Infiltrometer
D3468/D3468M	Standard Specification for Liquid-Applied Neoprene and Chlorosulfonated Polyethylene Used in Roofing and Waterproofing
D3516	Standard Test Methods for Ashing Cellulose
D3557	Standard Test Methods for Cadmium in Water
D3558	Standard Test Methods for Cobalt in Water
D3559	Standard Test Methods for Lead in Water
D3561	Standard Test Method for Lithium, Potassium, and Sodium Ions in Brackish Water, Seawater, and Brines by Atomic Absorption Spectrophotometry
D3590	Standard Test Methods for Total Kjeldahl Nitrogen in Water
D3633/D3633M	Standard Test Method for Electrical Resistivity of Membrane-Pavement Systems
D3645	Standard Test Methods for Beryllium in Water
D3651	Standard Test Method for Solvent Bearing Bituminous Compounds
D3665	Standard Practice for Random Sampling of Construction Materials
D3697	Standard Test Method for Antimony in Water
D3744/D3744M	Standard Test Method for Aggregate Durability Index
D3767	Standard Practice for Rubber – Measurement of Dimensions
D3847	Standard Practice for Rubber – Directions for Achieving Subnormal Test Temperatures
D3859	Standard Test Methods for Selenium in Water
D3866	Standard Test Methods for Silver in Water
D3867	Standard Test Methods for Nitrite-Nitrate in Water
D3868	Standard Test Method for Fluoride Ions in Brackish Water, Seawater, and Brines
D3869	Standard Test Methods for Iodide and Bromide Ions in Brackish Water, Seawater, and Brines
D3875	Standard Test Method for Alkalinity in Brackish Water, Seawater, and Brines
D3876	Standard Test Method for Methoxyl and Hydroxypropyl Substitution in Cellulose Ether Products by Gas Chromatography

D3919	Standard Practice for Measuring Trace Elements in Water by Graphite Furnace Atomic Absorption Spectrophotometry
D3920	Standard Test Method for Strontium in Water
D3963/D3963M	Standard Specification for Fabrication and Jobsite Handling of Epoxy-Coated Steel Reinforcing Bars
D3971	Standard Test Method for Dichloromethane-Soluble Matter in Cellulose
D3986	Standard Test Method for Barium in Brines, Seawater, and Brackish Water by Direct-Current Argon Plasma Atomic Emission Spectroscopy
D4014	Standard Specification for Plain and Steel-Laminated Elastomeric Bearings for Bridges
D4025	Standard Practice for Reporting Results of Examination and Analysis of Deposits Formed from Water for Subsurface Injection
D4071	Standard Practice for Use of Portland Cement Concrete Bridge Deck Water Barrier Membrane Systems
D4085	Standard Test Method for Metals in Cellulose by Atomic Absorption Spectrophotometry
D4127	Standard Terminology Used with Ion-Selective Electrodes
D4130	Standard Test Method for Sulfate Ion in Brackish Water, Seawater, and Brines
D4190	Standard Test Method for Elements in Water by Direct-Current Plasma Atomic Emission Spectroscopy
D4191	Standard Test Method for Sodium in Water by Atomic Absorption Spectrophotometry
D4192	Standard Test Method for Potassium in Water by Atomic Absorption Spectrophotometry
D4227	Standard Practice for Qualification of Coating Applicators for Application of Coatings to Concrete Surfaces
D4228	Standard Practice for Qualification of Coating Applicators for Application of Coatings to Steel Surfaces
D4258	Standard Practice for Surface Cleaning Concrete for Coating
D4259	Standard Practice for Preparation of Concrete by Abrasion Prior to Coating Application
D4260	Standard Practice for Liquid and Gelled Acid Etching of Concrete
D4261	Standard Practice for Surface Cleaning Concrete Masonry Units for Coating
D4262	Standard Test Method for pH of Chemically Cleaned or Etched Concrete Surfaces
D4263	Standard Test Method for Indicating Moisture in Concrete by the Plastic Sheet Method
D4285	Standard Test Method for Indicating Oil or Water in Compressed Air
D4286	Standard Practice for Determining Coating Contractor Qualifications for Nuclear Powered Electric Generation Facilities
D4309	Standard Practice for Sample Digestion Using Closed Vessel Microwave Heating Technique for the Determination of Total Metals in Water
D4318	Standard Test Methods for Liquid Limit, Plastic Limit, Plasticity Index of Soils
D4327	Standard Test Method for Anions in Water by Suppressed Ion Chromatography
D4328	Standard Practice for Calculation of Supersaturation of Barium Sulfate, Strontium Sulfate, and Calcium Sulfate Dihydrate (Gypsum) in Brackish Water, Seawater, and Brines
D4380	Standard Test Method for Density of Bentonitic Slurries
D4381/D4381M	Standard Test Method for Sand Content by Volume of Bentonitic Slurries
D4382	Standard Test Method for Barium in Water, Atomic Absorption Spectrophotometry, Graphite Furnace
D4434/D4434M	Standard Specification for Poly(Vinyl Chloride) Sheet Roofing
D4458	Standard Test Method for Chloride Ions in Brackish Water, Seawater, and Brines
D4469	Standard Practice for Calculating Percent Asphalt Absorption by the Aggregate in Asphalt Mixtures

D4511	Standard Test Method for Hydraulic Conductivity of Essentially Saturated Peat
D4520	Standard Practice for Determining Water Injectivity Through the Use of On-Site Floods
D4580/D4580M	Standard Practice for Measuring Delaminations in Concrete Bridge Decks by Sounding
D4637/D4637M	Standard Specification for EPDM Sheet Used In Single-Ply Roof Membrane
D4638	Standard Guide for Preparation of Biological Samples for Inorganic Chemical Analysis
D4658	Standard Test Method for Sulfide Ion in Water
D4691	Standard Practice for Measuring Elements in Water by Flame Atomic Absorption Spectrophotometry
D4748	Standard Test Method for Determining the Thickness of Bound Pavement Layers Using Short-Pulse Radar
D4788	Standard Test Method for Detecting Delaminations in Bridge Decks Using Infrared Thermography
D4791	Standard Test Method for Flat Particles, Elongated Particles, or Flat and Elongated Particles in Coarse Aggregate
D4792/D4792M	Standard Test Method for Potential Expansion of Aggregates from Hydration Reactions
D4794	Standard Test Method for Determination of Ethoxyl or Hydroxyethoxyl Substitution in Cellulose Ether Products by Gas Chromatography
D4811/D4811M	Standard Specification for Nonvulcanized (Uncured) Rubber Sheet Used as Roof Flashing
D5084	Standard Test Methods for Measurement of Hydraulic Conductivity of Saturated Porous Materials Using a Flexible Wall Permeameter
D5093	Standard Test Method for Field Measurement of Infiltration Rate Using Double-Ring Infiltrometer with Sealed-Inner Ring
D5106	Standard Specification for Steel Slag Aggregates for Bituminous Paving Mixtures
D5257	Standard Test Method for Dissolved Hexavalent Chromium in Water by Ion Chromatography
D5298	Standard Test Method for Measurement of Soil Potential (Suction) Using Filter Paper
D5361/D5361M	Standard Practice for Sampling Compacted Asphalt Mixtures for Laboratory Testing
D5367	Standard Practice for Evaluating Coatings Applied Over Surfaces Treated With Inhibitors Used to Prevent Flash Rusting of Steel When Water or Water/Abrasive Blasted
D5400	Standard Test Methods for Hydroxypropylcellulose
D5405/D5405M	Standard Test Method for Conducting Time-to-Failure (Creep-Rupture) Tests of Joints Fabricated from Nonbituminous Organic Roof Membrane Material
D5444	Standard Test Method for Mechanical Size Analysis of Extracted Aggregate
D5463	Standard Guide for Use of Test Kits to Measure Inorganic Constituents in Water
D5673	Standard Test Method for Elements in Water by Inductively Coupled Plasma – Mass Spectrometry
D5821	Standard Test Method for Determining the Percentage of Fractured Particles in Coarse Aggregate
D5856	Standard Test Method for Measurement of Hydraulic Conductivity of Porous Material Using a Rigid- Wall, Compaction-Mold Permeameter
D5896	Standard Test Method for Carbohydrate Distribution of Cellulosic Materials
D5897	Standard Test Method for Determination of Percent Hydroxyl on Cellulose Esters by Potentiometric Titration – Alternative Method
D5907	Standard Test Methods for Filterable Matter (Total Dissolved Solids) and Nonfilterable Matter (Total Suspended Solids) in Water
D5977	Standard Specification for High Load Rotational Spherical Bearings for Bridges and Structures

D5992	Standard Guide for Dynamic Testing of Vulcanized Rubber and Rubber-Like Materials Using Vibratory Methods
D6087	Standard Test Method for Evaluating Asphalt-Covered Concrete Bridge Decks Using Ground Penetrating Radar
D6134	Standard Specification for Vulcanized Rubber Sheets Used in Waterproofing Systems
D6147	Standard Test Method for Vulcanized Rubber and Thermoplastic Elastomer – Determination of Force Decay (Stress Relaxation) in Compression
D6153	Standard Specification for Materials for Bridge Deck Waterproofing Membrane Systems
D6155	Standard Specification for Nontraditional Coarse Aggregates for Asphalt Paving Mixtures
D6188	Standard Test Method for Viscosity of Cellulose by Cuprammonium Ball Fall
D6297	Standard Specification for Asphaltic Plug Joints for Bridges
D6383/D6383M	Standard Practice for Time-to-Failure (Creep-Rupture) of Adhesive Joints Fabricated from EPDM Roof Membrane Material
D6391	Standard Test Method for Field Measurement of Hydraulic Conductivity Using Borehole Infiltration
D6501	Standard Test Method for Phosphonate in Brines
D6508	Standard Test Method for Determination of Dissolved Inorganic Anions in Aqueous Matrices Using Capillary Ion Electrophoresis and Chromate Electrolyte
D6539	Standard Test Method for Measurement of the Permeability of Unsaturated Porous Materials by Flowing Air
D6581	Standard Test Methods for Bromate, Bromide, Chlorate, and Chlorite in Drinking Water by Suppressed Ion Chromatography
D6601	Standard Test Method for Rubber Properties – Measurement of Cure and After-Cure Dynamic Properties Using a Rotorless Shear Rheometer
D6754/D6754M	Standard Specification for Ketone Ethylene Ester Based Sheet Roofing
D6800	Standard Practice for Preparation of Water Samples Using Reductive Precipitation Preconcentration Technique for ICP-MS Analysis of Trace Metals
D6836	Standard Test Methods for Determination of the Soil Water Characteristic Curve for Desorption Using Hanging Column, Pressure Extractor, Chilled Mirror Hygrometer, and/or Centrifuge
D6850	Standard Guide for QC of Screening Methods in Water
D6878/D6878M	Standard Specification for Thermoplastic Polyolefin Based Sheet Roofing
D6910/D6910M	Standard Test Method for Marsh Funnel Viscosity of Construction Slurries
D6919	Standard Test Method for Determination of Dissolved Alkali and Alkaline Earth Cations and Ammonium in Water and Wastewater by Ion Chromatography
D6928	Standard Test Method for Resistance of Coarse Aggregate to Degradation by Abrasion in the Micro-Deval Apparatus
D6942	Standard Test Method for Stability of Cellulose Fibers in Alkaline Environments
D6994	Standard Test Method for Determination of Metal Cyanide Complexes in Wastewater, Surface Water, Groundwater and Drinking Water Using Anion Exchange Chromatography with UV Detection
D7100	Standard Test Method for Hydraulic Conductivity Compatibility Testing of Soils with Aqueous Solutions
D7121	Standard Test Method for Rubber Property – Resilience Using Schob Type Rebound Pendulum
D7172	Standard Test Method for Determining the Relative Density (Specific Gravity) and Absorption of Fine Aggregates Using Infrared
D7243	Standard Guide for Measuring the Saturated Hydraulic Conductivity of Paper Industry Sludges
D7357	Standard Specification for Cellulose Fibers for Fiber-Reinforced Concrete
D7370/D7370M	Standard Test Method for Determination of Relative Density and Absorption of Fine, Coarse and Blended Aggregate Using Combined Vacuum Saturation and Rapid Submersion

D7426	Standard Test Method for Assignment of the DSC Procedure for Determining Tg of a Polymer or an Elastomeric Compound
D7428	Standard Test Method for Resistance of Fine Aggregate to Degradation by Abrasion in the Micro-Deval Apparatus
D7503	Standard Test Method for Measuring the Exchange Complex and Cation Exchange Capacity of Inorganic Fine-Grained Soil
D7635/D7635M	Standard Test Method for Measurement of Thickness of Coatings Over Fabric Reinforcement
D7664	Standard Test Methods for Measurement of Hydraulic Conductivity of Unsaturated Soils
D7781	Standard Test Method for Nitrite-Nitrate in Water by Nitrate Reductase
D7980	Standard Guide for Ion-Chromatographic Analysis of Anions in Grab Samples of Ultrapure Water (UPW) in the Semiconductor Industry
E11	Standard Specification for Woven Wire Test Sieve Cloth and Test Sieves
E154/E154M	Standard Test Methods for Water Vapor Retarders Used in Contact with Earth Under Concrete Slabs, on Walls, or as Ground Cover
E605	Standard Test Methods for Thickness and Density of Sprayed Fire-Resistive Material (SFRM) Applied to Structural Members
E736	Standard Test Method for Cohesion/Adhesion of Sprayed Fire-Resistive Materials Applied to Structural Members
E759	Standard Test Method for Effect of Deflection on Sprayed Fire-Resistive Material Applied to Structural Members
E760	Standard Test Method for Effect of Impact on Bonding of Sprayed Fire-Resistive Material Applied to Structural Members
E761	Standard Test Method for Compressive Strength of Sprayed Fire-Resistive Material Applied to Structural Members
E859	Standard Test Method for Air Erosion of Sprayed Fire-Resistive Materials (SFRMs) Applied to Structural Members
E937	Standard Test Method for Corrosion of Steel by Sprayed Fire-Resistive Material (SFRM) Applied to Structural Members
E1155/ E1155M	Standard Test Method for Determining FF Floor Flatness and FL Floor Levelness Numbers
E1399/E1399M	Standard Test Method for Cyclic Movement and Measuring the Minimum and Maximum Joint Widths of Architectural Joint Systems
E1486/E1486M	Standard Test Method for Determining Floor Tolerances Using Waviness, Wheel Path and Levelness Criteria
E1513	Standard Practice for Application of Sprayed Fire-Resistive Materials (SFRMs)
E1612/E1612M	Standard Specification for Preformed Architectural Compression Seals for Buildings and Parking Structures
E1643	Standard Practice for Selection, Design, Installation, and Inspection of Water Vapor Retarders Used in Contact with Earth or Granular Fill Under Concrete Slabs
E1745	Standard Specification for Plastic Water Vapor Retarders Used in Contact with Soil or Granular Fill under Concrete Slabs
E1783/E1783M	Standard Specification for Preformed Architectural Strip Seals for Buildings and Parking Structures
E1918	Standard Test Method for Measuring Solar Reflectance of Horizontal and Low-Sloped Surfaces in the Field
E1980	Standard Practice for Calculating Solar Reflectance Index of Horizontal and Low-Sloped Opaque Surfaces
E1993/E1993M	Standard Specification for Bituminous Water Vapor Retarders Used in Contact with Soil or Granular Fill Under Concrete Slabs
E2174	Standard Practice for On-Site Inspection of Installed Firestops

E2393	Standard Practice for On-Site Inspection of Installed Fire Resistive Joint Systems and Perimeter Fire Barriers
E2634	Standard Specification for Flat Wall Insulating Concrete Form (ICF) Systems
E2785	Standard Test Method for Exposure of Firestop Materials to Severe Environmental Conditions
E2786	Standard Test Methods for Measuring Expansion of Intumescent Materials Used in Firestop and Joint Systems
E2923	Standard Practice for Longevity Assessment of Firestop Materials Using Differential Scanning Calorimetry
E2924	Standard Practice for Intumescent Coatings
F693	Standard Practice for Sealing Seams of Resilient Sheet Flooring Products by Use of Liquid Seam Sealers
F710	Standard Practice for Preparing Concrete Floors to Receive Resilient Flooring
F1482	Standard Practice for Installation and Preparation of Panel Type Underlayments to Receive Resilient Flooring
F1516	Standard Practice for Sealing Seams of Resilient Flooring Products by the Heat Weld Method (when Recommended)
F1869	Standard Test Method for Measuring Moisture Vapor Emission Rate of Concrete Subfloor Using Anhydrous Calcium Chloride
F2170	Standard Test Method for Determining Relative Humidity in Concrete Floor Slabs Using in situ Probes
F2419	Standard Practice for Installation of Thick Poured Gypsum Concrete Underlayments and Preparation of the Surface to Receive Resilient Flooring
F2471	Standard Practice for Installation of Thick Poured Lightweight Cellular Concrete Underlayments and Preparation of the Surface to Receive Resilient Flooring
F2659	Standard Guide for Preliminary Evaluation of Comparative Moisture Condition of Concrete, Gypsum Cement and Other Floor Slabs and Screeds Using a Non-Destructive Electronic Moisture Meter
F2678	Standard Practice for Preparing Panel Underlayments, Thick Poured Gypsum Concrete Underlayments, Thick Poured Lightweight Cellular Concrete Underlayments, and Concrete Subfloors with Underlayment Patching Compounds to Receive Resilient Flooring
F2753	Standard Practice to Evaluate the Effect of Dynamic Rolling Load over Resilient Floor Covering System
F2873	Standard Practice for the Installation of Self-Leveling Underlayment and the Preparation of Surface to Receive Resilient Flooring
F3010	Standard Practice for Two-Component Resin Based Membrane-Forming Moisture Mitigation Systems for Use Under Resilient Floor Coverings
G3	Standard Practice for Conventions Applicable to Electrochemical Measurements in Corrosion Testing
G4	Standard Guide for Conducting Corrosion Tests in Field Applications
G5	Standard Reference Test Method for Making Potentiodynamic Anodic Polarization Measurements
G59	Standard Test Method for Conducting Potentiodynamic Polarization Resistance Measurements
G61	Standard Test Method for Conducting Cyclic Potentiodynamic Polarization Measurements for Localized Corrosion Susceptibility of Iron-, Nickel-, or Cobalt-Based Alloys
G69	Standard Test Method for Measurement of Corrosion Potentials of Aluminum Alloys
G71	Standard Guide for Conducting and Evaluating Galvanic Corrosion Tests in Electrolytes
G82	Standard Guide for Development and Use of a Galvanic Series for Predicting Galvanic Corrosion Performance
G96	Standard Guide for Online Monitoring of Corrosion in Plant Equipment (Electrical and Electrochemical Methods)
G100	Standard Test Method for Conducting Cyclic Galvanostaircase Polarization
G102	Standard Practice for Calculation of Corrosion Rates and Related Information from Electrochemical Measurements
G106	Standard Practice for Verification of Algorithm and Equipment for Electrochemical Impedance Measurements

G108	Standard Test Method for Electrochemical Reactivation (EPR) for Detecting Sensitization of AISI Type 304 and 304L Stainless Steels
G109	Standard Test Method for Determining Effects of Chemical Admixtures on Corrosion of Embedded Steel Reinforcement in Concrete Exposed to Chloride Environments
G148	Standard Practice for Evaluation of Hydrogen Uptake, Permeation, and Transport in Metals by an Electrochemical Technique
G150	Standard Test Method for Electrochemical Critical Pitting Temperature Testing of Stainless Steels and Related Alloys
G180	Standard Test Method for Corrosion Inhibiting Admixtures for Steel in Concrete by Polarization Resistance in Cementitious Slurries
G189	Standard Guide for Laboratory Simulation of Corrosion Under Insulation
G192	Standard Test Method for Determining the Crevice Repassivation Potential of Corrosion-Resistant Alloys Using a Potentiodynamic-Galvanostatic-Potentiostatic Technique
G198	Standard Test Method for Determining the Relative Corrosion Performance of Driven Fasteners in Contact with Treated Wood
G199	Standard Guide for Electrochemical Noise Measurement

AASHTO STANDARDS

American Association of State Highway and Transportation Officials (AASHTO) documents related to aggregates, cement, and concrete that are relevant to or referred to in the text are listed as follows and can be obtained at transportation.org.

M 6	Standard Specification for Fine Aggregate for Hydraulic Cement Concrete
M 17	Standard Specification for Mineral Filler for Bituminous Paving Mixtures
M 29	Standard Specification for Fine Aggregate for Bituminous Paving Mixtures
M 31	Standard Specification for Deformed and Plain Carbon and Low-Alloy Steel Bars for Concrete Reinforcement
M 43	Standard Specification for Sizes of Aggregate for Road and Bridge Construction
M 45	Standard Specification for Aggregate for Masonry Mortar
M 54	Standard Specification for Welded Deformed Steel Bar Mats for Concrete Reinforcement
M 80	Standard Specification for Coarse Aggregate for Hydraulic Cement Concrete
M 85	Standard Specification for Portland Cement
M 92	Standard Specification for Wire-Cloth Sieves for Testing Purposes
M 143	Standard Specification for Sodium Chloride
M 144	Standard Specification for Calcium Chloride
M 152	Standard Specification for Flow Table for Use in Tests of Hydraulic Cement
M 154	Standard Specification for Air-Entraining Admixtures for Concrete
M 157	Standard Specification for Ready-Mixed Concrete
M 182	Standard Specification for Burlap Cloth Made from Jute or Kenaf and Cotton Mats
M 194	Standard Specification for Chemical Admixtures for Concrete
M 195	Standard Specification for Lightweight Aggregates for Structural Concrete
M 201	Standard Specification for Mixing Rooms, Moist Cabinets, Moist Rooms, and Water Storage Tanks Used in the Testing of Hydraulic Cements and Concretes
M 203	Standard Specification for Steel Strand, Uncoated Seven-Wire for Concrete Reinforcement
M 204	Standard Specification for Uncoated Stress-Relieved Steel Wire for Prestressed Concrete
M 205	Standard Specification for Molds for Forming Concrete Test Cylinders Vertically
M 210	Standard Specification for Use of Apparatus for the Determination of Length Change of Hardened Cement Paste, Mortar, and Concrete
M 224	Standard Specification for Use of Protective Sealers for Portland Cement Concrete
M 231	Standard Specification for Weighing Devices Used in the Testing of Materials
M 233	Standard Specification for Boiled Linseed Oil Mixture for Treatment of Portland Cement Concrete
M 240	Standard Specification for Blended Hydraulic Cement
M 241	Standard Specification for Concrete Made by Volumetric Batching and Continuous Mixing
M 261	Standard Specification for Rib-Tread Standard Tire for Special-Purpose Pavement Frictional Property Tests
M 286	Standard Specification for Smooth-Tread Standard Tire for Special-Purpose Pavement Frictional Property Tests
M 295	Standard Specification for Coal Fly Ash and Raw or Calcined Natural Pozzolan for Use in Concrete
M 302	Standard Specification for Ground Granulated Blast-Furnace Slag for Use in Concrete and Mortars
M 307	Standard Specification for Silica Fume Used in Cementitious Mixtures

M 321	Standard Specification for High-Reactivity Pozzolans for Use in Hydraulic-Cement Concrete, Mortar, and Grout
M 322	Standard Specification for Rail-Steel and Axle-Steel Deformed Bars for Concrete Reinforcement
M 327	Standard Specification for Processing Additions for Use in the Manufacture of Hydraulic Cements
M 230	Standard Specification for Expanded and Extruded Foam Board (Polystyrene)
M 235	Standard Specification for Epoxy Resin Adhesives
PP 58	Standard Practice for Static Segregation of Hardened Self-Consolidating Concrete (SCC) Cylinders
PP 84	Standard Practice for Developing Performance Engineered Concrete Pavement Mixtures
R 8	Standard Practice for Evaluation of Transportation-Related Earthborne Vibrations
R 10	Standard Practice for Definition of Terms Related to Quality and Statistics as Used in Highway Construction
R 16	Standard Practice for Regulatory Information for Chemicals Used in AASHTO Tests
R 25	Standard Practice for Technician Training and Qualification Programs
R 32	Standard Practice for Calibrating the Load Cell and Deflection Sensors for a Falling Weight Deflectometer
R 33	Standard Practice for Calibrating the Reference Load Cell Used for Reference Calibrations for a Falling Weight Deflectometer
R 34	Standard Practice for Evaluating Deicing Chemicals
R 39	Standard Practice for Making and Curing Concrete Test Specimens in the Laboratory
R 44	Standard Practice for Independent Assurance (IA) Programs
R 45	Standard Practice for Installing, Monitoring, and Processing Data of the Traveling Type Slope Inclinometer
R 60	Standard Practice for Sampling Freshly Mixed Concrete
R 64	Standard Practice for Sampling and Fabrication of 50-mm (2-in) Cube Specimens Using Grout (Non Shrink) or Mortar
R 70	Standard Specification for Use of Apparatus for the Determination of Length Change of Hardened Cement Paste, Mortar, and Concrete
R 71	Standard Method of Test for Sampling and Amount of Testing of Hydraulic Cement
R 72	Standard Practice for Match Curing of Concrete Test Specimens
R 76	Standard Method of Test for Reducing Samples of Aggregate to Testing Size
R 80	Standard Practice for Determining the Reactivity of Concrete Aggregates and Selecting Appropriate Measures for Preventing Deleterious Expansion in New Concrete Construction
R 81	Standard Practice for Static Segregation of Hardened Self-Consolidating Concrete (SCC) Cylinders
R 91	Standard Practice for Determining Aggregate Source Shape Values from Digital Image Analysis Shape Properties
T 2	Standard Method of Test for Sampling of Aggregates
T 11	Standard Method of Test for Materials Finer Than 75-μm (No. 200) Sieve in Mineral Aggregates by Washing
T 19	Standard Method of Test for Bulk Density ("Unit Weight") and Voids in Aggregate
T 21	Standard Method of Test for Organic Impurities in Fine Aggregates for Concrete
T 22	Standard Method of Test for Compressive Strength of Cylindrical Concrete Specimens
T 23	Standard Method of Test for Making and Curing Concrete Test Specimens in the Field
T 24	Standard Method of Test for Obtaining and Testing Drilled Cores and Sawed Beams of Concrete
T 26	Standard Method of Test for Quality of Water to Be Used in Concrete

T 27	Standard Method of Test for Sieve Analysis of Fine and Coarse Aggregates
T 30	Standard Method of Test for Mechanical Analysis of Extracted Aggregate
T 37	Standard Method of Test for Sieve Analysis of Mineral Filler for Hot Mix Asphalt (HMA)
T 71	Standard Method of Test for Effect of Organic Impurities in Fine Aggregate on Strength of Mortar
T 84	Standard Method of Test for Specific Gravity and Absorption of Fine Aggregate
T 85	Standard Method of Test for Specific Gravity and Absorption of Coarse Aggregate
T 96	Standard Method of Test for Resistance to Degradation of Small-Size Coarse Aggregate by Abrasion and Impact in the Los Angeles Machine
T 97	Standard Method of Test for Flexural Strength of Concrete (Using Simple Beam with Third-Point)
T 98	Standard Method of Test for Fineness of Portland Cement by the Turbidimeter
T 103	Standard Method of Test for Soundness of Aggregates by Freezing and Thawing
T 104	Standard Method of Test for Soundness of Aggregate by Use of Sodium Sulfate or Magnesium Sulfate
T 105	Standard Method of Test for Chemical Analysis of Hydraulic Cement
T 106	Standard Method of Test for Compressive Strength of Hydraulic Cement Mortar (Using 50-mm or 2 in. Cube Specimens)
T 107	Standard Method of Test for Autoclave Expansion of Hydraulic Cement
T 112	Standard Method of Test for Clay Lumps and Friable Particles in Aggregate
T 113	Standard Method of Test for Lightweight Pieces in Aggregate
T 119	Standard Method of Test for Slump of Hydraulic Cement Concrete
T 121	Standard Method of Test for Density (Unit Weight), Yield, and Air Content (Gravimetric) of Concrete
T 127	Standard Method of Test for Sampling and Amount of Testing of Hydraulic Cement
T 129	Standard Method of Test for Normal Consistency of Hydraulic Cement
T 131	Standard Method of Test for Time of Setting of Hydraulic Cement by Vicat Needle
T 132	Standard Method of Test for Tensile Strength of Hydraulic Cement Mortars
T 133	Standard Method of Test for Density of Hydraulic Cement
T 134	Standard Method of Test for Moisture-Density Relations of Soil-Cement Mixtures
T 137	Standard Method of Test for Air Content of Hydraulic Cement Mortar
T 140	Standard Method of Test for Compressive Strength of Concrete Using Portions of Beams Broken in Flexure
T 148	Standard Method of Test for Measuring Length of Drilled Concrete Cores
T 152	Standard Method of Test for Air Content of Freshly Mixed Concrete by the Pressure Method
T 153	Standard Method of Test for Fineness of Hydraulic Cement by Air Permeability Apparatus
T 154	Standard Method of Test for Time of Setting of Hydraulic Cement Paste by Gillmore Needles
T 155	Standard Method of Test for Water Retention by Liquid Membrane-Forming Curing Compounds for Concrete
T 157	Standard Method of Test for Air-Entraining Admixtures for Concrete
T 158	Standard Method of Test for Bleeding of Concrete
T 160	Standard Method of Test for Length Change of Hardened Hydraulic Cement Mortar and Concrete
T 161	Standard Method of Test for Resistance of Concrete to Rapid Freezing and Thawing
T 162	Standard Method of Test for Mechanical Mixing of Hydraulic Cement Pastes and Mortars of Plastic Consistency

T 168	Standard Method of Test for Sampling Bituminous Paving Mixtures
T 177	Standard Method of Test for Flexural Strength of Concrete (Using Simple Beam with Center-Point Loading)
T 178	Standard Method of Test for Portland-Cement Content of Hardened Hydraulic-Cement Concrete
T 180	Standard Method of Test for Moisture – Density Relations of Soils Using a 4.54-kg (10-lb) Rammer and a 457-mm (18-in.) Drop
T 185	Standard Method of Test for Early Stiffening of Hydraulic Cement (Mortar Method)
T 186	Standard Method of Test for Early Stiffening of Hydraulic Cement (Paste Method)
T 188	Standard Method of Test for Evaluation by Freezing and Thawing of Air-Entraining Additions to Hydraulic Cement
T 192	Standard Method of Test for Fineness of Hydraulic Cement by the 45-µm (No. 325) Sieve
T 196	Standard Method of Test for Air Content of Freshly Mixed Concrete by the Volumetric Method
T 197	Standard Method of Test for Time of Setting of Concrete Mixtures by Penetration Resistance
T 198	Standard Method of Test for Splitting Tensile Strength of Cylindrical Concrete Specimens
T 199	Standard Method of Test for Air Content of Freshly Mixed Concrete by the Chace Indicator
T 210	Standard Method of Test for Aggregate Durability Index
T 231	Standard Practice for Capping Cylindrical Concrete Specimens
T 248	Standard Method of Test for Reducing Samples of Aggregate to Testing Size
T 253	Standard Method of Test for Coated Dowel Bars
T 255	Standard Method of Test for Total Evaporable Moisture Content of Aggregate by Drying
T 259	Standard Method of Test for Resistance of Concrete to Chloride Ion Penetration
T 260	Standard Method of Test for Sampling and Testing for Chloride Ion in Concrete and Concrete Raw Materials
T 276	Standard Method of Test for Measuring Early-Age Compression Strength and Projecting Later-Age Strength
T 277	Standard Method of Test for Electrical Indication of Concrete's Ability to Resist Chloride Ion Penetration
T 279	Standard Method of Test for Accelerated Polishing of Aggregates Using the British Wheel
T 285	Standard Method of Test for Bend Test for Bars for Concrete Reinforcement
T 290	Standard Method of Test for Determining Water-Soluble Sulfate Ion Content in Soil
T 303	Standard Method of Test for Accelerated Detection of Potentially Deleterious Expansion of Mortar Bars Due to Alkali-Silica Reaction
T 304	Standard Method of Test for Uncompacted Void Content of Fine Aggregate
T 309	Standard Method of Test for Temperature of Freshly Mixed Portland Cement Concrete
T 318	Standard Method of Test for Water Content of Freshly Mixed Concrete Using Microwave Oven Drying
T 323	Standard Method of Test for Determining the Shear Strength at the Interface of Bonded Layers of Portland Cement Concrete
T 325	Standard Method of Test for Estimating the Strength of Concrete in Transportation Construction by Maturity Tests
T 336	Standard Method of Test for Coefficient of Thermal Expansion of Hydraulic Cement Concrete
T 326	Standard Method of Test for Uncompacted Void Content of Coarse Aggregate (As Influenced by Particle Shape, Surface Textures, and Grading)
T 327	Standard Method of Test for Resistance of Coarse Aggregate to Degradation by Abrasion in the Micro Deval Apparatus

T 330	Standard Method of Test for The Qualitative Detection of Harmful Clays of the Smectite Group in Aggregates Using Methylene Blue
T 332	Standard Method of Test for Determining Chloride Ions in Concrete and Concrete Materials by Specific Ion Probe
T 334	Standard Method of Test for Estimating the Cracking Tendency of Concrete
T 335	Standard Method of Test for Determining the Percentage of Fracture in Coarse Aggregate
T 336	Standard Method of Test for Coefficient of Thermal Expansion of Hydraulic Cement Concrete
T 345	Standard Method of Test for Passing Ability of Self-Consolidating Concrete (SCC) by J-Ring
T 353	Standard Method of Test for Particle Size Analysis of Hydraulic Cement and Related Materials by Light Scattering
T 358	Standard Method of Test for Surface Resistivity Indication of Concrete's Ability to Resist Chloride Ion Penetration
T 384	Standard Air-Void Characteristics of Freshly Mixed Concrete by Buoyancy Change
TP 59	Standard Method of Test for Determining Air Content of Hardened Portland Cement Concrete by High-Pressure Air Meter
TP 64	Standard Method of Test for Predicting Chloride Penetration of Hydraulic Cement Concrete by the Rapid Migration Procedure
TP 77	Standard Method of Test for Specific Gravity and Absorption of Aggregate by Volumetric Immersion Method
TP 110	Standard Method of Test for Potential Alkali Reactivity of Aggregates and Effectiveness of ASR Mitigation Measures (Miniature Concrete Prism Test, MCPT)
TP 381	Standard Method of Test for Determining Aggregate Shape Properties by Means of Digital Image Analysis

ASTM-AASHTO EQUIVALENT STANDARDS

ASTM STANDARDS		AASHTO EQUIVALENT	
A184	Standard Specification for Welded Deformed Steel Bar Mats for Concrete Reinforcement	M 54	Standard Specification for Welded Deformed Steel Bar Mats for Concrete Reinforcement
A416	Standard Specification for Low-Relaxation, Seven-Wire Steel Strand for Prestressed Concrete	M 203	Standard Specification for Steel Strand, Uncoated Seven-Wire for Concrete Reinforcement
A421	Standard Specification for Stress-Relieved Steel Wire for Prestressed Concrete	M 204	Standard Specification for Uncoated Stress-Relieved Steel Wire for Prestressed Concrete
A615	Standard Specification for Deformed and Plain Carbon-Steel Bars for Concrete Reinforcement	M 31	Standard Specification for Deformed and Plain Carbon and Low-Alloy Steel Bars for Concrete Reinforcement
A996	Standard Specification for Rail-Steel and Axle-Steel Deformed Bars for Concrete Reinforcement	M 322	Standard Specification for Rail-Steel and Axle-Steel Deformed Bars for Concrete Reinforcement
C29	Standard Test Method for Bulk Density ("Unit Weight") and Voids in Aggregate	T 19	Standard Method of Test for Bulk Density ("Unit Weight") and Voids in Aggregate
C31	Standard Practice for Making and Curing Concrete Test Specimens in the Field	T 23	Standard Method of Test for Making and Curing Concrete Test Specimens in the Field
C39	Standard Test Method for Compressive Strength of Cylindrical Concrete Specimens	T 22	Standard Method of Test for Compressive Strength of Cylindrical Concrete Specimens
C40	Standard Test Method for Organic Impurities in Fine Aggregates for Concrete	T 21	Standard Method of Test for Organic Impurities in Fine Aggregates for Concrete
C42	Standard Test Method for Obtaining and Testing Drilled Cores and Sawed Beams of Concrete	T 24	Standard Method of Test for Obtaining and Testing Drilled Cores and Sawed Beams of Concrete
C78	Standard Test Method for Flexural Strength of Concrete (Using Simple Beam with Third-Point Loading)	T 97	Standard Method of Test for Flexural Strength of Concrete (Using Simple Beam with Third-Point)
C87	Standard Test Method for Effect of Organic Impurities in Fine Aggregate on Strength of Mortar	T 71	Standard Method of Test for Effect of Organic Impurities in Fine Aggregate on Strength of Mortar
C94	Standard Specification for Ready-Mixed Concrete	M 157	Standard Specification for Ready-Mixed Concrete
C109	Standard Test Method for Compressive Strength of Hydraulic Cement Mortars (Using 2-in. or [50-mm] Cube Specimens)	T 106	Standard Method of Test for Compressive Strength of Hydraulic Cement Mortar (Using 50-mm or 2 in. Cube Specimens)
C114	Standard Test Methods for Chemical Analysis of Hydraulic Cement	T 105	Standard Method of Test for Chemical Analysis of Hydraulic Cement
C117	Standard Test Method for Materials Finer than 75-μm (No. 200) Sieve in Mineral Aggregates by Washing	T 11	Standard Method of Test for Materials Finer Than 75-μm (No. 200) Sieve in Mineral Aggregates by Washing
C123	Standard Test Method for Lightweight Particles in Aggregate	T 113	Standard Method of Test for Lightweight Pieces in Aggregate
C127	Standard Test Method for Relative Density (Specific Gravity) and Absorption of Coarse Aggregate	T 85	Standard Method of Test for Specific Gravity and Absorption of Coarse Aggregate
C128	Standard Test Method for Relative Density (Specific Gravity) and Absorption of Fine Aggregate	T 84	Standard Method of Test for Specific Gravity and Absorption of Fine Aggregate
C131	Standard Test Method for Resistance to Degradation of Small-Size Coarse Aggregate by Abrasion and Impact in the Los Angeles Machine	T 96	Standard Method of Test for Resistance to Degradation of Small-Size Coarse Aggregate by Abrasion and Impact in the Los Angeles Machine
C136	Standard Test Method for Sieve Analysis of Fine and Coarse Aggregates	T 27	Standard Method of Test for Sieve Analysis of Fine and Coarse Aggregates
C138	Standard Test Method for Density (Unit Weight), Yield, and Air Content (Gravimetric) of Concrete	T 121	Standard Method of Test for Density (Unit Weight), Yield, and Air Content (Gravimetric) of Concrete

ASTM STANDARDS		AASHTO EQUIVALENT	
C142	Standard Test Method for Clay Lumps and Friable Particles in Aggregates	T 112	Standard Method of Test for Clay Lumps and Friable Particles in Aggregate
C143	Standard Test Method for Slump of Hydraulic-Cement Concrete	T 119	Standard Method of Test for Slump of Hydraulic Cement Concrete
C144	Standard Specification for Aggregate for Masonry Mortar	M 45	Standard Specification for Aggregate for Masonry Mortar
C150	Standard Specification for Portland Cement	M 85	Standard Specification for Portland Cement
C151	Standard Test Method for Autoclave Expansion of Hydraulic Cement	T 107	Standard Method of Test for Autoclave Expansion of Hydraulic Cement
C156	Standard Test Method for Water Loss [from a Mortar Specimen] Through Liquid Membrane-Forming Curing Compounds for Concrete	T 155	Standard Method of Test for Water Retention by Liquid Membrane-Forming Curing Compounds for Concrete
C157	Standard Test Method for Length Change of Hardened Hydraulic-Cement Mortar and Concrete	T 160	Standard Method of Test for Length Change of Hardened Hydraulic Cement Mortar and Concrete
C172	Standard Practice for Sampling Freshly Mixed Concrete	M 241	Standard Specification for Concrete Made by Volumetric Batching and Continuous Mixing
C173	Standard Test Method for Air Content of Freshly Mixed Concrete by the Volumetric Method	T 196	Standard Method of Test for Air Content of Freshly Mixed Concrete by the Volumetric Method
C174	Standard Test Method for Measuring Thickness of Concrete Elements Using Drilled Concrete Cores	T 148	Standard Method of Test for Measuring Length of Drilled Concrete Cores
C183	Standard Practice for Sampling and the Amount of Testing of Hydraulic Cement	R 71	Standard Method of Test for Sampling and Amount of Testing of Hydraulic Cement
C185	Standard Test Method for Air Content of Hydraulic Cement Mortar	T 137	Standard Method of Test for Air Content of Hydraulic Cement Mortar
C187	Standard Test Method for Amount of Water Required for Normal Consistency of Hydraulic Cement Paste	T 129	Standard Method of Test for Normal Consistency of Hydraulic Cement
C188	Standard Test Method for Density of Hydraulic Cement	T 133	Standard Method of Test for Density of Hydraulic Cement
C191	Standard Test Methods for Time of Setting of Hydraulic Cement by Vicat Needle	T 131	Standard Method of Test for Time of Setting of Hydraulic Cement by Vicat Needle
C192	Standard Practice for Making and Curing Concrete Test Specimens in the Laboratory	R 39	Standard Practice for Making and Curing Concrete Test Specimens in the Laboratory
C204	Standard Test Methods for Fineness of Hydraulic Cement by Air-Permeability Apparatus	T 153	Standard Method of Test for Fineness of Hydraulic Cement by Air Permeability Apparatus
C230	Standard Specification for Flow Table for Use in Tests of Hydraulic Cement	M 152	Standard Specification for Flow Table for Use in Tests of Hydraulic Cement
C231	Standard Test Method for Air Content of Freshly Mixed Concrete by the Pressure Method	T 152	Standard Method of Test for Air Content of Freshly Mixed Concrete by the Pressure Method
C233	Standard Test Method for Air-Entraining Admixtures for Concrete	R 60	Standard Practice for Sampling Freshly Mixed Concrete
C260	Standard Specification for Air-Entraining Admixtures for Concrete	M 154	Standard Specification for Air-Entraining Admixtures for Concrete
C266	Standard Test Method for Time of Setting of Hydraulic-Cement Paste by Gillmore Needles	T 154	Standard Method of Test for Time of Setting of Hydraulic Cement Paste by Gillmore Needles
C293	Standard Test Method for Flexural Strength of Concrete (Using Simple Beam With Center-Point Loading)	T 177	Standard Method of Test for Flexural Strength of Concrete (Using Simple Beam with Center-Point Loading)

ASTM STANDARDS		AASHTO EQUIVALENT	
C305	Standard Practice for Mechanical Mixing of Hydraulic Cement Pastes and Mortars of Plastic Consistency	T 162	Standard Method of Test for Mechanical Mixing of Hydraulic Cement Pastes and Mortars of Plastic Consistency
C330	Standard Specification for Lightweight Aggregates for Structural Concrete	M 195	Standard Specification for Lightweight Aggregates for Structural Concrete
C359	Standard Test Method for Early Stiffening of Hydraulic Cement (Mortar Method)	T 185	Standard Method of Test for Early Stiffening of Hydraulic Cement (Mortar Method)
C403	Standard Test Method for Time of Setting of Concrete Mixtures by Penetration Resistance	T 197	Standard Method of Test for Time of Setting of Concrete Mixtures by Penetration Resistance
C430	Standard Test Method for Fineness of Hydraulic Cement by the 45-μm (No. 325) Sieve	T 192	Standard Method of Test for Fineness of Hydraulic Cement by the 45-μm (No. 325) Sieve
C451	Standard Test Method for Early Stiffening of Hydraulic Cement (Paste Method)	T 186	Standard Method of Test for Early Stiffening of Hydraulic Cement (Paste Method)
C465	Standard Specification for Processing Additions for Use in the Manufacture of Hydraulic Cements	M 327	Standard Specification for Processing Additions for Use in the Manufacture of Hydraulic Cements
C470	Standard Specification for Molds for Forming Concrete Test Cylinders Vertically	M 205	Standard Specification for Molds for Forming Concrete Test Cylinders Vertically
C490	Standard Practice for Use of Apparatus for the Determination of Length Change of Hardened Cement Paste, Mortar, and Concrete	R 70	Standard Specification for Use of Apparatus for the Determination of Length Change of Hardened Cement Paste, Mortar, and Concrete
C494	Standard Specification for Chemical Admixtures for Concrete	M 194	Standard Specification for Chemical Admixtures for Concrete
C496	Standard Test Method for Splitting Tensile Strength of Cylindrical Concrete Specimens	T 198	Standard Method of Test for Splitting Tensile Strength of Cylindrical Concrete Specimens
C511	Standard Specification for Mixing Rooms, Moist Cabinets, Moist Rooms, and Water Storage Tanks Used in the Testing of Hydraulic Cements and Concretes	M 201	Standard Specification for Mixing Rooms, Moist Cabinets, Moist Rooms, and Water Storage Tanks Used in the Testing of Hydraulic Cements and Concretes
C566	Standard Test Method for Total Evaporable Moisture Content of Aggregate by Drying	T 255	Standard Method of Test for Total Evaporable Moisture Content of Aggregate by Drying
C595	Standard Specification for Blended Hydraulic Cements	M 240	Standard Specification for Blended Hydraulic Cement
C617	Standard Practice for Capping Cylindrical Concrete Specimens	T 231	Standard Practice for Capping Cylindrical Concrete Specimens
C618	Standard Specification for Coal Ash and Raw or Calcined Natural Pozzolan for Use in Concrete	T 157	Standard Method of Test for Air-Entraining Admixtures for Concrete
C666	Standard Test Method for Resistance of Concrete to Rapid Freezing and Thawing	T 161	Standard Method of Test for Resistance of Concrete to Rapid Freezing and Thawing
C685	Standard Specification for Concrete Made by Volumetric Batching and Continuous Mixing	M 241	Standard Specification for Concrete Made by Volumetric Batching and Continuous Mixing
C702	Standard Practice for Reducing Samples of Aggregate to Testing Size	R 76	Standard Method of Test for Reducing Samples of Aggregate to Testing Size
C881	Standard Specification for Epoxy-Resin-Base Bonding Systems for Concrete	M 235	Standard Specification for Epoxy Resin Adhesives
C918	Standard Test Method for Measuring Early-Age Compressive Strength and Projecting Later-Age Strength	T 276	Standard Method of Test for Measuring Early-Age Compression Strength and Projecting Later-Age Strength
C989	Standard Specification for Slag Cement for Use in Concrete and Mortars	M 302	Standard Specification for Ground Granulated Blast-Furnace Slag for Use in Concrete and Mortars

ASTM STANDARDS		AASHTO EQUIVALENT	
C1084	Standard Test Method for Portland-Cement Content of Hardened Hydraulic-Cement Concrete	T 178	Standard Method of Test for Portland-Cement Content of Hardened Hydraulic-Cement Concrete
C1202	Standard Test Method for Electrical Indication of Concrete's Ability to Resist Chloride Ion Penetration	T 277	Standard Method of Test for Electrical Indication of Concrete's Ability to Resist Chloride Ion Penetration
C1240	Standard Specification for Silica Fume Used in Cementitious Mixtures	M 307	Standard Specification for Silica Fume Used in Cementitious Mixtures
D98	Standard Specification for Calcium Chloride	M 144	Standard Specification for Calcium Chloride
D448	Standard Classification for Sizes of Aggregate for Road and Bridge Construction	M 43	Standard Specification for Sizes of Aggregate for Road and Bridge Construction
D546	Standard Test Method for Sieve Analysis of Mineral Filler for Asphalt Paving Mixtures	T 37	Standard Method of Test for Sieve Analysis of Mineral Filler for Hot Mix Asphalt (HMA)
D979	Standard Practice for Sampling Bituminous Paving Mixtures	T 168	Standard Method of Test for Sampling Bituminous Paving Mixtures
D1073	Standard Specification for Fine Aggregate for Asphalt Paving Mixtures	M 29	Standard Specification for Fine Aggregate for Bituminous Paving Mixtures
D3319	Standard Practice for the Accelerated Polishing of Aggregates Using the British Wheel	T 279	Standard Method of Test for Accelerated Polishing of Aggregates Using the British Wheel
D3744	Standard Test Method for Aggregate Durability Index	T 210	Standard Method of Test for Aggregate Durability Index
D6928	Standard Test Method for Resistance of Coarse Aggregate to Degradation by Abrasion in the Micro-Deval Apparatus	T 327	Standard Method of Test for Resistance of Coarse Aggregate to Degradation by Abrasion in the Micro Deval Apparatus

CRD STANDARDS

United States Army Corps of Engineers (USACE) Handbook for Concrete and Cement can be obtained at wbdg.org/ccb.

C6	Method of Test for Remolding Effort of Freshly Mixed Concrete
C24	Standard Test Method for Comparing Concretes on Basis of the Bond Developed with Reinforcing Steel
C32	Standard Test Method for Flow Under Water of Hydraulic-Cement Concrete
C36	Method of Test for Thermal Diffusivity of Concrete
C37	Method of Test for Thermal Diffusivity of Mass Concrete
C38	Method of Test for Temperature Rise in Concrete
C39	Test Method for Coefficient of Linear Thermal Expansion of Concrete
C44	Method for Calculation of Thermal Conductivity of Concrete
C45	Method of Test for Thermal Conductivity of Lightweight Insulating Concrete
C48	Standard Test Method for Water Permeability of Concrete
C49	Supplementary Standard Methods of Making and Curing Concrete Test Specimens in the Laboratory
C50	Method of Test for Effect of Grinding During Concrete Mixing on the Grading of Aggregates
C53	Test Method for Consistency of No-Slump Concrete Using the Modified Vebe Apparatus
C55	Test Method for Within-Batch Uniformity of Freshly Mixed Concrete
C61	Test Method for Determining the Resistance of Freshly Mixed Concrete to Washing Out in Water
C62	Method of Testing Cylindrical Test Specimens for Planeness and Parallelism of Ends and Perpendicularity of Sides
C71	Standard Test Method for Ultimate Tensile Strain Capacity of Concrete
C72	Standard Test Methods for Determining the Cement Content of Freshly Mixed Concrete
C82	Standard Test Methods for Determining the Water Content of Freshly Mixed Concrete
C89	Method of Test for Longitudinal Shear Strength, Unconfined, Single Plane
C90	Method of Test for Transverse Shear Strength, Confined, Single Plane or Double Plane
C94	Corps of Engineers Specification for Surface Retarders
C100	Method of Sampling Concrete Aggregate and Aggregate Sources, and Selection of Material for Testing
C104	Method of Calculation of the Fineness Modulus of Aggregate
C112	Method of Test for Surface Moisture in Aggregate by Water Displacement
C114	Test Method for Soundness of Aggregates by Freezing and Thawing of Concrete Specimens
C120	Test Method for Flat and Elongated Particles in Fine Aggregate
C124	Method of Test for Specific Heat of Aggregates, Concrete, and Other Materials (Method of Mixtures)
C125	Method of Test for Coefficient of Linear Thermal Expansion of Coarse Aggregates (Strain-Gage Method)
C126	Method of Test for Coefficient of Thermal Expansion of Mortar
C130	Standard Recommended Practice for Estimating Scratch Hardness of Coarse Aggregate Particles
C143	Specifications for Meters for Automatic Indication of Moisture in Fine Aggregates
C144	Standard Test Method for Resistance of Rock to Freezing and Thawing
C148	Method of Testing Stone for Expansive Breakdown on Soaking in Ethylene Glycol

C154	Standard Test Method for Determination of Moisture in Fine Aggregate for Concrete by Means of a Calcium-Carbide Gas Pressure Meter
C161	Standard Practice for Selecting Proportions for Roller-Compacted Concrete (RCC) Pavement Mixtures Using Soil Compaction Concepts
C163	Test Method for Water Permeability of Concrete Using Triaxial Cell
C164	Standard Test Method for Direct Tensile Strength of Cylindrical Concrete or Mortar Specimens
C166	Standard Test Method for Static Modulus of Elasticity of Concrete in Tension
C169	Standard Test Method for Resistance of Rock to Wetting and Drying
C171	Standard Test Method for Determining Percentage of Crushed Particles in Aggregate
C213	Test Method for the Presence of Sugar in Cement, Mortar, Concrete, or Aggregates
C231	Method of Test for Volume Changes in Neat Cement Bars
C260	Standard Test Method for Tensile Strength of Hydraulic Cement Mortars
C261	Method of Test to Determine Whether Portland Cement is Contaminated with Fly Ash
C300	Corp of Engineers Specifications for Membrane-Forming Compounds for Curing Concrete
C301	Methods for Sampling, Packaging, Marking, and Delivery of Membrane-Forming Compounds for Curing Concrete
C302	Test Method for Sprayability and Unit Moisture Loss Through the Membrane Formed by a Concrete Curing Compound
C307	Method for Calculation of Amount of Ice Needed to Produce Mixed Concrete of a Specified Temperature
C311	Method of Test for Drying Time and Reflectance by the Membrane Formed by a Concrete Curing Compound
C316	Federal Specification: Thinner, Paint, Mineral Spirits, Regular and Odorless (TT-T-291F with Interim Amendment 1 and Notice 1)
C318	Federal Specification (for) Cloth, Burlap, Jute (or Kenaf) (CCC-C-467C with Notice 1)
C400	Requirements for Water for Use in Mixing or Curing Concrete
C401	Method of Test for the Straining Properties of Water
C403	Test Method for Determination of Sulfate Ion in Soils and Water
C406	Test Method for Compressive Strength of Mortar for Use in Evaluating Water for Mixing Concrete
C513	Corps of Engineers Specifications for Rubber Waterstops
C514	Concrete Plant Standards – Ninth Revision
C515	Method of Preparation of Test Specimens from Rubber of Plastic Waterstop
C516	Method of Test for Verification of Compression Testing Machines Using Calibrated Proving Rings
C517	Method of Test for Verifying Machines for Use in Flexural Strength Tests of Concrete Mattresses
C521	Standard Test Method for Frequency and Amplitude of Vibrators for Concrete
C522	Standard Methods of Testing Joint Sealants, Cold-Applied, Non-Jet-Fuel-Resistant for Rigid and Flexible Pavements
C525	Corps of Engineers Test Method for Evaluation of Hot-Applied Joint Sealants for Bubbling Due to Heating
C526	Federal Specification: Sealants, Joint, Two-Component, Jet-Blast-Resistant, Cold-Applied for Portland Cement Concrete Pavement (SS-S-200E)
C527	Standard Specification for Joint Sealants, Cold-Applied, Non-Jet-Fuel-Resistant, for Rigid and Flexible Pavements

C529	Federal Specification, Sealants, Joint, Jet-Fuel-Resistant, Hot-Applied, for Portland Cement and Tar Concrete Pavements (SS-S-1614A with Amendment 1 and Notices 1 & 2)
C530	Federal Specification, Sealant, Joint, Not-Jet-Fuel-Resistant, Hot-Applied, for Portland Cement and Asphalt Concrete Pavements (SS-S-1401C with Amendment 1 and Notices 1 & 2)
C540	Standard Specification for Nonbituminous Inserts for Contraction Joints in Portland Cement Concrete Airfield Pavements, Sawable Type
C541	Sealing Compound, Elastomeric Type, Single Component (for Calking, Sealing and Glazing in Buildings and Other Structures) (TT-S-00230c with Amendment 2 and Notice 1)
C542	Sealing Compound: Silicone Rubber Base (for Calking, Sealing and Glazing in Buildings and Other Structures) (TT-S-001543A)
C543	Sealing Compound: Single Component, Butyl Rubber Based, Solvent Release Type (for Buildings and Other Types of Construction) (TT-S-001657 with Notice 1)
C544	Insulation, Blankets, Thermal (Mineral Fiber, Industrial Type) (Inch-Pound) (HH-I-558C)
C547	Standard Methods of Testing for Jet-Fuel and Heat Resistance of Preformed Polychloroprene Elastomeric Joint Seals for Rigid Pavements
C548	Standard Specification for Jet-Fuel and Heat-Resistant Preformed Polychloroprene Elastomeric Joint Seals for Rigid Pavements
C572	Corps of Engineers Specifications for Polyvinylchloride Waterstops
C575	Method of Test for Change in Weight of Rubber on Immersion in Water
C620	Standard Method of Sampling Fresh Grout
C643	Standard Test Methods for Compressive Strength of Masonry Prisms
C649	Standard Test Method for Unit Weight, Marshall Stability, and Flow of Bituminous Mixtures
C650	Standard Test Method for Density and Percent Voids in Compacted Bituminous Paving Mixtures
C651	Standard Gyratory Testing Machine Method for Design of Hot-Mix Bituminous Pavement Mixtures
C652	Standard Test Method for Measurement of Reduction in Marshall Stability of Bituminous Mixtures Caused by Immersion in Water
C653	Standard Test Method for Determination of Moisture-Density Relation of Soils
C654	Standard Test Method for Determining the California Bearing Ratio of Soils
C655	Standard Test Method for Determining the Modulus of Soil Reaction
C656	Standard Test Method for Determining the California Bearing Ratio and for Sampling Pavement by the Small-Aperture Procedure
C661	Specification for Antiwashout Admixtures for Concrete
C662	Determining the Potential Alkali-Silica Reactivity of Combinations of Cementitious Materials, Lithium Nitrate Admixture and Aggregate (Accelerated Mortar-Bar Method)

CONVERSION FACTORS

The following list provides the conversion relationship between U.S. customary units (Inch-pound) and International System units (Metric, or SI).

The proper conversion procedure is to multiply the specified value on the left by the conversion factor exactly as given below and then round to the appropriate number of significant digits desired.

For example, to convert 11.4 ft to meters:
$11.4 \times 0.3048 = 3.47472$, which rounds to 3.47 meters.

Do not round either value before performing the multiplication, as accuracy would be reduced. A complete guide to the SI system and its use can be found in IEEE / ASTM SI 10, *American National Standard for Metric Practice*.

TO CONVERT FROM	TO	MULTIPLY BY	
LENGTH			
inch (in.)	micrometer (μm)	25,400	E*
inch (in.)	millimeter (mm)	25.4	E
inch (in.)	meter (m)	0.0254	E
foot (ft)	meter (m)	0.3048	E
yard (yd)	meter (m)	0.9144	
AREA			
square foot (ft^2)	square meter (m^2)	0.09290304	E
square inch (in.^{2})	square millimeter (mm^2)	645.2	E
square inch (in.^{2})	square meter (m^2)	0.00064516	E
square yard (yd^2)	square meter (m^2)	0.8361274	
VOLUME			
cubic inch (in.^{3})	cubic millimeter (mm^3)	16.387064	
cubic inch (in.^{3})	cubic meter (m^3)	0.00001639	
cubic foot (ft^3)	cubic meter (m^3)	0.02831685	
cubic yard (yd^3)	cubic meter (m^3)	0.7645549	
gallon (gal) U.S. liquid	liter (L)	3.7854118	
gallon (gal) U.S. liquid	cubic meter (m^3)	0.00378541	
gallon (gal) U.S. liquid	gallon (gal) Canadian	0.8327	
fluid ounce (fl oz)	milliliters (mL)	29.57353	
fluid ounce (fl oz)	cubic meter (m^3)	0.00002957	

TO CONVERT FROM	TO	MULTIPLY BY	
FORCE			
kip (1000 lb)	kilogram (kg)	453.6	
kip (1000 lb)	newton (N)	4448.222	
pound (lb) avoirdupois	kilogram (kg)	0.4535924	
pound (lb)	newton (N)	4.448222	
PRESSURE OR STRESS			
pound per square foot (psf)	kilogram per square meter (kg/m^2)	4.8824	
pound per square foot (psf)	pascal (Pa)†	47.88	
pound per square inch (psi)	kilogram per square centimeter (kg/cm^2)	0.07031	
pound per square inch (psi)	pascal (Pa) †	6894.757	
pound per square inch (psi)	megapascal (MPa)	0.00689476	
pound per square inch (psi)	milimeters of mercury (mmHg)	51.71507548	
MASS (WEIGHT)			
pound (lb), avoirdupois	kilogram (kg)	0.4535924	
ton, 2000 lb	kilogram (kg)	907.1848	
MASS (WEIGHT) PER LENGTH			
kip per linear foot (klf)	kilogram per meter (kg/m)	0.001488	
pound per linear foot (plf)	kilogram per meter (kg/m)	1.488	
MASS PER VOLUME (DENSITY)			
pound per cubic foot (lb/ft^3)	kilogram per cubic meter (kg/m^3)	16.01846	
pound per cubic yard (lb/yd^3)	kilogram per cubic meter (kg/m^3)	0.5933	
TEMPERATURE			
degree Fahrenheit (°F)	degree Celsius (°C)	$t_C = (t_F - 32)/1.8$	
degree Rankine (°R)	degree Kelvin (K)	$T_K = T_R/1.8$	
degree Kelvin (K)	degree Celsius (°C)	$t_C = T_K - 273.15$	
degree Rankine (°R)	degree Fahrenheit (°F)	$t_F = T_R - 450.67$	

TO CONVERT FROM	TO	MULTIPLY BY	
ENERGY AND HEAT			
British thermal unit (Btu)	joule (J)	1055.056	
calorie (cal)	joule (J)	4.1868	E
Btu/ °F hr • ft²	W/m² • K	5.678263	
kilowatt-hour (kwh)	joule (J)	3,600,000	E
British thermal unit per pound (Btu/lb)	calories per gram (cal/g)	0.55556	
British thermal unit per hour (Btu/hr)	watt (W)	0.2930711	
PERMEABILITY			
darcy	centimeter per second (cm/s)	0.000968	
feet per day (ft/day)	centimeter per second (cm/s)	0.000352	
CONCENTRATION			
pounds per cubic foot (lb/ft²)	parts per million (ppm)	16018.46337	
pounds per cubic yard (lb/yd³)	parts per million (ppm)	593.276421	
pounds per gallon (lb/gal)	kilogram per liter (kg/L)	8.345404	
parts per million (ppm)	milligram per liter (mg/L)	1	
parts per million (ppm)	gram per liter (g/L)	0.001	

*E indicates that the factor given is exact
†A pascal equals 1.000 newton per square meter

Note:

One U.S. gallon of water weighs 8.34 pounds (U.S.) at 60°F.

One cubic foot of water weighs 62.4 pounds (U.S.).

One milliliter of water has a mass of 1 gram and has a volume of one cubic centimeter.

One U.S. bag of cement weighs 94 lb.

The prefixes and symbols listed below are commonly used to form names and symbols of the decimal multiples and submultiples of the SI units.

MULTIPLICATION FACTOR	PREFIX	SYMBOL
$1,000,000,000 = 10^9$	giga	G
$1,000,000 = 10^6$	mega	M
$1,000 = 10^3$	kilo	k
$1 = 10^0$	–	–
$0.01 = 10^{-2}$	centi	c
$0.001 = 10^{-3}$	milli	m
$0.000001 = 10^{-6}$	micro	µ
$0.000000001 = 10^{-9}$	nano	n

CEMENT AND CONCRETE RESOURCES

AASHTO	American Association of State Highway and Transportation Officials transportation.org.
ACAA	American Coal Ash Association acaa-usa.org
ACI	American Concrete Institute concrete.org
ACPA	American Concrete Pavement Association pavement.com
ACPA	American Concrete Pipe Association concrete-pipe.org
ACPA	American Concrete Pumping Association concretepumpers.com
ACPPA	American Concrete Pressure Pipe Association acppa.org
ACS	American Ceramic Society ceramics.org
AIT	International Ferrocement Society ferrocement-ifs.com
APA	Architectural Precast Association archprecast.org
API	American Petroleum Institute api.org
ANSI	American National Standards Institute ansi.org
ASCC	American Society of Concrete Contractors ascconline.org
ASCE	American Society of Civil Engineers asce.org
ASHRAE	American Society of Heating, Refrigerating, and Air-Conditioning Engineers, Inc. ashrae.org
ASI	American Shotcrete Association shotcrete.org
ASTM	ASTM International astm.org
BRE	Building Research Establishment Group (UK) bregroup.com
BSI	British Standards Institution bsigroup.com
CAC	Cement Association of Canada cement.ca
CCAA	Cement Concrete & Aggregates Australia concrete.net.au
CDRA	Construction and Demolition Recycling Association cdrecycling.org
CEMBUREAU	European Cement (Industry) Association cembureau.eu
CEN	European Committee for Standardization cen.eu
CFA	Concrete Foundations Association cfawalls.org
CP Tech	National Concrete Pavement Technology Center cptechcenter.org
CPMB	Concrete Plant Manufacturers Bureau cpmb.org
CRSI	Concrete Reinforcing Steel Institute crsi.org
CSA	Canadian Standards Association Group csagroup.org
CSCE	Canadian Society for Civil Engineering csce.ca
CSDA	Concrete Sawing & Drilling Association csda.org
CSI	Construction Specifications Institute csiresources.org
CSI	Cast Stone Institute caststone.org
DIN	Deutsches Institut für Normung e.V. (German Standards Institution) din.de
ECRA	European Cement Research Academy ecra-online.org
EPRI	Electric Power Research Institute epri.com

ERMCO	European Ready Mixed Concrete Organization ermco.eu
ESCSI	Expanded Shale, Clay and Slate Institute escsi.org
FHWA	Federal Highway Administration highways.dot.gov
FICEM	Federacion Interamericana del Cemento (Inter-American Cement Federation) ficem.org
GCCA	Global Cement and Concrete Association gccassociation.org
ICC	International Code Council iccsafe.org
ICCC	International Cement Chemistry Congress Secretariat iccc-online.org
ICFMA	Insulating Concrete Forms Manufacturers Association icf-ma.org
ICMA	International Cement Microscopy Association cemmicro.org
ICPI	Interlocking Concrete Pavement Institute icpi.org
ICRI	International Concrete Repair Institute icri.org
IEEE	Institute of Electrical and Electronics Engineers ieee.org
IGGA	International Grooving & Grinding Association igga.net
IMCYC	Instituto Mexicano del Cemento y del Concreto A.C. (Mexican Institute of Cement and Concrete) imcyc.net
ISO	International Standards Organization iso.org
JCI	Japan Concrete Institute jci-net.or.jp/
MIT CSHub	MIT Concrete Sustainability Hub cshub.mit.edu
MRS	Materials Research Society mrs.org
NACE	NACE International, National Association of Corrosion Engineers nace.org
NAHB	National Association of Home Builders nahb.com
NCMA	National Concrete Masonry Association ncma.org
NIBS	National Institute of Building Sciences nibs.org
NIOSH	National Institute for Occupational Safety and Health cdc.gov/niosh
NIST	National Institute of Standards and Technology concrete.nist.gov
NPCA	National Precast Concrete Association precast.org
NRCC	National Research Council of Canada nrc.canada.ca
NRMCA	National Ready Mixed Concrete Association nrmca.org
NSA	National Slag Association nationalslag.org
NSSGA	National Stone, Sand & Gravel Association nssga.org
NTMA	National Terrazzo and Mosaic Association, Inc. ntma.com
OPCMIA	Operative Plasterers' and Cement Masons' International Association opcmia.org
OSHA	Occupational Safety and Health Administration osha.gov
PCA	Portland Cement Association cement.org
PCI	Precast/Prestressed Concrete Institute www.pci.org
PI	Perlite Institute perlite.org
PTI	Post-Tensioning Institute post-tensioning.org
RILEM	International Union of Laboratories and Experts in Construction Materials, Systems and Structures rilem.net

SCA	Slag Cement Association slagcement.org
SFA	Silica Fume Association silicafume.org
TAC	Transportation Association of Canada tac-atc.ca
TCA	Tilt-Up Concrete Association tilt-up.org
TMMB	Truck Mixer Manufacturers Bureau tmmb.org
TRB	Transportation Research Board trb.org
TVA	The Vermiculite Association vermiculite.org
USACE	U.S. Army Corps of Engineers usace.army.mil
USBR	U.S. Bureau of Reclamation usbr.gov/tsc
USGS	US Geological Survey usgs.gov/centers/nmic/cement-statistics-and-information
VCCTL	Virtual Cement and Concrete Testing Laboratory nist.gov/services-resources/software/vcctl-software
VMMB	Volumetric Mixer Manufacturers Bureau vmmb.org
VDZ	Verein Deutscher Zementwerke (German Cement Works Association) vdz-online.de
WRI	Wire Reinforcement Institute wirereinforcementinstitute.org

PCA CORPORATE MEMBERS

Argos USA Corporation

Buzzi Unicem USA

CalPortland Company

Capitol Aggregates, Inc.

Cemex USA

Continental Cement Company

CRH Americas

Drake Cement LLC

Federal White Cement, Ltd.

GCC of America, Inc.

GCHI - Giant Cement Holding, Inc.

Heidelberg Materials

Martin Marietta Materials, Inc.

Mitsubishi Cement Corporation

The Monarch Cement Company

National Cement Company, Inc.

Salt River Materials Group

Titan America LLC

Votorantim Cimentos North America

For a complete up to date list of PCA Corporate Members visit cement.org/members

PCA ASSOCIATE MEMBERS

Airgas

Arcosa Specialty Materials

BEUMER

ClimeCo.

DCL, Inc.

Eco Material Technologies

ES2, Engineering System Solutions

ESAB Welding & Cutting Products

Euclid Chemical

Evonik Corporation

ExxonMobil

FCT Combustion, Inc.

FLSmidth Inc.

ICR Staffing Services

Industrial Accessories Company

International Materials

North American Mining

Omaha Track, Inc.

Quantum IR Technologies

Ramco Systems Corporation

Refratechnik North America Inc.

RHI Magnesita

Solidia Technologies

thyssenkrupp Industrial Solutions (USA), Inc.

VLS Environmental Solutions

W.L. Gore

For a complete up to date list of PCA Associate Members visit cement.org/associatemembers

B

C

D

E

F

Q

T

U

V

W

X

Y

Z